BEGINNING ALGEBRA

Third Edition

John Tobey
Jeffrey Slater
North Shore Community College
Beverly, Massachusetts

Prentice Hall, Englewood Cliffs, NJ 07632

Library of Congress Cataloging-in-Publication Data

Tobey, John
Beginning algebra / John Tobey, Jeffrey Slater.—3rd ed.
p. cm.
Includes index.
ISBN 0-13-299900-5
1. Algebra. I. Slater, Jeffrey. II. Title.
QA152.2.T6 1995
512.9—dc20 94-21954
CIP

Editor-in-chief: *Jerome Grant*
Acquisitions editor: *Melissa Acuña*
Development editor: *Phyllis Dunsay*
Project manager: *Robert C. Walters*
Interior design: *Lisa A. Jones*
Cover design: *DeFranco Design, Inc.*
Production coordinator: *Trudy Pisciotti*
Photo editor: *Lorinda Morris-Nantz*
Photo research: *Rhoda Sidney*
Cover photo: *Nance Trueworthy/Liaison International*

Illustration credits: 1, Mark E. Gibson/The Stock Market • 20, © Time-Life Books, Inc., used by permission • 21, © Time-Life Books, Inc., used by permission • 29, Wesley Bocxs/Photo Researchers • 40, Steve Elmore/The Stock Market • 48, Mike Yamashita/Woodfin Camp & Associates • 67, Wolfgang Kaehler • 80, George Diebold/The Stock Market • 110, John Coletti/Stock Boston • 113, David Stoecklein/The Stock Market • 125, Gordon Garradd/Science Photo Library/Photo Researchers • 157, J. Messerschmidt/The Stock Market • 166, Richard Pasley/Stock Boston • 179, Randy Duchaine/The Stock Market • 196, Gilbert Grant/Photo Researchers • 213, Jon Feingersh/The Stock Market • 219, Donald C. Johnson/The Stock Market • 227, Bill Ross/Woodfin Camp & Associates • 239, Julian Baum/Science Photo Library/Photo Researchers • 258, Brock May/Photo Researchers • 291, Thomas Stewart/The Stock Market • 330, Hutchings/Photoedit • 334, Hans Halberstadt/Photo Researchers • 339, John Elk III/Stock Boston • 363, Renee Lynn/Photo Researchers • 383, Phyllis Picardi/Stock Boston • 393, Stephen Frink/The Stock Market • 457, Jose Palaez/The Stock Market • 478, Chris Jones/The Stock Market • 503, Goodwin/Monkmeyer Press • 557, Chuck O'Rear/West Light.

A Simon & Schuster Company
Englewood Cliffs, New Jersey 07632

Printed in the United States of America

10 9 8 7 6 5 4 3 2 1

ISBN 0-13-299900-5

Prentice-Hall International (UK) Limited, *London*
Prentice-Hall of Australia Pty, Limited, *Sydney*
Prentice-Hall Canada, Inc., *Toronto*
Prentice-Hall Hispanoamericana, S.A., *Mexico*
Prentice-Hall of India Private Limited, *New Delhi*
Prentice-Hall of Japan, Inc., *Tokyo*
Simon & Schuster Asia Pte. Ltd., *Singapore*
Editora Prentice-Hall do Brasil, Ltda., *Rio de Janeiro*

This book is dedicated to the memory of
Lexie Tobey and John Tobey, Sr.
They have left a legacy of love, a memory
of four decades of faithful teaching, and a sense
of helping others that will influence generations to come.

Contents

2

Equations and Inequalities *125*

3

Solving Applied Problems *179*

4

Exponents and Polynomials *239*

5

Factoring *291*

9

Radicals *503*

10

Quadratic Equations *557*

Preface to the Student

This book was written with your needs and interests in mind. The original manuscript and the first and second editions of this book have been class tested with students all across the country. Based on the suggestions of many students, the book has been refined and improved to maximize your learning while using this text.

We realize that students who enter college have sometimes never enjoyed math or never done well in a math course. You may find that you are anxious about taking this course. We want you to know that this book has been written to help you overcome those difficulties. Literally thousands of students across the country have found an amazing ability to learn math as they have used previous editions of this book. We have incorporated several learning tools and various types of exercises and examples to assist you in learning this material.

It helps to know that learning mathematics is going to help you in life. Perhaps you have entered this course feeling that little or no mathematics will be necessary for you in your future job. However, farmers, elementary school teachers, bus drivers, laboratory technicians, nurses, telephone operators, cable TV repair personnel, photographers, pharmacists, salespeople, doctors, architects, inspectors, counselors, and custodians who once believed that they needed little if any mathematics are finding that the mathematical skills presented in this course can help them. Mathematics, you will find, can help you too. In this book, a great number of examples and problems come from everyday life. You will be amazed at the number of ways mathematics is used in the world each day.

Our greatest wish is that you will find success and personal satisfaction in the mathematics course. Here are a few suggestions to help you on your way:

1. Be sure to take the time to read the boxes marked "Developing Your Study Skills." These ideas come from faculty and students throughout the United States. They have found ways to succeed in mathematics and they want to share those ideas with you.
2. Read through each section of the book that is assigned by your instructor. You will be amazed at how much information you will learn as you read. It will "fill in the pieces" of mathematical knowledge. Reading a math book is a key part of learning.
3. Study carefully each sample example. Then work out the related practice problem. Direct involvement in doing problems and not just thinking about them is one of the greatest guarantees of success in this course.
4. Work out all assigned homework problems. Verify your answers in the back of the book. Ask questions when you don't understand or when you are not sure of an answer.
5. At the end of a chapter take advantage of items that will help you. Study the Chapter Organizers. Do the Chapter Review Problems and the Practice Test problems.
6. If you need help, remember there are teachers and tutors who can assist you. There is a STUDENT SOLUTIONS MANUAL for this Textbook that you will find most valuable. This Manual shows worked out solutions for all the odd-numbered exercises as well as all Diagnostic Pretests, Chapter Reviews, Chapter Tests, and Cumulative Test Problems in this book. (If your college bookstore does not carry the STUDENT SOLUTIONS MANUAL, ask them to order a copy for you.) There are Videos and MathPro (a computerized tutorial) that can be used to help you with each section of this text.

7. Remember, learning mathematics takes time. However, the time spent is well worth the effort. Take the time to study, do homework, review, ask questions, and just reflect over what you have learned. The time you invest in learning mathematics will reap dividends in your future courses and your future life.

You may find the following *walk through* a helpful tool to understanding the organization of the text.

- Diagnostic Pretest—covers the entire course. For those students who have some background in Algebra, the diagnostic pretest will help you to determine your strengths and your weaknesses, and will tell you where to begin.
- Chapter Pretest—allows students to check their knowledge of the chapter material before the text addresses it formally.
- Section Objectives—provides a preview of the section and helps you to focus on the skills to be covered.
- Examples—provide a detailed solution to a mathematical problem. An explanation for each step can be found to the right of each step in italic type.
- Practice Problems—give you an opportunity to solve an exercise similar to the preceding example, on your own. Solutions to the practice problems appear at the end of the text.
- Exercise Sets—are paired and graded to ease you into the material. Answers to the odd-numbered problems are provided in the back of the text.
- Calculator icons are used to alert you to a problem that should be solved using a calculator.
- Verbal and Writing Skills—provide you with an opportunity to clarify your thinking and to become familiar with mathematical terms.
- Applications—provide you with a wide range of problems that demonstrate the use of mathematics in real-life.
- To Think About—provides you with an opportunity to stretch your mind, to look for patterns, and to draw conclusions based on your previous experience. Do not be timid. Some of these problems may appear to be difficult to solve. You may not see the solution immediately. Be patient. Give yourself time. You may find that with time the solution becomes apparent.
- Cumulative Review Problems—a brief review of material covered thus far.
- Putting Your Skills to Work—provides a more comprehensive application of a mathematical skill. These are problems you can sink your teeth into. We suggest that you work with one or more fellow students on the *Cooperative Group Activity.*
- Chapter Organizer—This great learning tool furnishes you with a ready made outline of the chapter, including examples.
- Review Problems—Grouped by sections, these problems can be used to test your knowledge as you complete sections, or at the end of the chapter as a quiz before taking the chapter test.
- Chapter Test—is a representative test of the chapter to help you gauge what sections to relearn before taking the actual test.
- Cumulative Test—one-half the content of these tests are from previous chapters.
- Practice Final Examination—provides you with a test that covers all the material in the text to help you to gauge your level of understanding and skill before taking the actual final.
- Glossary—is found at the end of the book and presents both a definition and an example of the term.

ACKNOWLEDGMENTS

This book is the product of many years of work and many contributions from faculty and students across the country. We would like to thank the many reviewers and participants in focus groups and special meetings with the authors.

Our deep appreciation to each of the following:

George J. Apostolopoulos, DeVry Institute of Technology

Katherine Barringer, Central Virginia Community College

Jamie Blair, Orange Coast College

Larry Blevins, Tyler Junior College

Robert Christie, Miami-Dade Community College

Mike Contino, California State University at Heyward

Judy Dechene, Fitchburg State University

Floyd L. Downs, Arizona State University

Barbara Edwards, Portland State University

Janice F. Gahan-Rech, University of Nebraska at Omaha

Colin Godfrey, University of Massachusetts, Boston

Carl Mancuso, William Paterson College

Janet McLaughlin, Montclair State College

Gloria Mills, Tarrant County Junior College

Norman Mittman, Northeastern Illinois University

Elizabeth A. Polen, County College of Morris

Ronald Ruemmler, Middlesex County College

Sally Search, Tallahassee Community College

Ara B. Sullenberger, Tarrant County Junior College

Michael Trappuzanno, Arizona State University

Jerry Wisnieski, Des Moines Community College

We have been greatly helped by a supportive group of colleagues who not only teach at North Shore Community College but who have provided a number of ideas as well as extensive help on all of our mathematics books. Also, a special word of thanks to Hank Harmeling, Tom Rourke, Wally Hersey, Bob McDonald, Judy Carter, SueEllen Robinson, Kathy LeBlanc, Russ Sullivan, and Rick Ponticelli and Nancy Tufo. Joan Peabody has done an excellent job of typing various materials for the manuscript and her help is gratefully acknowledged.

Behind every project is a quiet hero who does an overwhelming number of tasks. For this book our Developmental Editor, Phyllis Dunsay, has proved to be a virtual hero editor with many talents. Each of her talents have been most appreciated as this book has been improved, modified, refined and shaped. Peter Dunsay, P.E., was a significant help as he designed a number of real world problems and gave excellent suggestions for Applied topics throughout the text. Our Senior Editor, Melissa Acuña has provided words of encouragement, ideas for assistance, helpful and honest suggestions and personal support when it was most needed. Jerome Grant, the Editor-in-Chief, has been a wise and supportive leader in the development of this project.

Book writing is impossible for us without the loyal support of our families. Our deepest thanks and love to Nancy, Johnny, Melissa, Marcia, Shelly, Rusty, and Abby. Your understanding, your love and help, and your patience have been a source of great encouragement. Finally, we thank God for the strength and energy to write and the opportunity to help others through this textbook.

We have spent more than 20 years teaching mathematics. Each teaching day we find our greatest joy is helping students learn. We take a personal interest that each student has a good learning experience in taking this course. If you have some personal comments, suggestions, or ideas for future editions of this textbook, please write to us at:

Prof. John Tobey and Prof. Jeff Slater
Prentice Hall Publishing
Office of the College Mathematics Editor
113 Sylvan Avenue
Englewood Cliffs, NJ 07632

We wish you success in this course and in your future life!

John Tobey
Jeffrey Slater

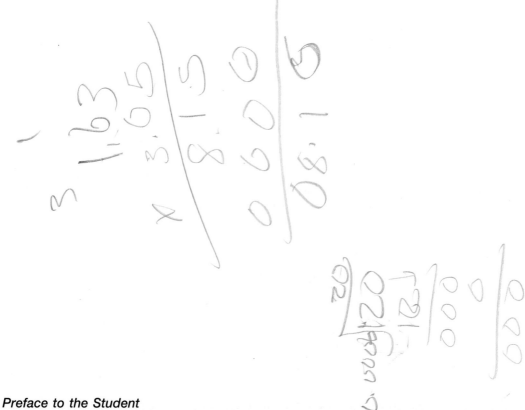

Diagnostic Pretest: Beginning Algebra

Follow the directions for each problem. Simplify each answer.

Chapter 0

1. Add. $3\frac{1}{4} + 2\frac{3}{5}$

2. Multiply. $\left(1\frac{1}{6}\right)\left(2\frac{2}{3}\right)$ 7 8

3. Divide. $\frac{15}{4} \div \frac{3}{8}$

4. Multiply. $(1.63)(3.05)$

5. Divide. $120 \div 0.0006$

6. Find 7% of 64,000.

Chapter 1

7. Add. $(-3) + (-4) + (+12)$

8. Subtract. $-20 - (-23)$

9. Combine. $5x - 6xy - 12x - 8xy$

10. Evaluate $2x^2 - 3x - 4$ when $x = -3$.

11. Remove the grouping symbols.
$2 - 3\{5 + 2[x - 4(3 - x)]\}$

12. Evaluate.
$-3(2 - 6)^2 + (-12) \div (-4)$

Chapter 2

Solve each equation for x.

13. $40 + 2x = 60 - 3x$

14. $7(3x - 1) = 5 + 4(x - 3)$

15. $\frac{2}{3}x - \frac{3}{4} = \frac{1}{6}x + \frac{21}{4}$

16. $\frac{4}{5}(3x + 4) = 20$

17. Solve for p.
$A = \frac{1}{2}(3p - 4f)$

18. Solve for x and graph the result.
$42 - 18x < 48x - 24$

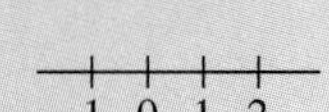

Chapter 3

19. The length of a rectangle is 7 meters longer than the width. The perimeter is 46 meters. Find the dimensions.

20. One side of a triangle is triple the second side. The third side is 3 meters longer than double the second side. Find each side of the triangle if the perimeter of the triangle is 63 meters.

21. Hector has four test scores of 80, 90, 83, and 92. What does he need to score on the fifth test to have an average of 86 on the five tests?

22. Marcia invested \$6000 in two accounts. One earned 5% interest, while the other earned 7% interest. After one year she earned \$394 in interest. How much did she invest in each account?

23. Melissa has 3 more dimes than nickels. She has twice as many quarters as nickels. The value of the coins is \$4.20. How many of each coin does she have?

24. The drama club put on a play for Thursday, Friday, and Saturday nights. The total attendance for the three nights was 6210. Thursday night had 300 fewer people than Friday night. Saturday night had 510 more people than Friday night. How many people came each night?

1. 13/9
2. 15/9
3. 5/2
4. 04.15
5. 020
6.
7.
8.
9.
10.
11.
12.
13.
14.
15.
16.
17.
18.
19.
20.
21.
22.
23.
24.

25. ______

26. ______

27. ______

28. ______

29. ______

30. ______

31. ______

32. ______

33. ______

34. ______

35. ______

36. ______

37. ______

38. ______

39. ______

40. ______

41. ______

42. ______

43. ______

44. ______

45. ______

46. ______

Chapter 4

25. Multiply. $(-2xy^2)(-4x^3y^4)$

26. Divide. $\dfrac{36x^5y^6}{-18x^3y^{10}}$

27. Raise to the indicated power. $(-2x^3y^4)^5$

28. Evaluate. $(-3)^{-4}$

29. Multiply. $(3x^2 + 2x - 5)(4x - 1)$

30. Divide. $(x^3 + 6x^2 - x - 30) \div (x - 2)$

Chapter 5

Factor completely.

31. $5x^2 - 5$

32. $x^2 - 12x + 32$

33. $8x^2 - 2x - 3$

34. $3ax - 8b - 6a + 4bx$

Solve for x.

35. $x^3 + 7x^2 + 12x = 0$

36. $16x^2 - 24x + 9 = 0$

Chapter 6

37. Simplify. $\dfrac{x^2 + 3x - 18}{2x - 6}$

38. Multiply.
$$\frac{6x^2 - 14x - 12}{6x + 4} \cdot \frac{x + 3}{2x^2 - 2x - 12}$$

39. Divide and simplify.
$$\frac{x^2}{x^2 - 4} \div \frac{x^2 - 3x}{x^2 - 5x + 6}$$

40. Add.
$$\frac{3}{x^2 - 7x + 12} + \frac{4}{x^2 - 9x + 20}$$

41. Solve for x.
$$2 - \frac{5}{2x} = \frac{2x}{x + 1}$$

42. Simplify.
$$\frac{3 + \dfrac{1}{x}}{\dfrac{9}{x} + \dfrac{3}{x^2}}$$

Chapter 7

43. Graph $y = 2x - 4$.

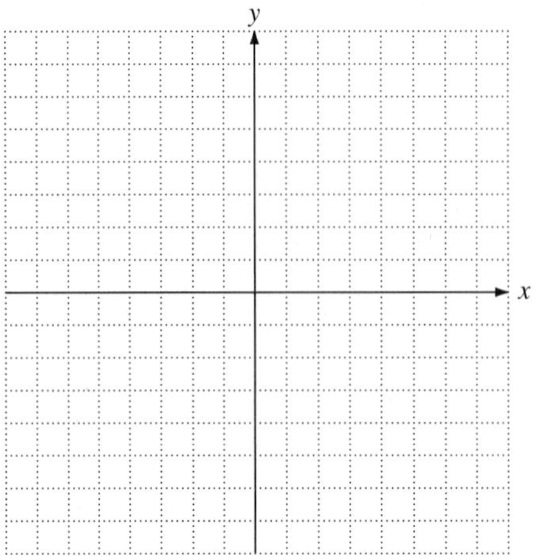

44. Graph $3x + 4y = -12$.

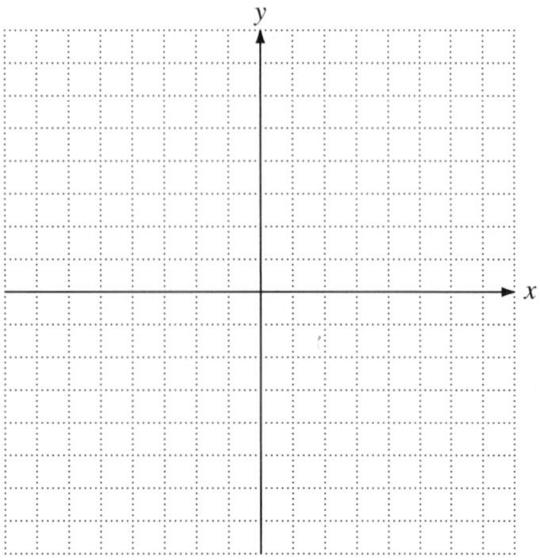

45. What is the slope of a line passing through $(6, -2)$ and $(-3, 4)$?

46. If $f(x) = 2x^2 - 3x + 1$, find $f(3)$.

47. Graph the region
$$y \geq -\frac{1}{3}x + 2$$

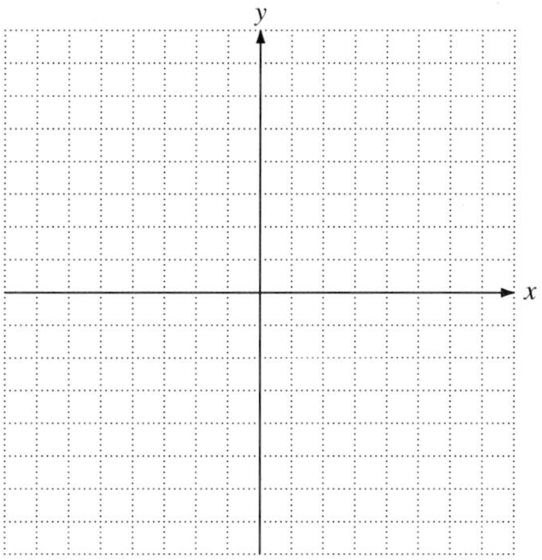

48. Find the equation of a line with a slope of $\frac{3}{5}$ that passes through the point $(-1, 3)$.

Chapter 8

Solve each system by the appropriate method.

49. Substitution method.
$$x + y = 17$$
$$2x - y = -5$$

50. Addition method.
$$-5x + 4y = 8$$
$$2x + 3y = 6$$

51. Any method.
$$2(x - 2) = 3y$$
$$6x = -3(4 + y)$$

52. Any method.
$$x + \frac{1}{3}y = \frac{10}{3}$$
$$\frac{3}{2}x + y = 8$$

53. Is $(2, -3)$ a solution for the system
$$3x + 5y = -9$$
$$2x - 3y = 13?$$

54. A man bought 3 pairs of gloves and 4 scarves for \$53. A woman bought 2 pairs of the same priced gloves and 3 scarves of the same priced scarves for \$38. How much did each item cost?

Chapter 9

55. Evaluate. $\sqrt{121}$

56. Simplify. $\sqrt{125x^3y^5}$

57. Multiply and simplify.
$$(\sqrt{2} + \sqrt{6})(2\sqrt{2} - 3\sqrt{6})$$

58. Rationalize the denominator.
$$\frac{\sqrt{5} - \sqrt{3}}{\sqrt{6}}$$

59. In the right triangle with sides a, b, and c, find side c if side $a = 4$ and side $b = 6$.

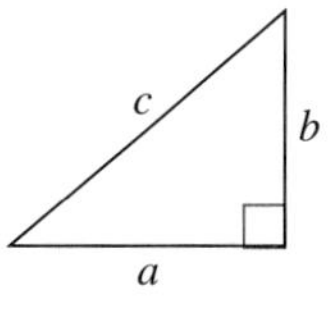

60. y varies directly with x. When $y = 56$, then $x = 8$. Find y when $x = 11$.

47. ____________

48. ____________

49. ____________

50. ____________

51. ____________

52. ____________

53. ____________

54. ____________

55. ____________

56. ____________

57. ____________

58. ____________

59. ____________

60. ____________

61.

62.

63.

64.

65.

66.

Chapter 10

Solve for x.

61. $14x^2 + 21x = 0$

62. $2x^2 + 1 = 19$

63. $2x^2 - 4x - 5 = 0$

64. $x^2 - x + 8 = 5 + 6x$

65. Graph the equation.
$y = x^2 + 8x + 15$

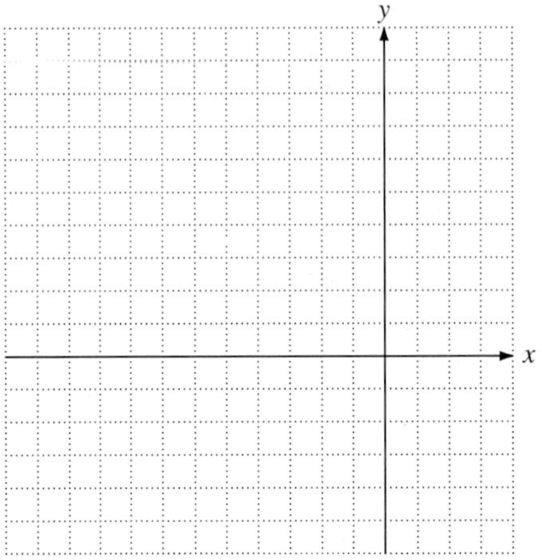

66. A rectangular box is 4 inches longer in length than in width. The area of the box is 96 square inches. Find the length and the width of the box.

0

A Brief Review of Arithmetic Skills

People who work in construction do calculations with fractions, with decimals, and with percents. You will be able to do the same kind of calculations after you have completed this chapter. Carpenters use fractions, estimators use decimals, and project managers use percents. To see how these professionals use mathematics, see Putting Your Skills to Work on pages 21, 40, and 48.

Pretest Chapter 0

1. ________	18. ________
2. ________	19. ________
3. ________	20. ________
4. ________	21. ________
5. ________	22. ________
6. ________	23. ________
7. ________	24. ________
8. ________	25. ________
9. ________	26. ________
10. ________	27. ________
11. ________	28. ________
12. ________	29. ________
13. ________	30. ________
14. ________	31. ________
15. ________	
16. ________	
17. ________	

Take this pretest for Chapter 0 and compare your answers to the solutions for the pretest listed in the Answer Key at the end of the book. If you obtain the correct answer for 25 or more problems, you are ready to begin Chapter 1. If you got more than one wrong in any one section, you should review the necessary sections of Chapter 0 to improve your arithmetic skills. If you take the time to master these arithmetic skills now, you will definitely find Chapter 1 and the remaining chapters of this book easier to complete.

Section 0.1

Simplify each fraction. **1.** $\frac{21}{27}$ **2.** $\frac{26}{39}$

3. Write $\frac{15}{4}$ as a mixed number. **4.** Change $4\frac{5}{7}$ to an improper fraction.

Find the missing number. **5.** $\frac{3}{7} = \frac{?}{14}$ **6.** $\frac{7}{4} = \frac{?}{20}$

Section 0.2

7. Find the LCD, but do not add. $\frac{3}{8}, \frac{5}{6},$ and $\frac{7}{15}$

Perform the calculation indicated. Write the answer in simplest form.

8. $\frac{8}{9} + \frac{5}{9}$ **9.** $\frac{5}{48} + \frac{5}{16}$ **10.** $2\frac{3}{4} + 5\frac{2}{3}$ **11.** $3\frac{1}{5} - 1\frac{3}{8}$

Section 0.3

Perform the calculations indicated. Write the answer in simplest form.

12. $\frac{25}{7} \times \frac{14}{45}$ **13.** $2\frac{4}{5} \times 3\frac{3}{4}$ **14.** $4 \div \frac{8}{7}$ **15.** $2\frac{1}{3} \div 3\frac{1}{4}$

Section 0.4

Change to a decimal. **16.** $\frac{5}{8}$ **17.** $\frac{5}{9}$

Perform the calculations indicated.

18. 15.23 + 3.6 + 0.821 **19.** 3.28 × 0.63

20. 3.015 ÷ 6.7 **21.** 12.13 − 9.884

Section 0.5

Change to a percent. **22.** 0.7 **23.** 0.006

Change to a decimal. **24.** 327% **25.** 2%

26. What is 25% of 1630? **27.** What percent of 500 is 36?

Section 0.6

Estimate. **28.** 8456 + 1749 **29.** 7386 × 2856

30. Estimate the number of miles per gallon achieved by Marcia's car if she drove for 235.8 miles and used 10.7 gallons of gas.

Section 0.7

Solve. You may want to use the Mathematical Blueprint for Problem Solving.

31. Susan drove 870 miles. She started with a full tank of gas. During the trip, she made the following gasoline purchases: 3.7 gal, 10 gal, 12 gal, and 4.9 gal. How many miles per gallon did her car achieve? (Round to the nearest tenth.)

0.1 Simplifying and Finding Equivalent Fractions

After studying this section, you will be able to:

1. *Understand and use some basic mathematical definitions.*
2. *Simplify fractions to lowest terms.*
3. *Change forms between improper fractions and mixed numbers.*
4. *Change a fraction into an equivalent one with a different denominator.*

Chapter 0 is designed to give you a mental "warm-up." In this chapter you'll be able to step back a bit and tone up your math skills. This brief review of arithmetic will increase your math flexibility and give you a good running start into algebra.

1 *Basic Definitions*

Whole numbers are the set of numbers 0, 1, 2, 3, 4, 5, 6, 7, ... They are used to describe whole objects, or entire quantities.

Fractions are a set of numbers that are used to describe parts of whole quantities. In the object shown on the right there are four equal parts. The *three* of the *four* parts that are shaded are represented by the fraction $\frac{3}{4}$. In the fraction $\frac{3}{4}$ the number 3 is called the **numerator** and the number 4, the **denominator.**

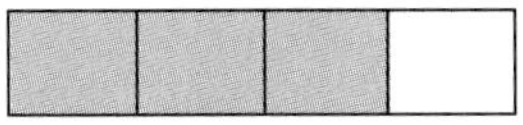

$\frac{3}{4}$

3 ⟵ *Numerator* is on the top
4 ⟵ *Denominator* is on the bottom

The *denominator* of a fraction shows the number of equal parts in the whole and the *numerator* shows the number of these parts being talked about or being used.

Numerals are symbols we use to name numbers. There are many different numerals that can be used to describe the same number. We know that $\frac{1}{2} = \frac{2}{4}$. The fraction $\frac{1}{2}$ and $\frac{2}{4}$ both describe the same number.

Usually, we find it more useful to use fractions that are simplified. A fraction is considered to be in simplest form when the numerator (top) and the denominator (bottom) can both be divided exactly by no number other than 1.

$\frac{1}{2}$ is in simplest form.

$\frac{2}{4}$ is *not* in simplest form since the numerator and the denominator can both be divided by 2.

If you get the answer $\frac{2}{4}$ to a problem, you should state it in simplest form, $\frac{1}{2}$. The process of changing $\frac{2}{4}$ to $\frac{1}{2}$ is called simplifying the fraction.

2 *Simplifying Fractions*

Natural numbers or **counting numbers** are the set of whole numbers excluding 0. Thus the natural numbers are the numbers 1, 2, 3, 4, 5, 6, ...

When two or more numbers are multiplied, each number that is multiplied is called a **factor.** For example, when we write $3 \times 7 \times 5$, each of the numbers 3, 7, and 5 is called a factor.

Prime numbers are all natural numbers greater than 1 whose only natural number factors are 1 and itself. The number 5 is prime. The only natural number factors of 5 are 5 and 1.

$$5 = 5 \times 1$$

The number 6 is not prime. The natural number factors of 6 are 3 and 2 or 6 and 1.

$$6 = 3 \times 2 \qquad 6 = 6 \times 1$$

The first 15 prime numbers are

2, 3, 5, 7, 11, 13, 17, 19, 23, 29, 31, 37, 41, 43, 47

Any natural number greater than 1 is either prime or can be written as the product of prime numbers. For example we can take each of the numbers 12, 30, 14, 19, and 29 and either indicate that they are prime or, if they are not prime, write them as the product of prime numbers. We write as follows:

$12 = 2 \times 2 \times 3$ $\quad 30 = 2 \times 3 \times 5$ $\quad$ 29 is a prime number.

$14 = 2 \times 7$ $\quad$ 19 is a prime number.

To reduce a fraction, we use prime numbers to factor numerator and denominator. Write each part of the fraction (numerator and denominator) as a product of prime numbers. Note any *factor* that appears in both the *numerator* (top) and *denominator* (bottom) of the fraction. If we divide numerator and denominator by that value, we will obtain an equivalent fraction in *simplest form*. When the new fraction is simplified, it is said to be in *lowest terms*. Throughout this text, when we say *simplify* a fraction, we always mean to lowest terms.

EXAMPLE 1 Simplify each fraction. **(a)** $\frac{14}{21}$ **(b)** $\frac{15}{35}$ **(c)** $\frac{20}{70}$

(a) $\frac{14}{21} = \frac{\cancel{7} \times 2}{\cancel{7} \times 3} = \frac{2}{3}$ *We factor* 14 *and factor* 21. *Then we divide numerator and denominator by* 7.

(b) $\frac{15}{35} = \frac{\cancel{5} \times 3}{\cancel{5} \times 7} = \frac{3}{7}$ *We factor* 15 *and factor* 35. *Then we divide numerator and denominator by* 5.

(c) $\frac{20}{70} = \frac{2 \times \cancel{2} \times \cancel{5}}{7 \times \cancel{2} \times \cancel{5}} = \frac{2}{7}$ *We factor* 20 *and factor* 70. *Then we divide numerator and denominator by both* 2 *and* 5.

Practice Problem 1 Simplify.

(a) $\frac{10}{16}$ **(b)** $\frac{24}{36}$ **(c)** $\frac{42}{36}$ ■

Sometimes when we simplify a fraction, all the prime factors in the top (numerator) are divided out. When this happens, we put a 1 in the numerator. If we did not put the 1 in the numerator, we would not realize that the answer was a fraction.

EXAMPLE 2 Simplify each fraction. **(a)** $\frac{7}{21}$ **(b)** $\frac{15}{105}$

(a) $\frac{7}{21} = \frac{\cancel{7} \times 1}{\cancel{7} \times 3} = \frac{1}{3}$ **(b)** $\frac{15}{105} = \frac{\cancel{5} \times \cancel{3} \times 1}{7 \times \cancel{5} \times \cancel{3}} = \frac{1}{7}$

Practice Problem 2 Simplify.

(a) $\frac{4}{12}$ **(b)** $\frac{25}{125}$ **(c)** $\frac{73}{146}$ ■

If all the prime numbers in the bottom (denominator) are divided out, we do not need to leave a 1 in the denominator, since we do not need to express the answer as a fraction. The answer is then a whole number and is not usually expressed as a fraction.

EXAMPLE 3 Simplify each fraction. **(a)** $\frac{35}{7}$ **(b)** $\frac{70}{10}$

(a) $\frac{35}{7} = \frac{5 \times \cancel{7}}{\cancel{7}} = 5$ **(b)** $\frac{70}{10} = \frac{7 \times \cancel{5} \times \cancel{2}}{\cancel{5} \times \cancel{2}} = 7$

Practice Problem 3 Simplify.

(a) $\frac{18}{6}$ **(b)** $\frac{146}{73}$ **(c)** $\frac{28}{7}$ ■

Sometimes the fraction we use represents how many of a certain thing are successful. For example, if a major league baseball player was at bat 30 times and achieved 12 hits, we could say that he had a hit $\frac{12}{30}$ of the time. If we reduce the fraction, we could say he had a hit $\frac{2}{5}$ of the time.

EXAMPLE 4 Cindy got 48 out of 56 questions correct on a test. Write this as a fraction.

Express as a fraction in simplest form the number of correct responses out of the total number of questions on the test.

$$48 \text{ out of } 56 \longrightarrow \frac{48}{56} = \frac{6 \times \cancel{8}}{7 \times \cancel{8}} = \frac{6}{7}$$

Cindy answered the questions correctly $\frac{6}{7}$ of the time.

Practice Problem 4

The major league pennant winner in 1917 won 56 games out of 154 games played. Express as a fraction in simplest form the number of games won in relation to the number of games played. ■

The number *one* can be expressed as 1, $\frac{1}{1}$, $\frac{2}{2}$, $\frac{6}{6}$, $\frac{8}{8}$, and so on, since

$$1 = \frac{1}{1} = \frac{2}{2} = \frac{6}{6} = \frac{8}{8}$$

We say that these numerals are *equivalent ways* of writing the number *one* because they all express the same quantity even though they appear to be different.

SIDELIGHT When we simplify fractions, we are actually using the fact that we can multiply any number by 1 without changing the value of that number. (Mathematicians call the number 1 the multiplicative identity because it leaves any number it multiplies with the same identical value as before.)

Let's look again at one of the previous problems.

$$\frac{14}{21} = \frac{7 \times 2}{7 \times 3} = \frac{7}{7} \times \frac{2}{3} = 1 \times \frac{2}{3} = \frac{2}{3}$$

So we see that

$$\frac{14}{21} = \frac{2}{3}$$

When we simplify fractions, we are using this property of multiplying by 1.

3 *Improper Fractions and Mixed Numbers*

If the numerator is less than the denominator the fraction is a **proper fraction.** A proper fraction is used to describe a quantity smaller than a whole.

Fractions can also be used to describe quantities larger than a whole. The figure below shows two bars that are equal in size. Each bar is divided into 5 equal pieces. The first bar is shaded in completely. The second bar has 2 of the 5 pieces shaded in.

The shaded-in region can be represented by $\frac{7}{5}$ since 7 of the pieces (each of which is $\frac{1}{5}$ of a whole box) are shaded. The fraction $\frac{7}{5}$ is called an improper fraction. An **improper fraction** is one in which the numerator is larger than or equal to the denominator.

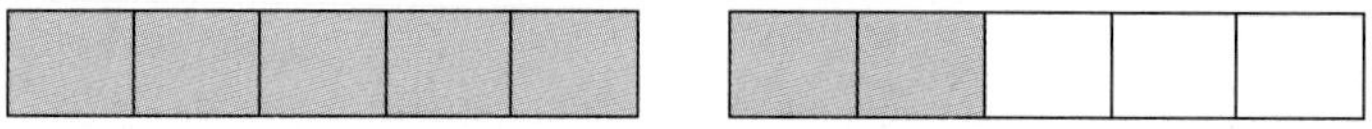

The shaded-in region can also be represented by 1 whole added to $\frac{2}{5}$ of a whole, or $1 + \frac{2}{5}$. This is written as $1\frac{2}{5}$. The fraction $1\frac{2}{5}$ is called a mixed number. A **mixed number** consists of a whole number added to a proper fraction (numerator is smaller than the denominator). The addition is understood but not written. When we write $1\frac{2}{5}$, it represents $1 + \frac{2}{5}$. The numbers $1\frac{7}{8}$, $2\frac{3}{4}$, $8\frac{1}{3}$, and $126\frac{1}{10}$ are all mixed numbers. From the figure above it seems clear that $\frac{7}{5} = 1\frac{2}{5}$. This suggests that we can change from one form to the other without changing the value of the fraction.

From a picture it is easy to see how to *change improper fractions to mixed numbers.* For example, if we start with the fraction $\frac{11}{3}$ and represent it by the figure below (where 11 of the pieces that are $\frac{1}{3}$ of a box are shaded), we see that $\frac{11}{3} = 3\frac{2}{3}$, since 3 whole boxes and $\frac{2}{3}$ of a box are shaded.

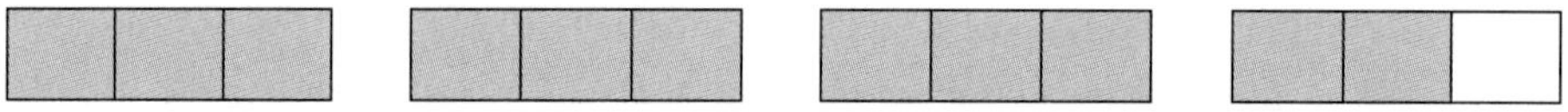

Changing Improper Fractions to Mixed Numbers

You can do the same procedure without a picture. For example, to change $\frac{11}{3}$ to a mixed number, we can do the following:

$$\frac{11}{3} = \frac{3}{3} + \frac{3}{3} + \frac{3}{3} + \frac{2}{3}$$ *By the rule for adding fractions (which is discussed in detail in Section 0.2).*

$$= 1 + 1 + 1 + \frac{2}{3}$$ *Write 1 in place of $\frac{3}{3}$, since $\frac{3}{3} = 1$.*

$$= 3 + \frac{2}{3}$$ *Write 3 in place of $1 + 1 + 1$.*

$$= 3\frac{2}{3}$$ *Use the notation for mixed numbers.*

Now that you know how to perform the change and why it works, here is a shorter method.

> **To Change an Improper Fraction to a Mixed Number**
>
> **1.** Divide the denominator into the numerator.
>
> **2.** The result is the whole-number part of the mixed number.
>
> **3.** The remainder from the division will be the numerator of the fraction. The denominator of the fraction remains unchanged.

We can write the fraction as a division example and divide. The arrows show how to write the mixed number.

$$\frac{7}{5} = 5\overline{)7}\quad \begin{array}{r} 1 \\ 5\overline{)7} \\ \underline{5} \\ 2 \end{array}$$

1 Whole-number part; 2 Remainder → $1\frac{2}{5}$ (2 ← Numerator of fraction)

Thus, $\frac{7}{5} = 1\frac{2}{5}$.

$$\frac{11}{3} = \begin{array}{r} 3 \\ 3\overline{)11} \\ \underline{9} \\ 2 \end{array}$$

3 Whole-number part ⟶ $3\frac{2}{3}$; 2 Remainder → Numerator of fraction

Thus, $\frac{11}{3} = 3\frac{2}{3}$.

Sometimes the remainder is 0. The improper fraction changes to a whole number.

EXAMPLE 5 Change to a mixed number or to a whole number.

(a) $\frac{7}{4}$ **(b)** $\frac{15}{3}$

(a) $\frac{7}{4} = 7 \div 4 = \begin{array}{r} 1 \\ 4\overline{)7} \\ \underline{4} \\ 3 \end{array}$ Remainder

Thus $\frac{7}{4} = 1\frac{3}{4}$.

(b) $\frac{15}{3} = 15 \div 3 = \begin{array}{r} 5 \\ 3\overline{)15} \\ \underline{15} \\ 0 \end{array}$ Remainder

Thus $\frac{15}{3} = 5$.

Practice Problem 5 Change to a mixed number or to a whole number.

(a) $\frac{12}{7}$ **(b)** $\frac{20}{5}$ ■

Changing Mixed Numbers to Improper Fractions

It is not difficult to see how to change mixed numbers to improper fractions. Suppose that you wanted to write $2\frac{2}{3}$ as an improper fraction:

$$2\frac{2}{3} = 2 + \frac{2}{3} \quad \textit{The meaning of mixed number notation.}$$

$$= 1 + 1 + \frac{2}{3} \quad \textit{Since } 1 + 1 = 2.$$

$$= \frac{3}{3} + \frac{3}{3} + \frac{2}{3} \quad \textit{Since } 1 = \tfrac{3}{3}.$$

When we draw a picture of $\frac{3}{3} + \frac{3}{3} + \frac{2}{3}$, we have this figure:

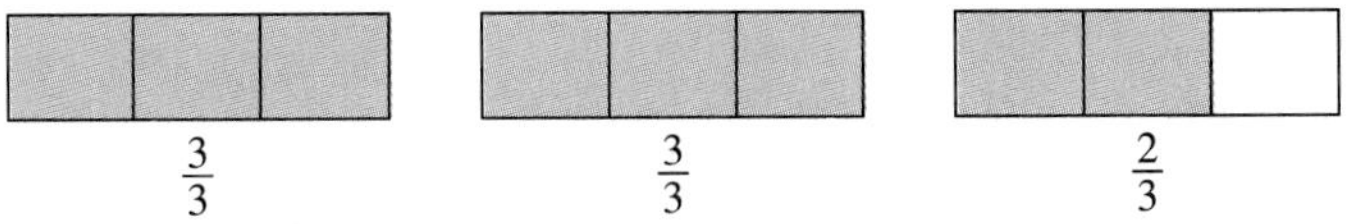

If we count the shaded parts, we see that

$$\frac{3}{3} + \frac{3}{3} + \frac{2}{3} = \frac{8}{3} \quad \text{Thus } 2\frac{2}{3} = \frac{8}{3}$$

Now that you can see how this change can be done, here is a shorter method.

> **To Change a Mixed Number to an Improper Fraction**
>
> **1.** Multiply the whole number by the denominator.
>
> **2.** Add this to the numerator. The result is the new numerator. The denominator does not change.

EXAMPLE 6 Change to an improper fraction. **(a)** $3\frac{1}{7}$ **(b)** $5\frac{4}{5}$

(a) $3\frac{1}{7} = \frac{(3 \times 7) + 1}{7} = \frac{21 + 1}{7} = \frac{22}{7}$ **(b)** $5\frac{4}{5} = \frac{(5 \times 5) + 4}{5} = \frac{25 + 4}{5} = \frac{29}{5}$

Practice Problem 6 Change to an improper fraction.

(a) $3\frac{2}{5}$ **(b)** $1\frac{3}{7}$ **(c)** $2\frac{6}{11}$ **(d)** $4\frac{2}{3}$ ■

4 *Changing a Fraction to an Equivalent Fraction with a Different Denominator*

Fractions can be changed to an equivalent fraction with a different denominator by multiplying both numerator and denominator by the same number.

$$\frac{5}{6} = \frac{5 \times 2}{6 \times 2} = \frac{10}{12} \qquad \frac{3}{7} = \frac{3 \times 3}{7 \times 3} = \frac{9}{21}$$

So $\frac{5}{6}$ is equivalent to $\frac{10}{12}$. $\frac{3}{7}$ is equivalent to $\frac{9}{21}$.

We often want to get a *particular denominator.*

EXAMPLE 7 Find the missing number.

(a) $\frac{3}{5} = \frac{?}{25}$ **(b)** $\frac{3}{7} = \frac{?}{21}$ **(c)** $\frac{2}{9} = \frac{?}{36}$

(a) $\frac{3}{5} = \frac{?}{25}$ *Observe that we need to multiply the denominator by 5 to obtain 25. So we multiply the numerator 3 by 5 also.*

$\frac{3 \times 5}{5 \times 5} = \frac{15}{25}$ *The desired numerator is 15.*

(b) $\frac{3}{7} = \frac{?}{21}$ *Observe that 7 × 3 = 21. We need to multiply by 3 to get the new numerator.*

$\frac{3 \times 3}{7 \times 3} = \frac{9}{21}$ *The desired numerator is 9.*

(c) $\frac{2}{9} = \frac{?}{36}$ *Observe that 9 × 4 = 36. We need to multiply by 4 to get the new numerator.*

$\frac{2 \times 4}{9 \times 4} = \frac{8}{36}$ *The desired numerator is 8.*

Practice Problem 7 Find the missing number.

(a) $\frac{3}{8} = \frac{?}{24}$ **(b)** $\frac{5}{6} = \frac{?}{30}$ **(c)** $\frac{12}{13} = \frac{?}{26}$ **(d)** $\frac{2}{7} = \frac{?}{56}$

(e) $\frac{5}{9} = \frac{?}{27}$ **(f)** $\frac{3}{10} = \frac{?}{60}$ **(g)** $\frac{3}{4} = \frac{?}{28}$ **(h)** $\frac{8}{11} = \frac{?}{55}$ ■

0.1 Exercises

Verbal and Writing Skills

1. In the fraction $\frac{12}{13}$, what number is the numerator?

2. In the fraction $\frac{17}{20}$, what number is the denominator?

3. What is a factor? Give an example.

4. Give some examples of the number 1 written as a fraction.

5. Draw a diagram to illustrate $2\frac{3}{5}$.

Simplify each fraction.

6. $\dfrac{18}{24}$ **7.** $\dfrac{35}{49}$ **8.** $\dfrac{16}{48}$ **9.** $\dfrac{12}{60}$ **10.** $\dfrac{32}{8}$ **11.** $\dfrac{75}{15}$

Change to a mixed number.

12. $\dfrac{13}{8}$ **13.** $\dfrac{39}{7}$ **14.** $\dfrac{21}{9}$ **15.** $\dfrac{125}{4}$ **16.** $\dfrac{46}{5}$ **17.** $\dfrac{89}{6}$

Change to an improper fraction.

18. $3\dfrac{1}{5}$ **19.** $2\dfrac{6}{7}$ **20.** $5\dfrac{2}{3}$ **21.** $4\dfrac{3}{4}$ **22.** $8\dfrac{8}{9}$ **23.** $13\dfrac{1}{6}$

Find the missing numerator.

24. $\dfrac{3}{11} = \dfrac{?}{44}$ **25.** $\dfrac{5}{7} = \dfrac{?}{28}$ **26.** $\dfrac{3}{5} = \dfrac{?}{35}$ **27.** $\dfrac{2}{7} = \dfrac{?}{21}$ **28.** $\dfrac{9}{11} = \dfrac{?}{55}$ **29.** $\dfrac{13}{17} = \dfrac{?}{51}$

Applications

Solve each word problem.

30. A baseball player obtained a hit 342 times out of 620 times at bat. Express as a fraction in simplified form how much of the time he obtained a hit.

31. A developer purchased 105 acres of land. Only 75 acres of land are buildable. Write a fraction in simplified form that indicates the fractional part of the land that is buildable.

32. Ted and Carol last year had a combined income of \$32,000. They paid \$6600 in federal income taxes. Write a fraction in simplified form that indicates what fractional part of their income is spent on federal income tax.

★ **33.** Juan got 7 out of 8 problems correct on a math quiz. The professor stated that there would be 136 problems all semester on tests and quizzes. How many of the 136 problems will Juan need to work correctly on tests and quizzes to maintain the same fractional ratio of correct problems?

The following chart gives the recipe for a trail mix.

34. What part of the Premium Blend is nuts?

35. What part of the High Energy Blend is raisins?

36. What part of the Premium Blend is not sunflower seeds?

THE ROCKING B TRAIL MIX			
	Nuts	Raisins	Sunflower Seeds
Premium Blend	1	1	2
High Energy Blend	1	3	1

The following chart provides some statistics for the Hometown Team.

37. Determine how often each player hit the ball based on the number of times at bat. Write each answer as a reduced fraction.

38. Which player was the best hitter?

BASEBALL STATS FOR HOMETOWN TEAM		
	Hits	Times at Bat
Aaron Dunsay	15	25
Paul Banks	12	30
Tom Re	28	35

Developing Your Study Skills

WHY IS HOMEWORK NECESSARY?

Mathematics is a set of skills that you learn by doing, not by watching someone else do it. Your instructor may make solving a mathematics problem look very easy, but for you to learn the necessary skills, you must practice them over and over again, just as your instructor once had to do. There is no other way. Learning mathematics is like learning to play a musical instrument, to type, or to play a sport. No matter how much you watch someone else do it, how many books you may read on "how to" do it, or how easy it may seem to be, the key to success is practice on a regular basis.

Homework provides this practice. The amount of practice varies for each individual, but usually students need to do most or all of the exercises provided at the end of each section in the text. The more problems you do, the better you get. Some problems in a set are more difficult than others, and some stress different concepts. Only by working all the problems will you cover the full range of difficulty and coverage.

0.2 Addition and Subtraction of Fractions

After studying this section, you will be able to:

1. *Add or subtract fractions with a common denominator.*
2. *Find the least common denominator of any two or more fractions.*
3. *Add or subtract fractions that do not have a common denominator.*
4. *Add or subtract mixed numbers.*

1 *Adding or Subtracting Fractions with a Common Denominator*

If fractions have the same denominator, the numerators may be added or subtracted. The denominator remains the same.

To Add or Subtract Two Fractions with the Same Denominator

1. Add or subtract the numerators.
2. Keep the same denominator.
3. Simplify the answer whenever possible.

EXAMPLE 1 Add the fractions. Simplify your answer whenever possible.

(a) $\frac{5}{7}+\frac{1}{7}$ **(b)** $\frac{2}{3}+\frac{1}{3}$ **(c)** $\frac{1}{8}+\frac{3}{8}+\frac{2}{8}$ **(d)** $\frac{3}{5}+\frac{4}{5}$

(a) $\frac{5}{7}+\frac{1}{7}=\frac{5+1}{7}=\frac{6}{7}$

(b) $\frac{2}{3}+\frac{1}{3}=\frac{2+1}{3}=\frac{3}{3}=1$

(c) $\frac{1}{8}+\frac{3}{8}+\frac{2}{8}=\frac{1+3+2}{8}=\frac{6}{8}=\frac{3}{4}$

(d) $\frac{3}{5}+\frac{4}{5}=\frac{3+4}{5}=\frac{7}{5}=1\frac{2}{5}$

Practice Problem 1 Add.

(a) $\frac{3}{6}+\frac{2}{6}$ **(b)** $\frac{3}{11}+\frac{2}{11}+\frac{6}{11}$ **(c)** $\frac{5}{8}+\frac{2}{8}+\frac{7}{8}$ ■

EXAMPLE 2 Subtract the fractions. Simplify your answer whenever possible.

(a) $\frac{9}{11}-\frac{2}{11}$ **(b)** $\frac{5}{6}-\frac{1}{6}$

(a) $\frac{9}{11}-\frac{2}{11}=\frac{9-2}{11}=\frac{7}{11}$ **(b)** $\frac{5}{6}-\frac{1}{6}=\frac{5-1}{6}=\frac{4}{6}=\frac{2}{3}$

Practice Problem 2 Subtract. **(a)** $\frac{11}{13}-\frac{6}{13}$ **(b)** $\frac{8}{9}-\frac{2}{9}$ ■

Our ability to add and subtract fractions with a common denominator allows us to solve simple problems. Most problems, however, involve fractions that do not have a common denominator. Fractions and mixed numbers such as halves, fourths, and eighths are commonly used. To add or subtract such fractions, we begin by finding a common denominator.

2 *Finding the Least Common Denominator*

Before you can add or subtract, fractions must have the same denominator. To save work, we select the smallest possible common denominator. This is called the **least common denominator** or LCD (also known as the *lowest common denominator*).

> The LCD of two or more fractions is the smallest whole number that is exactly divisible by each denominator of the fractions.

EXAMPLE 3 Find the LCD. $\frac{2}{3}$ and $\frac{1}{4}$

The numbers are small enough to find the LCD by inspection. The LCD is 12, since 12 is exactly divisible by 4 and by 3. There is no smaller number that is exactly divisible by 4 and 3.

Practice Problem 3 Find the LCD. $\frac{1}{8}$ and $\frac{5}{12}$ ■

In some cases, the LCD cannot easily be determined by inspection. If we write each denominator as the product of prime factors, we will be able to find the LCD. We will use (·) to indicate multiplication. For example, $30 = 2 \cdot 3 \cdot 5$. This means $30 = 2 \times 3 \times 5$.

> **Procedure to Find the LCD Using Prime Factors**
>
> 1. Write each denominator as the product of prime factors.
> 2. The LCD is the product of each different factor.
> 3. If a factor occurs more than once in any one denominator, the LCD will contain that factor repeated the greatest number of times that it occurs in any one denominator.

EXAMPLE 4 Find the LCD of $\frac{5}{6}$ and $\frac{1}{15}$ by this new procedure.

$6 = 2 \cdot 3$ *Write each denominator as the product of prime factors.*

$15 = \quad 3 \cdot 5$

$\downarrow \quad \downarrow \quad \downarrow$

$\text{LCD} = 2 \cdot 3 \cdot 5$

$\text{LCD} = 2 \cdot 3 \cdot 5 = 30$ *The LCD is the product of each different prime factor. The different factors are 2, 3, and 5.*

Practice Problem 4 Use the prime factors to find the LCD of $\frac{8}{35}$ and $\frac{6}{15}$. ■

Great care should be used to determine the LCD in the case of repeated factors.

EXAMPLE 5 Find the LCD of $\frac{4}{27}$ and $\frac{5}{18}$.

$27 = 3 \cdot 3 \cdot 3$

$18 = \quad 3 \cdot 3 \cdot 2$

$\text{LCD} = 3 \cdot 3 \cdot 3 \cdot 2$

Write each denominator as the product of prime factors. We observe that the factor 3 *occurs three times in the factorization of* 27.

$\text{LCD} = 3 \cdot 3 \cdot 3 \cdot 2 = 54$

The LCD is the product of each different factor. The factor 3 *occurred most* in the factorization of 27, where it occurred *three* times. Thus the LCD will be the product of *three* 3's and *one* 2.

Practice Problem 5 Find the LCD of $\frac{5}{12}$ and $\frac{7}{30}$. ■

EXAMPLE 6 Find the LCD of $\frac{5}{12}$, $\frac{1}{15}$, and $\frac{7}{30}$.

$12 = 2 \cdot 2 \cdot 3$

$15 = \quad 3 \cdot 5$

$30 = \quad 2 \cdot 3 \cdot 5$

$\text{LCD} = 2 \cdot 2 \cdot 3 \cdot 5$

Write each denominator as the product of prime factors. Notice that the repeated factor is 2, *which* occurs twice *in the factorization of* 12.

$\text{LCD} = 2 \cdot 2 \cdot 3 \cdot 5 = 60$

The LCD is the product of each different factor with the factor 2 repeated since it occurred twice in one denominator.

Practice Problem 6 Find the LCD of $\frac{2}{27}$, $\frac{1}{18}$, and $\frac{5}{12}$. ■

3 *Adding or Subtracting Fractions That Do Not Have a Common Denominator*

Before you can add or subtract them, fractions must have the same denominator. Using the LCD will make your work easier. First you must find the LCD. Then change each fraction to a fraction that has the LCD as the denominator. Sometimes one of the fractions will already have the LCD as the denominator. Once all the fractions have the same denominator, you can add or subtract. Be sure to simplify the fraction in your answer if this is possible.

To Add or Subtract Fractions That Do Not Have a Common Denominator

1. Find the LCD of the fractions.
2. Change each fraction to an equivalent fraction with the LCD for a denominator.
3. Add or subtract the fractions.
4. Simplify the answer whenever possible.

Let us return to the two fractions of Example 3. We have previously found that the LCD is 12.

EXAMPLE 7 Add. $\frac{2}{3} + \frac{1}{4}$

We must change $\frac{2}{3}$ and $\frac{1}{4}$ to fractions with the same denominator. We change each fraction to an equivalent fraction with a common denominator of 12, the LCD.

$$\frac{2}{3} = \frac{?}{12} \qquad \frac{2}{3} \times \frac{4}{4} = \frac{8}{12} \quad \text{so} \quad \frac{2}{3} = \frac{8}{12}$$

$$\frac{1}{4} = \frac{?}{12} \qquad \frac{1}{4} \times \frac{3}{3} = \frac{3}{12} \quad \text{so} \quad \frac{1}{4} = \frac{3}{12}$$

Then we rewrite the problem with common denominators and then add.

$$\frac{2}{3} + \frac{1}{4} = \frac{8}{12} + \frac{3}{12} = \frac{8+3}{12} = \frac{11}{12}$$

Practice Problem 7 Add. $\frac{1}{8} + \frac{5}{12}$ ■

Sometimes one of the denominators is the LCD. In such cases the fraction that has the LCD for the denominator will not need to be changed. If every other denominator divides into the largest denominator, the largest denominator is the LCD.

EXAMPLE 8 Find the LCD and then add. $\frac{3}{5} + \frac{7}{20} + \frac{1}{2}$

We can see by inspection that both 5 and 2 divide exactly into 20. Thus 20 is the LCD. Now add:

$$\frac{3}{5} + \frac{7}{20} + \frac{1}{2}$$

We change $\frac{3}{5}$ and $\frac{1}{2}$ to equivalent fractions with a common denominator of 20, the LCD.

$$\frac{3}{5} = \frac{?}{20} \qquad \frac{3}{5} \times \frac{4}{4} = \frac{12}{20} \quad \text{so} \quad \frac{3}{5} = \frac{12}{20}$$

$$\frac{1}{2} = \frac{?}{20} \qquad \frac{1}{2} \times \frac{10}{10} = \frac{10}{20} \quad \text{so} \quad \frac{1}{2} = \frac{10}{20}$$

Then we rewrite the problem with common denominators and add:

$$\frac{3}{5} + \frac{7}{20} + \frac{1}{2} = \frac{12}{20} + \frac{7}{20} + \frac{10}{20} = \frac{12+7+10}{20} = \frac{29}{20} \quad \text{or} \quad 1\frac{9}{20}$$

Practice Problem 8 Find the LCD and add. $\frac{3}{5} + \frac{4}{25} + \frac{1}{10}$ ■

Now we turn to examples where the selection of the LCD is not so obvious. In Examples 9 through 11 we will use the prime factorization method to find the LCD.

EXAMPLE 9 Add. $\frac{7}{18}+\frac{5}{12}$

First we find the LCD

$$\begin{aligned} 18 &= 3 \cdot 3 \cdot 2 \\ 12 &= \phantom{3 \cdot{}} 3 \cdot 2 \cdot 2 \\ \text{LCD} &= 3 \cdot 3 \cdot 2 \cdot 2 = 9 \cdot 4 = 36 \end{aligned}$$

Now we change $\frac{7}{18}$ and $\frac{5}{12}$ to equivalent fractions that have the LCD.

$$\frac{7}{18}=\frac{?}{36} \qquad \frac{7}{18}\times\frac{2}{2}=\frac{14}{36}$$

$$\frac{5}{12}=\frac{?}{36} \qquad \frac{5}{12}\times\frac{3}{3}=\frac{15}{36}$$

Now we add the fractions.

$$\frac{7}{18}+\frac{5}{12}=\frac{14}{36}+\frac{15}{36}=\frac{29}{36}$$ *This fraction cannot be simplified.*

Practice Problem 9 Add. $\frac{1}{49}+\frac{3}{14}$ ■

EXAMPLE 10 Subtract. $\frac{25}{48}-\frac{5}{36}$

First we find the LCD.

$$\begin{aligned} 48 &= 2 \cdot 2 \cdot 2 \cdot 2 \cdot 3 \\ 36 &= \phantom{2 \cdot 2 \cdot{}} 2 \cdot 2 \cdot 3 \cdot 3 \\ \text{LCD} &= 2 \cdot 2 \cdot 2 \cdot 2 \cdot 3 \cdot 3 = 16 \cdot 9 = 144 \end{aligned}$$

Now we change $\frac{25}{48}$ and $\frac{5}{36}$ to equivalent fractions that have the LCD for a denominator.

$$\frac{25}{48}=\frac{?}{144} \qquad \frac{25}{48}\times\frac{3}{3}=\frac{75}{144}$$

$$\frac{5}{36}=\frac{?}{144} \qquad \frac{5}{36}\times\frac{4}{4}=\frac{20}{144}$$

Now we subtract the fractions.

$$\frac{25}{48}-\frac{5}{36}=\frac{75}{144}-\frac{20}{144}=\frac{55}{144}$$ *This fraction cannot be simplified.*

Practice Problem 10 Subtract. $\frac{1}{12}-\frac{1}{30}$ ■

EXAMPLE 11 Add. $\frac{1}{5} + \frac{1}{6} + \frac{3}{10}$

First we find the LCD.

$$\begin{aligned} 5 &= 5 \\ 6 &= \phantom{5 \cdot{}} 2 \cdot 3 \\ 10 &= 5 \cdot 2 \\ \text{LCD} &= 5 \cdot 2 \cdot 3 = 10 \cdot 3 = 30 \end{aligned}$$

Now we change $\frac{1}{5}$, $\frac{1}{6}$, and $\frac{3}{10}$ to equivalent fractions that have the LCD for a denominator.

$$\frac{1}{5} = \frac{?}{30} \qquad \frac{1}{5} \times \frac{6}{6} = \frac{6}{30}$$

$$\frac{1}{6} = \frac{?}{30} \qquad \frac{1}{6} \times \frac{5}{5} = \frac{5}{30}$$

$$\frac{3}{10} = \frac{?}{30} \qquad \frac{3}{10} \times \frac{3}{3} = \frac{9}{30}$$

Now we add the three fractions.

$$\frac{1}{5} + \frac{1}{6} + \frac{3}{10} = \frac{6}{30} + \frac{5}{30} + \frac{9}{30} = \frac{20}{30} = \frac{2}{3}$$

Note the important step of simplifying the fraction to obtain the final answer.

Practice Problem 11 Combine. $\frac{2}{3} + \frac{3}{4} - \frac{3}{8}$ ■

4 *Adding or Subtracting Mixed Numbers*

If the problem you are adding or subtracting has mixed numbers, change them to improper fractions first and then combine (addition or subtraction). Often the final answer is changed to a mixed number. Here is a good rule to follow:

> If the original problem contains mixed numbers, express the result as a mixed number rather than as an improper fraction.

EXAMPLE 12 Combine. Simplify your answer whenever possible.

(a) $5\frac{1}{2} + 2\frac{1}{3}$ **(b)** $2\frac{1}{5} - 1\frac{3}{4}$ **(c)** $1\frac{5}{12} + \frac{7}{30}$

(a) First we change the mixed numbers to improper fractions.

$$5\frac{1}{2} = \frac{5 \times 2 + 1}{2} = \frac{11}{2} \qquad 2\frac{1}{3} = \frac{2 \times 3 + 1}{3} = \frac{7}{3}$$

Next we change each fraction to an equivalent form with a common denominator of 6.

$$\frac{11}{2} = \frac{?}{6} \qquad \frac{11}{2} \times \frac{3}{3} = \frac{33}{6}$$

$$\frac{7}{3} = \frac{?}{6} \qquad \frac{7}{3} \times \frac{2}{2} = \frac{14}{6}$$

Finally, we add the two fractions and change our answer to a mixed number.

$$\frac{33}{6} + \frac{14}{6} = \frac{47}{6} = 7\frac{5}{6}$$

Thus $5\frac{1}{2} + 2\frac{1}{3} = 7\frac{5}{6}$.

(b) First we change mixed numbers to improper fractions.

$$2\frac{1}{5} = \frac{2 \times 5 + 1}{5} = \frac{11}{5} \qquad 1\frac{3}{4} = \frac{1 \times 4 + 3}{4} = \frac{7}{4}$$

Next we change each fraction to an equivalent form with a common denominator of 20.

$$\frac{11}{5} = \frac{?}{20} \qquad \frac{11}{5} \times \frac{4}{4} = \frac{44}{20}$$

$$\frac{7}{4} = \frac{?}{20} \qquad \frac{7}{4} \times \frac{5}{5} = \frac{35}{20}$$

Now we subtract the two fractions.

$$\frac{44}{20} - \frac{35}{20} = \frac{9}{20}$$

Thus $2\frac{1}{5} - 1\frac{3}{4} = \frac{9}{20}$.

Note: It is not necessary to use these exact steps to add and subtract mixed numbers. If you know another method and can use it to obtain the correct answers, it is all right to continue to use that method throughout this chapter.

(c) Now we add $1\frac{5}{12} + \frac{7}{30}$.

The LCD of 12 and 30 is 60. Why? Change the mixed number to an improper fraction. Then change each fraction to an equivalent form with a common denominator.

$$1\frac{5}{12} = \frac{17}{12} \times \frac{5}{5} = \frac{85}{60} \qquad \frac{7}{30} \times \frac{2}{2} = \frac{14}{60}$$

Then add the fractions, simplify, and write the answer as a mixed number.

$$\frac{85}{60} + \frac{14}{60} = \frac{99}{60} = \frac{33}{20} = 1\frac{13}{20}$$

Thus $1\frac{5}{12} + \frac{7}{30} = 1\frac{13}{20}$.

Practice Problem 12 Combine. **(a)** $1\frac{2}{3} + 2\frac{4}{5}$ **(b)** $5\frac{1}{4} - 2\frac{2}{3}$ ■

EXAMPLE 13 Manuel is enclosing a triangular-shaped exercise yard for his new dog. He wants to determine how much fencing he will need. The sides of the yard measure $20\frac{3}{4}$ feet, $15\frac{1}{2}$ feet, and $18\frac{1}{8}$ feet. What is the perimeter of (total distance around) the triangle?

Understand the problem. Begin by drawing a picture.

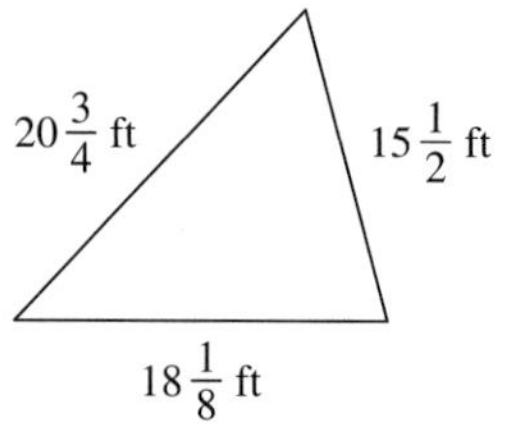

We want to add up the lengths of all three sides of the triangle. This distance around the triangle is called the **perimeter.**

$$20\frac{3}{4} + 15\frac{1}{2} + 18\frac{1}{8} = \frac{83}{4} + \frac{31}{2} + \frac{145}{8}$$

$$= \frac{166}{8} + \frac{124}{8} + \frac{145}{8} = \frac{435}{8} = 54\frac{3}{8} \text{ feet}$$

Practice Problem 13

Find the perimeter of a rectangle with sides of $4\frac{1}{5}$ cm and $6\frac{1}{2}$ cm. Begin by drawing a picture. Label the picture by including the measure of *each* side. ■

Developing Your Study Skills

CLASS ATTENDANCE

A student of mathematics needs to get started in the right direction by choosing to attend class every day, beginning with the first day of class. Statistics show that class attendance and good grades go together. Classroom activities are designed to enhance learning and, therefore, you must be in class to benefit from them. Vital information and explanations are given each day that can help you in understanding concepts. Do not be deceived into thinking that you can just find out from a friend what went on in class. There is no good substitute for firsthand experience. Give yourself a push in the right direction by developing the habit of going to class every day.

CLASS PARTICIPATION

People learn mathematics through active participation, not through observation from the sidelines. If you want to do well in this course, be involved in classroom activities. Sit near the front where you can see and hear well, where your focus is on the instruction process and not on those students around you. Ask questions, be ready to contribute toward solutions, and take part in all classroom activities. Your contributions are valuable to the class and to yourself. Class participation requires an investment of yourself in the learning process, which you will find pays huge dividends.

0.2 Exercises

Verbal and Writing Skills

1. Explain why the denominator 8 is the least common denominator of $\frac{3}{4}$ and $\frac{5}{8}$.

2. What must you do before you add or subtract fractions that do not have a common denominator?

Find the LCD (least common denominator) of each of the following pairs of fractions. Do not combine the fractions; only find the LCD.

3. $\frac{1}{12}$ and $\frac{5}{9}$ **4.** $\frac{11}{12}$ and $\frac{5}{18}$ **5.** $\frac{7}{10}$ and $\frac{1}{4}$ **6.** $\frac{3}{16}$ and $\frac{1}{24}$

7. $\frac{5}{63}$ and $\frac{5}{21}$ **8.** $\frac{7}{36}$ and $\frac{5}{54}$ **9.** $\frac{1}{18}$ and $\frac{13}{30}$ **10.** $\frac{3}{14}$ and $\frac{11}{16}$

Combine. Be sure to simplify your answer whenever possible.

11. $\frac{3}{11}+\frac{2}{11}+\frac{4}{11}$ **12.** $\frac{9}{13}+\frac{2}{13}+\frac{1}{13}$ **13.** $\frac{8}{17}-\frac{5}{17}$ **14.** $\frac{16}{19}-\frac{13}{19}$

15. $\frac{5}{6}+\frac{8}{15}$ **16.** $\frac{1}{12}+\frac{2}{9}$ **17.** $\frac{5}{14}-\frac{1}{4}$ **18.** $\frac{9}{20}-\frac{1}{15}$ **19.** $\frac{2}{9}+\frac{5}{6}$

20. $\frac{3}{4}+\frac{3}{10}$ **21.** $\frac{4}{5}-\frac{3}{8}$ **22.** $\frac{3}{5}-\frac{1}{3}$ **23.** $\frac{9}{8}+\frac{7}{12}$ **24.** $\frac{5}{6}+\frac{7}{15}$

25. $\frac{7}{18}+\frac{1}{12}$ **26.** $\frac{4}{7}+\frac{7}{9}$ **27.** $\frac{2}{3}+\frac{7}{12}$ **28.** $\frac{3}{35}+\frac{4}{7}$ **29.** $\frac{1}{15}+\frac{7}{12}$

30. $\frac{15}{16}-\frac{2}{3}$ **31.** $\frac{23}{25}-\frac{13}{15}$ **32.** $3\frac{1}{5}+2\frac{2}{3}$ **33.** $5\frac{1}{6}+3\frac{1}{4}$ **34.** $1\frac{5}{24}+\frac{5}{18}$

35. $6\frac{2}{3}+\frac{3}{4}$ **36.** $7\frac{1}{6}-2\frac{1}{4}$ **37.** $6\frac{3}{8}-2\frac{3}{4}$ **38.** $9\frac{4}{5}-3\frac{1}{2}$ **39.** $4\frac{7}{12}-1\frac{5}{6}$

40. $2\frac{1}{8}+3\frac{2}{3}$ **41.** $3\frac{1}{7}+4\frac{1}{3}$ **42.** $12\frac{1}{8}-9\frac{3}{8}$ **43.** $14\frac{1}{9}-12\frac{5}{9}$ **44.** $2\frac{1}{8}+6\frac{3}{4}$

Applications

Solve each word problem.

45. Saul jogged $3\frac{1}{8}$ miles on Monday, $2\frac{2}{3}$ miles on Tuesday, and $4\frac{1}{2}$ miles on Wednesday. What was his total jogging distance for the three days?

46. From a fish tank holding $9\frac{7}{10}$ gallons of water, Won Lin drained $3\frac{4}{5}$ gallons. How many gallons of water remained in the tank?

47. Alice has $8\frac{1}{2}$ hours to study this weekend. She estimates that it will take $2\frac{2}{3}$ hours to type a paper and $1\frac{3}{4}$ hours to read a book for literature class. How much study time will she have left for math?

★**48.** Adena is enclosing a triangular-shaped yard for her puppy. The sides of the yard measure $10\frac{1}{2}$ feet, $16\frac{1}{4}$ feet, and $24\frac{7}{8}$ feet. How much fencing will she need? How much more or less fencing does she need than Manuel on page 18?

Carpenters use fractions in their work. The diagram at the right is a diagram of a spice cabinet. The symbol ″ means inches.

49. To determine where to place the cabinet, calculate the height, A, and the width B. Don't forget to include the $\frac{1}{2}$-inch thickness of the wood where needed.

50. Look at the close-up of the drawer. The width is $4\frac{9}{16}$″. In the diagram, the same width is $4\frac{5}{8}$″. What is the difference?
Why do you think the drawer is smaller than the opening?

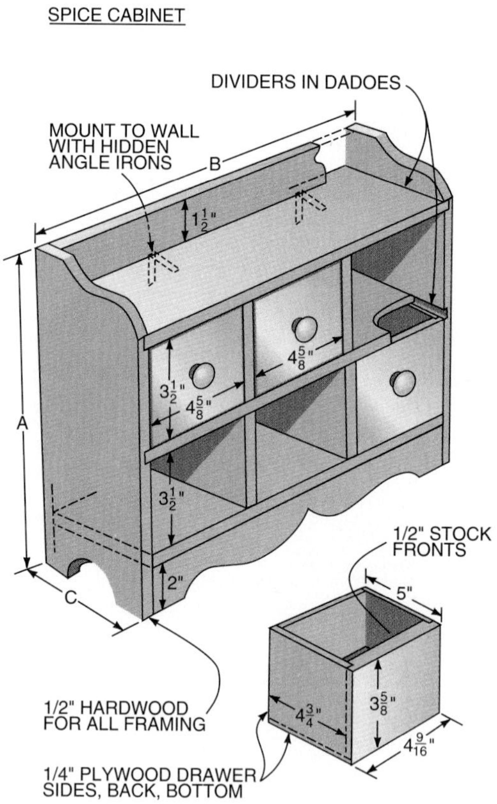

Cumulative Review Problems

51. Simplify. $\dfrac{36}{44}$.

52. Change to an improper fraction. $26\dfrac{3}{5}$.

53. Change to a mixed number. $\dfrac{136}{7}$.

54. Find the missing numerator. $\dfrac{3}{14} = \dfrac{?}{56}$.

Putting Your Skills to Work

CARPENTRY

Carpenters provide the framing for a house upon which the walls are placed. This framing may be done on site or it may come in prefabricated panels of various sizes. The basic panel is 8 feet by 7 feet $4\frac{1}{2}$ inches. The studs are butt-nailed on 16-inch centers. The basic panel can be modified for doors and windows and for corner pieces. A basic panel 4 feet wide serves as a filler so that walls can be made in multiples of 4 feet.

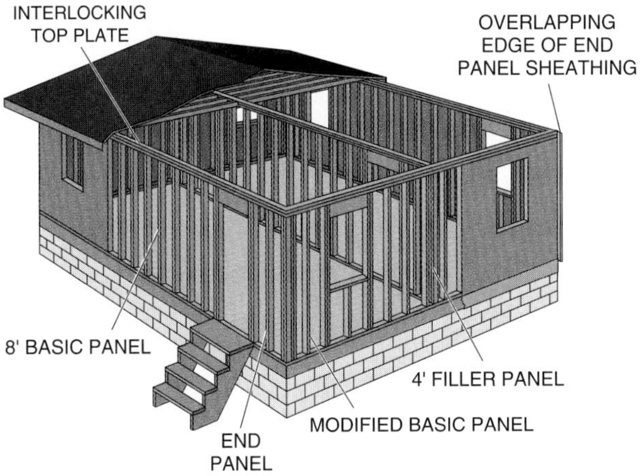

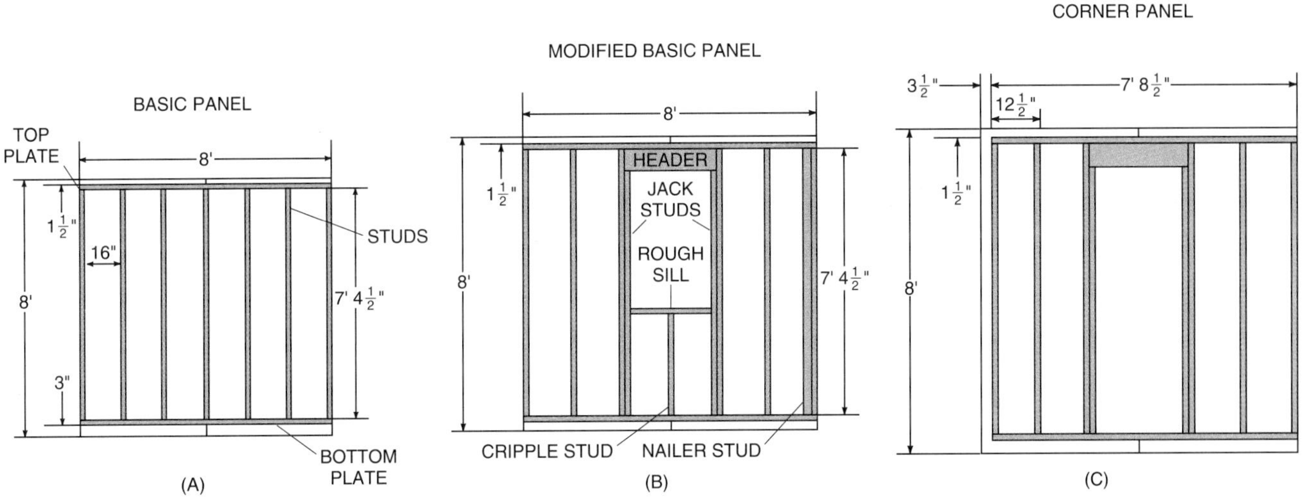

COOPERATIVE GROUP ACTIVITY

1. Look at the three panels above.

(a) Determine the thickness of the 2-by-4 top plate and the bottom plate in the basic panel.

(b) About how wide is the window in the modified basic panel?

(c) About how wide is the door in the corner panel?

(d) Design a room with two windows and one door. Decide how many panels and what kind of panels you will need to frame the room. Provide the information in a chart and in a diagram.

(e) Determine the number of 8-foot-long 2-by-4's you will need to frame your room.

0.3 Multiplication and Division of Fractions

After studying this section, you will be able to:

1 *Multiply fractions, whole numbers, and mixed numbers.*
2 *Divide fractions, whole numbers, and mixed numbers.*

1 *Multiplying Fractions, Whole Numbers, and Mixed Numbers*

Multiplication of Fractions

The multiplication rule for fractions states that to multiply two fractions we multiply the two numerators and multiply the two denominators.

> **To Multiply any Two Fractions**
>
> **1.** Multiply the numerators.
>
> **2.** Multiply the denominators.

Consider these examples.

EXAMPLE 1 Multiply.

(a) $\frac{3}{5} \times \frac{2}{7}$ **(b)** $\frac{1}{3} \times \frac{5}{4}$ **(c)** $\frac{7}{3} \times \frac{1}{5}$ **(d)** $\frac{6}{5} \times \frac{2}{3}$

(a) $\frac{3}{5} \times \frac{2}{7} = \frac{3 \cdot 2}{5 \cdot 7} = \frac{6}{35}$

(b) $\frac{1}{3} \times \frac{5}{4} = \frac{1 \cdot 5}{3 \cdot 4} = \frac{5}{12}$

(c) $\frac{7}{3} \times \frac{1}{5} = \frac{7 \cdot 1}{3 \cdot 5} = \frac{7}{15}$

(d) $\frac{6}{5} \times \frac{2}{3} = \frac{6 \cdot 2}{5 \cdot 3} = \frac{12}{15} = \frac{4}{5}$

Note that we must simplify.

Practice Problem 1 Multiply. **(a)** $\frac{2}{7} \times \frac{5}{11}$ **(b)** $\frac{8}{9} \times \frac{3}{10}$ ■

It is possible to avoid having to simplify a fraction at the last step. In many cases we can divide by a value that appears as a factor in both a numerator and a denominator. Often it is helpful to write a number (as a product of prime factors) in order to do this.

EXAMPLE 2 Multiply.

(a) $\frac{3}{5} \times \frac{5}{7}$ **(b)** $\frac{4}{11} \times \frac{5}{2}$ **(c)** $\frac{15}{8} \times \frac{10}{27}$ **(d)** $\frac{8}{7} \times \frac{5}{12}$

(a) $\frac{3}{5} \times \frac{5}{7} = \frac{3 \cdot 5}{5 \cdot 7} = \frac{3 \cdot \cancel{5}}{7 \cdot \cancel{5}} = \frac{3}{7}$ *Note here we divided numerator and denominator by 5.*

If we factor each number, we can see the common factors.

(b) $\frac{4}{11} \times \frac{5}{2} = \frac{2 \cdot \cancel{2}}{11} \times \frac{5}{\cancel{2}} = \frac{10}{11}$

(c) $\frac{15}{8} \times \frac{10}{27} = \frac{\cancel{3} \cdot 5}{2 \cdot 2 \cdot \cancel{2}} \times \frac{5 \cdot \cancel{2}}{\cancel{3} \cdot 3 \cdot 3} = \frac{25}{36}$

(d) $\frac{8}{7} \times \frac{5}{12} = \frac{2 \cdot \cancel{2} \cdot \cancel{2}}{7} \times \frac{5}{\cancel{2} \cdot \cancel{2} \cdot 3} = \frac{10}{21}$

By dividing out common factors, the multiplication involves smaller numbers and the answers are in simplified form.

Practice Problem 2 Multiply. **(a)** $\frac{3}{5} \times \frac{4}{3}$ **(b)** $\frac{9}{10} \times \frac{5}{12}$ ■

SIDELIGHT Why does this method of dividing out a value that appears as a factor in both numerator and denominator work? Let's reexamine one of the examples we have solved previously.

$$\frac{3}{5} \times \frac{5}{7} = \frac{3}{\cancel{5}} \times \frac{\cancel{5}}{7} = \frac{3}{7}$$

Consider the following steps and reasons.

$\frac{3}{5} \times \frac{5}{7} = \frac{3 \cdot 5}{5 \cdot 7}$ *Definition of multiplication of fractions.*

$= \frac{5 \cdot 3}{5 \cdot 7}$ *Change the order of the factors in the numerator, since $3 \cdot 5 = 5 \cdot 3$. This is called the commutative property of multiplication.*

$= \frac{5}{5} \cdot \frac{3}{7}$ *Definition of multiplication of fractions.*

$= 1 \cdot \frac{3}{7}$ *Write 1 in place of $\frac{5}{5}$ since 1 is another name for $\frac{5}{5}$.*

$= \frac{3}{7}$ *$1 \cdot \frac{3}{7} = \frac{3}{7}$ since any number can be multiplied by 1 without changing the value of the number.*

Think about this concept. It is an important one that we will use again when we discuss rational expressions.

Multiplication of a Fraction by a Whole Number

Whole numbers can be named using fractional notation. 3, $\frac{9}{3}$, $\frac{6}{2}$, and $\frac{3}{1}$ are ways of expressing the number *three*. Therefore,

$$3 = \frac{9}{3} = \frac{6}{2} = \frac{3}{1}$$

When we multiply a fraction by a whole number, we merely express the whole number as a fraction whose denominator is 1 and follow the multiplication rule for fractions.

EXAMPLE 3 Multiply. **(a)** $7 \times \frac{3}{5}$ **(b)** $\frac{3}{16} \times 4$

(a) $7 \times \frac{3}{5} = \frac{7}{1} \times \frac{3}{5} = \frac{21}{5} = 4\frac{1}{5}$

(b) $\frac{3}{16} \times 4 = \frac{3}{16} \times \frac{4}{1} = \frac{3}{4 \cdot \cancel{4}} \times \frac{\cancel{4}}{1} = \frac{3}{4}$

Notice that in (b) we did not use *prime* factors to factor 16. We recognized that $16 = 4 \cdot 4$. This is a more convenient factorization of 16 for this problem. Choose the factorization that works best for each problem. If you cannot decide what is best, factor into primes.

Practice Problem 3 Multiply. **(a)** $4 \times \frac{2}{7}$ **(b)** $12 \times \frac{3}{4}$ ■

Multiplication of Mixed Numbers

When multiplying mixed numbers, we first change them to improper fractions and then follow the multiplication rule for fractions.

EXAMPLE 4 Multiply. **(a)** $3\frac{1}{3} \times 2\frac{1}{2}$ **(b)** $1\frac{2}{5} \times 2\frac{1}{3}$

(a) $3\frac{1}{3} \times 2\frac{1}{2} = \frac{10}{3} \times \frac{5}{2} = \frac{\not{2} \cdot 5}{3} \times \frac{5}{\not{2}} = \frac{25}{3} = 8\frac{1}{3}$

(b) $1\frac{2}{5} \times 2\frac{1}{3} = \frac{7}{5} \times \frac{7}{3} = \frac{49}{15} = 3\frac{4}{15}$

Practice Problem 4 Multiply.

(a) $2\frac{1}{5} \times \frac{3}{7}$ **(b)** $3\frac{1}{3} \times 1\frac{2}{5}$ ■

EXAMPLE 5 Multiply. $2\frac{2}{3} \times \frac{1}{4} \times 6$

$$2\frac{2}{3} \times \frac{1}{4} \times 6 = \frac{8}{3} \times \frac{1}{4} \times \frac{6}{1} = \frac{\not{4} \cdot 2}{\not{3}} \times \frac{1}{\not{4}} \times \frac{2 \cdot \not{3}}{1} = \frac{4}{1} = 4$$

Practice Problem 5 Multiply. $3\frac{1}{2} \times \frac{1}{14} \times 4$ ■

2 *Dividing Fractions, Whole Numbers, and Mixed Numbers*

Division of Fractions

To divide two fractions, we invert the second fraction and then multiply the two fractions.

To Divide Two Fractions

1. Invert the second fraction.
2. Now multiply the two fractions.

EXAMPLE 6 Divide. **(a)** $\frac{1}{3} \div \frac{1}{2}$ **(b)** $\frac{2}{5} \div \frac{3}{10}$ **(c)** $\frac{2}{3} \div \frac{7}{5}$

(a) $\frac{1}{3} \div \frac{1}{2} = \frac{1}{3} \times \frac{2}{1} = \frac{2}{3}$

(b) $\frac{2}{5} \div \frac{3}{10} = \frac{2}{5} \times \frac{10}{3} = \frac{2}{\not{5}} \times \frac{\not{5} \cdot 2}{3} = \frac{4}{3} = 1\frac{1}{3}$

(c) $\frac{2}{3} \div \frac{7}{5} = \frac{2}{3} \times \frac{5}{7} = \frac{10}{21}$

Practice Problem 6 Divide.

(a) $\frac{2}{5} \div \frac{1}{3}$ **(b)** $\frac{12}{13} \div \frac{4}{3}$ ■

Division of a Fraction and a Whole Number

The process of inverting the second fraction and then multiplying the two fractions should be done very carefully when one of the original values is a whole number. Remember, a whole number such as 2 is equivalent to $\frac{2}{1}$.

EXAMPLE 7 Divide. **(a)** $\dfrac{1}{3} \div 2$ **(b)** $5 \div \dfrac{1}{3}$

(a) $\dfrac{1}{3} \div 2 = \dfrac{1}{3} \div \dfrac{2}{1} = \dfrac{1}{3} \times \dfrac{1}{2} = \dfrac{1}{6}$

(b) $5 \div \dfrac{1}{3} = \dfrac{5}{1} \div \dfrac{1}{3} = \dfrac{5}{1} \times \dfrac{3}{1} = \dfrac{15}{1} = 15$

Practice Problem 7 Divide.

(a) $\dfrac{3}{7} \div 6$ **(b)** $8 \div \dfrac{2}{3}$ ■

SIDELIGHT ***Number Sense*** Look at the answers to the problems in Example 7. In part (a), you will notice that $\frac{1}{6}$ is less than the original number $\frac{1}{3}$. Does this seem reasonable? Let's see. If $\frac{1}{3}$ is divided by 2, it means that $\frac{1}{3}$ will be divided into 2 equal parts. We would expect that each part would be less than $\frac{1}{3}$. $\frac{1}{6}$ is a reasonable answer to this division problem.

In part (b), 15 is greater than the original number 5. Does this seem reasonable? Think of what $5 \div \frac{1}{3}$ means. It means that 5 will be divided into thirds. Let's think of an easier problem. What happens when we divide 1 into thirds? We get *three* thirds. We would expect, therefore, that when we divide 5 by thirds, we would get 5×3 or 15 thirds. 15 is a reasonable answer to this division problem.

Complex Fraction

Sometimes division is written in the form of a complex fraction with one fraction in the numerator and one fraction in the denominator. It is best to write this in standard division notation first; then complete the problem using the rule for division.

EXAMPLE 8 Divide. **(a)** $\dfrac{\frac{3}{7}}{\frac{3}{5}}$ **(b)** $\dfrac{\frac{2}{9}}{\frac{5}{7}}$

(a) $\dfrac{\frac{3}{7}}{\frac{3}{5}} = \dfrac{3}{7} \div \dfrac{3}{5} = \dfrac{\cancel{3}}{7} \times \dfrac{5}{\cancel{3}} = \dfrac{5}{7}$

(b) $\dfrac{\frac{2}{9}}{\frac{5}{7}} = \dfrac{2}{9} \div \dfrac{5}{7} = \dfrac{2}{9} \times \dfrac{7}{5} = \dfrac{14}{45}$

Practice Problem 8 Divide.

(a) $\dfrac{\frac{3}{11}}{\frac{5}{7}}$ **(b)** $\dfrac{\frac{12}{5}}{\frac{8}{15}}$ ■

SIDELIGHT Why does the method of "invert and multiply" work? The division rule really depends on the property that any number can be multiplied by 1 without changing the value of the number. Let's look carefully at an example of division of fractions:

$$\frac{2}{5} \div \frac{3}{7} = \frac{\frac{2}{5}}{\frac{3}{7}}$$ *We can write the problem using a complex fraction.*

$$= \frac{\frac{2}{5}}{\frac{3}{7}} \times 1$$ *We can multiply by 1, since any number can be multiplied by 1 without changing the value of the number.*

$$= \frac{\frac{2}{5}}{\frac{3}{7}} \times \frac{\frac{7}{3}}{\frac{7}{3}}$$ *We write 1 in the form $\frac{\frac{7}{3}}{\frac{7}{3}}$, since any nonzero number divided by itself equals 1. We choose this value as a multiplier because it will help simplify the denominator.*

$$= \frac{\frac{2}{5} \times \frac{7}{3}}{\frac{3}{7} \times \frac{7}{3}}$$ *Definition of multiplication of fractions.*

$$= \frac{\frac{2}{5} \times \frac{7}{3}}{1}$$ *The product in the denominator equals 1.*

$$= \frac{2}{5} \times \frac{7}{3}$$

Thus we have shown that $\frac{2}{5} \div \frac{3}{7}$ is equivalent to $\frac{2}{5} \times \frac{7}{3}$ and have shown some justification for the "invert and multiply rule."

Division of Mixed Numbers

This method for division of fractions can be used with mixed numbers. However, we first must change the mixed numbers to improper fractions and then use the rule for dividing fractions.

EXAMPLE 9 Divide. **(a)** $2\frac{1}{3} \div 3\frac{2}{3}$ **(b)** $4\frac{1}{2} \div 1\frac{5}{7}$ **(c)** $\frac{2}{3\frac{1}{2}}$

(a) $2\frac{1}{3} \div 3\frac{2}{3} = \frac{7}{3} \div \frac{11}{3} = \frac{7}{\not{3}} \times \frac{\not{3}}{11} = \frac{7}{11}$

(b) $4\frac{1}{2} \div 1\frac{5}{7} = \frac{9}{2} \div \frac{12}{7} = \frac{9}{2} \times \frac{7}{12} = \frac{3 \cdot \not{3}}{2} \times \frac{7}{4 \cdot \not{3}} = \frac{21}{8} = 2\frac{5}{8}$

(c) $\frac{2}{3\frac{1}{2}} = 2 \div 3\frac{1}{2} = \frac{2}{1} \div \frac{7}{2} = \frac{2}{1} \times \frac{2}{7} = \frac{4}{7}$

Practice Problem 9 Divide.

(a) $1\frac{2}{5} \div 2\frac{1}{3}$ **(b)** $4\frac{2}{3} \div 7$ **(c)** $\frac{1\frac{1}{5}}{1\frac{2}{7}}$ ■

EXAMPLE 10 A chemist has 96 fluid ounces of a solution. She pours the solution into test tubes. Each test tube holds $\frac{3}{4}$ fluid ounce. How many test tubes can she fill?

We need to divide the total number of ounces, 96, by the number of ounces in each test tube, $\frac{3}{4}$.

$$96 \div \frac{3}{4} = \frac{96}{1} \div \frac{3}{4} = \frac{96}{1} \times \frac{4}{3} = \frac{\cancel{3} \cdot 32}{1} \times \frac{4}{\cancel{3}} = \frac{128}{1} = 128$$

She will be able to fill 128 test tubes.

Pause for a moment to think about the answer. Does 128 test tubes filled with solution seem like a reasonable answer? Did you perform the correct operation?

Practice Problem 10

A chemist has 64 fluid ounces of a solution. He wishes to fill several jars each holding $5\frac{1}{3}$ fluid ounces. How many jars can he fill? ■

Sometimes when solving word problems involving fractions, or mixed numbers, it is helpful to solve the problem using simpler numbers. Once you understand what operation is involved, you can go back and solve using the original numbers in the word problem.

EXAMPLE 11 A car traveled 301 miles on $10\frac{3}{4}$ gallons of gas. How many miles per gallon did it get?

Use simpler numbers: 300 miles on 10 gallons of gas. We want to find out how many miles the car traveled on 1 gallon of gas. You may want to draw a picture.

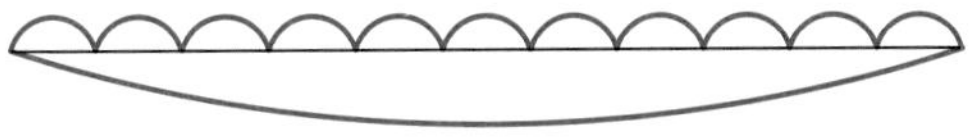

Divide. $300 \div 10 = 30$.

Now use the original numbers given.

$$301 \div 10\frac{3}{4} = \frac{301}{1} \div \frac{43}{4} = \frac{301}{1} \times \frac{4}{43} = \frac{1204}{43} = 28$$

The car got 28 miles to the gallon.

Practice Problem 11

A car can travel $25\frac{1}{2}$ miles on 1 gallon of gas. How many miles can a car travel on $5\frac{1}{4}$ gallons of gas? Check your answer to see if it is reasonable. ■

0.3 Exercises

Multiply. Simplify your answer whenever possible.

1. $\frac{3}{5} \times \frac{2}{11}$ **2.** $\frac{5}{7} \times \frac{3}{4}$ **3.** $\frac{25}{3} \times \frac{6}{5}$ **4.** $\frac{7}{9} \times \frac{18}{5}$ **5.** $8 \times \frac{3}{7}$

6. $\frac{7}{13} \times 4$ **7.** $1\frac{1}{3} \times 2\frac{1}{2}$ **8.** $2\frac{3}{4} \times 1\frac{1}{2}$ **9.** $3\frac{1}{3} \times \frac{3}{4} \times 2$ **10.** $4\frac{1}{5} \times \frac{2}{3} \times 5$

Divide. Simplify your answer whenever possible.

11. $\frac{5}{3} \div \frac{5}{2}$ **12.** $\frac{7}{8} \div \frac{7}{5}$ **13.** $\frac{3}{7} \div 3$ **14.** $\frac{7}{8} \div 4$ **15.** $10 \div \frac{5}{7}$ **16.** $18 \div \frac{2}{9}$

17. $\dfrac{\frac{5}{2}}{\frac{2}{3}}$ **18.** $\dfrac{\frac{3}{4}}{\frac{3}{8}}$ **19.** $1\frac{3}{7} \div 6\frac{1}{4}$ **20.** $4\frac{1}{2} \div 3\frac{3}{8}$ **21.** $\dfrac{2}{1\frac{1}{5}}$ **22.** $\dfrac{\frac{5}{8}}{1\frac{3}{4}}$

Mixed Practice

Perform the proper calculation. Reduce your answer whenever possible.

23. $\frac{6}{5} \times \frac{10}{12}$ **24.** $\frac{5}{24} \times \frac{18}{15}$ **25.** $\frac{5}{16} \div \frac{1}{8}$ **26.** $\frac{2}{11} \div 4$ **27.** $10\frac{3}{7} \times 5\frac{1}{4}$ **28.** $10\frac{2}{9} \div 2\frac{1}{3}$

29. $2\frac{1}{8} \div \frac{1}{4}$ **30.** $4 \div 1\frac{7}{9}$ **31.** $6 \times 2\frac{1}{2}$ **32.** $\frac{1}{3} \div 9$ **33.** $2\frac{1}{2} \times \frac{1}{10} \times \frac{3}{4}$ **34.** $3\frac{1}{3} \times \frac{1}{5} \times \frac{2}{3}$

Applications

Solve each word problem.

35. A certain size of man's shirt requires $2\frac{3}{4}$ yards of material. How many yards would be needed to make 26 shirts?

36. Martha jogged for $4\frac{1}{5}$ miles on Monday. Two-thirds of the distance was over a cross-country trail. How long is the cross-country trail?

Putting Your Skills to Work

THE STOCK MARKET

The price of stocks bought and sold on the stock exchange is calculated in eighths of a dollar. The following is a sample of a listing you might find in the newspaper on a typical day. Note that the High and Low columns that appear after the Sales column refer to the highest and the lowest price the stock traded for on that day. The Change is the change from the closing price from the day before. The name of each stock is abbreviated.

STOCK EXCHANGE									CONSOLIDATED TRADING	
365-DAY HIGH	LOW	STOCK	DIV	YLD %	P/E	SALES 100S	HIGH	LOW	LAST	CHANGE
$10\frac{1}{8}$	9	ACMOP	.79	8.0	Q	96	$9\frac{7}{8}$	$9\frac{3}{4}$	$9\frac{7}{8}$	+1/8
$58\frac{7}{8}$	$46\frac{3}{4}$	BELLSO	2.76	5.1	17	3448	$54\frac{3}{4}$	$54\frac{1}{8}$	$54\frac{1}{8}$	−5/8
20	16	CENTEN	1.60	9.0	11	2181	$17\frac{7}{8}$	$17\frac{5}{8}$	$17\frac{3}{4}$	—
$40\frac{1}{8}$	$34\frac{1}{2}$	FPLGX	2.48	6.4	14	4046	$39\frac{1}{8}$	$38\frac{1}{2}$	$38\frac{1}{2}$	$-1\frac{1}{8}$
$33\frac{3}{8}$	$18\frac{3}{32}$	QKREILY	.38a	1.1	10	454	$33\frac{3}{4}$	33	$33\frac{1}{2}$	+1/2
$56\frac{5}{8}$	13	TIMBCo	—	—	41	964	59	$56\frac{1}{4}$	$58\frac{3}{4}$	$+2\frac{1}{2}$

Use the stock market chart above to solve each problem. Note that transactions on the stock market involve a commission for the stock broker. Commissions are not taken into account in the following problems.

1. Maria bought 100 shares of ACMOp at $9\frac{3}{4}$. How much did she pay for the shares?

2. Joseph sold 50 shares of CentEn at its high for the day. How much did he receive?

3. What is the difference between the high and the low for TimbCo for the day?

4. What was the closing price for FPLGx yesterday?

5. Ashe has \$5025. How many shares of QKReily can she buy if she buys it at the last price of the day?

6. Alex bought 100 shares of FPLGx at its low and sold them at its high. How much did Alex receive for the day?

COOPERATIVE GROUP ACTIVITY

7. You have \$5000 to buy stock. You but at the low and sell at the high on that day.
(a) Which stock would be the best stock to buy? Why?
(b) How many shares can you buy?
(c) What will your profit be?
(d) Is it better to put all of your money into one stock? Why or why not?

8. Pick an actual stock(s) from the newspaper. Invest \$5000 on paper. Determine how much you can buy and chart the course of the stock(s) for a month. You may buy or sell during that time, but keep a record of your transactions. Then write a report that includes a brief description of the company(ies), the reason for the investment, your record of transactions, and a graph.

0.4 Use of Decimals

After studying this section, you will be able to:

1. *Change from a fraction to a decimal.*
2. *Change from a decimal to a fraction.*
3. *Add and subtract decimals.*
4. *Multiply decimals.*
5. *Divide decimals.*
6. *Multiply or divide a number by a multiple of 10.*

The Basic Concept of Decimals

We can express a part of a whole as a fraction or as a decimal. A decimal is another way of writing a fraction whose denominator is 10, 100, 1000, and so on.

$$\frac{3}{10} = 0.3 \qquad \frac{5}{100} = 0.05 \qquad \frac{172}{1000} = 0.172 \qquad \frac{58}{10{,}000} = 0.0058$$

The period in decimal notation is known as the **decimal point.** The number of digits in a number to the right of the decimal point is known as the number of **decimal places** of the number. The place value of decimals are shown below.

Hundred-thousands	Ten-thousands	Thousands	Hundreds	Tens	Ones	← Decimal point	Tenths	Hundredths	Thousandths	Ten-thousandths	Hundred-thousandths
100,000	10,000	1000	100	10	1	.	$\frac{1}{10}$	$\frac{1}{100}$	$\frac{1}{1000}$	$\frac{1}{10{,}000}$	$\frac{1}{100{,}000}$

EXAMPLE 1 Write the following decimals as a fraction. Give the number of decimal places. Write out in words the way the number would be spoken.

(a) 0.6 **(b)** 0.29 **(c)** 0.527
(d) 1.38 **(e)** 0.00007

	Decimal Form	Fraction Form	Number of Decimal Places	The Words Used to Describe the Number
(a)	0.6	$\frac{6}{10}$	one	six-tenths
(b)	0.29	$\frac{29}{100}$	two	twenty-nine hundredths
(c)	0.527	$\frac{527}{1000}$	three	five hundred twenty-seven thousandths
(d)	1.38	$1\frac{38}{100}$	two	one and thirty-eight hundredths
(e)	0.00007	$\frac{7}{100{,}000}$	five	seven hundred-thousandths

Practice Problem 1 Write each decimal as a fraction and in words.

(a) 1.371 **(b)** 0.09 ■

You have seen that a given fraction can be written in several different but equivalent ways. There are also several different equivalent ways of writing the decimal form of fractions. The decimal 0.18 can be written in the following equivalent ways:

$$\text{Fractional form: } \frac{18}{100} = \frac{180}{1000} = \frac{1800}{10{,}000} = \frac{18{,}000}{100{,}000}$$

$$\text{Decimal form: } 0.18 = 0.180 = 0.1800 = 0.18000$$

Thus we see that any number of *terminal zeros may be added onto the right-hand side* of a decimal without changing its value.

$$0.13 = 0.1300 \qquad 0.162 = 0.162000$$

Similarly, any number of *terminal zeros may be removed* from the right-hand side of a decimal without changing its value.

1 *Changing a Fraction to a Decimal*

A fraction can be changed to a decimal by dividing the denominator into the numerator.

EXAMPLE 2 Write each of the following fractions as a decimal.

(a) $\frac{3}{4}$ **(b)** $\frac{21}{20}$ **(c)** $\frac{1}{8}$ **(d)** $\frac{3}{200}$

(a) $\frac{3}{4} = 0.75$ since
$$\begin{array}{r} 0.75 \\ 4\overline{)3.00} \\ \underline{2\,8} \\ 20 \\ \underline{20} \end{array}$$

(b) $\frac{21}{20} = 1.05$ since
$$\begin{array}{r} 1.05 \\ 20\overline{)21.00} \\ \underline{20} \\ 100 \\ \underline{100} \end{array}$$

(c) $\frac{1}{8} = 0.125$ since
$$\begin{array}{r} 0.125 \\ 8\overline{)1.000} \\ \underline{8} \\ 20 \\ \underline{16} \\ 40 \\ \underline{40} \end{array}$$

(d) $\frac{3}{200} = 0.015$ since
$$\begin{array}{r} 0.015 \\ 200\overline{)3.000} \\ \underline{200} \\ 1000 \\ \underline{1000} \end{array}$$

CALCULATOR

Fraction to Decimal

You can use a calculator to change $\frac{3}{5}$ to a decimal.

Enter:

3 [÷] 5 [=]

The display should read:

[0.6]

Try **(a)** $\frac{17}{25}$ **(b)** $\frac{2}{9}$ **(c)** $\frac{13}{10}$ **(d)** $\frac{15}{19}$

Note: 0.7894737 is an approximation for $\frac{15}{19}$.

Practice Problem 2 Write as decimals.

(a) $\frac{3}{8}$ **(b)** $\frac{7}{200}$ **(c)** $\frac{33}{20}$ ■

Sometimes division yields an infinite repeating decimal. We use three dots to indicate that the pattern continues forever. For example:

$$\frac{1}{3} = 0.3333\ldots \qquad \begin{array}{r} 0.333 \\ 3\overline{)1.000} \\ \underline{9} \\ 10 \\ \underline{9} \\ 10 \\ \underline{9} \\ 1 \end{array}$$

An alternative notation is to place a bar over the repeating digits:

$$0.3333\ldots = 0.\overline{3} \qquad 0.575757\ldots = 0.\overline{57}$$

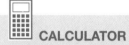

EXAMPLE 3 Write each fraction as a decimal. **(a)** $\frac{2}{11}$ **(b)** $\frac{5}{6}$

(a) $\frac{2}{11} = 0.181818\ldots$ or $0.\overline{18}$ **(b)** $\frac{5}{6} = 0.8333\ldots$ or $0.8\overline{3}$

$$\begin{array}{r} 0.1818 \\ 11\overline{)2.0000} \\ \underline{11} \\ 90 \\ \underline{88} \\ 20 \\ \underline{11} \\ 90 \\ \underline{88} \\ 2 \end{array} \qquad \begin{array}{r} 0.8333 \\ 6\overline{)5.0000} \\ \underline{48} \\ 20 \\ \underline{18} \\ 20 \\ \underline{18} \\ 20 \\ \underline{18} \\ 2 \end{array}$$

Note that the 8 does not repeat. Only the digit 3 is repeating.

Practice Problem 3 Write each fraction as a decimal.

(a) $\frac{1}{6}$ **(b)** $\frac{5}{11}$ ■

Sometimes division must be carried out to many places in order to observe the repeating pattern. This is true in the following example:

$\frac{2}{7} = 0.285714285714285714\ldots$ This can also be written as $\frac{2}{7} = 0.\overline{285714}$.

It can be shown that the denominator determines the maximum number of decimal places that might repeat. So $\frac{2}{7}$ must repeat in the seventh decimal place or sooner.

2 *Changing a Decimal to a Simple Fraction*

To convert from a decimal to a fraction, merely write the decimal as a fraction with a denominator of 10, 100, 1000, 10,000, and so on, and simplify the result when possible.

EXAMPLE 4 Write each decimal as a fraction.

(a) 0.2 **(b)** 0.35 **(c)** 0.516 **(d)** 0.74 **(e)** 0.138 **(f)** 0.008

(a) $0.2 = \frac{2}{10} = \frac{1}{5}$ **(b)** $0.35 = \frac{35}{100} = \frac{7}{20}$

(c) $0.516 = \frac{516}{1000} = \frac{129}{250}$ **(d)** $0.74 = \frac{74}{100} = \frac{37}{50}$

(e) $0.138 = \frac{138}{1000} = \frac{69}{500}$ **(f)** $0.008 = \frac{8}{1000} = \frac{1}{125}$

Practice Problem 4

Write each decimal as a fraction and simplify whenever possible.

(a) 0.8 **(b)** 0.88 **(c)** 0.45 **(d)** 0.148 **(e)** 0.612 **(f)** 0.016 ■

All repeating decimals can also be converted to fractional form. In practice, however, repeating decimals are usually rounded to a few places. It will not be necessary, therefore, to learn how to convert $0.\overline{033}$ to $\frac{11}{333}$ for this course.

3 *Adding and Subtracting Decimals*

Adding or subtracting decimals is similar to adding and subtracting whole numbers except that it is necessary to line up decimal points. To perform the operation 19.8 + 24.7 we line up the numbers in column form and add the digits

$$\begin{array}{r} 19.8 \\ +\ 24.7 \\ \hline 44.5 \end{array}$$

> **Addition and Subtraction of Decimals**
>
> **1.** Write in column form and line up decimal points.
>
> **2.** Add or subtract the digits.

EXAMPLE 5 Perform the following operations.

(a) 3.6 + 2.3 **(b)** 127.32 − 38.48

(c) 3.1 + 42.36 + 9.034 **(d)** 5.0006 − 3.1248

(a) $\begin{array}{r} 3.6 \\ +\ 2.3 \\ \hline 5.9 \end{array}$ **(b)** $\begin{array}{r} 127.32 \\ -\ 38.48 \\ \hline 88.84 \end{array}$ **(c)** $\begin{array}{r} 3.1 \\ 42.36 \\ +\ 9.034 \\ \hline 54.494 \end{array}$ **(d)** $\begin{array}{r} 5.0006 \\ -\ 3.1248 \\ \hline 1.8758 \end{array}$

Practice Problem 5 Add or subtract.

(a) 3.12 + 5.08 + 1.42 **(b)** 152.003 − 136.118

(c) 1.1 + 3.16 + 5.123 **(d)** 1.0052 − 0.1234 ■

SIDELIGHT When we added fractions, we had to have common denominators. Since decimals are really fractions, why can we add them without having common denominators? Actually, we have to have common denominators to add any fractions, whether they are in decimal form or fraction form. However, sometimes the notation does not show this. Let's examine Example 5(c), worked above.

Original Problem

$\begin{array}{r} 3.1 \\ 42.36 \\ +\ 9.034 \\ \hline 54.494 \end{array}$

We are adding the three numbers:

$3\frac{1}{10} + 42\frac{36}{100} + 9\frac{34}{1000}$

$3\frac{100}{1000} + 42\frac{360}{1000} + 9\frac{34}{1000}$

3.100 + 42.360 + 9.034 *This is the New Problem.*

New Problem	Original Problem	
$\begin{array}{r} 3.100 \\ 42.360 \\ +\ 9.034 \\ \hline 54.494 \end{array}$	$\begin{array}{r} 3.1 \\ 42.36 \\ +\ 9.034 \\ \hline 54.494 \end{array}$	*We notice that the results are the same. The only difference is the notation. We are using the property that any number of zeros may be added to the right-hand side of a decimal without changing its value.*

This shows the convenience of adding and subtracting fractions in the decimal form. Little work is needed to change the decimals to a common denominator. All that is required is to add zeros to the right-hand side of the decimal (and we usually do not even write out that step except when subtracting).

As long as we line up the decimal points, we can add or subtract any decimal fractions.

In the following example we will find it useful to add zeros to the right-hand side of the decimal.

EXAMPLE 6 Perform the following operations.

(a) $1.0003 + 0.02 + 3.4$ **(b)** $12 - 0.057$

We will add zeros so that each number shows the same number of decimal places.

(a)
$$\begin{array}{r} 1.0003 \\ 0.0200 \\ +\ 3.4000 \\ \hline 4.4203 \end{array}$$

(b)
$$\begin{array}{r} 12.000 \\ -\ \ 0.057 \\ \hline 11.943 \end{array}$$

Practice Problem 6 Perform the following operations.

(a) $0.061 + 5.0008 + 1.3$ **(b)** $18 - 0.126$ ■

4 *Multiplying Decimals*

> **Multiplication of Decimals**
>
> To multiply decimals, you first multiply as with whole numbers. To determine the position of the decimal point, you count the total number of decimal places in the two numbers being multiplied. This will determine the number of decimal places that should appear in the answer.

EXAMPLE 7 Multiply. 0.8×0.4

$$\begin{array}{rl} 0.8 & \text{(one decimal place)} \\ \times\ 0.4 & \text{(one decimal place)} \\ \hline 0.32 & \text{(two decimal places)} \end{array}$$

Practice Problem 7 Multiply. 0.5×0.3 ■

Note that you will often have to add zeros to the left of the digits obtained in the product so that you obtain the necessary number of decimal places.

EXAMPLE 8 Multiply. 0.123×0.5

$$\begin{array}{rl} 0.123 & \text{(three decimal places)} \\ \times\ \ \ \ 0.5 & \text{(one decimal place)} \\ \hline 0.0615 & \text{(four decimal places)} \end{array}$$

Practice Problem 8 Multiply. 0.12×0.4 ■

Here are some examples that involve more decimal places.

EXAMPLE 9 Multiply. **(a)** 2.56×0.003 **(b)** 0.0036×0.008

(a)

$$\begin{array}{rl} 2.56 & \text{(two decimal places)} \\ \times\ 0.003 & \text{(three decimal places)} \\ \hline 0.00768 & \text{(five decimal places)} \end{array}$$

(b)

$$\begin{array}{rl} 0.0036 & \text{(four decimal places)} \\ \times\ 0.008 & \text{(three decimal places)} \\ \hline 0.0000288 & \text{(seven decimal places)} \end{array}$$

Practice Problem 9 Multiply.

(a) 1.23×0.005 **(b)** 0.003×0.00002 ■

SIDELIGHT Why do we count the number of decimal places? The rule really comes from the properties of fractions. If we write the problem in Example 8 in fraction form, we have

$$(0.123) \times (0.5) = \frac{123}{1000} \times \frac{5}{10} = \frac{615}{10{,}000} = 0.0615$$

5 *Dividing Decimals*

When discussing division of decimals, we frequently refer to the three primary parts of a division problem. Be sure you know the meaning of each term.

The **divisor** is the number you divide into another.

The **dividend** is the number to be divided.

The **quotient** is the result of dividing one number by another.

In the problem $6 \div 2 = 3$ we represent each of these terms as follows:

$$\text{Divisor} \longrightarrow 2\overline{)6} \quad (3 \leftarrow \text{Quotient}; \ 6 \leftarrow \text{Dividend})$$

$$\begin{array}{r} \text{Quotient} \\ \text{Divisor}\overline{)\text{Dividend}} \end{array}$$

> When dividing two decimals, count the *number of decimal places* in the divisor. Then move the *decimal point to the right* that *same number of places* in both *the divisor* and *the dividend.* Mark that position with a caret (‸). Finally, perform the division. Be sure to line up the decimal point in the quotient with the position indicated by the caret in the dividend.

EXAMPLE 10 Divide. $32.68 \div 4$

$$\begin{array}{r} 8.17 \\ 4\overline{)32.68} \\ \underline{32} \\ 6 \\ \underline{4} \\ 28 \\ \underline{28} \end{array}$$

Since there are no decimal places in the divisor, we do not need to move the decimal point. We must be careful, however, to place the decimal point in the quotient directly above the decimal point in the dividend.

Thus $32.68 \div 4 = 8.17$.

Practice Problem 10 Divide. $0.1116 \div 0.18$ ■

EXAMPLE 11 Divide. 5.75 ÷ 0.5

$$0.5_{\wedge}\overline{)5.7_{\wedge}5}$$

One decimal place

*There is **one** decimal place in the divisor, so we move the decimal point **one place** to **the right** in the **divisor** and **dividend** and we mark that new position by a caret ($_\wedge$).*

$$\begin{array}{r} 11.5 \\ 0.5_{\wedge}\overline{)5.7_{\wedge}5} \\ \underline{5} \\ 7 \\ \underline{5} \\ 25 \\ \underline{25} \end{array}$$

Now we perform the division as with whole numbers. The decimal point in the answer is directly above the caret in the dividend.

Thus 5.75 ÷ 0.5 = 11.5

Practice Problem 11 Divide. 0.09 ÷ 0.3 ■

Note that sometimes we will need to place extra zeros in the dividend in order to move the decimal point the required number of places.

EXAMPLE 12 Divide. 16.2 ÷ 0.027

$$0.027_{\wedge}\overline{)16.200_{\wedge}}$$

Three decimal places

*There are **three** decimal places in the divisor, so we move the decimal point **three places** to **the right** in the divisor and **dividend** and mark the new position by a caret. Note that we must add two zeros to 16.2 in order to do this.*

$$\begin{array}{r} 600. \\ 0.027_{\wedge}\overline{)16.200_{\wedge}} \\ \underline{16\,2} \\ 000 \end{array}$$

Now perform the division as with whole numbers. The decimal point in the answer is directly above the caret in the dividend.

Thus 16.2 ÷ 0.027 = 600.

Practice Problem 12 Divide. 1800 ÷ 0.06 ■

Special care must be taken to line up the digits in the quotient. Note that sometimes we will need to place zeros in the quotient after the decimal point.

EXAMPLE 13 Divide. 0.04288 ÷ 3.2

$$3.2_{\wedge}\overline{)0.0_{\wedge}4288}$$

One decimal place

*There is **one** decimal place in the divisor, so we move the decimal point **one place** to **the right** in the **divisor** and **dividend** and mark the new position by a caret.*

$$\begin{array}{r} 0.0134 \\ 3.2_{\wedge}\overline{)0.0_{\wedge}4288} \\ \underline{32} \\ 108 \\ \underline{96} \\ 128 \\ \underline{128} \\ 0 \end{array}$$

Now perform the division as for whole numbers. The decimal point in the answer is directly above the caret in the dividend. Note the need for the initial zero after the decimal point in the answer.

Thus 0.04288 ÷ 3.2 = 0.0134.

Practice Problem 13 Divide. 0.01764 ÷ 4.9 ■

SIDELIGHT Why does this method of dividing decimals work? Essentially, we are using the steps we used in Section 0.1 to change a fraction to an equivalent fraction by multiplying both the numerator and denominator by the same number. Let's reexamine Example 13.

$0.04288 \div 3.2 = \dfrac{0.04288}{3.2}$ *Write the original problem using fraction notation.*

$= \dfrac{0.04288 \times 10}{3.2 \quad \times 10}$ *Multiply the numerator and denominator by 10. Since this is the same as multiplying by 1, we are not changing the fraction.*

$= \dfrac{0.4288}{32}$ *Write the result of multiplication by 10.*

$= 0.4288 \div 32$ *Rewrite the fraction as an equivalent problem with division notation.*

Notice that we have obtained a new problem that is the same as the problem in Example 13 when we moved the decimal one place to the right in the divisor and dividend. We see that the reason we can move the decimal point so many places to the right in divisor and dividend is that the numerator and denominator of a fraction are both being multiplied by 10, 100, 1000, and so on, to obtain an equivalent fraction.

6 *Multiplying and Dividing by a Multiple of 10*

When multiplying by 10, 100, 1000, and so on, a simple rule may be used to obtain the answer. For every zero in the multiplier, move the decimal point one place to the right.

EXAMPLE 14 Multiply. **(a)** 3.24×10 **(b)** 15.6×100 **(c)** 0.0026×1000

(a) $3.24 \times 10 = 32.4$ *One zero—move decimal point one place to the right.*

(b) $15.6 \times 100 = 1560$ *Two zeros—move decimal point two places to the right.*

(c) $0.0026 \times 1000 = 2.6$ *Three zeros—move decimal point three places to the right.*

Practice Problem 14 Multiply.

(a) 0.0016×100 **(b)** 2.34×1000 **(c)** $56.75 \times 10{,}000$ ■

The reverse rule is true for division. When dividing by 10, 100, 1000, 10,000, and so on, move the decimal point one place to the left for every zero in the divisor.

EXAMPLE 15 Divide. **(a)** $52.6 \div 10$ **(b)** $0.0038 \div 100$ **(c)** $5936.2 \div 1000$

(a) $\dfrac{52.6}{10} = 5.26$ *Move one place to the left.*

(b) $\dfrac{0.0038}{100} = 0.000038$ *Move two places to the left.*

(c) $\dfrac{5936.2}{1000} = 5.9362$ *Move three places to the left.*

Practice Problem 15 Divide.

(a) $\dfrac{5.82}{10}$ **(b)** $123.4 \div 1000$ **(c)** $\dfrac{0.00614}{10{,}000}$ ■

0.4 Exercises

Verbal and Writing Skills

1. A decimal is another way of writing a fraction whose denominator is ____________________.

2. To describe 0.03 in words we write ____________________.

3. When multiplying 0.059 by 10,000, move the decimal point ____ places to the ______.

Write each fraction as a decimal.

4. $\frac{5}{8}$

5. $\frac{3}{20}$

6. $\frac{7}{11}$

7. $\frac{15}{16}$

8. $\frac{9}{20}$

9. $\frac{6}{20}$

Write each decimal as a fraction in simplified form.

10. 0.8

11. 0.15

12. 1.125

13. 0.08

14. 0.75

15. 0.625

Add or subtract.

16. $3.6 + 1.28 + 4.5$

17. $1.2 + 3.9 + 2.62$

18. $0.00381 - 0.00228$

19. $1.0076 - 0.0982$

20. $156.0 + 0.03 + 3.621$

21. $22.0 + 0.42 + 1.936$

22. $158.23 - 39.67$

23. $121.98 - 34.78$

Multiply.

24. 2.18×1.3

25. 7.12×2.6

26. 5.26×0.0015

27. 7.21×0.071

28. 0.0062×0.018

29. 0.17×0.0084

30. $368{,}000 \times 0.00021$

31. $169{,}000 \times 0.0013$

Divide.

32. $0.0455 \div 0.13$

33. $0.5535 \div 0.15$

34. $0.7208 \div 2.12$

35. $27.808 \div 3.16$

36. $0.5230 \div 0.002$

37. $0.031 \div 0.005$

38. $186.16 \div 5.2$

39. $46.62 \div 7.4$

Mixed Practice

Multiply or divide by moving the decimal point.

40. 1.36×1000

41. $0.76 \div 100$

42. $164{,}320 \div 10{,}000$

43. 0.02×100

44. $3.52 \div 1000$

45. $0.00243 \times 100{,}000$

46. $7.36 \times 10{,}000$

47. $73{,}892 \div 100{,}000$

Perform the following calculations.

48. 26.13×0.04

49. 1.936×0.003

50. $1.62 + 2.005 + 8.1007$

51. $28 - 3.64$

52. $0.04296 \div 0.012$

53. $126.93 \div 0.0004$

54. $0.9281 \times 100{,}000$

55. $34.72 \div 10{,}000$

Applications

Solve each problem.

56. Fred's car usually gets 28.5 miles per gallon when driven on the highway. The gas tank holds 23.4 gallons. What is the driving range of his car when driven on the highway?

57. Alice drove 478.6 miles last week. She used 26.8 gallons of gas. Find how many miles per gallon she averaged in her car last week. (Round your answer to the nearest tenth.)

Cumulative Review Problems

Perform each calculation.

58. $3\frac{1}{2} \div 5\frac{1}{4}$

59. $\frac{3}{8} \cdot \frac{12}{27}$

60. $\frac{12}{25} + \frac{9}{20}$

61. $1\frac{3}{5} - \frac{1}{2}$

Developing Your Study Skills

MAKING A FRIEND IN THE CLASS

Attempt to make a friend in your class. You may find that you enjoy sitting together and drawing support and encouragement from one another. Exchange phone numbers so you can call each other whenever you get stuck in your study. Set up convenient times to study together on a regular basis, to do homework, and to review for exams.

You must not depend on a friend or fellow student to tutor you, do your work for you, or in any way be responsible for your learning. However, you will learn from one another as you seek to master the course. Studying with a friend and comparing notes, methods, and solutions can be very helpful. And it can make learning mathematics a lot more fun!

Putting Your Skills to Work

CONSTRUCTION

Estimators use decimals in their work to determine the cost of a job. They need to know the quantity and the unit price. Some of these abbreviations may be new to you. CY means cubic yards; CBF means hundred board feet; SF means square feet.

1. Calculate the total for each item shown in the table. Then calculate the G.C. (General Contractor's) total.

COST ESTIMATE ACE CONSTRUCTION CO.			
Item	Quantity	Cost per Unit	Total
Excavation	220 CY	\$50/CY	
Concrete Foundation	20 CY	\$400/CY	
Framing	50 CBF	\$200/CBF	
Roofing	300 SF	\$2/SF	
Finishing	300 SF	\$16/SF	
		G.C. Total	

2. Bry Construction Co. submitted a bid of \$38,000. Which company submitted the lower bid?
   How much lower?

3. Ace guarantees to finish the job in 120 days or by July 20. What is the cost per day?

COOPERATIVE GROUP ACTIVITY

4. The company needs to have the job completed by June 30. Labor is 60% of the total cost of the project. Ace will charge time and a half to work Saturdays and holidays. How much will the job cost if work begins on March 22?

5. What will the affect on the cost be if work begins on April 1? A regular day is 7 hours, overtime is at time and a half.

0.5 Use of Percent

After studying this section, you will be able to:

1 *Change a decimal to a percent.*
2 *Change a percent to a decimal.*
3 *Find the percent of a number.*
4 *Find the missing percent when given two numbers.*

The Basic Concept of Percents

A **percent** is a fraction that has a denominator of 100. When you say "sixty-seven percent" or write 67%, you are just using another way of expressing the fraction $\frac{67}{100}$. The word *percent* is a shortened form of the Latin words *per centum,* which means "by the hundred." In everyday use, percent means per one hundred.

It is important to see that 49% means 49 parts out of 100 parts. It can also be written as a fraction, $\frac{49}{100}$, or as a decimal: 0.49. Understanding the meaning of the notation allows you to change from one notation to another. For example,

$$49\% = 49 \text{ out of } 100 \text{ parts} = \frac{49}{100} = 0.49$$

Similarly, you can express a fraction with denominator 100 as a percent or a decimal.

$$\frac{11}{100} \text{ means 11 parts out of 100 or } 11\%; \qquad \frac{11}{100} \text{ as a decimal is } 0.11$$

So $\frac{11}{100} = 11\% = 0.11$.

1 *Changing a Decimal to a Percent*

Now that we understand the concept, we can use some quick procedures to change from decimals to percent, and vice versa.

> **Changing a Decimal to Percent**
>
> **1.** Move the decimal point two places to the right.
>
> **2.** Add the % symbol.

EXAMPLE 1 Change to a percent. **(a)** 0.23 **(b)** 0.461 **(c)** 0.4

We move the decimal point two places to the right and add the % symbol.

(a) 0.23 = 23% **(b)** 0.461 = 46.1% **(c)** 0.4 = 0.40 = 40%

Practice Problem 1 Change to a percent. **(a)** 0.92 **(b)** 0.418 **(c)** 0.7 ■

Be sure to follow the same procedure for percents that are less than 1%. Remember 0.01 is 1%. Thus we would expect 0.001 to be less than 1%. 0.001 = 0.1% or one-tenth (0.1) of a percent.

EXAMPLE 2 Change to a percent. **(a)** 0.0364 **(b)** 0.0026 **(c)** 0.0008

We move the decimal point two places to the right and add the % symbol.

(a) 0.0364 = 3.64% **(b)** 0.0026 = 0.26% **(c)** 0.0008 = 0.08%

Practice Problem 2 Change to a percent.

(a) 0.0019 **(b)** 0.0736 **(c)** 0.0003 ■

Be sure to follow the same procedure for percents that are greater than 100%. Remember 1 is 100%. Thus we would expect 1.5 to be greater than 100%. 1.5 = 150%

EXAMPLE 3 Change to a percent. **(a)** 1.48 **(b)** 2.938 **(c)** 4.5

We move the decimal point two places to the right and add the percent symbol.

(a) 1.48 = 148% **(b)** 2.938 = 293.8% **(c)** 4.5 = 4.50 = 450%

Practice Problem 3 Change to a percent.

(a) 3.04 **(b)** 5.186 **(c)** 2.1 ■

2 *Changing from a Percent to a Decimal*

In this procedure we move the decimal point to the left and remove the % symbol.

CALCULATOR

Percent to Decimal

You can use a calculator to change 52% to a decimal. If your calculator has a [%] key,

Enter: 52 [%]

The display should read:

[0.52]

Try **(a)** 46% **(b)** 137% **(c)** 9.3% **(d)** 6%

Note: The calculator divides by 100 when the percent key is pressed.

Changing a Percent to a Decimal

1. Move the decimal point two places to the left.
2. Remove the % symbol.

EXAMPLE 4 Change to a decimal. **(a)** 16% **(b)** 143%

First we move the decimal point two places to the left. Observe in each case that the decimal point is not written but is understood to be to the right of the last digit in the percent. Then we remove the % symbol.

(a) 16% = 16.% = 0.16

↑ The unwritten decimal point is understood to be here.

(b) 143% = 143.% = 1.43

Practice Problem 4 Change to a decimal. **(a)** 82% **(b)** 347% ■

EXAMPLE 5 Change to a decimal. **(a)** 4% **(b)** 3.2% **(c)** 0.6%

First we move the decimal point two places to the left. Then we remove the % symbol.

(a) 4% = 4.% = 0.04

↑ The unwritten decimal point is understood to be here.

(b) 3.2% = 0.032 **(c)** 0.6% = 0.006

Practice Problem 5 Change to a decimal. **(a)** 7% **(b)** 9.3% **(c)** 0.2% ■

EXAMPLE 6 Change to a decimal. **(a)** 192% **(b)** 254.8% **(c)** 0.027%

First we move the decimal point two places to the left. Then we remove the % symbol.

(a) 192% = 192.% = 1.92

↑ The unwritten decimal point is understood to be here.

(b) 254.8% = 2.548 **(c)** 0.027% = 0.00027

Practice Problem 6 Change to a decimal.

(a) 131% **(b)** 301.6% **(c)** 0.04% ■

3 *Finding the Percent of a Number*

How do we find 60% of 20? Let us relate it to a problem we did in Section 0.2.

Consider the problem

$$\text{What is } \frac{3}{5} \text{ of } 20?$$

$$\downarrow \quad \downarrow \quad \downarrow \quad \downarrow \quad \downarrow$$

$$\boxed{?} = \frac{3}{5} \times 20$$

$$\boxed{?} = \frac{3}{\cancel{5}} \times \overset{4}{\cancel{20}} = 12 \quad \text{The answer is 12.}$$

Since a percent is really a fraction, a percent problem is solved in a similar way to solving a fraction problem. Since $\frac{3}{5} = \frac{6}{10} = 60\%$, we could write the problem as

$$\text{What is } 60\% \text{ of } 20?$$

$$\downarrow \quad \downarrow \quad \downarrow \quad \downarrow \quad \downarrow$$

$$\boxed{?} = 60\% \times 20$$

$$\boxed{?} = 0.60 \times 20$$

$$\boxed{?} = 12.0 \quad \text{The answer is 12.}$$

Thus we have developed a rule.

> **Finding the Percent of a Number**
>
> To find the percent of a number, merely change the percent to a decimal and multiply by the decimal.

EXAMPLE 7 Find.

(a) 10% of 36 **(b)** 2% of 350 **(c)** 182% of 12 **(d)** 0.3% of 42

(a) 10% of 36 = 0.10 × 36 = 3.6 **(b)** 2% of 350 = 0.02 × 350 = 7

(c) 182% of 12 = 1.82 × 12 = 21.84 **(d)** 0.3% of 42 = 0.003 × 42 = 0.126

Practice Problem 7 Find.

(a) 18% of 50 **(b)** 4% of 64 **(c)** 156% of 35

(d) 0.8% of 60 **(e)** 1.3% of 82 **(f)** 0.002% of 564 ■

There are many real-life applications for finding the percent of a number. When you go shopping in a store, you may find sale merchandise marked 35% off. This means that the sale price is 35% off the regular price. That is, 35% of the regular price is subtracted from the regular price to get the sale price.

> **CALCULATOR**
>
> **Using Percents**
>
> You can use a calculator to find 12% of 48.
>
> Enter:
>
> 12 [%] [×] 48 [=]
>
> The display should read:
>
> [5.76]
>
> If your calculator does not have a [%] key,
>
> Enter: 0.12 [×] 48 [=]
>
> *Try* What is 54% of 450?

EXAMPLE 8 A store is having a sale of 35% off the retail price of all sofas. Melissa wants to buy a particular sofa that normally sells for \$595.

(a) How much will Melissa save if she buys the sofa on sale?

(b) What is the purchase price if Melissa buys the sofa on sale?

(a) To find 35% of \$595 we will need to multiply 0.35×595.

$$\begin{array}{r} \$595 \\ \times \quad 0.35 \\ \hline 2975 \\ 1785 \\ \hline \$208.25 \end{array}$$

Thus Melissa would save \$208.25 if she buys the sofa on sale.

(b) The purchase price is the difference between the original price and the amount saved.

$$\begin{array}{r} \$595.00 \\ -\quad 208.25 \\ \hline \$386.75 \end{array}$$

Thus Melissa bought the sofa on sale for \$386.75.

Practice Problem 8

John received a 4.2% pay raise at work this year. He previously earned \$18,000 per year.

(a) What is the amount of his pay raise in dollars?

(b) What will his new salary be? ■

4 *Finding the Missing Percent When Given Two Numbers*

Recall that we can write $\frac{3}{4}$ as $\frac{75}{100}$ or 75%. If we were asked the question, "What percent is 3 of 4?," we would say 75%. This gives us a procedure for finding what percent one number is of a second number.

> **Finding the Missing Percent**
>
> **1.** Write a fraction with the two numbers. The number *after* the word "of" is always the denominator, and the other number is the numerator.
>
> **2.** Simplify the fraction (if possible).
>
> **3.** Change the fraction to a decimal.
>
> **4.** Express the decimal as a percent.

A very common problem is the following:

EXAMPLE 9 What percent of 24 is 15?

This can be quickly solved as follows:

Step 1 $\dfrac{15}{24}$ *Write the relationship as a fraction. The number after "of" is 24, so the 24 is in the denominator.*

Step 2 $= \dfrac{5}{8}$ *Simplify the fraction (when possible).*

Step 3 $= 0.625$ *Change the fraction to a decimal.*

Step 4 $= 62.5\%$ *Change the decimal to percent.*

Practice Problem 9 What percent of 148 is 37? ■

The question in Example 9 can also be written as "15 is what percent of 24?" To answer the question, we begin by writing the relationship as $\frac{15}{24}$. Remember "of" 24 means 24 will be in the denominator.

EXAMPLE 10 **(a)** 82 is what percent of 200? **(b)** What percent of 16 is 3.8? **(c)** $150 is what percent of $120?

(a) 82 is what percent of 200?

$\frac{82}{200}$ *Write the relationship as a fraction with 200 in the denominator.*

$$\frac{82}{200} = \frac{41}{100} = 0.41 = 41\%$$

(b) What percent of 16 is 3.8?

$\frac{3.8}{16}$ *Write the relationship as a fraction.*

You can divide to change the fraction to a decimal and then change the decimal to a percent.

$$16\overline{)3.8000} = 0.2375 \rightarrow 23.75\%$$

(c) $150 is what percent of $120?

$\frac{150}{120} = \frac{5}{4}$ *Reduce the fraction whenever possible to make the division easier.*

$$4\overline{)5.00} = 1.25 \rightarrow 125\%$$

Practice Problem 10

(a) What percent of 48 is 24?
(b) 4 is what percent of 25? ■

EXAMPLE 11 Marcia had 29 shots on the goal during the last high school field hockey season. She actually scored a goal 8 times. What percent of her total shots were goals? Round your answer to the nearest whole percent.

Marcia scored a goal 8 times out of 29 tries. We want to know what percent of 29 is 8.

Step 1 $\frac{8}{29}$ *Express the relationship as a fraction. The number after the word "of" is 29, so the 29 appears in the denominator.*

Step 2 *Note that this fraction cannot be reduced.*

Step 3 $= 0.2758\ldots$ *The decimal equivalent of the fraction has many digits.*

Step 4 $= 27.58\ldots\%$ $\approx 28\%$ *We change the decimal to a percent, which we round to the nearest whole percent.*

Therefore, Marcia scored a goal approximately 28% of the time she made a shot on the goal.

Practice Problem 11

Roberto scored a basket 430 times out of 1256 attempts during his high school basketball career. What percent of the time did he score a basket? Round your answer to the nearest whole percent. ■

CALCULATOR

Using Percents

You can use a calculator to find a missing percent.

What percent of 95 is 19?

1. Enter as a fraction.

19 [÷] 95

2. Change to a percent.

19 [÷] 95 [×] 100 [=]

The display should read:

[20]

This means 20%.

Try What percent of 625 is 250?

0.5 Exercises

Verbal and Writing Skills

1. When you write 19% what do you really mean? Describe in your own words.

2. When you try to solve a problem like ''What percent of 80 is 30?,'' how do you know if you should write the fraction as $\frac{80}{30}$ or as $\frac{30}{80}$?

Change to a percent.

3. 0.624

4. 0.078

5. 0.003

6. 0.007

7. 1.56

8. 1.6

Change to decimal.

9. 4%

10. 70%

11. 0.2%

12. 0.62%

13. 250%

14. 175%

Find the following.

15. What is 35.8% of 1000?

16. What is 7.7% of 450?

17. What is 0.5% of 54?

18. What is 7% of 69?

19. What is 140% of 212?

20. What is 176% of 340?

21. What is 80% of 220?

22. What is 0.06% of 410?

23. What percent of 600 is 30?

24. What percent of 800 is 50?

25. 6 is what percent of 120?

26. 33 is what percent of 330?

27. What percent of 56 is 84?

28. 49 is what percent of 28?

29. What percent of 500 is 2?

30. 36 is what percent of 300?

31. What percent of 16 is 22?

32. 45 is what percent of 18?

Applications

Solve each word problem.

33. In a town of 30,000 people, 47% are registered as Democrats. How many people are Democrats in this town?

★ **34.** Melinda took an exam with 80 questions. She had 12 wrong. What was her grade for the exam? Write the grade as a percent.

35. In a local college survey it was found that 97 out of 180 students smoked. What percent of the students in the sample said that they smoked? (Round your answer to the nearest hundredth of a percent.)

36. Susan is earning $4.20 an hour. She will get an 8% pay raise this year. What will her raise be, and what will be her new hourly pay rate? (Round your answer to the nearest cent.)

37. Defects were found in 1.8% of the products manufactured by Deemo Industries last week. If 36,000 products were manufactured last week, how many of them were defected?

38. The Watson family has a combined monthly income of $1850. Their food budget is $380 per month. What percent of their monthly income is budgeted for food? Round your answer to the nearest whole percent.

39. Eastern Shores charter airline had 2390 flights last year. Of these flights, 560 did not land on time. What percent of the flights landed on time? Round your answer to the nearest whole percent.

40. Enrique and Rosa went out to dinner. The cost of the meal was $41.00. They want to leave a tip of 15%. How large a tip should they leave?

41. Juan traveled 18,600 miles in his car last year. As a traveling salesperson, 65% of his mileage was for business purposes. (a) How many miles did he travel on business last year? (b) If his company reimburses him $0.275 or 27.5¢ for each mile traveled, what is the amount that his company should pay him for his travel?

42. Abdul sells computers for a local computer outlet. He gets paid $450 per month plus a commission of 3.8% on all the computer hardware he sells. Last year he sold $780,000 worth of computer hardware. (a) What was his sales commission on the $780,000 worth of hardware he sold? (b) What was his annual salary if the monthly pay and the commission are added together?

Putting Your Skills to Work

CONSTRUCTION

At different times during the construction, project managers need to know what percent of the job is complete.

1. On May 18, the concrete for the Central Post Office had been poured. If this concrete cost \$27,000 out of a total of \$540,000 to build the post office, what percent of the job was complete?

2. By June 5, the framing had been erected at a cost of \$43,200. What percent of the job was complete by June 5?

3. If ACE Construction pays for materials within 10 days, they get a 2% discount. How much does ACE save by paying a \$12,500 bill within 10 days?

4. Lumber costs 80¢ per board foot (BF). What is the savings on 1000 BF for a bill paid during the discount period?

COOPERATIVE GROUP ACTIVITY

5. Complete the chart below.

Item	Calendar Days	Total Cost	Cost per Day	Work Complete	Percent Complete
Excavation	20	\$11,000		\$11,000	
Concrete	10	8,000		19,000	
Framing	10	10,000			
Roof	6	600			
Finishing	40	4,800			
Total					

6. After 30 calendar days, what would be the percent of the job complete?

7. After 60 calendar days, what would be the percent of the job complete?

0.6 Estimation

After studying this section, you will be able to:

1 *Use rounding to estimate.*

1 *Using Rounding to Estimate*

Estimation is the process of finding an approximate answer. It is not designed to provide an exact answer. Estimation will give you a rough idea of what the answer might be. For any given problem, you may choose to estimate in many different ways.

Estimation by Rounding

1. Round each number to an appropriate place.
2. Perform the calculation with the rounded numbers.

EXAMPLE 1 Find an estimate of the product 5368×2864.

Step 1 Round 5368 to 5000.
Round 2864 to 3000.

Step 2 Multiply.

$$5000 \times 3000 = 15{,}000{,}000$$

An estimate of the product is 15,000,000.

Practice Problem 1

Find an estimate of the product $128{,}621 \times 378$. ■

EXAMPLE 2 The four walls of a college classroom are $22\frac{1}{4}$ feet long and $8\frac{3}{4}$ feet high. A painter needs to know the area of these four walls in square feet. Since paint is sold in gallons, an estimate will do. Estimate the area of the four walls.

Step 1 Round $22\frac{1}{4}$ feet to 20 feet.
Round $8\frac{3}{4}$ feet to 9 feet.

Step 2 Multiply 20×9 to obtain an estimate of the area of one wall.
Multiply $20 \times 9 \times 4$ to obtain an estimate of the area of all four walls.

$$20 \times 9 \times 4 = 720 \text{ square feet}$$

Our estimate for the painter is 720 square feet of wall space.

Practice Problem 2

Mr. and Mrs. Ramirez need to carpet two rooms of their house. One room measures $12\frac{1}{2}$ feet by $9\frac{3}{4}$ feet. The other room measures $14\frac{1}{4}$ feet by $18\frac{1}{2}$ feet. Estimate the number of square feet (square footage) in these two rooms. ■

EXAMPLE 3 Won Lin has a small compact car. He drove 396.8 miles in his car and used 8.4 gallons of gas.

(a) Estimate the number of miles he gets per gallon.

(b) Estimate how much it will cost him for fuel to drive on a cross-country trip of 2764 miles if gasoline usually costs $\$1.39\frac{9}{10}$ per gallon.

(a) Round 396.8 miles to 400 miles. Round 8.4 gallons to 8 gallons. Now divide.

$$8\overline{)400}\quad 50$$

Won Lin's car gets about 50 miles per gallon.

(b) We will need to use the information we found in part (a) to determine how many gallons of gasoline Won Lin will use on his trip. Round 2764 miles to 3000 miles and divide 3000 miles by 50 gallons.

$$50\overline{)3000}\quad 60$$

Won Lin will use about 60 gallons of gas for his cross-country trip.

To estimate the cost, we need to ask ourselves, "What kind of an estimate are we looking for?" It may be sufficient to round $\$1.39\frac{9}{10}$ to \$1.00 and multiply.

$$60 \times \$1.00 = \$60.00$$

Keep in mind that this is a broad estimate. You may want an estimate that will be closer to the exact answer. In that case round $\$1.39\frac{9}{10}$ to \$1.40 and multiply.

$$60 \times \$1.40 = \$84.00$$

Caution: An estimate is only a rough guess. If we are estimating the cost of something, we may want to round the unit cost so that we get closer to the actual amount. It is a good idea to round above the unit cost to make sure that we will have enough money for the expenditure. The actual cost of the cross-country trip to the nearest penny is \$81.86.

Practice Problem 3

Roberta drove 422.8 miles in her truck and used 19.3 gallons of gas. Assume that gasoline costs $\$1.29\frac{9}{10}$ per gallon.

(a) Estimate the number of miles she gets per gallon.

(b) Estimate how much it will cost her to drive to Chicago and back, a distance of 3862 miles. ■

Other words that indicate estimation in word problems are *about* and *approximate*.

EXAMPLE 4 A local manufacturing company releases 1684 pounds of sulfur dioxide per day from its plant. The company operates 294 days per year and has been operating at this level for 32 years. Approximately how many pounds of sulfur dioxide has been released into the air by this company over the last 32 years.

The word *approximately* means we are looking for an estimate. We begin by rounding the given data so that there is only one nonzero digit in each number.

Round 1684 tons to 2000 tons.

Round 294 days to 300 days.

Round 32 years to 30 years.

Now determine the calculation you would perform to find the answer and calculate using the rounded numbers. We multiply

amount released in one day × the number of days in an operating year × years

$$2000 \times 300 \times 30 = 18{,}000{,}000$$

Thus 18,000,000 tons is an approximation of the sulfur dioxide released by the company during 32 years of operation.

Practice Problem 4

A space probe is sent from Earth at 43,300 miles per hour toward the planet Pluto, which has an average distance from earth of 3,580,000,000 miles.

(a) About how many *hours* will it take for the space probe to travel from Earth to Pluto?

(b) About how many *days* will it take for the space probe to travel from Earth to Pluto? ■

EXAMPLE 5 Find an estimate for 2.68% of $54,361.92.

Begin by writing 2.68% as a decimal.

$$2.68\% = 0.0268$$

Next round 0.0268 to the nearest hundredth.

0.0268 is 0.03 rounded to the nearest hundredth.

Finally, round $54,361.92 to $50,000.

Now multiply.

$0.03 \times 50{,}000 = 1500$ *Remember that to find a percent of a number you multiply.*

Thus we estimate the answer to be $1500.

To get a better estimate of 2.68% of $54,361.92, you may wish to round $54,361.92 to the nearest thousand. Try it and compare this estimate to the one above. Remember, an estimate is only an approximation and how close you wish to get to the exact answer depends on what you need the estimate for.

Practice Problem 5

Find an estimate for 56.93% of $293,567.12. ■

0.6 Exercises

Follow the principles of estimation to find an estimated value for each of the following. Do not find the exact value. There are many acceptable answers, depending on your rounding method.

1. 878×203

2. 329×773

3. $58{,}126 \times 1786$

4. $193{,}562 \times 38$

5. $93 + 87 + 56 + 22 + 34$

6. $127 + 341 + 778 + 567$

7. $23\overline{)578{,}962}$

8. $82\overline{)4{,}320{,}760}$

9. $\dfrac{0.002714}{0.0315}$

10. $\dfrac{0.5361}{0.00786}$

11. Find 3.6% of \$58,921.63.

12. Find 4.9% of \$9321.88.

Applications

Determine an estimate of the exact answer. Use estimation by rounding. Do not find the exact value.

13. Estimate the floor area of a room that measures $18\frac{3}{4}$ feet long and $12\frac{1}{8}$ feet wide.

14. Estimate the floor area of a room that measures $31\frac{1}{2}$ feet long and $17\frac{3}{4}$ feet wide.

15. Garcia drove for 234.8 miles and used 12.6 gallons of gas. Estimate the number of miles this car gets per gallon.

16. Yoshiko drove for 183.4 miles and used 9.8 gallons of gas. Estimate the number of miles his car gets per gallon.

17. A typical customer at a local supermarket spends approximately \$82 at the checkout register. The store keeps four registers open and each handles 22 customers per hour. Estimate the amount of money the store receives in one hour.

18. A local bank has found on Fridays that the average customer withdraws \$85 from the cash teller machine. Each machine averages 19 customers per hour. There are five machines. Estimate the amount of money withdrawn by customers in one hour.

19. Angela and Ramon went to a restaurant and ordered 3 hamburgers at \$3.95 each, 2 salads at \$2.59 each, and 2 soft drinks at \$1.25 each. Estimate the cost of the meal.

20. Wendy and George went to a Tex-Mex restaurant and ordered 3 tacos at \$2.95 each, 2 burritos at \$3.95 each, and 2 soft drinks at \$1.75 each. Estimate the cost of the meal.

21. A 6% sales tax is added to the cost of a \$13,587 car. Estimate the amount of the sales tax.

22. The city's population 10 years ago was 287,356. The population has grown by 18%. Estimate the increase in population over the last 10 years.

23. The governor of a midwestern state estimates that 8.3% of her budget of \$19,364,282,153.18 goes toward education. Estimate the amount of money spent on education.

24. A recent medical survey estimates that 6.23% of all operations performed in hospitals are not necessary. Assume that 28,364,122 operations were performed last year. Estimate the number of unnecessary operations.

To Think About

25. Carlos usually averages 43 long-distance calls per month. The average cost of his long-distance calls is \$3.24 per call. A new long-distance telephone company has promised Carlos a 23% *reduction* in the cost of his phone bills. If that is true, estimate how much he will save in *one year*.

26. Maria usually drives to work in her car. Her car costs her 8 cents per mile to drive. She drives 22 miles round trip each day when driving to work. She drives to work about 280 days per year. She can buy an unlimited-use bus pass for one year for \$300. Estimate how much she would save by taking the bus to work.

Developing Your Study Skills

READING THE TEXTBOOK

Your homework time each day should begin with the careful reading of the section(s) assigned in your textbook. Usually, much time and effort have gone into the selection of a particular text, and your instructor has decided that this is the book that will help you to become successful in this mathematics class. Textbooks are expensive, but they can be a wise investment if you take advantage of them by reading them.

Reading a mathematics textbook is unlike reading many other types of books that you may find in your literature, history, psychology, or sociology courses. Mathematics texts are technical books that provide you with exercises to practice on. Reading a mathematics text requires slow and careful reading of each word, which takes time and effort.

Begin reading your textbook with a paper and pencil in hand. As you come across a new definition, or concept, underline it in the text and/or write it down in your notebook. Whenever you encounter an unfamiliar term, look it up and make a note of it. When you come to an example, work through it step by step. Be sure to read each word and to follow directions carefully.

Notice the helpful hints that the author provides. They guide you to correct solutions and prevent you from making errors. Take advantage of these pieces of expert advice.

Be sure that you understand what you are reading. Make a note of any of those things that you do not understand and ask your instructor about them. Do not hurry through the material. Learning mathematics takes time.

0.7 Mathematics Blueprint for Problem Solving

After studying this section, you will be able to:

1 *Use the mathematical blueprint to solve applied problems.*

1 *Using the Mathematical Blueprint to Solve Applied Problems*

When a builder constructs a new home or office building, he often has a blueprint. This accurate drawing shows the basic form of the building. It also shows the dimensions of the structure to be built. This blueprint serves as a useful reference throughout the construction process.

Similarly, when solving applied problems, it is helpful to have a "mathematical blueprint." This is a simple way to organize the information provided in the word problem, in a chart, or in a graph. You can record the facts you need to use. You can determine what it is you are trying to find and how you can go about actually finding it. You can record other information that you think will be helpful as you work through the problem.

As we solve applied problems, we will utilize three steps.

Step 1 *Understand the problem.* Here we will read through the problem and use the mathematical blueprint as a guide to assist us in thinking through the steps needed to solve the problem.

Step 2 *Solve the problem and state the answer.* We will use arithmetic or algebraic procedures along with problem-solving strategies to find a solution.

Step 3 *Check.* We will use a variety of techniques to see if the answer in step 2 is the solution to the word problem.

EXAMPLE 1 Nancy and John want to install wall-to-wall carpeting in their living room. The floor of the rectangular living room is $11\frac{2}{3}$ feet wide and $19\frac{1}{2}$ feet long. How much will it cost if the carpet is \$18.00 per square yard?

1. *Understand the problem.*
First, read the problem carefully. Drawing a sketch of the living room may help you to see what is required. The carpet will cover the floor of the living room. This is area. Now we fill in the mathematics blueprint.

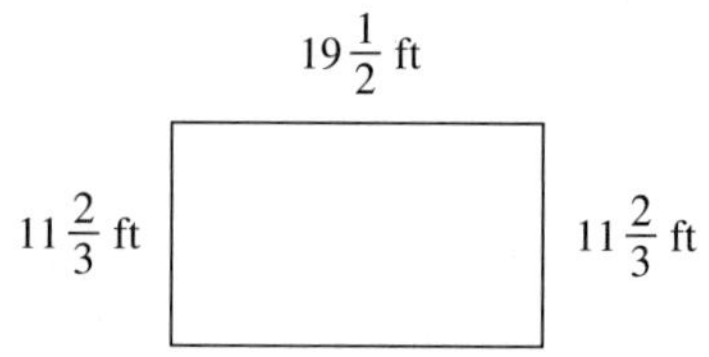

MATHEMATICS BLUEPRINT FOR PROBLEM SOLVING			
Gather the facts.	What am I solving for?	What must I calculate?	Key points to remember.
The living room measures $11\frac{2}{3}$ ft by $19\frac{1}{2}$ ft. The carpet costs \$18.00 per square yard.	(a) The area of the room in square feet (b) The area of the room in square yards (c) The cost of the carpet	(a) Multiply $11\frac{2}{3}$ ft by $19\frac{1}{2}$ ft to get area in square feet. (b) Divide the number of square feet by 9 to get the number of square yards. (c) Multiply the number of square yards by \$18.00.	There are 9 square feet, 3 feet × 3 feet, in 1 square yard; therefore, we must divide the number of square feet by 9 to obtain square yards.

2. *Solve and state the answer.*

(a) To find the area of a rectangle, we multiply the length times the width.

$$11\frac{2}{3} \times 19\frac{1}{2} = \frac{35}{3} \times \frac{39}{2}$$

$$= \frac{455}{2} = 227\frac{1}{2} \text{ sq ft}$$

A minimum of $227\frac{1}{2}$ square feet of carpet will be needed. We say a minimum because some carpet may be wasted in cutting. Carpet is sold by the square yard. We will want to know the amount of carpet needed in square yards.

(b) To determine the area in square yards, we divide $227\frac{1}{2}$ by 9. 9 sq ft = 1 sq yd.

$$227\frac{1}{2} \div 9 = \frac{455}{2} \div \frac{9}{1}$$

$$= \frac{455}{2} \times \frac{1}{9} = \frac{455}{18} = 25\frac{5}{18} \text{ sq yd}$$

A minimum of $25\frac{5}{18}$ square yards of carpet will be needed.

(c) Since the carpet costs \$18.00 per square yard, we will multiply the number of square yards needed by \$18.00.

$$25\frac{5}{18} \times 18 = \frac{455}{18} \times \frac{18}{1} = \$455$$

The carpet will cost a minimum of \$455.00 for this room.

3. *Check.*

We will estimate to see if our answers are reasonable.

(a) We will estimate by rounding each number to the nearest 10.

$$11\frac{2}{3} \times 19\frac{1}{2} \longrightarrow 10 \times 20 = 200 \text{ sq ft}$$

This is close to our answer to $227\frac{1}{2}$ sq ft. Our answer is reasonable. ✓

(b) We will estimate by rounding to one significant digit.

$$227\frac{1}{2} \div 9 \longrightarrow 200 \div 10 = 20 \text{ sq yd}$$

This is close to our answer of $25\frac{5}{18}$ sq yd. Our answer is reasonable. ✓

(c) We will estimate by rounding each number to the nearest 10.

$$25\frac{5}{18} \times 18 \longrightarrow 30 \times 20 = \$600$$

This is close to our answer of \$450. Our answer seems reasonable. ✓

Practice Problem 1

Jeff went to help Abby pick out carpet for her apartment. Her rectangular living room measures $16\frac{1}{2}$ feet by $10\frac{1}{2}$ feet. How much will it cost to carpet the room if the carpet costs \$20 per square yard?

MATHEMATICS BLUEPRINT FOR PROBLEM SOLVING			
Gather the facts.	What am I solving for?	What must I calculate?	Key points to remember.

To Think About Assume that the carpet in Example 1 comes in a standard width of 12 feet wide. How much wasted carpet will there be if the carpet is laid out for a 12-foot width and a length of $19\frac{1}{2}$ feet? How much wasted carpet would there be if the carpet is laid in two sections side to side that are each $11\frac{2}{3}$ feet long? Assuming you have to pay for wasted carpet, what is the minimum cost to carpet the room?

EXAMPLE 2 Below is a chart showing the 1995 sales of Micropower Computer Software for each of the four regions of the United States. Use the chart to answer the following questions (round all answers to nearest whole percent):

(a) What percent of the sales personnel are assigned to the Northeast?

(b) What percent of the volume of sales is attributed to the Northeast?

(c) What percent of the sales personnel are assigned to the Southeast?

(d) What percent of the volume of sales is attributed to the Southeast?

(e) Which of these two regions of the country has sales personnel that appear to be more effective in terms of the volume of sales?

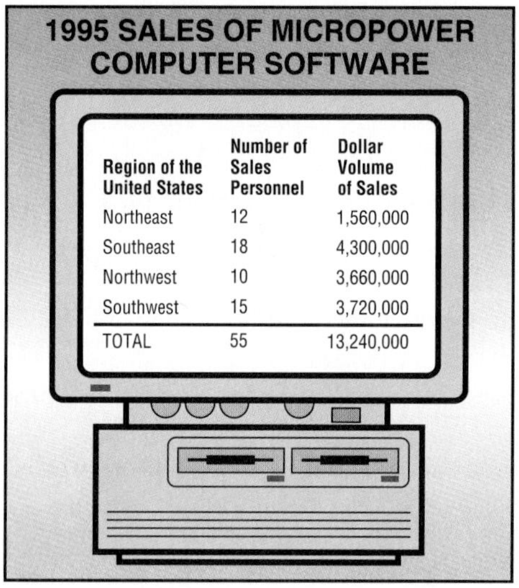

Region of the United States	Number of Sales Personnel	Dollar Volume of Sales
Northeast	12	1,560,000
Southeast	18	4,300,000
Northwest	10	3,660,000
Southwest	15	3,720,000
TOTAL	55	13,240,000

1. *Understand the problem*
We will only need to deal with figures from the Northeast region and the Southeast region.

MATHEMATICS BLUEPRINT FOR PROBLEM SOLVING			
Gather the facts.	**What am I solving for?**	**What must I calculate?**	**Key points to remember.**
Personnel: 12 Northeast 18 Southeast 55 total Sales Volume: 1,560,000 NE 4,300,000 SE 13,240,000 Total	(a) The percent of the total personnel that is in the Northeast (b) The percent of the total sales made in the Northeast (c) The percent of the total personnel that is in the Southeast (d) The percent of the total sales made in the Southeast (e) Compare the percentages from the two regions	(a) 12 of 55 is what percent? Divide. $12 \div 55$ (b) 1,560,000 of 13,240,000 is what percent? $1{,}560{,}000 \div 13{,}240{,}000$ (c) $18 \div 55$ (d) $4{,}300{,}000 \div 13{,}240{,}000$	We do not need to use the numbers relating to the Northwest or the Southwest in this problem.

2. *Solve and state the answer.*

(a) $\dfrac{12}{55} = 0.21818 \approx 22\%$ **(b)** $\dfrac{1{,}560{,}000}{13{,}240{,}000} = \dfrac{156}{1324} = 0.1178 \approx 12\%$

(c) $\dfrac{18}{55} = 0.32727 \approx 33\%$ **(d)** $\dfrac{4{,}300{,}000}{13{,}240{,}000} = \dfrac{430}{1324} \approx 0.3248 \approx 32\%$

(e) We notice that 22% of the sales force in the Northeast made 12% of the sales. The percent of the sales compared to the percent of the sales force is about half (12% of 24% would be half) or 50%. 33% of the sales force in the Southeast made 32% of the sales. The percent of sales compared to the percent of the sales force is close to 100%. We must be cautious here. *If there are no other significant factors,* it would appear that the Southeast sales force is more effective. (There may be other significant factors affecting sales, such as a recession in the Northeast, new and inexperienced sales personnel, or fewer competing companies in the Southeast.)

3. *Check.*
You may want to use a calculator to check the division in step 2, or you may use estimation.

(a) $\dfrac{12}{55} \rightarrow \dfrac{10}{60} \approx 0.17 = 17\%$ ✓ **(b)** $\dfrac{1{,}560{,}000}{13{,}240{,}000} \rightarrow \dfrac{1.6}{13} \approx 0.12 = 12\%$ ✓

(c) $\dfrac{18}{55} \rightarrow \dfrac{20}{60} \approx 0.33 = 33\%$ ✓ **(d)** $\dfrac{4{,}300{,}000}{13{,}240{,}000} \rightarrow \dfrac{4.3}{13} \approx 0.33 = 33\%$ ✓

Practice Problem 2

Using the chart for Example 2, answer the following questions (round all answers to nearest whole percent):

(a) What percent of the sales personnel are assigned to the Northwest?

(b) What percent of the sales volume is attributed to the Northwest?

(c) What percent of the sales personnel are assigned to the Southwest?

(d) What percent of the sales volume is attributed to the Southwest?

(e) Which of these two regions of the country has sales personnel that appear to be more effective in terms of volume of sales?

MATHEMATICS BLUEPRINT FOR PROBLEM SOLVING			
Gather the facts.	What am I solving for?	What must I calculate?	Key points to remember.

To Think About Suppose in 1996 the number of sales personnel (55) increases by 60%. What would the new number of sales personnel be? Suppose in 1997 that the number of sales personnel decreases by 60% from the number of sales personnel in 1996. What would the new number be? Why is this number not 55 since we have increased by 60% and decreased by 60%? Explain.

0.7 Exercises

Use the mathematical blueprint for problem solving to assist you in solving each of the following problems.

1. Carlos and Maria want to install wall-to-wall carpeting in their family room. The floor of the rectangular room is $11\frac{1}{4}$ feet long and $18\frac{1}{2}$ feet wide. How much will it cost to carpet the room if the carpet costs \$20.00 per square yard?

2. Michael and Barbara want to install wall-to-wall carpeting in their bedroom. The floor of the rectangular room is $13\frac{1}{2}$ feet long and $8\frac{3}{4}$ feet wide. How much will it cost to carpet the room if the carpet costs \$24.00 per square yard?

3. Fred found out his parents want to install a concrete driveway in front of their house. The driveway will measure $58\frac{1}{2}$ feet long and $18\frac{1}{2}$ feet wide. The driveway will be $\frac{1}{2}$ foot thick. How many cubic yards of cement will be needed for the driveway? (There are 27 cubic feet in 1 cubic yard. Round your answer to the nearest cubic yard.)

4. Linda found out her parents want to install a concrete driveway in front of their house. The driveway will measure $62\frac{1}{3}$ feet long and 19 feet wide. The driveway will be $\frac{2}{3}$ foot thick. How many cubic yards of cement will be needed for the driveway? (There are 27 cubic feet in one cubic yard. Round your answer to the nearest cubic yard.)

The following directions are posted on the wall at the gym.

5. Betty's athletic trainer told her to follow the beginning exercise training schedule on day 1. On day 2 she is to increase all distances and times by $\frac{1}{3}$ of day 1. On day 3 she is to increase all distances and times by $\frac{1}{3}$ of day 2. What will be her training schedule on day 3?

6. Melinda's athletic trainer told her to follow the beginning exercise training schedule on day 1. On day 2 she is to increase all distances and times by $\frac{1}{3}$ of day 1. On day 3 she is to once again increase all distances and times by $\frac{1}{3}$ of day 1. What will be her training schedule on day 3?

Beginning Exercise Training Schedule

On day 1 each athlete will begin the morning as follows:

Jog................$1\frac{1}{5}$ miles
Walk for...........$1\frac{3}{4}$ miles
Rest for........... $2\frac{1}{2}$ minutes
Walk for...........1 mile

To Think About

7. Who will have a more demanding schedule on day 3, Betty or Melinda? Why?

8. If Betty kept up the same type of increase day after day, how many miles would she be jogging on day 5?

9. If Melinda kept up the same type of increase day after day, how many miles would she be jogging on day 7?

10. Which athletic trainer would appear to have the best plan for training athletes if they used this plan for 14 days? Why?

11. Roberto drove out to visit his grandparents. His odometer read 45,678.2 when he started his trip and 47,001.2 at the end. He started and ended the trip on a full tank of gas. He made gas purchases of 11 gallons, 12.8 gallons, 13.2 gallons, and 12 gallons for the car during the trip. How many miles per gallon did the car get on the trip?

12. Evita has applied to attend a college in the Midwest and drove out to visit it. Her odometer reader 37,375.3 when she started the trip and 38,363.3 at the end. She started and ended the trip on a full tank of gas. She made gas purchases of 7.9 gallons, 10.3 gallons, 12.8 gallons, and 7 gallons for the car during the trip. How many miles per gallon did the car get on the trip?

13. In 1985, Springfield had a population of 120,000. Between 1985 and 1990, the city increased its population by 10%. Between 1990 and 1995, the city increased its population by 8%. What was the population in 1995? What might be some of the reasons for an increase in population in Springfield?

14. In Center City in 1986, 450 newborn infants died in City Hospital within 4 days of their birth. Soon the city began a massive health awareness campaign for pregnant mothers. Between 1986 and 1990, the number of newborn deaths decreased by 50%. Between 1990 and 1994, the number of newborn deaths decreased by 20%. How many newborn infants died within 4 days of their birth in City Hospital in 1994?

North Shore Community College students recently conducted a survey of 1000 passengers at Logan Airport. The passengers were classified as follows:

	Men Age 21 or Older	Women Age 21 or Older	Young People Age 20 or Younger	Total
Business	380	210	30	620
Pleasure	110	150	120	380
Total	490	360	150	1000

15. What percent of the total travelers were men age 21 or older? What percent of the business travelers were men age 21 or older? Which percentage is greater? Why do you suppose this is the case?

16. What percent of the business travelers were men age 21 or older? What percent of the business travelers were women age 21 or older? What percent of the business travelers were young people age 20 or younger? Why do you suppose this percentage is so small?

Use the following information from a pay check stub to solve Problems 17 through 20.

TOBEY & SLATER, INC.
5000 STILLWELL AVENUE
QUEENS, NY 10001

Check Number	Payroll Period From Date	Payroll Period To Date	Pay Date
495885		11-30-95	12-01-95

Name	Social Security No.	I.D. Number	File No.	Rate/Salary	Department	MS	Dep	Res
Fred J. Gilliani	012-34-5678	01	1379	1150.00	0100	M	5	NY

	Current	Year To Date		Current	Year To Date
GROSS	1 , 150 . 00	6 , 670 . 00	STATE	67 . 76	388 . 45
FEDERAL	138 . 97	781 . 07	LOCAL	5 . 18	30 . 04
FICA	87 . 98	510 . 28	DIS-SUI	. 00	. 00
W-2 GROSS		6 , 670 . 00	NET	790 . 47	4 , 960 . 16

Earnings No.	Type	Hours	Rate	Amount	Dept/Job No.	Deductions/Specials No.	Description	Amount
96	REGULAR			1 , 150 . 00	0100	82	Retirement	12 . 56
						75	Medical	36 . 28
						56	Union Dues	10 . 80

Gross pay is the pay an employee receives for his or her services before deductions. The Net pay is the pay the employee actually gets to take home. You may round each amount to the nearest whole percent for Problems 17 through 20.

17. What percent of Fred's gross pay is deducted for federal, state, and local taxes?

18. What percent of Fred's gross pay is deducted for retirement and medical?

19. What percent of Fred's gross pay does he actually get to take home?

20. What percent of Fred's deductions are special deductions?

Chapter Organizer for Fractions

Topic	Procedure	Examples
Simplifying fractions, p. 3.	**1.** Write the **numerator** and **denominator** as a product of prime factors. **2.** Use the basic rule of fractions that $\frac{a \times c}{b \times c} = \frac{a}{b}$ for any factor that appears in both the numerator and the denominator. **3.** Multiply the remaining factors for the numerator and separately for the denominator.	$\frac{15}{25} = \frac{\cancel{5} \cdot 3}{\cancel{5} \cdot 5} = \frac{3}{5}$ $\frac{36}{48} = \frac{\cancel{2} \cdot \cancel{2} \cdot 3 \cdot \cancel{3}}{\cancel{2} \cdot \cancel{2} \cdot 2 \cdot 2 \cdot \cancel{3}} = \frac{3}{4}$ $\frac{26}{39} = \frac{2 \cdot \cancel{13}}{3 \cdot \cancel{13}} = \frac{2}{3}$
Changing improper fractions to mixed numbers, p. 6.	**1.** Divide the denominator into the numerator to obtain the whole-number part of the mixed fraction. **2.** The remainder from the division will be the numerator of the fraction. **3.** The denominator remains unchanged.	$\frac{14}{3} = 4\frac{2}{3}$ since $\begin{array}{r} 4 \\ 3\overline{)14} \\ \underline{12} \\ 2 \end{array}$ $\frac{19}{8} = 2\frac{3}{8}$ since $\begin{array}{r} 2 \\ 8\overline{)19} \\ \underline{16} \\ 3 \end{array}$
Changing mixed numbers to improper fractions, p. 8.	**1.** Multiply the whole number by the denominator and add the result to the numerator. This will yield the new numerator. **2.** The denominator does not change.	$4\frac{5}{6} = \frac{(4 \times 6) + 5}{6} = \frac{24 + 5}{6} = \frac{29}{6}$ $3\frac{1}{7} = \frac{(3 \times 7) + 1}{7} = \frac{21 + 1}{7} = \frac{22}{7}$
Changing fractions to equivalent fractions with a given denominator, p. 14.	**1.** Divide the original denominator into the new denominator. This result is the value that we use for multiplication. **2.** Multiply numerator and denominator of the original fraction by that value.	$\frac{4}{7} = \frac{?}{21}$ $\begin{array}{r} 3 \\ 7\overline{)21} \end{array} \leftarrow$ Use this to multiply $\frac{4 \times 3}{7 \times 3} = \frac{12}{21}$
Addition and subtraction of fractions with the same denominator, p. 11.	**1.** Add or subtract the numerators. **2.** Leave the denominator unchanged. **3.** Simplify the answer if possible.	$\frac{1}{12} + \frac{5}{12} = \frac{6}{12} = \frac{1}{2}$ $\frac{13}{17} - \frac{5}{17} = \frac{8}{17}$
Finding the LCD (least common denominator) of two or more fractions, p. 12.	**1.** Write each denominator as the product of prime factors. **2.** The LCD is the product of each different factor. **3.** If a factor occurs more than once in any one denominator, the LCD will contain that factor repeated the greatest number of times that it occurs in any one denominator.	Find the LCD of $\frac{4}{15}$ and $\frac{3}{35}$. $15 = 5 \cdot 3$ $35 = 5 \cdot 7$ $\text{LCD} = 3 \cdot 5 \cdot 7 = 105$ Find the LCD of $\frac{11}{18}$ and $\frac{7}{45}$. $18 = 3 \cdot 3 \cdot 2$ (factor 3 twice) $45 = 3 \cdot 3 \cdot 5$ (factor 3 twice) $\text{LCD} = 2 \cdot 3 \cdot 3 \cdot 5 = 90$
Adding and subtracting fractions that do not have a common denominator, p. 13.	**1.** Find the LCD. **2.** Change each fraction to an equivalent fraction with the LCD for a denominator. **3.** Add or subtract the fractions, simplifying the answer if possible.	$\frac{3}{8} + \frac{1}{3} = \frac{3 \cdot 3}{8 \cdot 3} + \frac{1 \cdot 8}{3 \cdot 8}$ $= \frac{9}{24} + \frac{8}{24} = \frac{17}{24}$ $\frac{11}{12} - \frac{1}{4} = \frac{11}{12} - \frac{1 \cdot 3}{4 \cdot 3}$ $= \frac{11}{12} - \frac{3}{12} = \frac{8}{12} = \frac{2}{3}$
Adding and subtracting mixed numbers, p. 16.	**1.** Change the mixed numbers to improper fractions. **2.** Follow the rules for adding and subtracting fractions. **3.** If necessary, change your answer to a mixed number.	$1\frac{2}{3} + 1\frac{3}{4} = \frac{5}{3} + \frac{7}{4} = \frac{5 \cdot 4}{3 \cdot 4} + \frac{7 \cdot 3}{4 \cdot 3}$ $= \frac{20}{12} + \frac{21}{12} = \frac{41}{12} = 3\frac{5}{12}$ $2\frac{1}{4} - 1\frac{3}{4} = \frac{9}{4} - \frac{7}{4} = \frac{2}{4} = \frac{1}{2}$

Chapter Organizer for Decimals and Percents

Topic	Procedure	Examples
Multiplying fractions, p. 22.	**1.** If there are no common factors, multiply the numerators. Then multiply the denominators. **2.** If possible, write numerators and denominators as the product of prime factors. Use the basic rule of fractions for any value that appears in both a numerator and a denominator. Multiply the remaining factors in the numerator. Multiply the remaining factors in the denominator.	$\frac{3}{7}\times\frac{2}{13}=\frac{6}{91}$ $\frac{6}{15}\times\frac{35}{91}=\frac{2\cdot\cancel{3}}{\cancel{3}\cdot\cancel{5}}\times\frac{\cancel{5}\cdot\cancel{7}}{\cancel{7}\cdot 13}=\frac{2}{13}$ $3\times\frac{5}{8}=\frac{3}{1}\times\frac{5}{8}=\frac{15}{8}$ or $1\frac{7}{8}$
Dividing fractions, p. 24.	**1.** Change the division sign to multiplication. **2.** Invert the second fraction. **3.** Multiply the fractions.	$\frac{4}{7}\div\frac{11}{3}=\frac{4}{7}\times\frac{3}{11}=\frac{12}{77}$ $\frac{5}{9}\div\frac{5}{7}=\frac{\cancel{5}}{9}\times\frac{7}{\cancel{5}}=\frac{7}{9}$
Multiplying and dividing mixed numbers, p. 24.	**1.** Change each mixed number to an improper fraction. **2.** Use the rules for multiplying or dividing fractions. **3.** Change any improper fractions in the answer to mixed numbers.	$2\frac{1}{4}\times 3\frac{3}{5}=\frac{9}{4}\times\frac{18}{5}$ $=\frac{3\cdot 3}{2\cdot\cancel{2}}\times\frac{\cancel{2}\cdot 3\cdot 3}{5}$ $=\frac{3\cdot 3\cdot 3\cdot 3}{2\cdot 5}=\frac{81}{10}=8\frac{1}{10}$ $1\frac{1}{4}\div 1\frac{1}{2}=\frac{5}{4}\div\frac{3}{2}$ $=\frac{5}{4}\times\frac{2}{3}=\frac{5}{2\cdot 2}\times\frac{2}{3}=\frac{5}{6}$
Changing fractional form to decimal form, p. 31.	Divide the **denominator** into the **numerator.**	$\frac{5}{8}=0.625$ Since $8\overline{)5.000}$ = 0.625 48 20 16 40 40
Changing decimal form to fractional form, p. 32.	**1.** Write the decimal as a fraction with a denominator of 10, 100, 1000, and so on. **2.** Simplify the fraction, if possible.	$0.37=\frac{37}{100}$ $\quad 0.375=\frac{375}{1000}=\frac{3}{8}$ $0.4=\frac{4}{10}=\frac{2}{5}$
Adding and subtracting decimals, p. 33.	**1.** Carefully line up the decimal points as an indicated addition or subtraction. (Extra zeros may be added if desired.) **2.** Add or subtract the appropriate digits.	Add. 1.236 + 7.825 1.236 + 7.825 9.061 Subtract. 2 − 1.32 2.00 − 1.32 0.68
Multiplying decimals, p. 34.	**1.** First multiply the appropriate digits. **2.** Count the total number of decimal places in each number being multiplied. **3.** Find the sum of the two decimal places. **4.** Place the decimal point so that that number of decimal places appears in the answer.	0.9 (one place) × 0.7 (one place) 0.63 (two places) 0.009 (three places) × 0.07 (two places) 0.00063 (five places)
Dividing decimals, p. 35.	**1.** Count the number of decimal places in the **divisor.** **2.** Move the decimal point to the right the same number of places in both the **divisor** and **dividend.** **3.** Mark that position with a caret ($_\wedge$). **4.** Perform the division. Line up the decimal point in the **quotient** with the position indicated by the caret in the **dividend.**	Divide. 7.5 ÷ 0.6. Move decimal point one place to the right. $0.6_\wedge\overline{)7.5_\wedge 0}$ = 12.5 6 15 12 30 30 Therefore, 7.5 ÷ 0.6 = 12.5.

Chapter Organizer for Decimals and Percents

Topic	Procedure	Examples
Multiplying by a multiple of 10, p. 37.	For every zero in the multiplier, move the decimal point one place to the right.	$3.86 \times 100 = 386$ $0.0072 \times 1000 = 7.2$
Dividing by a multiple of 10, p. 37.	For every zero in the **divisor,** move the decimal point one place to the left.	$12.36 \div 10 = 1.236$ $1.97 \div 1000 = 0.00197$
Changing a decimal to percent, p. 41.	1. Move the decimal point two places to the right. 2. Add the % symbol.	$0.46 = 46\%$ $0.002 = 0.2\%$ $1.59 = 159\%$ $0.013 = 1.3\%$ $0.0007 = 0.07\%$
Changing a percent to a decimal, p. 42.	1. Remove the % symbol. 2. Move the decimal point two places to the left.	$49\% = 0.49$ $180\% = 1.8$ $59.8\% = 0.598$ $0.13\% = 0.0013$
Finding a percent of a number, p. 43.	1. Convert the **percent** to a decimal. 2. Multiply the decimal by the number.	Find 12% of 86. $12\% = 0.12$ $\begin{array}{r} 86 \\ \times\, 0.12 \\ \hline 1\,72 \\ 8\,6 \\ \hline 10.32 \end{array}$ Therefore, 12% of 86 = 10.32.
Finding what percent one number is of another number, p. 44.	1. Place the number after the word *of* in the denominator. 2. Place the other number in the **numerator.** 3. If possible, simplify the fraction. 4. Change the **fraction** to a decimal. 5. Express the decimal as a **percent.**	What percent of 8 is 7? $\frac{7}{8} = 0.875 = 87.5\%$ 42 is what percent of 12? $\frac{42}{12} = \frac{7}{2} = 3.5 = 350\%$
Estimation, p. 49.	1. Round each number to an appropriate place. 2. Perform the calculation with the rounded numbers.	Estimate the number of square feet in a room that is 22 feet long and 13 feet wide. Assume that the room is rectangular in shape. 1. We round 22 to 20. We round 13 to 10. 2. To find the area of a rectangle, we multiply length times width. We multiply $20 \times 10 = 200.$ We estimate that there are 200 square feet in the room.
Problem Solving, p. 54.	In solving an applied problem, you may find it helpful to complete the following steps. You will not use all of the steps all of the time. Choose the steps that best fit the conditions of the problem. 1. *Understand the problem.* (a) Read the problem carefully. (b) Draw a picture if this helps you. (c) Use the Mathematics Blueprint for Problem Solving. 2. *Solve and state the answer.* 3. *Check.* (a) Estimate to see if your answer is reasonable. (b) Repeat your calculation. (c) Work backwards from your answer. Do you arrive at the original conditions of the problem?	Susan is installing wall to wall carpeting in her $10\frac{1}{2}$ ft by 12 ft bedroom. How much will it cost at \$20 a square yard? 1. *Understand the problem.* (a) We need to find the area of the room in square yards. (b) Then we can find the cost. 2. *Solve and state the answer.* (a) $10\frac{1}{2} \times 12 = \frac{21}{2} \times \frac{12}{1} = 84$ sq ft (b) $84 \div 9 = 14$ sq yd (c) $14 \times 20 = \$280$ The carpeting will cost \$280. 3. *Check.* (a) $10 \times 12 = 120$ sq ft $120 \div 9 = 13\frac{1}{3}$ sq yd (b) $13 \times 20 = \$260$ Our answer is reasonable. ✓

Chapter 0 Review Problems

0.1 *In Problems 1 through 4, simplify.*

1. $\frac{36}{48}$ **2.** $\frac{15}{50}$ **3.** $\frac{24}{72}$ **4.** $\frac{18}{30}$

5. Write as an improper fraction $2\frac{5}{7}$. **6.** Write as a mixed number $\frac{34}{5}$. **7.** Write as a mixed fraction $\frac{27}{4}$.

Change the given fraction to an equivalent fraction with the specified denominator.

8. $\frac{5}{8} = \frac{?}{24}$ **9.** $\frac{1}{3} = \frac{?}{45}$ **10.** $\frac{4}{7} = \frac{?}{21}$ **11.** $\frac{2}{5} = \frac{?}{55}$

0.2 *Combine.*

12. $\frac{2}{3} + \frac{1}{4}$ **13.** $\frac{1}{12} + \frac{3}{8}$ **14.** $\frac{7}{20} - \frac{1}{12}$ **15.** $\frac{5}{12} - \frac{1}{16}$

16. $3\frac{1}{6} + 2\frac{3}{5}$ **17.** $1\frac{1}{4} + 2\frac{7}{10}$ **18.** $2\frac{1}{5} - 1\frac{2}{3}$ **19.** $3\frac{1}{15} - 1\frac{3}{20}$

0.3 *Multiply.*

20. $\frac{2}{3} \times \frac{9}{10}$ **21.** $\frac{3}{5} \times \frac{10}{21}$ **22.** $6 \times \frac{5}{11}$ **23.** $2\frac{1}{3} \times 4\frac{1}{2}$ **24.** $1\frac{1}{8} \times 2\frac{1}{9}$ **25.** $\frac{4}{7} \times 5$

Divide.

26. $\frac{1}{3} \div \frac{1}{6}$ **27.** $\frac{1}{2} \div \frac{1}{8}$ **28.** $\frac{2}{5} \div \frac{4}{3}$ **29.** $\frac{5}{7} \div \frac{15}{14}$ **30.** $\frac{15}{16} \div 6\frac{1}{4}$ **31.** $2\frac{6}{7} \div \frac{10}{21}$

0.4 *Combine.*

32. 1.324 + 2.008 + 1.130 **33.** 24.831 − 17.094 **34.** 14.037 − 2.61

35. 1.6 + 3.21 + 0.004 **36.** 1.3 + 1.8 + 2.6 + 7.2 + 0.8 **37.** 100.01 + 10.001 + 1.1011

Multiply.

38. 0.002×4.31

39. 362.341×1000

40. $2.6 \times 0.03 \times 1.02$

41. $1.08 \times 0.06 \times 160$

Divide.

42. $0.186 \div 100$

43. $71.32 \div 1000$

44. $0.186 \div 93$

45. $1.35 \div 0.015$

46. $0.147 \div 2.1$

47. $0.19 \div 0.38$

48. Write as a percent. $\frac{3}{8}$

49. Write as a simplified fraction. 24%

0.5 *In Problems 50 through 53, write the percentages in decimal form.*

50. 1.4%

51. 36.1%

52. 0.02%

53. 125.3%

54. What is 65% of 400?

55. Find 250% of 36.

56. What is 1.8% of 1000?

57. What percent of 120 is 15?

58. What percent of 1250 is 750?

59. 36% of Americans have blood type B. If there are 270,000,000 Americans, how many of these have blood type B?

60. In a given university, 720 of the 960 freshmen had a math deficiency. What percentage of the class had a math deficiency?

0.6 *Estimate each of the following. Do not find an exact value.*

61. $234{,}897 \times 1{,}936{,}112$

62. $357 + 923 + 768 + 417$

63. $780{,}000 - 198{,}000$

64. $7\frac{1}{3} + 3\frac{5}{6} + 8\frac{3}{7}$

65. Find 18% of $56,297

66. $12{,}482 \div 389$

67. Estimate the cost to have a chimney repaired by a stone mason who charges $18.00 per hour and says the job will take her $3\frac{1}{2}$ days. Assume that she works 8 hours in one day.

68. Estimate the monthly cost for each of three roommates who want to share an apartment that costs $923.50 per month.

0.7 *Solve. You may use the mathematical blueprint for problem solving.*

69. Mr. and Mrs. Carr are installing wall-to-wall carpeting in a room that measures $12\frac{1}{2}$ ft by $9\frac{2}{3}$ ft. How much will it cost if the carpet is $26.00 per square yard?

70. The population of Calais was 34,000 in 1980 and 36,040 in 1995. What was the average percent increase per year in population?

Chapter 0 Test

In Problems 1 and 2, simplify.

1. $\frac{16}{18}$

2. $\frac{35}{40}$

3. Write as an improper fraction. $6\frac{3}{7}$

4. Write as a mixed number. $\frac{108}{33}$

5. Change the given fraction to an equivalent fraction with the specified denominator.
$\frac{11}{16} = \frac{?}{48}$

Perform the operations indicated. Simplify answers whenever possible.

6. $\frac{5}{9} + \frac{1}{6}$

7. $\frac{1}{3} + \frac{1}{4} + \frac{5}{6}$

8. $\frac{2}{3} - \frac{1}{6}$

9. $1\frac{1}{8} + 3\frac{3}{4}$

10. $3\frac{2}{3} - 2\frac{5}{6}$

11. $6 - 4\frac{3}{4}$

12. $\frac{5}{7} \times \frac{28}{15}$

13. $\frac{3}{11} \times \frac{7}{5}$

14. $\frac{7}{4} \div \frac{1}{2}$

15. $2\frac{1}{2} \times 3\frac{1}{4}$

16. $8\frac{5}{6} \div 4\frac{1}{9}$

17. Write as a fraction. 0.72

18. Write as a decimal. $\frac{7}{16}$

1. ____________
2. ____________
3. ____________
4. ____________
5. ____________
6. ____________
7. ____________
8. ____________
9. ____________
10. ____________
11. ____________
12. ____________
13. ____________
14. ____________
15. ____________
16. ____________
17. ____________
18. ____________

19.

20.

21.

22.

23.

24.

25.

26.

27.

28.

29.

30.

31.

32.

33.

34.

35.

36.

Perform the calculations indicated.

19. $1.6 + 3.24 + 9.8$

20. $2003.42 - 196.8$

21. 32.8×0.04

22. 0.1632×100

23. $12.88 \div 0.056$

24. $19{,}365.8 \div 1000$

25. Write as a percent. 0.073

26. Write as a decimal. 196.5%

27. What is 18% of 350?

28. What is 2% of 16.8?

29. What percent of 120 is 15?

30. What percent of 460 is 138?

31. At Western College, 416 of 800 freshmen are women. What percent are women?

32. A 4-inch stack of computer chips is on the table. Each computer chip is $\frac{2}{9}$ of an inch thick. How many computer chips are in the stack?

Estimate each of the following.

33. $52{,}344\overline{)4{,}678{,}987}$

34. $18\frac{1}{3} + 22\frac{6}{7} + 57\frac{1}{2}$

35. Estimate the amount of money it would cost to buy the following four items: a computer for \$1867.85, a computer stand for \$98.87, a computer printer for \$397.49, and a VGA monitor for \$278.59.

Solve. You may use the mathematical blueprint for problem solving.

36. Allison is paid \$14,000 per year plus a sales commission of 3% of the value of her sales. Last year she sold \$870,000 worth of products. What percent of her total income was her commission?

1
Real Numbers and Variables

Scientists on a research team at the South Pole are using a computer that can be exposed to temperatures as high as 60°C and as low as −50°C. A formula will help them determine the Fahrenheit temperature range to which the computer can be exposed. See Section 1.8.

Pretest Chapter 1

1. ______ 18. ______
2. ______ 19. ______
3. ______ 20. ______
4. ______ 21. ______
5. ______ 22. ______
6. ______ 23. ______
7. ______ 24. ______
8. ______ 25. ______
9. ______ 26. ______
10. ______ 27. ______
11. ______
12. ______
13. ______
14. ______
15. ______
16. ______
17. ______

If you are familiar with the topics in this chapter, take this test now. Check your answers with those in the back of the book. If you obtained an incorrect answer or could not do a problem, study the appropriate section of the chapter.

If you are not familiar with the topics in this chapter, do not take the test now. Instead, study the examples, work the practice problems, and then take the test.

This test will help you to identify the concepts that you have mastered and the concepts that you must study more.

Perform the necessary operation and simplify your answer.

Sections 1.1–1.2

1. $(-3) - (-6)$ **2.** $-5 + 6 - 3 - 2$

3. $(-7) + (-11)$ **4.** $7 - (-13)$

Section 1.3

5. $(-7)(-2)(+3)(-1)$ **6.** $\dfrac{-\frac{2}{3}}{-\frac{1}{4}}$

7. $(-7)(-4)$ **8.** $20 \div (-5)$

Section 1.4

9. $(-2)^4$ **10.** 4^3 **11.** $\left(-\dfrac{2}{3}\right)^3$ **12.** -4^4

Section 1.5

13. $-2x(3x - 2xy + z)$ **14.** $-(3x - 4y - 12)$

Section 1.6

Combine like terms.

15. $5x^2 - 3xy - 6x^2y - 8xy$ **16.** $11x - 3y + 7 - 4x + y$

17. $3(2x - 5y) - (x - 8y)$ **18.** $2y(x^2 + y) - 3(x^2y - 5y^2)$

Section 1.7

Simplify.

19. $6 \cdot 2 + 8 \cdot 3 - 4 \cdot 2$ **20.** $3(4)^3 + 18 \div 3 - 1$

21. $-\dfrac{1}{6}\left(\dfrac{1}{2}\right) + \dfrac{3}{4} \div \dfrac{1}{12}$ **22.** $1.22 - 4.1(1.4) + (-3.3)^2$

Section 1.8

23. Evaluate $3x^2 - 5x - 4$ when $x = -2$.

24. Evaluate $ab + 3a^2 - 5b$ when $a = 3$ and $b = -1$.

25. Determine the Celsius temperature when the Fahrenheit temperature is 77°. Use the formula $C = \dfrac{5}{9}(F - 32)$.

Section 1.9

Simplify.

26. $3[(x - y) - (2x - y)] - 3(2x + y)$

27. $2x^2 - 3x[2x - (x + 2y)]$

1.1 Addition of Signed Numbers

After studying this section, you will be able to:

1. *Understand the names of different types of numbers.*
2. *Recognize real-life situations for signed numbers.*
3. *Add signed numbers with the same sign.*
4. *Add signed numbers with opposite signs.*

1 *Different Types of Numbers*

Let's review some of the basic terms we use to talk about numbers.

Integers are numbers such as . . . , -3, -2, -1, 0, 1, 2, 3, ...

Rational numbers are numbers like $\frac{3}{2}$, $\frac{5}{7}$, $-\frac{3}{8}$, $-\frac{4}{13}$, $\frac{6}{1}$, and $-\frac{8}{2}$.

Rational numbers can be written as one integer divided by another integer (as long as the denominator is not zero!). Integers can be written as fractions ($3 = \frac{3}{1}$, for example), so we can see that all integers are rational numbers. Rational numbers can be expressed in decimal form. For example, $\frac{3}{2} = 1.5$, $-\frac{3}{8} = -0.375$, and $\frac{1}{3} = 0.333\ldots$ or $\frac{1}{3}$. It is important to note that rational numbers in decimal form are either terminating decimals or repeating decimals.

Irrational numbers are numbers that cannot be expressed as one integer divided by another integer. The numbers π, $\sqrt{2}$, and $\sqrt[3]{7}$ are irrational numbers.

Irrational numbers can be expressed in decimal form. The decimal form of an irrational number is a nonterminating, nonrepeating decimal. For example, $\sqrt{2} = 1.414213\ldots$ can be carried out to an infinite number of decimal places with no repeating of digits.

EXAMPLE 1 Classify each of the following numbers as an integer, as a rational number, or as an irrational number.

(a) 5 **(b)** $-\dfrac{1}{3}$ **(c)** 2.85 **(d)** $\sqrt{2}$ **(e)** 0.777 . . .

Make a table. Check off the description of the number that applies.

Number	Integer	Rational Number	Irrational Number
a. 5	✓	✓	
b. $-\frac{1}{3}$		✓	
c. 2.85		✓	
d. $\sqrt{2}$			✓
e. 0.777 . . .		✓	

Practice Problem 1 Classify each of the following numbers.

(a) $-\dfrac{2}{5}$ **(b)** 1.515151 . . . **(c)** -8 **(d)** π ■

Integers and rational numbers can be pictured on a number line.

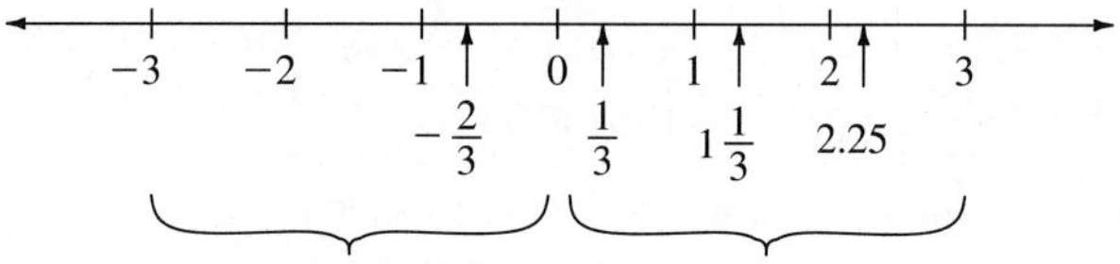

Positive numbers are to the right of 0 on the number line.

Negative numbers are to the left of 0 on the number line.

The **signed numbers** include the positive numbers, the negative numbers, and zero.

2 *Real-life Situations for Signed Numbers*

We often encounter practical examples of a number line that include positive and negative rational numbers. For example, we can tell by reading the thermometer below that the temperature is 20° below 0. From the stock market report, we see that the stock opened at 36, closed at $34\frac{1}{2}$, and the net change for the day was $-1\frac{1}{2}$.

Temperature in degrees Fahrenheit

The temperature is 20° below 0

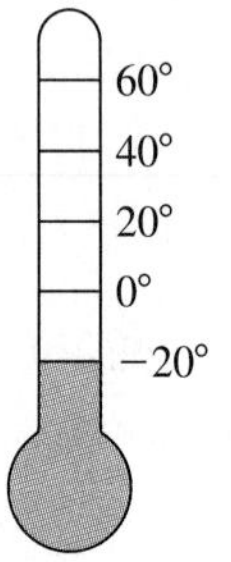

A thermometer

Stock value in dollars

The stock opened at 36

The stock closed at $34\frac{1}{2}$

Net change for the day $-1\frac{1}{2}$

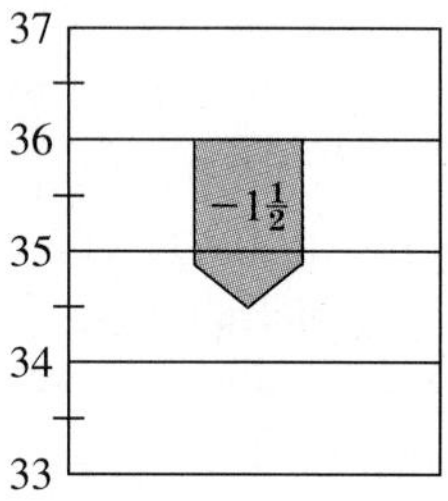

A stock market report

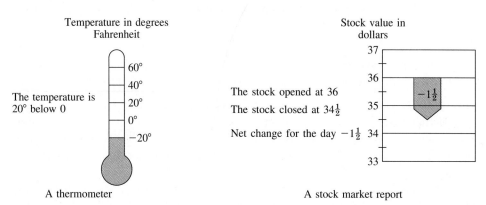

In the following example we use signed numbers to represent real-life situations.

EXAMPLE 2 Use a signed number to represent each situation.

(a) A temperature of 128.6°F below zero is recorded at Vostok Antarctica.

(b) The Himalayan peak K-2 rises 29,064 feet above sea level.

(c) The Dow gains 10.24 points.

(d) A withdrawal of $316 is made from a savings account.

A key word can help you to decide if a number is positive or negative.

(a) 128.6°F *below* zero is −128.6.

(b) 29,064 feet *above* sea level is +29,064.

(c) A *gain* of 10.24 points is +10.24.

(d) A *withdrawal* of $316 is −316.

Practice Problem 2 Use a signed number to represent each situation.

(a) A population growth of 1,259

(b) A depreciation of $763

(c) A wind chill factor of minus 10 ■

In everyday life we consider positive numbers to be the opposite of negative numbers. For example, a gain of 3 yards in a football game is the opposite of a loss of 3 yards; a check written for \$2.16 on a checking account is the opposite of a deposit of \$2.16.

Each positive number has an opposite negative number. Similarly, each negative number has an opposite positive number. **Opposite numbers** have the same magnitude but different signs and can be represented on the number line.

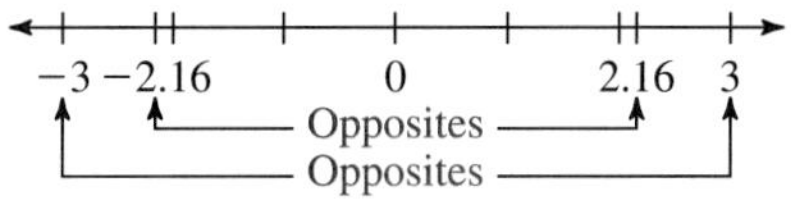

EXAMPLE 3 Find the opposite of each number. **(a)** -7 **(b)** $\frac{1}{4}$

(a) The opposite of -7 is $+7$. **(b)** The opposite of $\frac{1}{4}$ is $-\frac{1}{4}$.

Practice Problem 3 Find the opposite of each number.

(a) $+\frac{2}{5}$ **(b)** -1.92 ■

3 *Adding Signed Numbers with the Same Sign*

To use signed numbers, we need to be clear about the sign part of a signed number. When we write the number three as $+3$, the sign indicates that it is a positive number. The positive sign can be omitted. If someone writes three (3) it is understood that it is a positive three ($+3$). To write a negative number such as negative three (-3), we must include the sign.

A concept that will help us to add and subtract signed numbers is the idea of absolute value. The **absolute value** of a number is the distance between that number and zero on the number line.

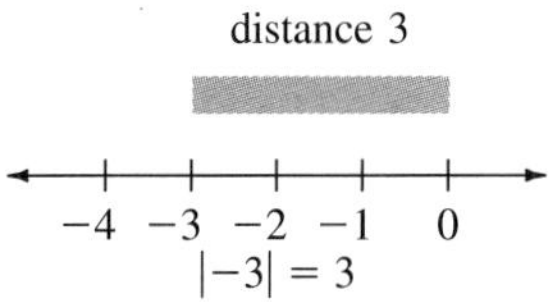

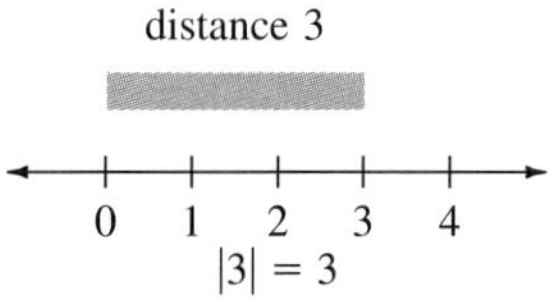

The distance is always a positive number regardless of the direction we travel. This means that the absolute value of any number will be a positive value or zero. We place the symbol "| |" around a number to mean we find the absolute value of the number.

The distance from 0 to 3 is 3, so $|3| = 3$.

This is read "the absolute value of 3 is 3."

The distance from 0 to -3 is 3, so $|-3| = 3$.

This is read "the absolute value of -3 is 3."

Some other examples are

$$|-22| = 22, \qquad |5.6| = 5.6, \qquad |0| = 0$$

Thus, the absolute value of a number can be thought of as the magnitude of the number, without regard to sign.

Now let's look at addition of signed numbers when the two numbers have the same sign. Suppose that you are keeping track of your checking account at a local bank. When you make a deposit of 5 dollars, you record it as (+5). When you write a check for 4 dollars, you record it as (−4), as a debit. Consider two situations.

Situation 1

You made a deposit of 20 dollars on one day and a deposit of 17 dollars the next day. You want to know the total value of your deposits.

Your record for situation 1.

(+20)	+	(+17)	=	(+37)
The amount of the deposit of the first day	added to	the amount of the deposit of the second day	is	the total of deposits made over the two days.

Situation 2

You write a check for 36 dollars to pay one bill and two days later write a check for 5 dollars. You want to know the total value of debits to your account for the two checks.

Your record for situation 2.

(−36)	+	(−5)	=	(−41)
The value of the first check	added to	the value of the second check	is	the total debit to your account.

In each situation we found that we added the absolute value of each number. (That is, we added the numbers without regarding their sign.) The answer always contained the sign that was common to both numbers.

We will now state these results as a formal rule.

Addition Rule for Two Numbers with Same Sign

To add two numbers with the same sign, add the absolute value of the numbers and use the common sign in the answer.

EXAMPLE 4 Add. **(a)** (+14) + (+16) **(b)** (−8) + (−7)

(a) (+14) + (+16)

$14 + 16 = 30$ *Add the absolute value of the numbers.*

$(+14) + (+16) = +30$ *Use the common sign in the answer. Here the common sign is the + sign.*

(b) (−8) + (−7)

$8 + 7 = 15$ *Add the absolute value of the numbers.*

$(-8) + (-7) = -15$ *Use the common sign in the answer. Here the common sign is the −.*

Practice Problem 4 Add. (−23) + (−35) ■

EXAMPLE 5 Add. $\left(+\frac{2}{3}\right) + \left(+\frac{1}{7}\right)$

$\left(+\frac{2}{3}\right) + \left(+\frac{1}{7}\right)$

$\left(+\frac{14}{21}\right) + \left(+\frac{3}{21}\right)$ *Change each fraction to an equivalent fraction with a common denominator of 21.*

$\frac{14}{21} + \frac{3}{21} = \frac{17}{21}$ *Add the absolute value of the numbers.*

$\left(+\frac{14}{21}\right) + \left(+\frac{3}{21}\right) = +\frac{17}{21}$ *or* $\frac{17}{21}$ *Use the common sign in the answer. Note that if no sign is written the number is understood to be positive.*

Practice Problem 5 Add. $\left(-\frac{3}{5}\right) + \left(-\frac{4}{7}\right)$ ■

EXAMPLE 6 Add. $(-4.2) + (-3.9)$

$(-4.2) + (-3.9)$

$4.2 + 3.9 = 8.1$ *Add the absolute value of the numbers.*

$(-4.2) + (-3.9) = -8.1$ *Use the common sign in the answer.*

Practice Problem 6 Add. $(-12.7) + (-9.38)$ ■

The rule for adding two numbers with the same signs can be extended to more than two numbers. If we add more than two numbers with the same sign, the answer will have the sign common to all.

EXAMPLE 7 Add. $(-7) + (-2) + (-5)$

$(-7) + (-2) + (-5)$ *We are adding three signed numbers all with the same sign. We will first add the first two numbers.*

$= (-9) + (-5)$ *Add* $(-7) + (-2) = (-9)$.

$= -14$ *Add* $(-9) + (-5) = -14$.

Of course, this can be shortened by adding the three numbers without regard to sign and then using the common sign for the answer.

Practice Problem 7 Add. $(-7) + (-11) + (-33)$ ■

4 *Adding Signed Numbers with Opposite Signs*

What if the signs of the numbers you are adding are different? Let's consider our checking account again to see how such a situation might occur.

Situation 3

You made a deposit of 30 dollars on one day. On the next day you write a check for 25 dollars. You want to know the result of your two transactions.

Your record for situation 3.

(+30)	+	(−25)	=	(+5)
A positive 30, for the deposit,	added to	a negative 25, for the check, which is a debit,	gives a result of	a net increase of 5 dollars (+5) in the account.

Situation 4

You made a deposit of 10 dollars on one day. The next day you write a check for 40 dollars. You want to know the result of your two transactions.

Your record for situation 4.

+10	+	(−40)	=	(−30)
A positive 10, for the deposit,	added to	a negative 40, for the check, which is a debit,	gives a result of	a net decrease of 30 dollars (−30) in the account

CALCULATOR

Negative Numbers

To enter a negative number on a calculator, find the key marked [+/−]. To enter the number −2, press the key 2 and then the key +/−. The display should read

[−2]

To find (−32) + (−46), enter

32 [+/−] [+] 46 [+/−] [=]

The display should read

[−78]

Try

(a) (−256) + 184

(b) 94 + (−51)

(c) (−18) − (−24)

(d) (−6) + (−10) − (−15)

Note: The [+/−] key changes the sign of a number from + to − or − to +.

The result is a negative thirty (−30), because the check was larger than the deposit. If you do not have at least 30 dollars in your account at the start of Situation 4, you have overdrawn your account.

What do we observe from Situations 3 and 4? In each case we found the difference of the absolute value of the two numbers. The sign of the result was always the sign of the number with the greater absolute value. Thus, in Situation 3, 30 is larger than 25. The sign of 30 is positive. The sign of the answer (+5) is positive. In Situation 4, 40 is larger than 10. The sign of 40 is negative. The sign of the answer (−30) is negative.

We will now state these results as a formal rule.

Addition Rule for Two Numbers with Different Signs

1. Find the difference of the larger absolute value and the smaller.
2. Give the answer the sign of the number having the larger absolute value.

EXAMPLE 8 Add. (+8) + (−7)

(+8) + (−7)	*We are to add two signed numbers with opposite signs.*
8 − 7 = 1	*Find the difference between the two absolute values, which is* 1.
(+8) + (−7) = +1 *or* 1	*The answer will have the sign of the number with the larger absolute value. That number is* +8. *Its sign is* ***positive,*** *so the answer will be* ***positive*** 1.

Practice Problem 8 Add. (−9) + (+3) ■

EXAMPLE 9 Add. $(-20) + (+13)$

$(-20) + (+13)$	*We are adding two signed numbers with different signs.*
$20 \quad - \quad 13 \quad = 7$	*Take the difference between the two absolute values.*
$(-20) + (+13) = -7$	*The answer will have the sign of the number with the larger absolute value. That number is −20, which is negative, so the answer will be negative 7.*

Practice Problem 9 Add. $(+37) + (-40)$ ■

It is useful to know the following three properties of signed numbers.

1. *Addition is commutative.* This property states that if two numbers are added the result is the same if either number is written first. The order of the numbers does not affect the result.

$$(+3) + (+6) = (+6) + (+3) = 9$$

$$(-7) + (-8) = (-8) + (-7) = -15$$

$$(-15) + (+3) = (+3) + (-15) = -12$$

2. *Addition of zero to any given number will result in that given number again.*

$$0 + (+5) = 5$$

$$(-8) + (0) = -8$$

3. *Addition is associative.* This property states that if three numbers are added it does not matter which two numbers are grouped by parentheses.

$3 + (5 + 7) = (3 + 5) + 7$ $3 + (12) = (8) + 7$ $15 = 15$	*First combine numbers inside parentheses; then combine the remaining numbers. The results are the same no matter which numbers are grouped first.*

We can use these properties along with the rules we have for adding signed numbers to add three or more numbers. We go from left to right, adding two numbers at a time.

EXAMPLE 10 Add. $\left(+\frac{3}{17}\right) + \left(-\frac{8}{17}\right) + \left(+\frac{4}{17}\right)$

$\left(+\frac{3}{17}\right) + \left(-\frac{8}{17}\right) + \left(+\frac{4}{17}\right)$	*We observe that all the fractions have a common denominator. We will add the first two numbers first.*
$= \left(-\frac{5}{17}\right) + \left(+\frac{4}{17}\right)$	*Add* $(\frac{3}{17}) + (-\frac{8}{17}) = (-\frac{5}{17})$. *The answer is negative since the larger of the two absolute values is negative.*
$= -\frac{1}{17}$	*Add* $(-\frac{5}{17}) + (+\frac{4}{17}) = (-\frac{1}{17})$. *The answer is negative since the larger of the two absolute values is negative.*

Practice Problem 10 Add. $\left(-\frac{5}{12}\right) + \left(+\frac{7}{12}\right) + \left(-\frac{11}{12}\right)$ ■

Sometimes the numbers being added have the same signs; sometimes they are different. To add three or more numbers, you may need to use rules for both.

EXAMPLE 11 Add. $(-1.8) + (+1.4) + (-2.6)$

$(-1.8) + (1.4) + (-2.6)$	*Add the first two numbers, which are opposite in sign.*
$= (-0.4) + (-2.6)$	*Add* $(-1.8) + (1.4) = (-0.4)$. *We take the difference of* 1.8 *and* 1.4 *and use the sign of the number with the larger absolute value.*
$= -3.0$	*Add* $(-0.4) + (-2.6) = -3.0$. *The signs are the same; we add the absolute values of the numbers and use the common sign.*

Practice Problem 11 Add. $(-6) + (-8) + (+2)$ ■

If many signed numbers are added, it is often easier to add numbers with like signs in a column format. Remember that addition is commutative; therefore, signed numbers can be *added in any order*. You do *not* need to combine the first two numbers as your first step.

EXAMPLE 12 Add. $(-8) + (+3) + (-5) + (-2) + (+6) + (+5)$

$$\begin{array}{r} -8 \\ -5 \\ -2 \\ \hline -15 \end{array} \qquad\qquad \begin{array}{r} +3 \\ +6 \\ +5 \\ \hline +14 \end{array}$$

All the signs are the same. Add the three negative numbers to obtain -15.

All the signs are the same. Add the three positive numbers to obtain $+14$.

Add the two results.

$$(-15) + (+14) = -1$$

The answer is negative because the larger absolute value is negative.

Practice Problem 12 Add. $(-6) + (+5) + (-7) + (-2) + (+5) + (+3)$ ■

In actual practice the positive sign and the parentheses are often omitted if it will not change the meaning of the mathematical expression. The only time we really need to show a sign of a number is when the number is negative. For example, -3. The only time we need to show parentheses when we add signed numbers is when we have two different signs preceding a number. For example, $-5 + (-6)$.

EXAMPLE 13 Add each of the following.

(a) $2.8 + (-1.3)$ **(b)** $-\frac{2}{5} + \left(-\frac{3}{5}\right)$ **(c)** $-8 + 4 + (-2)$

(a) $2.8 + (-1.3)$ is equivalent to $(+2.8) + (-1.3) = 1.5$

(b) $-\frac{2}{5} + \left(-\frac{3}{5}\right)$ is equivalent to $\left(-\frac{2}{5}\right) + \left(-\frac{3}{5}\right) = -\frac{5}{5} = -1$

(c) $-8 + 4 + (-2)$ is equivalent to $(-8) + (+4) + (-2) = (-4) + (-2) = -6$

Practice Problem 13 Add each of the following.

(a) $-2.9 + (-5.7)$ **(b)** $\frac{2}{3} + \left(-\frac{1}{4}\right)$ **(c)** $-10 + (-3) + 15 + 4$ ■

1.1 Exercises

Verbal and Writing Skills

Check off the description of the number that applies.

Number	Rational Number	Irrational Number
1. 23		
2. $-\frac{4}{5}$		
3. π		
4. 2.34		
5. $-6.666\ldots$		

Number	Rational Number	Irrational Number
6. $-\frac{7}{9}$		
7. $-2.3434\ldots$		
8. 14		
9. $\sqrt{2}$		
10. $3.232232223\ldots$		

Use a signed number to represent each situation.

11. Aaron writes a check for \$53.

12. The house appreciates \$29,000 in value.

13. Cindy hikes down a gulley for 85 feet.

14. Jon withdraws \$102 from his account.

15. The temperature rises 7°F.

16. Maya won the game by 12 points.

Find the opposite of each number.

17. $\frac{3}{4}$

18. -2

19. $-\frac{5}{13}$

20. 128

Add.

21. $(-9) + (+5)$

22. $(-7) + (+2)$

23. $(-8) + (-5)$

24. $(-12) + (-6)$

25. $\left(-\frac{1}{3}\right) + \left(+\frac{2}{3}\right)$

26. $\left(-\frac{1}{5}\right) + \left(-\frac{3}{5}\right)$

27. $(+35) + (+20)$

28. $(+18) + (+40)$

29. $(-15) + (-16)$

30. $(-23) + (-13)$

31. $\left(-\frac{2}{13}\right) + \left(-\frac{5}{13}\right)$

32. $\left(-\frac{5}{14}\right) + \left(+\frac{2}{14}\right)$

33. $(-1.5) + (-2.3)$ **34.** $(-1.8) + (-1.4)$ **35.** $(0.6) + (-0.2)$ **36.** $(-0.8) + (+0.5)$

37. $(-12) + (-13)$ **38.** $(-17) + (-21)$ **39.** $\left(\frac{1}{3}\right) + \left(-\frac{1}{4}\right)$ **40.** $\left(-\frac{2}{7}\right) + \left(+\frac{3}{14}\right)$

41. $(+2) + (-7) + (-6)$ **42.** $-5 + 3 + (-7)$ **43.** $(-3) + (+8) + (+5) + (-7)$

44. $(-2) + (-6) + (+7) + (+3)$ **45.** $(-2) + (+8) + (-3) + (-5)$ **46.** $(+9) + (+5) + (-3) + (-12)$

47. $(+31) + (-16) + (+15) + (-17)$ **48.** $(-22) + (-36) + (+19) + (+23)$ **49.** $(+0.5) + (-3.2) + (-2) + (+1.5)$

In Problems 50 to 65 the addition problems have been written without the use of extra parentheses and positive signs. If you find these difficult to do, you may insert the extra parentheses and positive signs before doing the calculation. (See Example 12)

Add.

50. $2 + (-17)$ **51.** $21 + (-4)$ **52.** $-77 + 96$ **53.** $-98 + 103$

54. $-\frac{3}{4} + \left(-\frac{2}{5}\right)$ **55.** $-\frac{7}{8} + \frac{1}{2}$ **56.** $-\frac{1}{10} + \frac{1}{2}$ **57.** $-\frac{2}{3} + \left(-\frac{1}{4}\right)$

58. $4.3 + (-3.6)$ **59.** $5.7 + (-9.1)$ **60.** $19 + (-6) + 9$ **61.** $-14 + (-23) + 5$

62. $-27 + 9 + (-54) + 30$ **63.** $18 + (-39) + 25 + (-3)$ **64.** $34.56 + (-3.14) + 0.76$ **65.** $12.59 + 1.98 + (-0.84)$

Applications

Solve each word problem.

66. Marcia has \$25 in her checking account. She makes out a check for \$32. What is her final balance?

67. The highest temperature yesterday in Fairbanks was $-18°$F. The highest temperature today is up $12°$F. What is the highest temperature today?

68. Melissa is standing at a point 126 feet below sea level. She hikes down a gulley to a point 43 feet lower. Represent her distance below sea level as a signed number.

69. A submarine is 85 feet below sea level. It dives to a point 180 feet lower. Represent her distance below sea level as a signed number.

70. On three successive running plays, Jon gained 9 yards, lost 11 yards, and gained 5 yards. What was his gain or loss?

71. The temperature at noon on a May day was 75. It dropped 18 degrees in the next 5 hours. What was the temperature at 5:00 P.M.?

72. An airplane is flying at an altitude of 8000 feet. It then descends 3000 feet and a half-hour later descends another 1500 feet. What is its new altitude?

73. Aaron owes \$258 to a credit card company. He makes a purchase of \$32 with the card and then makes a payment of \$150 on the account. How much does he still owe?

To Think About

74. What number must be added to -13 to get 5?

75. What number must be added to -18 to get 10?

76. $(-1.2) + 5.6 + (-2.8) + 1.3 + 2.8 + (-1.3) + 1.2 + (-5.6)$

77. For what values of x and y is $x + y$ negative?

78. For what values of x and y is $x + (-y) = 0$?

79. Glen makes a deposit of \$50 into his checking account and then writes out a check for \$89. What is the least amount of money he needs in his checking account to cover the transactions?

Putting Your Skills to Work

BALANCING A CHECKBOOK

Almost all college students have or will have a checking account. When you get your checking account statement from the bank, you should:

A. Write down the number that the bank statement lists as your ENDING BALANCE.

B. Add as POSITIVE NUMBERS any deposits that you have recently made but that have not yet been recorded on your bank statement.

C. Add as NEGATIVE NUMBERS any recent checks that you have made or cash withdrawals that you have taken that have not yet appeared as transactions on your bank statement.

D. Combine these numbers. The answer should agree with your checkbook balance.

Account Reconciliation

(C) Outstanding Checks/Cash Withdrawals

No. or Date	Dollars	Cents
Total (C)		

	Dollars	Cents
A) Closing Balance Shown On This Statement		
B) PLUS Deposits Made After Last Entry On This Statement		
Total (A Plus B)		
C) Minus Total Outstanding Checks		
D) YOUR CHECKBOOK BALANCE		

Solve each problem. You may use the chart on the right.

1. Brenda just received a bank statement with an ending balance of \$334.56. The statement does not include a deposit she made last week for \$46.78 nor does it include three checks she made out for \$12.39, \$67.99, and \$35. What should be the balance in her checkbook?

2. Carlos just received a bank statement with an ending balance of \$299.59. The statement does not include a deposit he made two weeks ago for \$200, nor does it include a check he made out for \$67.99, nor does it include a cash withdrawal he made from an ATM machine for \$170. What should be his balance?

COOPERATIVE GROUP ACTIVITY

3. Mr. Abington tried to balance his checkbook using the four-step procedure above. The ending balance on his bank statement was \$569.98. He found the total of his outstanding checks to be \$268.00 for the following checks: \$125.38, \$72.50, and \$98.12. He made two deposits after the last entry on the statement for \$110.00 and \$50.25. The checkbook balance in part (D) of the account reconciliation and the actual balance in his checkbook differed by \$28.25. What were the possible error(s)?

1.2 Subtraction of Signed Numbers

After studying this section, you will be able to:

1 *Subtract signed numbers.*

1 *Subtracting Signed Numbers*

So far we have developed the rules for adding signed numbers. We can use these rules to subtract signed numbers. Let's look at a checkbook situation to see how.

Situation 5

You have a balance of 20 dollars in your checking account. The bank calls you and says that a deposit of 5 dollars that belongs to another account was erroneously added to your account. They say they will correct the account balance to 15 dollars. The bank tells you that since they cannot take away the erroneous credit, they will add a debit to your account. You want to keep track of what's happening to your account.

Your record for situation 5.

(+20)	−	(+5)	=	15
From your present balance	subtract	the deposit	to give	the new balance

This equation shows what needs to be done to your account. The bank tells you that because the error happened in the past they cannot "take it away." However, they can add to your account a debit of 5 dollars. Here is the equivalent addition.

(+20)	+	(−5)	=	15
To your present balance	add	a debit	to give	the new balance

Subtracting a positive 5 has the same effect as adding a negative 5.

> **Subtraction of Signed Numbers**
>
> To subtract signed numbers, add the opposite of the second number to the first.

The rule tells us to do three things when we subtract signed numbers. First, change subtraction to addition. Second, replace the second number by its opposite. Third, add the two numbers using the rules for addition of signed numbers.

EXAMPLE 1 Subtract. $(+6) - (-2)$

$(+6) \quad - \quad (-2)$

Change subtraction to addition. *Write the opposite of the second number.*

$= (+6) \quad + \quad (+2)$

Add the two signed numbers with the same sign.

$= 8$

Practice Problem 1 Subtract. $(-9) - (+3)$ ■

EXAMPLE 2 Subtract. $(-8) - (-6)$

$$(-8) \quad - \quad (-6)$$

Change subtraction to addition. *Write the opposite of the second number.*

$$= (-8) \quad + \quad (+6)$$

Add the two signed numbers with opposite signs.

$$= -2$$

Practice Problem 2 Subtract. $(+9) - (+12)$ ■

EXAMPLE 3 Subtract. **(a)** $\left(+\frac{3}{7}\right) - \left(+\frac{6}{7}\right)$ **(b)** $\left(-\frac{7}{18}\right) - \left(-\frac{1}{9}\right)$

(a) $\left(+\frac{3}{7}\right) + \left(-\frac{6}{7}\right)$ *Change the subtraction problem to one of adding the opposite of the second number. We note that the problem has two fractions with the same denominator.*

$= -\frac{3}{7}$ *Add two numbers with different signs.*

(b) $\left(-\frac{7}{18}\right) + \left(+\frac{1}{9}\right)$ *Change subtracting to adding the opposite.*

$= \left(-\frac{7}{18}\right) + \left(+\frac{2}{18}\right)$ *Change $\frac{1}{9}$ to $\frac{2}{18}$ since* $LCD = 18$.

$= -\frac{5}{18}$ *Add two numbers with different signs.*

Practice Problem 3 Subtract. $\left(-\frac{1}{5}\right) - \left(+\frac{1}{4}\right)$ ■

EXAMPLE 4 Subtract. $(-5.2) - (-5.2)$

$(-5.2) + (+5.2)$ *Change the subtraction problem to one of adding the opposite of the second number.*

$= 0$ *Add two numbers with different signs.*

Example 4 illustrates what is sometimes called the **property of adding opposites.** When you add two real numbers that are opposites of each other, you will obtain zero. Examples of this are the following:

$$(+5) + (-5) = 0 \qquad (-186) + (186) = 0 \qquad \left(-\frac{1}{8}\right) + \left(+\frac{1}{8}\right) = 0$$

Practice Problem 4 Subtract. $\left(-\frac{3}{17}\right) - \left(-\frac{3}{17}\right)$ ■

EXAMPLE 5 Subtract -3 from -20.

Note that the -3 is subtracted FROM the -20. Thus we write the -20 first and then subtract -3 from it. Therefore, we have

$$(-20) - (-3) = (-20) + (+3) = -17$$

Practice Problem 5 Subtract -18 from 30. ■

Extra parentheses and extra positive signs within parentheses are often omitted in the writing of subtraction problems in algebra. When such problems are encountered, we want to be sure which number is subtracted from which. If it helps you to insert the optional parentheses and plus signs to make the problem clearer, then you should do so.

EXAMPLE 6 Perform the following.

(a) $-8 - 2$ **(b)** $23 - 28$ **(c)** $5 - (-3)$ **(d)** $\frac{1}{4} - 8$

(a) $-8 - 2$ *Notice that we are subtracting a positive* 2.
$= -8 - (+2)$ *Rewrite using parentheses.*
$= -8 + (-2)$ *Change to addition.*
$= -10$ *Add.*

In a similar fashion we have

(b) $23 - 28 = (+23) - (+28) = (+23) + (-28) = -5$

(c) $5 - (-3) = (+5) - (-3) = (+5) + (+3) = 8$

(d) $\frac{1}{4} - 8 = \left(+\frac{1}{4}\right) - (+8) = \left(+\frac{1}{4}\right) + (-8) = \left(+\frac{1}{4}\right) + \left(-\frac{32}{4}\right) = -\frac{31}{4}$

Practice Problem 6 Perform the following. **(a)** $-21 - 9$ **(b)** $17 - 36$ ■

EXAMPLE 7 A satellite is recording radioactive emissions from nuclear waste buried 3 miles below sea level. The satellite orbits the earth at 98 miles above sea level. How far is the satellite from the nuclear waste?

We want to find the difference between $+98$ miles and -3 miles.

$98 - (-3)$
$= 98 + (+3)$
$= 101$

The satellite is 101 miles from the nuclear waste.

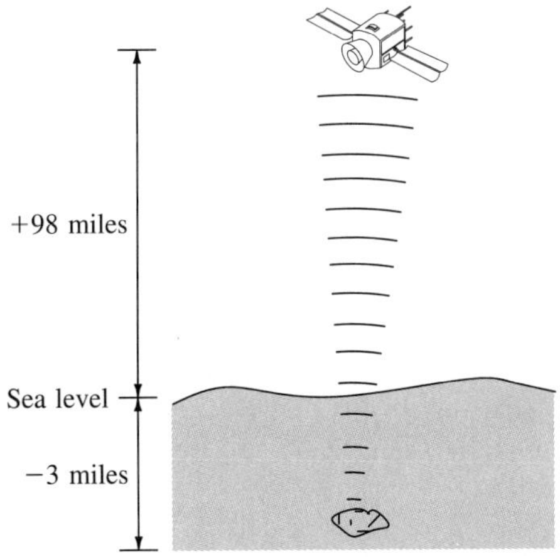

Practice Problem 7 A helicopter is directly over a sunken vessel. The helicopter is 350 feet above sea level. The vessel lies 186 feet below sea level. How far is the helicopter from the sunken vessel? ■

1.2 Exercises

Subtract by adding the opposite.

1. $(+8) - (+5)$ **2.** $(+12) - (+6)$ **3.** $(+7) - (-3)$ **4.** $(-12) - (-8)$ **5.** $(-12) - (-4)$

6. $(-17) - (-6)$ **7.** $(+20) - (+46)$ **8.** $(+15) - (+28)$ **9.** $(-32) - (-6)$ **10.** $(+18) - (-8)$

11. $(-52) - (-60)$ **12.** $(-48) - (-80)$ **13.** $(0) - (-5)$ **14.** $(0) - (-7)$ **15.** $(+15) - (+20)$

16. $(+18) - (+24)$ **17.** $(-18) - (-18)$ **18.** $(+24) - (-24)$ **19.** $(-17) - (-13)$ **20.** $(-26) - (-7)$

21. $(-0.6) - (+0.3)$ **22.** $(-0.9) - (+0.5)$ **23.** $(+2.64) - (-1.83)$ **24.** $(+0.07) - (-0.09)$ **25.** $(-1.5) - (-3.5)$

26. $\left(+\frac{1}{3}\right) - \left(-\frac{2}{5}\right)$ **27.** $\left(+\frac{3}{4}\right) - \left(-\frac{3}{5}\right)$ **28.** $\left(-\frac{2}{3}\right) - \left(+\frac{1}{4}\right)$ **29.** $\left(-\frac{1}{5}\right) - \left(+\frac{3}{8}\right)$ **30.** $\left(-\frac{7}{10}\right) - \left(+\frac{10}{15}\right)$

Perform each calculation. You may want to insert the extra parentheses and positive signs before changing the subtraction problem to an equivalent addition problem. (See Example 6.)

31. $34 - 87$ **32.** $19 - 76$ **33.** $-67 - 32$ **34.** $-98 - 34$ **35.** $2.3 - (-4.8)$

36. $8.4 - (-2.7)$ **37.** $3 - \frac{1}{5}$ **38.** $12 - \left(-\frac{2}{3}\right)$ **39.** $\frac{4}{9} - (-14)$ **40.** $\frac{5}{6} - 4$

41. $-\frac{3}{10} - \frac{3}{4}$ **42.** $-\frac{11}{12} - \frac{5}{18}$ **43.** $-135 - (-126.5)$ **44.** $-97.6 - (-146)$ **45.** $0.0067 - (-0.0432)$

46. $0.0762 - (-0.0094)$ **47.** $\frac{1}{5} - 6$ **48.** $\frac{2}{7} - (-3)$ **49.** $-\frac{3}{5} - \left(-\frac{3}{10}\right)$ **50.** $-1 - 0.0059$

51. Subtract -9 from -2. **52.** Subtract -12 from 20. **53.** Subtract 13 from -35.

Change each subtraction operation to "adding the opposite." Then combine the numbers.

54. $9 + 6 - (-5)$ **55.** $7 + (-6) - (+3)$ **56.** $8 + (-4) - (+10)$ **57.** $-10 + 6 - (-15)$

58. $18 - (-15) - 3$ **59.** $7 + (-42) - 27$ **60.** $-43 - (-17) + 25$ **61.** $-31 - (-14) - 17$

62. $-40 - (-6) + 34 - (-5)$ **63.** $-84 + 12 - (-45) - (+14)$

64. $-5 - (-30) + 20 + 40 - (-12)$ **65.** $42 - (-30) - 65 - (-11) + 20$

Applications

Solve each word problem.

66. A helicopter is 300 feet above sea level. A submerged rock directly below the helicopter is 126 feet below sea level. How far is the helicopter from the rock?

67. Yesterday Jackie had \$112 in her checking account. Today her account reads "balance $-$\$37." Find the difference in these two amounts.

68. On January 6, 1971, the temperature at Hawley Lake, Arizona, was -23°F. On January 7, 1971, the temperature at the same place was -40°F. What was the change in temperature from January 6 to January 7?

69. On January 19, 1937, the temperature at Boca, California, was -29°F. On January 20 the temperature at the same place was -45°F. What was the change in temperature from January 19 to January 20?

Cumulative Review Problems

70. $-37 + 16$ **71.** $-37 + (-14)$ **72.** $-3 + (-6) + (-10)$

73. What is the temperature after a rise of 13°C from a start of -21°C?

1.3 Multiplication and Division of Signed Numbers

After studying this section, you will be able to:

1. *Multiply signed numbers.*
2. *Divide signed numbers.*

1 *Multiplying Signed Numbers*

We are familiar with the meaning of multiplication for positive numbers. For example, $5 \times 40 = 200$ might mean that you receive five weekly checks of 40 dollars each and you gain \$200. Let's look at a situation that corresponds to $5 \times (-40)$. What might that mean?

Situation 6

You write a check for five weeks in a row to pay your weekly room rent of 40 dollars. You want to know the total impact on your checking account balance.

Your record for situation 6.

$(+5)$	(-40)	$=$	-200
The number of checks you have written	negative 40, the value of each check that was a debit to your account.	gives a result of	negative 200 dollars, a net debit to your account

Note that a multiplication symbol is not needed between the $(+5)$ and the (-40) because the two sets of parentheses indicate multiplication. In this case the multiplication $(5)(-40)$ is the same as repeated addition of five (-40)'s.

$$\underbrace{(-40) + (-40) + (-40) + (-40) + (-40)}_{\text{repeated addition of five }(-40)\text{'s}} = -200$$

This example seems to show that a positive number multiplied by a negative number is negative.

What if the negative number came first? If $(5)(-40) = -200$, then $(-40)(5) = -200$, by the commutative property of multiplication. This is an example showing *that when two numbers with opposite signs* (one positive, one negative) *are multiplied, the result is negative.*

But what if both numbers are negative? Consider the following situation.

Situation 7

Last year at college you rented a room at 40 dollars per week for 36 weeks, which included two semesters and summer school. This year you will not attend the summer session, so you will be renting the room for only 30 weeks. Thus the number of weekly rental checks will be six less than last year. You are making out your budget for this year. You want to know the financial impact of renting the room for six fewer weeks.

Your record for situation 7.

(-6)	(-40)	$=$	240
The number of checks this year compared to last is -6, which is negative to show a decrease,	the value of each check paid out, -40	gives a result of	$+240$ dollars. The product is positive, because your financial situation will be 240 dollars better this year.

You could check that the answer is positive by calculation of the total rental expenses.

	Dollars in rent last year	(36)(40) = 1440
(subtract)	Dollars in rent this year	−(30)(40) = −1200
	Extra dollars available this year	= +240

This agrees with our answer above: $(-6)(-40) = +240$.

In this situation it seems reasonable that a negative number times a negative number yields a positive answer. We have already known from arithmetic that a positive number times a positive number should yield a positive answer. Thus we might see the general rule that *when two numbers with the same sign* (both positive or both negative) *are multiplied, the result is positive.*

We will now state our rule.

Multiplication of Signed Numbers

To multiply two numbers with the **same sign,** multiply the absolute values. The sign of the result is **positive.**

To multiply two signed numbers with **opposite signs,** multiply the absolute values. The sign of the result is **negative.**

EXAMPLE 1 Multiply.

(a) $(3)(6)$ **(b)** $\left(-\frac{5}{7}\right)\left(-\frac{2}{9}\right)$ **(c)** $(-4)(8)$ **(d)** $\left(\frac{2}{7}\right)(-3)$

(a) $(3)(6) = 18$

(b) $\left(-\frac{5}{7}\right)\left(-\frac{2}{9}\right) = \frac{10}{63}$

When multiplying two numbers with the same sign, the result is a positive number.

(c) $(-4)(8) = -32$

(d) $\left(\frac{2}{7}\right)(-3) = \left(\frac{2}{7}\right)\left(-\frac{3}{1}\right) = -\frac{6}{7}$

When multiplying two numbers with opposite signs, the result is a negative number.

Practice Problem 1 Multiply.

(a) $(-6)(-2)$ **(b)** $(7)(9)$ **(c)** $\left(-\frac{3}{5}\right)\left(\frac{2}{7}\right)$ **(d)** $(40)(-20)$ ■

To multiply more than two numbers, take two numbers at a time.

EXAMPLE 2 Multiply. $(-4)(-3)(-2)$

$= (-4)(-3)(-2)$ *We first multiply the first two numbers.*

$= (+12)(-2)$ *Multiply two numbers,* (−4) *and* (−3). *The signs are the same. The answer is positive* 12.

$= -24$ *Multiply two numbers,* (+12) *and* (−2). *The signs are different. The answer is negative* 24.

Practice Problem 2 Multiply $(-3)(-4)(-3)$ ■

EXAMPLE 3 Multiply. **(a)** $(-3)(-8)$ **(b)** $\left(-\frac{1}{2}\right)(-1)(-4)$

(c) $(-2)(-2)(-2)(-2)$

Multiply two numbers at a time. See if you find a pattern.

(a) $(-3)(-8) = +24$ or 24

(b) $\left(-\frac{1}{2}\right)(-1)(-4) = +\frac{1}{2}(-4) = -2$

(c) $(-2)(-2)(-2)(-2) = +4(-2)(-2) = -8(-2) = +16$ or 16

What do you think would happen if we multiplied five negative numbers? If you guessed "negative," you probably see the pattern.

Practice Problem 3 Determine the sign of the product. Then multiply to check.

(a) $(-2)(-3)(-4)(-1)$ **(b)** $(-1)(-3)(-2)$ **(c)** $(-4)\left(-\frac{1}{4}\right)(-2)(-6)$ ■

> When you multiply two or more signed numbers:
>
> 1. The result is always **positive** if there are an **even** number of negative signs.
> 2. The result is always **negative** if there are an **odd** number of negative signs.

EXAMPLE 4 Multiply. $(-3)(+2)(-1)(+6)(-3)$

$(-3)(+2)(-1)(+6)(-3) = -$ *The answer is negative since there are three negative signs and 3 is* ***odd.***

$3 \times 2 \times 1 \times 6 \times 3 = 108$ *Multiply the numerical values.*

$(-3)(+2)(-1)(+6)(-3) = -108$ *The answer is negative* 108.

Practice Problem 4 Multiply. $(-2)(3)(-4)(-1)(-2)$ ■

For convenience, we will list the properties of multiplication.

> 1. *Multiplication is commutative.* This property states that if two numbers are multiplied the order of the numbers does not affect the result. The result is the same if either number is written first.
>
> $$(5)(7) = (7)(5) = 35, \quad \left(\frac{1}{3}\right)\left(\frac{2}{7}\right) = \left(\frac{2}{7}\right)\left(\frac{1}{3}\right) = \frac{2}{21}$$
>
> 2. *Multiplication of any number by zero will result in zero.*
>
> $$(5)(0) = 0, \quad (-5)(0) = 0, \quad (0)\left(\frac{3}{8}\right) = 0, \quad (0)(0) = 0$$
>
> 3. *Multiplication of any number by 1 will result in that same number.*
>
> $$(5)(1) = 5, \quad (1)(-7) = -7, \quad (1)\left(-\frac{5}{3}\right) = -\frac{5}{3}$$
>
> 4. *Multiplication is associative.* This property states that if three numbers are multiplied it does not matter which two numbers are grouped by parentheses.
>
> $2 \times (3 \times 4) = (2 \times 3) \times 4$ *First multiply numbers in parentheses. Then multiply the remaining numbers.*
>
> $2 \times (12) = (6) \times 4$ *The results are the same no matter which*
>
> $24 = 24$ *numbers are grouped first.*

2 Dividing Signed Numbers

What about division? Any division problem can be rewritten as a multiplication problem.

We know that $(+20) \div (+4) = +5$ because $(+4)(+5) = +20$.

Similarly, $(-20) \div (-4) = +5$ because $(-4)(+5) = -20$.

In both division problems the answer is positive 5. Thus we see that *when you divide two numbers with the same sign* (both positive or both negative) *the answer is positive.* What if the signs are different?

We know that $(-20) \div (+4) = -5$ because $(+4)(-5) = -20$.

Similarly, $(+20) \div (-4) = -5$ because $(-4)(-5) = +20$.

In these two problems the answer is negative 5. So we have reasonable evidence to see that *when you divide two numbers with different signs* (one positive and one negative) *the answer is negative.*

We will now state our rule for division.

> **Division of Signed Numbers**
>
> To divide two numbers with the **same sign,** divide the absolute values. The sign of the result is **positive.**
>
> To divide two numbers with **different signs,** divide the absolute values. The sign of the result is **negative.**

EXAMPLE 5 Divide.

(a) $12 \div 4$ **(b)** $(-25) \div (-5)$ **(c)** $\dfrac{-36}{18}$ **(d)** $\dfrac{42}{-7}$

(a) $12 \div 4 = 3$

(b) $(-25) \div (-5) = 5$

When dividing two numbers with the same sign, the result is a positive number.

(c) $\dfrac{-36}{18} = -2$

(d) $\dfrac{42}{-7} = -6$

When dividing two numbers with diffferent signs, the result is a negative number

Practice Problem 5 Divide.

(a) $(-36) \div (-2)$ **(b)** $18 \div 9$ **(c)** $\dfrac{50}{-10}$ **(d)** $(-49) \div 7$ ■

EXAMPLE 6 Divide. **(a)** $(-36) \div (0.12)$ **(b)** $(-2.4) \div (-0.6)$

(a) $(-36) \div (0.12)$ *Look at the problem to determine the sign. When dividing two numbers with different signs, the result will be a negative number.*

We then divide the absolute values.

$$\begin{array}{r} 3\,00. \\ 0.12_\wedge\overline{)36.00_\wedge} \\ \underline{36} \\ 00 \end{array}$$

Thus $(-36) \div (0.12) = -300$. The answer is a negative number.

(b) $(-2.4) \div (-0.6)$ *Look at the problem to determine the sign. When dividing two numbers with the same sign, the result will be positive.*

We then divide the absolute values.

$$0.6_{\wedge}\overline{)2.4_{\wedge}} = 4.$$
$$24$$

Thus $(-2.4) \div (-0.6) = 4$. The answer is a positive number.

Practice Problem 6 Divide.

(a) $(-1.242) \div (-1.8)$ **(b)** $(0.235) \div (-0.0025)$ ■

Note how similar the rules for multiplication and division are. When you **multiply** or **divide** two numbers with the **same** sign, you obtain a **positive** number. When you **multiply** or **divide** two numbers with **different** signs, you obtain a **negative** number.

EXAMPLE 7 Divide. $\left(-\frac{12}{5}\right) \div \left(\frac{2}{3}\right)$

$$\left(-\frac{12}{5}\right) \div \left(\frac{2}{3}\right) = \left(-\frac{12}{5}\right)\left(\frac{3}{2}\right)$$ *Divide two fractions. We invert the second fraction and multiply by the first fraction.*

$$\left(-\frac{\overset{6}{\cancel{12}}}{5}\right)\left(\frac{3}{\underset{1}{\cancel{2}}}\right) = -\frac{18}{5} \quad or \quad -3\frac{3}{5}$$ *The answer is negative since the two numbers divided have different signs.*

Practice Problem 7 Divide. $\left(-\frac{5}{16}\right) \div \left(-\frac{10}{13}\right)$ ■

Note that division can be indicated by the symbol ÷ or by the fraction bar: —. $\frac{2}{3}$ means $2 \div 3$.

EXAMPLE 8 Divide. **(a)** $\dfrac{\frac{7}{8}}{-21}$ **(b)** $\dfrac{-\frac{2}{3}}{-\frac{7}{13}}$

(a) $\dfrac{\frac{7}{8}}{-21}$ means $\frac{7}{8} \div (-21)$

$$= \left(\frac{7}{8}\right) \div \left(-\frac{21}{1}\right)$$ *Change −21 to a fraction.* $-21 = -\frac{21}{1}$

$$= \left(\frac{\overset{1}{\cancel{7}}}{8}\right)\left(-\frac{1}{\underset{3}{\cancel{21}}}\right)$$ *Change the division to multiplication. Cancel where possible.*

$$= -\frac{1}{24}$$ *Simplify.*

(b) $\dfrac{-\frac{2}{3}}{-\frac{7}{13}}$ means $\left(-\frac{2}{3}\right) \div \left(-\frac{7}{13}\right) = \left(-\frac{2}{3}\right)\left(-\frac{13}{7}\right) = \frac{26}{21}$ *or* $1\frac{5}{21}$

Practice Problem 8 Divide. **(a)** $\dfrac{-12}{-\frac{4}{5}}$ **(b)** $\dfrac{-\frac{2}{9}}{\frac{8}{13}}$ ■

There are two properties of division that are important to understand.

1. *Division of 0 by any nonzero number gives 0 as a result.*

$$0 \div 5 = 0, \quad 0 \div \frac{2}{3} = 0, \quad \frac{0}{5.6} = 0, \quad \frac{0}{1000} = 0$$

You can divide zero by 5, 3, 5.6, 1000, or any number (except 0).

2. *Division of any number by 0 is not allowed.*

$$7 \div 0 \qquad \frac{64}{0} \qquad \frac{0}{0}$$

$$\uparrow \qquad \uparrow \qquad \uparrow$$

None of these operations is possible. **You cannot divide by 0!**

When combining two numbers, it is important to be sure you know which rule applies. Think about the concepts in the following chart. See if you agree with each example.

Operation	Two Numbers with the Same Sign	Two Numbers with Different Signs
Addition	Result may be positive or negative. $9 + 2 = 11$ $-5 + (-6) = -11$	Result may be positive or negative. $-3 + 7 = 4$ $4 + (-12) = -8$
Subtraction	Result may be positive or negative. $15 - (+6) = 15 + (-6) = 9$ $-12 - (-3) = -12 + (+3) = -9$	Result may be positive or negative. $-12 - (+3) = -12 + (-3) = -15$ $5 - (-6) = 5 + (+6) = 11$
Multiplication	Result is always positive. $(9)(3) = 27$ $(-8)(-5) = 40$	Result is always negative. $(-6)(12) = -72$ $(8)(-3) = -24$
Division	Result is always positive. $150 \div 6 = 25$ $-72 \div (-2) = 36$	Result is always negative. $-60 \div 10 = -6$ $30 \div (-6) = -5$

Developing Your Study Skills

READING THE TEXTBOOK

Begin reading your textbook with a paper and pencil in hand. As you come across a new definition, or concept, underline it in the text and/or write it down in your notebook. Whenever you encounter an unfamiliar term, look it up and make a note of it. When you come to an example, work through it step by step. Be sure to read each word and to follow directions carefully.

Notice the helpful hints the author provides. They guide you to correct solutions and prevent you from making errors. Take advantage of these pieces of expert advice.

Be sure that you understand what you are reading. Make a note of any of those things that you do not understand and ask your instructor about them. Do not hurry through the material. Learning mathematics takes time.

1.3 Exercises

Multiply or divide. Be sure to write your answer in the simplest form.

1. $(3)(-12)$ **2.** $(5)(-4)$ **3.** $(-6)(-5)$ **4.** $(-8)(-7)$ **5.** $(-5)(-12)$

6. $9(-20)$ **7.** -8×3 **8.** -5×11 **9.** $0(-12)$ **10.** -3×150

11. 16×1.5 **12.** 24×2.5 **13.** $(-1.32)(-0.2)$ **14.** $0.6(-30)$ **15.** $(-1.8)(-0.03)$

16. $(-6)\left(\frac{3}{10}\right)$ **17.** $\left(\frac{12}{5}\right)(-10)$ **18.** $\left(\frac{2}{7}\right)\left(\frac{14}{3}\right)$ **19.** $\left(-\frac{3}{8}\right)\left(-\frac{5}{6}\right)$ **20.** $\left(-\frac{5}{21}\right)\left(\frac{3}{10}\right)$

21. $(12) \div (-4)$ **22.** $(-36) \div (-9)$ **23.** $0 \div (-15)$ **24.** $24 \div (-6)$ **25.** $-42 \div 7$

26. $(-52) \div (-2)$ **27.** $-360 \div (-10)$ **28.** $(-220) \div (-11)$ **29.** $105 \div (-7)$ **30.** $156 \div (-13)$

31. $(-0.6) \div 0.3$ **32.** $1.2 \div (-0.03)$ **33.** $(-2.4) \div (-0.6)$ **34.** $(0.36) \div (-0.9)$ **35.** $(-7.2) \div (8)$

36. $\frac{2}{7} \div \left(-\frac{3}{5}\right)$ **37.** $\left(-\frac{1}{5}\right) \div \left(\frac{2}{3}\right)$ **38.** $\left(\frac{5}{8}\right) \div \left(\frac{15}{16}\right)$ **39.** $\left(-\frac{3}{4}\right) \div \left(-\frac{7}{12}\right)$ **40.** $\frac{4}{9} \div \left(-\frac{8}{15}\right)$

41. $\dfrac{12}{-\frac{2}{5}}$ **42.** $\dfrac{-\frac{3}{7}}{-6}$ **43.** $\dfrac{-\frac{3}{8}}{-\frac{2}{3}}$ **44.** $\dfrac{-\frac{1}{7}}{\frac{3}{14}}$ **45.** $\dfrac{\frac{2}{5}}{-\frac{8}{15}}$

Multiply. You may want to determine the sign of the product before you multiply.

46. $(-6)(2)(-3)(4)$ **47.** $(-1)(-2)(-3)(4)$ **48.** $(-2)(-1)(3)(-2)(2)$ **49.** $(-1)(-3)(-2)(-2)(3)$

50. $(-3)(2)(-4)(0)(-2)$ **51.** $(-3)(-2)\left(\frac{1}{3}\right)(-4)(2)$ **52.** $(60)(-0.6)(-20)(0.5)$ **53.** $(-3)(-0.03)(100)(-2)$

54. $\frac{1}{9}\left(-\frac{2}{3}\right)\left(\frac{5}{7}\right)$ **55.** $\left(-\frac{5}{12}\right)\left(-\frac{4}{9}\right)\left(-\frac{3}{5}\right)$ **56.** $\left(-\frac{1}{2}\right)\left(\frac{4}{5}\right)\left(-\frac{7}{8}\right)\left(-\frac{2}{3}\right)$ **57.** $\left(-\frac{3}{4}\right)\left(-\frac{7}{15}\right)\left(-\frac{8}{21}\right)\left(-\frac{5}{9}\right)$

Mixed Review

Take a minute to review the chart following Example 8. Be sure that you can remember the sign rules for each operation. Then do problems 58 through 67.

58. $(-5) - (-2)$ **59.** $(-20) \div 5$ **60.** $(-3)(-9)$ **61.** $5 + (-7)$ **62.** $(-32) \div (-4)$

63. $8 - (-9)$ **64.** $(-6) + (-3)$ **65.** $6(-12)$ **66.** $18 \div (-18)$ **67.** $(-37) \div 37$

Applications

Solve each word problem.

68. You write a \$250 check each month for a car loan. What is the net debit to your account after 6 months?

69. After writing a check for \$100, you show a balance of $-\$70$. How much did you have in the account before you wrote the check?

70. The velocity (rate) of a projectile, $r = -30$ meters per second, indicates that it is moving to the left on a number line. Currently, it is at time $t = 0$ and at the zero mark on the number line. Find where it will be (distance) when $t = 3$ seconds. (Use $d = r \cdot t$.)

71. Referring to problem 70, where was the projectile 5 seconds ago ($t = -5$)?

Cumulative Review Problems

72. Add $(-17.4) + (8.31) + (2.40)$.

73. Add $(-13) + (-39) + (-20)$.

74. Subtract $(-3.7) - (-8.33)$.

75. Subtract $(-37) - (51)$.

Developing Your Study Skills

STEPS TOWARD SUCCESS IN MATHEMATICS

Mathematics is a building process, mastered one step at a time. The foundation of this process is built on a few basic requirements. Those who are successful in mathematics realize the absolute necessity for building a study of mathematics on the firm foundation of these six minimum requirements.

1. Attend class every day.
2. Read the textbook.
3. Take notes in class.
4. Do assigned homework every day.
5. Get help immediately when needed.
6. Review regularly.

1.4 Exponents

After studying this section, you will be able to:

1. *Write numbers in exponent form.*
2. *Evaluate numerical expressions that contain exponents.*

1 *Exponent Form*

In mathematics, we use exponents as a way to abbreviate repeated multiplication.

Long Notation		***Exponent Form***
$2 \cdot 2 \cdot 2 \cdot 2 \cdot 2 \cdot 2$	$=$	2^6
$3 \cdot 3 \cdot 3 \cdot 3$	$=$	3^4
$7 \cdot 7 \cdot 7 \cdot 7 \cdot 7$	$=$	7^5

There are two parts to exponent notation: (1) the **base** and (2) the **exponent.** The **base** tells you what number is being multiplied and the **exponent** tells you how many times this number is used as a factor. (A *factor,* you recall, is a number being multiplied.) In the first example above,

$$2 \cdot 2 \cdot 2 \cdot 2 \cdot 2 \cdot 2 = 2^{6}$$

The *base* is 2 (the number being multiplied)

the *exponent* is 6 (the number of times 2 is used as a factor)

If the base is a *positive integer,* the exponent appears to the right and slightly above the level of the number, for example, 5^6 and 8^3. If the base is a *negative integer,* then parentheses are used around the number and the exponent appears outside the parentheses. For example, $(-2)(-2)(-2) = (-2)^3$.

EXAMPLE 1 Write each of the following in exponent form.

(a) $(9)(9)(9)$ **(b)** $(13)(13)(13)(13)$

(c) $(-7)(-7)(-7)(-7)(-7)$ **(d)** $(-4)(-4)(-4)(-4)(-4)(-4)$

(a) The 9 is used as a factor three times. Thus $(9)(9)(9) = 9^3$.

(b) The 13 is used as a factor four times. Thus $(13)(13)(13)(13) = 13^4$.

(c) The -7 is used as a factor five times. The answer must contain parentheses. Thus $(-7)(-7)(-7)(-7)(-7) = (-7)^5$.

(d) The -4 is used as a factor six times. The answer must contain parentheses. Thus $(-4)(-4)(-4)(-4)(-4)(-4) = (-4)^6$.

Practice Problem 1 Write each of the following in exponent form.

(a) $(6)(6)(6)(6)$ **(b)** $(-2)(-2)(-2)(-2)(-2)$

(c) $(108)(108)(108)$ **(d)** $(-11)(-11)(-11)(-11)(-11)(-11)$ ■

How Do You Say It?

If the value has an exponent of 2, we say the value is ''squared.''

If the value has an exponent of 3, we say the value is ''cubed.''

If the value has an exponent greater than 3, we say ''to the (exponent)-th power.''

Thus

5^2 is read ''five squared.''

2^3 is read ''two cubed.''

3^6 is read ''three to the sixth power'' or simply ''three to the sixth.''

2 *Evaluating Numerical Expressions That Contain Exponents*

EXAMPLE 2 Evaluate. **(a)** 4^3 **(b)** 2^5 **(c)** 7^4

(a) $4^3 = (4)(4)(4) = 64$

(b) $2^5 = (2)(2)(2)(2)(2) = 32$

(c) $7^4 = (7)(7)(7)(7) = 2401$

Practice Problem 2 Evaluate. **(a)** 5^3 **(b)** 3^5 **(c)** 9^4 ■

If the base is negative, be especially careful in determining the sign. Notice the following:

$$(-3)^2 = (-3)(-3) = +9$$

$$(-3)^3 = (-3)(-3)(-3) = -27$$

$$(-3)^4 = (-3)(-3)(-3)(-3) = +81$$

From Section 1.2 we know that when you multiply two or more signed numbers you multiply their absolute values.

- The result is positive if there are an even number of negative signs.
- The result is negative if there are an odd number of negative signs.

This can be generalized as follows.

> **CALCULATOR**
>
> **Exponents**
>
> You can use a calculator to evaluate 3^5. Press the following keys:
>
> 3 y^x 5 =
>
> The display should read
>
> 243
>
> Try **(a)** 4^6 **(b)** $(0.2)^5$
> **(c)** 18^6 **(d)** 3^{12}

Sign Rule for Exponents

Suppose that a number is written in exponent form and the base is negative. The result is **positive** if the exponent is **even.** The result is **negative** if the exponent is **odd.**

Be careful how you read expressions with exponents and negative signs.

$$(-3)^4 \text{ means } (-3)(-3)(-3)(-3) \text{ or } +81$$

$$-3^4 \text{ means } -(3)(3)(3)(3) \text{ or } -81$$

EXAMPLE 3 Evaluate. **(a)** $(-2)^3$ **(b)** $(-4)^6$ **(c)** -3^6

(a) $(-2)^3 = -8$ *The answer is negative since the exponent 3 is odd.*

(b) $(-4)^6 = +4096$ *The answer is positive since the exponent 6 is even.*

(c) $-3^6 = -729$ *The negative sign is not contained in the parentheses. -3^6 means the negative of 3 raised to the sixth power.*

Practice Problem 3 Evaluate. **(a)** $(-3)^3$ **(b)** -2^4 **(c)** $(-2)^6$ **(d)** $(-1)^{10}$ ■

EXAMPLE 4 Evaluate. **(a)** $\left(\frac{1}{2}\right)^4$ **(b)** $(0.2)^4$ **(c)** $\left(\frac{2}{5}\right)^3$

(a) $\left(\frac{1}{2}\right)^4 = \left(\frac{1}{2}\right)\left(\frac{1}{2}\right)\left(\frac{1}{2}\right)\left(\frac{1}{2}\right) = \frac{1}{16}$ **(b)** $(0.2)^4 = (0.2)(0.2)(0.2)(0.2) = 0.0016$

(c) $\left(\frac{2}{5}\right)^3 = \left(\frac{2}{5}\right)\left(\frac{2}{5}\right)\left(\frac{2}{5}\right) = \frac{8}{125}$

Notice $\frac{1}{16}$ and 0.0016 are not the same number. 0.0016 as a fraction is $\frac{16}{10{,}000}$.

Practice Problem 4 Evaluate. **(a)** $\left(\frac{1}{3}\right)^3$ **(b)** $(0.3)^4$ **(c)** $\left(\frac{3}{2}\right)^4$ ■

1.4 Exercises 1.4

Verbal and Writing Skills

Write an expression in exponent form, given the base and the exponent. Then evaluate the expression.

1. Base = 4, exponent = 3

2. Base = 7, exponent = 2

3. Base = −5, exponent = 4

4. Base = −4, exponent = 4

5. Base = −7, exponent = 3

6. Base = −6, exponent = 3

Evaluate.

7. 2^3

8. 2^4

9. 3^4

10. 3^3

11. 7^3

12. 5^4

13. $(-3)^3$

14. $(-2)^3$

15. $(-2)^6$

16. $(-3)^4$

17. $(-6)^3$

18. $(-7)^3$

19. $\left(\frac{1}{4}\right)^2$

20. $\left(\frac{1}{2}\right)^3$

21. $\left(\frac{2}{5}\right)^3$

22. $\left(\frac{2}{3}\right)^4$

23. $(0.3)^2$

24. $(0.4)^2$

25. $(0.5)^3$

26. $(0.6)^3$

27. $(-8)^4$

28. $(-7)^4$

29. -8^4

30. -7^4

31. $(-0.6)^5$

32. $(-0.8)^4$

33. $(0.1)^6$

34. $8^2(-2)^3$

35. $(-5)^2(3)$

36. $3^3(-2)$

37. $2^3(3)^2(-10)^3$

38. $-4^2(-1)^7(2)^4$

39. $(-7)4^2(3)^2$

40. $(-2)2^3(-3)^2$

41. $\left(\frac{2}{3}\right)^2\left(\frac{1}{4}\right)3^3$

42. $\left(\frac{2}{5}\right)^2(0.1)^2(-10)^3$

43. $3^3 - 2^4$

44. $(-2)^3 - (-5)^4$

45. $-7 - (-3)^3$

To Think About

★**46.** Compare $\frac{1}{3}$ with $(\frac{1}{3})^3$. Which is larger?

★**47.** What number can be raised to the sixth power to obtain 64?

Cumulative Review Problems

Evaluate.

48. $(-11) + (-13) + 6 + (-9) + 8$

49. $\frac{3}{4} \div \left(-\frac{9}{20}\right)$

50. $-17 - (-9)$

51. $(-2.1)(-1.2)$

1.5 Use the Distributive Property to Simplify Expressions

After studying this section, you will be able to:

1 *Recognize polynomials.*
2 *Use the distributive properties to multiply a polynomial by a monomial.*

1 *Polynomials*

In mathematics and in problem solving, we often do not know the value of a number. When this happens, we use a letter to represent the unnknown number. We call the number a **variable.** Almost all problems in algebra use variables in one way or another.

> **Definition of Variable**
>
> A **variable** is a symbol used to represent an unknown number (or numbers) until such time as its value can be determined.

Note that if a number is multiplied by a variable we do not need any symbol in between the number and variable. Thus, to indicate $(2)(x)$, we write $2x$. To indicate $3 \cdot y$, we write $3y$. If one variable is multiplied by another variable, we place the variables next to each other. Thus, $(a)(b)$ is written ab. We use exponent form if an unknown number (a variable) is used several times as a factor. Thus, $x \cdot x \cdot x = x^3$. Similarly, $(y)(y)(y)(y) = y^4$. All these ideas are combined in the expression $5ab^2$ below.

Thus, you can combine variables and numbers to write algebraic expressions such as

$$(a + b), \qquad 2x - 3, \qquad \frac{x+1}{x^2-1}, \qquad 5ab^2$$

The kinds of algebraic expressions that we often deal with are called **polynomials.** Polynomials are variable expressions that contain terms with nonnegative integer exponents. The following three expressions are all polynomials.

$$5x^2y + 1, \qquad 3a + 2b^2, \qquad 5a + 6b + 3c$$

There are special names for polynomials with one, two, or three terms. (A **term** is a number, a variable, or a product of numbers and variables.)

A **monomial** has *one* term:

$$5a, \qquad 3x^2yz^6, \qquad 10xy$$

A **binomial** has *two* terms:

$$5x^3 + 2y^2, \qquad 3x + 1, \qquad 10x + y$$

A **trinomial** has *three* terms:

$$5x^2 - 6x + 3, \qquad 2ab^3 - 6ab^2 - 8ab$$

2 *The Distributive Property*

An important property of algebra is the **distributive property.** We can state it in an equation as follows:

$$a(b + c) = ab + ac$$

A numerical example shows that it does seem reasonable.

$$5(3 + 6) = 5(3) + 5(6)$$
$$5(9) = 15 + 30$$
$$45 = 45$$

We can use the distributive property to multiply any polynomial by a monomial. In Section 1.4, we defined the word *factor*. We can use the word factor with polynomials and with monomials. Two or more algebraic expressions joined by multiplication are called **factors.** Consider the following examples of multiplying a polynomial by a monomial.

EXAMPLE 1 Multiply. **(a)** $5(a + b)$ **(b)** $-1(3x + 2y)$ **(c)** $6x(x + 2y)$

(a) $5(a + b) = 5a + 5b$ *Multiply the factor $(a + b)$ by the factor 5.*

(b) $-1(3x + 2y) = -1(3x) + -1(2y) = -3x - 2y$ *Multiply the factor $(3x + 2y)$ by the factor -1.*

(c) $6x(x + 2y) = 6x^2 + 12xy$ *Multiply the factor $(x + 2y)$ by the factor $6x$.*

Practice Problem 1 Multiply.

(a) $-3(x + 2y)$ **(b)** $-a(a - 3b)$ **(c)** $5x(x + 4y)$ ■

If the parentheses are preceded by a negative sign, we consider this to be the product of (-1) and the polynomial.

EXAMPLE 2 Multiply. **(a)** $-(a - 2b)$ **(b)** $-(3x^2y - 2xy^2 + 6x)$

(a) $-(a - 2b) = (-1)(a - 2b) = (-1)(a) + (-1)(-2b) = -a + 2b$

(b)
$$\begin{aligned} -(3x^2y - 2xy^2 + 6x) &= (-1)(3x^2y - 2xy^2 + 6x) \\ &= (-1)(3x^2y) + (-1)(-2xy^2) + (-1)(6x) \\ &= -3x^2y + 2xy^2 - 6x \end{aligned}$$

Practice Problem 2 Multiply. **(a)** $-(-3x + y)$ **(b)** $-(5r - 3s - 1)$ ■

In general, we see that in all these examples we have multiplied each term of the polynomial in the parentheses by the monomial in front of the parentheses.

EXAMPLE 3 Multiply. **(a)** $\frac{2}{3}(x^2 - 6x + 8)$ **(b)** $1.4(a^2 + 2.5a + 1.8)$

(a)
$$\begin{aligned} \frac{2}{3}(x^2 - 6x + 8) &= \left(\frac{2}{3}\right)(1x^2) + \left(\frac{2}{3}\right)(-6x) + \left(\frac{2}{3}\right)(8) \\ &= \frac{2}{3}x^2 + (-4x) + \frac{16}{3} \\ &= \frac{2}{3}x^2 - 4x + \frac{16}{3} \end{aligned}$$

(b)
$$\begin{aligned} &1.4(a^2 + 2.5a + 1.8) \\ &= 1.4(1a^2) + (1.4)(2.5a) + (1.4)(1.8) \\ &= 1.4a^2 + 3.5a + 2.52 \end{aligned}$$

Practice Problem 3 Multiply.

(a) $\frac{3}{5}(a^2 - 5a + 25)$ **(b)** $2.5(x^2 - 3.5x + 1.2)$ ■

The distributive property can also be presented with the a on the right.

$$(b + c)a = ba + ca$$

The a is "distributed" over the b and c inside the parentheses.

EXAMPLE 4 Multiply. **(a)** $(2x^2 - x)(-3)$ **(b)** $(5x + 3y - 2)(2y)$

(a) $(2x^2 - x)(-3) = 2x^2(-3) + (-x)(-3)$

$= -6x^2 + 3x$

(b) $(5x + 3y - 2)(2y) = 5x(2y) + 3y(2y) + (-2)(2y)$

$= 10xy + 6y^2 - 4y$

Practice Problem 4 Multiply.

(a) $(3x^2 - 2x)(-4)$ **(b)** $(4y)(7x^2 + 1)$ ■

Developing Your Study Skills

HOW TO DO HOMEWORK

Set aside time each day for your homework assignments. Do not attempt to do a whole week's worth on the weekend. Two hours spent studying outside of class for each hour in class is usual for college courses. You may need more than that for mathematics.

Before beginning to solve your homework problems, read your texbook very carefully. Expect to spend much more time reading a few pages of a mathematics textbook than several pages of another text. Read for complete understanding, not just for the general idea.

As you begin your homework assignments, read the directions carefully. You need to understand what is being asked for. Concentrate on each exercise or problem, taking time to solve it accurately. Rushing through your work usually causes errors. Check your answers with those given in the back of the textbook. If your answer is incorrect, check to see that you are doing the right problem. Redo the problem, watching for little errors. If it is still wrong, check with a friend. Perhaps the two of you can figure it out.

Work on your assignments every day and do as many problems as it takes for you to know what you are doing. Begin by doing all the problems that have been assigned. If there are more available in the section in your text, then do more. When you think you have done enough problems to understand fully the topic at hand, do a few more to be sure. This may mean that you do many more problems than the instructor assigns, but you can never practice too much. Practice improves your skills and increases your accuracy, speed, competence, and confidence.

If a problem is difficult to solve, check the examples in the textbook or in your notes for a similar example or problem. Can this one be solved in the same way? Give it some thought. You may want to leave it for a while by taking a break or doing a different one. But come back later and try again. If you are still unable to figure it out, ask your instructor for help during office hours or in class.

1.5 Exercises

Verbal and Writing Skills

Complete the sentence by filling in the blank.

1. A ______________ is a symbol used to represent an unknown number.

2. A trinomial has ________ terms.

3. $3x^2yz^6$ has one term and is called a __________________.

4. Explain why you think the property $a(b + c) = ab + ac$ is called the distributive property. What does distributive mean?

★ **5.** Does the following distributive property work? $a(b - c) = ab - ac$ Why or why not? Give an example.

Multiply. Use the distributive property.

6. $a(a - 2b)$

7. $x(2x - y)$

8. $4(-2a - 6b)$

9. $-2(3x^2 - 2)$

10. $-2x(x - y)$

11. $5a(-3a - 2b^2)$

12. $-(a + 2b - c^2)$

13. $-3x(x + 2y - 1)$

14. $4(2x - 8y - 19)$

15. $-(5x^2y - 2xy^2 - xy)$

16. $-9(9x - 5y + 8)$

17. $-5(3x + 9 - 7y)$

18. $\frac{1}{3}(3x^2 + 2x - 1)$

19. $\frac{1}{4}(x^2 + 2x - 8)$

20. $\frac{3}{5}(-5x^2 - 10x + 4)$

21. $\frac{4}{7}(7x^2 - 21x - 3)$

22. $\frac{x}{5}(x + 10y - 4)$

23. $\frac{y}{3}(3y - 4x - 6)$

24. $5x(x + 2y + z - 1)$

25. $3a(2a + b - c - 4)$

26. $(8 - x)(-3)$

27. $(4 - 2x)(-3)$

28. $(3x^2 + 5x)(-2)$

29. $(-5x^3 + x)(-5)$

30. $2x(3x + y - 4)$

31. $3x(4x - 5y - 6)$

32. $-3x(x + y - 2)$

33. $-4x(x + y - 5)$

34. $(3x + 2y - 1)(-xy)$

35. $(4a - 2b - 1)(-ab)$

36. $(2x + 3y - 2)3xy$

37. $(-2x + y - 3)4xy$

38. $1.5(2.8x^2 + 3.0x - 2.5)$

39. $2.5(1.5a^2 - 3.5a + 2.0)$

40. $-0.5x(-0.3y + 1.0xy + 0.2)$

41. $-0.4a(0.3a - 0.2b + 0.02b^2)$

42. $0.5x(0.6x + 0.8y - 5)$

Cumulative Review Problems

Add.

43. $-18 + (-20) + 36 + (-14)$.

Evaluate.

44. $(-2)^6$

45. $(-1)^{200}$

46. $-27 - (-41)$

1.6 Combine Like Terms

After studying this section, you will be able to:

1. *Identify like terms.*
2. *Combine like terms.*

1 *Identifying Like Terms*

We can add or subtract quantities that are like quantities. This is called **combining** like quantities.

$$5 \text{ inches} + 6 \text{ inches} = 11 \text{ inches}$$

$$20 \text{ square inches} - 16 \text{ square inches} = 4 \text{ square inches}$$

However, we cannot combine things that are not the same.

$$16 \text{ square inches} - 4 \text{ inches} \quad \text{(cannot be done!)}$$

Similarly, in algebra we can *combine like terms.* This means to add or subtract like terms. Remember, we cannot combine terms that are not the same. A **term** is a number, a variable, or a product of numbers and variables. **Like terms** are terms that have identical variables and exponents. In other words, like terms must have exactly the same letter parts.

EXAMPLE 1 List the like terms of each expression.

(a) $5x - 2y + 6x$ **(b)** $2x^2 - 3x - 5x^2 - 8x$

(a) $5x$ and $6x$ are like terms. These are the only like terms.

(b) $2x^2$ and $-5x^2$ are like terms.
$-3x$ and $-8x$ are like terms.
Note that x^2 and x are not like terms.

Practice Problem 1 List the like terms of each expression.

(a) $5a + 2b + 8a - 4b$ **(b)** $x^2 + y^2 + 3x - 7y^2$ ■

Do you really understand what a term is? A **term** is a number, a variable, or a product of numbers and variables. Terms are the parts of an algebraic expression separated by plus or minus signs. The sign in front of the product is considered part of the term.

EXAMPLE 2

(a) List the terms of this expression. $3a - 2b + 5a^2 + 6a - 8b - 12a^2$

(b) Identify which terms are like terms.

(a) There are six terms. They are $3a$, $-2b$, $5a^2$, $6a$, $-8b$, and $-12a^2$.

(b) $3a$ and $6a$ are like terms.
$-2b$ and $-8b$ are like terms.
$5a^2$ and $-12a^2$ are like terms.

Practice Problem 2

(a) List the terms of this expression. $5x^2 - 6x + 8 - 4x - 9x^2 + 2$

(b) Identify which terms are like terms. ■

2 *Combining Like Terms*

It is important to know how to combine like terms. Since

$$4 \text{ inches} + 5 \text{ inches} = 9 \text{ inches}$$

we would expect in algebra that

$$4x + 5x = 9x$$

Why is this true? Let's take a look at the distributive property.

> Like terms may be added or subtracted by using the distributive property:
>
> $$ab + ac = a(b + c) \quad \text{and} \quad ba + ca = (b + c)a$$

For example,

$$-7x + 9x = (-7 + 9)x = 2x$$

$$5x^2 + 12x^2 = (5 + 12)x^2 = 17x^2$$

EXAMPLE 3 Combine like terms. **(a)** $-4x^2 + 8x^2$ **(b)** $5x + 3x + 2x$

(a) Notice that each term contains the factor x^2. Using the distributive property,

$$-4x^2 + 8x^2 = (-4 + 8)x^2 = 4x^2$$

(b) Note that each term contains the factor x. Using the distributive property,

$$5x + 3x + 2x = (5 + 3 + 2)x = 10x$$

Practice Problem 3 Combine like terms.

(a) $5a + 7a + 4a$ **(b)** $16y^3 + 9y^3$ ■

In this section, the direction *simplify* means to combine like terms.

EXAMPLE 4 Simplify. $5a^2 - 2a^2 + 6a^2$

$$5a^2 - 2a^2 + 6a^2 = (5 - 2 + 6)a^2 = 9a^2$$

Practice Problem 4 Simplify. $-8y^2 - 9y^2 + 4y^2$ ■

After doing a few problems, you will find that it is not necessary to write out the step of using the distributive property. We will omit this step for the remaining examples in this section.

EXAMPLE 5 Simplify. **(a)** $5a + 2b + 7a - 6b$

(b) $3x^2y - 2xy^2 + 6x^2y$ **(c)** $2a^2b + 3ab^2 - 6a^2b^2 - 8ab$

(a) $5a + 2b + 7a - 6b = 12a - 4b$ *We combine the a terms and the b terms separately.*

(b) $3x^2y - 2xy^2 + 6x^2y = 9x^2y - 2xy^2$ ***Note:*** *x^2y and xy^2 are not like terms because of different powers.*

(c) $2a^2b + 3ab^2 - 6a^2b^2 - 8ab$ *These terms cannot be combined; there are no like terms in this expression.*

Practice Problem 5 Simplify. **(a)** $-x + 3a - 9x + 2a$

(b) $5ab - 2ab^2 - 3a^2b + 6ab$ **(c)** $7x^2y - 2xy^2 - 3x^2y - 4xy^2 + 5x^2y$ ■

The two skills in this section that a student must practice are identifying like terms and correctly adding or subtracting like terms. If a problem involves many terms, you may find it helpful to rearrange the terms so that like terms are together.

EXAMPLE 6 Simplify. $3a - 2b + 5a^2 + 6a - 8b - 12a^2$

There are three pairs of like terms.

$= 3a + 6a - 2b - 8b + 5a^2 - 12a^2$ *You can rearrange the terms so that like terms are together, making it easier to combine them.*

a terms, b terms, a^2 terms

$= 9a - 10b - 7a^2$ *Combine like terms.*

The order of terms in an answer in these problems is not significant. These three terms can be rearranged in a different order. $-10b + 9a - 7a^2$ and $-7a^2 + 9a - 10b$ are also correct. However, we usually write polynomials in order of descending powers. $-7a^2 + 9a - 10b$ would be the preferred way to write the answer.

Practice Problem 6 Simplify. $5xy - 2x^2y + 6xy^2 - xy - 3xy^2 - 7x^2y$ ■

EXAMPLE 7 Simplify. $6(2x + 3xy) - 8x(3 - 4y)$

First remove the parentheses; then collect like terms.

$6(2x + 3xy) - 8x(3 - 4y) = 12x + 18xy - 24x + 32xy$ *Use the distributive property.*

$= -12x + 50xy$ *Combine like terms.*

Practice Problem 7 Simplify. $5a(2 - 3b) - 4(6a + 2ab)$ ■

Use extra care with fractional coefficients.

EXAMPLE 8 Simplify. $\frac{3}{4}x^2 - 5y - \frac{1}{8}x^2 + \frac{1}{3}y$

We need a least common denominator for the x^2 terms, which is 8. Change $\frac{3}{4}$ to eighths by multiplying the numerator and denominator by 2.

$$\frac{3}{4}x^2 - \frac{1}{8}x^2 = \frac{3 \cdot 2}{4 \cdot 2}x^2 - \frac{1}{8}x^2 = \frac{6}{8}x^2 - \frac{1}{8}x^2 = \frac{5}{8}x^2$$

The least common denominator for the y terms is 3. Change 5 to thirds.

$$-\frac{5}{1}y + \frac{1}{3}y = \frac{-5 \cdot 3}{1 \cdot 3}y + \frac{1}{3}y = \frac{-15}{3}y + \frac{1}{3}y = -\frac{14}{3}y$$

Thus, our solution is $\frac{5}{8}x^2 - \frac{14}{3}y$.

Practice Problem 8 Simplify. $\frac{1}{7}a^2 - \frac{5}{12}b + 2a^2 - \frac{1}{3}b$ ■

1.6 Exercises

Verbal and Writing Skills

Identify the like terms in each problem. Do not combine.

1. 5 miles + 7 inches + 3 miles

2. 2 quarts + 4 gallons + 9 quarts

3. $5a - 2b - 12a$

4. $3x + 15x - 26$

5. $3xy - 5x + 6x + 2xy$

6. $9x^2 - 3x + 5 - 7x - 2x^2$

Simplify.

7. $7x + 3x$

8. $2a - 8a$

9. $-12x^3 - 16x^3$

10. $-8a^2 + 4a^2$

11. $10x^4 + 8x^4 + 7x^2$

12. $3xy - 2y + 6x$

13. $5a + 2b - 7a^2$

14. $3a^3 - 6a^2 + 5a^3$

15. $2ab + 1 - 6ab - 8$

16. $2x^2 + 3x^2 - 7 - 5x^2$

17. $1.3x - 2.6y + 5.8x - 0.9y$

18. $3.1ab - 0.2b - 0.8ab + 5.3b$

19. $1.6x - 2.8y - 3.6x - 5.9y$

20. $1.9x - 2.4b - 3.8x - 8.2b$

21. $\frac{1}{2}x^2 - 3y - \frac{1}{3}y + \frac{1}{4}x^2$

22. $\frac{1}{5}a^2 - 2b - \frac{1}{2}a^2 - 3b$

23. $\frac{1}{2}x - \frac{1}{4}y - \frac{3}{4}x - \frac{1}{8}y$

24. $\frac{1}{5}a - \frac{1}{8}b + \frac{3}{5}a - \frac{1}{16}b$

25. $3x + 2y - 6 - 8x - 9y - 14$

26. $-a + 2b + 8 - 7a - b - 1$

27. $5x^2y - 10xy^6 + 6xy^2 - 7xy^2$

28. $5bcd - 8cd - 12bcd + cd$

29. $2ab + 5bc - 6ac - 2ab$

30. $5x^2y + 12xy^2 - 8x^2 - 12xy^2$

31. $x^3 - 3x^2 + 1 - 2x + 6x^2 - x^3$

32. $x^2 - 2x + 1 - 3x^2 - 5x + 6$

33. $2y^2 - 8y + 9 - 12y^2 - 8y + 3$

34. $5 - 2y^2 + 3y - 8y - 9y^2 - 12$

35. $ab + 3a - 4ab + 2a - 8b$

Simplify. Use the distributive property to remove parentheses; then combine like terms.

36. $5(a - 3b) + 2(-b - 4a)$

37. $3(x + y) - 5(-2y + 3x)$

38. $6x(x + y) - 3(x^2 - 2xy)$

39. $5a(a + 3b) - 2(a^2 - 6ab)$

40. $-3(x^2 + 3y) + 5(-6y - x^2)$

41. $-7(a^2 + 3b) + 4(2b - 8a^2)$

42. $4(2 - x) - 3(-5 - 12x)$

43. $7(3 - x) - 6(8 - 13x)$

Cumulative Review Problems

44. $-\frac{1}{3} - \left(-\frac{1}{5}\right)$

45. $\left(-\frac{5}{3}\right)\left(\frac{1}{2}\right)$

46. $\frac{4}{5} + \left(-\frac{1}{25}\right) + \left(-\frac{3}{10}\right)$

47. $\left(\frac{5}{7}\right) \div \left(-\frac{14}{3}\right)$

1.7 Order of Arithmetic Operations

After studying this section, you will be able to:

1 *Use the order of operations to simplify numerical expressions involving addition, subtraction, multiplication, division, and exponents.*

1 *Order of Operations*

It is important to know *when* to do certain operations as well as how to do them. For example, for the expression $2 - 4 \cdot 3$, should we subtract first or multiply first? The following list will assist you. It tells which operations to do first: the correct **order of operations.** We call this a *list of priorities.*

Order of Operations for Numbers

Follow this order of operations:

Do first	**1.** Combine numbers inside parentheses.
	2. Raise numbers to a power.
	3. Multiply and divide numbers from left to right.
Do last	**4.** Add and subtract numbers from left to right.

Let's return to the problem: $2 - 4 \cdot 3$. There are no parentheses or numbers raised to a power, so multiplication comes next. We do that first. Then we subtract since this comes last on our list.

$2 - 4 \cdot 3$ *Follow the order of operations by first multiplying* $4 \cdot 3 = 12$.

$= 2 - 12$

$= -10$ *Combine* $2 - 12 = -10$.

EXAMPLE 1 Evaluate. $8 \div 2 \cdot 3 + 4^2$

$= 8 \div 2 \cdot 3 + 16$ *Evaluate* $4^2 = 16$ *because the highest priority in this problem is raising to a power.*

$= 4 \cdot 3 + 16$ *Next multiply and divide from left to right, so* $8 \div 2 = 4$ *and* $4 \cdot 3 = 12$.

$= 12 + 16$ *Finally, add.*

$= 28$

Practice Problem 1 Evaluate. $18 - (-3)^3$ ■

Note: Multiplication and division have equal priority. We do not do multiplication problems first. Rather, we work from left to right, doing any multiplication or division that we encounter. Similarly, addition and subtraction have equal priority.

EXAMPLE 2 Evaluate. $(-3)^3 - 2^4$

The highest priority is to raise the expressions to the appropriate power.

$(-3)^3 - 2^4 = -27 - 16$ *In* $(-3)^3$ *we are cubing the number* -3 *to obtain* -27. *Be careful,* -2^4 *is not* $(-2)^4$! *Raise 2 to the fourth power and take the negative.*

$= -43$ *The last step is to add and subtract from left to right.*

Practice Problem 2 Evaluate. $12 \div (-2)(-3) - 2^4$ ■

CALCULATOR

Order of Operations

Use your calculator to evaluate $3 + 4 \cdot 5$. Enter

3 [+] 4 [×] 5

If the display is [23], the correct order of operations is built in. If the display is not 23, you will need to modify the way you enter the problem. You should use

4 [×] 5 [+] 3 [=]

Try $6 + 3 \cdot 4 - 8 \div 2$.

EXAMPLE 3 Evaluate. $2 \cdot (2 - 3)^3 + 6 \div 3 + (8 - 5)^2$

$2 \cdot (2 - 3)^3 + 6 \div 3 + (8 - 5)^2$ *Combine the numbers inside the parentheses.*

$= 2 \cdot (-1)^3 + 6 \div 3 + 3^2$

$= 2 \cdot (-1) + 6 \div 3 + 9$ *Next, raise to a power. Note that we need parentheses for -1 because of the negative sign, but they are not needed for 3.*

$= -2 + 2 + 9$ *Next, multiply and divide from left to right.*

$= 9$ *Finally, add and subtract from left to right.*

Practice Problem 3 Evaluate. $6 - (8 - 12)^2 + 8 \div 2$ ■

EXAMPLE 4 Evaluate. $\left(-\frac{1}{5}\right)\left(\frac{1}{2}\right) - \left(\frac{3}{2}\right)^2$

The highest priority is to raise $\frac{3}{2}$ to the second power.

$$\left(\frac{3}{2}\right)^2 = \left(\frac{3}{2}\right)\left(\frac{3}{2}\right) = \frac{9}{4}$$

$= \left(-\frac{1}{5}\right)\left(\frac{1}{2}\right) - \frac{9}{4}$ *Next we multiply.*

$= -\frac{1}{10} - \frac{9}{4}$ *We need to write each fraction as an equivalent fraction with the LCD of 20.*

$= -\frac{1 \cdot 2}{10 \cdot 2} - \frac{9 \cdot 5}{4 \cdot 5}$

$= -\frac{2}{20} - \frac{45}{20}$ *Add.*

$= -\frac{47}{20}$

Practice Problem 4 Evaluate. $\left(-\frac{1}{7}\right)\left(\frac{-14}{5}\right) + \left(-\frac{1}{2}\right) \div \left(\frac{3}{4}\right)$ ■

Developing Your Study Skills

PREVIEWING NEW MATERIAL

Part of your study time each day should consist of looking over the sections in your text that are to be covered the following day. You do not necessarily need to study and learn the material on your own, but a survey of the concepts, terminology, diagrams, and examples will help the new ideas seem more familiar as the instructor presents them. You can look for concepts that appear confusing or difficult and be ready to listen carefully for your instructor's explanations. You can be prepared to ask the questions that will increase your understanding. Previewing new material enables you to see what is coming and prepares you to be ready to absorb it.

1.7 Exercises

Verbal and Writing Skills

A friend is counting some money and announces "I have three twos plus four fives."

1. Write this as a number expression.

2. How much money does your friend have?

3. What answer would you get to our number expression if
(a) you used left-to-right entry?
(b) you used order-of-operation entry?

4. Which order best fits this problem?

Evaluate.

5. $4^2 + 2(4)$

6. $2 \div 2 \cdot 3 + 1$

7. $5 + 6 \cdot 2 \div 4 - 1$

8. $(2 - 8)^2 \div 6$

9. $(3 - 5)^2 \cdot 6 \div 4$

10. $2(3 - 5 + 6) + 5$

11. $8 - 2^3 \cdot 5 + 3$

12. $1 - 16 \cdot 2 \div 4 + 5$

13. $4 + 27 \div 3 \cdot 2 - 8$

14. $3 \cdot 5 + 7 \cdot 3 - 5 \cdot 3$

15. $5(3)^3 - 20 \div (-2)$

16. $4(2)^4 + 15 \div (-3)$

17. $3(5 - 7)^2 - 6(3)$

18. $-2(3 - 6)^2 - (-2)$

19. $5 \cdot 6 - (3 - 5)^2 + 8 \cdot 2$

20. $(-3)^2 \cdot 6 \div 9 + 4 \cdot 2$

21. $\frac{1}{2} \div \frac{2}{3} + 6 \cdot \frac{1}{4}$

22. $2 + \left(\frac{1}{3}\right)^2 - \left(2 - \frac{1}{2}\right)$

23. $0.8 + 0.3(0.6 - 0.2)^2$

24. $0.05 + 1.4 - (0.5 - 0.7)^3$

25. $\frac{3}{4}\left(-\frac{2}{5}\right) - \left(-\frac{3}{5}\right)$

26. $-\frac{2}{3}\left(\frac{3}{5}\right) + \frac{5}{7} \div \frac{5}{3}$

27. $7.2 - (-3.8) + (5.5)^2$

28. $1.62 - 3.2(1.4) + (-0.5)^2$

29. $\left(\frac{1}{2}\right)^3 + \frac{1}{4} - \left(\frac{1}{6} - \frac{1}{12}\right) - \frac{2}{3} \cdot \left(\frac{1}{4}\right)^2$

30. $(2.4 \cdot 1.2)^2 - 1.6 \cdot 2.2 \div 4.0 - 3.6$

Cumulative Review Problems

Simplify.

31. $(2x - 3y) - (x - 3y)$

32. $-\frac{3}{4} - \frac{5}{6}$

33. -1^{2086}

34. $3\frac{3}{5} \div 6\frac{1}{4}$

1.8 Use Substitution to Evaluate Expressions

After studying this section, you will be able to:

1 *Evaluate a variable expression for a specified value.*
2 *Evaluate a formula by substitution.*

1 *Evaluating a Variable Expression*

You will use the order of operations to evaluate variable expressions.

Evaluate $6 + 3x$ for $x = -4$

In general, x represents some unknown number. Here we are told x has the value -4. We can replace x by -4. Use parentheses around -4. Note that we always put replacement values in parentheses.

$$6 + 3(-4) = 6 + (-12) \quad \textit{Order of operations}$$
$$= -6$$

> When we replace a variable by a particular value, we say we have *substituted* the value for the variable. We then evaluate the expression (find a value for it).

EXAMPLE 1 Evaluate. $\frac{2}{3}x - 5$ for $x = -6$

$\frac{2}{3}(-6) - 5$ — *Substitute $x = -6$. Be sure to enclose the -6 in parentheses.*

Multiply $\left(\frac{2}{3}\right)\left(-\frac{6}{1}\right) = -4.$

$-4 \quad -5$ — *Combine.*

-9

Practice Problem 1 Evaluate. $4 - \frac{1}{2}x$ for $x = -8$ ■

Compare parts (a) and (b) in the next example. The two parts illustrate that you must be careful what value you raise to a power. *Note:* In part (b) we will need parentheses within parentheses. To avoid confusion, we use brackets [ ] to represent the outside parentheses.

EXAMPLE 2 **(a)** Evaluate $2x^2$ for $x = -3$. **(b)** Evaluate $(2x)^2$ for $x = -3$.

(a) Here the value x is squared.

$2x^2 = 2(-3)^2$

$= 2(9)$ — *First square -3.*

$= 18$ — *Then multiply.*

(b) Here the value $(2x)$ is squared.

$(2x)^2 = [(2)(-3)]^2$

$= (-6)^2$ — *First multiply the numbers inside the parentheses.*

$= 36$ — *Then square -6.*

Practice Problem 2 Evaluate each expression for $x = -3$.

(a) $-x^4$ **(b)** $(-x)^4$ ■

Carefully study the solutions to Example 3(a) and Example 3(b). You will find that taking the time to see *how* and *why* they are different is a good investment of study time.

EXAMPLE 3 Evaluate $x^2 + x$ for $x = -4$.

$x^2 + x = (-4)^2 + (-4)$ *Replace x by −4 in the original expression.*

$= 16 + (-4)$ *Raise to a power.*

$= 12$ *Finally, add.*

Practice Problem 3 Evaluate $(5x)^3 + 2x$ for $x = -2$. ■

2 *Evaluating Formulas*

We can *evaluate a formula* by substituting values for the variables. For example, the area of a triangle can be found using the formula $A = \frac{1}{2}bh$, where b is the length of the base of a triangle and h is the height (or the altitude) of the triangle (see figure). If we know values for b and h, we can substitute those values into the formula to find the area. The units for area are *square units.*

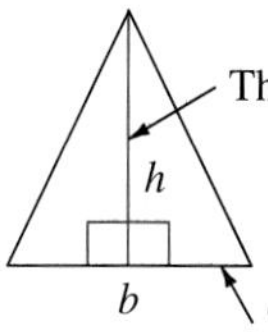

EXAMPLE 4 Find the area of a triangle with a base of 16 centimeters (cm) and a height of 12 centimeters (cm).

Use the formula

$$A = \frac{1}{2}bh$$

Substitute 16 centimeters for b and 12 centimeters for h.

$A = \frac{1}{2}(16 \text{ centimeters})(12 \text{ centimeters})$

$= \frac{1}{2}(16)(12)(\text{cm})(\text{cm})$ *If you take $\frac{1}{2}$ of 16 first, it will make your calculation easier.*

$= (8)(12)(\text{cm})^2 = 96$ square centimeters

The area of the triangle is 96 square centimeters.

Practice Problem 4 Find the area of a triangle with a height of 3 meters and a base of 7 meters. ■

The area of a circle is given by

$$A = \pi r^2$$

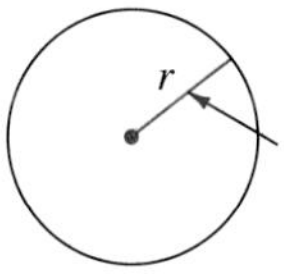

We will use 3.14 as an approximation for the *irrational number* π.

EXAMPLE 5 Find the area of a circle if the radius is 2 inches.

$A = \pi r^2 = (3.14)(2 \text{ inches})^2$ *Write the formula and substitute the given values for the variable.*

$= (3.14)(4)(\text{in.})^2$ *Raise to a power. Then multiply.*

$= 12.56$ square inches

Practice Problem 5 Find the area of a circle if the radius is 3 meters. ■

The formula $C = \frac{5}{9}(F - 32)$ allows us to find the Celsius temperature if we know the Fahrenheit temperature. That is, we can substitute a value for F in degrees Fahrenheit into the formula to obtain a temperature C in degrees Celsius.

EXAMPLE 6 What is the Celsius temperature when the Fahrenheit temperature is $F = -22°$?

Use the formula.

$$C = \frac{5}{9}(F - 32)$$

$$= \frac{5}{9}(-22 - 32) \quad \textit{Substitute } -22 \textit{ for } F \textit{ in the formula.}$$

$$= \frac{5}{9}(-54) \quad \textit{Combine the numbers inside the parentheses.}$$

$$= (5)(-6) \quad \textit{Simplify.}$$

$$= -30 \quad \textit{Multiply.}$$

The temperature is $-30°$ Celsius.

Practice Problem 6 What is the Celsius temperature when the Fahrenheit temperature is $F = 68°$? Use the formula $C = \frac{5}{9}(F - 32)$. ■

When driving in Canada or Mexico, we must observe speed limits posted in kilometers per hour. A formula that converts r (miles per hour) or k (kilometers per hour) is $k = 1.61r$. Note that this is an approximation.

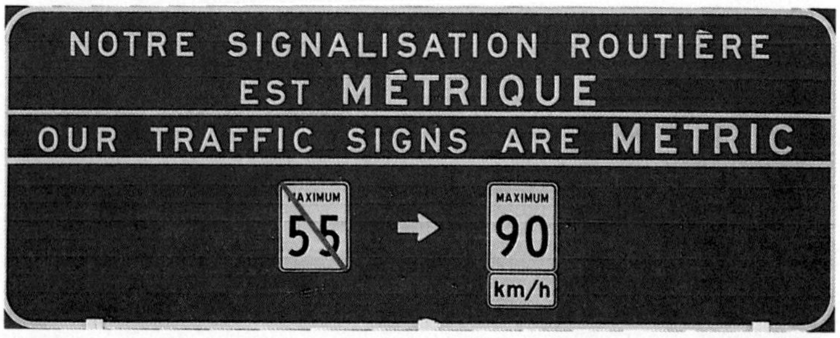

EXAMPLE 7 You are driving on a highway in Mexico. It is posted for a maximum speed of 100 kilometers per hour. You are driving at 61 miles per hour. Are you exceeding the speed limit?

Use the formula.

$$k = 1.61r$$

$$= (1.61)(61) \quad \textit{Replace } r \textit{ by } 61.$$

$$= 98.21 \quad \textit{Multiply the numbers.}$$

You are driving at approximately 98 kilometers per hour. You are not exceeding the speed limit.

Practice Problem 7 You are driving behind a heavily loaded truck on a Canadian highway. The highway has a posted minimum speed of 65 kilometers per hour. When you travel at exactly the same speed as the truck ahead of you, you observe that the speedometer reads 35 miles per hour. Assuming that your speedometer is accurate, determine if the truck is violating the minimum speed law. ■

1.8 Exercises

Evaluate.

1. $-2x + 1$ for $x = 3$

2. $-4x - 2$ for $x = 5$

3. $\frac{1}{2}x + 2$ for $x = -8$

4. $\frac{1}{3}x - 6$ for $x = 9$

5. $5x + 10$ for $x = \frac{1}{2}$

6. $7x + 20$ for $x = -\frac{1}{2}$

7. $2 - 4x$ for $x = 7$

8. $3 - 5x$ for $x = 8$

9. $x^2 + 2x$ for $x = 3$

10. $x^2 + 3x$ for $x = 4$

11. $3x^2$ for $x = -1$

12. $4x^2$ for $x = -1$

13. $-2x^2$ for $x = 5$

14. $-5x^2$ for $x = 3$

15. $9x + 13$ for $x = -2$

16. $5x + 7$ for $x = -2$

17. $2x^2 + 3x$ for $x = -3$

18. $18 - 5x$ for $x = -3$

19. $(2x)^2 + x$ for $x = 3$

20. $2 - x^2$ for $x = -2$

21. $2 - (-x)^2$ for $x = -2$

22. $2x - 3x^2$ for $x = -4$

23. $16 - 7x + x^2$ for $x = -2$

24. $5x + (3x)^2$ for $x = -2$

25. $2x^2 + 3x + 6$ for $x = 2$

26. $3x^2 + 2x - 4$ for $x = 3$

27. $\frac{1}{2}x^2 - 3x + 9$ for $x = -4$

Evaluate each expression for different values of the variable.

28. $x^2 + 2x + 3$
   (a) for $x = 4$
   (b) for $x = 0$
   (c) for $x = -2$

29. $x^2 - 3x + 4$
   (a) for $x = 4$
   (b) for $x = 0$
   (c) for $x = -2$

30. $5 - 6x - 2x^2$
   (a) for $x = 2$
   (b) for $x = 0$
   (c) for $x = -1$

31. $6 - 2x - 3x^2$
   (a) for $x = 2$
   (b) for $x = 0$
   (c) for $x = -1$

★32. $x^2 - 2y + 3y^2$ for $x = -3$ and $y = 4$

★33. $2x^2 - 3xy + 2y$ for $x = 4$ and $y = -1$

★34. $a^3 + 2abc - 3c^2$ for $a = 5$, $b = 9$, and $c = -1$

★35. $a^2 - 2ab + 2c^2$ for $a = 3$, $b = 2$, and $c = -4$

Applications

For Problems 36 and 37, draw and label a diagram. Use the formula $A = \frac{1}{2}bh$.

36. Find the area of a triangle with a base of 14 feet and a height of 17 feet.

37. Find the area of a triangle with a base of 13 inches and a height of 14 inches.

For Problems 38 and 39, draw and label a diagram. Use the formula $A = \pi r^2$. Use $\pi = 3.14$.

38. Find the area of a circle whose radius is 4 cm.

39. Find the area of a circle whose radius is 5 meters.

For Problems 40 and 41, use the formula $F = \frac{9}{5}C + 32$.

40. Find the Fahrenheit temperature F when the Celsius temperature is 25°C.

41. Find the Fahrenheit temperature F when the Celsius temperature is −10°C.

Solve each word problem. Round to the nearest tenth.

42. Find the cost of making a circle of thin gold leaf on an outdoor sign at a cost of \$290 per square centimeter if the radius of the circle is 9 cm. (Use $\pi = 3.14$.)

43. Find the cost of making a circle of solid aluminum for an outdoor sign at a cost of \$350 per square meter if the radius of the circle is 7 meters. (Use $\pi = 3.14$.)

44. Scientists on a research team are using a computer that can be exposed to temperatures as high as 60°C and as low as −50°C. What Fahrenheit temperature range can the computer be exposed to? (Use the formula $F = \frac{9}{5}C + 32$.)

45. The heated living unit of the scientists at the South Pole is built to provide comfortable temperatures inside for an outside temperature of −60 degrees Celsius with no wind blowing or −30 degrees Celsius with wind gusts of up to 50 miles per hour. What is the corresponding Fahrenheit temperature range? (Use the formula $F = \frac{9}{5}C + 32$.)

46. You are excessively tired on a bicycle trip in Canada. You see on the map that the elevation of the town is 2.3 kilometers above sea level. How many miles above sea level is the town? Why do you think you are so tired? (Use the formula $k = 1.61\,r$.)

47. While taking a bicycle trip in Canada you see on the map that it is 20 kilometers to the nearest town. Approximately how many miles is it to the nearest town? (Use the formula $k = 1.61r$.)

Cumulative Review Problems

Simplify.

48. $(-2)^4 - 4 \div 2 - (-2)$

49. $3(x - 2y) - (x^2 - y) - (x - y)$

50. $-12 - 6 + 8 - (-19) + (-13) - 14 + (-16)$

51. $(-0.3) \times (2.4)$

Putting Your Skills to Work

RETIREMENT PLANS

Retirement plans are based on how much a worker earns and on the number of years the worker is employed by a company. The following is one formula for monthly retirement income (R).

$$R = \frac{3}{5}S\left(1 + \frac{2n}{100}\right)$$

where S = the average monthly salary
n = the number of years employed

1. Mrs. Terpak worked for Begen Steel Corporation for 32 years at an average monthly salary of \$1575. What is her monthly retirement income?

2. Mrs. Bounds worked for the McNally Construction Company for 27 years at an average monthly salary of \$1295. What is her monthly retirement income?

3. If Pete stays at his present position, he expects to earn an average monthly income of \$1700 over 25 years. If Pete goes for additional training, he has an opportunity to advance himself, and his average monthly income would be \$2100 over 20 years. How much more will he receive each month for his retirement if he goes for additional training?

4. Paul plans to retire early but would like to have a retirement income of at least \$1200 a month. He expects to have an average salary of \$1650 regardless of the number of years of employment. Can Paul retire in 10 years? 15 years? or 20 years?

COOPERATIVE GROUP ACTIVITY

5. Determine what your retirement needs will be in the years to come. This should include housing, food, clothing, medical, and personal wants. Incorporate price increases due to inflation. Investigate your possible sources of income from your employer, savings in CDs (Certificates of Deposit) or stocks, an IRA (Individual Retirement Account), or an annuity. Assign each member of your group a specific task and pool your findings to draw up a model retirement plan. Include the number of years you need to work and/or save to achieve your retirement goals.

1.9 Grouping Symbols

After studying this section, you will be able to:

1 *Simplify variable expressions with several grouping symbols.*

1 *Simplifying Variable Expressions*

Many expressions in algebra use **grouping symbols** such as parentheses, brackets, and braces. Sometimes expressions are inside other expressions. Because it can be confusing to have more than one set of parentheses, brackets and braces are also used. How do we know what to do first when we see an expression like $2[5 - 4(a + b)]$?

To simplify the expression, we start with the innermost grouping symbols. Here it is a set of parentheses. We first use the distributive law to multiply

$$2[5 - 4(a + b)] = 2[5 - 4a - 4b]$$

We use the distributive law again.

$$= 10 - 8a - 8b$$

There are no like terms, so this is our final answer.

Notice that we started with two sets of grouping symbols, but our final answer had none. So we can say we *removed* the grouping symbols. Of course, we didn't just take them away: We used the distributive law and the rules for signed numbers to simplify as much as possible. Although we are doing many steps, we sometimes say ''remove parentheses'' as a shorthand direction. Sometimes we say ''simplify.''

Remember to remove the innermost parentheses first. Keep working from the inside out.

EXAMPLE 1 Simplify. $3[6 - 2(x + y)]$

We want to remove the innermost parentheses first. Therefore, we first use the distributive property to simplify $-2(x + y)$.

$3[6 - 2(x + y) = 3[6 - 2x - 2y]$ *Use the distributive property.*

$= 18 - 6x - 6y$ *Use the distributive property again.*

Practice Problem 1 Simplify. $5[4x - 3(x - 2)]$ ■

You recall that a negative sign in front of parentheses is equivalent to having a coefficient of negative 1. You can write the -1 and then multiply by -1 using the distributive property.

$$-(x + 2y) = -1(x + 2y) = -x - 2y$$

Notice that this has the effect of removing the parentheses. Each term in the result now has its sign changed from that of the original.

Similarly, a positive sign in front of parentheses can be viewed as multiplication by $+1$.

$$+(5x - 6y) = +1(5x - 6y) = 5x - 6y$$

If a grouping symbol has a positive or negative sign in front, we mentally multiply by $+1$ or -1.

EXAMPLE 2 Simplify. $-2[3a - (b + 2c) + (d - 3e)]$

$= -2[3a - b - 2c + d - 3e]$ — *Remove the two innermost sets of parentheses. Since one is not inside the other, we remove both sets at once.*

$= -6a + 2b + 4c - 2d + 6e$ — *Now we remove the brackets by multiplying each term by -2.*

Practice Problem 2 Simplify. $3ab - [2ab - (2 - a)]$ ■

EXAMPLE 3 Simplify. $2[3x - (y + w)] - 3[2x + 2(3y - 2w)]$

$= 2[3x - y - w] - 3[2x + 6y - 4w]$ — *In each set of brackets, remove the inner parentheses.*

$= 6x - 2y - 2w - 6x - 18y + 12w$ — *Remove each set of brackets by multiplying by the appropriate number.*

$= -20y + 10w$ or $10w - 20y$ — *Combine like terms. (Note that $6x - 6x = 0x = 0$.)*

Practice Problem 3 Simplify. $3[4x - 2(1 - x)] - [3x + (x - 2)]$ ■

You can always simplify problems with many sets of grouping symbols by the method shown above. Essentially, you just keep removing one level of grouping symbols at each step. Finally, at the end you add up the like terms if possible.

Sometimes it is possible to collect like terms at each step.

EXAMPLE 4 Simplify. $-3\{7x - 2[x - (2x - 1)]\}$

$= -3\{7x - 2[x - 2x + 1]\}$ — *Remove the inner parentheses by multiplying each term within the parentheses by negative 1.*

$= -3\{7x - 2[-x + 1]\}$ — *Collect like terms by combining $+x - 2x$.*

$= -3\{7x + 2x - 2\}$ — *Remove the brackets by multiplying each term within it by -2.*

$= -3\{9x - 2\}$ — *Combine the x terms.*

$= -27x + 6$ — *Remove the braces by multiplying each term by -3.*

Practice Problem 4 Simplify. $-2\{5x - 3x[2x - (x^2 - 4x)]\}$ ■

Developing Your Study Skills

EXAM TIME: THE NIGHT BEFORE

With adequate preparation, you can spend the night before the exam pulling together the final details.

1. Look over each section to be covered in the exam. Review the steps needed to solve each type of problem.
2. Review your list of terms, rules, and formulas that you are expected to know for the exam.
3. Take the Practice Test at the end of the chapter just as though you were taking the actual exam. Do not look in your text or get help in any way. Time yourself so that you know how long it takes you to complete the test.
4. Check the Practice Test. Redo the problems you missed.
5. Be sure you have ready the necessary supplies for taking your exam.

1.9 Exercises

Verbal and Writing Skills

1. Rewrite the expression $-3x - 2y$ using a negative sign and parentheses.
2. Rewrite the expression $-x + 5y$ using a negative sign and parentheses.
3. To simplify expressions with grouping symbols, we use the ________________ property.
4. When an expression contains many grouping symbols, remove the ________________ parentheses first.

Simplify. Remove grouping symbols and combine like terms.

5. $x - 4(x + y)$
6. $y - 3(y + z)$
7. $2(a + 3b) - 3(b - a)$
8. $4(x - y) - 2(3x + y)$
9. $x^2(x^2 - 3y^2) - 2(x^4 - x^2y^2)$
10. $ab(ab - 6b) - 8(a^2b^2 - 2ab^2)$
11. $5[3 + 2(x - 26) + 3x]$
12. $-4[-(x + 3y) - 2(y - x)]$
13. $2x + [4x^2 - 2(x - 3)]$
14. $2x[4x^2 - 2(x - 3)]$
15. $4y[-3y^2 + 2(4 - y)]$
16. $4y + [-3y^2 + 2(4 - y)]$
17. $2(x - 2y) - [3 - 2(x - y)]$
18. $5[a - b(a - b) - b^2]$
19. $2[a - 3b(a + b) - b^2]$
20. $3(x + 2y) - [4 - 2(x + y)]$
21. $x(x^2 + 2x - 3) - 2(x^3 + 6)$
22. $5x^2(x + 6) - 2[x - 2(1 + 2x^2)]$
23. $3a^2 - 4[2b - 3b(b + 2)]$
24. $x - \{2y - 3[x - 2(x + y)]\}$
25. $5x - \{4y - 2[2x + 3(x - y)]\}$
26. $2b^2 - 3[5b + 2b(2 - b)]$
27. $-\{4a^2 - 2[3a^2 - (b - a^2)]\}$
28. $3x - x[4x + 2(1 - x)]\}$
29. $-4\{3a^2 - 2[4a^2 - (b + a^2)]\}$
30. $-2\{x^2 - 3[x - (x - 2x^2)]\}$

Cumulative Review Problems

31. Use $F = 1.8C + 32$ to find the Fahrenheit temperature equivalent to 36.4° Celsius.
32. Use 3.14 as an approximation for π to compute the area covered by a circular irrigation system with radial arm of length 385 feet. $A = \pi r^2$.
33. Simplify. $3 - 4(-2)^3$
34. Evaluate. $4x^2 - x + 11$ for $x = -3$

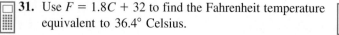

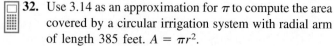

Putting Your Skills to Work

COMPOUND GROWTH

City planners need to be able to predict what the population will be in the future to determine needs for services such as schools, housing, and transportation. In general, for any town or city experiencing continuous growth at a steady rate over a period of time, the population (P) can be approximately determined by the formula

$$P = n(1 + r)^t$$

where r = the annual rate of growth

t = the number of years

To evaluate this formula, you can use either a compound growth table or a scientific calculator.

COMPOUND GROWTH TABLE

n	$(1.04)^n$	$(1.08)^n$	$(1.12)^n$
1	1.040000	1.080000	1.120000
2	1.081600	1.166400	1.254400
3	1.124864	1.259712	1.404928
4	1.169859	1.360489	1.573519
5	1.216653	1.469328	1.762342
6	1.265319	1.586874	1.973823
7	1.315932	1.713824	2.210681
8	1.368569	1.850930	2.475963
9	1.423312	1.999005	2.773079
10	1.480244	2.158925	3.105848
11	1.539454	2.331639	3.478550
12	1.601032	2.518170	3.895976
13	1.665074	2.719624	4.363493
14	1.731676	2.937194	4.887112
15	1.800944	3.172169	5.473566

EXAMPLE Determine the population in 5 years for a town of 4000 people experiencing a 4% growth rate per year.

$P = n(1 + r)^t$

$= 4000(1 + 0.04)^5$

Use the following keystrokes.

4000 [×] 1.04 [y^x] 5 [=]

The display reads

[4866.6116]

or *Find* $(1.04)^n$ *for* $n = 5$ *in the Compound Growth Table.*

$(1.04)^5 = 1.216653$.

Now use your calculator to multiply 4000(1.216653).

$= 4867$ rounded to the nearest person

1. A town has a population of 7000. For 8 years it grows at 12% per year. Then for the next 5 years it grows at only 8% per year. Finally, for the last 3 years it grows at only 4% per year. What will be the population of the town after the 16 years? Round your answer to nearest whole unit.

2. A town has a population of 15,000. For 14 years it grows at 12% per year. After 14 years it must build a school. The cost of the new school is a total of \$4,000,000 plus \$100 per person for any population above 50,000 people. What is the estimated cost to build the new school building?

COOPERATIVE GROUP ACTIVITY

3. Crestwood City has a population of 45,000. For the next 6 years it is projected to grow at the rate of 4% per year. Millville has a population of 38,000. For the next 6 years it is projected to grow at the rate of 8% per year. What will be the population of each city in 6 years? Which city will be larger? How much difference will there be in the population?

Chapter Organizer

Topic	Procedure	Examples
Absolute value, p. 71.	The absolute value of a number is the distance between that number and zero on the number line. The absolute value of any number will be positive or zero.	$\lvert 3\rvert = 3$ $\lvert -2\rvert = 2$ $\lvert 0\rvert = 0$ $\left\lvert -\frac{5}{6}\right\rvert = \frac{5}{6}$ $\lvert -1.38\rvert = 1.38$
Adding signed numbers with the same sign, p. 72.	If the signs are the same, add the absolute values of the numbers. Use the common sign in the answer.	$(-3) + (-7) = -10$
Adding signed numbers with opposite signs, p. 74.	If the signs are different: **1.** Find the difference of the larger absolute value and the smaller. **2.** Give the answer the sign of the number having the larger absolute value.	$(-7) + (+13) = 6$ $(7) + (-13) = -6$
Adding several signed numbers, p. 76.	When adding several signed numbers, separate them into two groups by common sign. Find the sum of all the positives and all the negatives. Combine these two subtotals by the method described above.	$(-7) + 6 + 8 + (-11) + (-13) + 22$ $-7 \quad +6$ $-11 \quad +8$ $-13 \quad +22$ $-31 \quad +36$ $(-31) + (+36) = +5$ positive since 36 is positive
Subtracting signed numbers, p. 81.	Change the sign of the second number and then add.	$(-3) - (-13) = (-3) + (+13) = 10$
Multiplying and dividing signed numbers, p. 87 and p. 89.	**1.** If the two numbers have the same sign, multiply (or divide). The result is positive. **2.** If the two numbers have different signs, multiply (or divide) as indicated. The result is negative.	$(-5)(-3) = +15$ $(-36) \div (-4) = +9$ $(28) \div (-7) = -4$ $(-6)(3) = -18$
Exponent form, p. 94.	The base tells you what number is being multiplied. The exponent tells you how many times this number is used as a factor.	$2^5 = 2 \cdot 2 \cdot 2 \cdot 2 \cdot 2 = 32$ $4^3 = 4 \cdot 4 \cdot 4 = 64$ $(-3)^4 = (-3)(-3)(-3)(-3) = 81$
Raising a negative number to a power, p. 95.	When the base is negative, the result is positive for even exponents, and negative for odd exponents.	$(-3)^3 = -27$ but $(-2)^4 = 16$
Removing parentheses, p. 98.	Use the distributive law to remove parentheses. $a(b + c) = ab + ac$	$3(5x + 2) = 15x + 6$
Combining like terms, p. 102.	Combine terms that have identical letters and exponents.	$7x^2 - 3x + 4y + 2x^2 - 8x - 9y = 9x^2 - 11x - 5y$
Order of operations, p. 105.	Remember the proper order of operations: **1.** Operations inside parentheses **2.** Powers **3.** Multiplication and division from left to right **4.** Addition and subtraction from left to right	$3(5 + 4)^2 - 2^2 \cdot 3 \div (9 - 2^3) = 3 \cdot 9^2 - 4 \cdot 3 \div 1$ $= 3 \cdot 81 - 12$ $= 243 - 12 = 231$
Substituting into variable expressions, p. 108.	**1.** Replace each letter by the numerical value given. **2.** Follow the order of operations in evaluating the expression.	Evaluate $2x^3 + 3xy + 4y^2$ for $x = -3$, $y = 2$. $2(-3)^3 + 3(-3)(2) + 4(2)^2$ $= 2(-27) + 3(-3)(2) + 4(4)$ $= -54 - 18 + 16$ $= -56$

Chapter Organizer

Topic	Procedure	Examples
Using formulas, p. 109.	1. Replace each variable in the formula by the given values. 2. Evaluate the expression. 3. Label units carefully.	Find the area of a circle with radius 4 feet. Use $A = \pi r^2$, with π as approximately 3.14. $A = (3.14)(4 \text{ feet})^2$ $= (3.14)(16 \text{ feet}^2)$ $= 50.24 \text{ feet}^2$ The area of the circle is approximately 50.24 square feet.
Removing grouping symbols, p. 114.	1. Remove innermost grouping symbols first. 2. Then remove remaining innermost grouping symbols. 3. Continue until all grouping symbols are removed. 4. Combine like terms.	$5\{3x - 2[4 + 3(x - 1)]\} = 5\{3x - 2[4 + 3x - 3]\}$ $= 5\{3x - 8 - 6x + 6\}$ $= 15x - 40 - 30x + 30$ $= -15x - 10$

Developing Your Study Skills

PROBLEMS WITH ACCURACY

Strive for accuracy. Mistakes are often made as a result of human error rather than by lack of understanding. Such mistakes are frustrating. A simple arithmetic or sign error can lead to an incorrect answer.

These five steps will help you cut down on errors.

1. Work carefully, and take your time. Do not rush through a problem just to get it done.
2. Concentrate on one problem at a time. Sometimes problems become mechanical, and your mind begins to wander. You become careless and make a mistake.
3. Check your problem. Be sure that you copied it correctly from the book.
4. Check your computations from step to step. Check the solution in the problem. Does it work? Does it make sense?
5. Keep practicing new skills. Remember the old saying, "Practice makes perfect." An increase in practice results in an increase in accuracy. Many errors are due simply to lack of practice.

There is no magic formula for eliminating all errors, but these five steps will be a tremendous help in reducing them.

Chapter 1 Review Problems

1.1 *Add the following.*

1. $(-6) + (-2)$ **2.** $(-12) + (+7.8)$ **3.** $(+5) + (-2) + (-12)$ **4.** $(+3.7) + (-1.8)$

5. $\left(+\frac{1}{2}\right) + \left(-\frac{5}{6}\right)$ **6.** $\left(-\frac{3}{11}\right) + \left(-\frac{1}{22}\right)$ **7.** $\left(+\frac{3}{4}\right) + \left(-\frac{1}{12}\right) + \left(-\frac{1}{2}\right)$ **8.** $\left(-\frac{4}{15}\right) + \left(+\frac{12}{5}\right) + \left(-\frac{2}{3}\right)$

9. $(-5) + (-15) + (+6) + (-9) + (+10)$ **10.** $-8 + 4 + (-3) + (-20) + 18 + 7$

1.2 *Subtract the following.*

11. $(+5) - (-3)$ **12.** $(-2) - (-15)$ **13.** $(-30) - (+3)$ **14.** $(+8) - (-1.2)$

15. $\left(-\frac{7}{8}\right) + \left(-\frac{3}{4}\right)$ **16.** $\left(-\frac{3}{14}\right) + \left(+\frac{5}{7}\right)$ **17.** $-20.8 - 1.9$ **18.** $-103 - (-76)$

Mixed Review 1.1–1.2 *Perform each operation indicated. Simplify all answers.*

19. $-5 + (-2) - (-3)$ **20.** $6 - (-4) + (-2) + (8)$ **21.** $(-16) + (-13)$ **22.** $(-11) - (-12)$

1.3

23. $(-7) \times 6$ **24.** $96 \div (-32)$ **25.** $(-3)(-2) + (-2)(4)$ **26.** $\dfrac{-18}{-\frac{2}{3}}$

27. $(-20) \div 4$ **28.** $(-100) \div (-50)$ **29.** $\left(-\frac{1}{2}\right) \div \left(\frac{3}{4}\right)$ **30.** $\left(\frac{5}{7}\right) \div \left(-\frac{5}{25}\right)$

31. $(-5)(-1)(2)$ **32.** $(-6)(3)(4)$ **33.** $(-1)(-2)(-3)(-5)$ **34.** $(-5)\left(-\frac{1}{2}\right)(4)(-3)$

Mixed Review 1.1–1.3

35. $-\frac{4}{3} + \frac{2}{3} + \frac{1}{6}$ **36.** $-\frac{6}{7} + \frac{1}{2} + \left(-\frac{3}{14}\right)$ **37.** $(-3)(-2)(-5)$ **38.** $-6 + (-2) - (-3)$

39. $3.5(-2.6)$ **40.** $(-5.4) \div (-6)$ **41.** $5 - (-3.5) + 1.6$ **42.** $-8 + 2 - (-4.8)$

43. $25 + 4.1 + (-26) + (-3.5)$ **44.** $29 + (-45) + 17.5 + (-3.8)$

Solve each word problem.

45. The football team had 3 plays in which they lost 8 yards each time. What was the total yardage lost?

46. The low temperature in Anchorage Alaska last night was $-34°F$. During the day the temperature rose $12°F$. What was the temperature during the day?

47. A mountain peak is 6895 feet above sea level. A location in Death Valley is 468 feet below sea level. What is the difference in height between these two locations?

48. IBM stock rose $1\frac{1}{2}$ points on Monday, dropped $3\frac{1}{4}$ points on Tuesday, rose 2 points on Wednesday, and dropped $2\frac{1}{2}$ points on Thursday. What was the total gain or loss on the value of the stock over this four-day period?

Evaluate.

49. $(-3)^4$

50. $(-2)^6$

51. $(-5)^3$

52. $\left(\frac{1}{2}\right)^3$

53. -8^2

54. $(0.7)^2$

55. $\left(\frac{5}{6}\right)^2$

56. $\left(\frac{3}{4}\right)^3$

Use the distributive property to multiply.

57. $5(3x - 7y)$

58. $2x(3x - 7y + 4)$

59. $-(7x^2 - 3x + 11)$

60. $(2xy + x - y)(-3y)$

Combine like terms.

61. $3a^2b - 2bc + 6bc^2 - 8a^2b - 6bc^2 + 5bc$

62. $3x + 7y + 11x - 9y$

63. $4x^2 - 13x + 7 - 9x^2 - 22x - 16$

$-5y^2 - 35x - 9$

64. $-x + \frac{1}{2} + 14x^2 - 7x - 1 - 4x^2$

Simplify using the order of operations.

65. $(5)(-4) + (3)(-2)^3$

66. $20 - (-10) - (-6) + (-5) - 1$

67. $(7 - 9)^3 + (-6)(-2) + (-3)$

1.8 *Evaluate each expression for the given value of the variable.*

68. $3x - 4$ for $x = -5$

69. $5 - \frac{2}{3}x$ for $x = 6$

70. $x^2 + 3x - 4$ for $x = -3$

71. $-3x^2 - 4x + 5$ for $x = 2$

72. $-3x^3 - 4x^2 + 2x + 6$ for $x = -2$

73. $vt - \frac{1}{2}at^2$ for $v = 24$, $t = 2$, and $a = 32$

74. $\frac{nRT}{V}$ for $n = 16$, $R = -2$, $T = 4$, and $V = -20$

Solve each word problem.

75. Find the simple interest on a loan of \$6000 at an annual interest rate of 18% per year for $\frac{3}{4}$ of a year. Use $I = prt$.

76. Find the Fahrenheit temperature if a radio announcer in Mexico City says that the high temperature today was 30°C. Use the formula $F = \frac{9C + 160}{5}$.

77. How much will it cost to paint a circular sign with a radius of 15 meters. The painter charges \$3 per square meter. Use $A = \pi r^2$, where π is approx 3.14.

78. Find the profit P if the initial cost $C = \$1200$, rent $R = \$300$, and sales $S = \$56$. Use the profit formula $P = 180S - R - C$.

1.9 *Simplify.*

79. $3x - 2(x + 4)$

80. $2(x - 1) - 7(3x + 4)$

81. $2[3 - (4 - 5x)]$

82. $-3x[x + 3(x - 7)]$

83. $2xy^3 - 6x^3y - 4x^2y^2 + 3(xy^3 - 2x^2y - 3x^2y^2)$

84. $-2y^2 - 8xy + 3x^2 - 4y^2 + 3xy - 8y^2 - 2x^2 + xy$

85. $-5(x + 2y - 7) + 3x(2 - 5y)$

86. $2\{x - 3(y - 2) + 4[x - 2(y + 3)]\}$

87. $8[a - b(3 - 4a)] - 6a[2 - a(3 - b)]$

88. $-5\{2a - b[5a - b(3 + 2a)]\}$

89. $-3\{2x - [x - 3y(x - 2y)]\}$

Chapter 1 Test

Simplify each answer.

1. $-3 + (-4) + 9 + 2$

2. $\frac{1}{2} + \left(-\frac{1}{6}\right)$

3. $-0.6 - (-0.8)$

4. $(-8)(-12)$

5. $(-5)(-2)(7)(-1)$

6. $(-12) \div (-3)$

7. $(-1.8) \div (0.6)$

8. $\dfrac{-5}{\frac{3}{2}}$

9. $(-4)^3$

10. $(1.2)^2$

11. -5^3

12. $\left(\frac{2}{3}\right)^4$

1. 4
2. ____
3. ____
4. ____
5. ____
6. ____
7. ____
8. ____
9. ____
10. ____
11. ____
12. ____

13. $-5x(x + 2y - 7)$

14. $-2ab^2(-3a - 2b + 7ab)$

15. $2xy - 5x^2y + 3xy^2 - 8xy - 7xy^2$

16. $6ab - \frac{1}{2}a^2b + \frac{3}{2}ab + \frac{5}{2}a^2b$

17. $12a(a + b) - 4(a^2 - 2ab)$

18. $3(2 - a) - 4(-6 - 2a)$

19. $5(3x - 2y) - (x + 6y)$

20. $7 - (6 - 9)^2 + 5(2)$

21. $3(4 - 6)^3 + 12 \div (-4) + 2$

Evaluate each expression for the value of the variable indicated.

22. $x^3 - 3x^2y + 2y - 5$ for $x = 3$ and $y = -4$

23. $3x^2 - 7x - 11$ for $x = -3$

24. $2a - 3b$ for $a = -4$ and $b = -3$

25. If you are traveling 60 miles per hour in Canada on a highway, how fast are you traveling in kilometers per hour? (Use $k = 1.61r$)

Simplify each expression.

26. $-3\{a + b[3a - b(1 - a)]\}$

27. $3\{[x - (5 - 2y)] - 4[3 + (6x - 7y)]\}$

13. ______

14. ______

15. ______

16. ______

17. ______

18. ______

19. ______

20. ______

21. ______

22. ______

23. ______

24. ______

25. ______

26. ______

27. ______

2

Equations and Inequalities

Suppose you are in an open field and you observe a very severe lightning storm approaching. You see a big bolt of lightning and hear a tremendous clap of thunder 3 seconds later. You would like to know how far away the storm is so that you can make a plan to take shelter from the storm. Using the techniques of this chapter, we will show you how to estimate the distance to the approaching storm. See page 157.

Pretest Chapter 2

1. ______

2. ______

3. ______

4. ______

5. ______

6. ______

7. ______

8. ______

9. (a) ______

(b) ______

10. (a) ______

(b) ______

11. ______

12. ______

13. ______

14. ______

15. ______

16. ______

If you are familiar with the topics in this chapter, take this test now. Check your answers with those in the back of the book. If an answer was wrong or you couldn't do a problem, study the appropriate section of the chapter.

If you are not familiar with the topics in this chapter, don't take this test now. Instead, study the examples, work the practice problems, and then take the test.

This test will help you to identify those concepts that you have mastered and those that need more study.

Sections 2.1–2.3

Solve for x.

1. $x - 9 = 49$ **2.** $\frac{1}{4}x = 12$

3. $-5x = 20$ **4.** $7x + 2 = 30$

5. $3x - 8 = 7x + 6$ **6.** $5x + 3 - 6x = 4 - 8x - 2$

Section 2.4

Solve for x.

7. $\frac{2}{5}x + \frac{1}{4} = \frac{1}{2}x$ **8.** $\frac{2}{3}(x - 2) + \frac{1}{2} = 5 - (3 + x)$

Section 2.5

9. (a) Solve for F. $C = \frac{5}{9}(F - 32)$.

(b) Find the temperature in degrees Fahrenheit if the Celsius temperature is $-15°$.

10. (a) Solve for r. $I = Prt$.

(b) Find the rate r if $P = \$2000$, $t = 2$ years, and $I = \$240$.

Sections 2.6–2.7

Replace the question mark with < or >.

11. 1 ? 10 **12.** 2 ? −3 **13.** −12 ? −13 **14.** 0 ? 0.9

Solve and graph.

15. $-2x + 5 \leq 4 - x + 3$ **16.** $\frac{1}{2}(x + 2) - 1 > \frac{x}{3} - 5 + x$

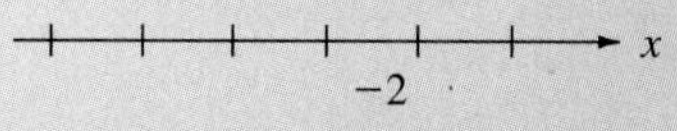

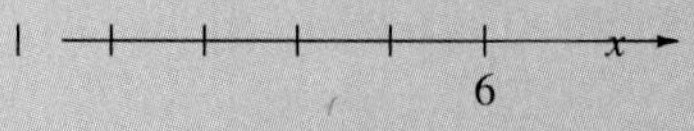

2.1 The Addition Principle

After studying this section, you will be able to:

1 *Solve equations of the form* $x + b = c$ using the addition principle.

1 *Using the Addition Principle*

When we use the equal sign (=), we indicate that two expressions are equal in value. This is called an **equation.** For example, $x + 5 = 23$ is an equation. By choosing certain procedures, you can go step by step from a given equation to the equation $x =$ some number. The number is the solution to the equation.

One of the first procedures used in solving equations has an application in our everyday world. Suppose that we place a 10-kilogram box on one side of a seesaw and a 10-kilogram stone on the other side. If the center of the box is the same distance from the balance point as the center of the stone, we would expect the seesaw to balance. The box and the stone do not look the same, but they have the same value in weight. If we add a 2-kilogram lead weight to the center of weight of each object at the same time, the seesaw should still balance. The results are equal.

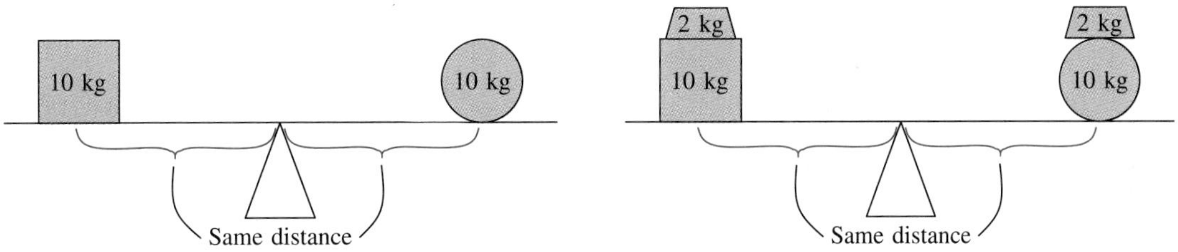

There is a similar principle in mathematics. We can state it in words like this.

> **The Addition Principle**
>
> If the same number is added to both sides of an equation, the results on each side are equal in value.

We can restate it in symbols this way.

> For real numbers a, b, c if $a = b$ then $a + c = b + c$

Here is an example.

$$\text{If } 3 = \frac{6}{2}, \qquad \text{then } 3 + 5 = \frac{6}{2} + 5$$

Since we added the same amount 5 to both sides, each side has an equivalent value.

$$3 + 5 = \frac{6}{2} + 5$$

$$8 = \frac{6}{2} + \frac{10}{2}$$

$$8 = \frac{16}{2}$$

$$8 = 8$$

We can use the addition principle to solve an equation.

EXAMPLE 1 Solve for x. $x + 16 = 20$

$x + 16 + (-16) = 20 + (-16)$	*Use the addition principle to add* -16 *to both sides.*
$x + 0 = 4$	*Simplify.*
$x = 4$	*The value of* x *is* 4.

We have just found the solution of the equation. The **solution** is a value for the variable that makes the equation true. We then say that the value, 4, in our example, **satisfies** the equation. We can easily verify that 4 is a solution by substituting this value in the original equation. This step is called **checking** the solution.

Check. $x + 16 = 20$

$4 + 16 \stackrel{?}{=} 20$

$20 = 20$ ✓

When the same value appears on both sides of the equals sign, we call the equation an **identity.** Because the two sides of the equation in our check have the same value, we know that the original equation has been correctly solved. We have found the solution.

Practice Problem 1 Solve for x and check your solution. $x + 0.3 = 1.2$ ■

When you are trying to solve these types of equations, you notice that you must add a particular number to both sides of the equation. What is the number to choose? Look at the number that is on the same side of the equation with x, that is, the number added to x. Then think of the number that is **opposite in sign.** This is called the **additive inverse** of the number. The additive inverse of 16 is -16. The additive inverse of -3 is 3. The number to add to both sides of the equation is precisely this additive inverse.

It does not matter which side of the equation contains the variable. The x term may be on the right or left. In the next example the x term will be on the right.

EXAMPLE 2 Solve for x. $14 = x - 3$

$14 + 3 = x - 3 + 3$	*Add 3 to both sides, since 3 is the additive inverse of* -3. *This will eliminate the* -3 *on the right and isolate* x.
$17 = x + 0$	*Simplify.*
$17 = x$	*The value of* x *is* 17.

Check. $14 = x - 3$

$14 \stackrel{?}{=} 17 - 3$ *Replace* x *by* 17.

$14 = 14$ ✓ *Simplify. It checks. The solution is* $x = 17$.

Practice Problem 2 Solve for x and check your solution. $17 = x - 5$ ■

Before you add a number to both sides, you should always simplify the equation. The following example shows how combining numbers by addition—separately, on both sides of the equation—simplifies the equation.

EXAMPLE 3 Solve for x. $15 + 2 = 3 + x + 2$

$17 = x + 5$	*Simplify by adding.*
$17 + (-5) = x + 5 + (-5)$	*Add the value -5 to both sides, since -5 is the additive inverse of 5.*
$12 = x$	*Simplify. The value of x is 12.*

Check. $15 + 2 = 3 + x + 2$

$15 + 2 \stackrel{?}{=} 3 + 12 + 2$	*Replace x by 12 in the original equation.*
$17 = 17$ ✓	*It checks.*

Practice Problem 3 Solve for x and check your solution. $5 - 12 = x - 3$ ■

In Example 3 we added -5 to each side. You could subtract 5 from each side and get the same result. In Chapter 1 we discussed how subtracting a 5 is the same as adding a negative 5. Do you see why?

We can determine if a value is a solution to an equation by following the same steps used to check an answer. Substitute the value to be tested for the variable in the original equation. We will obtain an identity if the value is a solution.

EXAMPLE 4 Is $x = 10$ the solution to the equation $-15 + 2 = x - 3$? If it is not, find the solution.

We substitute 10 for x in the equation and see if we obtain an identity.

$$-15 + 2 = x - 3$$

$$-15 + 2 = 10 - 3$$

$$-13 \neq 7 \qquad \textit{This is not true. It is not an identity.}$$

Thus, $x = 10$ is not a solution. Now we take the original equation and solve to find a solution.

$-15 + 2 = x - 3$	
$-13 = x - 3$	*Simplify by adding.*
$-13 + 3 = x - 3 + 3$	*Add 3 to both sides. 3 is the additive inverse of -3.*
$-10 = x$	

Check to see if $x = -10$ is a solution. The value $x = 10$ was incorrect because of a sign error. We must be especially careful to write the correct sign for each number when solving equations.

Practice Problem 4 Is $x = -2$ the solution to the equation $x + 8 = -22 + 6$? If it is not, find the solution. ■

EXAMPLE 5 Find the value of x that satisfies the equation $\frac{1}{5} + x = -\frac{1}{10} + \frac{1}{2}$.

To combine the fractions, the fractions must have common denominators. The least common denominator (LCD) of the fraction is 10.

$$\frac{1 \cdot 2}{5 \cdot 2} + x = -\frac{1}{10} + \frac{1 \cdot 5}{2 \cdot 5}$$ *Change each fraction to an equivalent fraction with a denominator of 10.*

$$\frac{2}{10} + x = -\frac{1}{10} + \frac{5}{10}$$ *This is an equivalent equation.*

$$\frac{2}{10} + x = \frac{4}{10}$$ *Simplify by adding.*

$$\frac{2}{10} + \left(-\frac{2}{10}\right) + x = \frac{4}{10} + \left(-\frac{2}{10}\right)$$ *Add the additive inverse of $\frac{2}{10}$ to each side.*

$$x = \frac{2}{10}$$ *Add the fractions.*

$$x = \frac{1}{5}$$ *Simplify the answer.*

Check. We substitute $\frac{1}{5}$ for x in the original equation and see if we obtain an identity.

$$\frac{1}{5} + x = -\frac{1}{10} + \frac{1}{2}$$

$$\frac{1}{5} + \frac{1}{5} = -\frac{1}{10} + \frac{1}{2}$$ *Substitute $\frac{1}{5}$ for x.*

$$\frac{2}{5} = -\frac{1}{10} + \frac{5}{10}$$

$$\frac{2}{5} = \frac{4}{10}$$

$$\frac{2}{5} = \frac{2}{5} \checkmark$$ *It checks.*

Practice Problem 5 Find the value of x that satisfies the following equation.

$$\frac{1}{20} - \frac{1}{2} = x + \frac{3}{5}$$ ■

Developing Your Study Skills

WHY STUDY MATHEMATICS?

In our present-day, technological world, it is easy to see mathematics at work. Many vocational and professional areas—such as the fields of business, statistics, economics, psychology, finance, computer science, chemistry, physics, engineering, electronics, nuclear energy, banking, quality control, and teaching—require a certain level of expertise in mathematics. Those who want to work in these fields must be able to function at a given mathematical level. Those who cannot will not make it. So if your field of study requires you to take higher-level mathematics courses, be sure to master the basics of this course. Then you will be ready for the next one.

2.1 Exercises

Verbal and Writing Skills

Fill in the blank with the appropriate word.

1. When we use the ______________ sign, we indicate two expressions are ______________ in value.

2. If the ______________ ______________ is added to both sides of an equation, the results on each side are equal in value.

3. The ______________ of an equation is a value of the variable that makes the equation true.

4. What is the additive inverse of -20?

5. Why do we add the additive inverse of a to each side of $x + a = b$ to solve for x?

Find the value of x that satisfies each equation. Check your answers for Problems 6 through 25.

6. $x + 5 = 9$

7. $x + 3 = 6$

8. $11 = 13 + x$

9. $6 = x + 7$

10. $x - 16 = 5$

11. $x - 18 = 2$

12. $0 = x + 5$

13. $0 = x - 7$

14. $3 + 5 = x - 7$

15. $8 - 2 = x + 5$

16. $7 + 3 + x = 5 + 5$

17. $18 - 2 + 3 = x + 19$

18. $x - 9 = -30$

19. $x - 5 = -26$

20. $-12 + x = 50$

21. $-18 + x = 48$

22. $18 - 11 = x - 5$

23. $23 - 8 = x - 12$

24. $5 - 19 + 3 = x + 1$

25. $x + 7 = 15 - 26 + 4$

In Problems 26 through 33, determine if the given solution is correct. If it is not, find the solution.

26. Is $x = 8$ the solution to $-12 + x = 4$?

27. Is $x = 12$ the solution to $-19 + x = 7$?

28. Is $x = -4$ the solution to $-12 + 5 = 3 + x$?

29. Is $x = -3$ the solution to $-10 - 3 = x - 20$?

30. Is $x = -33$ the solution to $x - 23 = -56$?

31. Is $x = -8$ the solution to $-39 = x - 47$?

32. Is $x = 35$ the solution to $15 - 3 + 20 = x - 3$?

33. Is $x = -12$ the solution to $x + 8 = 12 - 19 + 3$?

Find the value of x that satisfies each equation.

34. $-2 = -7 + x$

35. $-9 + x = -4$

36. $-16 = x + 25$

37. $27 = 5 + x$

38. $1.3 + x + 1.8 = 0.2$

39. $3.6 + 1.2 = x + 1.3$

40. $-1.5 + x = -3.8$

41. $-0.6 + x = -1.8$

42. $x - \frac{1}{4} = \frac{3}{4}$

43. $x + \frac{1}{3} = \frac{2}{3}$

44. $\frac{1}{2} + x = \frac{1}{4} + \frac{1}{3}$

45. $x - \frac{1}{5} = \frac{1}{2} + \frac{1}{10}$

46. $\frac{1}{18} - \frac{5}{9} = x - \frac{1}{2}$

47. $\frac{3}{22} - \frac{7}{11} = x - \frac{3}{2}$

48. $5\frac{1}{6} + x = 8$

49. $7\frac{1}{8} = -20 + x$

★50. $\frac{1}{2} + 3x = \frac{1}{4} + 2x - 3 - \frac{1}{2}$

51. $1.6 - 5x + 3.2 = -2x + 5.6 + 4x - 8x$

52. $x - 0.8613 = 2.1754$

53. $x + 9.3715 = -18.1261$

Cumulative Review Problems

Simplify by adding like terms.

54. $5x - 8 + 2x - 12$

55. $x + 3y - 5x - 7y + 2x$

56. $-3x^2 - 6x + 1 - 4x^2 + 2x - 8$

57. $y^2 + y - 12 - 3y^2 - 5y + 16$

2.2 The Multiplication Principle

After studying this section, you will be able to:

1 *Solve equations of the form* $\frac{1}{a}x = b$.

2 *Solve equations of the form* $ax = b$.

1 *Solving Equations of the Form* $\frac{1}{a}x = b$

The addition principle allows us to add the same number to both sides of an equation. What would happen if we multiplied each side of an equation by the same number? For example, what would happen if we multiplied each side of an equation by 3?

To answer this question, let's return to our simple example of the box and the stone on a balanced seesaw. If we triple the number of weights on each side (we are multiplying each side by 3), the seesaw should still balance. The ''weight value'' of each side remains equal.

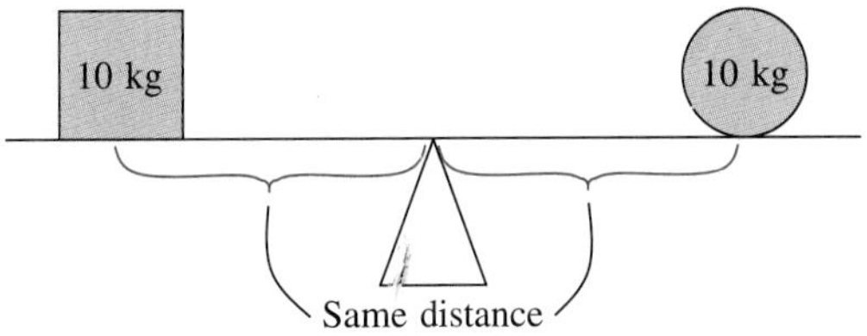

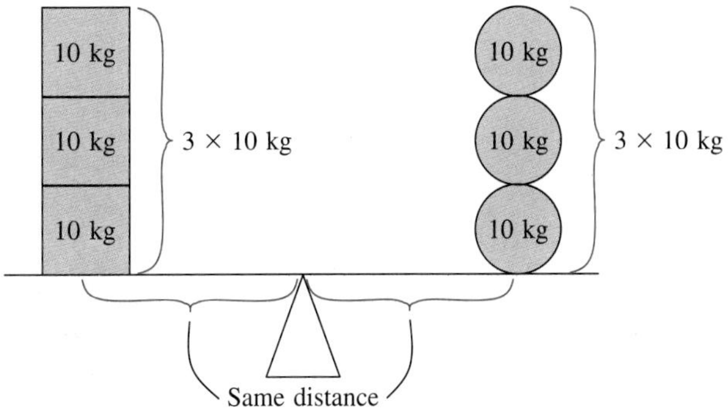

In words we can state this principle thus.

> **Multiplication Principle**
>
> If both sides of an equation are multiplied by the same number, the results on each side are equal in value.

In symbols we can restate the multiplication principle this way.

> For real numbers a, b, c with $c \neq 0$ if $a = b$ then $ca = cb$

Let us look at an equation where it would be helpful to multiply each side of the equation by 3.

EXAMPLE 1 Solve for x. $\frac{1}{3}x = -15$

We know that $(3)(\frac{1}{3}) = 1$. We will multiply each side of the equation by 3, because we want to isolate the variable x.

$$3\left(\frac{1}{3}x\right) = 3(-15) \qquad \textit{Multiply each side of the equation by 3 since } (3)(\tfrac{1}{3}) = 1.$$

$$\left(\frac{3}{1}\right)\left(\frac{1}{3}\right)(x) = -45$$

$$1x = -45 \qquad \textit{Simplify.}$$

$$x = -45$$

Check. $\frac{1}{3}(-45) \stackrel{?}{=} -15$ *Substitute* -45 *for* x *in the original equation.*

$-15 = -15$ ✓ *It checks.*

Practice Problem 1 Solve for x. $\frac{1}{8}x = -2$ ■

Note that $\frac{1}{5}x$ can be written as $\frac{x}{5}$. To solve the equation $\frac{x}{5} = 3$, we could multiply each side of the equation by 5. Try it. Then check your solution.

2 *Solving Equations of the Form $ax = b$*

We can see that using the multiplication principle to multiply each side of an equation by $\frac{1}{2}$ is the same as dividing each side of the equation by 2. Thus, it would seem that the multiplication principle would allow us to divide each side of the equation by any nonzero real number. Is there a real-world example of this idea?

Let's return to our simple example of the box and the stone on a balanced seesaw. Suppose that we were to cut the two objects in half (so that the amount of weight of each was divided by 2). We then return the objects to the same places on the seesaw. The seesaw would still balance. The ''weight value'' of each side remains equal.

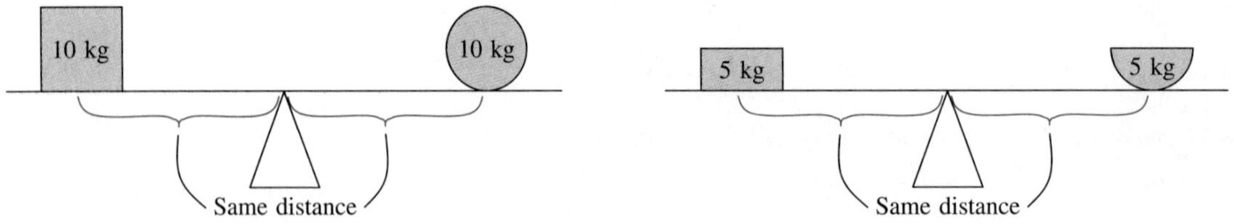

In words we can state this principle thus:

> **Division Principle**
>
> If both sides of an equation are divided by the same nonzero number, the results on each side are equal in value.

Note: We put a restriction on the number by which we are dividing. We *cannot divide* by zero. We say that expressions like $\frac{2}{0}$ are not defined. Thus we restrict our divisor to *nonzero* numbers. We can restate the division principle this way.

> For real numbers a, b, c where $c \neq 0$ if $a = b$ then $\dfrac{a}{c} = \dfrac{b}{c}$

EXAMPLE 2 Solve for x. $5x = 125$

$$\frac{\quad}{5} = \frac{\quad}{5} \qquad \textit{Divide both sides by 5.}$$

$$x = 25 \qquad \textit{Simplify. The solution is 25.}$$

Check. $5x = 125$

$5(25) \stackrel{?}{=} 125$ *Replace x* by 25.

$125 = 125$ ✓ *It checks.*

Practice Problem 2 Solve for x. $9x = 72$ ■

For equations of the form $ax = b$ (a number multiplied by x equals another number), we solve the equation by choosing to divide both sides by a particular number. What is the number to choose? We look at the side of the equation that has x. We notice the number that is multiplied by x. We divide by that number. The division principle tells us that we can still have a true equation provided that we divide by that number *on both sides* of the equation.

The solution to an equation may be a proper fraction or an improper fraction.

EXAMPLE 3 Solve for x. $4x = 38$

$$\frac{4x}{4} = \frac{38}{4} \qquad \textit{Divide both sides by 4.}$$

$$x = \frac{19}{2} \qquad \textit{Simplify. The solution is } \tfrac{19}{2}.$$

If you leave the solution as a fraction, it will be easier to check that solution in the original equation.

Check: $4x = 38$ *Replace x by* $\frac{19}{2}$.

$$\overset{2}{\cancel{4}}\left(\frac{19}{\cancel{2}}\right) \stackrel{?}{=} 38$$

$38 = 38$ ✓ *It checks.*

Practice Problem 3 Solve for x. $6x = 50$ ■

In Examples 2 and 3 we *divided by the number multiplied by x (the coefficient of x).* This procedure is followed regardless of whether the sign of that number is positive or negative.

EXAMPLE 4 Solve for x. $-3x = 48$

$$\frac{-3x}{-3} = \frac{48}{-3} \qquad \textit{Divide both sides by } -3.$$

$$x = -16 \qquad \textit{The solution is } -16.$$

Check. Can you check this solution?

Practice Problem 4 Solve for x. $-27x = 54$ ■

The coefficient of x may be 1 or -1. You may have to rewrite the equation so that the coefficient of 1 or -1 is obvious. With practice you may be able to "see" the coefficient without actually rewriting the equation.

EXAMPLE 5 Solve for x. $-x = -24$.

$$-1x = -24 \qquad \textit{Rewrite the equation. } -1x \textit{ is the same as } -x. \textit{ Now the coefficient of } -1 \textit{ is obvious.}$$

$$\frac{-1x}{-1} = \frac{-24}{-1} \qquad \textit{Divide both sides by } -1.$$

$$x = 24 \qquad \textit{The solution is } 24.$$

Check. Can you check this solution?

Practice Problem 5 Solve for x. $-x = 36$ ■

The variable can be on either side of the equation. The equation $-78 = -3x$ can be solved in exactly the same way as $-3x = -78$.

EXAMPLE 6 Solve for x. $-78 = -3x$

$$\frac{-78}{-3} = \frac{-3x}{-3} \qquad \textit{Divide both sides by } -3.$$

$$26 = x \qquad \textit{The solution is } 26.$$

Check:

$$-78 = -3x$$

$$-78 \stackrel{?}{=} -3(26) \qquad \textit{Replace } x \textit{ by } 26.$$

$$-78 = -78 \ \checkmark \qquad \textit{It checks.}$$

Practice Problem 6 Solve for x. $-51 = -6x$ ■

EXAMPLE 7 Solve for x. $31.2 = 6.0x - 0.8x$

$$31.2 = 6.0x - 0.8x \qquad \textit{There are like terms on the right side.}$$

$$31.2 = 5.2x \qquad \textit{Collect like terms.}$$

$$\frac{31.2}{5.2} = \frac{5.2x}{5.2} \qquad \textit{Divide both sides by } 5.2.$$

$$6 = x \qquad \textit{The solution is } 6.$$

Note. Be sure to place the decimal point in the quotient directly above the caret (‸) when performing the division.

$$\begin{array}{r} 6. \\ 5.2_{\wedge}\overline{)\,31.2_{\wedge}} \\ \underline{31\ 2} \\ 0 \end{array}$$

Check. The check is up to you.

Practice Problem 7 Solve for x. $21 = 4.2x$ ■

2.2 Exercises

Verbal and Writing Skills

1. To solve the equation $-5x = -30$, divide each side of the equation by ____.

2. To solve the equation $20 = -10x$, divide each side of the equation by ____.

3. To solve the equation $\frac{1}{7}x = -2$, multiply each side of the equation by ____.

4. To solve the equation $\frac{1}{9}x = 5$, multiply each side of the equation by ____.

In Problems 5 through 14, solve for x. Be sure to reduce your answers. Check your solution.

5. $\frac{1}{3}x = 21$

6. $\frac{1}{5}x = 30$

7. $\frac{1}{4}x = -2$

8. $\frac{1}{6}x = -3$

9. $\frac{x}{5} = 16$

10. $\frac{x}{10} = 8$

11. $-4 = \frac{x}{12}$

12. $-5 = \frac{x}{15}$

13. $7x = 28$

14. $9x = 36$

15. $-16 = 6x$

16. $-35 = 21x$

17. $1.5x = 75$

18. $2x = 0.36$

19. $-15 = -x$

20. $32 = -x$

21. $-84 = 12x$

22. $-72 = -9x$

23. $0.5x = 0.20$

24. $0.31x = 9.3$

Determine if the given solution is correct. If it is not, find the correct solution.

25. Is $x = 7$ the solution for $-3x = 21$?

26. Is $x = 8$ the solution for $5x = -40$?

27. Is $x = -6$ the solution for $-11x = 66$?

28. Is $x = -20$ the solution for $-x = 20$?

Find the value of the variable that satisfies each equation.

29. $-3y = 2.4$

30. $5z = -1.8$

31. $-27 = -12z$

32. $63 = -28y$

33. $-5.2y = 3.12$

34. $-3.6y = -1.8$

35. $4x + 3x = 21$

36. $5x + 4x = 36$

37. $2x - 7x = 20$

38. $3x - 9x = 18$

39. $-12y + y = -15$

$y = \frac{15}{11}$

40. $-14y + y = -17$

★**41.** $-\frac{2}{3} = -\frac{4}{7}x$

★**42.** $5.6 = -2.7x$

43. $3.6172x = -19.026472$

44. $-4.0518x = 14.505444$

Cumulative Review Problems

Evaluate using the correct order of operations. (Be careful to avoid sign errors.)

45. $5 - 3(2)$

46. $(-6)(-8) + (-3)(2)$

47. $(-3)^3 + (-20) \div 2$

48. $5 + (2 - 6)^2$

Developing Your Study Skills

GETTING HELP

Getting the right kind of help at the right time can be a key ingredient in being successful in mathematics. When you have gone to class on a regular basis, taken careful notes, methodically read your textbook, and diligently done your homework—all of which means making every effort possible to learn the mathematics—you may find that you are still having difficulty. If this is the case, then you need to seek help. Make an appointment with your instructor to find out what help is available to you. The instructor, tutoring services, a mathematics lab, video tapes, and computer software may be among resources you can draw on.

Once you discover the resources available in your school, you need to take advantage of them. Do not put it off, or you will find yourself getting behind. You cannot afford that. When studying mathematics, you must keep up with your work.

2.3 Use the Addition and Multiplication Principles Together

After studying this section, you will be able to:

1 *Solve equations of the form $ax + b = c$.*
2 *Solve equations in which the variable appears on both sides of the equation.*
3 *Solve equations with parentheses.*

1 *Solving Equations of the Form $ax + b = c$*

To solve many equations, we must use both the addition principle and the multiplication principle.

EXAMPLE 1 Solve for x and check your solution.

$$5x + 3 = 18$$

$5x + 3 + (-3) = 18 + (-3)$ *Add -3 to both sides, using the addition principle.*

$5x = 15$ *Simplify.*

$\dfrac{5x}{5} = \dfrac{15}{5}$ *Divide both sides by 5, using the division principle.*

$x = 3$ *The solution is 3.*

Check. $5(3) + 3 \stackrel{?}{=} 18$

$15 + 3 \stackrel{?}{=} 18$

$18 = 18$ ✓ *It checks.*

Practice Problem 1 Solve for x and check your solution. $9x + 2 = 38$ ■

2 *A Variable on Both Sides of the Equation*

In some cases the variable appears on both sides of the equation. We would like to rewrite the equation so that all the terms containing the variable appear on one side. To do this, we apply the addition principle to the variable term.

EXAMPLE 2 Solve for x. $9x = 6x + 15$

$9x + (-6x) = 6x + (-6x) + 15$ *Add $-6x$ to both sides. Notice $6x + (-6x)$ eliminates the variable on the right side.*

$3x = 15$ *Collect like terms.*

$\dfrac{3x}{3} = \dfrac{15}{3}$ *Divide both sides by 3.*

$x = 5$ *The solution is 5.*

Check. Left to the student.

Practice Problem 2 Solve for x. $13x = 2x - 66$ ■

Many problems have variable terms and constant terms on both sides of the equation. You will want to get all the variable terms on one side and all the constant terms on the other side.

EXAMPLE 3 Solve for x and check your solution. $9x + 3 = 7x - 2$.

$$9x + (-7x) + 3 = 7x + (-7x) - 2 \quad \textit{Add } -7x \textit{ to both sides of the equation.}$$

$$2x + 3 = -2 \quad \textit{Combine like terms.}$$

$$2x + 3 + (-3) = -2 + (-3) \quad \textit{Add } -3 \textit{ to both sides.}$$

$$2x = -5 \quad \textit{Simplify.}$$

$$\frac{2x}{2} = \frac{-5}{2} \quad \textit{Divide both sides by 2.}$$

$$x = -\frac{5}{2} \quad \textit{The solution is } -\tfrac{5}{2}.$$

Check. $9x + 3 = 7x - 2$

$$9\left(-\frac{5}{2}\right) + 3 \stackrel{?}{=} 7\left(-\frac{5}{2}\right) - 2 \quad \textit{Replace } x \textit{ by } -\tfrac{5}{2}.$$

$$-\frac{45}{2} + 3 \stackrel{?}{=} -\frac{35}{2} - 2 \quad \textit{Simplify.}$$

$$-\frac{45}{2} + \frac{6}{2} \stackrel{?}{=} -\frac{35}{2} - \frac{4}{2} \quad \textit{Change to equivalent fractions with a common denominator.}$$

$$-\frac{39}{2} = -\frac{39}{2} \ \checkmark \quad \textit{It checks. } x = -\tfrac{5}{2} \textit{ is the correct solution.}$$

Practice Problem 3 Solve for x and check your solution. $3x + 2 = 5x + 2$ ■

In our next example we will study equations that need simplifying before any other steps are taken. Where it is possible, you should first collect like terms on one or both sides of the equation. The variable terms can be collected on the right side or the left side. In this example we will collect all the x terms on the right side.

EXAMPLE 4 Solve for x. $5x + 26 - 6 = 9x + 12x$

$$5x + 20 = 21x \quad \textit{Combine like terms.}$$

$$5x + (-5x) + 20 = 21x + (-5x) \quad \textit{Add } -5x \textit{ to both sides.}$$

$$20 = 16x \quad \textit{Combine like terms.}$$

$$\frac{20}{16} = \frac{16x}{16} \quad \textit{Divide both sides by 16.}$$

$$\frac{5}{4} = x \quad \textit{(Don't forget to reduce the resulting fraction.)}$$

Check. Left to the student.

Practice Problem 4 Solve for z. $-z + 8 - z = 3z + 10 - 3$ ■

Do you really need all these steps? No. As you become more proficient you will be able to combine or eliminate some of these steps. However, it is best to write each step in its entirety until you are consistently obtaining the correct solution. It is much better to show every step than to take a lot of shortcuts but obtain a wrong "solution." This is a section of the algebra course where working neatly and accurately will help you—both now and as you progress through the course.

All the equations we have been studying so far are called first-degree equations. This means the variable terms are not squared (such as x^2 or y^2) or some higher power. It is possible to solve equations with x^2 and y^2 terms by the same methods we have used so far. If x^2 or y^2 terms appear, try to collect them on one side of the equation. If the squared term drops out, you may solve it as a first-degree equation using the methods discussed in this section.

EXAMPLE 5 Solve for y. $5y^2 + 6y - 2 = -y + 5y^2 + 12$

$5y^2 - 5y^2 + 6y - 2 = -y + 5y^2 - 5y^2 + 12$ *Subtract $5y^2$ from both sides.*

$6y - 2 = -y + 12$ *Combine, since $5y^2 - 5y^2 = 0$.*

$6y + y - 2 = -y + y + 12$ *Add y to each side.*

$7y - 2 = 12$ *Simplify.*

$7y - 2 + 2 = 12 + 2$ *Add 2 to each side.*

$7y = 14$ *Simplify.*

$\frac{7y}{7} = \frac{14}{7}$ *Divide each side by 7.*

$y = 2$ *Simplify. The solution is 2.*

Check. Left to the student.

Practice Problem 5 Solve for x. $2x^2 - 6x + 3 = -4x - 7 + 2x^2$ ■

3 *Solving Equations with Parentheses*

The equations that you solved in the preceding section are simpler versions of equations that we will now discuss. These equations contain parentheses. If the parentheses are first removed, the problems then become just like those encountered previously. We use the distributive property to remove the parentheses.

EXAMPLE 6 Solve for x and check your solution. $4(x + 1) - 3(x - 3) = 25$

$4(x + 1) - 3(x - 3) = 25$

$4x + 4 - 3x + 9 = 25$ *Multiply by 4 and -3 to remove parentheses. Be careful of the signs. Remember $(-3)(-3) = 9$.*

After removing the parentheses, it is important to collect like terms on each side of the equation. Do this before going on to isolate the variable.

$x + 13 = 25$ *Collect like terms.*

$x + 13 - 13 = 25 - 13$ *Add -13 to both sides to isolate the variable.*

$x = 12$ *The solution is 12.*

Check. $4(12 + 1) - 3(12 - 3) \stackrel{?}{=} 25$ *Replace x by 12.*

$4(13) - 3(9) \stackrel{?}{=} 25$ *Combine numbers inside parentheses.*

$52 - 27 \stackrel{?}{=} 25$ *Multiply.*

$25 = 25$ ✓ *Simplify. It checks.*

Practice Problem 6 Solve for x and check your solution.

$4x - (x + 3) = 12 - 3(x - 2)$ ■

EXAMPLE 7 Solve for x. $3(-x - 7) = -2(2x + 5)$

$-3x - 21 = -4x - 10$	*Remove parentheses. Watch the signs carefully.*
$-3x + 4x - 21 = -4x + 4x - 10$	*Add $4x$ to both sides.*
$x - 21 = -10$	*Simplify.*
$x - 21 + 21 = -10 + 21$	*Add 21 to both sides.*
$x = 11$	*The solution is 11.*

Check. Left to the student.

Practice Problem 7 Solve for x. $4(-2x - 3) = -5(x - 2) + 2$ ■

In problems that involve decimals, great care should be taken. In some steps you will be multiplying decimal quantities, and in other steps you will be adding them.

EXAMPLE 8 Solve for x. $0.3(1.2x - 3.6) = 4.2x - 16.44$

$0.36x - 1.08 = 4.2x - 16.44$	*Remove parentheses.*
$0.36x - 0.36x - 1.08 = 4.2x - 0.36x - 16.44$	*Subtract $0.36x$ from both sides.*
$-1.08 = 3.84x - 16.44$	*Collect like terms.*
$-1.08 + 16.44 = 3.84x - 16.44 + 16.44$	*Add 16.44 to both sides.*
$15.36 = 3.84x$	*Simplify.*
$\frac{15.36}{3.84} = \frac{3.84x}{3.84}$	*Divide both sides by 3.84.*
$4 = x$	*The solution is $x = 4$.*

Check. Left to the student.

Practice Problem 8 Solve for x. $0.3x - 2(x + 0.1) = 0.4(x - 3) - 1.1$ ■

EXAMPLE 9 Solve for z and check. $2(3z - 5) + 2 = 4z - 3(2z + 8)$

$6z - 10 + 2 = 4z - 6z - 24$	*Remove parentheses.*
$6z - 8 = -2z - 24$	*Collect like terms.*
$6z - 8 + 2z = -2z + 2z - 24$	*Add $2z$ to each side.*
$8z - 8 = -24$	*Simplify.*
$8z - 8 + 8 = -24 + 8$	*Add 8 to each side.*
$8z = -16$	*Simplify.*
$\frac{8z}{8} = \frac{-16}{8}$	*Divide each side by 8.*
$z = -2$	*Simplify. The solution is -2.*

Check. $2[3(-2) - 5] + 2 \stackrel{?}{=} 4(-2) - 3[2(-2) + 8]$	*Replace z by -2.*
$2[-6 - 5] + 2 \stackrel{?}{=} -8 - 3[-4 + 8]$	*Multiply.*
$2[-11] + 2 \stackrel{?}{=} -8 - 3[4]$	*Simplify.*
$-22 + 2 \stackrel{?}{=} -8 - 12$	
$-20 = -20$ ✓	*It checks.*

Practice Problem 9 Solve for z and check. $5(2z - 1) + 7 = 7z - 4(z + 3)$ ■

2.3 Exercises

Find the value of the variable that satisfies each equation. Check your solution.

1. $7x + 9 = 51$

2. $3x + 8 = 50$

3. $6x - 4 = 62$

4. $5x - 9 = 71$

5. $4x - 13 = -81$

6. $9x - 13 = -76$

7. $-5x + 14 = -41$

8. $-6x + 25 = -83$

9. $-15 + 2x = 15$

10. $-8 + 4x = 8$

11. $\frac{1}{2}x - 3 = 11$

12. $\frac{1}{3}x - 2 = 13$

13. $\frac{1}{6}x + 2 = -4$

14. $\frac{1}{5}x + 6 = -24$

15. $8x = 48 + 2x$

16. $5x = 22 + 3x$

17. $-6x = -27 + 3x$

18. $-7x = -26 + 6x$

19. $30 - x = 5x$

20. $42 - x = 6x$

21. $54 - 2x = -8x$

To Think About

22. Is $y = 2$ the solution for $2y + 3y = 12 - y$?

23. Is $y = 4$ the solution for $5y + 2 = 6y - 6 + y$?

24. Is $x = 11$ a solution for $7x + 6 - 3x = 2x - 5 + x$?

25. Is $x = -12$ a solution for $9x + 2 - 5x = -8 + 5x - 2$?

Solve for y by getting all the y terms on the left. Then solve for y by getting all the y terms on the right. Which approach is better?

26. $-3 + 10y + 6 = 15 + 12y - 18$

27. $7y + 21 - 5y = 5y - 7 + y$

Solve for the variable. You may move the variable terms to the right or to the left.

28. $15 - 3x = -x - 5$

29. $5x + 6 = -7x - 4$

30. $x - 6 = 8 - x$

31. $2x + 5 = 4x - 5$

32. $9y - 2 = 2y - 4$

33. $5y - 7 = 3y - 9$

34. $5x - 9 + 2x = 3x + 23 - 4x$

35. $9x - 5 + 4x = 7x + 43 - 2x$

Remove the parentheses. Solve for the variable. Check your solution.

36. $4(x + 15) = 16$

37. $3(2y - 4) = 12$

38. $6(3x + 2) - 8 = -2$

39. $4(2x + 1) - 7 = 6 - 5$

40. $7x - 3(5 - x) = 10$

41. $6(3 - 4x) + 17 = 8x - 3(2 - 3x)$

42. $0.3x - 0.2(x + 1) = 0.1$

43. $3(x + 0.2) = 2(x - 0.3) + 5.2$

44. $3(x + 1) + 3 = 7(x - 2)$

45. $5x - (2x - 3) = 4(x + 9)$

46. $2(4x - x) + 6 = 2(2x + x) + 8 - x$

47. $-3(y - 3y) + 4 = -4(3y - y) + 6 + 13y$

48. $3(x + 0.2) - (2x + 0.5) = 2(x + 0.3) - 0.5$

49. $0.2(x + 3) - (x - 1.5) = 0.3(x + 2) - 2.9$

50. $3(x - 1) + 2x + 8 - 4x - 2 = 3[(2x - 1) - 2(x + 3)] + 6 + 2x + 2$

Mixed Practice

Solve for the variable.

51. $1.2x + 4 = 3.2x - 8$

52. $9x + 1.9 = -7x - 3.1$

53. $5z + 7 - 2z = 32 - 2z$

54. $8 - 7z + 2z = 20 + 5z$

55. $-4w - 28 = -7 - w$

56. $-6w - 7 = -3 - 8w$

57. $9x - 16 + 6x = 11 + 4x - 5$

58. $7 - 9x - 12 = 3x + 5 - 8x$

59. $2x^2 - 3x - 8 = 2x^2 + 5x - 6$

60. $3x^2 + 4x - 7 = 3x^2 - 5x + 2$

61. $-3.5x + 1.3 = -2.7x + 1.5$

62. $2.8x - 0.9 = 5.2x - 3.3$

63. $5(4 + x) = 3(3x - 1) - 9$

64. $x - 0.8x + 4 = 2.6$

65. $17(y + 3) - 4(y - 10) = 13$

66. $3x + 2 - 1.7x = 0.6x + 31.4$

67. $10(x - 5) - 2(x - 9) = -28$

68. $2(5z + 4) + 19 = 4z - 3(2z + 11)$

Solve for x. Round your answer to the nearest hundredth.

69. $1.63x - 9.23 = 5.71x + 8.04$

70. $-2.21x + 8.65 = 3.69x - 7.78$

Cumulative Review Problems

Simplify.

71. $2x(3x - y) + 4(2x^2 - 3xy)$

72. $5[3(x + y) - 2(x - y)]$

73. $2\{3 - 2[x - 4(2x + 3)]\}$

74. $2\{x - 3[4 + 2(3 + x)]\}$

2.4 Equations with Fractions

After studying this section, you will be able to:

1 *Solve equations with fractions.*

1 *Solving Equations with Fractions*

Equations with fractions can be rather difficult to solve. This difficulty is simply due to the extra care we usually have to use when computing with fractions. The actual equation-solving procedures are the same, with fractions or without. To avoid unnecessary work, we transform the given equation with fractions to an equivalent equation that does not contain fractions. How do we do this? We multiply each side of the equation by the lowest common denominator of all the fractions contained in the equation. We then use the distributive property so that the LCD is multiplied by each term of the equation.

EXAMPLE 1 Solve for x. $\frac{1}{4}x - \frac{2}{3} = \frac{5}{12}x$

First we find that the LCD = 12.

$$12\left(\frac{1}{4}x - \frac{2}{3}\right) = 12\left(\frac{5}{12}x\right) \qquad \textit{Multiply each term by 12.}$$

$$\left(\frac{12}{1}\right)\left(\frac{1}{4}\right)(x) - \left(\frac{12}{1}\right)\left(\frac{2}{3}\right) = \left(\frac{12}{1}\right)\left(\frac{5}{12}\right)(x) \qquad \textit{Use the distributive property.}$$

$$3x - 8 = 5x \qquad \textit{Simplify.}$$

$$3x + (-3x) - 8 = 5x + (-3x) \qquad \textit{Add } -3x \textit{ to each side.}$$

$$-8 = 2x \qquad \textit{Simplify.}$$

$$-\frac{8}{2} = \frac{2x}{2} \qquad \textit{Divide each side by 2.}$$

$$-4 = x \qquad \textit{Simplify.}$$

Check. $\frac{1}{4}(-4) - \frac{2}{3} \stackrel{?}{=} \frac{5}{12}(-4)$

$$-1 - \frac{2}{3} \stackrel{?}{=} -\frac{5}{3}$$

$$-\frac{3}{3} - \frac{2}{3} \stackrel{?}{=} -\frac{5}{3}$$

$$-\frac{5}{3} = -\frac{5}{3} \quad \checkmark \qquad \textit{It checks.}$$

Practice Problem 1 Solve for x. $\frac{3}{8}x - \frac{3}{2} = \frac{1}{4}x$ ■

In Example 1 we multiply each side of the equation by the LCD. It is common practice to immediately go to the second step and multiply each term by the LCD, rather than to write out a separate step using the distributive property.

EXAMPLE 2 Solve for x and check your solution. $\frac{x}{3} + 3 = \frac{x}{5} - \frac{1}{3}$

$$15\left(\frac{x}{3}\right) + 15(3) = 15\left(\frac{x}{5}\right) - 15\left(\frac{1}{3}\right)$$ *The LCD is 15. Use the multiplication principle to multiply each term by 15.*

$5x + 45 = 3x - 5$ *Simplify.*

$5x - 3x + 45 = 3x - 3x - 5$ *Add $-3x$ to both sides.*

$2x + 45 = -5$ *Collect like terms.*

$2x + 45 - 45 = -5 - 45$ *Add -45 to both sides.*

$2x = -50$ *Simplify.*

$\frac{2x}{2} = \frac{-50}{2}$ *Divide both sides by 2.*

$x = -25$ *The solution is -25.*

Check. $\frac{-25}{3} + 3 \stackrel{?}{=} \frac{-25}{5} - \frac{1}{3}$

$$-\frac{25}{3} + \frac{9}{3} \stackrel{?}{=} -\frac{5}{1} - \frac{1}{3}$$

$$-\frac{16}{3} \stackrel{?}{=} -\frac{15}{3} - \frac{1}{3}$$

$$-\frac{16}{3} = -\frac{16}{3} \quad ✓$$

Practice Problem 2 Solve for x and check your solution. $\frac{5x}{4} - 1 = \frac{3x}{4} + \frac{1}{2}$ ■

EXAMPLE 3 Solve for x. $\frac{x+5}{7} = \frac{x}{4} + \frac{1}{2}$

$\frac{x}{7} + \frac{5}{7} = \frac{x}{4} + \frac{1}{2}$ *First we write as separate fractions.*

$$28\left(\frac{x}{7}\right) + 28\left(\frac{5}{7}\right) = 28\left(\frac{x}{4}\right) + 28\left(\frac{1}{2}\right)$$ *We observe that the LCD is 28, so we multiply each term by 28.*

$4x + 20 = 7x + 14$ *Simplify.*

$4x - 4x + 20 = 7x - 4x + 14$ *Add $-4x$ to both sides.*

$20 = 3x + 14$ *Collect like terms.*

$20 - 14 = 3x + 14 - 14$ *Add -14 to both sides.*

$6 = 3x$ *Collect like terms.*

$\frac{6}{3} = \frac{3x}{3}$ *Divide both sides by 3.*

$2 = x$ *The solution is $x = 2$.*

Check. Left to the student.

Practice Problem 3 Solve for x. $\frac{5x}{6} - \frac{5}{8} = \frac{3x}{4} - \frac{1}{3}$ ■

If a problem contains both parentheses and fractions, it is best to remove the parentheses first. Many students find it is helpful to have a written procedure to follow in solving these more involved equations.

Procedure to Solve Linear Equations

1. Remove any parentheses.
2. If fractions exist, multiply all terms on both sides by the lowest common denominator of all the fractions.
3. Collect like terms if possible. Simplify numerical work if possible.
4. Add or subtract terms on both sides of the equation to get all terms with the variable on one side of the equation.
5. Add or subtract a value on both sides of the equation to get all terms not containing the variable on the other side of the equation.
6. Divide both sides of the equation by the coefficient of the variable.
7. Simplify the solution (if possible).
8. Check your solution.

Let's use each step in solving this example.

EXAMPLE 4 Solve for x and check your solution. $\frac{1}{3}(x-2) = \frac{1}{5}(x+4) + 2$

Step 1 $\frac{x}{3} - \frac{2}{3} = \frac{x}{5} + \frac{4}{5} + 2$ *Remove parentheses.*

Step 2 $15\left(\frac{x}{3}\right) - 15\left(\frac{2}{3}\right) = 15\left(\frac{x}{5}\right) + 15\left(\frac{4}{5}\right) + 15(2)$ *Multiply by the LCD, 15.*

Step 3 $5x - 10 = 3x + 12 + 30$ *Simplify.*

$5x - 10 = 3x + 42$ *Simplify.*

Step 4 $5x - 3x - 10 = 3x - 3x + 42$ *Add $-3x$ to both sides.*

$2x - 10 = 42$ *Simplify.*

Step 5 $2x - 10 + 10 = 42 + 10$ *Add 10 to both sides.*

$2x = 52$ *Simplify.*

Step 6 $\frac{2x}{2} = \frac{52}{2}$ *Divide both sides by 2.*

Step 7 $x = 26$ *Simplify the solution.*

Step 8 *Check.* $\frac{1}{3}(26-2) \stackrel{?}{=} \frac{1}{5}(26+4) + 2$ *Replace x by 26.*

$\frac{1}{3}(24) \stackrel{?}{=} \frac{1}{5}(30) + 2$ *Combine values within parentheses.*

$8 \stackrel{?}{=} 6 + 2$ *Simplify.*

$8 = 8$ ✓ *$x = 26$ is the solution.*

Practice Problem 4 Solve for x and check your solution.

$\frac{1}{3}(x-2) = \frac{1}{4}(x+5) - \frac{5}{3}$ ■

It should be remembered that not every step will be needed in each problem. You can combine some steps as well, *as long as you are consistently obtaining the correct solution.* However, you are encouraged to write out every step as a way of helping you to avoid careless errors.

It is important to remember that when we write decimals these numbers are really fractions written in a special way. Thus, $0.3 = \frac{3}{10}$ and $0.07 = \frac{7}{100}$. It is possible to take a linear equation containing decimals and to multiply each term by the appropriate value to obtain integer coefficients.

EXAMPLE 5 Solve for x. $0.2(1 - 8x) + 1.1 = -5(0.4x - 0.3)$

$0.2 - 1.6x + 1.1 = -2.0x + 1.5$	*Remove parentheses.*
$10(0.2) + 10(1.6x) + 10(1.1) = 10(-2.0x) + 10(1.5)$	*Multiply each term by 10.*
$2 - 16x + 11 = -20x + 15$	*Multiplying by 10 moves the decimal point one place to the right.*
$-16x + 13 = -20x + 15$	*Simplify.*
$-16x + 20x + 13 = -20x + 20x + 15$	*Add 20x to each side.*
$4x + 13 = 15$	*Simplify.*
$4x + 13 + (-13) = 15 + (-13)$	*Add −13 to each side.*
$4x = 2$	*Simplify.*
$\frac{4x}{4} = \frac{2}{4}$	*Divide each side by 4.*
$x = \frac{1}{2}$ or 0.5	*Simplify.*

Check.

$$0.2[1 - 8(0.5)] + 1.1 \stackrel{?}{=} -5[0.4(0.5) - 0.3]$$
$$0.2[1 - 4] + 1.1 \stackrel{?}{=} -5[0.2 - 0.3]$$
$$0.2[-3] + 1.1 \stackrel{?}{=} -5[-0.1]$$
$$-0.6 + 1.1 \stackrel{?}{=} 0.5$$
$$0.5 = 0.5 \checkmark$$

Practice Problem 5 Solve for x. $2.8 = 0.3(x - 2) + 2(0.1x - 0.3)$ ■

Developing Your Study Skills

TAKING NOTES IN CLASS

An important part of studying mathematics is taking notes. To take meaningful notes, you must be an active listener. Keep your mind on what the instructor is saying, and be ready with questions whenever you do not understand something.

If you have previewed the lesson material, you will be prepared to take good notes. The important concepts will seem somewhat familiar. You will have a better idea of what needs to be written down. If you frantically try to write all that the instructor says or copy all the examples done in class, you may find your notes to be nearly worthless when you are home alone. You may find that you are unable to make sense of what you have written.

Write down *important* ideas and examples as the instructor lectures, making sure that you are listening and following the logic. Include any helpful hints or suggestions that your instructor gives you or refers to in your text. You will be amazed at how easily these are forgotten if they are not written down.

2.4 Exercises

In Problems 1 through 16, solve for the variable and check your answer.

1. $\frac{1}{5}x + \frac{1}{10} = \frac{1}{2}$

2. $\frac{1}{3}x - \frac{1}{9} = \frac{8}{9}$

3. $\frac{2}{3}x = \frac{1}{3}x + \frac{1}{6}$

4. $\frac{1}{2} = \frac{3}{10} - \frac{2}{5}x$

5. $\frac{y}{2} + \frac{y}{3} = \frac{5}{6}$

6. $\frac{x}{4} - 1 = \frac{x}{5}$

7. $20 - \frac{1}{3}x = \frac{1}{2}x$

8. $\frac{1}{5}x - \frac{1}{2} = \frac{1}{6}x$

9. $2 + \frac{y}{2} = \frac{3y}{4} - 3$

10. $\frac{x}{3} - 1 = -\frac{1}{2} - x$

11. $\frac{y-1}{2} = 4 - \frac{y}{7}$

12. $\frac{x-7}{6} = -\frac{1}{2}$

13. $0.5 - 0.3x = 3.8$

14. $1.2 + 5x = 3.2$

15. $5.9 = 2.5x - 1.6$

16. $0.09x + 2 = 6.5$

17. Is $y = 4$ a solution to $\frac{1}{2}(y-2) + 2 = \frac{3}{8}(3y-4)$?

18. Is $y = 2$ a solution to $\frac{1}{5}(y+2) = \frac{1}{10}y + \frac{3}{5}$?

19. Is $y = \frac{5}{8}$ a solution to $\frac{1}{2}\left(y - \frac{1}{5}\right) = \frac{1}{5}(y+2)$?

20. Is $y = \frac{13}{3}$ a solution to $\frac{y}{2} - \frac{7}{9} = \frac{y}{6} + \frac{2}{3}$?

Solve for the variable.

21. $4(3x-2) = \frac{1}{3}(x-1) + 4$

22. $\frac{2}{3}(x+2) = -2(1-3x)$

23. $0.7 - 0.3x - 0.6 = 0.1$

24. $-1.2 + 0.4x = 0.8 - 2$

25. $-0.3(x-1) + 2 = -0.4(2x+3)$

26. $0.7 - 0.3(6-2x) = -0.1 + 0.4x$

27. $-5(0.2x + 0.1) - 0.6 = 1.9$

28. $0.3x + 1.7 = 0.2x - 0.4(5x+1)$

Mixed Practice

Solve.

29. $\frac{1}{3}(y + 2) = 3y - 5(y - 2)$

30. $\frac{2}{5}(y + 3) - \frac{1}{2} = \frac{1}{3}(y - 2) + \frac{1}{2}$

31. $\frac{1 + 2x}{5} + \frac{4 - x}{3} = \frac{1}{15}$

32. $\frac{x + 3}{4} = 4x - 2(x - 3)$

33. $\frac{x}{5} - \frac{2}{3}x + \frac{16}{15} = \frac{1}{3}(x - 4)$

34. $3 + \frac{1}{2}(x - 1) = 2 - \frac{1}{3}x + 3$

35. $\frac{1}{3}(x - 2) = 3x - 2(x - 1) + \frac{16}{3}$

36. $\frac{3}{4}(x - 2) + \frac{3}{5} = \frac{1}{5}(x + 1)$

37. $\frac{1}{2}x + \frac{1}{9}x = \frac{3x - 2}{12}$

38. $1 - \frac{2}{5}x + \frac{1}{10}x - \frac{2}{5} = \frac{3}{10}$

39. $0.8(x - 3) = -5(2.1x + 0.4)$

40. $0.7(2x + 3) = -4(6.1x + 0.2)$

41. $\frac{1}{5}x - \frac{2}{3} + \frac{1}{15} = \frac{4}{3}x - \frac{7}{15}$

42. $\frac{1}{20}x - \frac{1}{4} + \frac{3}{5} = -\frac{1}{2}x + \frac{1}{4}x$

Solve for x in the following equations. Work carefully.

★ **43.** $7x + 2[(x + 2) + 3] + 2(x + 2) = 2[2(x + 2) + 2x] + 8$

★ **44.** $\frac{7}{2} - \frac{2x - 3}{2} - (3x + 5) = \frac{3}{2} - \frac{5x}{2} - x$

Cumulative Review Problems

45. Add. $\frac{3}{7} + 1\frac{5}{10}$

46. Subtract. $3\frac{1}{5} - 2\frac{1}{4}$

47. Multiply. $\left(2\frac{1}{5}\right)\left(6\frac{1}{8}\right)$

48. Divide. $5\frac{1}{2} \div 1\frac{1}{4}$

2.5 Formulas

After studying this section, you will be able to:

1 *Solve formulas for a specified variable.*

1 *Solving for a Specified Variable in a Formula*

Formulas are equations with one or more variables that are used to describe real-world situations. The formula describes the relationship that exists among the variables. For example, in the formula $d = rt$, distance (d) is related to the rate of speed (r) and to time (t). We can use this formula to find distance if we know the rate and time. Sometimes, however, we are given the distance and the rate, and we are asked to find the time.

EXAMPLE 1 Joseph drove a distance of 156 miles at an average speed of 52 miles per hour. How long did it take Joseph to make the trip?

$d = rt$ *Use the distance formula.*

$156 = 52t$ *Substitute the known values for the variables.*

$\frac{156}{52} = \frac{52}{52}t$ *Divide both sides of the equation by 52 to solve for t.*

$3 = t$ *We have solved for t.*

It took Joseph 3 hours to drive 156 miles at 52 miles per hour.

Practice Problem 1 Find the rate for $d = 240$ miles and $t = 5$ hours. ■

If we have many problems that ask us to find the time given the distance and rate, it may be worthwhile to rewrite the formula in terms of time.

EXAMPLE 2 Solve for t. $d = rt$

$\frac{d}{r} = \frac{rt}{r}$ *We want to isolate t. Therefore, we are dividing both sides of the equation by the coefficient of t, which is r.*

$\frac{d}{r} = t$ *You have solved for the variable indicated.*

Practice Problem 2 Solve for m. $E = mc^2$ ■

A simple first degree equation with two variables can be thought of as the equation of a line. It is often useful to solve for y in order to make graphing the line easier.

EXAMPLE 3 Solve for y. $3x - 2y = 6$

$-2y = 6 - 3x$ *We want to isolate the term containing y, so we subtract 3x from both sides.*

$\frac{-2y}{-2} = \frac{6 - 3x}{-2}$ *Divide both sides by the coefficient of y.*

$y = \frac{6}{-2} + \frac{-3x}{-2}$ *Rewrite the fraction.*

$= \frac{3}{2}x - 3$ *Simplify and regroup.*

This is known as the slope–intercept form of the equation of a line.

Practice Problem 3 Solve for y. $8 - 2y + 3x = 0$ ■

Our procedure for solving a first degree equation can be rewritten to give us a procedure for solving a formula for a specified variable.

Procedure to Solve a Formula for a Specified Variable

1. Remove any parentheses.
2. If fractions exist, multiply all terms on both sides by the LCD of all the fractions.
3. Collect like terms or simplify if possible.
4. Add or subtract terms on both sides of the equation to get all terms with the desired variable on one side of the equation.
5. Add or subtract the appropriate quantity to get all terms that do *not* have the desired variable on the other side of the equation.
6. Divide both sides of the equation by the coefficient of the desired variable.
7. Simplify if possible.

EXAMPLE 4 A trapezoid is a four-sided figure with two parallel sides. If the parallel sides are a and b and the altitude is h, the area is given by

$$A = \frac{h}{2}(a + b)$$

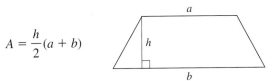

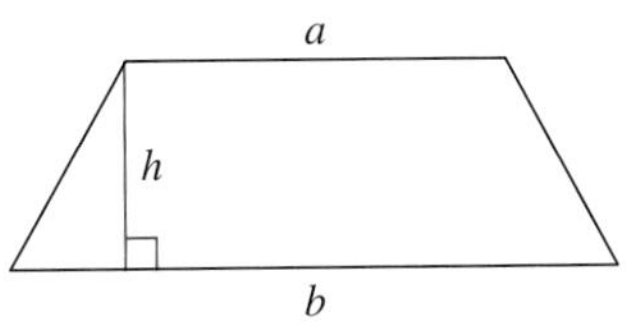

Solve this equation for a.

$$A = \frac{h}{2}(a + b)$$

$$A = \frac{ha}{2} + \frac{hb}{2} \qquad \textit{Remove the parentheses.}$$

$$2(A) = 2\left(\frac{ha}{2}\right) + 2\left(\frac{hb}{2}\right) \qquad \textit{Multiply all terms by LCD of 2.}$$

$$2A = ha + hb \qquad \textit{Simplify.}$$

$$2A - hb = ha$$ *We want to isolate the term containing a. Therefore, we subtract hb from both sides.*

$$\frac{2A - hb}{h} = \frac{ha}{h} \qquad \textit{Divide both sides by h (the coefficient of a).}$$

$$\frac{2A - hb}{h} = a \qquad \textit{The solution is obtained.}$$

Note: Although the solution is in simple form, it could be written in an alternative way. Since

$$\frac{2A - hb}{h} = \frac{2A}{h} - \frac{hb}{h} = \frac{2A}{h} - b$$

we could have $\frac{2A}{h} - b = a$ as an alternative way of writing the answer.

Practice Problem 4 Solve for d. $c = \pi d$ ■

2.5 Exercises

Applications

1. Use the formula $d = rt$ to find the rate of speed of a car that covers 450 miles in 9 hours.

2. The formula for calculating the temperature in degrees Fahrenheit when you know the temperature in degrees Celsius is $F = \frac{9}{5}C + 32$. Use this formula to find the Celsius temperature when the Fahrenheit temperature is $-49°$.

3. The formula for the area of a triangle is $A = \frac{1}{2}bh$
where b is the *base* of the triangle
h is the *height* or *altitude* of the triangle

(a) Use this formula to find the base of a triangle that has an area of 60 square meters and a height of 12 meters.

(b) Use this formula to find the altitude of a triangle that has an area of 88 square meters and a base of 11 meters.

4. The formula for calculating simple interest is $I = Prt$
where P is the *principal* (amount of money invested)
r is the *rate* at which the money is invested
t is the *time*

(a) Use this formula to find how long it would take to earn \$720 in interest on an investment of \$3000 at the rate of 6%.

(b) Use this formula to find the rate of interest if \$5000 earns \$400 interest in 2 years.

(c) Use this formula to find the amount of money invested if the interest earned was \$120 and the rate of interest was 5% over 3 years.

5. The standard form of a line is $3x - 5y = 15$.

(a) Solve for the variable y.

(b) Use this result to find y with $x = -5$.

In each formula or equation, solve for the variable indicated.

Area of a triangle

6. $A = \frac{1}{2}bh$ Solve for b.

7. $A = \frac{1}{2}bh$ Solve for h.

Simple interest formula

8. $I = Prt$ Solve for P.

9. $I = Prt$ Solve for r.

Slope–intercept form of a line

10. $y = mx + b$ Solve for m.

11. $y = mx + b$ Solve for b.

Simple interest formula

12. $A = P(1 + rt)$ Solve for t.

13. $A = P(1 + rt)$ Solve for r.

Standard form of a line

14. $5x - 6y = 6$ Solve for y.

15. $5x + 9y = -18$ Solve for y.

Slope–intercept form of a line

16. $y = -\frac{3}{4}x + 9$ Solve for x.

17. $y = \frac{6}{7}x - 12$ Solve for x.

Standard form of a line

18. $ax + by = c$ Solve for y.

19. $ax + by = c$ Solve for x.

Area of a circle

20. $A = \pi r^2$ Solve for r^2.

Surface area of a sphere

21. $s = 4\pi r^2$ Solve for r^2.

Energy equation

22. $E = mc^2$ Solve for c^2.

23. $E = mc^2$ Solve for m.

Distance of falling object

24. $S = \frac{1}{2}gt^2$ Solve for g.

25. $S = \frac{1}{2}gt^2$ Solve for t^2.

Electronics equation

26. $E = IR$ Solve for R.

27. $E = IR$ Solve for I.

Volume of a rectangular prism

28. $V = LWH$ Solve for L.

29. $V = LWH$ Solve for H.

Volume of pyramid

30. $V = \frac{1}{3}Bh$ Solve for h.

31. $V = \frac{1}{3}Bh$ Solve for B.

Perimeter of a rectangle

32. $P = 2L + 2W$ Solve for W.

33. $P = 2L + 2W$ Solve for L.

Pythagorean theorem

34. $c^2 = a^2 + b^2$ Solve for a^2.

35. $c^2 = a^2 + b^2$ Solve for b^2.

Temperature conversion formulas

36. $F = \frac{9}{5}C + 32$ Solve for C.

37. $C = \frac{5}{9}(F - 32)$ Solve for F.

Volume of a sphere

38. $V = \frac{4}{3}\pi r^3$ Solve for r^3.

Distance of a projectile

39. $d = 16t^2 + vt$ Solve for v.

Area of a sector of a circle

40. $A = \dfrac{\pi r^2 S}{360}$

(a) Solve for S (angle measure).

(b) Solve for r^2 (radius).

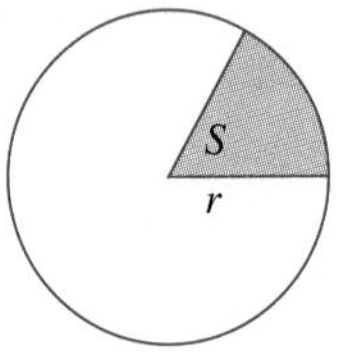

To Think About

41. In $I = Prt$, if t doubles, what is the effect on I?

42. In $I = Prt$, if both r and t double, what is the effect on I?

43. In $A = \pi r^2$, if r doubles, what is the effect on A?

44. In $A = \pi r^2$, if r is halved, what is the effect on A?

Cumulative Review Problems

45. Find 12% of 260.

46. What is 0.2% of 48?

47. Twelve out of 30 attempts are successful. What percent is successful?

48. Four defective parts are found in a shipment of 160 parts. What percent of the shipment is defective?

Putting Your Skills to Work

FINDING THE DISTANCE FROM A LIGHTNING STRIKE

Suppose that you see a flash of lightning and a few seconds later hear the rumble of thunder. Do you know how far away the lightning strike is? Using the formula $d = rt$, you can approximate the distance by substituting $r = 1100$ feet per second, the speed of sound in air, and $t =$ the time in seconds between the moment you see the flash of lightning and the moment you first hear the clap of thunder.

For example, suppose you see a lightning flash and in 5 seconds you hear the rumble of thunder that accompanied that flash. We will use $d = rt$.

$d = (1100)(5)$ *Substitute the known rate and the known time.*

$d = 5500$ feet *Simplify.*

The distance to the lightning strike is approximately 5500 feet, or slightly greater than a mile.

EXERCISES

1. You hear a thunderclap 1.4 seconds after seeing the flash. How far away did the lightning strike?

2. You hear a thunderclap 2.5 seconds after seeing the flash. How far away did the lightning strike?

3. You are talking on a car telephone to a friend who is on the same road driving toward the same city and is using her car telephone. You both mention that a severe thunderstorm appears to be headed toward you. Her driving companion counted 3.5 seconds between a flash of lightning and the thunderclap. Your driving companion counted 3.0 seconds between the same flash of lightning and the thunderclap. Is she closer to the storm or farther from the storm than you are? How far apart are the two cars? (Assume that the road is a straight line.)

COOPERATIVE GROUP ACTIVITY

4. The speed of sound in air that has a temperature range of 50°F to 90°F is approximately 1100 feet per second. If it were hotter than that temperature range, would you expect the speed of sound to be faster or slower? Why?

5. Light travels at 186,282 miles per second. Yet we did not use that figure in our calculations. Why did we not calculate the time it takes for the light from the lightning strike to reach our eyes? By ignoring this fact, what is the effect on our estimates for the distance to the lightning?

2.6 Write and Graph Inequalities

After studying this section, you will be able to:

1 *Interpret an inequality statement.*
2 *Graph an inequality on a number line.*

1 *Inequality Statements*

We frequently speak of one value being greater than or less than another value. We say that "5 is less than 7" or "9 is greater than 4." These relationships are called **inequalities.** We can write inequalities in mathematics by using symbols. We use the symbol $<$ to represent the words "is less than." We use the symbol $>$ to represent the words "is greater than."

Statement in Words	*Statement in Algebra*
5 is less than 7.	$5 < 7$
9 is greater than 4.	$9 > 4$

Note. "5 is less than 7" and "7 is greater than 5" have the same meaning. Similarly, $5 < 7$ and $7 > 5$ have the same meaning. They represent two equivalent ways of describing the same relationship between the two numbers 5 and 7.

We can illustrate the concept of inequality graphically if we examine a number line.

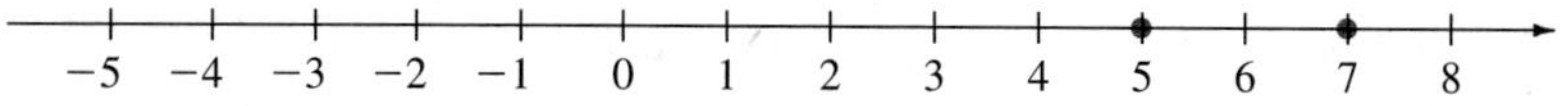

We say that **one number is greater than another** if it is to the right of the other on the number line. Thus $7 > 5$, since 7 is to the right of 5.

What about negative numbers? We can say "-1 is greater than -3" and write it in symbols $-1 > -3$ because we know that -1 lies to the right of -3 on the number line.

EXAMPLE 1 Replace the question mark with the symbol $<$ or $>$ in each statement.

(a) $3 \,?\, -1$ **(b)** $-2 \,?\, 1$ **(c)** $-3 \,?\, -4$ **(d)** $0 \,?\, 3$ **(e)** $-3 \,?\, 0$

(a) $3 > -1$ *Use $>$, since 3 is to the right of -1.*
(b) $-2 < 1$ *Use $<$, since -2 is to the left of 1.*
(Or equivalently, we could say that 1 is to the right of -2.)
(c) $-3 > -4$ *Since -3 is to the right of -4.*
(d) $0 < 3$
(e) $-3 < 0$

Practice Problem 1 Replace the question mark with the symbol $<$ or $>$ in each statement.

(a) $7 \,?\, 2$ **(b)** $-3 \,?\, -4$ **(c)** $-1 \,?\, 2$

(d) $-8 \,?\, -5$ **(e)** $0 \,?\, -2$ **(f)** $\frac{2}{5} \,?\, \frac{3}{8}$ ■

2 Graphing an Inequality on a Number Line

Sometimes we will use an inequality to express the relationship between a variable and a number. $x > 3$ means that x could have the value of *any number* greater than 3. This can be pictured on the number line in a graph as follows:

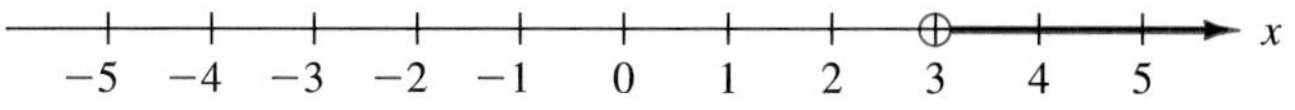

Note that the open circle at 3 suggests that we do not include the point for the number 3.

Similarly, we could represent graphically $x < -2$ as follows:

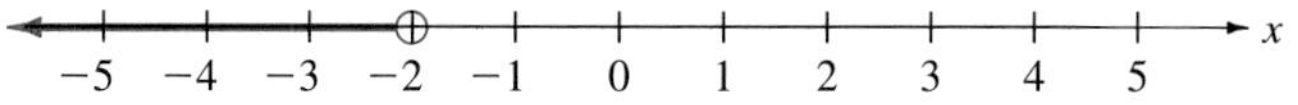

Sometimes a variable will be either greater than or equal to a certain number. In the statement "x is greater than or equal to 3," we are implying that x could have the value of 3 or any number greater than 3. We write this as $x \geq 3$. We represent it graphically as follows:

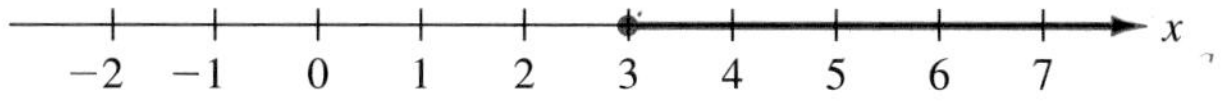

Note that the closed circle at 3 suggests that *we do include* the point for the number 3.

Similarly, we could represent graphically $x \leq -2$ as follows:

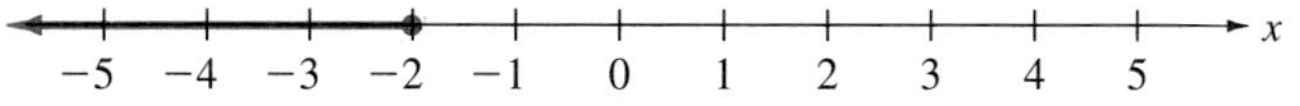

EXAMPLE 2 For parts (a) through (d), state each mathematical relationship in words and then illustrate it graphically.

(a) $x < -2$ **(b)** $-3 < x$ **(c)** $x \geq -2$ **(d)** $x \leq -6$

(a) We state that "x is less than -2."

$x < -2$

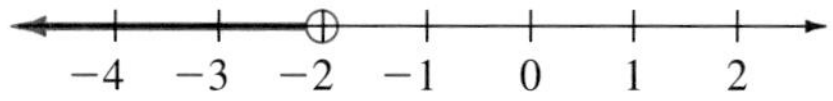

(b) We can state that "-3 is less than x" or, an equivalent statement, "x is greater than -3." Be sure you see that $-3 < x$ is equivalent to $x > -3$. Although both ways are correct, we *usually write the variable first* in a simple linear inequality containing a variable and a numerical value.

$x > -3$

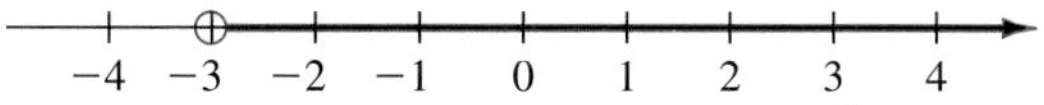

(c) We state that "x is greater than or equal to -2."

$x \geq -2$

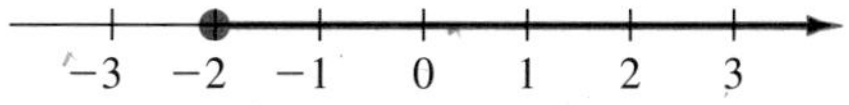

(d) We state that "x is less than or equal to 6."

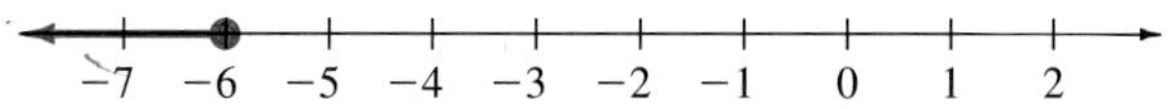

Practice Problem 2 State each mathematical relationship in words and then illustrate it graphically on the number lines in the margin.

(a) $x > 5$ **(b)** $x \leq -2$ **(c)** $3 > x$ **(d)** $x \geq -\frac{3}{2}$ ■

PRACTICE PROBLEM 2

(a)

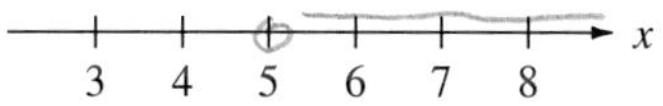

(b)

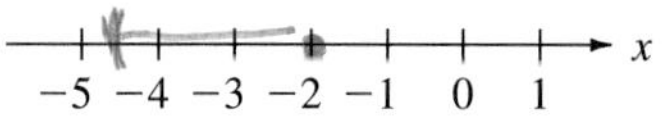

(c)

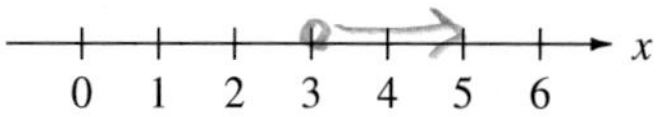

(d)

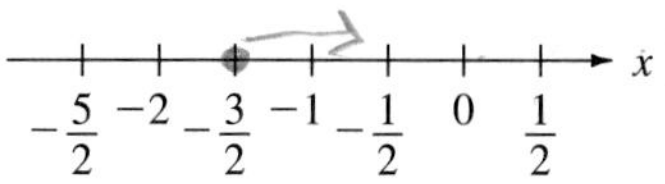

There are many everyday situations involving an unknown value and an inequality. We can translate these situations into algebraic statements. This is the first step in solving word problems using inequalities.

EXAMPLE 3 Translate each English statement into an algebraic statement.

(a) The police on the scene said that the car was traveling greater than 80 miles per hour (use the variable s for speed).

(b) The owner of the trucking company said that the payload of a truck must never exceed 4500 pounds (use the variable p for payload).

(a) Since the speed must be greater than 80 we have $s > 80$.

(b) If the payload of the truck can never exceed 4500 pounds, then the payload must be always less than or equal to 4500 pounds. Thus we write $p \leq 4500$.

Practice Problem 3 Translate the English statement into an inequality.

(a) During the drying cycle, the temperature inside the clothes dryer must never exceed 180 degrees Fahrenheit (use the variable t for temperature).

(b) The bank loan officer said that the total consumer debt incurred by Wally and Mary must be less than $15,000 if they want to qualify for a mortgage to buy their first home (use the variable d for debt). ■

Developing Your Study Skills

KEEP TRYING

You may be one of those students who have had much difficulty with mathematics in the past and who are sure that you cannot do well in this course. Perhaps you are thinking, "I have never been any good at mathematics," or "I have always hated mathematics," or "Math always scares me," or "I have not had any math for so long that I have forgotten it all." You may even have picked up on the label "math anxiety" and attached it to yourself. That is most unfortunate, and it is time for you to reprogram your thinking. Replace those negative thoughts with more positive ones. You need to say things like, "I will give this math class my best shot," or "I can learn mathematics if I work at it," or "I will try to do better than I have done in previous math classes." You will be pleasantly surprised at the difference this more positive attitude makes!

We live in a highly technical world, and you cannot afford to give up on the study of mathematics. Dropping mathematics may prevent you from entering certain career fields that you may find interesting. You may not have to take math courses as high-level as calculus, but such courses as intermediate algebra, finite math, college algebra, and trigonometry may be necessary. Learning mathematics can open new doors for you.

Learning mathematics is a process that takes time and effort. You will find that regular study and daily practice are necessary to strengthen your skills and to help you to grow academically. This process will lead you toward success in mathematics. Then, as you become more successful, your confidence in your ability to do mathematics will grow.

2.6 Exercises

Verbal and Writing Skills

1. Is the statement $5 > -6$ equivalent to the statement $-6 < 5$? Why?

2. Is the statement $-8 < -3$ equivalent to the statement $-3 > -8$? Why?

Replace the ? by < or >.

3. $9 ? -3$ **4.** $-2 ? 5$ **5.** $-4 ? -2$ **6.** $-3 ? -6$

7. $\frac{2}{3} ? \frac{3}{4}$ **8.** $\frac{5}{7} ? \frac{6}{8}$ **9.** $-1.2 ? +2.1$ **10.** $-3.6 ? +2.4$

11. $-\frac{13}{3} ? -4$ **12.** $-3 ? -\frac{15}{4}$ **13.** $-6.9 ? -7.2$ **14.** $-3\frac{1}{2} ? -3\frac{1}{4}$

Which is greater?

15. $\frac{123}{4986}$ or 0.0247?   **16.** $\frac{997}{6384}$ or 0.15613?

Give a graphical representation on the number line.

17. $x \geq -6$

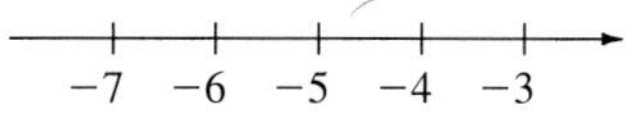

18. $x \leq -2$

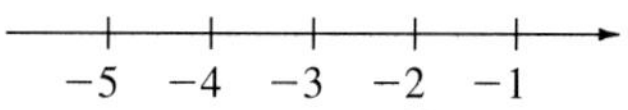

19. $x < 3$

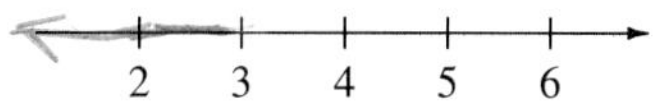

20. $x > 5$

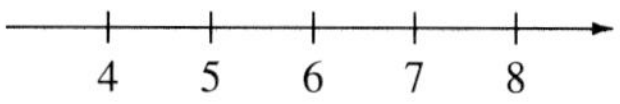

21. $x > \frac{3}{4}$

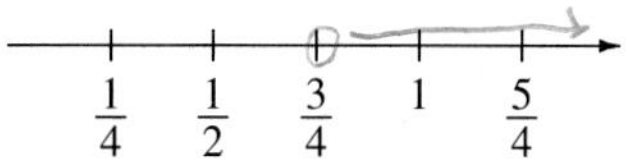

22. $x \geq -\frac{5}{2}$

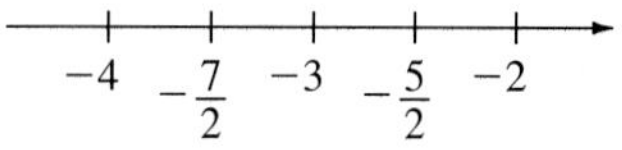

23. $x \leq -3.6$

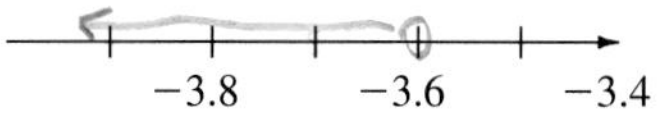

24. $x < -2.2$

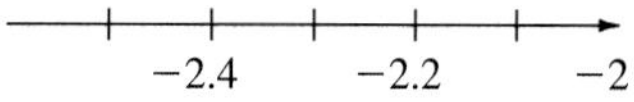

25. $x > 35$

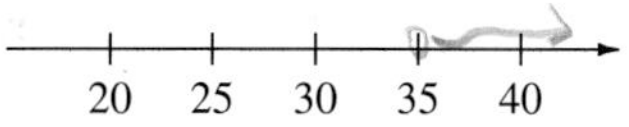

26. $x > 65$

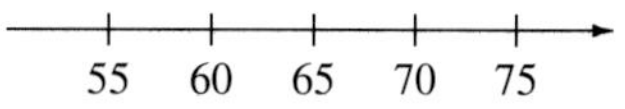

Translate the graphical representation to an inequality using the variable x.

27.

X is Greater than Equal to −1

28.

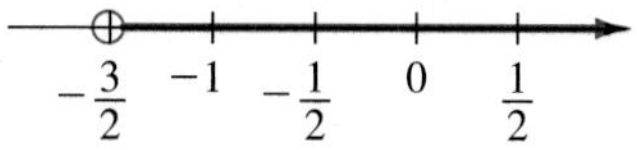

X is greater than −3/2

29.

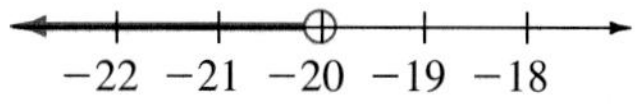

X is Less than −20

30.

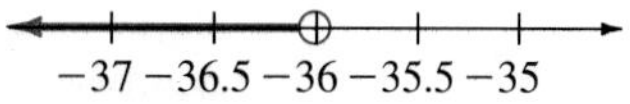

X is less than −36

31.

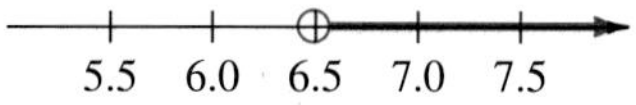

X is greater than 6.5

32.

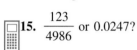

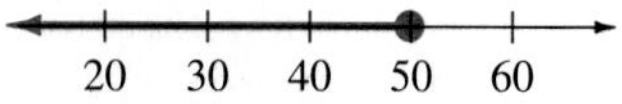

X is less than or equal to 50

Translate the English statement into an inequality.

33. The cost must be less than \$56. (Use the variable c for cost.)

34. The speed was greater than 580 kilometers per hour. (Use the variable V for speed.)

35. The number of hours worked in any week cannot be less than 37. (Use the variable h for hours.)

36. The number of nurses on duty on the floor can never exceed 6. (Use the variable n for the number of nurses.)

To Think About

37. Suppose that the variable x must satisfy *all* of these conditions.

(a) $x \leq 2$ **(b)** $x > -3$ **(c)** $x < \frac{5}{2}$ **(d)** $x \geq -\frac{5}{2}$

Graph on a number line the region that satisfies all conditions.

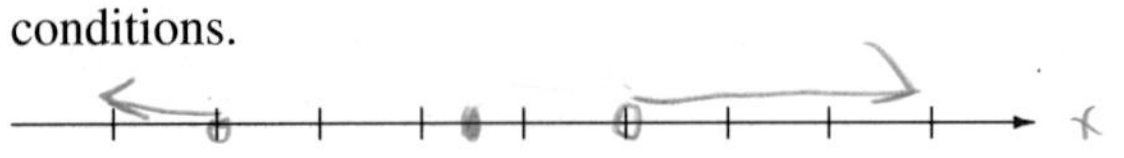

38. Suppose that the variable x must satisfy *all* of these conditions.

(a) $x < 3$ **(b)** $x > -4$ **(c)** $x \leq \frac{7}{2}$ **(d)** $x \geq -\frac{7}{2}$

Graph on a number line the region that satisfies all conditions.

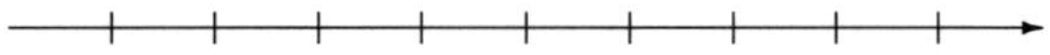

Cumulative Review Problems

39. Find 16% of 38.

40. Find 1.3% of 1250.

41. Sixteen out of 800 engines are defective. What percent are defective?

42. Write the fraction $\frac{3}{8}$ as a percent.

Developing Your Study Skills

EXAM TIME: GETTING ORGANIZED

Studying adequately for an exam requires careful preparation. Begin early so that you will be able to spread your review over several days. Even though you may still be learning new material at this time, you can be reviewing concepts previously learned in the chapter. Giving yourself plenty of time for review will take the pressure off. You need this time to process what you have learned and to tie concepts together.

Adequate preparation enables you to feel confident and to think clearly with less tension and anxiety.

2.7 Solve Inequalities

After studying this section, you will be able to:

1 *Solve an inequality.*

1 *Solving Inequalities*

The possible values that make an inequality true are called its **solutions.** Thus, when we **solve an inequality,** we are finding *all* the values that make it true. To solve an inequality, we simplify it to the point where we can clearly see the possible values for the variable. We've solved equations by adding, subtracting, multiplying, and dividing a particular value on both sides of the equation. Here we do similar operations with inequalities, with one important exception. We'll show some examples so that you can see the operations we can do with inequalities just as with equations.

We will first examine the pattern that takes place when we perform a given operation on both sides of an inequality.

EXAMPLE 1

	Original Inequality				*New Inequality*
(a)	$3 < 5$	$\longrightarrow$	Multiply both sides by 2.	$\longrightarrow$	$6 < 10$
(b)	$-2 < -1$	$\longrightarrow$	Add -3 to both sides.	$\longrightarrow$	$-5 < -4$
(c)	$0 > -4$	$\longrightarrow$	Divide both sides by 2.	$\longrightarrow$	$0 > -2$
(d)	$8 > 4$	$\longrightarrow$	Subtract 6 from both sides.	$\longrightarrow$	$2 > -2$

Note that *we avoided multiplying or dividing by a negative number!*

Practice Problem 1 Perform the given operation and write a new inequality.

(a) $9 > 6$ Divide both sides by 3.

(b) $2 < 5$ Multiply both sides by 2.

(c) $-8 < -3$ Add 4 to each side.

(d) $-3 < 7$ Add -7 to each side. ■

Now let us examine what would happen if we did multiply or divide by a negative number. We start with an original, true inequality. We want to get a new, also true inequality.

Original Inequality				*New Inequality*
$3 < 5$	$\longrightarrow$	Multiply by -2.	$\longrightarrow$	$-6 \,?\, -10$

What is the correct inequality sign? Since -6 is to the right of -10, we know the new inequality should be $-6 > -10$, if we wish the statement to remain true. Notice how we reverse the direction of the inequality from $<$ (less than) to $>$ (greater than). We would thus obtain the new inequality $-6 > -10$. Thus

$$3 < 5 \quad \longrightarrow \quad \text{Multiply by } -2. \quad \longrightarrow \quad -6 > -10$$

The $<$ sign we started with $(3 < 5)$ is reversed to $>$ $(-6 > -10)$. A similar reversal takes place in the following example.

EXAMPLE 2

	Original Inequality				*New Inequality*
(a)	$-2 < -1$	$\longrightarrow$	Multiply by -3.	$\longrightarrow$	$6 > 3$
(b)	$0 > -4$	$\longrightarrow$	Divide both sides by -2.	$\longrightarrow$	$0 < 2$
(c)	$8 > 4$	$\longrightarrow$	Divide both sides by -4.	$\longrightarrow$	$-2 < -1$

Notice that we do the arithmetic with signed numbers just as we always do. But the new inequality has its inequality sign reversed (from that of the original inequality). *Whenever both sides of an inequality are multiplied or divided by a negative quantity, the direction of the inequality is reversed.*

Practice Problem 2

(a) $7 > 2$ Multiply each side by -2.

(b) $-3 < -1$ Multiply each side by -1.

(c) $-10 > -20$ Divide each side by -10.

(d) $-15 < -5$ Divide each side by -5. ■

> **Procedure for Solving Inequalities**
>
> You may use the same procedures to solve inequalities that you did to solve equations *except* that the direction of an inequality is *reversed* if you *multiply* or *divide* both sides *by a negative number.*

EXAMPLE 3 Solve and graph $3x + 7 \geq 13$.

$3x + 7 - 7 \geq 13 - 7$ *Subtract 7 from both sides.*

$3x \geq 6$ *Simplify.*

$\frac{3x}{3} \geq \frac{6}{3}$ *Divide both sides by 3.*

$x \geq 2$ *Simplify. Note the direction of the inequality is not changed, since we have divided by a positive number.*

The graphical representation is

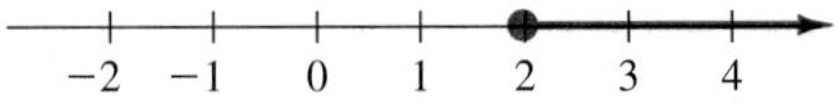

PRACTICE PROBLEM 3

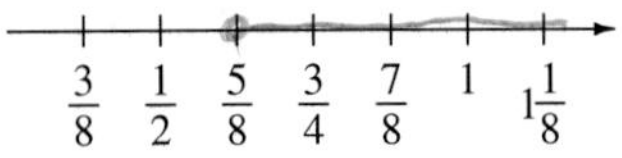

Practice Problem 3 Solve and graph $8x - 2 < 3$. ■

EXAMPLE 4 Solve and graph $5 - 3x > 7$.

$5 - 5 - 3x > 7 - 5$ *Subtract 5 from both sides.*

$-3x > 2$ *Simplify.*

$\frac{-3x}{-3} < \frac{2}{-3}$ *Divide by -3, and **reverse the inequality,** since both sides are divided by negative 3.*

$x < -\frac{2}{3}$ *Note the direction of the inequality.*

The graphical representation is

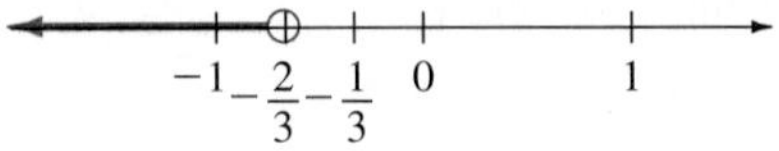

PRACTICE PROBLEM 4

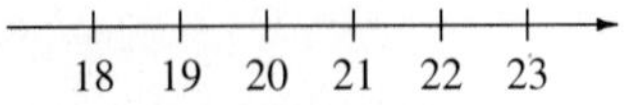

Practice Problem 4 Solve and graph $7 + x < 28$. ■

Just like equations, some inequalities contain parentheses and fractions. The initial steps to solve these inequalities will be the same as those used to solve equations with parentheses and fractions. When the variable appears on both sides of the inequality, it is advisable to collect the x terms on the left side of the inequality symbol.

EXAMPLE 5 Solve and graph $-\frac{13x}{2} \leq \frac{x}{2} - \frac{15}{8}$.

$8\left(\frac{-13x}{2}\right) \leq 8\left(\frac{x}{2}\right) - 8\left(\frac{15}{8}\right)$ *Multiply all terms by LCD = 8. We do **not** reverse the direction of the inequality symbol since we are multiplying by a positive number, 8.*

$-52x \leq 4x - 15$ *Simplify.*

$-52x - 4x \leq 4x - 15 - 4x$ *Add $-4x$ to both sides.*

$-56x \leq -15$ *Combine like terms.*

$\frac{-56x}{-56} \geq \frac{-15}{-56}$ *Divide both sides by -56. We **reverse** the direction of the inequality when we divide both sides by a negative number.*

$x \geq \frac{15}{56}$

The graphical representation is

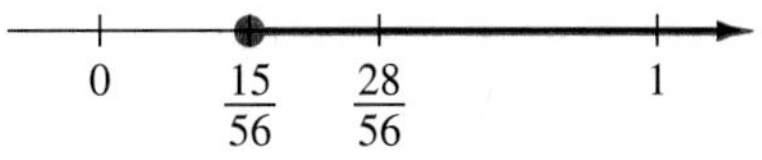

PRACTICE PROBLEM 5

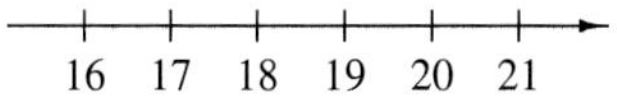

Practice Problem 5 Solve and graph $\frac{1}{2}x + 3 < \frac{2}{3}x$. ■

EXAMPLE 6 Solve and graph $\frac{1}{3}(3 - 2x) \leq -4(x + 1)$.

$1 - \frac{2x}{3} \leq -4x - 4$ *Remove parentheses.*

$3(1) - 3\left(\frac{2x}{3}\right) \leq 3(-4x) - 3(4)$ *Multiply all terms by LCD = 3.*

$3 - 2x \leq -12x - 12$ *Simplify.*

$3 - 2x + 12x \leq -12x + 12x - 12$ *Add 12x to both sides.*

$3 + 10x \leq -12$ *Combine like terms.*

$3 - 3 + 10x \leq -12 - 3$ *Subtract 3 from both sides.*

$10x \leq -15$ *Simplify.*

$\frac{10x}{10} \leq \frac{-15}{10}$ *Divide both sides by 10. Since we are dividing by a **positive** number, the inequality is **not** reversed.*

$x \leq -\frac{3}{2}$

The graphical representation is

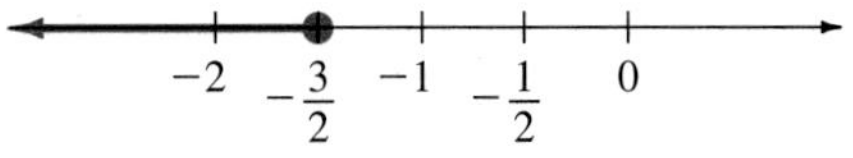

PRACTICE PROBLEM 6

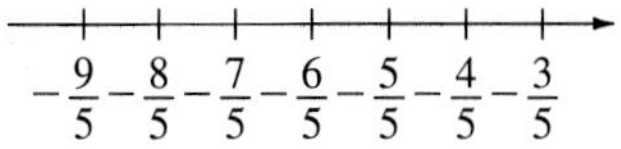

Practice Problem 6 Solve and graph $\frac{1}{2}(3 - x) \leq 2x + 5$. ■

The most common error students make in solving inequalities is forgetting to reverse the direction of the inequality symbol when multiplying or dividing by a negative number.

CAUTION When each side of an inequality is multiplied or divided by a negative number, the direction of the inequality symbol is reversed.

EXAMPLE 7 A hospital director has determined that the costs of operating one floor of the hospital for an 8-hour shift must never exceed \$2370. An expression for the cost of operating one floor of the hospital is $130n + 1200$, where n is the number of nurses. This expression is determined by \$1200 in fixed costs and \$130 per nurse for an 8-hour shift. Set up and solve an inequality to determine the number of nurses that may be on duty on this floor for an 8-hour shift if this cost factor is to be followed.

The cost expression must never exceed \$2370. This means the cost expression is always less than or equal to \$2370.

$$130n + 1200 \leq 2370 \qquad \textit{Write the inequality.}$$

$$130n + 1200 + (-1200) \leq 2370 + (-1200) \qquad \textit{Add } (-1200) \textit{ to each side.}$$

$$130n \leq 1170 \qquad \textit{Simplify.}$$

$$\frac{130n}{130} \leq \frac{1170}{130} \qquad \textit{Divide each side by } 130.$$

$$n \leq 9$$

The number of nurses on duty on this floor for an 8-hour shift must always be less than or equal to 9.

Practice Problem 7 The company president of Staywell, Inc., wants the monthly profits to never be less than \$2,500,000. He has determined that an expression for monthly profit for the company is $2200n - 700{,}000$. In the expression, n is the number of exercise machines manufactured each month. The profit on each machine is \$2000 and the −\$700,000 represents fixed costs to run the manufacturing division. How many machines must be made and sold each month to satisfy these financial goals? ■

2.7 Exercises

Solve each inequality. Graph the results.

1. $x - 5 > 3$

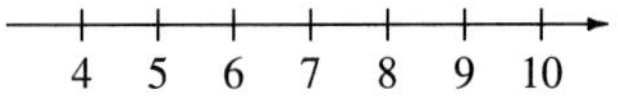

2. $x + 9 \leq 2$

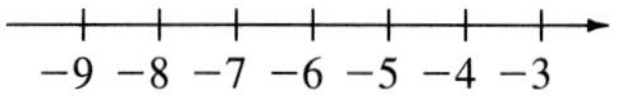

3. $4x < 12$

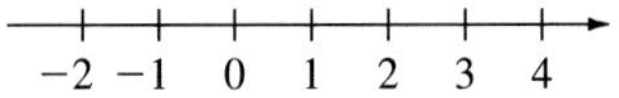

4. $-3x > 15$

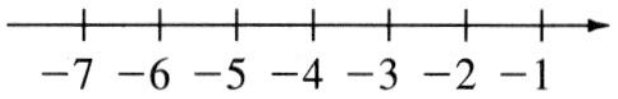

5. $\frac{1}{2}x \geq 4$

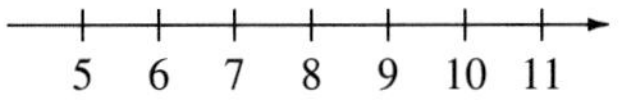

6. $-\frac{1}{5}x < 10$

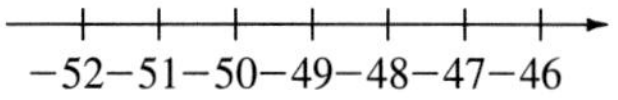

7. $2x - 3 < 4$

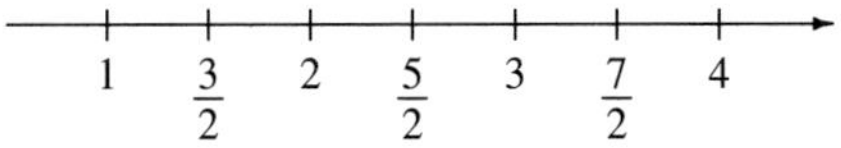

8. $3 - 3x > 12$

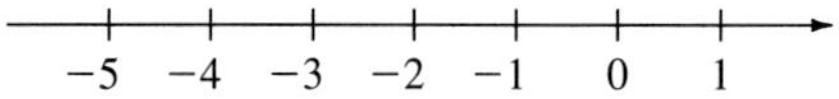

9. $6 - 5x \geq 3x + 14$

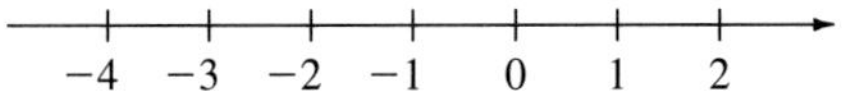

10. $7 - 8x \leq -6x - 5$

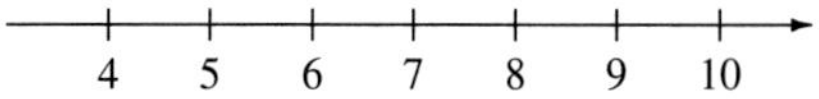

11. $\frac{5x}{6} - 5 > \frac{x}{6} - 9$

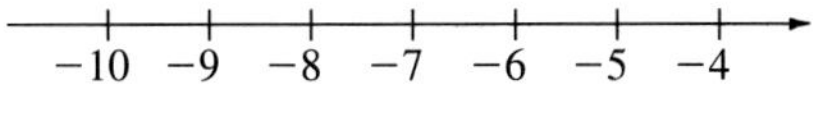

12. $\frac{x}{4} - 2 < \frac{3x}{4} + 5$

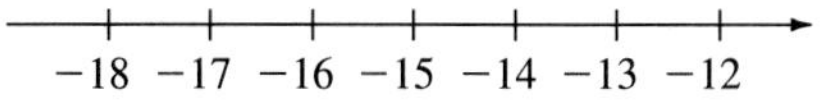

13. $3(x + 2) \leq 2x + 4$

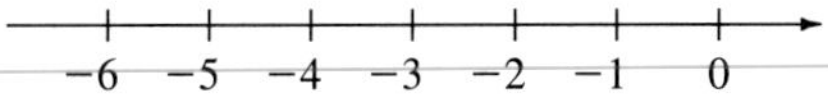

14. $4(x + 5) \geq 3x + 7$

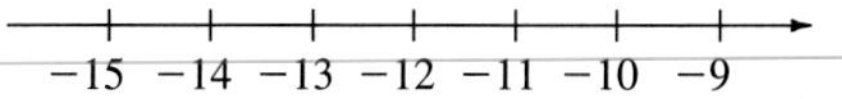

Verbal and Writing Skills

15. Add -2 to both sides of the inequality $5 > 3$. What is the result? Why was the direction of the inequality not reversed?

16. Divide -3 into both sides of the inequality $-21 > -29$. What is the result? Why was the direction of the inequality reversed?

Mixed Practice

Solve each inequality. Collect the variable terms on the left side of the inequality.

17. $2x - 5 < 5x - 11$

18. $4x - 7 > 9x - 2$

19. $6x - 2 \geq 4x + 6$

20. $5x - 5 \leq 2x + 10$

21. $0.3(x - 1) < 0.1x - 0.5$

22. $0.2(3 - x) + 0.1 > 0.1(x - 2)$

23. $5 - 2(3 - x) \leq 2(2x + 5) + 1$

24. $5(x + 1) + 2 \geq x - 3(2x + 1)$

25. $2x - \left(\frac{x}{2} - 6\right) \geq 1 + \frac{2x}{3}$

26. $\frac{2}{3}(x - 1) - \frac{3}{4} \leq \frac{1}{2}(x + 2)$

★**27.** $3(0.3 + 0.1x) + 0.1 < 0.5(x + 2)$

★**28.** $0.3 + 0.4(2 - x) \leq 6(-0.2 + 0.1x) + 0.3$

★**29.** $\frac{1}{6} - \frac{1}{2}(3x + 2) < \frac{1}{3}\left(x - \frac{1}{2}\right)$

★**30.** $\frac{2}{3}(2x - 5) + 3 \geq \frac{1}{4}(3x + 1) - 5$

Solve for x. Round your answer to the nearest hundredth.

31. $1.96x - 2.58 < 9.36x + 8.21$

32. $3.5(1.7x - 2.8) \leq 7.96x - 5.38$

Applications

33. To pass a course with a B grade, a student must have an average of 80 or greater. A student's grades on three tests are 75, 83, and 86. What score must the student get on the next test to get a B average or better?

34. A salesperson may choose to receive \$300 or 6% of his sales as payment for his work. How much does he need to sell to make the 6% offer a better deal?

35. The supervisor of a computer chip manufacturing factory has determined that her monthly profit factor is given by the expression $12.5n - 300{,}000$. Here n represents the number of computer chips manufactured each month. Each chip manufactured produces a profit of \$12.50. The fixed costs of the factory are \$300,000 per month.
(a) How many chips must be manufactured to break even?
(b) How many chips must be manufactured monthly to ensure a profit of \$650,000? a profit of \$470,000?

Putting Your Skills to Work

OVERSEAS TRAVEL

Maria lives in Dallas. Her family lives in Boston. She works for an American company that produces computers. Her employer often sends her to Europe on business trips, so she keeps these charts handy in order to prepare her travel budget.

CORPORATE RATES FLYING OVERSEAS AIRLINES IN BUSINESS CLASS TO AND FROM BOSTON		
Round-trip Fares between Boston and:	Stay at Least One Week Including a Saturday	Not Staying One Week Including a Saturday
London	\$460	\$680
Paris	\$490	\$710
Geneva	\$500	\$730

CORPORATE RATES FLYING DOMESTIC AIRLINES BETWEEN BOSTON AND DALLAS		
Round-trip Fares between Boston and Dallas on:	Spend at Least One Week Including a Saturday	Not Staying One Week Including a Saturday
First class	\$640	\$900
Business class	\$500	\$680
Regular coach class	\$450	\$570
14-Day advance coach class	\$310	\$460

Her company allows her to take vacation time to stay with her family for a day or two before and after overseas business trips. Use an inequality to solve the following problems.

1. In three weeks she must travel from Dallas to Paris. She will be in Paris for two weeks. What is the least amount that her airfare could be?

2. Her company has notified her that she must fly from Dallas to London in three days. They want her to travel first class on the domestic part of the trip in order to prepare for a company presentation. She will only be in London for 4 days. What is the least amount that her airfare could be?

COOPERATIVE GROUP ACTIVITY

3. Soon her company will send her to Geneva for 7 days. She does not know how soon she will be asked to leave. What is the least amount that her airfare could be? If she flies regular coach class, what is the greatest amount that her airfare could be?

Chapter Organizer

Topic	Procedure	Examples
Solving equations without parentheses or fractions, p. 139.	1. On each side of the equation, collect like terms if possible. 2. Add or subtract terms on both sides of the equation in order to get all terms with the variable on one side of the equation. 3. Add or subtract a value on both sides of the equation to get all terms not containing the variable on the other side of the equation. 4. Divide both sides of the equation by the coefficient of the variable. 5. If possible, simplify solution. 6. Check your solution by substituting the obtained value into the original equation.	Solve for x: $5x + 2 + 2x = -10 + 4x + 3$ $7x + 2 = -7 + 4x$ $7x + (-4x) + 2 = -7 + 4x + (-4x)$ $3x + 2 = -7$ $3x + 2 + (-2) = -7 + (-2)$ $3x = -9$ $\frac{3x}{3} = \frac{-9}{3}$ $x = -3$ *Check:* Is $x = -3$ a solution for $5x + 2 + 2x = -10 + 4x + 3$? $5(-3) + 2 + 2(-3) \stackrel{?}{=} -10 + 4(-3) + 3$ $-15 + 2 - 6 \stackrel{?}{=} -10 + (-12) + 3$ $-13 - 6 \stackrel{?}{=} -22 + 3$ $-19 = -19$
Solving equations with parentheses and/or fractions, p. 141 and p. 148.	1. Remove any parentheses. 2. Simplify, if possible. 3. If fractions exist, multiply all terms on both sides by the lowest common denominator of all the fractions. 4. Now follow the remaining steps of solving an equation without parentheses or fractions.	Solve for y: $5(3y - 4) = \frac{1}{4}(6y + 4) - 48$ $15y - 20 = \frac{3}{2}y + 1 - 48$ $15y - 20 = \frac{3}{2}y - 47$ $2(15y) - 2(20) = 2\left(\frac{3}{2}y\right) - 2(47)$ $30y - 40 = 3y - 94$ $30y - 3y - 40 = 3y - 3y - 94$ $27y - 40 = -94$ $27y - 40 + 40 = -94 + 40$ $27y = -54$ $\frac{27y}{27} = \frac{-54}{27}$ $y = -2$

Chapter Organizer

Topic	Procedure	Examples
Solving formulas, p. 152.	**1.** Remove any parentheses and simplify if possible. **2.** If fractions exist, multiply all terms on both sides by the LCD, which may be a variable. **3.** Add or subtract terms on both sides of the equation in order to get all terms containing the *desired variable* on one side of the equation. **4.** Add or subtract terms on both sides of the equation in order to get all other terms on the opposite side of the equation. **5.** Divide both sides of the equation by the coefficient of the desired variable. This division may involve other variables. **6.** Simplify, if possible. **7.** (Optional) Check your solution by substituting the obtained expression into the original equation.	Solve for z: $B = \frac{1}{3}(hx + hz)$. First we remove parentheses. $B = \frac{1}{3}hx + \frac{1}{3}hz$ Now we multiply each term by 3. $3(B) = 3\left(\frac{1}{3}hx\right) + 3\left(\frac{1}{3}hz\right)$ $3B = hx + hz$ $3B - hx = hx - hx + hz$ $3B - hx = hz$ The coefficient of z is h, so we divide each side by h. $\frac{3B - hx}{h} = z$
Solving inequalities, p. 158.	**1.** Follow the steps for solving a first-degree equation up until the division step. **2.** If you divide both sides of the inequality by a positive number, the direction of the inequality is not reversed. **3.** If you divide both sides of the inequality by a *negative number,* the direction of the inequality is reversed.	Solve for x and graph your solution: $\frac{1}{2}(3x - 2) \leq -5 + 5x - 3$ First remove parentheses and simplify. $\frac{3}{2}x - 1 \leq -8 + 5x$ Now multiply each term by 2. $2\left(\frac{3}{2}x\right) - 2(1) \leq 2(-8) + 2(5x)$ $3x - 2 \leq -16 + 10x$ $3x - 10x - 2 \leq -16 + 10x - 10x$ $-7x - 2 \leq -16$ $-7x - 2 + 2 \leq -16 + 2$ $-7x \leq -14$ When we divide both sides by a negative number, the inequality is reversed. $\frac{-7x}{-7} \geq \frac{-14}{-7}$ $x \geq 2$ Graphical solution: 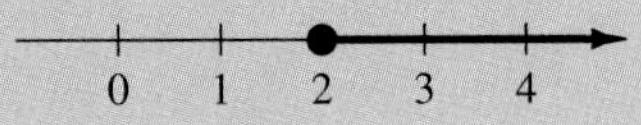

Chapter 2 Review Problems

2.1–2.3 *Solve for x.*

1. $5x + 20 = 3x$

2. $7x + 3 = 4x$

3. $7(x - 4) = x + 2$

4. $2(17 - x) = 6x + 2$

5. $4x - 3(x + 2) = 4$

6. $1 - 2(6 - x) = 3x + 2$

7. $x - (0.5x + 2.6) = 17.6$

8. $-0.2(x + 1) = 0.3(x + 11)$

9. $3(x - 2) = -4(5 + x)$

10. $\frac{1}{4}x = -16$

11. $x + 37 = 26$

12. $6(8x + 3) = 5(9x + 8)$

13. $3(x - 3) = 13x + 21$

14. $9x + 10 = 3x + 4$

15. $24 - 3x = 4(x - 1)$

16. $12 - x + 2 = 3x - 10 + 4x$

17. $36 = 9x - (3x - 18)$

18. $8 - 3x = -5x - 4$

19. $2(3 - x) = 1 - (x - 2)$

20. $4(x + 5) - 7 = 2(x + 3)$

21. $0.9x + 3 = 0.4x + 1.5$

22. $7x - 3.4 = 11.3$

23. $8(3x + 5) - 10 = 9(x - 2) + 13$

24. $8 - 3x + 5 = 13 + 4x + 2$

25. $3 = 2x + 5 - 3(x - 1)$

26. $-2(x - 3) = -4x + 3(3x + 2)$

27. $2(5x - 1) - 7 = 3(x - 1) + 5 - 4x$

2.4 *Solve for the variable.*

28. $\frac{3}{4}x - 3 = \frac{1}{2}x + 2$

29. $1 = \frac{5x}{6} + \frac{2x}{3}$

30. $\frac{7x}{5} = 5 + \frac{2x}{5}$

31. $\frac{7x - 3}{2} - 4 = \frac{5x + 1}{3}$

32. $\frac{3x - 2}{2} + \frac{x}{4} = 2 + x$

33. $\frac{-3}{2}(x + 5) = 1 - x$

34. $\frac{-4}{3}(2x + 1) = -x - 2$

35. $\frac{1}{3}(x - 2) = \frac{x}{4} + 2$

36. $\frac{1}{5}(x - 3) = 2 - \frac{x}{2}$

37. $\frac{4}{5} + \frac{1}{2}x = \frac{1}{5}x + \frac{1}{2}$

38. $3x + \frac{6}{5} - x = \frac{6}{5}x - \frac{4}{5}$

39. $\frac{10}{3} - \frac{5}{3}x + x = \frac{2}{9} + \frac{1}{9}x$

40. $-\frac{8}{3}x - 8 + 2x - 5 = -\frac{5}{3}$

41. $\frac{1}{2} + \frac{5}{4}x = \frac{2}{5}x - \frac{1}{10} + 4$

42. $5 + 3x = \frac{7}{3}x + 5$

43. $\frac{1}{6}x - \frac{2}{3} = \frac{1}{3}(x - 4)$

44. $\frac{1}{2}(x - 3) = \frac{1}{4}(3x - 1)$

45. $\frac{7}{12}(x - 3) = -\frac{1}{3}x + 1$

46. $\frac{5}{6} + \frac{1}{3}(x - 3) = \frac{1}{2}(x + 9)$

47. $\frac{1}{7}(x + 5) - \frac{6}{14} = \frac{1}{2}(x + 3)$

48. $\frac{1}{6}(8x + 3) = \frac{1}{2}(2x + 7)$

49. $-\frac{1}{3}(2x - 6) = \frac{2}{3}(3 + x)$

50. $\frac{7}{9}x + \frac{2}{3} = 2 + \frac{1}{3}x$

2.5 *Solve for the variable indicated.*

51. Solve for y. $3x - y = 10$

52. Solve for y. $5x + 2y + 7 = 0$

53. Solve for r. $A = P(1 + rt)$

54. Solve for h. $A = 4\pi r^2 + 2\pi rh$

55. Solve for p. $H = \frac{1}{3}(a + 2p + 3)$

56. **(a)** Solve for T. $C = \frac{WRT}{1000}$.

(b) Use your result to find T if $C = 0.36$, $W = 30$, and $R = 0.002$.

57. **(a)** Solve for y. $5x - 3y = 12$

(b) Use your result to find y if $x = 9$.

58. **(a)** Solve for R. $I = \frac{E}{R}$

(b) Use your result to find R if $E = 100$ and $I = 20$.

2.6 *Solve the inequalities and graph the result.*

59. $7 - 2x \geq 4x - 5$

−2 −1 0 1 2 3

60. $2 - 3x \leq -5 + 4x$

−2 −1 0 1 2 3

61. $2x - 3 + x > 5(x + 1)$

−5 −4 −3 −2 −1 0

62. $-x + 4 < 3x + 16$

−5 −4 −3 −2 −1 0

63. $4x \geq 2(12 - 2x)$

−1 0 1 2 3 4

64. $5 - \frac{1}{2}x > 4$

−1 0 1 2 3 4

65. $2(x - 1) \geq 3(2 + x)$

−10 −9 −8 −7 −6 −5

66. $3x + 5 - 7x \leq -2x - 1$

0 1 2 3 4 5

67. $-4x - 14 < 4 - 2(3x - 1)$

7 8 9 10 11 12

68. $3(x - 2) + 8 < 7x + 14$

−5 −4 −3 −2 −1 0

69. $\frac{1}{2}(2x + 3) > 10$

$\frac{15}{2}$ 8 $\frac{17}{2}$ 9 $\frac{19}{2}$ 10

70. $\frac{1}{3}(x + 2) \leq \frac{1}{2}(3x - 5)$

$\frac{15}{7}$ $\frac{17}{7}$ $\frac{19}{7}$

71. $4(2 - x) - (-5x + 1) \geq -8$

−15 −13 −11

72. $5(1 - x) < 3(x - 1) - 2(3 - x)$

$\frac{7}{5}$ $\frac{8}{5}$ $\frac{9}{5}$ 2 $\frac{11}{5}$ $\frac{12}{5}$

Use inequalities to solve the following problems.

73. The cost of a substitute teacher for one day at Central High school is \$70. Let n = the number of substitute teachers. Set up an inequality to determine how many times a substitute teacher may be hired if the monthly budget for substitute teachers is \$3220. What is the maximum number of substitute teachers that may be hired during the month?

74. The cost of hiring a temporary secretary for the day is \$85. Let n = the number of temporary secretaries. Set up an inequality to determine how may times a temporary secretary may be hired if the company budget for temporary secretaries is \$1445 per month. What is the maximum number of days a temporary secretary may be hired during the month?

Solve for the variable.

1. $2x + 0.8 = 5.0$

2. $9x - 3 = -6x + 27$

3. $2(2y - 3) = 4(2y + 2)$

4. $\frac{1}{2}y + 5 = \frac{1}{3}y$

5. $2(6 - x) = 3x + 2$

6. $0.8x + 0.18 - 0.4x = 0.3(x + 0.2)$

7. $\frac{2y}{3} + \frac{1}{5} - \frac{3y}{5} + \frac{1}{3} = 1$

8. $3 - 2y = 2(3y - 2) - 5y$

Solve for x.

9. $5(20 - x) + 10x = 165$

10. $2(x + 75) + 5x = 1025$

11. $-2(2 - 3x) = 76 - 2x$

12. $20 - (2x + 6) = 5(2 - x) + 2x$

1. ______

2. ______

3. ______

4. ______

5. ______

6. ______

7. ______

8. ______

9. ______

10. ______

11. ______

12. ______

Solve for x.

13. $2x - 3 = 12 - 6x + 3(2x + 3)$

14. $\frac{1}{3}x - \frac{3}{4}x = \frac{1}{12}$

15. $\frac{3}{5}x + \frac{7}{10} = \frac{1}{3}x + \frac{3}{2}$

16. $\frac{15x - 2}{28} = \frac{5x - 3}{7}$

17. $\frac{1}{3}(7x - 1) + \frac{1}{4}(2 - 5x) = \frac{1}{3}(5 + 3x)$

18. Solve for w. $A = 3w + 2P$

19. Solve for w. $\frac{2w}{3} = 4 - \frac{1}{2}(x + 6)$

20. Solve for a. $A = \frac{1}{2}h(a + b)$

21. Solve for y. $5ax(2 - y) = 3axy + 5$

Solve and graph the inequalities.

22. $4(x - 1) \geq 12x$

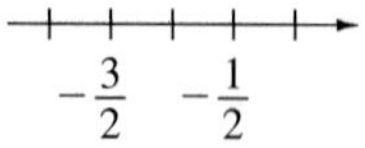

23. $2 - 7(x + 1) - 5(x + 2) < 0$

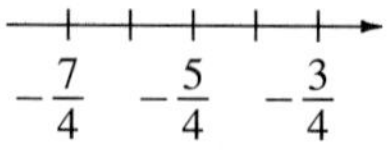

24. $5 + 8x - 4 < 2x + 13$

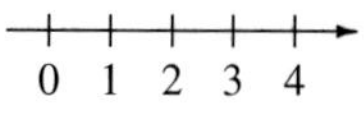

25. $\frac{1}{4}x + \frac{1}{16} \leq \frac{1}{8}(7x - 2)$

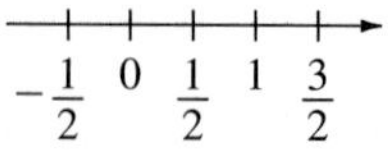

13. ____________

14. ____________

15. ____________

16. ____________

17. ____________

18. ____________

19. ____________

20. ____________

21. ____________

22. ____________

23. ____________

24. ____________

25. ____________

Cumulative Test for Chapters 0–2

Approximately one-half of this test covers the content of Chapters 0 and 1. The remainder covers the content of Chapter 2.

1. $\frac{6}{7} - \frac{2}{3}$

2. $1\frac{3}{4} + 2\frac{1}{5}$

3. $3\frac{1}{5} \div 1\frac{1}{2}$

4. Multiply. $\begin{array}{r} 1.23 \\ \times\ 0.56 \\ \hline \end{array}$

5. Divide. $0.144 \div 1.2$

6. What is 18% of 340?

7. Multiply. $(-3)(-5)(-1)(2)(-1)$

8. Collect like terms. $5ab - 7ab^2 - 3ab - 12ab^2 + 10ab - 9ab^2$

9. Simplify. $(5x^2y^3)^2$

10. Simplify. $2\{x + y[3 - 2x(1 - 4y)]\}$

11. Solve for x. $3(5 - x) = 2x - 10$

12. Solve for x. $\frac{1}{3}(x + 5) = 2x - 5$

13. Solve for y. $\frac{2y}{3} - \frac{1}{4} = \frac{1}{6} + \frac{y}{4}$

1. ______

2. ______

3. ______

4. ______

5. ______

6. ______

7. ______

8. ______

9. ______

10. ______

11. ______

12. ______

13. ______

14. ______

15. ______

16. ______

17. ______

18. ______

19. ______

20. ______

21. ______

22. ______

14. Solve for y. $3x - 7y + 2 = 0$

15. Solve for b. $H = \frac{2}{3}(b + 4a)$

16. Solve for t. $I = Prt$

17. Solve for a. $A = \frac{ha}{2} + \frac{hb}{2}$

In problems 18 through 22, solve and graph the inequalities.

18. $-6x - 3 < 2x - 10x + 7$

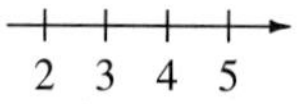

19. $\frac{1}{2}(x - 5) \geq x - 4$

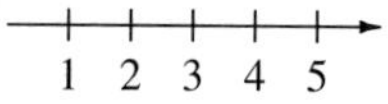

20. $4(2 - x) > 1 - 5x - 8$

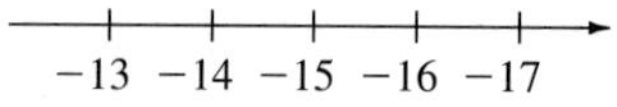

21. $x + \frac{5}{9} \leq \frac{1}{3} + \frac{7}{9}x$

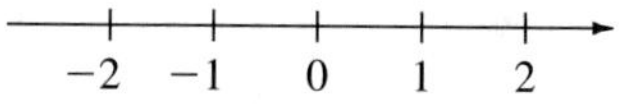

22. $4 - 16x \leq 6 - 5(3x - 2)$

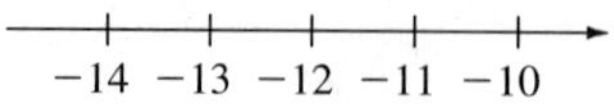

3
Solving Applied Problems

Making long range decisions about your future requires careful thought. Using the mathematics presented in this chapter, will enable you to make more well informed decisions about the financial implications of your future plans.

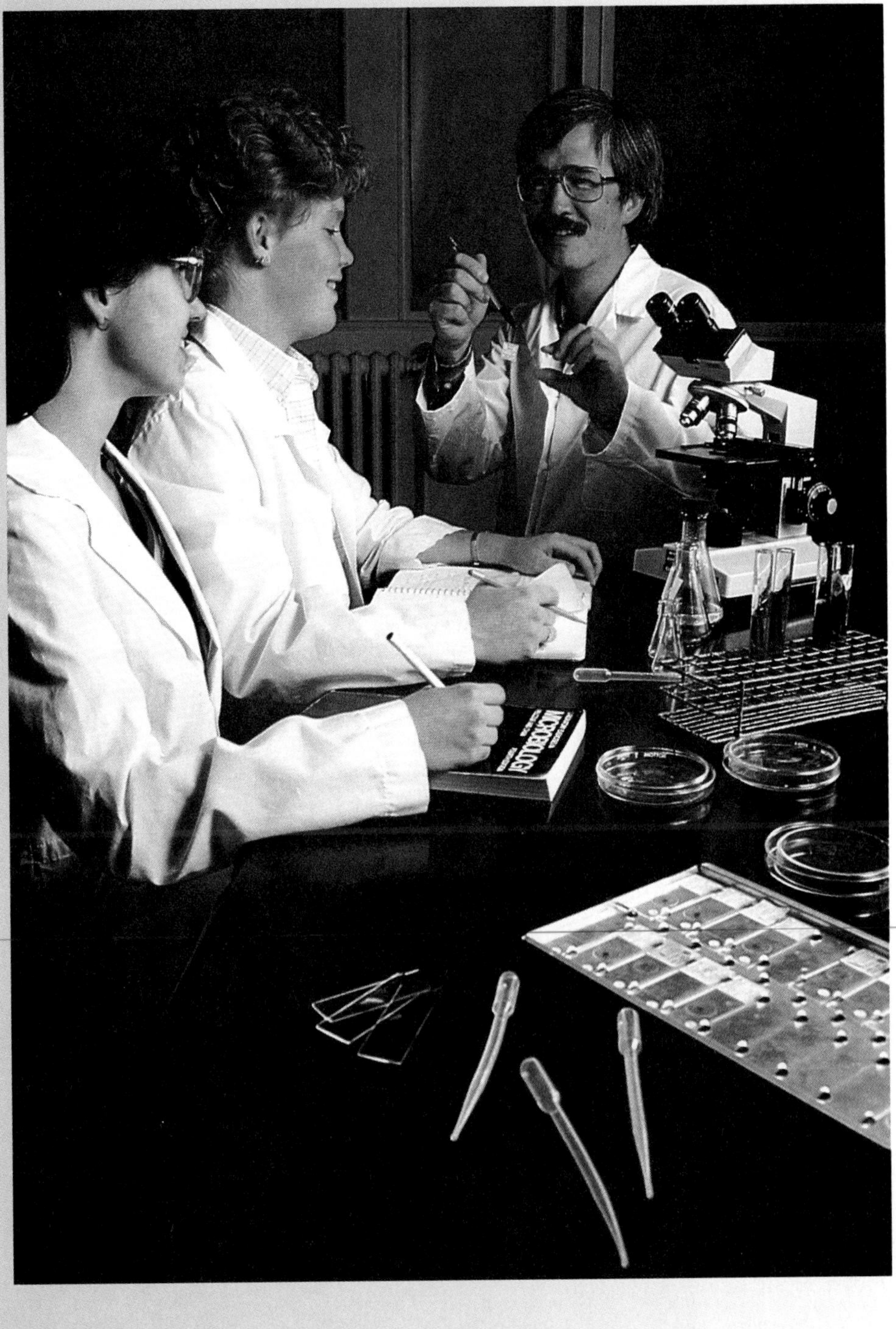

Pretest Chapter 3

1.

2.

3.

4.

5.

6.

7.

8.

If you are familiar with the topics in this chapter, take this test now. Check your answers with those in the back of the book. If an answer was wrong or you couldn't do a problem, study the appropriate section of the chapter.

If you are not familiar with the topics in this chapter, don't take this test now. Instead, study the examples, work the practice problems, and then take the test. This test will help you to identify those concepts that you have mastered and those that need more study.

Section 3.1

Write an algebraic expression. Use the variable x to represent the unknown number.

1. Twice a number is then decreased by thirty

2. Forty percent of a number is increased by 150

Section 3.2

Translate the information in each word problem into an equation, and solve.

3. A number is multiplied by seven and then is increased by nine. The result is forty-four. Find the original number.

4. There are three numbers with these properties: The first number is triple a second number. A third number is three more than the second number. The sum of the three numbers is twenty-three. Find the three numbers.

Section 3.3

5. One side of a triangular piece of land is 3 meters less than double the second side. The third side is 4 more meters in length than the first side. The perimeter of the triangle is 41 meters. Find the length of each side of the triangle.

6. Charlene took three packages to the post office to mail. The second package was $3\frac{1}{2}$ pounds less than the first package. The third package was 2 pounds less than the first package. The total weight of the three packages was 17 pounds. How much did each package weigh?

7. The length of a rectangle is 13 meters longer than twice the width. The perimeter of the rectangle is 74 meters. Find the dimensions of the rectangle.

Section 3.4

8. Two investments were made totaling \$1000. The investments gain interest once each year. The first investment yielded 12% interest after one year. The second investment yielded 9% interest. The total interest for both investments was \$102. How much was invested at each interest rate?

9. The population of Springville has grown 11% from the population level of five years ago. The town now has population of 24,420 people. What was the population five years ago?

9. ____________________

10. Enrique has $3.35 in change in his pocket. He has four more dimes than nickels. He has three more quarters than dimes. How many coins of each type does he have?

10. ____________________

Section 3.5

11. Find the area of a triangle whose base is 16 inches and whose altitude is 9 inches.

11. ____________________

12. Find the volume of a cylinder whose height is 12 meters and whose radius is 3 meters. (Use 3.14 for π.)

12. ____________________

13. Find the circumference of a circle whose radius is 4 centimeters. (Use 3.14 for π.)

13. ____________________

14. A stainless-steel plate in the shape of a trapezoid is to be manufactured according to the following diagram. The cost to manufacture the plate is $1.85 per square centimeter. What is the cost to manufacture this steel plate?

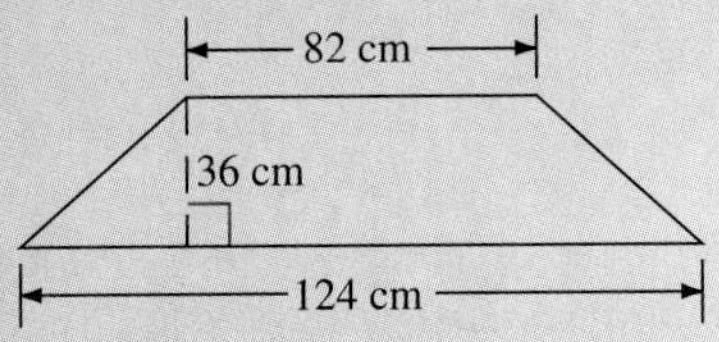

14. ____________________

15. ____________________

Section 3.6

Translate each phrase into an inequality.

15. The height must be no more than 72 inches.

16. ____________________

16. The rent will be at least $450 a month.

17. The average number must be less than 15.

17. ____________________

18. The cost of the car is more than $8900.

18. ____________________

Solve.

19. Bob makes 3% on the amount of his sales. If he wants to earn more than $900, how much will he need to sell?

19. ____________________

20. Lauren is making a rectangular pen. The width is 11 feet. The perimeter cannot be more than 54 feet. What is the length?

20. ____________________

3.1 Translate English Phrases into Algebraic Expressions

After studying this section, you will be able to:

1. *Write an algebraic expression for a simple English phrase.*
2. *Write an algebraic expression for two or more quantities that are being compared.*

1 *Translating English Phrases into Algebra*

One of the most useful applications of algebra is solving word problems. One of the first steps in solving word problems is translating the conditions of the problem into algebra. In this section we show you how to translate common English phrases into algebraic symbols. This process is similar to translating between languages like Spanish and French.

There are several English phrases to describe the operation of addition. If we represent an unknown number by the variable x, all of the following phrases can be translated into algebra as $x + 3$.

ENGLISH PHRASES DESCRIBING ADDITION	ALGEBRAIC EXPRESSION	DIAGRAM
Three *more than* a number The *sum of* a number and three A number *increased by* three Three is *added to* a number Three *greater than* a number A number *plus* three	$\mathbf{x + 3}$	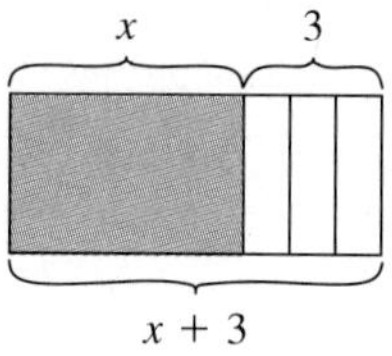

In a similar way we can use algebra to express English phrases that describe the operations of subtraction, multiplication, and division.

ENGLISH PHRASES DESCRIBING SUBTRACTION	ALGEBRAIC EXPRESSION	DIAGRAM
A number *decreased by* four Four *less than* a number Four is *subtracted from* a number Four *smaller than* a number Four *fewer than* a number A number *diminished by* four A number *minus* 4 The *difference between* a number and four	$\mathbf{x - 4}$	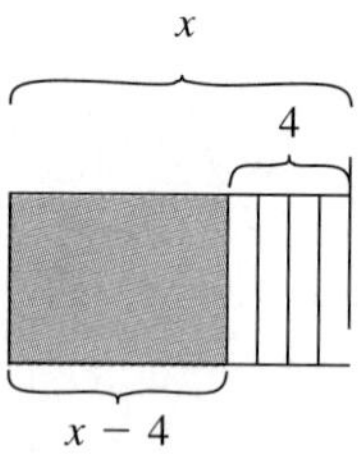

ENGLISH PHRASES DESCRIBING MULTIPLICATION	ALGEBRAIC EXPRESSION	DIAGRAM
Double a number *Twice* a number The *product* of two and a number Two *of* a number Two *times* a number	$\mathbf{2x}$	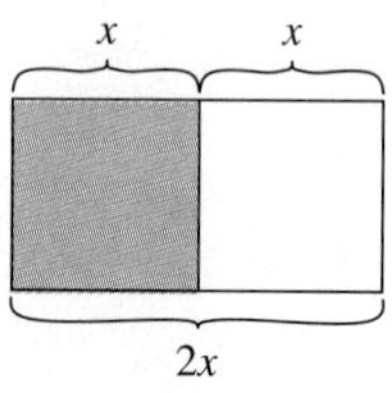

ENGLISH PHRASES DESCRIBING DIVISION	ALGEBRAIC EXPRESSION	DIAGRAM
A number *divided by* five One-*fifth* of a number	$\mathbf{\frac{x}{5}}$	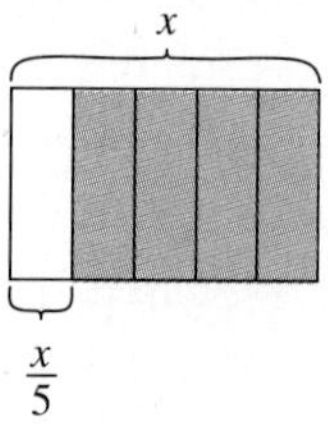

It is important to review these phrases and to see how they are represented as algebraic expressions. Often other words are used in English instead of the word "number." We can use a variable, such as x, here also.

EXAMPLE 1

ENGLISH PHRASE	ALGEBRAIC EXPRESSION
(a) A *quantity* is increased by five	$x + 5$
(b) Double the *value*	$2x$
(c) One-third of the *weight*	$\frac{x}{3}$ or $\frac{1}{3}x$
(d) Twelve *more than* a number	$x + 12$
(e) Seven *less than* a number	$x - 7$

Note that the algebraic expression for "seven less than a number" does not follow the order of the words in the English phrase. The variable or expression that follows the words *less than* always comes first.

seven less than x

$x - 7$

The variable or expression that follows the words *more than* always comes first.

Practice Problem 1 Write each English phrase as an algebraic expression.

(a) Four more than a number **(b)** Triple a value

(c) Eight less than a number **(d)** One-fourth of a height ■

Finally, we see that more than one operation can be described in an English phrase. Sometimes parentheses must be used to make clear which operation is done first.

EXAMPLE 2

ENGLISH PHRASE	ALGEBRAIC EXPRESSION
(a) Seven more than double a number	$2x + 7$
	*Note that these are **not** the same.*
(b) The value of the number is increased by seven and then doubled	$2(x + 7)$

Practice Problem 2 Write each English phrase as an algebraic expression.

(a) Eight more than triple a number

(b) A number is increased by eight and then it is tripled. ■

EXAMPLE 3 Write the following English phrase as an algebraic expression.

A number increased by five is subtracted from three times the same number.

We let x represent the number. We can visualize it this way:

$$\underbrace{\text{A number}}_{(x} \quad \underbrace{\text{increased by}}_{+} \quad \underbrace{5}_{5)}$$

Therefore, "a number increased by 5" is $(x + 5)$.

Now we must subtract the entire quantity $(x + 5)$ from three times the number.

A number increased by 5

$$3x \quad - \quad (x + 5)$$

is subtracted from

Three times the same number

Compare the English phrase with the algebraic expression. Notice that the part that follows the word *from* in the English phrase comes first in the algebraic expression. It is also important to note the use of parentheses here. $3x - x + 5$ would be wrong.

Practice Problem 3 Write each English phrase as an algebraic expression.

(a) Four less than three times a number

(b) Two-thirds of the sum of a number and five ■

2 *When Two or More Quantities are Compared*

Often in a word problem two or more quantities are described in terms of another. We will want to use a variable to represent one quantity and then write an algebraic expression using *the same variable* to represent the other quantity. Which quantity should we let the variable represent? We usually let the variable represent the quantity that is the basis of comparison: the quantity that the others are being *compared to.*

EXAMPLE 4 Use a variable and an algebraic expression to describe the two quantities in the English phrase "Mike's salary is \$2000 more than Fred's salary."

The two quantities that are being compared are Mike's salary and Fred's salary. Since Mike's salary is being *compared to* Fred's salary, we let the variable represent Fred's salary. The choice of the letter f helps us to remember that the variable represents Fred's salary.

Let f = Fred's salary

Then $f + \$2000$ = Mike's salary *Since Mike's salary is \$2000 more than Fred's.*

Practice Problem 4 Use a variable and an algebraic expression to describe the two quantities in the English phrase "Marie works 17 hours per week less than Ann." ■

EXAMPLE 5 The length of a rectangle is 3 meters shorter than twice the width. Use a variable and an algebraic expression to describe the length and the width. Draw a picture.

The length of the rectangle is being *compared to* the width. Use the letter w for width.

Let w = the width

3 meters shorter than twice the width

Then $2w - 3$ = the length

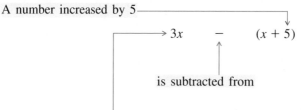

Draw a picture and use the algebraic expressions to identify the length and the width.

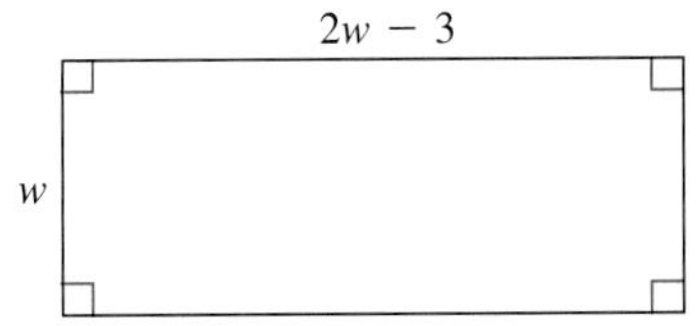

Practice Problem 5 The length of a rectangle is 5 meters longer than double the width. Use a variable and an algebraic expression to describe the length and the width. Draw a picture. ■

EXAMPLE 6 The first angle of a triangle is triple the second angle. The third angle of a triangle is 12° more than the second angle. Describe each angle algebraically. Draw a diagram.

Since the first and third angles are described in terms of the second angle, we let the variable represent the number of degrees in the second angle.

Let s = the number of degrees in the second angle

$3s$ = the number of degrees in the first angle

$s + 12$ = the number of degrees in the third angle

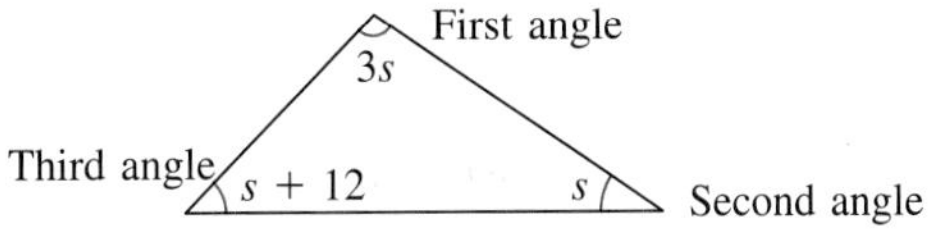

Practice Problem 6 The first angle of a triangle is 16° less than the second angle. The third angle is double the second angle. Describe each angle algebraically. Draw a diagram. ■

Some comparisons will involve fractions.

EXAMPLE 7 A theater manager was examining the records of attendance for last year. The number of people attending the theater in January was one-half of the number of people attending the theater in February. The number of people attending the theater in March was three-fifths of the number of people attending the theater in February. Use algebra to describe the attendance each month.

What are we looking for? The *number of people* who attended the theater *each month.* The basis of comparison is February. That is where we begin.

Let f = the number of people who attended in February

$\frac{1}{2}f$ = the number of people who attended in January

$\frac{3}{5}f$ = the number of people who attended in March

Practice Problem 7 The college dean noticed that in the spring the number of students on campus was two-thirds of the number of students on campus in the fall. She also noticed that in the summer the number of students on campus was one-fifth the number of students on campus in the fall. Use algebra to describe the number of students on campus in each of these three time periods. ■

3.1 Exercises

Verbal and Writing skills

Write an algebraic expression for each quantity. Let x represent the unknown value.

1. Seven more than a number

2. Nine greater than a number

3. Five less than a value

4. Twelve smaller than a quantity

5. One-third of a number

6. Three-fourths of a number

7. Triple a number

8. Double a number

9. One-eighth of a quantity

10. The product of twelve and a number

11. Five more than double a number

12. Eight less than triple a number

13. Two-sevenths of a value decreased by three

14. Double a quantity increased by four

15. Five times the sum of a number and twelve

16. One-fifth of the sum of a number and eight

17. Two-thirds of six less than a number

18. Seven less than one-fourth of a number

19. Three times a number added to one-half of the same number

20. Twice a number added to one-third of the same number

21. Four times a number decreased by one-half of the same number

Write an algebraic expression for each of the quantities that is being compared.

22. Wally's salary is \$1400 more than George's salary.

23. Alicia works 15 hours per week more than Barbara.

24. The cost of the chair was \$285 less than the cost of the sofa.

25. The cost of the printer was \$390 less than the cost of the computer.

26. The length of the rectangle is 7 inches more than double the width.

27. The length of the rectangle is 3 meters more than triple the width.

28. The attendance on Monday was 1185 people more than on Tuesday. The attendance on Wednesday was 365 people fewer than on Tuesday.

29. The attendance on Friday was 1600 people more than on Saturday. The attendance on Thursday was 783 people fewer than on Saturday.

30. The first angle of a triangle is 16 degrees less than the second angle. The third angle of a triangle is double the second angle.

31. The first angle of a triangle is 19 degrees more than the third angle. The second angle is triple the third angle.

32. The attendance on Tuesday was 350 people more than double the attendance on Monday.

33. The attendance on Friday was 200 people less than triple the attendance on Thursday.

34. The first angle of a triangle is triple the second angle. The third angle of a triangle is 14 degrees less than the second angle.

35. The second angle of a triangle is 6 degrees less than the first angle. The third angle of a triangle is double the first angle.

Applications

Use algebra to describe the situation in each word problem.

36. A local car dealer observed that car sales in August were one-half of the car sales in July. She also observed that the car sales in September were two-thirds of the car sales in July. Describe the number of cars sold in each of these three months.

37. Computer City found that the sales of computers in April were three-fourths of the sales of computers in March. The sales of computers in February were five-sixths of the sales of computers in March. Describe the number of computers sold in each of these three months.

38. The company president obtained this year's costs by increasing last year's costs by \$100 and by adding a surcharge of twenty percent of last year's costs. Describe the company's costs this year and last year.

39. The company president determined that the new profit this year was triple last year's profit diminished by \$400, further diminished by fifty percent of the last year's profit. Describe the company's profit this year and last year.

Cumulative Review Problems

Solve for the variable in each equation.

40. $x + 2(x + 2) = 55$

41. $2x + 2(3x - 4) = 72$

42. $5(x - 8) = 13 + x - 5$

43. $6(w - 1) - 3(2 + w) = 9$

3.2 Use Equations to Solve Word Problems

After studying this section, you will be able to:

1 *Solve number problems.*
2 *Solve applied word problems.*
3 *Use a formula to solve a word problem.*

The skills you have just developed in translating English phrases into algebraic expressions will enable you to solve a variety of word problems. Solving word problems can be challenging. You need to be organized. The following steps may help.

1. *Understand the problem.*
   - **(a)** Read the word problem carefully to get an overview.
   - **(b)** Determine what information you will need to solve the problem.
   - **(c)** Draw a sketch. Label it with the known information. Determine what needs to be found.
   - **(d)** Choose a variable to represent one unknown quantity.
   - **(e)** If necessary, represent other unknown quantities in terms of that very same variable.
2. *Write an equation.*
   - **(a)** Look for key words to help you to translate the words into algebraic symbols and expressions.
   - **(b)** Use a given relationship in the problem or an appropriate formula in order to write an equation.
3. *Solve the equation and state the answer.*
4. *Check.*
   - **(a)** Check the solution in the original equation.
   - **(b)** Be sure the solution to the equation answers the question in the word problem. You may need to do some additional calculations if it does not.

1 *Number Problems*

EXAMPLE 1 Two-thirds of a number is eighty-four. What is the number?

1. *Understand the problem.*
   Draw a sketch.

   Let x = the unknown number.

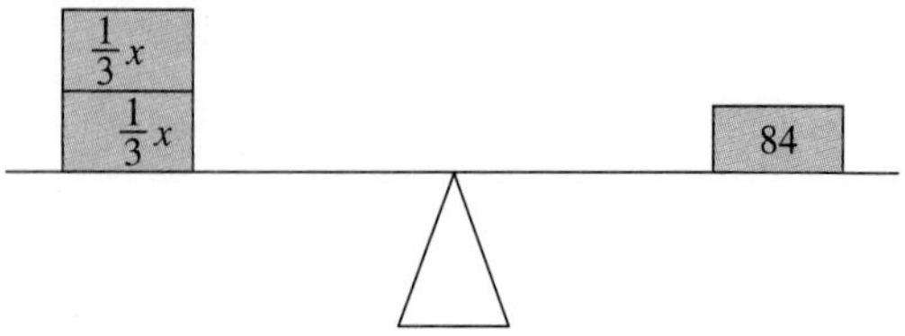

2. *Write an equation.*

$$\underbrace{\text{Two-thirds of a number}}_{\frac{2}{3}x} \ \underbrace{\text{is}}_{=} \ \underbrace{\text{eighty-four.}}_{84}$$

3. *Solve the equation and state the answer.*

$$\frac{2}{3}x = 84$$

$$3\left(\frac{2}{3}x\right) = 3(84) \qquad \textit{Multiply both sides of the equation by 3.}$$

$$2x = 252$$

$$\frac{2x}{2} = \frac{252}{2} \qquad \textit{Divide both sides by 2.}$$

$$x = 126$$

The number is 126.

4. *Check.* Is two-thirds of 126 eighty-four?

$$\frac{2}{3}(126) \stackrel{?}{=} 84$$

$$84 = 84 \checkmark$$

Practice Problem 1 Three-fourths of a number is negative eighty-one. What is the number? ■

EXAMPLE 2 Five more than six times a number is three hundred and five. Find the number.

1. *Understand the problem.*
Read the problem carefully. You may not need to draw a sketch.

Let $x =$ the unknown number.

2. *Write an equation.*

$$\underbrace{\text{Five more than}}_{5\ +}\ \underbrace{\text{six times a number}}_{6x}\ \underbrace{\text{is}}_{=}\ \underbrace{\text{three hundred and five.}}_{305}$$

3. *Solve the equation and state the answer.*
You may want to rewrite the equation to make it easier to solve.

$$6x + 5 = 305$$

$$6x + 5 - 5 = 305 - 5 \qquad \textit{Subtract 5 from both sides.}$$

$$6x = 300$$

$$\frac{6x}{6} = \frac{300}{6} \qquad \textit{Divide both sides by 6.}$$

$$x = 50$$

The number is 50.

4. *Check.* Is five more than six times 50 three hundred and five?

$$6(50) + 5 \stackrel{?}{=} 305$$

$$305 = 305$$

Practice Problem 2

Two less than triple a number is forty-nine. Find the number. ■

EXAMPLE 3 The larger of two numbers is three more than twice the smaller. The sum of the numbers is thirty-nine. Find each number.

1. *Understand the problem.*
   Read the problem carefully. The problem refers to *two* numbers. We must write an algebraic expression for *each number* before writing the equation. The larger number is being compared to the smaller number.

$$\text{Let } s = \text{the smaller number}$$

$$\underbrace{2s + 3}_{\text{three more than twice the smaller number}} = \text{the larger number}$$

2. *Write an equation.*
   Write an equation for the sum of the two numbers.

The sum of the numbers is thirty-nine.

$$s + (2s + 3) \quad = \quad 39$$

3. *Solve the equation.*

$$s + (2s + 3) = 39$$

$$3s + 3 = 39 \quad \textit{Collect like terms.}$$

$$3s = 36 \quad \textit{Subtract 3 from each side.}$$

$$s = 12 \quad \textit{Divide both sides by 3.}$$

4. *Check.*

$$12 + [2(12) + 3] \stackrel{?}{=} 39$$

$$39 = 39 \ \checkmark$$

The solution checks, but have we solved the word problem? We need to find *each* number. 12 is the smaller number. Substitute 12 into the expression $2s + 3$ to find the larger number.

$$2s + 3 = ?$$

$$2(12) + 3 = 27$$

The smaller number is 12. The larger number is 27.

Practice Problem 3 Consider two numbers. The second number is twelve less than triple the first number. The sum of the two numbers is twenty-four. Find each number. ■

2 *Applications*

You may never encounter problems like those in the previous examples in real life. They were presented to provide you with some of the basic skills needed to solve applied word problems. The following examples relate more directly to real-life situations.

EXAMPLE 4 The mean annual snowfall in Juneau, Alaska, is 105.8 inches. This is 20.2 inches less than three times the annual snowfall in Boston. What is the annual snowfall in Boston?

1. *Understand the problem.*
What information is given?

The annual snowfall in Juneau, Alaska.

The relationship between the snowfalls in Juneau and Boston.

What do you need to find?

The annual snowfall in Boston.

Choose a variable to represent the unknown quantity:

Let b = the annual snowfall in Boston

2. *Write an equation.*
Translate the words into algebraic symbols:

Juneau's snowfall is 20.2 less than three times Boston's snowfall.

$$105.8 = 3b - 20.2$$

3. *Solve the equation and state the answer.*
You may want to rewrite the equation to make it easier to solve.

$$3b - 20.2 = 105.8$$

$$3b = 126 \quad \textit{Add 20.2 to both sides.}$$

$$b = 42 \quad \textit{Divide both sides by 3.}$$

The annual snowfall in Boston is 42 inches.

4. *Check.* Reread the word problem. Work backward.

Three times 42 is 126.

126 less 20.2 is 105.8.

Is this the annual snowfall in Juneau? ✓

Practice Problem 4 The maximum 24-hour rainfall for Utah occurred at Bug Point on 9/5/70. It was 2 inches less than that recorded at Elk Point, South Dakota, on 9/10/70. If the combined rainfall was 14 inches, how much rainfall was recorded for each location? ■

Some word problems require a simple translation of the facts. Others require a little more detective work. You will not always need to use the four-step procedure to solve every word problem. As you gain confidence in problem solving, you will no doubt leave out some of the steps. We suggest that you use the procedure when you find yourself on unfamiliar ground. It is a powerful organizational tool.

3 *Formulas*

Sometimes the relationship between two quantities is so well understood that we have developed a formula to describe that relationship. We have already done some work with formulas in Section 2.5. The following examples show how you can use a formula to solve a word problem.

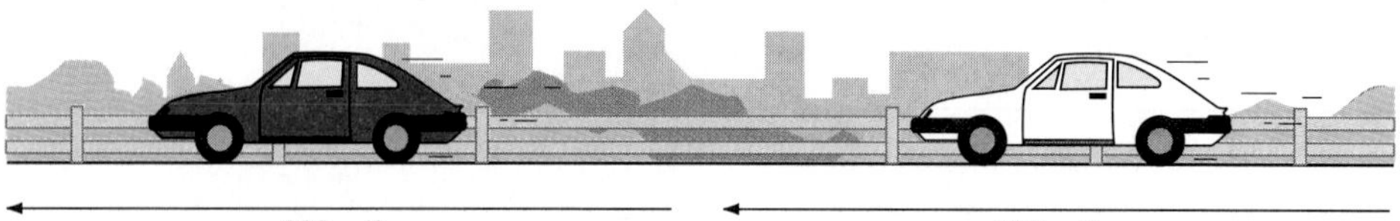

EXAMPLE 5 Two people travel in separate cars. They each travel a distance of 330 miles on an interstate highway. To maximize fuel economy, Fred travels at exactly 50 mph. Sam travels at exactly 55 mph. How much time did the trip take each person?

Here we use the formula

$$\text{distance} = (\text{rate}) \times (\text{time}) \quad \text{or} \quad d = rt$$

Time is the unknown quantity for each driver. We will use subscripts to denote the different values of t.

Let t_f = Fred's time   Let t_s = Sam's time

Substitute the known values into the formula and solve for t.

$d = rt$	$d = rt$
$330 = 50t_f$	$330 = 55t_s$
$6.6 = t_f$	$6 = t_s$

It took Fred 6.6 hours to drive 330 miles.   It took Sam 6 hours to drive 330 miles.

Check. Is this reasonable? Yes, you would expect Fred to take longer to drive the same distance because Fred is driving at a lower rate of speed.

Note: You may wish to express 6.6 hours in hours and minutes. To change 0.6 hours to minutes, proceed as follows:

$$0.6 \cancel{\text{hour}} \times \frac{60 \text{ minutes}}{1 \cancel{\text{hour}}} = (0.6)(60) \text{ minutes} = 36 \text{ minutes}$$

Thus, Fred drove for 6 hours and 36 minutes.

Practice Problem 5 Sarah left the city to visit her aunt and uncle, who live in a rural area north of the city. She traveled the 220-mile trip in 4 hours. When she went home she took a slightly longer route, which measured 225 miles on the car odometer. The return trip took 4.5 hours.

(a) What was her average speed on the trip leaving the city?

(b) What was her average speed on the return trip?

(c) On which trip did she travel faster and by how much? ■

EXAMPLE 6 A teacher told Melinda that she had a course average of 78 based on her six math tests. When she got home Melinda found five of her tests. She had scores of 87, 63, 79, 71, and 96 on the five tests. She could not find her sixth test. What score did she obtain on that test?

To find an average, we add all the test scores and divide by the number of tests. We will use this formula to solve the word problem. Read the problem again. What is the unknown quantity? We do not know the score Melinda received on her sixth test.

Let x = the score on the sixth test

We now write the equation you would use to find the average and then solve for x.

$$\frac{87 + 63 + 79 + 71 + 96 + x}{6} = 78$$

$$\frac{396 + x}{6} = 78 \qquad \textit{Add the numbers in the numerator.}$$

$$6\frac{396 + x}{6} = 6(78) \qquad \textit{Multiply both sides of the equation by 6 to remove the fraction.}$$

$$396 + x = 468 \qquad \textit{Simplify.}$$

$$x = 72 \qquad \textit{Subtract 396 from both sides to find } x.$$

Melinda's score on the sixth test was 72.

Check. To verify that this is correct, we see if the average of the 6 tests is 78.

$$\frac{87 + 63 + 79 + 71 + 96 + 72}{6} \stackrel{?}{=} 78$$

$$\frac{468}{6} \stackrel{?}{=} 78$$

$$78 = 78 \checkmark$$

The problem checks. We know that the score on the sixth test was 72.

Practice Problem 6 Barbara's math course has four tests and one final exam. The final exam counts as much as two tests. Barbara has test scores of 78, 80, 100, and 96 on her four tests. What grade does she need on the final exam if she wants to have a 90 average for the course? ■

3.2 Exercises

Solve each word problem. Check your solution.

1. What number added to 125 is 170?

2. Nine times what number is 756?

3. Two-fifths of a number is 288. What is the number?

4. Sixty-seven less than a number is 22. What is the number?

5. Fifteen greater than a number is 305. Find the number.

6. Twice a number is one. What is the number?

7. A number is doubled and then increased by three. The result is eighty-one. What is the original number?

8. Nine less than twelve times a number is one hundred seventeen. Find the original number.

9. When six is subtracted from one-third of a number, the result is 11. What is the original number?

10. A number is tripled and then increased by seven. The result is seventy-nine. What is the original number?

11. Four times a number is decreased by thirty to obtain seven times the original number. What is the number?

12. Thirty is added to five times a number. The result is double the original number. What is that number?

13. One number is 12 more than another number. The sum of the two numbers is 156. Find each number.

14. One number is 15 less than another number. The sum of the two numbers is 177. Find each number.

15. One number is 3 less than double another number. The sum of the two numbers is 48. Find each number.

16. One number is larger than another. The smaller number is one-third of the larger number. The sum of the smaller number and the larger number is 56. Find each number.

17. One-fourth of a number, one-eighth of that number, and one-fifth of that number are added. The result is forty-six. What is the original number?

18. One-eighth of a number, one-third of that number, and one-sixth of that number are added. The result is sixty. What is the original number?

19. One number is 41.8632 greater than another. The numbers have a sum of 196.0578. Find each number.

20. The sum of three numbers is 88.9635. The first number is double the second number. The third number is 6.3254 greater than the second number. Find each number.

Applications

Solve each word problem. Check to see if your answer is reasonable.

21. A furniture store maintains an inventory of four times as many chairs as tables. If there are 60 chairs, how many tables are now in stock?

22. Gloria is half as old as her aunt Sylvia. If Gloria is 19 years old, how old is Sylvia?

23. The monthly rental on a car is \$320. The cost of a new car is \$8000. In how many months will the rental equal the cost of the car?

24. The cost of renting a car is \$20 a day plus 25 cents per mile. How far can Jeff drive in one day if he has only \$75?

25. The cost to replace a transmission in a car came to \$950. This included \$540 for the transmission and \$45 per hour for labor. Find the number of hours of labor.

26. A city with a population of 15,900 is decreasing at a constant rate of 175 people per year. In how many years will the population drop to 11,000?

27. Rob is on a diet and can have 500 calories for lunch. A 3-ounce hamburger on whole-wheat bread has 315 calories and 12 fluid ounces of soda has 145 calories. How many French fries can he eat if there are 10 calories in one French fry?

28. Lauren Lee makes \$24 per hour for a 40-hour week and time and a half for every hour over 40 hours. If Lauren made \$1140 last week, how many overtime hours did she work?

29. Melissa traveled through the hill country for a distance of 270 miles. She averaged 45 miles per hour on the trip. How many hours did the trip take?

30. Mike traveled on the Ohio turnpike for a distance of 260 miles. The trip took four hours. What was his average rate of speed?

31. Alice jogs 3 miles in 30 minutes. What is her average rate of speed?

32. Paul drove 75 miles at 50 miles per hour. How long did the trip take?

33. Two joggers, Tom and Dave, start from the same point and run in the same direction. Tom jogs at 8 miles per hour and Dave jogs at 10 miles per hour. How far apart will they be in $2\frac{1}{2}$ hours?

34. Two trains leave a train station at the same time. One train travels east at 50 miles per hour. The other train travels west at 55 miles per hour. In how many hours will the two trains be 315 miles apart?

★ **35.** Roberto drove from the city to visit the Lost Caverns. He took 6 hours over the mountain road to travel 312 miles. He came home by a valley route. On the valley route he took 5.5 hours to travel 319 miles. How fast did he travel using the valley route? How fast did he travel using the mountain route? How much faster (in miles per hour) did he travel using the valley route?

★ **36.** Susan drives 30 miles per hour through the city and 55 miles per hour on the turnpike. She drove 90 miles from Battery Park to Peapack. How much of this was city driving if she spent 1.2 hours on the turnpike?

37. Juanita will have eight quizzes in her chemistry class. Her grades for her first seven quizzes are 75, 90, 85, 88, 92, 73, and 81. What score does she need to obtain on her last quiz to raise her quiz average to an 85?

38. In Won Sun's math class he will take four quizzes and two one-hour exams. Each exam will count as much as two quizzes. His four quiz grades are 91, 80, 86, and 83. He scored a 93 on his first one-hour exam. What score does he need on his last hour exam to have a 90 average for the course?

★ **39.** The Ramirez family has five licensed drivers and three cars. The smallest family car is a subcompact and gets excellent fuel mileage of 38 miles per gallon (mpg) in city driving. The second car is a compact and it gets 21 mpg in city driving. The third car is an ancient heavy station wagon. Dad Ramirez does not like to admit the mileage rating of the ''old beast.'' Rita calculated that if all three cars were driven the same number of miles in city driving each year, the *average* mpg rating of the three Ramirez cars would be $22\frac{2}{3}$ mph in city driving. What is the mpg rating of the old station wagon?

★ **40.** Four friends rented a house this semester while going to college. All four own cars and carry the same type of insurance coverage. Gregg pays \$350 per year to insure his car. Rich pays double that amount each year. Hector pays \$420 per year for his car insurance. Constanos purchased a new Corvette and won't tell anyone the huge amount he pays each year to insure the car. The four roommates have the same insurance agent. The agent said that if you took the average amount each of the four boys paid for car insurance it would be \$855. What is the yearly amount Constanos pays to insure his Corvette?

To Think About

41. In warmer climates, approximate temperature prediction can be made by counting the number of chirps of a cricket during a minute. The Fahrenheit temperature decreased by forty is equivalent to one-fourth of the number of cricket chirps.

(a) Write an equation for this relationship.

(b) Approximately how many chirps should be recorded if the temperature is 90°F?

(c) If a person recorded 148 cricket chirps in a minute, what would be the Fahrenheit temperature according to this formula?

Cumulative Review Problems

Simplify.

42. $5x(2x^2 - 6x - 3)$

43. $-2a(ab - 3b + 5a)$

Collect like terms.

44. $7x - 3y - 12x - 8y + 5y$

45. $5x^2y - 7xy^2 - 8xy - 9x^2y$

3.3 Solve Word Problems: Comparisons

After studying this section, you will be able to:

1 *Solve word problems involving comparisons.*

1 *Comparisons*

Many real-life problems involve comparisons. We often compare quantities such as length, height, or income. Sometimes not all the information is known about the quantities that are being compared. You need to identify each quantity and write an algebraic expression that describes the situation in the word problem.

EXAMPLE 1 Cindy and Aaron Dunsay have a combined income of \$32,500. If Cindy makes \$4000 more than Aaron, what is Aaron's income?

1. *Understand the problem.*
Read the problem: What information is given?
The combined income is \$32,500.
What is being compared?
Cindy's income is being compared to Aaron's income.
Express this algebraically:

$$\text{Let } a = \text{Aaron's income}$$
$$a + \$4000 = \text{Cindy's income}$$

2. *Write an equation.*
Reword the problem and translate:

Cindy's income combined with Aaron's income is \$32,500.

$$(a + \$4000) \quad + \quad a \quad = 32{,}500$$

3. *Solve the equation and state the answer.*

$$(a + 4000) + a = 32{,}500$$

$$2a + 4000 = 32{,}500 \qquad \textit{Combine like terms.}$$

$$2a + 4000 - 4000 = 32{,}500 - 4000 \qquad \textit{Subtract 4000 from both sides.}$$

$$2a = 28{,}500$$

$$\frac{2a}{2} = \frac{28{,}500}{2} \qquad \textit{Divide both sides by 2.}$$

$$a = 14{,}250$$

Aaron's income is \$14,250.

4. *Check.* $(14{,}250 + 4000) + 14{,}250 \stackrel{?}{=} 32{,}500$ *Substitute* 14,250 *for a in the equation.*

$$18{,}250 + 14{,}250 \stackrel{?}{=} 32{,}500$$

$$32{,}500 = 32{,}500 \checkmark$$

Practice Problem 1 A deck hand on a fishing boat is working with a rope that measures 89 feet. He needs to cut it into two pieces. The long piece must be 17 feet longer than the short piece. Find the length of each piece of rope. ■

If the word problem contains three unknown quantities, determine what is the basis of comparison for two of the quantities.

EXAMPLE 2 An airport filed a report showing the number of plane departures that took off from the airport during each month last year. The number of departures in March was 50 more than the number of departures in January. In July the number of departures was 150 less than triple the number of departures in January. In those three months, the airport had 2250 departures. How many departures were recorded for each month?

1. *Understand the problem.*
What is the basis of comparison?

The number of departures in March is compared to January.
The number of departures in July is compared to January.

Express this algebraically. It may help to underline the key phrases.

Let j = the departures in January

$$\underbrace{\text{March was 50 more than January}}$$

$j + 50$ = the departures in March

$$\underbrace{\text{July was 150 less than triple January}}$$

$3j - 150$ = the departures in July

2. *Write an equation.*

Number of departures in January	+	Number of departures in March	+	Number of departures in July	=	3 months' total departures
j	+	$(j + 50)$	+	$(3j - 150)$	=	2250

3. *Solve the equation and state the answer.*

$$j + (j + 50) + (3j - 150) = 2250$$

$$5j - 100 = 2250 \quad \textit{Collect like terms.}$$

$$5j = 2350 \quad \textit{Add 100 to each side.}$$

$$j = 470 \quad \textit{Divide both sides by 5.}$$

Now, if $j = 470$, then

$$j + 50 = 470 + 50 = 520$$

and

$$3j - 150 = 3(470) - 150 = 1410 - 150 = 1260$$

The number of departures in January was 470; the number of departures in March was 520; the number of departures in July was 1260.

4. *Check.* Do these answers seem reasonable? Yes. Do these answers verify all the statements in the word problem?

Is the number of departures in March 50 more than those in January?

$$520 \stackrel{?}{=} 50 + 470$$

$$520 = 520 \checkmark$$

Is the number of departures in July 150 less than triple those in January?

$$1260 \stackrel{?}{=} 3(470) - 150$$

$$1260 \stackrel{?}{=} 1410 - 150$$

$$1260 = 1260 \checkmark$$

Is the total number of departures in the 3 months equal to 2250?

$$470 + 520 + 1260 \stackrel{?}{=} 2250$$

$$2250 = 2250 \quad ✓$$

Yes, all conditions are satisfied. The three answers are correct.

Practice Problem 2 A social services worker was comparing the cost incurred by three families in heating their homes for the year. The first family has an annual heating bill of \$360 more than that of the second family. The third family has a heating bill of \$200 less than double the heating bill of the second family. The total annual heating bill for the three families was \$3960. What was the annual heating bill for each family? ■

EXAMPLE 3 A small plot of land is in the shape of a rectangle. The length is 7 meters longer than the width. The perimeter of the rectangle is 86 meters. Find the dimensions of the rectangle.

1. *Understand the problem.*
   Read the problem: What information is given?
   The perimeter of a rectangle is 86 *meters.*
   What is being compared?
   The length is being compared to the width.
   Express this algebraically, and draw a picture.

   Let w = the width
   $w + 7$ = the length

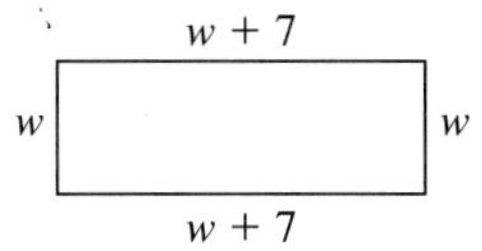

   Reread the problem: What are you being asked to do?
   Find the dimensions of the rectangle. The dimensions of the rectangle are the length and the width of the rectangle.

2. *Write an equation.*
   The perimeter is the total distance around the rectangle.

$$w + (w + 7) + w + (w + 7) = 86$$

3. *Solve the equation and state the answer.*

$$w + (w + 7) + w + (w + 7) = 86$$
$$4w + 14 = 86 \quad \textit{Combine like terms.}$$
$$4w = 72 \quad \textit{Subtract 14 from both sides.}$$
$$w = 18 \quad \textit{Divide both sides by 4.}$$

   The width of the rectangle is 18 meters. What is the length?

$$w + 7 = \text{the length}$$
$$18 + 7 = 25$$

   The length of the rectangle is 25 meters.

4. *Check.* Put in the actual dimensions in your drawing and add the length of the sides. Is the sum 86 meters? ✓

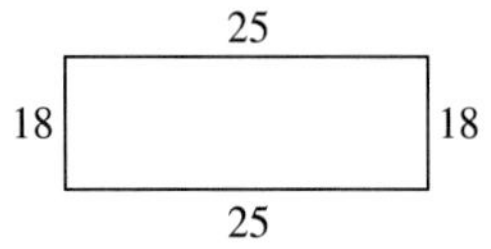

Practice Problem 3 A farmer purchased 720 meters of wire fencing to enclose a pasture. The pasture is in the shape of a triangle. One side of the triangle is 30 meters less than the longest side. The shortest side is one-half as long as the longest side. Find the dimensions of the triangle. ■

3.3 Exercises

Applications

Solve each word problem. Check to see if your answer is reasonable. Have you answered the question that was asked?

1. Heather earns \$3600 less per year than Melissa. Together the two girls earn \$29,100 per year. What is the annual salary of each girl?

2. Charlie walked 86 miles more than Robert last year in a local YMCA ''walk your way to health'' program. The two boys together walked a total of 1022 miles last year. How many miles did each person walk?

3. A drainage pipe measures 93 meters in length. It consists of two pieces. The long piece is 10 meters longer than the short piece. Find the length of each piece.

4. A copper conducting wire measures 84 centimeters in length. It is cut into two pieces. The shorter piece is 5 centimeters shorter than the long piece. Find the length of each piece.

5. The Topsfield Fair set records this year for total attendance at the fairgrounds. On Monday there were 350 more people attending than on Tuesday. The crowds on Wednesday were 3450 more people than on Tuesday. The total attendance for Monday, Tuesday, and Wednesday was 64,550. How many people attended each of those three days?

6. Alice bought a new car that cost \$1368 more than one purchased by Michael. Bettina's car cost \$852 less than the new car Michael bought. The three cars cost \$27,588. Can you determine how much each car cost?

7. Three overweight people went to a weight-reducing class for an entire year. During that time the total weight lost by all three people was 156 pounds. Rita lost two-thirds the number of pounds that Margaret lost. Tony lost 20 pounds less than Margaret. How many pounds did each lose?

8. An air sample was taken in an industrial city. It was found to contain 12.8 ppm (parts per million) of three harmful pollutants. The amount of the first pollutant was three times as great as that of the second pollutant. The amount of the third pollutant was twelve times as great as that of the second pollutant. How many parts per million of each pollutant were found?

9. The sum of the electric current in three transistors is 0.027 ampere (a measure of the amount of electric current). By performing a comparison test, a technician found that the electric current in the first transistor is half of that in the second. The technician also found that the amount of current in the third transistor is one and one-half times the amount of current in the second transistor. How much current is in each transistor?

10. The sum of the electric current in three transistors is 0.1445 ampere. A technician found that the electric current in the first transistor is one-fourth the amount of current in the second transistor. However, the amount of current in the third transistor is three times the amount in the second transistor. How much current is in each of the three transistors?

11. The length of a rectangular piece of tin is 3 centimeters more than double the width. The perimeter of the rectangle is 42 centimeters. Find the dimensions in centimeters of the width and length of the piece of tin.

12. A state park in Colorado has a perimeter of 92 miles. The park is in the shape of a rectangle. The length of the park is 30 miles less than triple the width of the park. What are the dimensions of the park?

13. Rashau made a rectangular picture frame for her latest oil painting. The length is 20 centimeters less than triple the width. The perimeter of the frame is 96 centimeters. Find the length and width of the frame.

14. A rectangular piece of land is being used by the college for an experimental student cooperative garden this semester. The garden will have a perimeter of 1570 feet. The length of the garden is 200 feet more than twice the width of the garden. Find the dimensions of the garden in feet.

15. A small region in front of a building is triangular in shape. The perimeter of the triangle is 37 meters. The second side is one-half of the first side in length. The third side is 3 meters less than the first side. Find the length in meters of each side of the triangular region.

16. An electronics firm needs to use a copper plate that is triangular in shape. The perimeter of the copper plate measures 78 centimeters. The longest side is 10 centimeters longer than the length of the second longest side. The shortest side is 4 centimeters shorter than the second longest side. Find the length in centimeters of each side.

17. A small triangular piece of metal is welded onto the hull of a whale watch boat. The perimeter of the metal is 46 inches. The shortest side of the triangle is four inches longer than one-half of the longest side. The second longest side is 3 inches shorter than the longest side. Find the length of each side.

18. Carmelina's uncle owns a triangular piece of land. The perimeter fence that surrounds the land measures 378 yards. The shortest side is 30 meters longer than one-half of the longest side. The second longest side is 2 meters shorter than the longest side. Find the length of each side.

19. The perimeter of a rectangle is 49.240 centimeters. The length of the rectangle is 1.334 centimeters larger than double the width. Find the length and width.

20. A triangle has a perimeter of 8.07844 meters. The first side is 0.20508 meter longer than the second side. The third side is double the second side. Find the length of each side.

To Think About

21. A small square is constructed. Then a new square is made by increasing each side by 2 meters. The perimeter of the new square is 3 meters shorter than five times the length of one side of the original square. Find the dimensions of the original square.

22. A rectangle is constructed. The length of this rectangle is double the width. Then a new rectangle is made by increasing each side by 3 meters. The perimeter of the new rectangle is 2 meters greater than 4 times the length of the old rectangle. Find the dimensions of the original rectangle.

Cumulative Review Problems

Simplify.

23. $-4x(2x^2 - 3x + 8)$

24. $5a(ab + 6b - 2a)$

Collect like terms.

25. $-7x + 10y - 12x - 8y - 2$

26. $3x^2y - 6xy^2 + 7xy + 6x^2y$

3.4 Solve Word Problems: The Value of Money and Percents

After studying this section, you will be able to:

1. *Solve applied problems with periodic rate charges.*
2. *Solve percent problems.*
3. *Solve investment problems involving simple interest.*
4. *Solve coin problems.*

The problems we now present are frequently encountered in business. They deal with money: buying, selling, and renting items; earning and borrowing money; the value of collections of stamps or coins. Many applications require an understanding of the use of percentage and decimals. Review Sections 0.4 and 0.5 if you are weak in these skills.

1 *Periodic Rate Charges*

EXAMPLE 1 A business executive rented a car. The Car Rental Agency charged \$39 per day and \$0.28 per mile. The executive rented the car for two days and the total rental cost was computed to be \$176. How many miles did the executive drive the rented car?

1. *Understand the problem.* How do you calculate the cost of renting a car?
   Total cost = Per day cost + Mileage cost
   What is known?
   It costs \$176 to rent the car for two days.
   What do you need to find?
   The number of miles the car was driven.

   Choose a variable:

   Let m = the number of miles driven in the rented car

2. *Write an equation.*
   Use the relationship for calculating the total cost.

   Per day cost + Mileage cost = Total cost

   $$(39)(2) \quad + \quad (0.28)m \quad = \quad 176$$

3. *Solve the equation and state the answer.*

   $$(39)(2) + (0.28)(m) = 176$$

   $$78 + 0.28m = 176 \qquad \textit{Simplify the equation.}$$

   $$0.28m = 98 \qquad \textit{Subtract 78 from both sides.}$$

   $$\frac{0.28m}{0.28} = \frac{98}{0.28} \qquad \textit{Divide both sides by 0.28.}$$

   $$m = 350 \qquad \textit{Simplify.}$$

   The executive drove 350 miles.

4. *Check.* Does this seem reasonable? If he drove the car 350 miles in two days, would it cost \$176?

   (Cost of \$39 per day for 2 days) + (cost of \$0.28 per mile for 350 miles) $\stackrel{?}{=}$ total cost of \$176

   $$(\$39)(2) + (350)(\$0.28) \stackrel{?}{=} \$176$$

   $$\$78 + \$98 \stackrel{?}{=} \$176$$

   $$\$176 = \$176 \checkmark$$

Practice Problem 1 Alfredo wanted to rent a truck to move to Florida. He determined that the cheapest rental rates for a truck of the correct size are from a local company that will charge him \$22 per day and \$0.18 per mile. He has not yet completed an estimate of the mileage of the trip, but he knows that he will need the truck for three days. He has allowed \$363 in his moving budget for the truck. How far can he travel for a rental cost of exactly \$363? ■

2 *Percents*

Many applied situations require finding a percent of an unknown number. If we want to find 23% of \$400, we multiply $0.23(400) = 92$. If we want to find 23% of an unknown number, we can express this using algebra by writing $0.23n$, where n represents the unknown number.

EXAMPLE 2 A sofa was marked with the following sign: "The price of this sofa has been reduced by 23%. You can save \$138 if you buy now." What was the original price of the sofa?

Understand the problem.

Let s = the original price of the sofa

Then $0.23s$ = the amount of the price reduction, which is \$138

Write an equation and solve.

$0.23s = 138$ *Write the equation.*

$\frac{0.23s}{0.23} = \frac{138}{0.23}$ *Divide each side of the equation by 0.23.*

$s = 600$ *Simplify.*

The original price of the sofa was \$600.

Check: Is \$600 a reasonable answer? ✓ Is 23% of \$600 = \$138? ✓

Practice Problem 2 John earns a commission of 38% of the cost of every set of encyclopedias that he sells. Last year he earned \$4560 in commissions. What was the cost of the encyclopedias that he sold last year? ■

EXAMPLE 3 Hector received a pay raise this year. The raise was 6% of last year's salary. This year he earns \$15,900 per year. What was his salary last year before the raise?

Understand the problem. What do we need to find?
Hector's salary last year
What do we know?
Hector received a 6% pay raise and now earns \$15,900.
What does this mean?

Reword the problem: *This year's salary of \$15,900 is 6% more than last year's salary.*

Choose a variable:

Let x = Hector's salary last year

Then $0.06x$ = the amount of the raise

Write an equation and solve.

Last year's salary	+	the amount of his raise	=	this year's salary

x	$+$	$0.06x$	$=$	$15{,}900$	*Write the equation.*
$1.00x$	$+$	$0.06x$	$=$	$15{,}900$	*Rewrite x as $1.00x$.*
		$1.06x$	$=$	$15{,}900$	*Collect like terms.*
		x	$=$	$\dfrac{15{,}900}{1.06}$	*Divide by 1.06.*
		x	$=$	$15{,}000$	*Simplify.*

Thus Hector's salary was $15,000 last year before the raise.

Check. Does it seem reasonable that Hector's salary last year was $15,000? The check is up to you.

Practice Problem 3 The price of Betsy's new car is 7% more than the price of a similar model last year. She paid $13,910 for her car this year. What would a similar model have cost last year? ■

3 *Interest Problems*

Interest rates affect our lives. They affect the national economy and they affect a consumers' ability to borrow money for big purchases. For these reasons, a student of mathematics should be able to solve problems involving interest. There are two basic types of interest: simple interest and compound interest.

Simple Interest

Simple interest is a charge for borrowing money or an income from investing money. It is computed by multiplying the amount of money borrowed or invested (which is called the *principal*) times the rate of interest times the period of time it is borrowed or invested (usually measured in years unless otherwise stated).

$$\text{Interest} = \text{principal} \times \text{rate} \times \text{time}$$

$$I = prt$$

Compound Interest

You often hear of banks offering a certain interest rate *compounded* quarterly, monthly, weekly, or daily. In **compound interest** the amount of interest is added to the amount of the original principal at the end of each time period, so future interest is based on the sum of both principal and previous interest. Most financial institutions use compound interest in their transactions.

Problems involving compound interest may be solved by:

1. Repeated calculations using the simple interest formula
2. Using a compound interest table
3. Using exponential functions, a topic that is usually covered in a higher-level college algebra course

All exercises discussed in this chapter will involve **simple interest.**

CALCULATOR

Interest

You can use a calculator to find simple interest. Find the interest on $450 invested at 6.5% for 15 months. Notice the time is in months. Since the interest formula is in years, $I = P \times R \times T$, you need to change 15 months to years by dividing 15 by 12.

Enter: 15 [÷] 12 [=]

Display: [1.25]

Leave this on the display and multiply as follows:

1.25 [×] 450 [×]
6.5 [%] [=]

The display should read

[36.5625]

which would round to $36.56.

Try **(a)** $9516 invested at 12% for 30 months.

(b) $593 borrowed at 8% for 5 months.

EXAMPLE 4 Find the interest on \$3000 borrowed at an interest rate of 18% for one year.

$I = prt$ *The simple interest formula.*

$I = (3000)(0.18)(1)$ *Substitute the value of the principal* = 3000; *the rate* = 18% = 0.18; *the time* = *one year.*

$I = 540$

Thus the interest charge for borrowing \$3000 for one year at a simple interest rate of 18% is \$540.

Practice Problem 4 Find the interest on \$7000 borrowed at an interest rate of 12% for one year. ■

Now we apply this concept to a word problem about investments.

EXAMPLE 5 A woman invested an amount of money in two places for one year. She invested some at 8% interest and the rest at 12% interest. Her total amount invested was \$1250. At the end of the year she had earned \$128.00 in interest. How much money did she invest in each place?

Understand the problem.
There are two amounts of money, some invested at 8% and the rest invested at 12%. The total amount invested is \$1250. Let us represent one unknown amount by x.

Let x = the amount invested at 8%

Then $1250 - x$ = the amount invested at 12%
(this is the amount that is left to invest at 12%)

Draw a diagram.

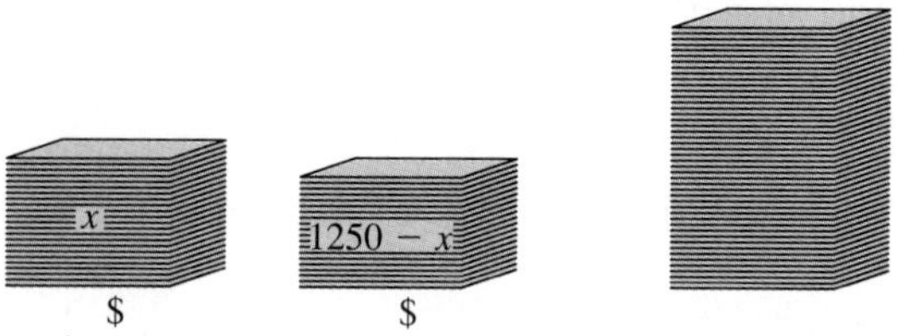

\$ Amount invested at 8% + \$ Amount invested at 12% = \$1250.00

Reread the problem. *At the end of the year she had earned \$128 in interest.*
Use the simple interest formula to represent the interest earned at 8% and at 12%.

INTEREST EARNED AT 8%

$I_1 = prt$

$= (x)(0.08)(1)$

$= 0.08x$

INTEREST EARNED AT 12%

$I_2 = prt$

$= (1250 - x)(0.12)(1)$

$= 0.12(1250 - x)$

Now we look at the total interest:

interest earned at 8%	+	interest earned at 12%	=	total interest earned during the year
I_1	+	I_2	=	\$128
$0.08x$	+	$0.12(1250 - x)$	=	128

In writing this equation, we substituted the two expressions for interest that we developed.

Solve the equation and state the answer.

$0.08x + 150 - 0.12x = 128$	*Remove parentheses.*
$-0.04x + 150 = 128$	*Collect like terms.*
$-0.04x = -22$	*Subtract* 150 *from both sides.*
$\dfrac{-0.04x}{-0.04} = \dfrac{-22}{-0.04}$	*Divide both sides by* -0.04
$x = 550$	*The amount invested at* 8% *interest is* $550.
$1250 - x = 1250 - 550 = 700$	*The amount invested at* 12% *interest is* $700.

Check. Are these values reasonable? Yes. Do the amounts equal $1250?

$$\$550 + \$700 \stackrel{?}{=} \$1250$$

$$\$1250 = \$1250 \checkmark$$

Would these amounts earn $128 interest in one year invested at the specified rates?

$$0.08(\$550) + 0.12(\$700) \stackrel{?}{=} \$128$$

$$\$44 + \$84 \stackrel{?}{=} \$128$$

$$\$128 = \$128 \checkmark$$

Practice Problem 5 A woman invested her savings of $8000 in two accounts that each calculate interest only once per year. She placed one amount in a special notice account that yields 15% annual interest. The remainder she placed in a tax-free All-Savers account that yields 12% annual interest. At the end of the year she earned $1155 in interest from the two accounts together. How much did she invest in each account? ■

4 Coin Problems

Coin problems provide an unmatched opportunity to use the concept of *value.* We must make a distinction between how many coins and the *value* of the coins.

Consider this example. Here we know *the value* of some coins, but do not know *how many* we have.

EXAMPLE 6 When Bob got out of math class, he had to make a long-distance call. He had exactly enough dimes and quarters to make a phone call that would cost $2.55. He had one less quarter than he had dimes. How many coins of each type did he have?

Let d = the number of dimes

Then $d - 1$ = the number of quarters

The total value of the coins is $2.55. How can we represent the value of the dimes and the value of the quarters? Think.

Each dime is worth $0.10.	Each quarter is worth $0.25.
5 dimes are worth $(5)(0.10) = 0.50$	8 quarters are worth $(8)(0.25) = 2.00$
d dimes are worth $(d)(0.10) = 0.10d$	$(d - 1)$ quarters are worth $(d - 1)(0.25) = 0.25(d - 1)$

Now we can write an equation for the total value.

(value of dimes)	+	(value of quarters)	=	\$2.55
$0.10d$	+	$0.25(d-1)$	=	2.55

$$0.10d + 0.25d - 0.25 = 2.55 \quad \textit{Remove parentheses.}$$

$$0.35d - 0.25 = 2.55 \quad \textit{Collect like terms.}$$

$$0.35d = 2.80 \quad \textit{Add 0.25 to both sides.}$$

$$\frac{0.35d}{0.35} = \frac{2.80}{0.35} \quad \textit{Divide both sides by 0.35.}$$

$$d = 8 \quad \textit{Simplify.}$$

$$d - 1 = 7$$

Thus Bob had 8 dimes and 7 quarters.

Check. Is this answer reasonable? Yes. Does Bob have one less quarter than he has dimes?

$$8 - 7 \stackrel{?}{=} 1$$

$$1 = 1 \checkmark$$

Are 8 dimes and 7 quarters worth \$2.55?

$$8(\$0.10) + 7(\$0.25) \stackrel{?}{=} \$2.55$$

$$\$0.80 + \$1.75 \stackrel{?}{=} \$2.55$$

$$\$2.55 = \$2.55 \checkmark$$

Practice Problem 6 Ginger has five more quarters than dimes. She has \$5.10 in change. If she has only quarters and dimes, how many coins of each type does she have? ■

EXAMPLE 7 Michele returned from the store with \$2.80 in change. She had twice as many quarters as nickels. She had two more dimes than nickels. How many nickels, dimes, and quarters did she have?

Let x = the number of nickels

$2x$ = the number of quarters

$x + 2$ = the number of dimes

Now we must find the *value* of the coins.

Since one nickel is worth \$0.05, we let $0.05x$ = *value* of the nickels.

Since one dime is worth \$0.10, we let $0.10(x + 2)$ = *value* of the dimes.

Since one quarter is worth \$0.25, we let $0.25(2x)$ = *value* of the quarters.

We can now write an equation for total value.

(value of nickels)	+	(value of dimes)	+	(value of quarters)	=	\$2.80
$0.05x$	+	$0.10(x+2)$	+	$0.25(2x)$	=	2.80

Solve the equation and check.

$$0.05x + 0.10x + 0.20 + 0.50x = 2.80 \quad \textit{Remove parentheses.}$$

$$0.65x + 0.20 = 2.80 \quad \textit{Collect like terms.}$$

$$0.65x = 2.60 \quad \textit{Subtract 0.20 from both sides.}$$

$$\frac{0.65x}{0.65} = \frac{2.60}{0.65} \quad \textit{Divide both sides by 0.65.}$$

$$x = 4 \quad \textit{Simplify. Michele had 4 nickels.}$$

$$2x = 8 \quad \textit{She had 8 quarters.}$$

$$x + 2 = 6 \quad \textit{She had 6 dimes.}$$

Check. Is the answer reasonable? Yes. Did Michele have twice as many quarters as nickels?

$$(4)(2) \stackrel{?}{=} 8 \qquad 8 = 8 \ \checkmark$$

Did she have two more dimes than nickels?

$$4 + 2 \stackrel{?}{=} 6 \qquad 6 = 6 \ \checkmark$$

Do 4 nickels, 8 quarters, and 6 dimes have a value of \$2.80?

$$4(\$0.05) + 8(\$0.25) + 6(\$0.10) \stackrel{?}{=} \$2.80$$

$$\$0.20 + \$2.00 + \$0.60 \stackrel{?}{=} \$2.80$$

$$\$2.80 = \$2.80 \ \checkmark$$

Practice Problem 7 A young boy told his friend that he had twice as many nickels as dimes in his pocket. He also said that he had four more quarters than dimes. He said that he had \$2.35 in change in his pocket. Can you determine how many nickels, dimes, and quarters he had? ■

Developing Your Study Skills

APPLICATIONS OR WORD PROBLEMS

Applications or word problems are the very life of mathematics! They are the reason for doing mathematics because they teach you how to put into use the mathematical skills you have developed. Learning mathematics without ever doing word problems is similar to learning all the skills of a sport without ever playing a game or learning all the notes on an instrument without ever playing a song.

The key to success is practice. Make yourself do as many problems as you can. You may not be able to do them all correctly at first, but keep trying. Do not give up whenever you reach a difficult one. If you cannot solve it, just try another one. Then come back and try it again later.

A misconception among students when they begin studying word problems is that each problem is different. At first the problems may seem this way, but as you practice more and more, you will begin to see the similarities, the different "types." You will see patterns in solving problems, which will enable you to solve problems of a given type more easily.

3.4 Exercises

Applications

Solve each word problem.

1. Barbara rented a car for $18 per day. She was charged a mileage rate of $0.08 per mile. She rented the car for four days. Her total bill to the rental agency was $100.80. How many miles did she drive?

2. Richard rented a truck for $38 per day. He used the truck for two days. The mileage charge was $0.20 per mile. He paid a total of $107.60 to the truck rental company. How many miles did he drive the truck?

3. In the summer Nancy has a sales job for which she is paid a salary plus a commission. She earns $10 per day plus $8 for every dictionary she sells. Last week she worked five days and earned $186. How many dictionaries did she sell?

4. Marybelle is contemplating a job as a waitress. She would be paid $2 per hour plus tips. The other waitresses have told her that an average tip at that restaurant is $3 per table served. If she works 20 hours per week, how many tables would she have to service in order to make $151 during the week?

5. Ramon has a summer job to earn his college tuition. He gets paid $6.00 per hour for the first 40 hours and $9.00 per hour for each hour in the week worked above the 40 hours. This summer his goal is to earn at least $303.00 per week. How many hours of overtime per week will he need to achieve his goal?

6. Maria has a summer job to earn her college tuition. She gets paid $6.50 per hour for the first 40 hours and $9.75 for each hour in the week worked above the 40 hours. This summer her goal is to earn at least $338.00 per week. How many hours of overtime per week will she need to achieve her goal?

7. A compact disc player has been reduced in price by 35%. The amount of the discount was $84. What was the original price of the compact disc player?

8. The tuition at Smithfield College went up this year by 14%. The amount of the increase is $784 per semester. What was the amount of tuition per semester last year?

9. The crime rate in Center City has increased by 7% from last year to this year. There were 56 more reported crimes this year in Center City than there were last year. How many reported crimes were there last year in Center City?

10. Due to a serious budget shortfall the state has been forced to lay off 9% of its employees. The number of employees laid off was 6030. How many employees did the state have before these employees were laid off?

11. When Harvey got promoted at work, he received a 12% pay raise. He now earns $16,240 per year. What did he earn annually before the pay raise?

12. Shirley borrowed some money from Easy Finance Company. They charged 18% interest on the loan. At the end of one year Shirley paid off the debt by paying back the amount of the original loan plus the 18% interest. She paid the company $2006. How much did she borrow originally?

13. A man invested some money at 14% interest. At the end of the year the total amount of his original principal and the interest was \$5700. How much did he originally invest?

14. The cost of living last year went up 13%. Fred was fortunate to get a 13% increase in salary. He now makes \$15,820 per year. How much did he make last year?

15. Linda has a job as a loan counselor in a bank. She advised a customer to invest part of her money in a 12% money market fund and the rest in a 14% investment fund. The customer has \$4000 to invest. If she earned \$508 in interest after one year, how much did she invest in each fund?

16. A retired couple earned \$656 last year from investments made in tax-free bonds. They invested some of the money at 8% and the rest at 10%. The total amount they invested was \$7200. If the couple invested the \$7200 for exactly one year, how much did they invest at each interest rate?

17. Suzanne invested \$6000 for one year. She invested part of it at 9% and the rest at 11%. At the end of the year she earned \$624 in interest. How much did she invest at each rate?

18. Eva invested \$5000 for one year. She invested part of it at 10% and part of it at 12%. At the end of the year she earned \$566. How much did she invest at each rate?

19. Ted and Linda invested \$9000 for one year. They invested part of it at 14% and the remainder at 9%. At the end of the year they had earned \$1135. How much did they invest at each rate?

20. Prior to going to college Carlos had saved up \$3000. He invested it for one year. Part of it was invested at 7% and the remainder at 9%. At the end of one year he had earned \$242 in interest. How much did he invest at each rate?

21. Little Melinda has nickels and quarters in her bank. She has four fewer nickels than quarters. She has \$3.70 in the bank. How many coins of each type does she have?

22. Fred's younger brother had several coins when he returned from his paper route. He had, in dimes and quarters, a total of \$5.35. He said he had six more quarters than he had dimes. How many of each coin did he have?

23. A newspaper carrier has \$3.75 in change. He has three more quarters than dimes but twice as many nickels as quarters. How many coins of each type does he have?

24. Alice collected a number of nickels, dimes, and quarters from her room. She found three more dimes than nickels but twice as many quarters as dimes. The value of her coins was \$5.05. How many coins of each type did she have?

25. Terry has two fewer quarters than nickels and one more half-dollar than twice the number of nickels. She has \$9.10. How many of each coin does she have?

26. Mike collected coins from a vending machine. He found three times as many nickels as dimes. He had eight more quarters than dimes. He collected \$4.50 from one machine. How many coins of each type did he collect?

★ **27.** The sum of the measures of the three interior angles of a triangle is always 180°. In a certain triangle the measure of one angle is double the measure of a second angle but is 5 degrees less than the measure of the third angle. What is the measure of each angle?

★ **28.** The length of a spring increases 1.32 centimeters for each kilogram it supports. The spring is 12.00 centimeters long when it is supporting 3.5 kilograms. Find the original length of the spring.

29. Walter invested \$2969 for one year and earned \$273.47 in interest. He invested part of the \$2969 at 7% and the remainder at 11%. How much did he invest at each rate?

30. Last year the town of Waterbury paid an interest payment of \$52,396.08 for a one-year note. This represented 11% of the amount the town borrowed. How much did they borrow?

To Think About

31. The Superior Car Rental Agency will rent a compact car for \$35 per day and an additional charge of \$0.24 per mile. The Peerless Car Rental Agency only charges \$0.16 per mile but charges \$41 per day. If a salesperson wanted to rent a car for three days, how many miles would that person have to drive to make the Peerless Car Rental Agency car a better bargain?

32. A pumping station pumps 2000 gallons per hour into a reservoir tank for a town's drinking water. The tank is initially empty. The station pumps for 3 hours. Then a leak in the reservoir tank is created by a large crack. Some water flows out of the reservoir tank at a constant rate. The pumping station continues pumping for 6 more hours while the leak is undetected. At the end of 9 hours the reservoir has 17,640 gallons. During the last 6 hours, how many gallons per hour were leaking from the reservoir?

Cumulative Review Problems

Do each operation in the proper order.

33. $5(3) + 6 \div (-2)$

34. $2 - (8 - 10)^3 + 5$

Evaluate when $x = -2$ *and* $y = 3$.

35. $2x^2 + 3xy - 2y^2$

36. $x^3 - 5x^2 + 3y - 6$

Putting Your Skills to Work

LOANS

Most interest today on loans and savings is compounded quarterly, monthly, and daily. Compound interest is interest paid not only on the principal but also on the accumulated interest. Amortization tables make it easier to determine the monthly payments on loans. For example, look at the table on the right. If you borrow \$4000 at 9% for five years (60 months), you would expect to pay \$83.04 each month for the five years to repay the loan with interest.

TYPICAL MONTHLY PAYMENTS

Months to Pay	Interest	Amount \$2,625	Amount \$4,000	Amount \$5,250
36 months	8%	\$82.26	\$125.35	\$164.52
	9%	83.48	127.20	166.95
	10%	—	—	—
48 months	8%	64.09	97.66	128.17
	9%	65.33	99.55	130.65
	10%	—	—	—
60 months	8%	53.23	81.11	106.46
	9%	54.50	83.04	108.99
	10%	53.83	82.00	107.54
72 months	8%	46.03	70.14	92.05
	9%	47.32	72.11	94.64
	10%	46.95	71.54	93.76

Use the amortization table to solve each problem.

1. Calculate the total interest you would pay on a \$4000 loan at 9% for five years?

2. What is the monthly payment on a loan for \$5250 at 8% for 48 months?

3. Calculate the total interest you would pay on a \$2625 loan at 10% for six years.

4. How much would you save if you borrow \$2625 at 8% for six years instead of at 10%?

5. How much would you save if you borrow \$2625 at 10% for five years instead of for six years?

6. Would it be better to borrow \$4000 at 9% for 36 months or to borrow \$4000 at 8% for 48 months?

COOPERATIVE GROUP ACTIVITY

7. Maria borrowed \$5250 at an interest rate of 8% for 4 years to pay for her attendance at a community college. At the end of community college she plans to work for 30 years. Her guidance counselor told her that in her part of the country community college graduates on average earn \$1800 per year more than high school graduates. How much more might Maria expect to earn in 30 years if she graduates from the community college than if she only graduated from high school? If Maria subtracts the \$5250 loan and the interest on that loan, how much more will she have earned in that 30-year period?

3.5 Solve Word Problems Using Geometric Formulas

After studying this section, you will be able to:

1 *Find the area and the perimeter of two-dimensional objects. Find the missing dimension given the area or perimeter and one dimension.*
2 *Find the volume and the surface area of three-dimensional objects. Find a missing dimension.*
3 *Solve more involved geometric problems.*

1 *Area and Perimeter*

We can use our knowledge of geometry to solve word problems. To do this, we first review some geometric facts and formulas. **Perimeter** is the distance around a plane figure. Perimeter measure is in linear units (inches, feet, centimeters, miles). **Area** is a measure of the amount of surface in a region. Area is measured in square units (square inches, square feet, square centimeters).

In our sketches we will show angles of 90° by using a small square (∟ ┘). This indicates that the two lines are at right angles. All angles that measure 90° are called *right angles.* An *altitude* is perpendicular to the base of a figure. That is, the altitude forms right angles with the base. The small corner square in a sketch helps us to identify the altitude of the figure.

The following box provides a handy guide to some facts and formulas you will need to know. Use it as a reference when solving word problems involving geometric figures.

Geometric Formulas: Two-dimensional Figures

A **parallelogram** is a four-sided figure with opposite sides parallel. In a parallelogram, opposite sides are equal and opposite angles are equal.

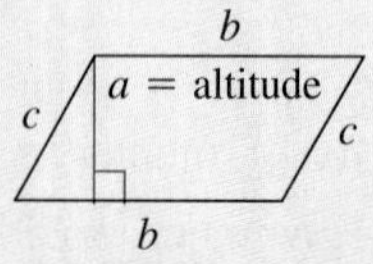

Perimeter = the sum of all four sides

Area = ab

A **rectangle** is a parallelogram with all interior angles measuring 90°.

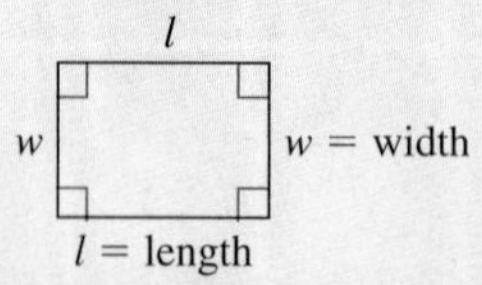

Perimeter = $2l + 2w$

Area = lw

A **square** is a rectangle with all four sides equal.

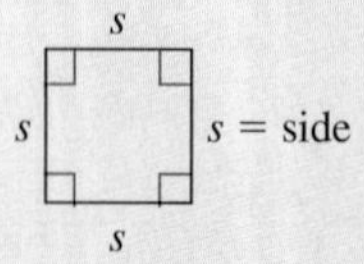

Perimeter = $4s$

Area = s^2

Geometric Formulas: Two-dimensional Figures

A **trapezoid** is a four-sided figure with two sides parallel. The parallel sides are called the bases of the trapezoid.

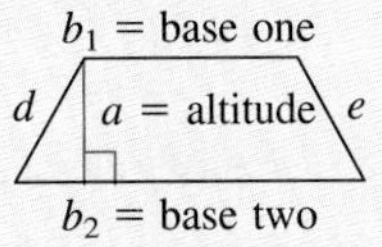

Perimeter = the sum of all four sides

$$\text{Area} = \frac{1}{2}a(b_1 + b_2)$$

A **triangle** is a closed plane figure with three sides.

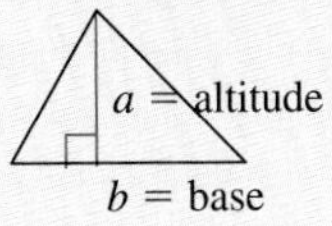

Perimeter = the sum of the three sides

$$\text{Area} = \frac{1}{2}ab$$

A **circle** is a plane curve consisting of all points at an equal distance from a given point called the center.

Circumference is the distance around a circle.

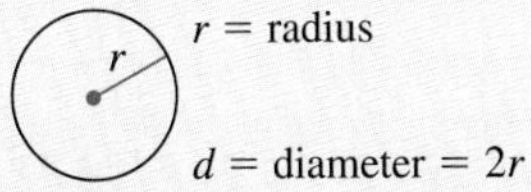

$$\text{Circumference} = 2\pi r$$

$$\text{Area} = \pi r^2$$

π is a constant associated with circles. It is an irrational number that is approximately 3.141592654. We usually use 3.14 as a sufficiently accurate approximation. Thus we write $\pi \approx 3.14$ for most of our calculations involving π.

We frequently encounter triangles in word problems. There are four important facts about triangles, which we list for convenient reference.

Triangle Facts

1. The sum of the interior angles of any triangle is 180°. That is,

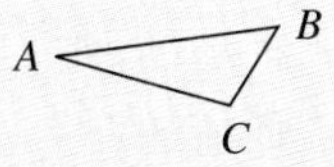

$$\text{measure of } \angle A + \text{measure of } \angle B + \text{measure of } \angle C = 180°$$

2. An **equilateral** triangle is a triangle with three sides equal in length and three angles that measure 60° each.

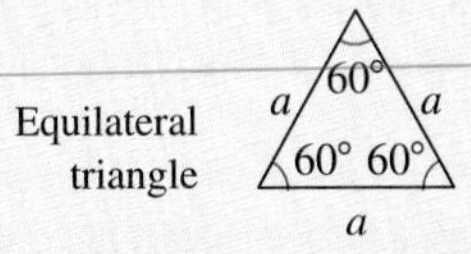

3. An **isosceles** triangle is a triangle with two equal sides. The two angles opposite the equal sides are also equal.

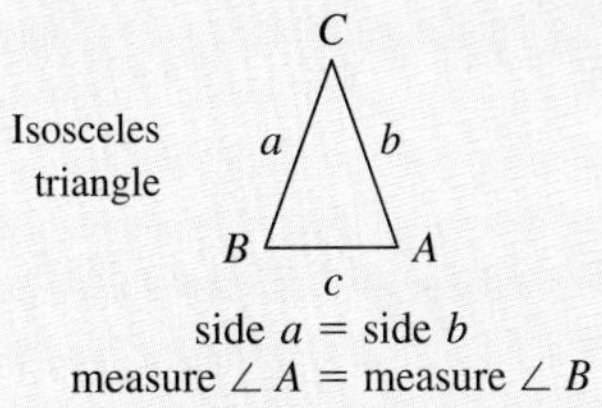

4. A **right** triangle is a triangle with one angle that measures 90°.

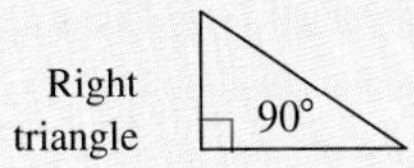

EXAMPLE 1 Find the area of a triangle whose base is 5 centimeters (cm) and whose altitude is 12 cm.

We can use the formula for the area of a triangle. Substitute the known values in the formula and solve for the unknown.

$$A = \frac{1}{2}ab \qquad \textit{Write the formula for the area of a triangle.}$$

$$= \frac{1}{2}(12 \text{ cm})(5 \text{ cm}) \qquad \textit{Substitute the known values in the formula.}$$

$$= \frac{1}{2}(12)(5)(\text{cm})(\text{cm}) \qquad \textit{Simplify.}$$

$$A = 30 \text{ cm}^2 \quad \text{or} \quad 30 \text{ square centimeters}$$

The area of the triangle is 30 square centimeters.

Practice Problem 1 Find the area of a triangle whose base is 14 inches and whose altitude is 20 inches. ■

EXAMPLE 2 The area of a rectangular field is 72 square meters (m^2). Find the width of the field if the length is 9 meters (m).

Draw a diagram.

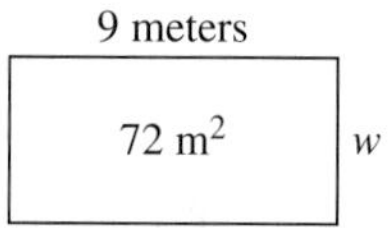

Write the formula for the area of a rectangle and solve for the unknown value.

$$A = lw$$

$$72(\text{m})^2 = (9 \text{ m})(w) \qquad \textit{Substitute the known values into the formula.}$$

$$\frac{72(\text{m})(\text{m})}{9 \text{ m}} = \frac{9 \text{ m}}{9 \text{ m}}w \qquad \textit{Divide both sides by 9 m.}$$

$$8 \text{ m} = w$$

The width of the rectangular field is 8 meters.

Practice Problem 2 The area of a rectangular field is 120 square yards. If the width of the field is 8 yards, what is the length? ■

EXAMPLE 3 The area of a trapezoid is 80 square inches (in.^{2}). The altitude is 20 inches and one of the bases is 3 inches. Find the length of the other base.

$$A = \frac{1}{2}a(b_1 + b_2) \qquad \textit{Write the formula for the area of a trapezoid.}$$

$$80(\text{in.}^2) = \frac{1}{2}(20 \text{ in.})(3 \text{ in.} + b_2) \qquad \textit{Substitute the known values.}$$

$$80(\text{in.})(\text{in.}) = 10 \text{ in.}(3 \text{ in.} + b_2) \qquad \textit{Simplify.}$$

$$= 30(\text{in.})(\text{in.}) + (10 \text{ in.})b_2 \qquad \textit{Subtract 30 in.}^2 \textit{ from both sides.}$$

$$50(\text{in.})(\text{in.}) = (10 \text{ in.})b_2 \qquad \textit{Divide both sides by 10 in.}$$

$$5 \text{ in.} = b_2$$

The other base is 5 inches long.

Practice Problem 3 The area of a trapezoid is 256 square feet. The bases are 12 feet and 20 feet, respectively. Find the length of the altitude. ■

EXAMPLE 4 Find the area of a circle whose diameter is 14 inches (use $\pi = 3.14$). Round to the nearest square inch.

$A = \pi r^2$ *Write the formula for the area of a circle.*

Note that the length of the diameter is given in the word problem. We need to know the length of the radius to use the formula. Since $d = 2r$, then 14 in. $= 2r$ and $r = 7$ in.

$A = \pi(7 \text{ in.})(7 \text{ in.})$ *Substitute known values into the formula.*

$= (3.14)(7 \text{ in.})(7 \text{ in.})$

$= (3.14)(49 \text{ in.}^2)$

$= 153.86 \text{ in.}^2$

Rounded to the nearest square inch, the area of the circle is approximately 159 square inches.

In this example, you were asked to find the area of the circle. Do not confuse the formula for area with the formula for circumference. $A = \pi r^2$, while $C = 2\pi r$. Remember that area involves square units; so it is only natural that in the area formula you would square the radius. $A = \pi r^2$.

Practice Problem 4 Find the circumference of a circle whose radius is 15 meters (use $\pi = 3.14$). Round your answer to the nearest meter. ■

EXAMPLE 5 Find the perimeter of a parallelogram whose longest side is 4 feet and whose shortest side is 2.6 feet.

Draw a picture.

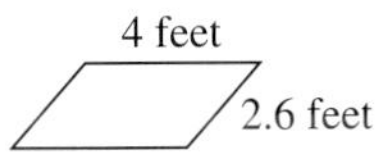

The perimeter is the distance around a figure. To find the perimeter, add the lengths of the sides. Since the opposite sides of a parallelogram are equal, we can write

$P = 2(4 \text{ feet}) + 2(2.6 \text{ feet})$

$= 8 \text{ feet} + 5.2 \text{ feet}$

$= 13.2 \text{ feet}$

The perimeter of the parallelogram is 13.2 feet.

Practice Problem 5 Find the perimeter of an equilateral triangle with a side that measures 15 centimeters. ■

2 *Volume and Surface Area*

Now let's examine three-dimensional figures. **Surface area** is the total area of the faces of the figure. You can find surface area by calculating the area of each face and then finding the sum. **Volume** is the measure of the amount of space inside the figure. Some formulas for the surface area and the volume of regular figures can be found in the table below.

Geometric Formulas: Three-dimensional Figures

Rectangular prism

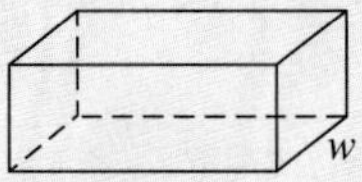

Surface area = $2lw + 2wh + 2lh$

Volume = lwh

Notice that all of the faces are rectangles.

Sphere

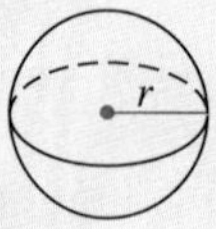

Surface area = $4\pi r^2$

Volume = $\frac{4}{3}\pi r^3$

Right circular cylinder

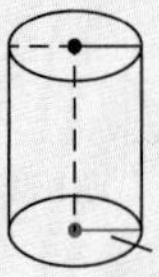

Surface area = $2\pi rh + 2\pi r^2$

Volume = $\pi r^2 h$

EXAMPLE 6 Find the volume of a sphere with radius 4 centimeters (use $\pi = 3.14$). Round your answer to the nearest cubic centimeter.

$V = \frac{4}{3}\pi r^3$ *Write the formula for the volume of a sphere.*

$= \frac{4}{3}(3.14)(4 \text{ cm})^3$ *Substitute the known values into the formula.*

$= \frac{4}{3}(3.14)(64)\text{cm}^3$

$= 267.946667 \text{ cm}^3$

Rounding to the nearest whole number, the volume of the sphere is 268 cubic centimeters.

Practice Problem 6 Find the surface area of a sphere with radius 5 meters. Round your answer to the nearest square meter. ■

EXAMPLE 7 A can is made of aluminum. It has a flat top and a flat bottom. The height is 5 inches and the radius is 2 inches. How much aluminum is needed to make the can? How much aluminum is needed to make 10,000 cans?

Understand the problem.
What do we need to find? Reword the problem.

We need to find the total surface area of a cylinder.

Draw a picture.

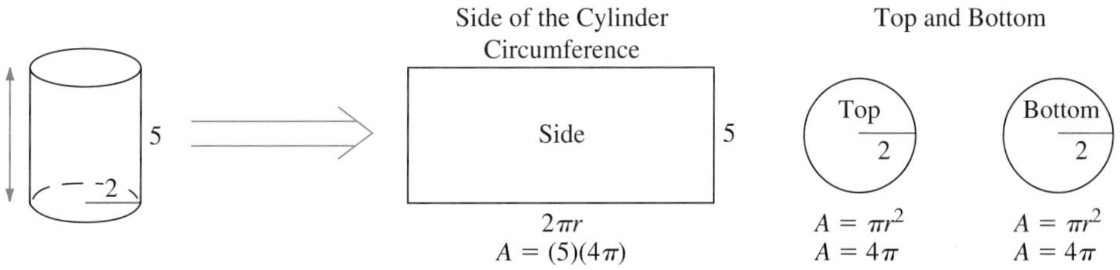

You may calculate the area of each piece and find the sum or you may use the formula.

$$\begin{aligned}\text{Surface area} &= 2\pi rh + 2\pi r^2 \\ &= 2(3.14)(2\text{ in.})(5\text{ in.}) + 2(3.14)(2\text{ in.})^2 \\ &= 62.8\text{ in.}^2 + 25.12\text{ in.}^2 \\ &= 87.92\text{ in.}^2\end{aligned}$$

87.92 square inches of aluminum is needed to make each can.

Have we answered all the questions in the word problem? Reread the problem. How much aluminum is needed to make 10,000 cans?

(aluminum for 1 can)(10,000)

$(87.92\text{ in.}^2)(10{,}000) = 879{,}200$ square inches

It would take 879,200 square inches of aluminum to make 10,000 cans.

Practice Problem 7 Sand is stored in a cylindrical drum that is 4 feet high and has a radius of 3 feet. How much sand can be stored in the drum? ■

3 More Involved Geometric Problems

Problems that at first seem complicated can sometimes be reduced to a combination of several simple problems.

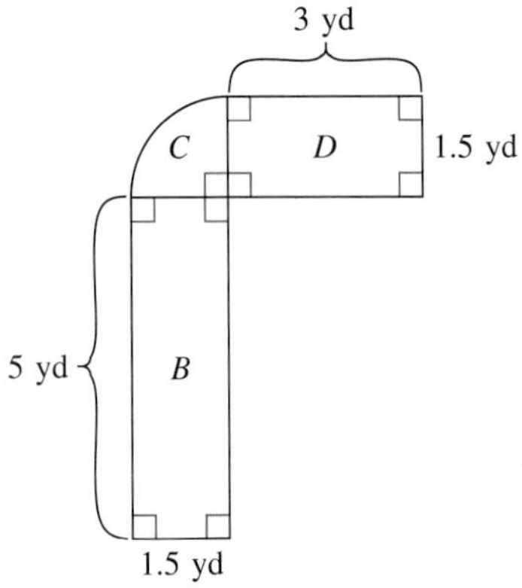

EXAMPLE 8 A quarter-circle (of radius 1.5 yards) is connected to two rectangles with dimensions labeled on the sketch in the margin. You need to lay a strip of carpet in your house according to this sketch. How many square yards of carpeting will be needed on the floor? (Use $\pi \approx 3.14$. Round your final answer to the nearest hundredth.)

The desired area is the sum of three areas, which we will call B, C, and D. Area B and area D are rectangular in shape. It is relatively easy to find the areas of these shapes.

$$A_B = (5 \text{ yd})(1.5 \text{ yd}) = 7.5 \text{ yd}^2 \qquad A_D = (3 \text{ yd})(1.5 \text{ yd}) = 4.5 \text{ yd}^2$$

Area C is one-fourth of a circle. The radius of the circle is 1.5 yd.

$$A_C = \pi\frac{r^2}{4} = \frac{(3.14)(1.5 \text{ yd})^2}{4} = \frac{(3.14)(2.25)\text{yd}^2}{4} = 1.76625 \text{ yd}^2 \approx 1.8 \text{ yd}^2$$

The total area is $A_B + A_D + A_C = 7.5 \text{ yd}^2 + 4.5 \text{ yd}^2 + 1.8 \text{ yd}^2 = 13.8 \text{ yd}^2$

13.8 square yards of carpeting will be needed to cover the floor.

Practice Problem 8 John has a swimming pool that measures 8 feet by 12 feet. He plans to make a concrete walkway around the pool that is 3 feet wide. What will be the total cost of the walkway at \$12 a square foot? ■

EXAMPLE 9 Find the weight of the water in a full cylinder containing water if the height of the cylinder is 5 feet and the radius is 2 feet. 1 cubic foot of water weighs 62.4 pounds. Round your answer to the nearest ten pounds.

Understand the problem.

Draw a diagram.

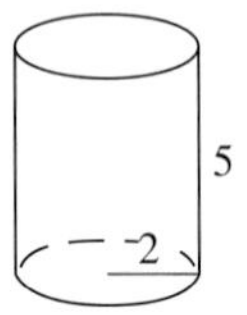

Write the formula for the volume of a cylinder.

$$\begin{aligned} V &= \pi r^2 h \\ &= (3.14)(2 \text{ ft})^2(5 \text{ ft}) \\ &= (3.14)(4 \text{ ft}^2)(5 \text{ ft}) \\ &= 62.8 \text{ ft}^3 \end{aligned}$$

The volume of the cylinder is 62.8 cubic feet.

Each cubic foot of water weighs 62.4 pounds. That is, water weighs 62.4 pounds per cubic foot. This can be written as 62.4 lb/ft³. Since the volume of the cylinder is 62.8 ft³, we multiply.

$$\begin{aligned} \text{Weight of the water} &= 62.8 \cancel{\text{ft}^3} \times \frac{62.4 \text{ lb}}{1 \cancel{\text{ft}^3}} \\ &= (62.8)(62.4) \text{ lb} \\ &= 3918.72 \text{ lb} \end{aligned}$$

The weight of the water in the cylinder is 3920 rounded to the nearest ten pounds.

Practice Problem 9 Find the weight of the water in a rectangular watertight container that is 5 feet wide, 6 feet long, and 8 feet high. Remember, 1 cubic foot of water weighs 62.4 pounds. Round your answer to the nearest 100 pounds. ■

3.5 Exercises

Verbal and Writing Skills

Write the word(s) to complete each sentence.

1. Perimeter is the ______________ a plane figure.

2. ____________ is the distance around a circle.

3. Area is a measure of the amount of ____________ in a region.

4. A ____________ is a four-sided figure with two sides parallel.

5. The sum of the interior angles of any triangle is __________.

Applications

6. Find the area of a triangle whose altitude is 12 inches and whose base is 7 inches.

7. Find the area of a parallelogram whose altitude is 14 inches and whose base is 5 inches.

8. The area of a parallelogram is 345 square meters. The altitude is 15 meters. What is the base?

9. The area of a triangle is 80 square feet. Find the altitude if the base is 16 feet.

10. Find the area of a rectangle whose length is 12 centimeters and whose width is 7 centimeters.

11. Find the length of a rectangle whose area is 225 square inches and whose width is 15 inches. What kind of a rectangle is this? Find the perimeter.

12. Find the area of a circle whose radius is 7.00 feet. (Use $\pi = 3.14$ as an approximate value.)

13. Find the area of a circle whose diameter is 6 meters. (Use $\pi = 3.14$ as an approximate value.)

14. Find the area of a trapezoid with an altitude of 12 meters and whose bases are 5 meters and 7 meters, respectively.

15. The area of a trapezoid is 600 square inches and the bases are 20 inches and 30 inches, respectively. Find the altitude.

16. Find the perimeter of a rectangle whose width is 12 feet and whose length is 18.5 feet.

17. Find the perimeter of a triangle whose sides are 12 meters, 27 meters, and 31 meters, respectively.

18. The perimeter of a rectangle is 88 meters. Find the length of the rectangle if the width is 20 ~~feet.~~ meters

19. Find the circumference of a circle whose diameter is 20 meters. (Use $\pi = 3.14$ as an approximate value.)

X = Lengths

88 = 2x + 2(20)

= 2x + 40

88 − 40 = 2x

48 = 2x

Perimeter: P = 2l + 2w

20. The circumference of a circle is 31.4 centimeters. Find the radius. (Use $\pi = 3.14$ as an approximate value.)

21. The perimeter of an equilateral triangle is 27 inches. Find the length of each side of the triangle.

22. What is the length of one side of a square whose perimeter is 124 centimeters?

23. The perimeter of an isosceles triangle is 32 feet. The base is $\frac{2}{3}$ as long as one of the equal sides. How long is each side of the triangle?

24. Two angles of a triangle measure 34 degrees and 59 degrees, respectively. What is the measure of the third angle?

25. A right triangle has one angle that measures 77 degrees. What does the other angle measure?

26. Each of the equal angles of an isosceles triangle is 4 times as large as the vertex angle. What is the measure of each angle?

★27. In a triangle, the measure of the first angle is twice the measure of the second angle. The measure of the third angle is 20 degrees less than the second angle. What is the measure of each angle?

28. What is the volume of a rectangular solid whose height is 12 centimeters, whose width is 5 centimeters, and whose length is 21 centimeters?

29. What is the volume of a cylinder whose height is 8 inches and whose radius is 10 inches. (Use $\pi = 3.14$ as an approximate value.)

30. The volume of a rectangular solid is 1320 cubic meters. Find the length if the height is 11 meters and the width is 10 meters.

31. The volume of a cylinder is 235.5 cubic feet. Find the height if the radius is 5 feet. (Use $\pi = 3.14$ as an approximate value.)

32. Find the surface area of a sphere with radius 3 feet. (Use $\pi = 3.14$ as an approximate value.)

33. The volume of a rectangular solid is 864 cubic inches. (a) Find the height of the solid if the length is 12 inches and the width is 9 inches. (b) What is the surface area?

34. A plastic cylinder is made with a solid top and bottom. The radius is 6.00 centimeters and the height is 4.00 centimeters. (a) Find the volume of the cylinder. (b) Find the total surface area of the cylinder.

35. A sphere is made to hold liquids in a science lab. The radius of the sphere is 3.00 centimeters. (a) Find the volume of the sphere. (b) Find the total surface area of the sphere.

36. Find the area of the object shown in the figure. The length of the rectangle is 4 inches. The width of the rectangle and the diameter of the small half-circle are both 2 inches. The radius of the large quarter-circle is 2 inches.

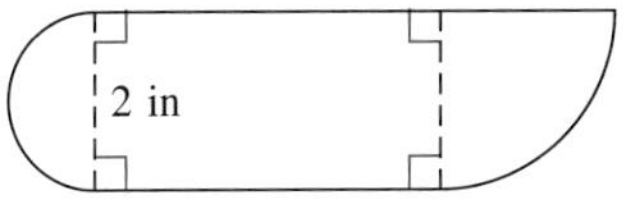

37. An oddly shaped field consists of a triangular region and a quarter-circle. This field must be fertilized. Green Lawn, Inc., will do the job for \$0.40 per square yard. How much will the job cost?

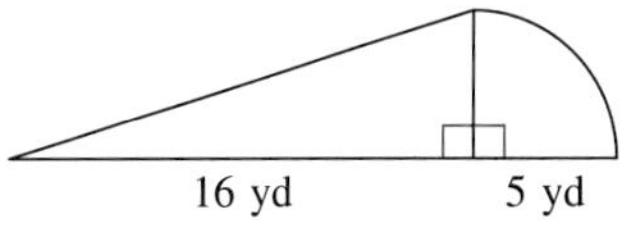

38. A cement walkway is to be poured. It consists of the two rectangles with dimensions as shown in the diagram and a quarter of a circle with radius 1.5 yards. (a) How many square yards will the walkway have? (b) If a painter paints it for \$2.50 per square yard, how much will the painting cost?

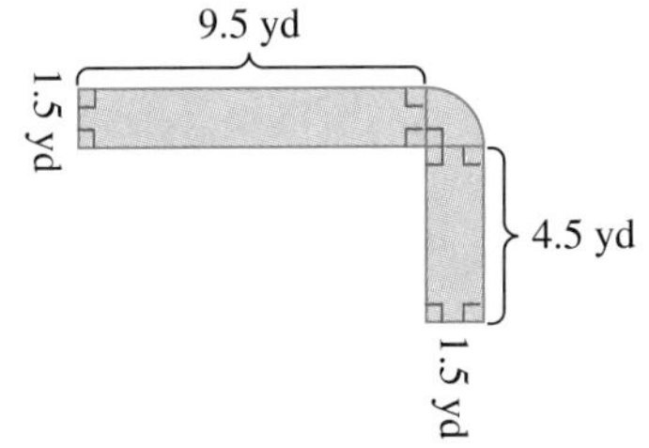

39. An aluminum plate is made according to the dimensions above. The outer radius is 10 inches and the inner radius is 3 inches. (a) How many square inches in area is the aluminum plate (the shaded region in the sketch)? (b) If the cost to construct the plate is \$7.50 per square inch, what is the construction cost?

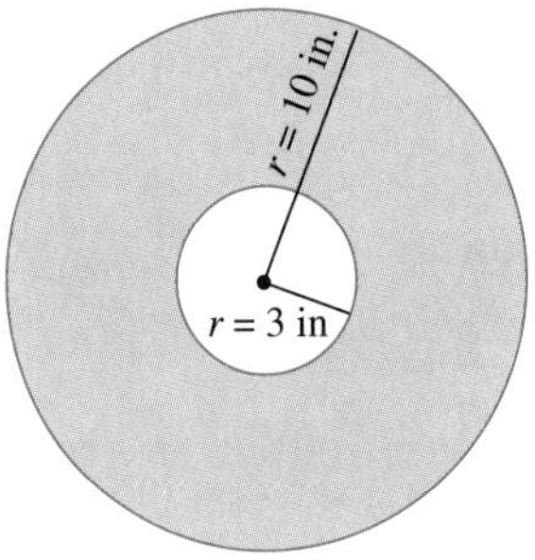

40. It cost \$120.93 to frame a painting with a flat aluminum strip. This was calculated at \$1.39 per foot for materials and labor. Find the dimensions of the painting if its length is 3 feet less than four times its width.

41. A concrete driveway measures 18 yards by 4 yards. The driveway will be 0.5 yards thick. (a) How many cubic yards of concrete will be needed? (b) How much will it cost if concrete is \$3.50 per cubic yard?

To Think About

42. An isosceles triangle has one side that measures 4 feet and another that measures 7.5 feet. The third side was not measured. Can you find one unique perimeter for this type of triangle? What are the two possible perimeters?

43. Find the total surface area of (a) a sphere of radius 3 inches and (b) a right circular cylinder with radius 2 inches and height 0.5 inch. Which object has a greater surface area?

44. Assume that a very long rope is stretched around the moon supported by poles that are 3 *feet tall.* (Neglect gravitational pull and assume that the rope takes the shape of a large circle, as shown in the figure.) The radius of the moon is approximately 1080 miles. How much longer would you need to make the rope if you wanted the rope to be supported by poles that are 4 *feet tall*?

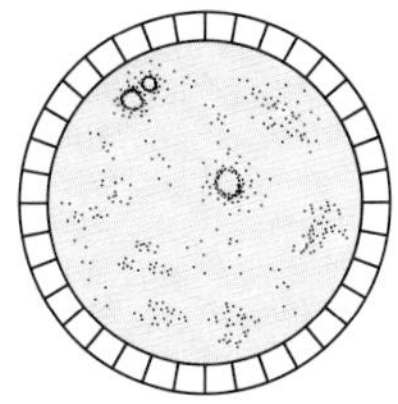

Cumulative Review

Simplify.

45. $2(x - 6) - 3[4 - x(x + 2)]$

46. $-5(x + 3) - 2[4 + 3(x - 1)]$

47. $-\{3 - 2[x - 3(x + 1)]\}$

48. $-2\{x + 3[2 - 4(x - 3)]\}$

Putting Your Skills to Work

PAINTING AND DECORATING

House painters need to be able to calculate surface area in order to determine how much material they will need to do a job and thus determine cost. Joe Lee wants to determine how much he should charge to paint the storage shed below. To help him to calculate the surface area, he draws a diagram of the *unfolded* shed.

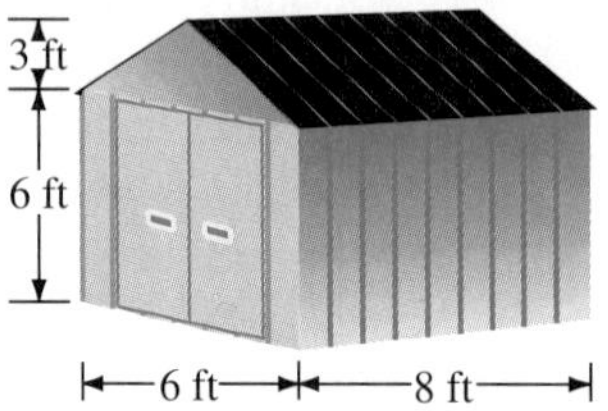

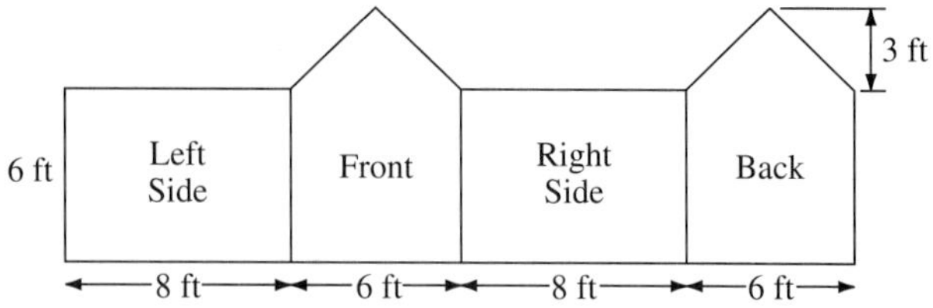

JOB ESTIMATE

The job consists of preparation (filling holes and cracks with putty, sanding the surface smooth, putting on the primer) putting on **two** finishing coats of paint, and clean up.

1. Calculate the total surface area to be painted.

2. The primer costs \$24 a gallon and will cover 300 square feet. The paint costs \$20 a gallon and will cover 320 square feet. Miscellaneous materials will come to \$18. Calculate the total cost in materials.

3. Joe knows that it will take a half a day for preparation and a half a day for each coat of paint and 2 hours for the cleanup. Joe charges \$100 a day or \$15 an hour. Calculate the cost in labor.

4. Joe will charge 20% of the total of items 2 and 3 and add this on to the bill to pay for the fixed expenses associated with running a business. Determine the estimate Joe should quote for this job.

5. Compare your estimate with the estimate of a fellow student. Why might there be differences?

COOPERATIVE GROUP ACTIVITY

Estimate the cost of decorating your classroom. Have one group wallpaper the room, another group may paint the room, and a third group may wallpaper two walls and paint the other two walls. Decide how you want to decorate the room.

6. Take measurements of the room and calculate the surface area. Remember to subtract windows and doors. You will want to treat these separately.

7. Determine the cost of materials. Primer for the paint is \$11 a gallon and will cover 310 square feet. Paint is \$21 a gallon and will cover 380 square feet. If your classroom has not been done for awhile, you may want to give it two coats. Sizing for the wallpaper is \$15 a gallon and will cover 550 square feet. Wallpaper is \$14 a roll, which will cover 25 square feet. Estimate miscellaneous materials at \$25.

8. Decide what coverings you want for the windows and calculate the cost based on their measurements.

3.6 Use Inequalities to Solve Word Problems

After studying this section, you will be able to:

1 *Translate English sentences into inequalities.*
2 *Solve applied problems using inequalities.*

1 *Translating English sentences into Inequalities*

Real-life situations often involve inequalities. For example, if a student wishes to maintain a certain grade point average (GPA), you may hear him or her say, "I have to get a grade of at least an 87 on the final exam." What does this mean? What grade does the student need to achieve? 87 would fit the condition, *at least an 87*. What about an 88? The student would be happy with an 88. In fact the student would be happy with any grade that *is greater than or equal to* an 87. We can write this as

$$\text{grade} \geq 87$$

Notice that the English phrase *at least an 87* translates mathematically as *greater than or equal to an 87*. Be careful when translating English phrases that involve inequalities into algebra. If you are not sure which symbol to use, try a few numbers to see if they fit the conditions of the problem and note the direction they take. Choose the mathematical symbol accordingly.

EXAMPLE 1 Translate each phrase into an inequality.

(a) The perimeter of the rectangle must be no more than 70 inches.

(b) The average number of errors must be less than 2 mistakes per page.

(c) His income must be at least \$950 per week.

(d) The new car will cost at most \$12,070.

(a) Perimeter must be *no more than* 70.

$$\text{Perimeter} \leq 70$$

Try 80. Does 80 fit the condition, no more than 70? No. Do 60 and 50 fit? Yes. Use the less than or equal to symbol.

(b) Average number must be *less than* 2.

$$\text{Average} < 2$$

(c) Income must be *at least* \$950.

$$\text{Income} \geq 950$$

"At least" means, "not less than." The income must be \$950 or more. Use ≥.

(d) The new car will cost *at most* \$12,070.

$$\text{Car} \leq 12{,}070$$

At most \$12,070 means it is the top price for the car. The cost of the car cannot be more than \$12,070. Use ≤.

Practice Problem 1 Translate each phrase into an inequality.

(a) The height of the person must be no greater than 6 feet.

(b) The speed of the car must have been greater than 65 miles per hour.

(c) The area of the room must be greater than or equal to 560 square feet.

(d) The profit margin cannot be less than \$50 per television set. ■

2 *Solving Applied Problems Using Inequalities*

We can use the same problem-solving techniques we used previously to solve word problems involving inequalities.

EXAMPLE 2 A manufacturing company makes 60-watt light bulbs. For every 1000 bulbs manufactured, a random sample of 5 bulbs is selected and tested. To meet quality control standards, the average number of hours these five bulbs last must be at least 900 hours. The test engineer tested 5 bulbs. Four of them lasted 850 hours, 1050 hours, 1000 hours, and 950 hours, respectively. How many hours must the fifth bulb last in order to meet quality control standards?

1. *Understand the problem.*

   Let n = the number of hours the fifth bulb must last

   The average of the five bulbs *must be at least* 900 hours. We could write

   $$\text{Average} \geq 900 \text{ hours}$$

2. *Write an inequality.*

   $$\text{Average of five bulbs} \geq 900 \text{ hours}$$

   $$\frac{850 + 1050 + 1000 + 950 + n}{5} \geq 900$$

3. *Solve the inequality and state the answer.*

   $$\frac{850 + 1050 + 1000 + 950 + n}{5} \geq 900$$

   $$\frac{3850 + n}{5} \geq 900 \quad \textit{Simplify the numerator.}$$

   $$3850 + n \geq 4500 \quad \textit{Multiply both sides by 5.}$$

   $$n \geq 650 \quad \textit{Subtract 3850 from each side.}$$

   If the fifth bulb burns for 650 hours or more, the quality control standards will be met.

4. *Check.* Any value of 650 or larger should satisfy the original inequality. Suppose we pick $n = 650$ and substitute it into the inequality.

   $$\frac{850 + 1050 + 1000 + 950 + 650}{5} \stackrel{?}{\geq} 900$$

   $$\frac{4500}{5} \stackrel{?}{\geq} 900$$

   $$900 \geq 900 \quad \checkmark$$

   If we try $n = 700$, the left side of the inequality will be 910. $910 \geq 900$. In fact, if we try any number greater than 650 for n, the left side of the inequality will be greater than 900. The inequality will be valid for $n = 650$ or any larger value.

Practice Problem 2 A manufacturing company makes coils of $\frac{1}{2}$-inch maritime rope. For every 300 coils of rope made, a random sample of four coils of rope is selected and tested. To meet quality control standards, the average number of pounds that each coil will hold must be at least 1100 pounds. The test engineer tested four coils of rope. The first three held 1050 pounds, 1250 pounds, and 950 pounds, respectively. How many pounds must the last coil hold if the rope is to meet quality control standards? ■

EXAMPLE 3 Juan is selling supplies for restaurants in the Southwest. He gets paid \$700 per month plus 8% of all the supplies he sells to restaurants. If he wants to earn more than \$3100 per month, what value of products will he need to sell?

1. *Understand the problem.*
   Juan earns a fixed salary and a salary that depends on how much he sells (commission salary) each month.

   Let x = the value of the supplies he sells each month

   Then $0.08x$ = the amount of the commission salary he earns each month

   He wants the total income to be more than \$3100 per month; thus

   Total income > 3100

2. *Write the inequality.*

   The fixed income added to the commission income must be greater than 3100

   $$700 \quad + \quad 0.08x \quad > \quad 3100$$

3. *Solve the inequality and state the answer.*

   $$700 + 0.08x > 3100$$

   $$0.08x > 2400 \qquad \textit{Subtract } 700 \textit{ from each side.}$$

   $$x > 30{,}000 \qquad \textit{Divide each side by } 0.08.$$

   Juan must sell more than \$30,000 worth of supplies each month.

4. *Check.* Any value greater than 30,000 should satisfy the original inequality. Suppose we pick $x = 30{,}001$ and substitute it into the inequality.

   $$700 + 0.08(30{,}001) \stackrel{?}{>} 3100$$

   $$700 + 2400.08 \stackrel{?}{>} 3100$$

   $$3100.09 > 3100 \checkmark$$

Practice Problem 3 Rita is selling commercial fire and theft alarm systems to local companies. She gets paid \$1400 per month plus 2% of the cost of all the systems she sells in one month. She wants to earn more than \$2200 per month. What value of products will she need to sell each month? ■

3.6 Exercises

Verbal and Writing Skills

Translate to an inequality.

1. The cost is greater than $67,000.

2. The clearance of the bridge is less than 12 feet.

3. The number of people is not more than 120.

4. The tax increase is more than $7890.

5. The perimeter is less than 34 miles.

6. The area is not greater than 550 square yards.

7. His earnings cannot be less than $3500 per month.

8. The number of passengers in the plane is not less than 26 people.

9. When you bake the cake, the temperature cannot exceed 350 degrees Fahrenheit.

10. The A student will never exceed four unexcused absences per semester.

Applications

Solve each of the following applied problems by using an inequality.

11. Two sides of a triagular field are 180 feet and 135 feet. A farmer has available 539 feet of fencing. If the farmer wants to place a fence around the outside of the field with his fencing, what are the possible values for the length of the third side of the field?

12. A triangular-shaped window on the side of a boat has two sides that measure 87 centimeters and 64 centimeters, respectively. The perimeter of the triangle must not exceed 291 centimeters. What are the possible values for the length of the third side of the window?

13. A student has scored 85, 77, and 68 on three tests in her biology course. She has one final test to take. She wants to obtain an average of 80 or more in the course. What possible values can she obtain on her final test and still achieve her goal?

14. A video rental store needs to average 48 or more customers per day for six days each week in order to make a profit. On the first five days the following number of customers came into the store: 60, 36, 40, 52, and 59. What possible number of customers can come into the rental store on the last day to achieve the customer goal for the week?

15. A businesswoman has a limit of $655 per month in travel and entertainment expenses. She spends monthly $230 in travel expenses. She often takes potential customers out to lunch. The average business lunch for two people is $25. How many possible clients can she take to lunch?

16. Carlos drives a cab part-time. He gets paid $100 per week plus tips. His average tip is $3 per taxi trip. He needs to earn an average of at least $217 per week to meet his college expenses. How many possible taxi trips can he take per week and meet his expense requirements?

17. A builder is constructing a house with a living room that is 12 feet wide. What are the possible lengths of the living room if the area of the room must be greater than or equal to 204 square feet.

18. A builder is constructing a house with a master bedroom that is 11 feet wide. What are the possible lengths of the bedroom if the area of the room must be greater than or equal to 176 square feet?

19. John is selling cars at Centerville Chrysler. He is getting paid \$850 per month plus \$130 for every car he sells during the month. He wants to earn at least \$2800 per month. How many cars will he need to sell each month?

20. Marcia is selling cars at the Beverly Jeep Center. She is getting paid \$920 per month plus \$120 for every car or wagon she sells during the month. she wants to earn at least \$2600 per month. How many cars and wagons will she need to sell to meet that goal?

21. Melissa is selling supplies for MidCentral Medical Service. She gets paid \$900 per month plus 3% of the sales price of all the medical supplies she sells. If she wants to earn more than \$2700 per month, what value of products will she need to sell?

22. Jerome is selling supplies for Commercial Photography Centers. He gets paid \$600 per month plus 4% of the sales price of all the supplies he sells. If he wants to earn more than \$2500 per month, what value of products will he need to sell?

23. The coolant in Tom's car is effective for radiator temperatures of less than 110 degrees Celsius. What would be the corresponding temperature range in degrees Fahrenheit? (Use $C = \frac{5}{9}(F - 32)$ in your inequality.)

24. A person's IQ is found by multiplying mental age (m), as indicated by standard tests, by 100 and dividing this result by the chronological age (c). This gives the formula IQ $= 100m/c$. If a group of 14-year-old children has an IQ of not less than 80, what is the mental age range of this group?

25. Dr. Slater is purchasing computer tables and chairs for the tutoring center at the college. The Dean has told him that he cannot spend more than \$850. Computer tables will cost him \$55 and chairs will cost him \$28. He has decided to purchase nine computer tables. What are the different possible numbers of chairs that he can buy?

26. Franklin and Roberto are planning a pizza party for their friends. They have decided they cannot spend more than \$88 for food. Each large pizza costs \$11.50 at their favorite pizza parlor. They purchase six-packs of cold soda for \$3.50 each. They have decided to buy six large pizzas. What are the different possible numbers of six-packs of cold soda that they can buy?

★ **27.** A compact disc recording company has a cost equation for the week of cost $= 5000 + 7n$, where n is the number of compact discs manufactured in one week. The income equation for the amount of income produced from selling these n discs is income $= 18n$. How many discs need to be manufactured and sold in one week for the income to be greater than the cost?

★ **28.** A company that manufactures VCRs has a cost equation for the week of cost $= 12{,}000 + 100n$, where n is the number of VCRs manufactured in one week. The income equation for the amount of income produced from selling these n VCRs is income $= 210n$. How many VCRs need to be manufactured and sold in one week for the income to be greater than the cost?

Cumulative Review Problems

Solve each inequality.

29. $10 - 3x > 14 - 2x$

30. $2x - 3 \geq 7x - 18$

31. $30 - 2(x + 1) \leq 4x$

32. $2(x + 3) - 22 < 4(x - 2)$

Chapter Organizer

Procedure for Solving Applied Problems

1. **Understand the problem.**
   (a) Read the word problem carefully to get an overview.
   (b) Determine what information you will need to solve the problem.
   (c) Draw a sketch. Label it with the known information. Determine what needs to be found.
   (d) Choose a variable to represent one unknown quantity.
   (e) If necessary, represent other unknown quantities in terms of that same variable.
2. **Write an equation.**
   (a) Look for key words to help you to translate the words into algebraic symbols.
   (b) Use a given relationship in the problems or an appropriate formula in order to write an equation.
3. **Solve the equation and state the answer.**
4. **Check.**
   (a) Check the solution in the original equation.
   (b) Be sure the solution to the equation answers the question in the word problem. You may need to do some additional calculations if it does not.

Example A

A. The perimeter of a rectangle is 126 meters. The length of the rectangle is 6 meters less than double the width. Find the dimensions of the rectangle.

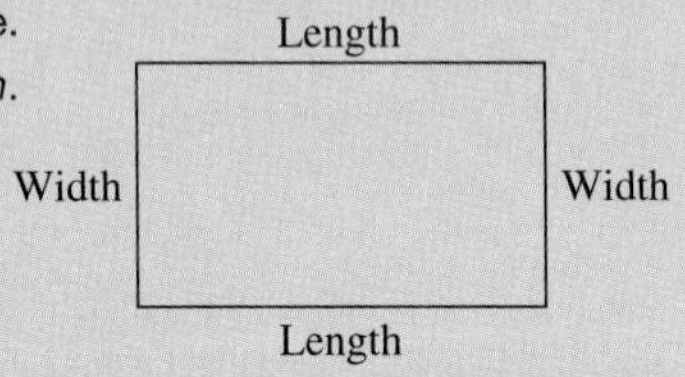

1. *Understand the problem.* We want to find the length and the width of a rectangle whose perimeter is 126 meters.

   The length is *compared to* the width, so we start with the width. — Let w = width

   The length is 6 meters less than double the width. — Let $l = 2w - 6$

2. *Write an equation.* The perimeter of a rectangle is $P = 2w + 2l$. — $126 = 2w + 2(2w - 6)$

3. *Solve the equation.*

$$126 = 2w + 4w - 12$$
$$126 = 6w - 12$$
$$138 = 6w$$
$$23 = w$$

The width is 23 meters.

$2w - 6 = 2(23) - 6 = 46 - 6 = 40$

The length is 40 meters.

4. *Check:* Is this reasonable? Yes. A rectangle 23 meters wide and 40 meters long seems to be about right for the perimeter to be 126 meters.

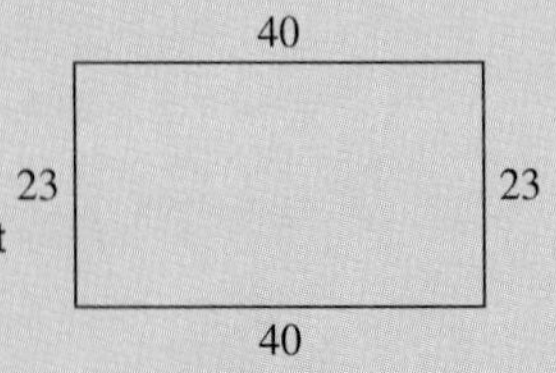

Is the perimeter exactly 126 meters?

$$2(23) + 2(40) \stackrel{?}{=} 126$$
$$46 + 80 \stackrel{?}{=} 126$$

Yes $126 = 126$ ✓

Is the length exactly 6 meters less than double the width?

$$40 \stackrel{?}{=} 2(23) - 6$$
$$40 \stackrel{?}{=} 46 - 6$$

Yes. $40 = 40$ ✓

Example B

B. Gina saved some money for college. For one year she invested $2400 and earned $225 in interest. She invested part of it at 12% and the rest of it at 9%. How much did she invest at each amount?

1. *Understand the problem.* We want to find each amount that was invested.

   Interest earned at 12% + interest earned at 9% = total interest of $225

   We let x represent one quantity of money. — Let x = amount of money invested at 12%

   We started with $2400. If we invest x at 12%, we still have $(2400 - x)$ left. — Let $2400 - x$ = the amount of money invested at 9%

   Interest = prt

   Interest at 12% — $I_1 = 0.12x$

   Interest at 9% — $I_2 = 0.09(2400 - x)$

2. *Now write the equation.* $I_1 + I_2 = 225$. — $0.12x + 0.09(2400 - x) = 225$

3. *Solve the equation.*

$$0.12x + 216 - 0.09x = 225$$
$$0.03x + 216 = 225$$
$$0.03x = 9$$
$$x = \frac{9.00}{0.03}$$
$$= 300$$

$300 was invested at 12%

$2400 - x = 2400 - 300 = 2100$

$2100 was invested at 9%

4. *Check:* Is this reasonable? Yes.

   Are the conditions of the problem satisfied? Does the total amount invested equal $2400?

$$2100 + 300 \stackrel{?}{=} 2400$$

Yes. $2400 = 2400$ ✓

Will $2100 at 9% and $300 at 12% yield $225 in interest?

$$0.09(2100) + 0.12(300) \stackrel{?}{=} 225$$
$$189 + 36 \stackrel{?}{=} 225$$

Yes. $225 = 225$ ✓

Topic	Procedure	Examples
Geometric formulas, p. 214.	Know the definitions of and formulas related to basic geometric shapes such as the triangle, rectangle, parallelogram, trapezoid, and circle. Review these formulas, from the beginning of Section 3.5 as needed. Formulas for the volume and total surface area of three solids are also given there.	The volume of a right circular cylinder found in a four-cylinder car engine is determined by the formula $V = \pi r^2 h$. Find V if $r = 6$ centimeters and $h = 5$ centimeters. $V = 3.14(6\text{ cm})^2(5\text{ cm})$ $= (3.14)(36)(5)\text{ cm}^3 = (113.04)(5)\text{ cm}^3$ $= 565.2$ cc (or cm^3) Four such cylinders would make this a 2260-cc engine.
Using inequalities to solve word problems, p. 225.	Be sure that the inequality you use fits the conditions of the word problem. **1.** At least an 87 means ≥ 87 **2.** No less than \$150 means $\geq \$150$ **3.** At most \$500 means $\leq \$500$ **4.** No more than 70 means ≤ 70 **5.** Less than 2 means <2 **6.** More than 10 means >10	Julie earns 2% on sales over \$1000. How much must Julie make in sales to earn at least \$300? Let x = amount of sales $(2\%)(x - \$1000) \geq \$300$ $0.02(x - 1000) \geq 300$ $x - 1000 \geq 15{,}000$ $x \geq 16{,}000$ Julie must make sales of \$16,000 or more (at least \$16,000).

Chapter 3 Review Problems

3.1 *Write an algebraic expression. Use the variable x to represent the unknown value.*

1. 19 more than a number

2. Two-thirds of a number

3. 56 less than a number

4. Triple a number

5. Double a number increased by seven

6. Twice a number decreased by three

7. Five times a number decreased by one-third of that same number

8. One-fourth of the sum of a number and six

Write an expression for each of the quantities compared. Use the letter specified.

9. The speed of the jet was 370 miles per hour faster than the speed of the helicopter. (Use the letter h.)

10. The cost of the truck was \$2000 less than the cost of the garage. (Use the letter g.)

11. The length of the rectangle is 5 meters more than triple the width. (Use the letter w.)

12. The number of working people is four times the number of retired people. The number of unemployed people is one-half the number of retired people. (Use the letter r.)

13. A triangle has three angles A, B, and C. The number of degrees in angle A is double the number of degrees in angle B. The number of degrees in angle C is 17 degrees less than the number in angle B. (Use the letter b.)

14. There are 29 more students in biology class than in algebra. There are one-half as many students in geology class as in algebra. (Use the letter a.)

3.2 *Solve each word problem.*

15. Twice a number is increased by five. The result is 29. What is the original number?

16. Triple a number is decreased by seven. The result is 47. What is the original number?

17. When one-fourth of a number is diminished by two the result is nine. Find the original number.

18. Seven times a number diminished by three is equal to four times the original number. Find the original number.

19. Jon is twice as old as his cousin David. If Jon is 32, how old is David?

20. Six chairs and a table cost \$450. If the table cost \$210, how much does one chair cost?

21. Two cars left Center City and drove to the shore 330 miles away. One car was driven at 50 miles an hour. The other car was driven at 55 miles an hour. How long did it take each car to make the trip?

22. A student took five math tests. She obtained scores of 85, 78, 65, 92, and 70. She needs to complete one more test. If she wants to get an average of 80 on all six tests, what score does she need on her last test?

3.3

23. A rectangle has a perimeter of 66 meters. The length of the rectangle is 3 meters longer than double the width. Find the dimensions of the rectangle.

24. The length of a rectangle is $2\frac{1}{2}$ times its width. The rectangle has a perimeter of 84 feet. Find the length and width of the rectangle.

25. The perimeter of a triangle is 40 yards. The second side is 1 yard shorter than double the length of the first side. The third side is 9 yards longer than the first side. Find the length of each side.

26. The measure of the second angle of a triangle is three times the measure of the first angle. The measure of the third angle is 12 degrees less than twice the measure of the first. Find the measure of each of the three angles.

27. A piece of rope 50 yards long is cut into two pieces. One piece is three-fifths as long as the other. Find the length of each piece.

28. Jon and Lauren have a combined income of \$48,000. Lauren earns \$3000 more than half of what Jon earns. How much do each earn?

3.4

29. Abel rented a car from Sunshine Car Rentals. He was charged \$39 per day and \$0.25 per mile. He rented the car for 3 days and paid a rental cost of \$187. How many miles did he drive the car?

30. The electric bill at Jane's house this month was \$71.50. The charge is based on a flat rate of \$25 per month plus a charge of \$0.15 per kilowatt-hour of electricity used. How many kilowatt-hours of electricity were used?

31. Jamie bought a new tape deck for her car. When she bought it, she found that the price had been decreased by 18%. She was able to buy the tape deck for \$36 less than if the item had not been on sale. What was the original price?

32. Alfred has worked for his company for 10 years as a store manager. Last year he received a 9% raise in salary. He now earns \$24,525 per year. What was his salary last year before the raise?

33. Peter and Shelly invested \$9000 for one year. Part of it was invested at 12% and the remainder was invested at 8%. At the end of one year the couple had earned exactly \$1000 in interest. How much did they invest at each rate?

34. A man invests \$5000 in a savings bank. He places part of it in a checking account that earns 4.5% and the rest in a regular savings account that earns 6%. His total annual income from this investment is \$270. How much was invested in each account?

35. Mary has \$3.75 in nickels, dimes, and quarters. She has three more quarters than dimes. She has twice as many nickels as quarters. How many of each coin does she have?

36. Tim brought in \$3.65 in coins. He had two more quarters than he had dimes. He had one fewer nickel than the number of dimes. How many coins of each type did he have?

3.5 *Solve each word problem.*

37. Find the diameter of a circle whose circumference is 25.12 meters. (Use $\pi \approx 3.14$.)

38. What is the perimeter of a rectangle whose length is 8 inches and whose width is $\frac{1}{4}$ inch?

39. Find the third angle of a triangle if one angle measures 62 degrees and a second angle measures 47 degrees.

40. Find the area of a triangle whose base is 8 miles and whose altitude is 10.5 miles.

41. Find the perimeter of a triangle whose smallest side is 8 inches and whose second side is double the first and whose third side is double the second.

42. Find the area of a parallelogram whose altitude is 8 centimeters and whose base is 21 centimeters.

43. Find the volume of a rectangular solid whose width is 8 feet, whose length is 12 feet, and whose height is 20 feet.

44. Find the volume of a sphere if the radius is 3 centimeters. (Use $\pi \approx 3.14$.)

45. The perimeter of an isosceles triangle is 46 inches. The two equal sides are each 17 inches long. How long is the other side?

46. Find the area of a circle whose diameter is 18.00 centimeters. (Use $\pi \approx 3.14$.)

47. Find the cost to build a cylinder out of aluminum if the radius is 5 meters and the height is 14 meters and the cylinder will have a flat top and flat bottom. The cost factor has been determined to be \$40.00 per square meter. (*Hint:* First find the total surface area.)

48. Fred wants to have the exterior of his house painted. Two sides of the house have a height of 18 feet and a length of 22 feet, and the other two sides of the house have a height of 18 feet and a length of 19 feet. The painter has quoted a price of \$2.50 per square foot. The windows and doors of the house take up 100 square feet. How much will the painter charge to paint the walls?

3.6 *Solve.*

49. Bruce scored 92, 85, 78, and 86 on his math tests. What must he score on his last test to obtain an 85 or more for the course?

50. Diane has more than \$10 in dimes and quarters. She has 10 times as many dimes as quarters. At least how many of each kind of coin does she have?

51. Chelsea rents a car for \$35 a day and 15 cents a mile. How far can she drive if she wants to rent the car for 2 days and does not want to spend more than \$100?

52. Philip earns \$12,000 per year in salary and 4% commission on his sales. How much were his sales if his annual income was more than \$24,000?

53. Michael plans to spend less than \$70 on shirts and ties. He bought three shirts for \$17.95 each. At most, how much can he spend for ties?

54. Two cars start from the same point going in opposite directions. One car travels at 50 miles per hour. The other car travels at 55 miles per hour. How long must they travel to be at least 315 miles apart?

Miscellaneous

Solve the following miscellaneous applied problems.

55. Fred timed a popular TV show that was scheduled for 30 minutes. The entertainment portion of the show lasted 4 minutes longer than four times the amount of time that was devoted to commercials. How many minutes long was the actual entertainment portion of the show?

56. Faye conducted an experiment. To verify the weight of four identical steel balls, she placed them on a balance. She placed three of the balls on one side of the balance. On the other side she placed an 8-ounce weight and one of the balls. She discovered that these two sides exactly balanced on the balance beam. How much does each steel ball weigh?

57. Susan borrowed \$333 from her uncle to help purchase a car. She promised to pay him back with an initial payment of \$45 and then \$18 per month until the loan is paid back. How many payments will it take to pay off the loan?

58. The college baseball team has won seven of their first nine games. They will play an additional 16 games. The coach desires that they win at least 84% of their games in order to be eligible for the regional playoffs. How many of the 16 games will they need to win?

Putting Your Skills to Work

During a winter storm, a fishing boat develops a leak and the ship's pumps fail. The sinking boat has made emergency radio contact with the Coast Guard. The captain reports that he is 100 miles from the coastal harbor station and that his ship is leaking. The boat is able to head directly toward the harbor, proceeding slowly at 5 miles per hour. The captain warns that he can stay afloat only for 1 more hour. A rescue helicopter leaves the Coast Guard harbor station 5 minutes ($\frac{1}{12}$ hour) later. The helicopter is flying at 120 miles per hour. Will the helicopter reach the boat before it sinks?

We use the formula $d = rt$ for both.

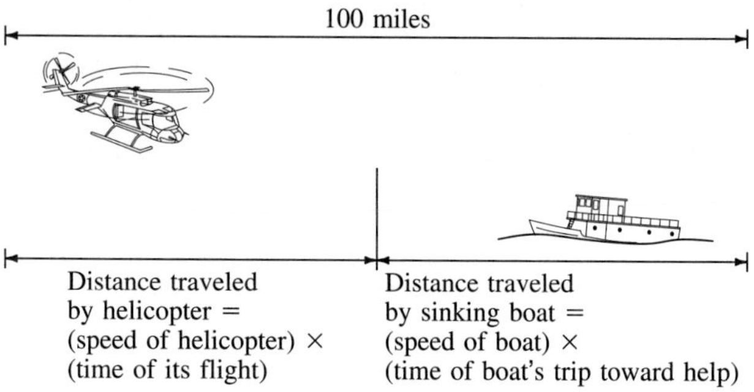

Let t = the time in hours of the boat traveling (starting with the time of the distress call)

$t - \dfrac{1}{12}$ = the time in hours that the helicopter flies to reach the boat (Since the helicopter left $\frac{1}{12}$ of an hour after the distress call the time is $t - \frac{1}{12}$ hours).

Distance traveled by helicopter + distance traveled by boat = 100 miles

$$(r_{\text{helicopter}})(t_{\text{helicopter}}) \quad + \quad (r_{\text{boat}})(t_{\text{boat}}) \quad = 100$$

$$120\left(t - \frac{1}{12}\right) + (5)(t) = 100 \qquad \textit{Substitute rate and time of both boat and helicopter.}$$

$$120t - 10 + 5t = 100 \qquad \textit{Simplify.}$$

$$125t - 10 = 100 \qquad \textit{Collect like terms.}$$

$$125t = 110 \qquad \textit{Add 10 to both sides.}$$

$$t = \frac{110}{125} = \frac{22}{25} \qquad \textit{Divide both sides by 125.}$$

The time the boat travels before the helicopter arrives is $\frac{22}{25}$ hour or about 52.8 minutes. The helicopter will reach the boat with approximately 7.2 minutes to spare!

1. A sinking boat is 76 miles from a Coast Guard station. The boat is traveling toward the Coast Guard station battling heavy seas. It is only able to move toward the station at 2 miles per hour. At the same time a rescue helicopter leaves the Coast Guard station and flies directly toward the boat at 150 miles per hour. If the boat does not sink, how long will it take the helicopter to reach the boat?

2. When the helicopter arrives, the boat has already sunk. No survivors or debris from the boat are visible to the pilot of the helicopter. The Coast Guard orders a full search by eight available rescue helicopters. The search area is defined by the trapezoid shown in the following sketch.

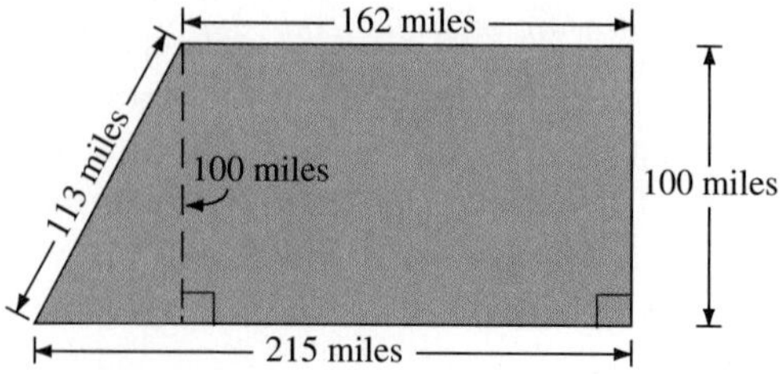

(a) How many square miles are in the search area?

(b) If a helicopter can search 15 square miles in an hour, how long will it take the eight helicopters to search the entire area for signs of the boat or its crew?

Solve each applied problem.

1. A number is doubled and then decreased by 11. The result is 59. What is the original number?

2. The sum of one-half of a number, one-ninth of a number, and one-twelfth of a number is twenty-five. Find the original number.

3. Triple a number is increased by six. The result is the same as when the original number was diminished by three and then doubled. Find the original number.

4. A triangular region has a perimeter of 66 meters. The first side is two-thirds of the second side. The third side is 14 meters shorter than the second side. What are the lengths of the three sides of the triangular region?

5. A rectangle has a length 7 meters longer than double the width. The perimeter is 134 meters. Find the dimensions of the rectangle.

6. Three harmful pollutants were measured by a consumer group in the city. The total sample contained 15 parts per million of three harmful pollutants. The amount of the first pollutant was double the second. The amount of the third pollutant was 75% of the second. How many parts per million of each pollutant were found?

7. Raymond has a budget of \$1940 to rent a computer for his company office. The computer company he wants to rent from charges \$200 for installation and service as a one-time fee. Then they charge \$116 per month rental for the computer. How many months will Raymond be able to rent a computer with this budget?

8. Last year the yearly tuition at King's College went up 8%. This year's charge for tuition for the year is \$9180. What was it last year before the increase went into effect?

1. ____________

2. ____________

3. ____________

4. ____________

5. ____________

6. ____________

7. ____________

8. ____________

9. ______

10. ______

11. ______

12. ______

13. ______

14. ______

15. ______

16. ______

17. ______

9. Franco invested \$4000 in money market funds. Part was invested at 14% interest, the rest at 11% interest. At the end of each year the fund company pays interest. After one year he earned \$482 in interest. How much was invested at each interest rate?

10. Mary has \$3.50 in change. She has twice as many nickels as quarters. She has one less dime than she has quarters. How many of each coin does she have?

11. Find the circumference of a circle with radius of 22.0 inches. Use $\pi = 3.14$ as an approximation.

12. Find the area of a trapezoid if the two bases are 10 inches and 14 inches, respectively, and the height is 16 inches.

13. Find the volume of a sphere with radius 10 inches. Use $\pi = 3.14$ as an approximation and round your answer to the nearest cubic inch.

14. Find the area of a parallelogram with a base of 10 centimeters and an altitude of 7 centimeters.

15. How much would it cost to carpet the area shown in the figure if carpeting will cost \$12 per square yard?

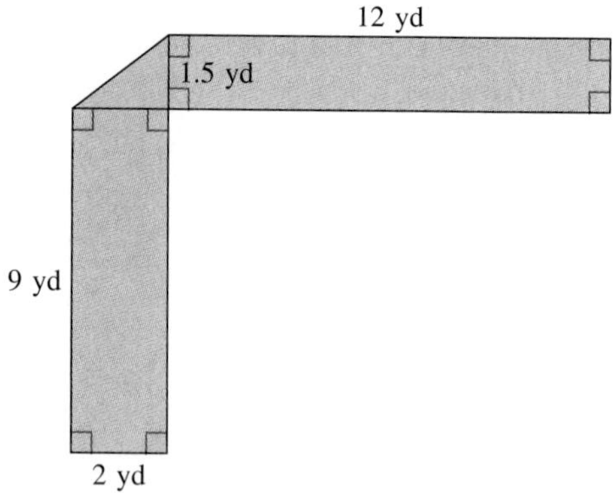

16. Cathy scored 76, 84, and 78 on three tests. How much must she score on the next test to have at least an 80 average?

17. Carol earns \$15,000 plus 5% commission on sales over \$10,000. How much must she make in sales to earn more than \$20,000?

Cumulative Test for Chapters 0–3

Approximately one-half of this test covers the content of Chapters 0–2. The remainder covers the content of Chapter 3.

1. Divide. $2.4\overline{)8.856}$

2. Add. $\frac{3}{8} + \frac{5}{12} + \frac{1}{2}$

3. Simplify. $3(2x - 3y + 6) + 2(-3x - 7y)$

4. Evaluate. $5 - 2(3 - 7) + 15$

5. Solve for b. $H = \frac{1}{2}(3a + 5b)$

6. Solve for x and graph your solution. $5x - 3 \leq 2(4x + 1) + 4$

7. A number is first tripled and then decreased by 17. The result is 103. Find the original number.

8. The number of students in Psychology class is 34 fewer than the number of students in World History. The total enrollment for the two classes is 134 students. How many students are in each class?

1. __________

2. __________

3. __________

4. __________

5. __________

6. __________

7. __________

8. __________

9. A rectangle has a perimeter of 78 centimeters. The length of the rectangle is 11 centimeters longer than triple the width. Find the dimensions of the rectangle.

10. This year the sales for Homegrown Video are 35% higher than last year. This year the sales are $182,250. What were the sales last year?

11. Hassan invested $7000 for one year. He invested part of it in a high-risk fund that pays 15% interest. He placed the rest in a safe, low-risk fund that pays 7% interest. At the end of one year he earned $730. How much did he invest in each of the two funds?

12. Linda has some dimes, nickels, and quarters in her purse. The total value of these coins is $2.55. She has three times as many dimes as quarters. She has three more nickels than quarters. How many coins of each type does she have?

13. Find the area of a triangle with an altitude of 13 meters and a base of 25 meters. What is the cost to construct such a triangle out of sheet metal that costs $4.50 per square meter?

14. Find the volume of a sphere that has a radius of 3.00 inches. How much will the contents of the sphere weigh if it is filled with a liquid that weighs 1.50 pounds per cubic inch?

9. ______

10. ______

11. ______

12. ______

13. ______

14. ______

4
Exponents and Polynomials

Our knowledge of exponents allows us to use scientific notation to describe very large or very small quantities. Calculating the time for the Voyager Space Probe to travel to the outer regions of the solar system requires calculations involving scientific notation. Calculations can be made for even longer distances in space using similar procedures. See Exercise 4.2, Problems 43 and 44.

Pretest Chapter 4

1.		16.	
2.		17.	
3.		18.	
4.		19.	
5.		20.	
6.		21.	
7.		22.	
8.		23.	
9.		24.	
10.		25.	
11.		26.	
12.		27.	
13.		28.	
14.		29.	
15.			

If you are familiar with the topics in this chapter, take this test now. Check your answers with those in the back of the book. If an answer was wrong or you couldn't do a problem, study the appropriate section of the chapter.

If you are not familiar with the topics in this chapter, don't take this test now. Instead, study the examples, work the practice problems, and then take the test.

This test will help you to identify those concepts that you have mastered and those that need more study.

Section 4.1

Multiply and leave your answer in exponent form.

1. $(3^6)(3^{10})$ **2.** $(-5x^3)(2x^4)$ **3.** $(-4ab^2)(2a^3b)(3ab)$

Simplify each fraction. Assume that all variables are nonzero.

4. $\dfrac{x^{26}}{x^3}$ **5.** $\dfrac{12xy^2}{-6x^3y^4}$ **6.** $\dfrac{-25a^0b^2c^3}{-15abc^2}$

Simplify. Assume that all variables are nonzero.

7. $(x^5)^{10}$ **8.** $(-2x^2y)^3$ **9.** $\left(\dfrac{3ab^2}{c^3}\right)^2$

Section 4.2

Express with positive exponents.

10. $3x^2y^{-3}z^{-4}$ **11.** $\dfrac{-2a^{-3}}{b^{-2}c^4}$

12. Evaluate. $(-3)^{-2}$

13. Write in scientific notation. 0.000638 **14.** Write in decimal notation. 1.894×10^{12}

Section 4.3

Combine the following polynomials.

15. $(3x^2 - 6x - 9) + (-5x^2 + 13x - 20)$ **16.** $(2x^3 - 5x^2 + 7x) - (x^3 - 4x + 12)$

Section 4.4

Multiply.

17. $-3x(2 - 7x)$ **18.** $2x^2(x^2 + 4x - 1)$ **19.** $(x^3 - 3xy + y^2)(5xy^2)$

20. $(x + 5)(x + 9)$ **21.** $(3x - 2y)(4x + 3y)$ **22.** $(2x^2 - y^2)(x^2 - 3y^2)$

Section 4.5

Multiply.

23. $(8x - 11y)(8x + 11y)$ **24.** $(6x + 3y)^2$ **25.** $(5ab - 6)^2$

26. $(x^2 - 3x + 2)(4x - 1)$ **27.** $(2x - 3)(2x + 3)(4x^2 + 9)$

Section 4.6

Divide.

28. $(42x^4 - 36x^3 + 66x^2) \div (6x)$ **29.** $(15x^3 - 28x^2 + 18x - 7) \div (3x - 2)$

4.1 The Rules of Exponents

After studying this section, you will be able to:

1. *Multiply exponential expressions with like bases.*
2. *Divide exponential expressions with like bases.*
3. *Raise exponential expressions to a power.*

1 *The Product Rule for Exponents*

Recall that x^2 means $x \cdot x$. That is, x appears as a factor two times. What happens when we multiply $x^2 \cdot x^2$? Is there a pattern that will help us to form a general rule?

Notice that

$$(2^2)(2^3) = \overbrace{(2 \cdot 2)(2 \cdot 2 \cdot 2)}^{5 \text{ twos}} = 2^5$$ *The exponent means 2 occurs 5 times as a factor.*

$$(3^3)(3^4) = \overbrace{(3 \cdot 3 \cdot 3)(3 \cdot 3 \cdot 3 \cdot 3)}^{7 \text{ threes}} = 3^7$$ *Notice that* $3 + 4 = 7$.

$$(x^3)(x^5) = \overbrace{(x \cdot x \cdot x)(x \cdot x \cdot x \cdot x \cdot x)}^{8\ x\text{'s}} = x^8$$ *The sum of the exponents* $3 + 5 = 8$.

$$(y^4)(y^2) = \overbrace{(y \cdot y \cdot y \cdot y)(y \cdot y)}^{6\ y\text{'s}} = y^6$$ *The sum of the exponents* $4 + 2 = 6$.

We can state the pattern in words and then use variables.

> **The Product Rule**
>
> To multiply two numbers or variables in exponent form that have the same base, *add the exponents* but keep the base unchanged.
>
> $$x^a \cdot x^b = x^{a+b}$$

Be sure to notice that this rule applies only to numbers or variables that have the *same base.* Here x represents the base (the same in both cases), while the letters a and b represent the exponents that are added.

It is important that you apply this rule when the exponent is 1. Every variable that does not have a written exponent is understood to have an exponent of 1. Thus $x^1 = x$, $y^1 = y$, and so on.

EXAMPLE 1 Multiply. **(a)** $x^3 \cdot x^6$ **(b)** $x \cdot x^5$

(a) $x^3 \cdot x^6 = x^{3+6} = x^9$

(b) $x \cdot x^5 = x^{1+5} = x^6$ *Note that the exponent of the first x is* 1.

Practice Problem 1 Multiply. **(a)** $a^7 \cdot a^5$ **(b)** $w^{10} \cdot w$ ■

EXAMPLE 2 Simplify. **(a)** $y^5 \cdot y^{11}$ **(b)** $2^3 \cdot 2^5$ **(c)** $x^6 \cdot y^8$

(a) $y^5 \cdot y^{11} = y^{5+11} = y^{16}$

(b) $2^3 \cdot 2^5 = 2^{3+5} = 2^8$ *Note that the base does not change! Only the exponent changes.*

(c) $x^6 \cdot y^8$ *The rule for multiplying numbers with exponents does not apply since the bases are not the same. This cannot be simplified.*

Practice Problem 2 Simplify if possible.

(a) $x^3 \cdot x^9$ **(b)** $3^7 \cdot 3^4$ **(c)** $a^3 \cdot b^2$ ■

We can now look at multiplying expressions such as

$$(2x^5)(3x^6)$$

The number 2 in $2x^5$ is called the **numerical coefficient.** A numerical coefficient is a number that is multiplied by a variable. What is the numerical coefficient in $3x^6$? When we multiply two expressions such as $2x^5$ and $3x^6$, we first multiply the numerical coefficients; we multiply the variables with their exponents separately.

EXAMPLE 3 Multiply. **(a)** $(2x^5)(3x^6)$ **(b)** $(5x^3)(x^6)$ **(c)** $(-6x)(-4x^5)$

(a) $(2x^5)(3x^6) = (2 \cdot 3)(x^5 \cdot x^6)$ *Multiply the numerical coefficients.*

$= 6(x^5 \cdot x^6)$ *Use the rule for multiplying expressions with exponents. Add the exponents.*

$= 6x^{11}$

(b) Every variable that does not have a visible numerical coefficient is understood to have a numerical coefficient of 1. Thus x^6 has a numerical coefficient of 1.

$$(5x^3)(x^6) = (5 \cdot 1)(x^3 \cdot x^6) = 5x^9$$

(c) $(-6x)(-4x^5) = (-6)(-4)(x^1 \cdot x^5) = 24x^6$ *Remember that x has an exponent of 1.*

Practice Problem 3 Multiply.

(a) $(-a^8)(a^4)$ **(b)** $(3y^2)(-2y^3)$ **(c)** $(-4x^3)(-5x^2)$ ■

Problems of this type may involve more than one variable or more than two factors.

EXAMPLE 4 Multiply. **(a)** $(-4x^2y^3)(-2x^5y^3)$ **(b)** $(5ab)(-\frac{1}{3}a)(9b^2)$

(a) $(-4x^2y^3)(-2x^5y^3) = (-4)(-2)(x^2 \cdot x^5)(y^3 \cdot y^3)$

$= 8x^7y^6$

(b) $(5ab)(-\frac{1}{3}a)(9b^2) = (5)(-\frac{1}{3})(9)(a \cdot a)(b \cdot b^2)$

$= -15a^2b^3$

As you do the following problems, keep in mind the rule for multiplying numbers with exponents and the rules for multiplying signed numbers.

Practice Problem 4 Multiply.

(a) $(-5a^3b)(-2ab^4)$ **(b)** $(2xy)(-\frac{1}{4}x^2y)(6xy^3)$ ■

2 *The Quotient Rule for Exponents*

Frequently, we must divide expressions like the ones we have just learned to multiply. Since division by zero is undefined, in all problems in this chapter we assume that the denominator of any variable expression is not zero. We'll look at division in three separate parts.

Suppose that we want to divide $x^5 \div x^2$. We could do the division in a long fashion.

$$\frac{x^5}{x^2} = \frac{(x)(x)(x)(\cancel{x})(\cancel{x})}{(\cancel{x})(\cancel{x})} = x^3$$

Here we are using the arithmetical property of reducing fractions (see Section 0.1). When the same factor appears in both numerator and denominator, that factor can be removed.

A simpler way is to *subtract the exponents.* Notice that the base remains the same.

The Quotient Rule (Part 1)

$$\frac{x^a}{x^b} = x^{a-b} \quad \text{if the larger exponent is in the numerator and } x \neq 0$$

EXAMPLE 5 Divide. **(a)** $\frac{2^{16}}{2^{11}}$ **(b)** $\frac{x^5}{x^3}$ **(c)** $\frac{y^{16}}{y^7}$

(a) $\frac{2^{16}}{2^{11}} = 2^{16-11} = 2^5$ Note that the base does *not* change.

(b) $\frac{x^5}{x^3} = x^{5-3} = x^2$ **(c)** $\frac{y^{16}}{y^7} = y^{16-7} = y^9$

Practice Problem 5 Divide. **(a)** $\frac{10^{13}}{10^7}$ **(b)** $\frac{x^{11}}{x}$ ■

Now we consider the situation where the larger exponent is in the denominator. Suppose that we want to divide $x^2 \div x^5$.

$$\frac{x^2}{x^5} = \frac{(\cancel{x})(\cancel{x})}{(\cancel{x})(\cancel{x})(x)(x)(x)} = \frac{1}{x^3}$$

The Quotient Rule (Part 2)

$$\frac{x^a}{x^b} = \frac{1}{x^{b-a}} \quad \text{if the larger exponent is in the denominator and } x \neq 0$$

EXAMPLE 6 Divide. **(a)** $\frac{12^{17}}{12^{20}}$ **(b)** $\frac{b^7}{b^9}$ **(c)** $\frac{x^{20}}{x^{24}}$

(a) $\frac{12^{17}}{12^{20}} = \frac{1}{12^{20-17}} = \frac{1}{12^3}$ Note that the base does *not* change.

(b) $\frac{b^7}{b^9} = \frac{1}{b^{9-7}} = \frac{1}{b^2}$ **(c)** $\frac{x^{20}}{x^{24}} = \frac{1}{x^{24-20}} = \frac{1}{x^4}$

Practice Problem 6 Simplify. **(a)** $\frac{c^3}{c^4}$ **(b)** $\frac{10^{31}}{10^{56}}$ ■

When there are numerical coefficients, use the rules for dividing signed numbers to reduce fractions to lowest terms.

EXAMPLE 7 Divide.

(a) $\dfrac{5x^5}{25x^7}$ **(b)** $\dfrac{-12x^8}{4x^3}$ **(c)** $\dfrac{2x^6}{6x^4}$ **(d)** $\dfrac{-16x^7}{-24x^8}$

(a) $\dfrac{5x^5}{25x^7} = \dfrac{1}{5x^{7-5}} = \dfrac{1}{5x^2}$ **(b)** $\dfrac{-12x^8}{4x^3} = -3x^{8-3} = -3x^5$

(c) $\dfrac{2x^6}{6x^4} = \dfrac{x^{6-4}}{3} = \dfrac{x^2}{3}$ **(d)** $\dfrac{-16x^7}{-24x^8} = \dfrac{2}{3x^{8-7}} = \dfrac{2}{3x}$

Practice Problem 7 Simplify. **(a)** $\dfrac{-7x^7}{-21x^9}$ **(b)** $\dfrac{15x^{11}}{-3x^4}$ ■

You have to work very carefully if two or more variables are involved. Treat the coefficients and each variable separately.

EXAMPLE 8 Divide. **(a)** $\dfrac{x^3y^2}{5xy^6}$ **(b)** $\dfrac{x^5y^3}{x^6w}$ **(c)** $\dfrac{-3x^2y^5}{12x^6y^8}$

(a) $\dfrac{x^3y^2}{5xy^6} = \dfrac{x^2}{5y^4}$ **(b)** $\dfrac{x^5y^3}{x^6w} = \dfrac{y^3}{xw}$ **(c)** $\dfrac{-3x^2y^5}{12x^6y^8} = \dfrac{-1}{4x^4y^3}$

Practice Problem 8 Simplify. **(a)** $\dfrac{x^7y^9}{y^{10}}$ **(b)** $\dfrac{12x^5y^6}{-24x^3y^8}$ ■

Suppose that, for a given base, the exponent is the same in numerator and denominator. In these cases we can use the fact that *any nonzero number divided by itself is* 1.

EXAMPLE 9 Divide. **(a)** $\dfrac{x^6}{x^6}$ **(b)** $\dfrac{3x^5}{x^5}$

(a) $\dfrac{x^6}{x^6} = 1$ **(b)** $\dfrac{3x^5}{x^5} = 3\left(\dfrac{x^5}{x^5}\right) = 3(1) = 3$

Practice Problem 9 Divide. **(a)** $\dfrac{10^7}{10^7}$ **(b)** $\dfrac{12a^4}{15a^4}$ ■

Do you see that if we had subtracted exponents for $\frac{x^6}{x^6}$ we would have obtained x^0 in Example 9? So we can surmise that any number (except 0) to the 0 power equals 1. We can write this fact as a separate rule.

The Quotient Rule (Part 3)

$$\frac{x^a}{x^a} = x^0 = 1 \quad \text{if } x \neq 0 \quad (0^0 \text{ remains undefined})$$

EXAMPLE 10 Divide. **(a)** $\dfrac{4x^0y^2}{8^0y^5z^3}$ **(b)** $\dfrac{5x^2y}{10x^2y^3}$

(a) $\dfrac{4x^0y^2}{8^0y^5z^3} = \dfrac{4(1)y^2}{(1)y^5z^3} = \dfrac{4y^2}{y^5z^3} = \dfrac{4}{y^3z^3}$ **(b)** $\dfrac{5x^2y}{10x^2y^3} = \dfrac{1x^0}{2y^2} = \dfrac{(1)(1)}{2y^2} = \dfrac{1}{2y^2}$

Practice Problem 10 Simplify. $\dfrac{-20a^3b^8c^4}{28a^3b^7c^5}$ ■

We can combine all three parts of the rule we have developed.

The Quotient Rule

$\dfrac{x^a}{x^b} = x^{a-b}$ if the larger exponent is in the numerator and $x \neq 0$

$\dfrac{x^a}{x^b} = \dfrac{1}{x^{b-a}}$ if the larger exponent is in the denominator and $x \neq 0$

$\dfrac{x^a}{x^a} = x^0 = 1$ if there is the same exponent in the numerator and denominator and $x \neq 0$

We can combine the product rule and the quotient rule to simplify algebraic expressions that involve both multiplication and division.

EXAMPLE 11 Simplify. $\dfrac{(8x^2y)(-3x^3y^2)}{-6x^4y^3}$

$$\frac{(8x^2y)(-3x^3y^2)}{-6x^4y^3} = \frac{-24x^5y^3}{-6x^4y^3} \quad \textit{Simplify the numerator. Then divide.}$$

$$= 4x$$

Practice Problem 11 Simplify. $\dfrac{(-6ab^5)(3a^2b^4)}{16a^5b^7}$ ■

3 *Raising a Power to a Power*

How do we simplify an expression such as $(x^4)^3$? $(x^4)^3$ is x^4 raised to the third power. For this type of problem we say that we are raising a power to a power. A problem such as $(x^4)^3$ could be done by writing

$$(x^4)^3 = x^4 \cdot x^4 \cdot x^4 \quad \text{By definition.}$$

$$= x^{12} \quad \text{By adding exponents.}$$

Notice that when we add the exponents we get $4 + 4 + 4 = 12$. This is the same as multiplying 4 by 3. That is, $4 \cdot 3 = 12$. This process can be summarized by the following rule.

Raising a Power to a Power

To raise a power to a power, keep the same base and multiply the exponents.

$$(x^a)^b = x^{ab}$$

Recall what happens when you raise a negative number to a power. $(-1)^2 = 1$. $(-1)^3 = -1$. In general,

$$(-1)^n = \begin{cases} +1, & n \text{ is even} \\ -1, & n \text{ is odd} \end{cases}$$

If an expression can be raised to a power, we sometimes give the direction *simplify*.

EXAMPLE 12 Simplify. **(a)** $(x^3)^5$ **(b)** $(2^7)^3$

(a) $(x^3)^5 = x^{3\cdot 5} = x^{15}$ **(b)** $(2^7)^3 = 2^{7\cdot 3} = 2^{21}$

Note that in both parts (a) and (b) the base does not change.

Practice Problem 12 Raise each expression to the power indicated.

(a) $(a^4)^3$ **(b)** $(10^5)^2$ ■

Here are two similar rules involving products and quotients that are very useful. We'll illustrate each with an example.

If a product in parentheses is raised to a power, the parentheses indicate that *each factor* must be raised to that power.

$$(xy)^2 = x^2y^2 \qquad (xy)^3 = x^3y^3$$

$$(xy)^a = x^ay^a$$

EXAMPLE 13 Simplify. **(a)** $(ab)^8$ **(b)** $(3x)^4$ **(c)** $(-2x^2)^3$

(a) $(ab)^8 = a^8b^8$ **(b)** $(3x)^4 = (3)^4x^4 = 81x^4$ **(c)** $(-2)^3 \cdot (x^2)^3 = -8x^6$

Practice Problem 13 Raise each expression to the power indicated.

(a) $(3xy)^3$ **(b)** $(yz)^{37}$ ■

If $\left(\frac{x}{y}\right)$ is raised to the second power, the parentheses indicate that numerator and denominator are each raised to that power.

$$\left(\frac{x}{y}\right)^2 = \frac{x^2}{y^2} \quad \text{if } y \neq 0$$

$$\left(\frac{x}{y}\right)^a = \frac{x^a}{y^a} \quad \text{if } y \neq 0$$

EXAMPLE 14 Simplify. **(a)** $\left(\frac{x}{y}\right)^5$ **(b)** $\left(\frac{7}{w}\right)^4$

(a) $\left(\frac{x}{y}\right)^5 = \frac{x^5}{y^5}$ **(b)** $\left(\frac{7}{w}\right)^4 = \frac{7^4}{w^4}$

Practice Problem 14 Raise to the power indicated. $\left(\frac{4a}{b}\right)^6$ ■

Many problems can be simplified by using the previous rules involving exponents. Be sure to take particular care to determine the correct sign, especially if the numerical coefficient of the term is a negative number.

EXAMPLE 15 Simplify. **(a)** $\left(\dfrac{-3x^2z^0}{y^3}\right)^4$ **(b)** $\left(\dfrac{-2xw^4}{x^2w}\right)^3$

(a) $\left(\dfrac{-3x^2z^0}{y^3}\right)^4 = \left(\dfrac{-3x^2}{y^3}\right)^4$ *Since $z^0 = 1$. Simplify inside the parentheses first.*

$= \dfrac{(-3)^4x^8}{y^{12}}$ *Apply the rules for raising a power to a power. Notice that we wrote $(-3)^4$ and not -3^4. We are raising -3 to the fourth power.*

$= \dfrac{81x^8}{y^{12}}$ *Simplify the coefficient. $(-3)^4 = +81$*

(b) $\left(\dfrac{-2xw^4}{x^3w}\right)^3 = \left(\dfrac{-2w^3}{x^2}\right)^3$ *Simplify the expression inside the parentheses first.*

$= \dfrac{(-2)^3w^9}{x^6}$ *Simplify the coefficient $(-2)^3 = -8$.*

$= \dfrac{-8w^9}{x^6}$

Practice Problem 15 Simplify. **(a)** $\left(\dfrac{-2x^3y^0z}{4xz^2}\right)^5$ **(b)** $\left(\dfrac{4x^8y^5}{-3x^4z}\right)^3$ ■

Developing Your Study Skills

WHY IS REVIEW NECESSARY?

You master a course in mathematics by learning the concepts one step at a time. There are basic concepts like addition, subtraction, multiplication, and division of whole numbers that are considered the foundation upon which all of mathematics is built. These must be mastered first. Then the study of mathematics is built step by step upon this foundation, each step supporting the next. The process is a carefully designed procedure, so no steps can be skipped. A student of mathematics needs to realize the importance of this building process to succeed.

Because learning new concepts depends on those previously learned, students often need to take time to review. The reviewing process will strengthen the understanding and application of concepts that are weak due to a lack of mastery or the passage of time. Review at the right time on the right concepts can strengthen previously learned skills and make progress possible.

Timely, periodic review of previously learned mathematical concepts is absolutely necessary in order to master new concepts. You may have forgotten a concept or grown a bit rusty in applying it. Reviewing is the answer. Make use of any review sections in your textbook, whether they are assigned or not. Look back to previous chapters whenever you have forgotten how to do something. Study the examples and practice some exercises to refresh your understanding.

Be sure that you understand and can perform the computations of each new concept. This will enable you to be able to move successfully on to the next ones.

Remember, mathematics is a step-by-step building process. Learn one concept at a time, skipping none, and reinforce and strengthen with review whenever necessary.

4.1 Exercises

Verbal and Writing Skills

1. Write in your own words the product rule for exponents.

2. To be able to use the rules of exponents, what must be true of the bases?

3. If the larger exponent is in the denominator, the quotient rule states that $\frac{x^a}{x^b} = \frac{1}{x^{b-a}}$. Provide an example to show why this is true.

Identify the numerical coefficient, the base(s), and the exponent(s).

4. $-8x^5y^2$

5. $6x^{11}y$

6. Evaluate. (a) $3x^0$ (b) $(3x)^0$ Why are the results different?

Write in simplest exponent form.

7. $3 \cdot x \cdot x \cdot x \cdot y \cdot y$

8. $4 \cdot a \cdot a \cdot a \cdot a \cdot b \cdot b \cdot b$

9. $(-3)(a)(a)(b)(c)(b)(c)(c)$

10. $(-7)(x)(y)(z)(y)(x)$

Multiply and leave your answer in exponent form.

11. $(3^8)(3^7)$

12. $(2^5)(2^8)$

13. $(5^{10})(5^{16})$

14. $(5^3)(2^6)$

15. $(3^5)(8^2)$

Multiply.

16. $-9x^2 \cdot 8x^3$

17. $-12x^2 \cdot 5x$

18. $(5x)(10x^2)$

19. $(-13x)(-2x^2)$

20. $(-8x^4)(-2x^6)$

21. $(4x^8)(-3x^3)$

22. $(-2ab)(3a^2b^3)$

23. $(5xy^2)(-2x^3y)$

24. $\left(\frac{1}{2}x^2y\right)\left(\frac{3}{4}xy^2\right)$

25. $\left(\frac{2}{3}x^3y\right)\left(\frac{5}{7}xy^4\right)$

26. $(1.1x^2z)(-2.5xy)$

27. $(2.3x^4w)(-3.5xy^4)$

28. $(-3x)(2y)(5x^3y)$

29. $(8a)(2a^3b)(0)$

30. $(-5ab)(2a^2)(0)$

31. $(-4x)(2x^2y)(3x)$

32. $(-20xy^3)(-4x^3y^2)$

33. $(-6xy^3)(-12x^5y^8)$

34. $(-3x^5)(20y^8)$

35. $(14a^5)(-2b^6)$

36. $(5a^3b^2)(-ab)$

37. $(-x^2y)(3x)$

38. $(-3x^2y)(0)(-5xy^3)$

39. $(3ab^3)(-2a^5b)(-2a^2b^2)$

40. $(5ab)(-2ab^2)(3a^3b)$

41. $(4x^3y)(0)(-12x^4y^2)$

42. $(5x^3y)(-2w^4z)$

43. $(6w^5z^6)(-4xy)$

44. $(3ab)(5a^2c)(-2b^2c^3)$

Divide. Leave your answer in exponent form. Assume that all variables in any denominator are nonzero.

45. $\frac{x^8}{x^{10}}$ **46.** $\frac{a^{15}}{a^{13}}$ **47.** $\frac{y^{12}}{y^5}$ **48.** $\frac{b^{20}}{b^{23}}$ **49.** $\frac{3^{45}}{3^{17}}$ **50.** $\frac{7^{20}}{7^{33}}$

51. $\frac{13^{20}}{13^{30}}$ **52.** $\frac{5^{19}}{5^{11}}$ **53.** $\frac{2^{20}}{2^7}$ **54.** $\frac{4^{15}}{4^{23}}$ **55.** $\frac{a^{13}}{4a^5}$ **56.** $\frac{4b^{16}}{b^{13}}$

57. $\frac{x^7}{y^9}$ **58.** $\frac{x^{20}}{y^3}$ **59.** $\frac{-10a^5b^2}{-20a^5b^3}$ **60.** $\frac{-36x^3y^6}{-18xy^8}$ **61.** $\frac{12x^5y^6}{-3xy}$ **62.** $\frac{-20a^3b^2}{4a^5b^2}$

63. $\frac{12ab}{-24a^3b^2}$ **64.** $\frac{a^5b^6}{a^5b^6}$ **65.** $\frac{-13x^5y^6}{13x^5y^6}$ **66.** $\frac{-15xy^3}{45x^6y^3}$ **67.** $\frac{-27x^5y^3z}{9x^3y^6z^2}$ **68.** $\frac{42m^6n^3}{4.2mn^6}$

69. $\frac{2.3x^8y^{10}}{23x^{12}y^5}$ **70.** $\frac{-51x^6y^8z^{12}}{17x^3y^8z^7}$ **71.** $\frac{30x^5y^4}{5x^3y^4}$ **72.** $\frac{27x^3y^2}{3xy^2}$ **73.** $\frac{8^0x^2y^3}{16x^5y}$ **74.** $\frac{3^2x^3y^7}{3^0x^5y^2}$

75. $\frac{18a^6b^3c^0}{24a^5b^3}$ **76.** $\frac{12a^7b^8}{16a^3b^8c^0}$ **77.** $\frac{25x^6}{35y^8}$ **78.** $\frac{39x^6y^7}{-3xy^5}$ **79.** $\frac{-32x^6y^3z}{64x^5y^2}$ **80.** $\frac{24y^5}{16x^3}$

Multiply the numerator **first.** *Then divide the two expressions.*

81. $\frac{(3x^2)(2x^3)}{3x^4}$ **82.** $\frac{(4a^5)(3a^6)}{2a^3}$ **83.** $\frac{(9a^2b)(2a^3b^6)}{-27a^8b^7}$

To Think About

84. What expression can be multiplied by $(-3x^3yz)$ to obtain $81x^8y^2z^4$?

85. $63a^5b^6$ is divided by an expression and the result is $-9a^4b$. What is this expression?

Find a value for the variable to show that the two expressions are **not** *equivalent.*

86. $(x^2)(x^3)$; x^6 **87.** $\frac{y^8}{y^4}$; y^2 **88.** $\frac{z^3}{z^6}$; z^3

Simplify.

89. $(a^{2b})(a^{3b})$ **90.** $\frac{b^{5x}}{b^{2x}}$ **91.** $\frac{(c^{3y})(c^4)}{c^{2y+2}}$

Raise each expression to the power indicated.

92. $(x^2)^6$ **93.** $(w^5)^8$ **94.** $(xy^2)^7$ **95.** $(a^3b^2)^5$

96. $(ab^6)^4$ **97.** $(a^3b^2c)^8$ **98.** $(2ab^2c^7)^4$ **99.** $(3^2xy^2)^4$

100. $(-3a^4)^2$ **101.** $(-2a^5)^4$ **102.** $\left(\frac{7}{w^2}\right)^8$ **103.** $\left(\frac{12x}{y^2}\right)^5$

104. $\left(\frac{3a}{2b}\right)^4$ **105.** $\left(\frac{6x}{5y^3}\right)^2$ **106.** $(-3a^2b^3c^0)^4$ **107.** $(-2a^5b^2c)^5$

108. $(-2x^3yz)^3$ **109.** $(-4xy^0z^4)^3$ **110.** $\frac{(2x)^4}{(2x)^5}$ **111.** $\frac{(4a^2b)^2}{(4ab^2)^3}$

112. $(3ab^2)^3(ab)$ **113.** $(-2a^2b^3)^3(ab^2)$ **114.** $\left(\frac{2x^2y^3}{3z^0}\right)^3$ **115.** $\left(\frac{4x^0y^4}{3z^3}\right)^2$

116. $\left(\frac{x^5}{3}\right)^3$ **117.** $\left(\frac{y^6}{2}\right)^3$ **118.** $\left(\frac{ab^2}{c^3d^4}\right)^4$ **119.** $\left(\frac{a^3b}{c^5d}\right)^5$

120. $(5xy^2)^3(xy^2)$ **121.** $(4x^3y)^2(x^3y)$ **122.** $\left(\frac{2ab^2}{c}\right)^4$ **123.** $\left(\frac{3a^3b}{c^2}\right)^4$

To Think About

124. What expression raised to the third power is $-27x^9y^{12}z^{21}$?

125. What expression raised to the fourth power is $16x^{20}y^{16}z^{28}$?

Find a value of the variable that shows that the two expressions are **not** *equivalent.*

126. $5x^3$; $(5x)^3$ **127.** $(y + 2)^2$; $y^2 + 2^2$ **128.** $(z - 3)^2$; $z^2 - 3^2$

129. Write $4^2 \cdot 8^3 \cdot 16$ as a power of 2. **130.** Write $9^3 \cdot 27 \cdot 81$ as a power of 3.

Cumulative Review Problems

Simplify.

131. $-3 - 8$ **132.** $-17 + (-32) + (-24) + (27)$

133. $\left(\frac{2}{3}\right)\left(-\frac{21}{8}\right)$ **134.** $\frac{-3}{4} \div \frac{-12}{80}$

4.2 Negative Exponents and Scientific Notation

After studying this section, you will be able to:

1. *Use negative exponents.*
2. *Write numbers in scientific notation.*

1 *Negative Exponents*

If n is a positive integer and $x \neq 0$, then x^{-n} is defined as follows:

Definition of a Negative Exponent

$$x^{-n} = \frac{1}{x^n}$$

EXAMPLE 1 Write with positive exponents.

(a) y^{-3} **(b)** z^{-6} **(c)** w^{-1} **(d)** x^{-12}

(a) $y^{-3} = \frac{1}{y^3}$ **(b)** $z^{-6} = \frac{1}{z^6}$

(c) $w^{-1} = \frac{1}{w^1} = \frac{1}{w}$ **(d)** $x^{-12} = \frac{1}{x^{12}}$

Practice Problem 1 Write with positive exponents.

(a) x^{-12} **(b)** w^{-5} **(c)** z^{-2} **(d)** y^{-7} ■

To evaluate a numerical expression with a negative exponent, first write the expression with a positive exponent. Then simplify.

EXAMPLE 2 Evaluate. **(a)** 2^{-5} **(b)** 3^{-4}

(a) $2^{-5} = \frac{1}{2^5} = \frac{1}{32}$ **(b)** $3^{-4} = \frac{1}{3^4} = \frac{1}{81}$

Practice Problem 2 Evaluate. **(a)** 4^{-3} **(b)** 6^{-2} ■

All the previously studied laws of exponents can be applied when the exponent is any integer. These laws are the summarized in the following box.

Laws of Exponents

The Product Rule

$$x^a \cdot x^b = x^{a+b}$$

The Quotient Rule

$$\frac{x^a}{x^b} = x^{a-b} \quad \text{if } a > b, \qquad \frac{x^a}{x^b} = \frac{1}{x^{b-a}} \quad \text{if } a < b$$

Raising a Power to a Power

$$(xy)^a = x^a y^a, \qquad (x^a)^b = x^{ab}, \qquad \left(\frac{x}{y}\right)^a = \frac{x^a}{y^a}$$

By using the definition of a negative exponent and the properties of fractions, we can derive two more helpful properties of exponents.

Properties of Negative Exponents

$$\frac{1}{x^{-n}} = x^n \qquad \frac{x^{-m}}{y^{-n}} = \frac{y^n}{x^m}$$

EXAMPLE 3 Simplify. Write the expression so that no negative exponents appear.

(a) $\frac{1}{x^{-6}}$ **(b)** $\frac{x^{-3}y^{-2}}{z^{-4}}$ **(c)** $x^{-2}y^3$

(a) $\frac{1}{x^{-6}} = x^6$ **(b)** $\frac{x^{-3}y^{-2}}{z^{-4}} = \frac{z^4}{x^3y^2}$ **(c)** $x^{-2}y^3 = \frac{y^3}{x^2}$

Practice Problem 3 Simplify by writing with no negative exponents.

(a) $\frac{3}{w^{-4}}$ **(b)** $\frac{x^{-6}y^4}{z^{-2}}$ **(c)** $x^{-6}y^{-5}$ ■

EXAMPLE 4 Simplify. Write each expression so that no negative exponents appear.

(a) $(3x^{-4}y^2)^{-3}$ **(b)** $\frac{x^2y^{-4}}{x^{-5}y^3}$

(a) $(3x^{-4}y^2)^{-3} = 3^{-3}x^{12}y^{-6} = \frac{x^{12}}{3^3y^6} = \frac{x^{12}}{27y^6}$

(b) $\frac{x^2y^{-4}}{x^{-5}y^3} = \frac{x^2x^5}{y^4y^3} = \frac{x^7}{y^7}$ *First rewrite the expression so that only positive exponents appear. Then simplify using the product rule.*

Practice Problem 4 Simplify by writing with no negative exponents.

(a) $(2x^4y^{-5})^{-2}$ **(b)** $\frac{y^{-3}z^{-4}}{y^2z^{-6}}$ ■

2 *Scientific Notation*

One common use of negative exponents is in writing numbers in scientific notation. Scientific notation is most useful in expressing very large and very small numbers.

Scientific Notation

A positive number is written in scientific notation if it is in the form $a \times 10^n$, where $1 \leq a < 10$ and n is an integer.

EXAMPLE 5 Write in scientific notation. **(a)** 4567 **(b)** 157,000,000

(a) $4567 = 4.567 \times 1000$ *To change 4567 to a number that is greater than 1 but less than 10, we move the decimal point three places to the left. We must then multiply the number by a power of 10 so that we do not change the value of the number. Use 1000.*

$= 4.567 \times 10^3$

(b) $157{,}000{,}000 = \underleftarrow{1.57000000}_{\text{8 places}} \times \underbrace{100000000}_{\text{8 zeros}}$

$= 1.57 \times 10^8$

Practice Problem 5 Write in scientific notation.

(a) 78,200 **(b)** 4,786,000 ■

Numbers that are smaller than 1 will have a negative power of 10 if they are written in scientific notation.

EXAMPLE 6 Write in scientific notation.

(a) 0.061 **(b)** 0.0034 **(c)** 0.000052

(a) We need to write 0.061 as a number that is greater than 1 but less than 10. In which direction do we move the decimal point?

$0.061 = 6.1 \times 10^{-2}$ *Move the decimal point 2 places to the right. Then multiply by* 10^{-2}.

(b) $0.0034 = 3.4 \times 10^{-3}$ *Move the decimal point 3 places to the right. Then multiply by* 10^{-3}.

(c) $0.000052 = 5.2 \times 10^{-5}$ *Why?*

Practice Problem 6 Write in scientific notation.

(a) 0.98 **(b)** 0.000763 **(c)** 0.000092 ■

The reverse procedure is used to go from scientific notation to ordinary decimal notation.

EXAMPLE 7 Write in decimal notation. **(a)** 1.568×10^2 **(b)** 7.432×10^{-3}

(a) $1.568 \times 10^2 = 1.568 \times 100$

$= 156.8$

Alternative method.

$1.568 \times 10^2 = 156.8$ *The exponent 2 tells us to move the decimal point 2 places to the right.*

(b) $7.432 \times 10^{-3} = 7.432 \times \frac{1}{1000}$

$= 0.007432$

Alternative method.

$7.432 \times 10^{-3} = 0.007432$ *The exponent −3 tells us to move the decimal point 3 places to the left.*

Practice Problem 7 Write in decimal notation.

(a) 2.96×10^3 **(b)** 1.93×10^6 **(c)** 5.43×10^{-2} **(d)** 8.562×10^{-5} ■

CALCULATOR

Scientific Notation

Most calculators can display only eight digits at one time. Numbers with more than eight digits are shown in scientific notation. 1.12 E 08 means 1.12×10^8. You can use the calculator to compute large numbers by entering the numbers using scientific notation. For example,

$(7.48 \times 10^{24}) \times (3.5 \times 10^8)$

Enter 7.48 [EXP]

24 [×] 3.5

[EXP] 8 [=]

Display: [2.618 E 33]

Compute on a calculator.

1. 35,000,000 + 77,000,000
2. 800,000,000 − 29,000,000
3. $(6.23 \times 10^{12})(4.9 \times 10^5)$
4. $(2.5 \times 10^7)^5$
5. $3.3284 \times 10^{32} \div 6.28 \times 10^{24}$
6. How many seconds are there in 1000 years?

The distance light travels in one year is called a *light-year.* A light-year is a convenient unit of measure to use when investigating the distances between stars.

EXAMPLE 8 A light-year is a distance of 9,460,000,000,000,000 meters. Write this in scientific notation.

$$9{,}460{,}000{,}000{,}000{,}000 = 9.46 \times 10^{15} \text{ meters}$$

Practice Problem 8 Astronomers measure distances to faraway galaxies in parsecs. A parsec is a distance of 30,900,000,000,000,000 meters. Write this in scientific notation. ■

To perform a calculation involving very large or very small numbers, it is usually helpful to write the numbers in scientific notation and then to use the laws of exponents to do the calculation.

EXAMPLE 9 Use scientific notation and the laws of exponents to find these answers. Leave your answer in scientific notation.

(a) (32,000,000)(1,500,000,000,000) **(b)** $\dfrac{0.00063}{0.021}$

(a) (32,000,000)(1,500,000,000,000)

$= (3.2 \times 10^7)(1.5 \times 10^{12})$ *Change each number to scientific notation.*

$= 3.2 \times 1.5 \times 10^7 \times 10^{12}$ *Rearrange the order. Remember multiplication is commutative.*

$= 4.8 \times 10^{19}$ *Multiply* 3.2×1.5. *Multiply* $10^7 \times 10^{12}$.

(b) $\dfrac{0.00063}{0.021} = \dfrac{6.3 \times 10^{-4}}{2.1 \times 10^{-2}}$ *Change each number to scientific notation.*

$= \dfrac{6.3}{2.1} \times \dfrac{10^{-4}}{10^{-2}}$ *Rearrange the order. We are actually using the definition of multiplication of fractions.*

$= \dfrac{6.3}{2.1} \times \dfrac{10^2}{10^4}$ *Rewrite with positive exponents.*

$= 3.0 \times 10^{-2}$

Practice Problem 9 Use scientific notation to find the following. Leave your answer in scientific notation.

(a) (56,000)(1,400,000,000) **(b)** $\dfrac{0.000111}{0.00000037}$ ■

EXAMPLE 10 The approximate distance from Earth to the star Polaris is 208 parsecs. A parsec is a distance of approximately 3.09×10^{13} kilometers. How long would it take a space probe traveling at 40,000 kilometers per hour to reach the star? Round your answer to three significant digits.

1. *Understand the problem.*
Recall that the distance formula is

$$\text{distance} = \text{rate} \times \text{time}$$

We are given the distance and the rate. We need to find the time.

Let's take a look at the distance. The distance is given in parsecs, but the rate is given in kilometers per hour. We need to change the distance to kilometers. We are told that a parsec is approximately 3.09×10^{13} kilometers. That is, there are 3.09×10^{13} kilometers per parsec. We use this information to change 208 parsecs to kilometers.

$$208 \text{ parsecs} = \frac{(208 \text{ parsecs})(3.09 \times 10^{13} \text{ kilometers})}{1 \text{ parsec}} = 642.72 \times 10^{13} \text{ kilometers}$$

2. *Write an equation.*
Use the distance formula.

$$d = r \times t$$

3. *Solve the equation and state the answer.*
Substitute the known values into the formula and solve for the unknown, time.

$$642.72 \times 10^{13} \text{ km} = \frac{40{,}000 \text{ km}}{1 \text{ hour}} \times t$$

$$6.4272 \times 10^{15} \text{ km} = \frac{4 \times 10^4 \text{ km}}{1 \text{ hr}} \times t \qquad \textit{Change the numbers to scientific notation.}$$

$$\frac{6.4272 \times 10^{15} \text{ km}}{\dfrac{4 \times 10^4 \text{ km}}{1 \text{ hr}}} = t \qquad \textit{Divide both sides by } \frac{4 \times 10^4 \text{ km}}{1 \text{ hr}}.$$

$$\frac{(6.4272 \times 10^{15} \cancel{\text{km}})(1 \text{ hr})}{4 \times 10^4 \cancel{\text{km}}} = t$$

$$1.6068 \times 10^{11} \text{ hr} = t$$

1.6068×10^{11} is 160.68×10^9 or 160.68 billion hours. The space probe will take approximately 160.68 billion hours to reach the star.

Reread the problem. Are we finished? What is left to do? We need to round the answer to three significant digits. Rounding to three significant digits, we have

$$160.68 \times 10^9 = 161 \times 10^9$$

This is approximately 161 billion hours or a little more than 18 million years.

4. *Check.* Unless you have had a great deal of experience working in astronomy, it would be difficult to determine if this is a reasonable answer. You may wish to reread your analysis and redo your calculations as a check.

Practice Problem 10 The average distance from Earth to the star Betelgeuse is 159 parsecs. How many hours would it take a space probe to travel from Earth to Betelgeuse at a speed of 50,000 kilometers per hour? Round your answer to three significant digits. ■

4.2 Exercises

Simplify each term and express your answer with positive exponents. Assume that all variables are nonzero.

1. $3x^{-2}$

2. $4xy^{-4}$

3. $(2xy)^{-1}$

4. $(3x^2y^3)^{-2}$

5. $\dfrac{3xy^{-2}}{z^{-3}}$

6. $\dfrac{4x^{-2}y^{-3}z^0}{y^4}$

7. $\dfrac{(3x)^{-2}}{(3x)^{-3}}$

8. $\dfrac{(2ab^2)^{-3}}{(2ab^2)^{-4}}$

9. $x^6y^{-2}z^{-3}w^{-10}$

10. $2x^2y^{-3}z^{-4}w^5$

11. $(4^{-3})(2^0)$

12. $(6^{-2})(3^2)$

13. $\left(\dfrac{3xy^2}{z^4}\right)^{-2}$

14. $\left(\dfrac{2a^3b^0}{c^2}\right)^{-3}$

15. $\dfrac{x^{-2}y^{-3}}{x^4y^{-2}}$

16. $\dfrac{a^{-6}b^3}{a^{-2}b^{-5}}$

17. $(-2x^3y^{-2})^{-3}$

18. $(-3x^{-4}y^2)^{-4}$

Write in scientific notation.

19. 123,780

20. 0.063

21. 0.000742

22. 889,610,000,000

23. 7,652,000,000

24. 0.00000001963

Write in decimal notation.

25. 5.63×10^4

26. 1.776×10^8

27. 3.3×10^{-5}

28. 1.99×10^{-1}

29. 9.83×10^5

30. 3.5×10^{-8}

31. The speed of light is 3.00×10^8 meters per second. Write this in decimal notation.

32. The mass of a proton is

0.00000000000000000000000000167 kilogram

Write this in scientific notation.

33. The mass of an electron is

0.000000000000000000000000000000911 kilogram

Write this in scientific notation.

34. Astronomers sometimes measure distances in megaparsecs. It is a distance of

30,800,000,000,000,000,000,000 meters

Write this in scientific notation.

Evaluate by using scientific notation and the laws of exponents. Leave your answer in scientific notation.

35. $\dfrac{(5{,}000{,}000)(16{,}000)}{8{,}000{,}000{,}000}$

36. (0.0075)(0.0000002)(0.001)

37. $(0.0002)^5$

38. $\dfrac{0.000000006}{0.000012}$

39. (150,000,000)(0.00005)(0.002)(30,000)

40. $(30{,}000{,}000)^4$

Applications

For fiscal year 1991 the national debt was determined to be approximately 3.502×10^{12} *dollars.*

41. The 1990 census records that the entire population of the United States was 2.49×10^8 people. If the national debt were evenly divided among every person in the country, how much debt would be assigned to each individual? Round your answer to three significant digits.

42. The 1990 census records that the number of people in the United States that were over the age of 17 was approximately 1.85×10^8 people. If the national debt were evenly divided among every person over the age of 17 in the country, how much debt would be assigned to each individual? Round your answer to three significant digits.

A parsec is a distance of approximately 3.09×10^{13} *kilometers.*

43. How long would it take a space probe to travel from Earth to the star Rigel, which is 276 parsecs from Earth? Assume that the space probe travels at 45,000 kilometers per hour. Round your answer to three significant digits.

44. How long would it take a space probe to travel from earth to the star Hadar, which is 150 parsecs from Earth? Assume that the space probe travels at 55,000 kilometers per hour. Round your answer to three significant digits.

45. The mass of an electron is approximately 9.11×10^{-28} gram. Find the mass of 125,000,000 electrons.

46. The mass of a proton is approximately 1.67×10^{-24} gram. Find the mass of 250,000,000 protons.

Cumulative Review Problems

Simplify.

47. $-2.7 - (-1.9)$

48. $(-1)^{2935}$

49. $\frac{-3}{4} + \frac{5}{7}$

Putting Your Skills to Work

GRAVITATIONAL FORCE

Many factors affect tides. But by far the greatest is the gravitational pull of the moon on Earth. Just how large is this force? We can use the following formula to calculate the gravitational force between two large objects such as Earth and the moon.

$$F = \frac{(6.67 \times 10^{-11})(M_1)(M_2)}{d^2}$$

where M_1 = the mass of one object

M_2 = the mass of the second object

d = the mean distance between the objects

The force is measured in newtons.

Use the formula to solve each problem. Round each answer to three significant digits.

1. Calculate the gravitational force between the Earth and the moon. The mass of Earth is 5.97×10^{24} kilograms. The mass of the moon is 7.35×10^{22} kilograms. The distance between them is 3.84×10^8 meters.

2. Calculate the gravitational force between Mars and its moon Phobos. The mass of Mars is 6.39×10^{23} kilograms. The mass of its moon Phobos is 1.93×10^{15} kilograms. The distance between them is 9.90×10^6 meters.

3. Calculate the force in newtons between Jupiter and its moon Europa. The mass of Jupiter is 1.90×10^{27} kilograms. The mass of its moon Europa is 4.85×10^{19} kilograms. The distance between them is 6.71×10^8 meters.

4. Since 1 newton is equivalent to 0.225 pound, we can find the gravitational force in pounds. Estimate the gravitational force between Earth and the moon in pounds.

COOPERATIVE GROUP ACTIVITY

5. Compare the forces in Problems 1 and 2. Which force is greater? Why?

6. What effect would moving two objects apart have on the gravitational force between them? How does the formula reflect this?

7. If the distance between two objects were kept the same, what effect would doubling the mass of both objects have on the gravitational force?

8. Information on how the gravitational forces of the moon and of the sun affect tides can be found in the World Almanac. What are spring tides and neap tides?

4.3 Addition and Subtraction of Polynomials

After studying this section, you will be able to:

1. *Add polynomials.*
2. *Subtract polynomials.*

1 *Adding Polynomials*

Recall that a polynomial in x is the sum of a finite number of terms of the form ax^n, where a is any real number and n is a whole number. Usually these polynomials are written in descending powers of the variable, such as

$$5x^3 + 3x^2 - 2x - 5 \quad \text{and} \quad 3.2x^2 - 1.4x + 5.6$$

You can add, subtract, multiply, and divide polynomials.

Let us take a look at addition. To add two polynomials, we add the respective like terms.

EXAMPLE 1 Add. $(5x^2 - 6x - 12) + (-3x^2 - 9x + 5)$

To begin we need to group like terms.

$$\begin{aligned}(5x^2 - 6x - 12) + (-3x^2 - 9x + 5) &= [5x^2 + (-3x^2)] + [-6x + (-9x)] + [-12 + 5]\\ &= [(5-3)x^2] + [(-6-9)x] + [-12+5]\\ &= 2x^2 + (-15x) + (-7)\\ &= 2x^2 - 15x - 7\end{aligned}$$

Alternative Method

We could write the problem in vertical form by lining up the like terms.

$$\begin{array}{r} 5x^2 - 6x - 12 \\ + \; -3x^2 - 9x + 5 \\ \hline 2x^2 - 15x - 7 \end{array}$$

You may use either horizontal form or vertical form to group the like terms, whichever form you find easier to use.

Practice Problem 1 Add.

$(-8x^3 + 3x^2 + 6) + (2x^3 - 7x^2 - 3)$ ■

The numerical coefficients of the polynomials may be any real number. Thus the polynomials may have numerical coefficients that are decimals or fractions.

EXAMPLE 2 Add. $(\frac{1}{2}x^2 - 6x + \frac{1}{3}) + (\frac{1}{5}x^2 - 2x - \frac{1}{2})$

$$\begin{aligned}(\tfrac{1}{2}x^2 - 6x + \tfrac{1}{3}) + (\tfrac{1}{5}x^2 - 2x - \tfrac{1}{2}) &= [\tfrac{1}{2}x^2 + \tfrac{1}{5}x^2] + [-6x + (-2x)] + [\tfrac{1}{3} + (-\tfrac{1}{2})]\\ &= [(\tfrac{1}{2} + \tfrac{1}{5})x^2] + [(-6-2)x] + [\tfrac{1}{3} + (-\tfrac{1}{2})]\\ &= [(\tfrac{5}{10} + \tfrac{2}{10})x^2] + [-8x] + [\tfrac{2}{6} - \tfrac{3}{6}]\\ &= \tfrac{7}{10}x^2 - 8x - \tfrac{1}{6}\end{aligned}$$

Practice Problem 2 Add. $(-\frac{1}{3}x^2 - 6x - \frac{1}{12}) + (\frac{1}{4}x^2 + 5x - \frac{1}{3})$ ■

EXAMPLE 3 Add. $(1.2x^3 - 5.6x^2 + 5) + (-3.4x^3 - 1.2x^2 + 4.5x - 7)$

Group like terms.

$$\begin{aligned}(1.2x^3 - 5.6x^2 + 5) + (-3.4x^3 - 1.2x^2 + 4.5x - 7) &= (1.2 - 3.4)x^3 + (-5.6 - 1.2)x^2 + 4.5x + (5 - 7)\\ &= -2.2x^3 - 6.8x^2 + 4.5x - 2\end{aligned}$$

Practice Problem 3

Add. $(3.5x^3 - 0.02x^2 + 1.56x - 3.5) + (-0.08x^2 - 1.98x + 4)$ ■

Polynomials may involve more than one variable.

EXAMPLE 4 Add. $(-5x^2 + 2xy - 15y^2) + (-30x^2 - 5xy + 2y^2)$

Group like terms.

$$\begin{aligned}(-5x^2 + 2xy - 15y^2) + (-30x^2 - 5xy + 2y^2) &= (-5 - 30)x^2 + (2 - 5)xy + (-15 + 2)y^2\\ &= -35x^2 - 3xy - 13y^2\end{aligned}$$

Practice Problem 4 Add. $(-3a^3 + 2ab + 7b) + (-4a^3 - 6ab + 3b^3)$ ■

2 *Subtracting Polynomials*

Recall that subtraction of real numbers can be defined as adding the opposite of the second number. Thus $(a) - (b) = (a) + (-b)$. That is, $3 - 5 = 3 + (-5)$. A similar method is used to subtract two polynomials.

To subtract two polynomials, change all the signs of the second polynomial and it becomes an addition problem.

EXAMPLE 5 Subtract. $(7x^2 - 6x + 3) - (5x^2 - 8x - 12)$

We change all the signs of the second polynomial and then add.

$$\begin{aligned}(7x^2 - 6x + 3) - (5x^2 - 8x - 12) &= (7x^2 - 6x + 3) + (-5x^2 + 8x + 12)\\ &= (7 - 5)x^2 + (-6 + 8)x + (3 + 12)\\ &= 2x^2 + 2x + 15\end{aligned}$$

Practice Problem 5

Subtract. $(5x^3 - 15x^2 + 6x - 3) - (-4x^3 - 10x^2 + 5x + 13)$ ■

EXAMPLE 6 Subtract. $(-6x^2y - 3xy + 7xy^2) - (5x^2y - 8xy - 15x^2y^2)$

Change the sign of each term in the second polynomial. Then look for like terms.

$$\begin{aligned}(-6x^2y - 3xy + 7xy^2) + (-5x^2y + 8xy + 15x^2y^2) &= (-6 - 5)x^2y + (-3 + 8)xy + 7xy^2 + 15x^2y^2\\ &= -11x^2y + 5xy + 7xy^2 + 15x^2y^2\end{aligned}$$

Nothing further can be done to combine these four terms.

Practice Problem 6 Subtract. $(x^3 - 7x^2y + 3xy^2 - 2y^3) - (2x^3 + 4xy - 6y^3)$ ■

4.3 Exercises

Add the following polynomials.

1. $(7x + 3) + (-5x - 20)$

2. $(-8x - 4) + (3x + 12)$

3. $(x^2 + 5x + 6) + (-2x^2 - 6x + 8)$

4. $(x^2 - 6x - 3) + (-4x^2 + x - 4)$

5. $(x^3 - 3x^2 + 7) + (x^2 + 5x - 8)$

6. $(x^2 - 7x - 8) + (x^3 + 3x^2 + 10)$

7. $(\frac{1}{2}x^2 + \frac{1}{3}x - 4) + (\frac{1}{3}x^2 + \frac{1}{6}x - 5)$

8. $(\frac{1}{4}x^2 - \frac{2}{3}x - 10) + (-\frac{1}{3}x^2 + \frac{1}{9}x + 2)$

9. $(3.4x^3 - 5.6x^2 - 7.1x + 3.4) + (-1.7x^3 + 2.2x^2 - 6.1x - 8.8)$

10. $(-4.6x^3 - 2.9x^2 + 5.6x - 0.3) + (-2.6x^3 + 9.8x^2 + 4.5x - 1.7)$

Subtract the following polynomials.

11. $(-8x + 2) - (3x + 4)$

12. $(12x + 9) - (8x - 6)$

13. $(5x^3 - 7x^2) - (-12x^3 - 8x^2)$

14. $(-9x^3 + 3x^2) - (5x^3 - 14x^2)$

15. $(\frac{1}{2}x^2 + 3x - \frac{1}{5}) - (\frac{2}{3}x^2 - 4x + \frac{1}{10})$

16. $(\frac{1}{8}x^2 - 5x + \frac{2}{7}) - (\frac{1}{4}x^2 - 6x + \frac{1}{14})$

17. $(-3x^2 + 5x) - (2x^3 - 3x^2 + 10)$

18. $(7x^3 - 3x^2 - 5x + 4) - (x^3 - 7x + 3)$

19. $(0.5x^4 - 0.7x^2 + 8.3) - (5.2x^4 + 1.6x + 7.9)$

20. $(1.3x^4 - 3.1x^3 + 6.3x) - (x^4 - 5.2x^2 + 6.5x)$

Perform the indicated operations.

21. $(7x - 8) - (5x - 6) + (8x + 12)$

22. $(-8x + 3) + (-6x - 20) - (-3x + 9)$

23. $(5x^2y - 6xy^2 + 2) + (-8x^2y + 12xy^2 - 6)$

24. $(7x^2y^2 - 6xy + 5) + (-15x^2y^2 - 6xy + 18)$

25. $(-7a^4 + 3a^2 - 2) - (4a^4 + 5a^3 + 3a^2)$

26. $(-9a^4 - 6a^3 + 2a^2) - (a^4 + 6a^2 - 3a)$

27. $(-2x^3 - 3x^2y + 4xy^2 - 7y^3) + (x^3 + x^2y - 9xy^2 + 2y^3)$

28. $(x^3 - 4x^2y - 6xy^2 - 10y^3) + (-5x^3 - 4x^2y + xy^2 + y^3)$

To Think About

29. When $5x^2 - 7xy - 9y^2$ is added to $-63x^2 + 34xy + y^2$, what is the result?

30. When $7x^2y^2 - 16xy - 45$ is added to $-8x^2y^2 + 23xy + 45$, what is the result?

31. When $5x^3 - 7x^2 - 28$ is subtracted from $12x^3 + 34x^2 - 6$, what is the result?

32. When $7x^3 + 4x - 18$ is subtracted from $-18x^3 - 16x + 3$, what is the result?

Cumulative Review Problems

33. Solve for x. $3y - 8x = 2$

34. Solve for B. $3A = 2BCD$

35. Solve for b. $A = \frac{1}{2}h(b + c)$

36. Solve for d. $B = \dfrac{5xy}{d}$

4.4 Multiplication of Polynomials

After studying this section, you will be able to:

1 *Multiply a monomial by a polynomial.*
2 *Multiply two binomials.*

1 *Multiplying a Monomial by a Polynomial*

We use the distributive property to multiply a monomial by a polynomial. Remember, the distributive property states that for real numbers a, b, c it is true that

$$a(b + c) = ab + ac$$

EXAMPLE 1 Multiply. $3x^2(5x - 2)$

$$3x^2(5x - 2) = 3x^2(5x) + 3x^2(-2) \qquad \textit{Use the distributive property.}$$
$$= (3 \cdot 5)(x^2 \cdot x) + (3)(-2)x^2$$
$$= 15x^3 - 6x^2$$

Practice Problem 1 Multiply. $4x^3(-2x^2 + 3x)$ ■

Try to do as much of the multiplication mentally as you can.

EXAMPLE 2 Multiply.

(a) $3(2 - y)$ **(b)** $2x(x^2 + 3x - 1)$ **(c)** $-2xy^2(x^2 - 2xy - 3y^2)$

(a) $3(2 - y) = 6 - 3y$

(b) $2x(x^2 + 3x - 1) = 2x^3 + 6x^2 - 2x$

(c) $-2xy^2(x^2 - 2xy - 3y^2) = -2x^3y^2 + 4x^2y^3 + 6xy^4$

Notice in part (c) that you are multiplying each term by the negative number $-2xy^2$. This will affect the sign of each term in the product.

Practice Problem 2 Multiply.

(a) $7(3 - 2x)$ **(b)** $-3x(x^2 + 2x - 4)$ **(c)** $6xy(x^3 + 2x^2y - y^2)$ ■

When we multiply by a monomial, the monomial may be on the right side.

EXAMPLE 3 Multiply. **(a)** $(x + 2y + z)3x$ **(b)** $(x^2 - 2x + 6)(-2xy)$

(a) $(x + 2y + z)3x = 3x^2 + 6xy + 3xz$

(b) $(x^2 - 2x + 6)(-2xy) = -2x^3y + 4x^2y - 12xy$

Practice Problem 3 Multiply.

(a) $(3x^2 - 7x + 1)4x$ **(b)** $(-6x^3 + 4x^2 - 2x)(-3xy)$ ■

2 *Multiplying Two Binomials*

We can build on our knowledge of the distributive property and our experience with multiplying monomials to learn how to multiply two binomials. Let's suppose that we want to multiply $(x + 2)(3x + 1)$. We can use the distributive property. Since a can represent any number, let $a = x + 2$. Then let $b = 3x$ and $c = 1$. We now have:

$$\begin{aligned} a(b + c) &= \quad ab \quad + \quad ac \\ (x + 2)(3x + 1) &= (x + 2)(3x) + (x + 2)(1) \\ &= 3x^2 + 6x + x + 2 \\ &= 3x^2 + 7x + 2 \end{aligned}$$

Let's take another look at the original problem, $(x + 2)(3x + 1)$. This time we will assign a letter to each term in the binomials. That is, let $a = x$, $b = 2$, $c = 3x$, and $d = 1$. Using substitution, we have

$$\begin{aligned} &(x + 2)(3x + 1) \\ &(a + b)(c + d) = (a + b)c + (a + b)d \\ &\qquad = ac + bc + ad + bd \\ &\qquad = (x)(3x) + (2)(3x) + (x)(1) + (2)(1) \qquad \textit{By substitution.} \\ &\qquad = 3x^2 + 6x + x + 2 \\ &\qquad = 3x^2 + 7x + 2 \end{aligned}$$

How does this compare with the result above?

The distributive property shows us *how* the problem can be done and *why* it can be done. In actual practice there is a somewhat easier approach to obtain the answer. It is often referred to as the FOIL method. The letters FOIL stand for

F	multiply the *F*irst terms
O	multiply the *O*uter terms
I	multiply the *I*nner terms
L	multiply the *L*ast terms

The FOIL letters are simply a way to remember the four terms in the final product and how they are obtained. Let's return to our original problem.

$(x + 2)(3x + 1)$	F	Multiply *first* terms and obtain $3x^2$.
$(x + 2)(3x + 1)$	O	Multiply the *outer* terms and obtain x.
$(x + 2)(3x + 1)$	I	Multiply the two *inner* terms and obtain $6x$.
$(x + 2)(3x + 1)$	L	Multiply the *last* two terms and obtain 2.

The result so far is $3x^2 + x + 6x + 2$. These four terms are the same four terms that we obtained when we multiplied using the distributive property. We can combine the like terms to obtain the final answer: $3x^2 + 7x + 2$. Our result is again the same as when we used the distributive property. Now let's study the use of the FOIL method in a few examples.

EXAMPLE 4 Multiply. $(2x - 1)(3x + 2)$

First Last

$(2x - 1)(3x + 2)$ First + Outer + Inner + Last

Inner

Outer

$$\begin{aligned} &\quad\ \, F \quad\ \ O \quad\ \ I \quad\ L \\ &= 6x^2 + 4x - 3x - 2 \\ &= 6x^2 + x - 2 \end{aligned}$$

Collect like terms.

Practice Problem 4 Multiply. $(5x - 1)(x - 2)$ ■

EXAMPLE 5 Multiply. $(3a + 2b)(4a - b)$.

First Last

$(3a + 2b)(4a - b)$

Inner

Outer

$$= 12a^2 - 3ab + 8ab - 2b^2$$

$$= 12a^2 + 5ab - 2b^2$$

Practice Problem 5 Multiply. $(8a - 5b)(3c - d)$ ■

After you have done several problems you may be able to combine the outer and inner products mentally.

EXAMPLE 6 Multiply. $(5x + 1)(x - 2)$

First Last

$(5x + 1)(x - 2)$

Inner

Outer

$$= 5x^2 - 9x - 2$$

(Mentally add $-10x + x$ and obtain $-9x$ for the middle term.)

Practice Problem 6 Multiply. $(3x + 2)(x - 1)$ ■

In some problems the inner and outer products cannot be combined.

EXAMPLE 7 Multiply. $(3x + 2y)(5x - 3z)$

First Last

$(3x + 2y)(5x - 3z)$

Inner

Outer

$$= 15x^2 - 9xz + 10xy - 6yz$$

Since there are no like terms, we cannot combine any terms.

Practice Problem 7 Multiply. $(3a + 2b)(2a - 3b)$ ■

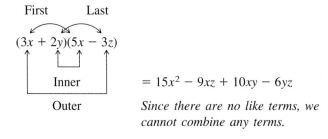

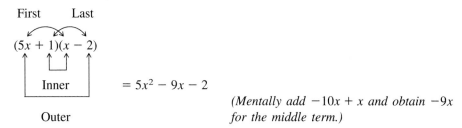

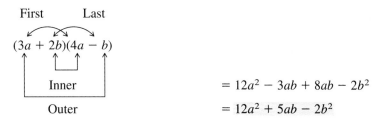

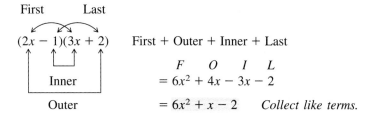

EXAMPLE 8 Multiply. $(7x - 2y)^2$

$(7x - 2y)(7x - 2y)$ *When we square a binomial, it is the same as multiplying a binomial by itself.*

$(7x - 2y)(7x - 2y)$

Inner

Outer

$$= 49x^2 - 14xy - 14xy + 4y^2$$
$$= 49x^2 - 28xy + 4y^2$$

Practice Problem 8 Multiply. $(3x - 2y)^2$ ■

The exponents of the variable may be higher than 1. We may multiply binomials containing x^2 or y^3, and so on.

EXAMPLE 9 Multiply. $(3x^2 + 4y^3)(2x^2 + 5y^3)$

$$(3x^2 + 4y^3)(2x^2 + 5y^3) = 6x^4 + 15x^2y^3 + 8x^2y^3 + 20y^6$$
$$= 6x^4 + 23x^2y^3 + 20y^6$$

Practice Problem 9 Multiply. $(2x^2 + 3y^2)(5x^2 + 6y^2)$ ■

EXAMPLE 10 The width of a living-room is $(x + 4)$ feet. The length of the room is $(3x + 5)$ feet. What is the area of the room in square feet?

$$A = (\text{length})(\text{width}) = (3x + 5)(x + 4)$$
$$= 3x^2 + 12x + 5x + 20$$
$$= 3x^2 + 17x + 20$$

There are $(3x^2 + 17x + 20)$ square feet in the room.

Practice Problem 10 What is the area in square feet of a room that is $(2x - 1)$ feet wide and $(7x + 3)$ feet long? ■

Developing Your Study Skills

EXAM TIME: HOW TO REVIEW

Reviewing adequately for an exam enables you to bring together the concepts you have learned over several sections. For your review, you will need to:

1. Reread your textbook. Make a list of any terms, rules, or formulas you need to know for the exam. Be sure you understand them all.
2. Reread your notes. Go over returned homework and quizzes. Redo the problems you missed.
3. Practice some of each type of problem covered in the chapter(s) you are to be tested on.
4. Use the end-of-chapter materials provided in your textbook. Read carefully through the Chapter Organizer. Do the Review Problems. Take the Chapter Test. When you are finished, check your answers. Redo any problems you missed.
5. Get help if any concepts give you difficulty.

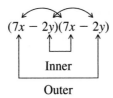

4.4 Exercises

Multiply.

1. $3x(5x^2 - x)$

2. $-6x(2x^2 + 3x)$

3. $-4x(x^2 - 3x - 6)$

4. $5x(-x^2 + 6x + 8)$

5. $2x(x^3 - 3x^2 + x - 6)$

6. $-3x(x^3 + 5x^2 - 6x - 7)$

7. $-5xy(x^2y^2 - 3xy + 5)$

8. $2xy(3x^2y + 9xy - 8)$

9. $3x^3(x^4 + 2x^2 - 3x + 7)$

10. $-7x^3(x^5 - 2x^3 + 3x^2 - 4)$

11. $2ab^2(3 - 2ab - 5b^3)$

12. $-2x^2y(x^2 - 8x + 2y^2)$

13. $(5x^3 - 2x^2 + 6x)(-3xy^2)$

14. $(3b^2 - 6b + 8ab)(5b^3)$

15. $(2b^2 + 3b - 4)(2b^2)$

16. $(2x^3 + x^2 - 6x)(2xy)$

17. $(x^3 - 3x^2 + 5x - 2)(3x)$

18. $(-2x^3 + 4x^2 - 7x + 3)(2x)$

19. $(x^2y^2 - 6xy + 8)(-2xy)$

20. $(x^2y^2 + 5xy - 9)(-3xy)$

21. $(-7x^3 + 3x^2 + 2x - 1)(4x^2y)$

22. $(-5x^3 - 6x^2 + x - 1)(5xy^2)$

23. $(5a^3 + a^2 - 7)(-3ab^2)$

24. $(6a^3 - 5a + 2)(-4a^2b)$

25. $6x^3(2x^4 - x^2 + 3x + 9)$

26. $8x^3(-2x^4 + 3x^2 - 5x - 14)$

27. $-2x^4(9x^3 + 2x - 4)$

28. $-3x^3(4x^4 - x^2 + 1)$

Multiply. Try to do most of the problems mentally without writing down intermediate steps.

29. $(x + 10)(x + 3)$

30. $(x + 3)(x + 4)$

31. $(x + 6)(x + 2)$

32. $(x - 8)(x + 2)$

33. $(x + 3)(x - 6)$

34. $(x - 5)(x - 4)$

35. $(x - 6)(x - 5)$

36. $(3x + 1)(-x - 4)$

37. $(5x + 1)(-x - 3)$

38. $(7x - 4)(x + 2y)$

39. $(3x - 5)(x + 3y)$

40. $(2y + 3)(5y - 2)$

41. $(3y + 2)(4y - 3)$

42. $(3y - 7)(5y - 1)$

43. $(6y - 5)(2y - 1)$

To Think About

44. What is *wrong* with this multiplication? $(x - 2)(-3) = 3x - 6$

45. What is *wrong* with this answer? $-(3x - 7) = -3x - 7$

46. What is the missing term? $(3x + 2)(3x + 2) = 9x^2 +$ ______ $+ 4$.

47. After multiplying these two binomials, write a brief description of what is special about your answer. $(7x + 3)(7x - 3)$

Multiply.

48. $(x + y)(-5x - 6y)$

49. $(4x - 2y)(-7x - 3y)$

50. $(5x - 3y)(-8x - 9y)$

51. $(3a - 2b^2)(5a - 6b^2)$

52. $(8a - 3b^2)(2a - 5b^2)$

53. $(2x^2 - 5y^2)(2x^2 + 5y^2)$

54. $(8x^2 - 3y^2)(8x^2 - 3y^2)$

55. $(8x - 2)^2$

56. $(5x - 3)^2$

57. $(5x^2 + 2y^2)^2$

58. $(3x^2 - 7y^2)^2$

59. $(5t - 9p)(3t + 7p)$

60. $(9w - 7z)(8w + 9z)$

61. $(3ab - 5d)(2ab - 7d)$

62. $(6xy - 2z)(5xy + 4z)$

63. $(5x + 8y)(6x - y)$

64. $(3x + 8y)(5x - 2y)$

65. $(7x - 2y)(5x + 3z)$

66. $(4a - 5b)(a + 3f)$

67. $(a - 8b)(4c - 3b)$

68. $(x - 3y)(z - 8y)$

★ **69.** $(\frac{1}{3}x^3y^4 - \frac{1}{7}xy^6)(\frac{1}{2}x^3y^4 + \frac{1}{3}xy^6)$

★ **70.** $(3.2a^3b - 0.9ab^3)(5.3a^3b - 1.1ab^3)$

Cumulative Review Problems

71. Solve for x. $3(x - 6) = -2(x + 4) + 6x$

72. Solve for w. $3(w - 7) - (4 - w) = 11w$

73. La Tanya has three more quarters than dimes, giving her \$3.55. How many of each coin does she have? (Use d for the number of dimes.)

74. A number decreased by three is seven more than three times the number. Find the number.

Multiply.

75. $(1.62x + 3.53y)(1.98x - 4.82y)$

76. $(7.56x - 8.65y)^2$

4.5 Multiplication: Special Cases

After studying this section, you will be able to:

1 *Multiply binomials of the type* $(a + b)(a - b)$.
2 *Multiply binomials of the type* $(a + b)^2$ *and* $(a - b)^2$.
3 *Multiply polynomials with more than two terms.*

1 *Multiplying Binomials of the Type* $(a + b)(a - b)$

The case when you multiply $(x + y)(x - y)$ is interesting and deserves special consideration. Using the FOIL method, we find

$$(x + y)(x - y) = x^2 - xy + xy - y^2 = x^2 - y^2$$

Notice that the sum of the inner product and the outer product is zero. We see that

$$(x + y)(x - y) = x^2 - y^2$$

This works in all cases when the binomials have the same two terms. In one factor the terms are added, while in the other factor the terms are subtracted.

$$(5a + 2b)(5a - 2b) = 25a^2 - 10ab + 10ab - 4b^2$$
$$= 25a^2 - 4b^2$$

The product is the difference of squares. That is, $(5a)^2 - (2b)^2$ or $25a^2 - 4b^2$.

Many students find it helpful to memorize this equation.

Multiplying Binomials: A Sum and a Difference

$$(a + b)(a - b) = a^2 - b^2$$

You may use this relationship to find the product quickly in cases where it applies. The terms must be the same and there must be a sum and a difference.

EXAMPLE 1 Multiply. $(7x + 2)(7x - 2)$

$$(7x + 2)(7x - 2) = (7x)^2 - (2)^2 = 49x^2 - 4$$

Practice Problem 1 Multiply. $(6x + 7)(6x - 7)$ ■

EXAMPLE 2 Multiply. $(5x - 8y)(5x + 8y)$

$$(5x - 8y)(5x + 8y) = (5x)^2 - (8y)^2 = 25x^2 - 64y^2$$

Practice Problem 2 Multiply. $(3x + 5y)(3x - 5y)$ ■

EXAMPLE 3 Multiply. $(2x^2 - 9)(2x^2 + 9)$

$$(2x^2 - 9)(2x^2 + 9) = (2x^2)^2 - (9)^2 = 4x^4 - 81$$

Practice Problem 3 Multiply. $(4x^2 - 1)(4x^2 + 1)$ ■

2 *Multiplying Binomials of the Type $(a + b)^2$ and $(a - b)^2$*

A second case that is worth special consideration is when a binomial is squared. Consider the problem

$$\begin{aligned}(3x + 2)^2 &= (3x + 2)(3x + 2)\\ &= 9x^2 + 6x + 6x + 4\\ &= 9x^2 + 12x + 4\end{aligned}$$

If you complete enough problems of this type, you will notice a pattern. The answer always contains the square of the first term added to double the product of the first and last terms added to the square of the last term. That is,

$3x$ is the first term		2 is the last term		Square the first term: $(3x)^2$		Double the product of the first and last terms: $2(3x)(2)$		Square the last term: $(2)^2$
↓		↓		↓		↓		↓
$(3x$	$+$	$2)^2$	$=$	$9x^2$	$+$	$12x$	$+$	4

We show the same steps using variables instead of words.

$$(a + b)^2 = a^2 + 2ab + b^2$$

This is also true for the square of a difference.

$$\begin{aligned}(2x - 3)^2 &= (2x)^2 - 2(2x)(3) + (3)^2\\ &= 4x^2 - 12x + 9\end{aligned}$$

You may wish to multiply this out using FOIL to verify. That is, multiply $(2x - 3)(2x - 3)$.

$$(a - b)^2 = a^2 - 2ab + b^2$$

The two types of multiplications, the square of a sum and the square of a difference, can be summarized as follows:

A Binomial Squared

$$\boldsymbol{(a + b)^2 = a^2 + 2ab + b^2}$$

$$\boldsymbol{(a - b)^2 = a^2 - 2ab + b^2}$$

EXAMPLE 4 Multiply. **(a)** $(5y - 2)^2$ **(b)** $(8x + 9y)^2$

(a) $(5y - 2)^2 = (5y)^2 - (2)(5y)(2) + (2)^2$

$= 25y^2 - 20y + 4$

(b) $(8x + 9y)^2 = (8x)^2 + (2)(8x)(9y) + (9y)^2$

$= 64x^2 + 144xy + 81y^2$

Practice Problem 4 Multiply. **(a)** $(5x + 4)^2$ **(b)** $(4a - 9b)^2$ ■

Warning

$(a + b)^2 \neq a^2 + b^2$! The two sides are not equal! Squaring a sum $(a + b)$ does not give $a^2 + b^2$! Beginning algebra students often make this error. Make sure you remember that when you square a binomial there is always a *middle term.*

$$(a + b)^2 = a^2 + 2ab + b^2$$

Sometimes a numerical example helps you to see this.

$$(3 + 4)^2 \neq 3^2 + 4^2$$
$$7^2 \neq 9 + 16$$
$$49 \neq 25$$

Notice that what is missing on the right is the $2ab = 2 \cdot 3 \cdot 4 = 24$.

3 *Multiplying Polynomials with More Than Two Terms*

We used the distributive property to multiply two binomials $(a + b)(c + d)$, and we obtained $ac + ad + bc + bd$. We could also use the distributive property to multiply the polynomials $(a + b)(c + d + e)$ and we would obtain $ac + ad + ae + bc + bd + be$. Let us see if we can find a direct way to multiply polynomials such as $(3x - 2)(x^2 - 2x + 3)$. It can be quickly done using an approach similar to that used in arithmetic for multiplying whole numbers. Consider the following arithmetic problem.

$$\begin{array}{rl} 128 & \\ \times\ \ 43 & \\ \hline 384 & \longleftarrow \text{The product of 128 and 3.} \\ 512\ \ & \longleftarrow \text{The product of 128 and 4 moved one space to the left.} \\ \hline 5504 & \longleftarrow \text{The sum of the two products.} \end{array}$$

Let us follow a similar format to multiply the two polynomials. For example, multiply $(x^2 - 2x + 3)(3x - 2)$.

$$\begin{array}{rl} x^2 - 2x + 3 & \text{We place the polynomial with the greatest number of} \\ 3x - 2 & \text{terms above the other polynomial.} \\ \hline -2x^2 + 4x - 6 & \longleftarrow \text{The product } (x^2 - 2x + 3)(-2). \\ 3x^3 - 6x^2 + 9x\qquad & \longleftarrow \text{The product } (x^2 - 2x + 3)(3x) \text{ moved one space to the} \\ \hline & \text{left so that like terms are underneath each other.} \\ 3x^3 - 8x^2 + 13x - 6 & \longleftarrow \text{The sum of the two products.} \end{array}$$

EXAMPLE 5 Multiply *vertically.* $(3x^3 + 2x^2 + x)(x^2 - 2x - 4)$

$$\begin{array}{rl} 3x^3 + 2x^2 + x & \text{We place one polynomial over the other.} \\ x^2 - 2x - 4 & \text{This is often called vertical multiplication.} \\ \hline -12x^3 - 8x^2 - 4x & \longleftarrow \text{The product } (3x^3 + 2x^2 + x)(-4). \\ -6x^4 - 4x^3 - 2x^2\qquad & \longleftarrow \text{The product } (3x^3 + 2x^2 + x)(-2x). \\ 3x^5 + 2x^4 + x^3\qquad\qquad & \longleftarrow \text{The product } (3x^3 + 2x^2 + x)(x^2). \\ \hline 3x^5 - 4x^4 - 15x^3 - 10x^2 - 4x & \longleftarrow \text{The sum of the three products.} \end{array}$$

Note that the answers for each product are placed so that similar terms are underneath each other.

Practice Problem 5 Multiply vertically the following polynomials.

$(3x^2 - 2xy + 4y^2)(x - 2y)$ ■

Alternative Method

Some students prefer to do this type of multiplication using a horizontal format similar to the FOIL method. The following example illustrates this approach.

EXAMPLE 6 Multiply horizontally. $(x^2 + 3x + 5)(x^2 - 2x - 6)$

We will use the distributive property repeatedly.

$$\begin{aligned}(x^2 + 3x + 5)(x^2 - 2x - 6) &= x^2(x^2 - 2x - 6) + 3x(x^2 - 2x - 6) + 5(x^2 - 2x - 6)\\ &= x^4 - 2x^3 - 6x^2 + 3x^3 - 6x^2 - 18x + 5x^2 - 10x - 30\\ &= x^4 + x^3 - 7x^2 - 28x - 30\end{aligned}$$

Practice Problem 6 Multiply horizontally. $(2x^2 - 5x + 3)(x^2 + 3x - 4)$ ■

Some problems may need to be done in two or more separate steps.

EXAMPLE 7 Multiply. $(2x - 3y)(x + 2y)(x + y)$

We first need to multiply any two binomials. Let us select the first pair.

$$\underbrace{(2x - 3y)(x + 2y)}_{\text{Do this product first.}}(x + y)$$

$$\begin{aligned}(2x - 3y)(x + 2y) &= 2x^2 + 4xy - 3xy - 6y^2\\ &= 2x^2 + xy - 6y^2\end{aligned}$$

Now we replace the first two sets of parentheses with the result obtained above.

$$\underbrace{(2x^2 + xy - 6y^2)}_{\text{Result of first product}}(x + y)$$

We then multiply.

$$\begin{aligned}(2x^2 + xy - 6y^2)(x + y) &= (2x^2 + xy - 6y^2)x + (2x^2 + xy - 6y^2)y\\ &= 2x^3 + x^2y - 6xy^2 + 2x^2y + xy^2 - 6y^3\\ &= 2x^3 + 3x^2y - 5xy^2 - 6y^3\end{aligned}$$

The vertical format of Example 5 is an alternative method for this type of problem.

$$\begin{array}{r} 2x^2 + xy - 6y^2\\ x + y\\ \hline 2x^2y + xy^2 - 6y^3\\ 2x^3 + x^2y - 6xy^2 \quad\\ \hline 2x^3 + 3x^2y - 5xy^2 - 6y^3 \end{array}$$

Be sure to use special care in writing the exponents correctly in problems with two different variables.

Thus we have

$$(2x - 3y)(x + 2y)(x + y) = 2x^3 + 3x^2y - 5xy^2 - 6y^3$$

Note that it does not matter which two binomials are multiplied first. For example, you could first multiply $(2x - 3y)(x + y)$ to obtain $2x^2 - xy - 3y^2$ and then multiply that result by $(x + 2y)$ and still obtain the same result.

Practice Problem 7 Multiply. $(3x - 2)(2x + 3)(3x + 2)$ (*Hint:* Rearrange.) ■

4.5 Exercises

Verbal and Writing Skills

1. In the special case of $(a + b)(a - b)$, a binomial times a binomial is a ______.

2. Identify which of the following could be the answer to a problem using the formula for $(a + b)(a - b)$.
(a) $9x^2 - 16$ **(b)** $4x^2 + 25$ **(c)** $9x^2 + 12x + 4$ **(d)** $x^4 - 1$

3. A student evaluated $(4x - 7)^2$ as $16x^2 + 49$. Is this correct? Why? Why not? What is missing? State the correct answer.

4. The square of a binomial, $(a - b)^2$, always produces a (choose):
(a) Binomial **(b)** Trinomial **(c)** Four-term polynomial

Use the relationship $(a + b)(a - b) = a^2 - b^2$ to multiply each of the following.

5. $(x + 4)(x - 4)$

6. $(y - 8)(y + 8)$

7. $(x - 9)(x + 9)$

8. $(x + 6)(x - 6)$

9. $(8x + 3)(8x - 3)$

10. $(6x + 5)(6x - 5)$

11. $(2x - 7)(2x + 7)$

12. $(3x - 10)(3x + 10)$

13. $(2x - 5y)(2x + 5y)$

14. $(9w - 4z)(9w + 4z)$

15. $(10p - 7q)(10p + 7q)$

16. $(11xy + 1)(11xy - 1)$

17. $(12x^2 + 7)(12x^2 - 7)$

18. $(1 - 8ab^2)(1 + 8ab^2)$

19. $(3x^2 - 4y^3)(3x^2 + 4y^3)$

Use the relationship for a binomial squared to multiply each of the following.

20. $(3y + 1)^2$

21. $(4x - 1)^2$

22. $(7x - 2)^2$

23. $(8x + 3)^2$

24. $(9x + 5)^2$

25. $(5x - 7)^2$

26. $(3x - 7)^2$

27. $(2x + 3y)^2$

28. $(3x + 4y)^2$

29. $(8x - 5y)^2$

30. $(3 - 4y)^2$

31. $(8 + 3x)^2$

32. $(4a + 5b)^2$

33. $(6w + 5z)^2$

34. $(5xy - 6z)^2$

Multiply. Use the special relationship that applies.

35. $(7x + 3y)(7x - 3y)$

36. $(12a - 5b)(12a + 5b)$

37. $(4x^2 - 7y)^2$

38. $(3a - 10b^2)^2$

Multiply out the polynomials.

39. $(x^2 - 2x + 1)(x - 2)$

40. $(x^2 + 3x - 1)(x + 3)$

41. $(3y + 2z)(3y^2 - yz - 2z^2)$

42. $(4w - 2z)(w^2 + 2wz + z^2)$

43. $(4x + 1)(x^3 - 2x^2 + x - 1)$

44. $(3x - 1)(x^3 + x^2 - 4x - 2)$

45. $(x + 2)(x - 3)(2x + 5)$

46. $(x + 6)(x - 1)(3x - 2)$

47. $(3x + 5)(x - 2)(x - 4)$

48. $(2x - 7)(x + 1)(x - 2)$

49. $(x + 4)(2x - 7)(x - 4)$

50. $(x - 2)(5x + 3)(x - 2)$

51. $(x^2 + 2x + 1)(x^2 - x - 3)$

52. $(y^2 - 3y + 1)(y^2 - 2y + 4)$

★ **53.** $(x^2 + 3y)(x^2 - 3y)(2x^2 + 5xy - 6y^2)$

★ **54.** $(x^3 + 3x^2 - 4x + 2)(2x^3 - 4x^2 + x + 5)$

Cumulative Review Problems

55. An executive invested \$18,000 in two accounts, one cash-on-reserve account that yielded 7% simple interest, the other a long-term certificate paying 11% simple interest. At the end of one year, the two accounts had earned her a total of \$1540. How much did she invest in each account?

56. The perimeter of a rectangular room measures 34 meters. The width is 2 meters more than half the length. Find the dimensions of the room.

Putting Your Skills to Work

AREA

You can use what you have learned about multiplying binomials to solve a geometry problem involving area. Look at the figure on the right. A small square is cut from a large square of cardboard. You can find the area of the piece that is left. That is, you can express the area of the shaded portion as a polynomial and then use this general formula for specific problems.

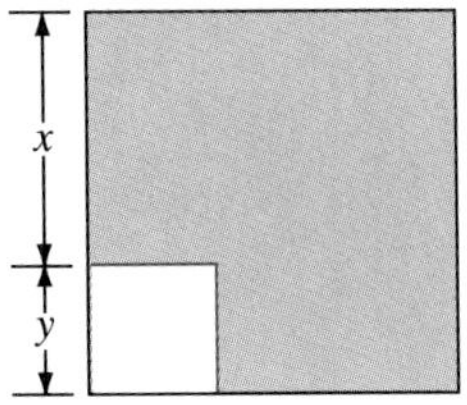

Use the figure on the right to solve the following problem.

1. Express the area of the shaded portion as a polynomial. (*Hint:* The side of the large square is $x + y$.)

Use the formula you derived in Problem 1 to solve Problems 2 and 3.

2. A small square with a side of 3 inches is cut from a large square with a side of 8 inches. What is the area of the portion that is left? (*Hint:* Draw and label a diagram.)

3. A small square with a side of 12 inches is cut from the middle of a large square with a side of 25 inches. What is the area of the portion that is left?

4. Let's take another look at the same problem. We could have labeled the sides as shown in the figure on the right. What formula does this picture suggest?

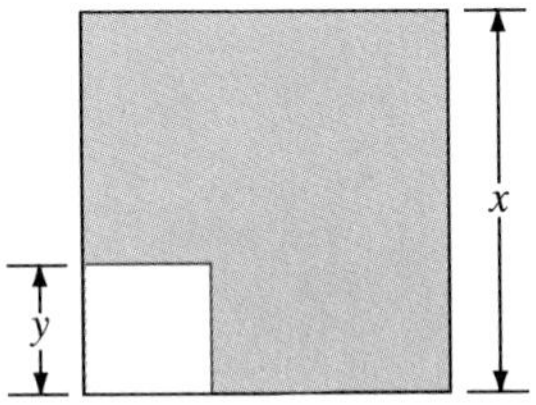

5. Use the formula in Problem 4 to do Problems 2 and 3. Keep in mind that x has been redefined by the new figure.

6. There is a third way of finding the area of the shaded region when a square is cut from a square. It is derived from the formula in Problem 4. What is this expression?

7. Find the area of the region that is left when a small square with a side of 6 inches is cut from a large square with a side of 15 inches. Use $(x + y)(x - y)$, where x is the side of the large square and y is the side of the small square.

COOPERATIVE GROUP ACTIVITY

8. What happens when the small square moves? That is, can you use the same formulas if the small square is cut from the corner of the large square, along the side of the large square, or from the middle of the large square? Is the area preserved? Construct a convincing argument to show that this is true.

4.6 Division of Polynomials

After studying this section, you will be able to:

1 *Divide a polynomial by a monomial.*
2 *Divide a polynomial by a binomial.*

1 *Dividing a Polynomial by a Monomial*

To divide a polynomial by a monomial, divide each term of the numerator by the denominator; then write the sum of the results. We are using the property of fractions that

$$\frac{a+b}{c} = \frac{a}{c} + \frac{b}{c}$$

> **Dividing a Polynomial by a Monomial**
>
> **1.** Divide each term of the polynomial by the monomial.
>
> **2.** When dividing variables, use the property $\frac{x^a}{x^b} = x^{a-b}$.

EXAMPLE 1 Divide. $\frac{25x^2 - 20x}{5x}$

$$\frac{25x^2 - 20x}{5x} = \frac{25x^2}{5x} - \frac{20x}{5x} = 5x - 4$$

Practice Problem 1 Divide. $\frac{14y^2 - 49y}{7y}$ ■

EXAMPLE 2 Divide. $\frac{8y^6 - 8y^4 + 24y^2}{8y^2}$

$$\frac{8y^6 - 8y^4 + 24y^2}{8y^2} = \frac{8y^6}{8y^2} - \frac{8y^4}{8y^2} + \frac{24y^2}{8y^2} = y^4 - y^2 + 3$$

Practice Problem 2 Divide. $\frac{15y^4 - 27y^3 - 21y^2}{3y^2}$ ■

2 *Dividing a Polynomial by a Binomial*

Division of a polynomial by a binomial is similar to long division in arithmetic. Notice the similarity in the following division problems.

Division of a Three-digit by a Two-digit Number	***Division of a Polynomial by a Binomial***
$\begin{array}{r} 32 \\ 21\overline{)672} \\ \underline{63} \\ 42 \\ \underline{42} \end{array}$	$\begin{array}{r} 3x + 2 \\ 2x + 1\overline{)6x^2 + 7x + 2} \\ \underline{6x^2 + 3x} \\ 4x + 2 \\ \underline{4x + 2} \end{array}$
To check, multiply. $(21)(32) = 672$	*To check, multiply.* $(2x + 1)(3x + 2) = 6x^2 + 7x + 2$

Dividing a Polynomial by a Binomial

1. Place the terms of the polynomial and binomial in descending order. Insert a 0 for any missing term.
2. Divide the first term of the polynomial by the first term of the binomial. The result is the first term of the answer.
3. Multiply the first term of the answer by the binomial and subtract the results from the first two terms of the polynomial. Bring down the next term to obtain a new polynomial.
4. Divide the new polynomial by the binomial using the process described in step 2.
5. Continue dividing, multiplying, and subtracting until the remainder is at a lower power than the variable in the first term of the binomial divisor.

EXAMPLE 3 Divide. $(x^3 + 5x^2 + 11x + 14) \div (x + 2)$

Step 1. The terms are arranged in descending order. No terms are missing.

Step 2. Divide the first term of the polynomial by the first term of the binomial. In this case, divide x^3 by x to get x^2.

$$\begin{array}{r} x^2 \\ x + 2\overline{)x^3 + 5x^2 + 11x + 14} \end{array}$$

Step 3. Multiply x^2 by $x + 2$ and subtract the results from the first two terms of the polynomial: $x^3 + 5x^2$, in this case.

$$\begin{array}{l} x^2 \\ x + 2\overline{)x^3 + 5x^2 + 11x + 14} \\ \underline{x^3 + 2x^2} \quad \downarrow \\ 3x^2 + 11x \end{array}$$

Bringing down the next term.

Step 4. Continue to use the step 2 process. Divide $3x^2 + 11x$ by $x + 2$. Write the resulting $3x$ as the next term of the answer.

$$\begin{array}{l} x^2 + 3x \\ x + 2\overline{)x^3 + 5x^2 + 11x + 14} \\ \underline{x^3 + 2x^2} \\ 3x^2 + 11x \end{array}$$

Step 5. Continue multiplying, dividing, and subtracting until the remainder is at a power lower than the variable in the binomial's first term. In this case, we stop when the remainder does not have an x.

$$\begin{array}{l} x^2 + 3x + 5 \\ x + 2\overline{)x^3 + 5x^2 + 11x + 14} \\ \underline{x^3 + 2x^2} \\ 3x^2 + 11x \\ \underline{3x^2 + 6x} \\ 5x + 14 \\ \underline{5x + 10} \\ 4 \longleftarrow \text{The remainder is 4.} \end{array}$$

Answer is $x^2 + 3x + 5 + \dfrac{4}{x + 2}$.

To check the answer, we multiply $(x + 2)(x^2 + 3x + 5)$ and add the remainder 4.

$$(x + 2)(x^2 + 3x + 5) = x^3 + 5x^2 + 11x + 10$$

Now add 4. We get $x^3 + 5x^2 + 11x + 14$. This is the original polynomial. It checks.

Practice Problem 3 Divide. $(x^3 + 10x^2 + 31x + 35) \div (x + 4)$ ■

Take great care with the subtraction step when negative numbers are involved.

EXAMPLE 4 Divide. $(5x^3 - 24x^2 + 9) \div (5x + 1)$

We must first put in a $0x$ to represent the missing term. Then we divide $5x^3 - 24x^2$ by $5x + 1$.

$$\begin{array}{r}
x^2 \phantom{{}+0x+9} \\
5x + 1 \overline{)\, 5x^3 - 24x^2 + 0x + 9} \\
\underline{5x^3 + x^2} \phantom{{}+0x+9} \\
-25x^2 \phantom{{}+0x+9}
\end{array}$$

Note that we are subtracting:
$-24x^2 - (+1x^2)$
$= -24x^2 - 1x^2 = -25x^2$

Next we divide $-25x^2 + 0x$ by $5x + 1$.

$$\begin{array}{r}
x^2 - 5x \phantom{{}+9} \\
5x + 1 \overline{)\, 5x^3 - 24x^2 + 0x + 9} \\
\underline{5x^3 + x^2} \phantom{{}+0x+9} \\
-25x^2 + 0x \phantom{{}+9} \\
\underline{-25x^2 - 5x} \phantom{{}+9} \\
5x \phantom{{}+9}
\end{array}$$

Note we are subtracting:
$0x - (-5x)$
$= 0x + (5x) = 5x$

Finally, we divide $5x + 9$ by $5x + 1$.

$$\begin{array}{r}
x^2 - 5x + 1 \\
5x + 1 \overline{)\, 5x^3 - 24x^2 + 0x + 9} \\
\underline{5x^3 + x^2} \phantom{{}+0x+9} \\
-25x^2 + 0x \phantom{{}+9} \\
\underline{-25x^2 - 5x} \phantom{{}+9} \\
5x + 9 \\
\underline{5x + 1} \\
8
\end{array} \longleftarrow \text{The remainder is 8.}$$

The answer is $x^2 - 5x + 1 + \dfrac{8}{5x + 1}$.

To check, multiply $(5x + 1)(x^2 - 5x + 1)$ and add the remainder 8.

$$(5x + 1)(x^2 - 5x + 1) = 5x^3 - 24x^2 + 1$$

Now add the remainder. $5x^3 - 24x^2 + 1 + 8 = 5x^3 - 24x^2 + 9$. This is the original polynomial. Our answer is correct.

Practice Problem 4 Divide. $(2x^3 - x^2 + 1) \div (x - 1)$ ■

Now we will perform the division by writing a minimum of steps. See if you can follow each step.

EXAMPLE 5 Divide and check. $(12x^3 - 11x^2 + 8x - 4) \div (3x - 2)$

$$
\begin{array}{r}
4x^2 - x + 2 \\
3x - 2\overline{)12x^3 - 11x^2 + 8x - 4} \\
\underline{12x^3 - 8x^2} \phantom{{}+ 8x - 4} \\
- 3x^2 + 8x \phantom{{}- 4} \\
\underline{- 3x^2 + 2x} \phantom{{}- 4} \\
6x - 4 \\
\underline{6x - 4}
\end{array}
$$

Check. $(3x - 2)(4x^2 - x + 2) = 12x^3 - 3x^2 + 6x - 8x^2 + 2x - 4$

$= 12x^3 - 11x^2 + 8x - 4$ *Our answer is correct.*

Practice Problem 5 Divide and check. $(20x^3 - 11x^2 - 11x + 6) \div (4x - 3)$ ■

Developing Your Study Skills

EXAM TIME: TAKING THE EXAM

Allow yourself plenty of time to get to your exam. You may even find it helpful to arrive a little early in order to collect your thoughts and ready yourself. This will help you feel more relaxed.

After you get your exam, you will find it helpful to do the following.

1. Take two or three moderately deep breaths. Inhale, then exhale slowly. You will feel your entire body begin to relax.
2. Write down on the back of the exam any formulas or ideas that you need to remember.
3. Look over the entire test quickly in order to pace yourself and use your time wisely. Notice how many points each problem is worth. Spend more time on items of greater worth.
4. Read directions carefully, and be sure to answer all questions clearly. Keep your work neat and easy to read.
5. Ask your instructor about anything that is not clear to you.
6. Work the problems and answer the questions that are easiest for you first. Then come back to the more difficult ones.
7. Do not get bogged down on one problem too long because it may jeopardize your chances of finishing other problems. Leave the tough problem and come back to it when you have time later.
8. Check your work. This will help you to catch minor errors.
9. Stay calm if others leave before you do. You are entitled to use the full amount of allotted time.

4.6 Exercises

Divide.

1. $\dfrac{8x^3 - 20x^2 + 4x}{2x}$

2. $\dfrac{21x^3 - 6x^2 + 9x}{3x}$

3. $\dfrac{8y^4 - 12y^3 - 4y^2}{4y^2}$

4. $\dfrac{10y^4 - 35y^3 + 5y^2}{5y^2}$

5. $\dfrac{49x^6 - 21x^4 + 56x^2}{7x^2}$

6. $\dfrac{36x^6 + 54x^4 - 6x^2}{6x^2}$

7. $(16x^5 - 72x^3 + 24x^2) \div (8x^2)$

8. $(18x^4 - 27x^3 + 36x^2) \div (9x^2)$

9. $(30x^4 - 21x^3 - 3x^2 + 15x) \div (3x)$

10. $(100x^4 - 16x^3 - 28x^2 + 56x) \div (2x)$

Divide. Check your answers for Exercises 11 through 18 by multiplication.

11. $\dfrac{6x^2 + 13x + 5}{2x + 1}$

12. $\dfrac{12x^2 + 19x + 5}{3x + 1}$

13. $\dfrac{x^2 - 9x - 6}{x - 7}$

14. $\dfrac{x^2 - 8x - 4}{x - 6}$

15. $\dfrac{3x^3 - x^2 + 4x - 2}{x + 1}$

16. $\dfrac{2x^3 - 3x^2 - 3x + 6}{x - 1}$

17. $\dfrac{2x^3 - x^2 + 3x + 2}{2x + 1}$

18. $\dfrac{2x^3 - 3x^2 + 4x + 4}{2x + 1}$

19. $\dfrac{6x^3 + x^2 - 7x + 2}{3x - 1}$

20. $\dfrac{3x^3 + 8x^2 - 6x + 2}{3x - 1}$

21. $\dfrac{8y^3 - 38y^2 + 49y + 15}{4y - 1}$

22. $\dfrac{4y^3 - 5y^2 + 2y - 18}{y - 2}$

23. $(y^3 - y^2 - 13y - 12) \div (y + 3)$

24. $(4y^3 - 17y^2 + 7y + 10) \div (4y - 5)$

25. $(4y^3 - 1 + 3y) \div (2y - 1)$

26. $(y^3 - 7y^2 - 8y - 9) \div (y + 3)$

27. $(y^4 - 3y^2 + 2y - 1) \div (y + 3)$

28. $(y^4 - 2y^3 + y^2 - 1) \div (y - 2)$

29. $(y^4 - 9y^2 - 5) \div (y - 2)$

30. $(2y^4 + 3y^3 - 3y + 5) \div (2y - 1)$

Cumulative Review Problems

Solve and graph each inequality.

31. $x - 6 \leq 4(x - 3) - x$

32. $3 + 4x < 3x + 4 - 2x - 7$

33. $3 + 2x > 6x - 9 - 7x + 5x$

34. $11 + 3(x + 1) \geq 5x + 16$

Chapter Organizer

Topic	Procedure	Examples
Multiplying monomials, p. 241.	$x^a \cdot x^b = x^{a+b}$ **1.** Multiply the numerical coefficients. **2.** Add the exponents of a given base.	$3^{12} \cdot 3^{15} = 3^{27}$ $x^3 \cdot x^4 = x^7$ $(-3x^2)(6x^3) = -18x^5$ $(2ab)(4a^2b^3) = 8a^3b^4$
Dividing monomials, p. 243.	$\frac{x^a}{x^b} = \begin{cases} x^{a-b} & \text{if } a \text{ is greater than } b \\ \frac{1}{x^{b-a}} & \text{if } b \text{ is greater than } a \end{cases}$ **1.** Divide or reduce the fraction created by the quotient of the numerical coefficients. **2.** Subtract the exponents of a given base.	$\frac{16x^7}{8x^3} = 2x^4$ $\frac{5x^3}{25x^5} = \frac{1}{5x^2}$ $\frac{-12x^5y^7}{18x^3y^{10}} = \frac{-2x^2}{3y^3}$
Exponent of zero, p. 244.	$x^0 = 1 \quad \text{if } x \neq 0$	$5^0 = 1 \quad \frac{x^6}{x^6} = 1$ $w^0 = 1 \quad 3x^0y = 3y$
Raising a power to a power, p. 245.	$(x^a)^b = x^{ab}$ $(xy)^a = x^ay^a$ $\left(\frac{x}{y}\right)^a = \frac{x^a}{y^a} \quad (y \neq 0)$ **1.** Raise the numerical coefficient to the power outside the parentheses. **2.** Multiply the exponent outside the parentheses times the exponent inside the parentheses.	$(x^9)^3 = x^{27}$ $(3x^2)^3 = 27x^6$ $\left(\frac{2x^2}{y^3}\right)^3 = \frac{8x^6}{y^9}$ $(-3x^4y^5)^4 = 81x^{16}y^{20}$ $(-5ab)^3 = -125a^3b^3$
Negative exponents, p. 251.	If $x \neq 0$, $y \neq 0$ Then $x^{-n} = \frac{1}{x^n}$ $\frac{1}{x^{-n}} = x^n$ $\frac{x^{-m}}{y^{-n}} = \frac{y^n}{x^m}$	Write with positive exponents. $3^{-4} = \frac{1}{3^4} = \frac{1}{81}$ $x^{-6} = \frac{1}{x^6}$ $\frac{1}{w^{-3}} = w^3$ $\frac{w^{-12}}{z^{-5}} = \frac{z^5}{w^{12}}$ $(2x^2)^{-3} = 2^{-3}x^{-6} = \frac{1}{2^3x^6}$ $= \frac{1}{8x^6}$
Scientific notation, p. 252.	A positive number is written in scientific notation if it is in the form $a \times 10^n$, where $1 \leq a < 10$ and n is an integer.	$128 = 1.28 \times 10^2$ $2{,}568{,}000 = 2.568 \times 10^6$ $13{,}200{,}000{,}000 = 1.32 \times 10^{10}$ $0.16 = 1.6 \times 10^{-1}$ $0.00079 = 7.9 \times 10^{-4}$ $0.0000034 = 3.4 \times 10^{-6}$
Add polynomials, p. 259.	To add two polynomials, we add the respective like terms.	$(-7x^3 + 2x^2 + 5) + (x^3 + 3x^2 + x)$ $= -6x^3 + 5x^2 + x + 5$
Subtracting polynomials, p. 260.	To subtract the polynomials, change all signs of the second polynomial and add the result to the first polynomial. $(a) - (b) = (a) + (-b)$	$(5x^2 - 6) - (-3x^2 + 2)$ $= (5x^2 - 6) + (+3x^2 - 2)$ $= 8x^2 - 8$
Multiplying a monomial by a polynomial, p. 263.	Use the distributive property. $a(b + c) = ab + ac$ $(b + c)a = ba + ca$	Multiply. $-5x(2x^2 + 3x - 4) = -10x^3 - 15x^2 + 20x$ $(6x^3 - 5xy - 2y^2)(3xy) = 18x^4y - 15x^2y^2 - 6xy^3$

Chapter Organizer

Topic	Procedure	Examples
Multiplying two binomials, p. 264.	**1.** The product of the sum and difference of the same two values yields the difference of their squares. $(a + b)(a - b) = a^2 - b^2$ **2.** The square of a binomial yields a trinomial: the square of the first, plus twice the product of first and second, plus the square of the second. $(a + b)^2 = a^2 + 2ab + b^2$ $(a - b)^2 = a^2 - 2ab + b^2$ **3.** Use FOIL for other binomial multiplication. The middle terms can often be combined, giving a trinomial answer.	$(3x + 7y)(3x - 7y) = 9x^2 - 49y^2$ $(3x + 7y)^2 = 9x^2 + 42xy + 49y^2$ $(3x - 7y)^2 = 9x^2 - 42xy + 49y^2$ $(3x - 5)(2x + 7) = 6x^2 + 21x - 10x - 35$ $= 6x^2 + 11x - 35$
Multiplying two polynomials, p. 271.	To multiply two polynomials, multiply each term of one by each term of the other. This method is similar to the multiplication of many-digit numbers.	Vertical method $(3x^2 - 7x + 4)$ $\times \quad (3x - 1)$ $-3x^2 + 7x - 4$ $9x^3 - 21x^2 + 12x$ $9x^3 - 24x^2 + 19x - 4$ Horizontal method $(5x + 2)(2x^2 - x + 3)$ $= 10x^3 - 5x^2 + 15x + 4x^2 - 2x + 6$ $= 10x^3 - x^2 + 13x + 6$
Multiplying three or more polynomials, p. 272.	**1.** Multiply any two polynomials. **2.** Multiply the result by any remaining polynomials.	$(2x + 1)(x - 3)(x + 4) = (2x^2 - 5x - 3)(x + 4)$ $2x^2 - 5x - 3$ $x + 4$ $8x^2 - 20x - 12$ $2x^3 - 5x^2 - 3x$ $2x^3 + 3x^2 - 23x - 12$
Dividing a polynomial by a monomial, p. 276.	**1.** Divide each term of the polynomial by the monomial. **2.** When dividing variables, use the property $\frac{x^a}{x^b} = x^{a-b}$	Divide. $(15x^3 + 20x^2 - 30x) \div (5x)$ $= \frac{15x^3}{5x} + \frac{20x^2}{5x} + \frac{-30x}{5x}$ $= 3x^2 + 4x - 6$
Dividing a polynomial by a binomial, p. 277.	**1.** Place the terms of the polynomial and binomial in descending order. Insert a 0 for any missing term. **2.** Divide the first term of the polynomial by the first term of the binomial. **3.** Multiply the partial answer by the binomial, and subtract the results from the first two terms of the polynomial. Bring down the next term to obtain a new polynomial. **4.** Divide the new polynomial by the binomial using the process described in step 2. **5.** Continue dividing, multiplying, and subtracting until the remainder is at a lower power than the variable in the first term of the binomial divisor.	Divide. $(8x^3 - 13x + 2x^2 + 7) \div (4x - 1)$ We rearrange the terms. $2x^2 + x - 3$ $4x - 1\overline{)8x^3 + 2x^2 - 13x + 7}$ $8x^3 - 2x^2$ $4x^2 - 13x$ $4x^2 - x$ $-12x + 7$ $-12x + 3$ 4 The answer is $2x^2 + x - 3 + \frac{4}{4x - 1}$

Chapter 4 Review Problems

4.1 *Simplify and leave your answer in exponent form.*

1. $(-6a^2)(3a^5)$ **2.** $(5^{10})(5^{13})$ **3.** $(3xy^2)(2x^3y^4)$ **4.** $5xy(-2x^3)(-4xy^4)$

5. $\frac{8^{20}}{8^3}$ **6.** $\frac{7^{15}}{7^{27}}$ **7.** $\frac{x^8}{x^{12}}$ **8.** $\frac{w^{20}}{w^{15}}$

9. $\frac{3x^8y^0}{9x^4}$ **10.** $\frac{-15xy^2}{25x^6y^6}$ **11.** $\frac{-12a^3b^6}{18a^2b^{12}}$ **12.** $\frac{-20a^8b^6c^2}{-12a^5b^5c^3}$

13. $(x^3)^8$ **14.** $(5xy^2)^3$ **15.** $(-3a^3b^2)^2$ **16.** $[3(ab)^2]^3$

17. $\frac{2x^4}{3y^2}$ **18.** $\left(\frac{5ab^2}{c^3}\right)^2$ **19.** $\frac{(2xy^2)^3}{(2xy^2)^4}$ **20.** $\left(\frac{x^0y^3}{4w^5z^2}\right)^3$

4.2 *Simplify. Write with positive exponents.*

21. x^{-3} **22.** $x^{-6}y^{-8}$ **23.** $\frac{2x^{-6}}{y^{-3}}$ **24.** $2^{-1}x^5y^{-6}$

25. $(2x^3)^{-2}$ **26.** $\frac{3x^{-3}}{y^{-2}}$ **27.** $\frac{4x^{-5}y^{-6}}{w^{-2}z^8}$ **28.** $\frac{3^{-3}a^{-2}b^5}{c^{-3}d^{-4}}$

Write in scientific notation.

29. 156,340,200,000 **30.** 88,367 **31.** 179,632 **32.** 0.0078 **33.** 0.00006173 **34.** 0.13

Write in decimal notation.

35. 1.2×10^5 **36.** 8.367×10^{10} **37.** 3×10^6

38. 2.5×10^{-1} **39.** 5.708×10^{-8} **40.** 6×10^{-9}

Perform the calculation and leave the answer in exponent form.

41. $\frac{(28{,}000{,}000)(5{,}000{,}000{,}000)}{7000}$ **42.** $(3.12 \times 10^5)(2.0 \times 10^6)(1.5 \times 10^8)$

43. $(1.6 \times 10^{-3})(3.0 \times 10^{-5})(2.0 \times 10^{-2})$ **44.** $\frac{(0.00078)(0.000005)(0.00004)}{0.002}$

45. If a space probe travels at 40,000 kilometers per hour for 1 year, how far will it travel? (Assume that 1 year = 365 days.)

46. An atomic clock is based on the fact that cesion emits 9,192,631,770 cycles of radiation in one second. How many of these cycles occur in one day? Round your answer to 3 significant digits.

47. Today's fastest modern computers can perform one operation in 1×10^{-8} second. How many operations can such a computer perform in 1 minute?

4.3 *Combine the following polynomials.*

48. $(2x^2 - 3x + 5) + (-7x^2 - 8x - 23)$

49. $(-7x^2 - 6x + 3) + (2x^2 - 10x - 15)$

50. $(x^3 + x^2 - 6x + 2) - (2x^3 - x^2 - 5x - 6)$

51. $(4x^3 - x^2 - x + 3) - (-3x^3 + 2x^2 + 5x - 1)$

52. $(-6x^2 - 3xy + y^2) + (-5x^2 + 10xy - 21y^2)$

53. $(8x^2 + 2y + 3xy) + (-5x^2 - 8y + 17xy)$

54. $(9x^3y^3 + 3xy - 4) - (4x^3y^3 + 2x^2y^2 - 7xy)$

55. $(4x^2y^2 + 6xy - 2y) - (-7x^2y^2 + 5xy + 3y^2)$

56. $(5x^2 + 3x) + (-6x^2 + 2) - (5x - 8)$

57. $(2x^2 - 7) - (3x^2 - 4) + (-5x^2 - 6x)$

4.4 *Multiply.*

58. $(3x + 1)(5x - 1)$

59. $(8x - 2)(4x - 3)$

60. $(2x + 3)(10x + 9)$

61. $5x(2x^2 - 6x + 3)$

62. $(3x^2y^2 - 5xy + 6)(-2xy)$

63. $(x^3 - 3x^2 + 5x - 2)(4x)$

64. $(5a + 7b)(a - 3b)$

65. $(2x^2 - 3)(4x^2 - 5y)$

66. $-3x^2y(5x^4y + 3x^2 - 2)$

4.5 *Multiply.*

67. $(3x - 2)^2$

68. $(5x + 3)(5x - 3)$

69. $(7x + 6y)(7x - 6y)$

70. $(5a - 2b)^2$

71. $(8x + 9y)^2$

72. $(x^2 + 7x + 3)(4x - 1)$

73. $(x^3 + 2x^2 - x - 4)(2x - 3)$

74. $(x - 6)(2x - 3)(x + 4)$

75. $(2x + 3)^3$

4.6 *Divide.*

76. $(12y^3 + 18y^2 + 24y) \div (6y)$

77. $(30x^5 + 35x^4 - 90x^3) \div (5x^2)$

78. $(16x^3y^2 - 24x^2y + 32xy^2) \div (4xy)$

79. $(106x^6 - 24x^5 + 38x^4 + 26x^3) \div (2x^3)$

80. $(15x^2 + 41x + 14) \div (5x + 2)$

81. $(16x^2 - 8x - 3) \div (4x - 3)$

82. $(2x^3 - x^2 + 3x - 1) \div (x + 2)$

83. $(6x^3 + x^2 + 6x + 5) \div (2x - 1)$

84. $(8x^2 - 6x + 6) \div (2x + 1)$

85. $(12x^2 + 11x + 2) \div (3x + 2)$

86. $(x^3 - x - 24) \div (x - 3)$

87. $(2x^3 - 3x + 1) \div (x - 2)$

88. $(3x^4 + 2x^3 - 4x - 5) \div (x + 1)$

89. $(12x^3 + 11x^2 - 2x - 1) \div (3x - 1)$

Chapter 4 Test

Simplify each problem. Leave your answer in exponent form.

1. $(3^{10})(3^{24})$ **2.** $\frac{3^{35}}{3^7}$ **3.** $(8^4)^6$

Simplify.

4. $(-2x^2y)(-3x^5y^6)$

5. $\frac{-35x^8y^{10}}{25x^5y^{10}}$

6. $(-5xy^6)^3$

7. $\left(\frac{7a^7b^2}{3c^0}\right)^2$

8. $\frac{4a^5b^6}{16a^{10}b^{12}}$

9. Evaluate. 5^{-3}

Write with only positive exponents.

10. $6a^{-4}b^{-3}c^5$

11. $\frac{2x^{-3}y^{-4}}{w^{-6}z^8}$

12. Write in scientific notation. 0.0000001765

13. Write in decimal notation. 5.82×10^8

14. Multiply and leave your answer in scientific notation.

$(4.0 \times 10^{-3})(3.0 \times 10^{-8})(2.0 \times 10^4)$

1. ______

2. ______

3. ______

4. ______

5. ______

6. ______

7. ______

8. ______

9. ______

10. ______

11. ______

12. ______

13. ______

14. ______

15.

16.

17.

18.

19.

20.

21.

22.

23.

24.

25.

26.

Combine the following polynomials.

15. $(2x^2 - 3x - 6) + (-4x^2 + 8x + 6)$

16. $(5x^2 + 6xy - 7y^2) - (2x^2 + 3xy + 6y)$

Multiply.

17. $-7x^2(3x^3 - 4x^2 + 6x - 2)$

18. $(5x^2y^2 - 6xy + 2)(3x^2y)$

19. $(5a - 4b)(2a + 3b)$

20. $(3x + 2)(2x + 1)(x - 3)$

21. $(7x^2 + 2y^2)^2$

22. $(9x - 2y)(9x + 2y)$

23. $(3x - 2)(4x^3 - 2x^2 + 7x - 5)$

24. $(3x^2 - 5xy)(x^2 + 3xy)$

Divide.

25. $15x^6 - 5x^4 + 25x^3 \div 5x^3$

26. $(4y^4 + 6y^3 + 3y - 1) \div (2y^2 + 1)$

Cumulative Test for Chapters 0–4

Approximately one-half of this test covers the content of Chapters 0 through 3. The remainder covers the content of Chapter 4.

Simplify.

1. $\frac{5}{12} - \frac{7}{8}$

2. $(-3.7) \times (0.2)$

3. $\left(-4\frac{1}{2}\right) \div \left(5\frac{1}{4}\right)$

4. 7% of \$128.00

5. $7x(3x - 4) - 5x(2x - 3) - (3x)^2$

6. Evaluate $2x^2 - 3xy + y^2$ when $x = -2$ and $y = 3$.

Solve.

7. $7x - 3(4 - 2x) = 14x - (3 - x)$

8. $\frac{x - 5}{3} = \frac{1}{4}x + 2$

9. $4 - 7x < 11$

10. Solve for f: $B = \frac{1}{2}a(c + 3f)$

11. A national walkout of 11,904 employees of the VBM Corp. occurred last month. This was 96% of the number of employees. How many employees does VBM have?

12. How much interest would \$3320 earn in one year if invested at 6%?

1. ______
2. ______
3. ______
4. ______
5. ______
6. ______
7. ______
8. ______
9. ______
10. ______
11. ______
12. ______

13. ______

14. ______

15. ______

16. ______

17. ______

18. ______

19. ______

20. ______

21. ______

22. ______

23. ______

24. ______

13. Multiply. $(3x - 7)(5x - 4)$

14. Multiply. $(3x - 5)^2$

15. $(2x + 1)^3$ *Hint:* Write it in the form $(2x + 1)(2x + 1)(2x + 1)$ and then multiply.

Simplify.

16. $(-4x^4y^5)(5xy^3)$

17. $\dfrac{14x^8y^3}{-21x^5y^{12}}$

18. $(-3xy^4z^2)^3$

19. Write with only positive exponents. $\dfrac{9x^{-3}y^{-4}}{w^2z^{-8}}$

20. Write in scientific notation. 1,360,000,000,000,000

21. Perform the calculation and leave your answer in scientific notation.

$$\frac{(2.0 \times 10^{-12})(8.0 \times 10^{-20})}{4.0 \times 10^3}$$

22. Subtract the polynomials. $(x^3 - 3x^2 - 5x + 20) - (-4x^3 - 10x^2 + x - 30)$

Multiply.

23. $-6xy^2(6x^2 - 3xy + 8y^2)$

24. $(x^2 - 6x + 3)(2x^2 - 3x + 4)$

5
Factoring

If you are in a fishing boat out in the ocean over 100 miles from the shore, you depend on the reliability of your motor and your generator. They must run efficiently and correctly or your own safety is at risk. How can a boat generator produce the required amount of power with the minimum amount of current? These kinds of problems can be solved by factoring quadratic equations. See Exercises 5.7, Problem 42.

Pretest Chapter 5

1. ________	13. ________
2. ________	14. ________
3. ________	15. ________
4. ________	16. ________
5. ________	17. ________
6. ________	18. ________
7. ________	19. ________
8. ________	20. ________
9. ________	21. ________
10. ________	22. ________
11. ________	23. ________
12. ________	

If you are familiar with the topics in this chapter, take this test now. Check your answers with those in the back of the book. If an answer was wrong or you couldn't do a problem, study the appropriate section of the chapter.

If you are not familiar with the topics in this chapter, don't take this test now. Instead, study the examples, work the practice problems, and then take the test.

This test will help you to identify those concepts that you have mastered and those that need more study.

Section 5.1

Remove the greatest common factor.

1. $2x^2 - 6xy + 12xy^2$ **2.** $3x(a - 2b) + 4y(a - 2b)$ **3.** $36ab^2 - 18ab$

Section 5.2

Factor.

4. $5a - 10b - 3ax + 6xb$ **5.** $3x^2 - 4y + 3xy - 4x$ **6.** $21x^2 - 14x - 9x + 6$

Section 5.3

Factor.

7. $x^2 - 22x - 48$ **8.** $x^2 - 8x + 15$ **9.** $x^2 + 6x + 5$

10. $2x^2 + 8x - 24$ **11.** $3x^2 - 6x - 189$

Section 5.4

Factor.

12. $15x^2 - 16x + 4$ **13.** $6y^2 + 5yz - 6z^2$ **14.** $12x^2 + 44x + 40$

Section 5.5

Factor completely.

15. $81x^4 - 16$ **16.** $49x^2 - 28xy + 4y^2$ **17.** $25x^2 + 80x + 64$

Section 5.6

Factor completely. If not possible, so state.

18. $6x^3 + 15x^2 - 9x$ **19.** $32x^2y^2 - 48xy^2 + 18y^2$ **20.** $36x^2 + 49$

Section 5.7

Solve for the root of each quadratic equation.

21. $2x^2 + x - 3 = 0$ **22.** $x^2 + 30 = 11x$ **23.** $\dfrac{3x^2 - 7x}{2} = 3$

5.1 Introduction to Factoring

After studying this section, you will be able to:

1 *Factor polynomials containing a common factor in each term.*

1 *Common Factors*

Recall that when two or more numbers, variables, or algebraic expressions are multiplied, each is called a **factor.**

$$3 \cdot 2 \qquad 3x^2 \cdot 5x^3 \qquad (2x-3)(x+4)$$

factor factor factor factor factor factor

When you are asked to factor a number or an algebraic expression, you are being asked, ''What, when multiplied, will give that number or expression?'' For example, you can factor 6 as $3 \cdot 2$ since $3 \cdot 2 = 6$. You can factor $15x^5$ as $3x^2 \cdot 5x^3$ since $3x^2 \cdot 5x^3 = 15x^5$. Factoring is simply the reverse of multiplying. 6 and $15x^5$ are simple expressions to factor and can be factored in different ways. The factors of the polynomial $2x^2 + x - 12$ are not so easy to recognize. In this chapter we will be learning techniques to find the factors of a polynomial. We will begin with common factors.

EXAMPLE 1 Factor. **(a)** $3x - 6y$ **(b)** $9x + 2xy$

Begin by looking for a common factor, the factor that both terms have in common. Then rewrite the expression as a product.

(a) $3x - 6y = 3(x - 2y)$ *This is true because* $3(x - 2y) = 3x - 6y$.

(b) $9x + 2xy = x(9 + 2y)$ *This is true because* $x(9 + 2y) = 9x + 2xy$.

Some people find it helpful to think of factoring as the distributive property in reverse. When we write $3x - 6y = 3(x - 2y)$, we are ''undistributing'' the 3.

Practice Problem 1 Factor. **(a)** $21a - 7b$ **(b)** $p + prt$ ■

When we factor, we are looking for the largest common factor. For example, in the polynomial $48x - 16y$, a common factor is 2. We could factor $48 - 16y$ as $2(24 - 8y)$. However, this is not complete. To factor $48 - 16y$ completely, we look for the greatest common factor of 48 and of 16.

$$48x - 16y = 16(3x - y)$$

EXAMPLE 2 Factor. $24xy + 12x^2 + 36x^3$

Find the greatest common factor of 24, of 12, and of 36. You may want to factor each number or you may notice that 12 is a common factor. 12 is the greatest common factor. Notice also that x is a factor of each term.

$$24xy + 12x^2 + 36x^3 = 12x(2y + x + 3x^2)$$

Practice Problem 2 Factor. $12a^2 + 16ab^2 - 12a^2b$ ■

Common Factors of a Polynomial

1. You can determine the numerical common factor by asking, "What is the largest integer that will divide into the coefficient of each term?"
2. You can determine the variable common factor by asking, "What variables are common to each term, and what is the largest exponent of those variables that is common?"

EXAMPLE 3 Factor. **(a)** $12x^2 + 18y^2$ **(b)** $x^2y^2 + 3xy^2 + y^3$

(a) Note that the largest integer that is common to both terms is 6 (not 3 or 2).

$$12x^2 + 18y^2 = 6(2x^2 + 3y^2)$$

(b) Although y is common to all terms, we factor out y^2 since 2 is the largest exponent of y that is common to all terms. We do not factor out x since x is not common to all terms.

$$x^2y^2 + 3xy^2 + y^3 = y^2(x^2 + 3x + y)$$

Practice Problem 3 Factor. **(a)** $16a^3 - 24b^3$ **(b)** $r^3s^2 - 4r^4s + 7r^5$ ∎

Checking

You can check any factoring problem by multiplying the factors you obtain. The result should be the same as the original polynomial.

EXAMPLE 4 Factor. $8x^3y + 16x^2y^2 + 24x^3y^3$

We see that 8 is the largest integer that will divide evenly into the three numerical coefficients. We can factor an x^2 out of each term (not just x). For the other variable we factor y out of each term.

$$8x^3y + 16x^2y^2 + 24x^3y^3 = 8x^2y(x + 2y + 3xy^2)$$

Check

$$8x^2y(x + 2y + 3xy^2) = 8x^3y + 16x^2y^2 + 24x^3y^3 \quad ✓$$

Practice Problem 4 Factor. $18a^3b^2c - 27ab^3c^2 - 45a^2b^2c^2$ ∎

EXAMPLE 5 Factor. $9a^3b^2 + 9a^2b^2$

We observe that both terms contain a common factor of 9. We can remove a^2 and b^2 from each term.

$$9a^3b^2 + 9a^2b^2 = 9a^2b^2(a + 1)$$

WARNING Don't forget to include the 1 inside the parentheses in Example 5. The solution is wrong without it. You will see why if you try to check a result written without the 1.

Practice Problem 5 Factor and check. $30x^3y^2 - 24x^2y^2 + 6xy^2$ ∎

EXAMPLE 6 Factor. $3x(x - 4y) + 2(x - 4y)$

Be sure you understand what are terms and what are factors in this example. There are two *terms.* The expression $3x(x - 4y)$ is one term. The expression $2(x - 4y)$ is the second term. Each term is made up of two factors. Observe that the binomial $(x - 4y)$ is a common *factor* in each term. A common factor may be a set of parentheses containing any type of polynomial. Thus we can remove the common factor $(x - 4y)$.

$$3x(x - 4y) + 2(x - 4y) = (x - 4y)(3x + 2)$$

Practice Problem 6 Factor. $3(a + b) + x(a + b)$ ■

EXAMPLE 7 Factor. $7x^2(2x - 3y) - (2x - 3y)$

The common factor for each term is $(2x - 3y)$. What happens when we factor out $(2x - 3y)$? What are we left with in the second term? Recall that $(2x - 3y) = 1(2x - 3y)$. Thus

$7x^2(2x - 3y) - (2x - 3y) = 7x^2(2x - 3y) - 1(2x - 3y)$ *Rewrite the original expression.*

$= (7x^2 - 1)(2x - 3y)$ *Factor out $(2x - 3y)$.*

Practice Problem 7 Factor. $8y(9y^2 - 2) - (9y^2 - 2)$ ■

EXAMPLE 8 A computer programmer is writing a program to find the area of 4 circles. She uses the formula $A = \pi r^2$. The radii of the circles are a, b, c, and d, respectively. She wants the final answer to be in factored form with the value of π occurring only once, in order to minimize round-off area. Write the total area with a formula that has π occurring only once.

For each circle $A = \pi r^2$, where $r = a, b, c$, or d.

The total area is $\pi a^2 + \pi b^2 + \pi c^2 + \pi d^2$.

In factored form the total area $= \pi(a^2 + b^2 + c^2 + d^2)$

Practice Problem 8 Use $A = \pi r^2$ to find the shaded area. The radius of the larger circle is b. The radius of the smaller circle is a. Write the total area formula in factored form so that π only appears once.

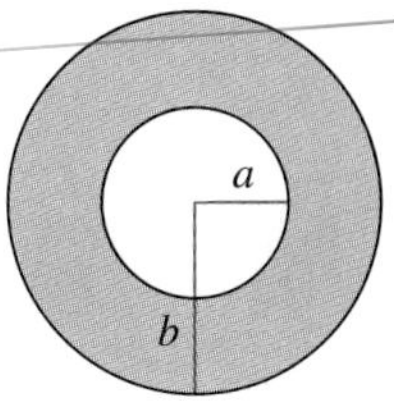

■

5.1 Exercises

Verbal and Writing Skills

Write a word or words to complete each sentence.

1. In the expression $3x^2 \cdot 5x^3$, $3x^2$ and $5x^3$ are called ______________ .

2. In the expression $3x^2 + 5x^3$, $3x^2$ and $5x^3$ are called ______________.

3. We can factor $30a^4 + 15a^3 - 45a^2$ as $5a(6a^3 + 3a^2 - 9a)$. Is the factoring complete? Why or why not?

Remove the largest possible common factor. Check your answers for Problems 4 through 15 by multiplication.

4. $6x^2 - 6x$

5. $5ab - 5b$

6. $21abc - 14ab^2$

7. $18wz - 27w^2z$

8. $5x^3 + 25x^2 - 15x$

9. $8x^3 - 10x^2 - 14x$

10. $18abc + 12d - 6ab$

11. $9pqr - 12p^2qr - 14rx$

12. $ax^2 - 2bx^2 - cx^2 + dx$

13. $a^4y - a^3y^2 + a^2y^3 - 2a$

14. $60x^3 - 50x^2 + 25x$

15. $6x^9 - 8x^7 + 4x^5$

16. $2\pi rh + 2\pi r^2$

17. $9a^2b^2 - 36ab$

18. $14x^2y - 35xy - 63x$

19. $40a^2 - 16ab - 24a$

20. $39x^2 - 26xy - 13x$

21. $34x^2 - 51xy - 17x$

22. $20a^2b - 4ab^2 + 12ab + 8a^2b^2$

23. $18xy^2 + 6x^2y + 24xy - 12x^2y^2$

Hint: In Problems 24 through 37, refer to Examples 6 and 7.

24. $2(a + b) + 3x(a + b)$

25. $5(x - 2y) - z(x - 2y)$

26. $3x(x - 4) - 2(x - 4)$

27. $5x(x - 7) + 3(x - 7)$

28. $7a(x - 3y) - 4(x - 3y)$

29. $5b(c + 2d) - 2a(c + 2d)$

30. $3(x^2 + 1) + 2y(x^2 + 1) + w(x^2 + 1)$

31. $5a(bc - 1) + b(bc - 1) + c(bc - 1)$

32. $2a(xy - 3) - 4(xy - 3) - z(xy - 3)$

33. $4(x^2 + 8) - 2y(x^2 + 8) - w(x^2 + 8)$

34. $4y(x + 2y) + (x + 2y)$

35. $3x^2(x - 2y) - (x - 2y)$

36. $9a(5a - b) - (5a - b)$

37. $2ab(9x + 5y) + (9x + 5y)$

Factor. Be sure to remove the greatest common factor.

★ **38.** $100x^6 - 25x^5 + 36x^4 - 72x^3 - 68x^2 + 120x$

★ **39.** $3x^2y(5a - 3b + 1) - 19xy(5a - 3b + 1) - y(5a - 3b + 1)$

40. Remove a common factor of 1.6845 from $5.0535x^2 - 1.6845w^2 + 15.1605z^2$.

41. Remove a common factor of 57,982 from $173{,}946x^2 - 347{,}892y^2 - 637{,}802z^2$.

Applications

42. Find a formula for the area of 4 rectangles of width 2.786 inches. The length of each rectangle is a, b, c, or d inches. Write the formula in factored form.

43. Find a formula for the total cost of all purchases by four people. Each person spent \$25.95 per item. Frank bought a items, Sam bought b items, Marcia bought c items, and Melissa bought d items. Write the formula in factored form.

Cumulative Review Problems

44. $(2x - 3)(2x + 3)$

45. $(4a - b)^2$

46. The sum of three consecutive odd numbers is 40 more than the smallest. Find the largest of the three numbers. (*Hint:* Consecutive odd integers are integers such as $-5, -3, -1, 1, 3, 5, 7, \dots, x, x + 2, \dots$)

47. Tuition, room, and board at Ivy Hall U cost Rosa \$13,284. This was an 8% increase over her costs last year. What did Rosa pay last year for tuition, room, and board?

5.2 Factor by Grouping

After studying this section, you will be able to:

1 *Factor problems with four terms by grouping.*

1 *Factoring by Grouping*

A common factor can be a number, a variable, or an algebraic expression. Sometimes the polynomial is written so that it is easy to recognize the common factor. This is especially true when the common factor is enclosed by parentheses.

EXAMPLE 1 Factor. $x(x-3)+2(x-3)$

Observe each term:

$$\underbrace{x(x-3)}_{\text{first term}}+\underbrace{2(x-3)}_{\text{second term}}$$

The common factor for both the first and second terms is the quantity $(x-3)$, so we have

$$x(\underline{x-3})+2(\underline{x-3})=(\underline{x-3})(x+2)$$

Practice Problem 1 Factor. $3y(2x-7)-8(2x-7)$ ■

In many cases these types of problems do not have any parentheses and we are asked to factor the polynomial. In such cases we remove a common factor from the first two terms and a different common factor from the second two terms. If the two parentheses contain the same expressions at this step, the problem may then be completed as we did in Example 1. This procedure of factoring is often called *factoring by grouping.*

EXAMPLE 2 Factor. $2x^2+3x+6x+9$

$2x^2+3x$	+	$6x+9$
Remove a common factor of x from the first two terms.		Remove a common factor of 3 from the second two terms.
$x(2x+3)$		$3(2x+3)$

Note that the two sets of parentheses contain the same expression at this step. The expression in parentheses is now a common factor.

Now we finish the factoring.

$$x(2x+3)+3(2x+3)=(2x+3)(x+3)$$

Practice Problem 2 Factor. $6x^2-15x+4x-10$ ■

EXAMPLE 3 Factor. $4x + 8y + ax + 2ay$

Remove a common factor of 4 from the first two terms.

$$4x + 8y + ax + 2ay = 4(x + 2y) + a(x + 2y)$$

Remove a common factor of a from the second two terms.

$4(\underline{x + 2y}) + a(\underline{x + 2y}) = (\underline{x + 2y})(4 + a)$ *Both sets of parentheses contain the same expression. The common factor of each term is the expression in parentheses,* $(x + 2y)$.

Practice Problem 3 Factor by grouping. $ax + 2a + 4bx + 8b$ ■

In practice, these problems are done in just two steps.

EXAMPLE 4 Factor. $cx + cy + 4x + 4y$

$cx + cy + 4x + 4y = c(x + y) + 4(x + y)$ *Remove the common factor of c from the first two terms. Remove the common factor of 4 from the second two terms.*

$= (x + y)(c + 4)$ *Since the two parentheses above are the same, we complete the factoring.*

Practice Problem 4 Factor. $12x + 4y + 21ax + 7ay$ ■

In some problems the terms are out of order. We have to rearrange the order of the terms first so that the first two terms have a common factor.

EXAMPLE 5 Factor. $bx + 4y + 4b + xy$

$= bx + 4b + xy + 4y$ *Rearrange the terms so that the first terms have a common factor.*

$= b(x + 4) + y(x + 4)$ *Remove the common factor of b from the first two terms.*
Remove the common factor of y from the second two terms.

$= (x + 4)(b + y)$ *Since the two sets of parentheses above contain the same expression, we can complete the factoring.*

Practice Problem 5 Factor. $6a^2 + 5bc + 10ab + 3ac$ ■

Sometimes you will need to remove a negative common factor from the second two terms to obtain two sets of parentheses that contain the same expression.

EXAMPLE 6 Factor. $2x^2 + 5x - 4x - 10$

$= x(2x + 5) - 2(2x + 5)$ *Remove the common factor of x from the first two terms and a common factor of −2 from the second two terms.*

$= (2x + 5)(x - 2)$ *Since the two sets of parentheses above contain the same expression, we can complete the factoring.*

Notice that if you removed a common factor of +2 in the first step, the two sets of parentheses would not contain the same expression. If the expressions inside the two sets of parentheses are not exactly the same, you cannot express the polynomial as a product of two factors!

Practice Problem 6 Factor. $6xy + 14x - 15y - 35$ ■

EXAMPLE 7 Factor. $2ax - a - 2bx + b$

$= a(2x - 1) - b(2x - 1)$ *Remove the common factor of a from the first two terms.*
Remove the common factor of −b from the second two terms.

$= (2x - 1)(a - b)$ *Since the two sets of parentheses contain the same expression, we can complete the factoring.*

Practice Problem 7 Factor. $3x - 10ay + 6y - 5ax$ ■

WARNING Many students find that they make a factoring error in the first step of problems like Example 7. Multiplying back the results of the first step (even if you do it in your head) will usually help you to detect any error you may have made.

EXAMPLE 8 Factor and check your answer. $8ad + 21bc - 6bd - 28ac$

We observe that the first two terms do not have a common factor.

$= 8ad - 6bd - 28ac + 21bc$ *Rearrange the order using the commutative and associate properties of addition.*

$= 2d(4a - 3b) - 7c(4a - 3b)$ *Remove the common factor of 2d from first terms.*
Remove the common factor of −7c from last terms.

$= (4a - 3b)(2d - 7c)$ *Remove the common factor of (4a − 3b).*

To check, we multiply out the two binomials using the FOIL procedure.

$(4a - 3b)(2d - 7c) = 8ad - 28ac - 6bd + 21bc$

$= 8ad + 21bc - 6bd - 28ac$ ✓ *Rearrange the order of the terms. This is the original problem. Thus it checks.*

Practice Problem 8 Factor and check your answer.
$10ab + 27bc - 6bd - 45ac$ ■

5.2 Exercises

Factor by grouping. Check your answers for Problems 1 through 12.

1. $3x - 6 + xy - 2y$
2. $2x - 10 + xy - 5y$
3. $ax + ay - 2bx - 2by$
4. $4x + 8y - 3wx - 6wy$
5. $x^3 - 4x^2 + 3x - 12$
6. $x^3 - 6x^2 + 2x - 12$
7. $3ax + bx - 6a - 2b$
8. $4ax + bx - 28a - 7b$
9. $5a + 12bc + 10b + 6ac$
10. $2x + 15yz + 6y + 5xz$
11. $5a - 5b - 2ax + 2xb$
12. $xy - 4x - 3y + 12$
13. $y^2 - 2y - 3y + 6$
14. $x^2 + 4x - 5x - 20$
15. $3ax - y + 3ay - x$
16. $xa + 2bx - a - 2b$
17. $6ax - y + 2ay - 3x$
18. $4ab - c + 2ac - 2b$
19. $x^2 + 6x - 8x - 48$
20. $x^2 - 7x - 9x + 63$
21. $28x^2 + 8xy^2 + 21xw + 6y^2w$
22. $8xw + 10x^2 + 35xy^2 + 28y^2w$
23. $8a^2 - 6ab + 20ae - 15be$
24. $21a^2 - 6ab - 28ae + 8be$

★ 25. $44x^3 - 63y^2w - 36xy^2 + 77x^2w$

★ 26. $50x^7 - 40x^3y^5 - 15x^4w^2 + 12y^5w^2$

Verbal and Writing Skills

27. Although $6a^2 - 12bd - 8ad + 9ab = 6(a^2 - 2bd) - a(8d - 9b)$ is true, it is not correct as the solution to the problem "Factor $6a^2 - 12bd - 8ad + 9ab$." Explain. Can this problem be factored?

Cumulative Review Problems

28. Multiply. $(11x - 7y)(11x + 7y)$.

29. A polynomial of three terms is called a ______________.

30. We lost count of how many tickets of each type we sold to the school jazz concert, but we know that 120 tickets were sold altogether. Some were to students at \$1.75 each, and the rest were general admission at \$3.25 each. The total collected from all the sales was \$264. How many tickets of each type were sold?

31. Thu has scored 79, 95, 82, and 89 on his four tests. Tomorrow is his final, which counts double (that is, worth 200 points compared to 100 for each test). He would like to score well enough on this final to have an 88 grade average. What must he score on the final?

5.3 Factoring Trinomials of the Form $x^2 + bx + c$

After studying this section, you will be able to:

1. *Factor polynomials of the form $x^2 + bx + c$.*
2. *Factor polynomials that have a common factor and a factor of the form $x^2 + bx + c$.*

1 *Factoring Polynomials of the Form $x^2 + bx + c$*

Suppose that you wanted to factor $x^2 + 5x + 6$. After some trial and error you *might* obtain $(x + 2)(x + 3)$ or you might get discouraged and not get an answer. If you did get these factors, you could check this answer by the FOIL method.

$$(x + 2)(x + 3) = x^2 + 3x + 2x + 6$$
$$= x^2 + 5x + 6$$

But trial and error can be a long process. There is another way.

Look at the original polynomial $x^2 + 5x + 6$. We know immediately that the answer will be of the form ( )( ). Since the first term of the polynomial is x^2, we can place an x in each of the parentheses, $(x \quad)(x \quad)$. Next look at the sign of the last term. The sign is positive. This means that the sign of each factor must *be the same.* Since the sign of the middle term is positive, the sign of each factor must be positive, and we can write $(x + \quad)(x + \quad)$. Look again at the last term. Think of factors of 6. Try 2 and 3 since $2 + 3 = 5$, the coefficient of the middle term. We write $(x + 2)(x + 3)$ and our factoring is complete. We can check this using FOIL as shown above. To summarize:

The coefficient of x is the *sum* of these two numbers.

Factor. $x^2 + 5x + 6$ Solution: $(x + 2)(x + 3)$

The last term is the *product* of these two numbers.

Let's write the procedure we have observed and try a few examples.

> **Factoring Trinomials of the Form $x^2 + bx + c$**
>
> **1.** The answer will be of the form $(x + \_\_)(x + \_\_)$.
>
> **2.** The two numbers at the end of each set of parentheses are numbers such that:
> **(a)** When you multiply them, you get the last term, which is c.
> **(b)** When you add them, you get the coefficient of x, which is b.

EXAMPLE 1 Factor. $x^2 + 7x + 12$

The answer is of the form $(x + \_\_)(x + \_\_)$. We want to find the two numbers you can multiply to get 12 but add to get 7. The numbers are 3 and 4.

$$x^2 + 7x + 12 = (x + 3)(x + 4)$$

Practice Problem 1 Factor. $x^2 + 8x + 12$ ■

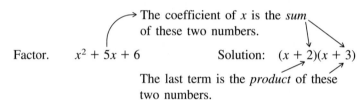

EXAMPLE 2 Factor. $x^2 + 12x + 20$

We want two numbers that have a product of 20 but a sum of 12. The numbers are 10 and 2.

$$x^2 + 12x + 20 = (x + \underline{10})(x + \underline{2})$$

Note: If you cannot think of the numbers in your head, write down the possible factors whose product is 20.

Product	*Sum*
$1 \cdot 20 = 20$	$1 + 20 = 21$
$2 \cdot 10 = 20$	$2 + 10 = 12$ ←
$4 \cdot 5 = 20$	$4 + 5 = 9$

Then select the pair whose sum is 12. Select this pair.

Practice Problem 2 Factor. $x^2 + 17x + 30$ You may find that it is helpful to list all the factors whose product is 30 first. ■

EXAMPLE 3 Factor. $x^2 + 11x + 10$

Two numbers whose product is 10 but whose sum is 11 are the numbers 10 and 1.

$$x^2 + 11x + 10 = (x + 10)(x + 1)$$

Note: The order of the parenthetical factors is not important. You can write the answer as $(x + 1)(x + 10)$. This concept applies to all factoring problems. Remember, multiplication is commutative.

Practice Problem 3 Factor. $x^2 + 7x + 6$ ■

So far we have only factored trinomials of the form $x^2 + bx + c$, where b and c are positive numbers. The same procedure applies if b is a negative number and c is positive. Remember, if c is positive, the sign of both factors will *be the same.*

EXAMPLE 4 Factor. $x^2 - 8x + 15$

We want two numbers that have a product of $+15$ but a sum of -8. They must be negative numbers since the sign of the middle term is negative.

the sum of $-5 + (-3)$

$$x^2 - 8x + 15 = (x - 5)(x - 3)$$

the product of $(-5)(-3)$

Think: $(-5)(-3) = +15$ *and* $-5 + -3 = -8$.

Multiply using FOIL to check.

Practice Problem 4 Factor. $x^2 - 11x + 18$ ■

EXAMPLE 5 Factor. $x^2 - 9x + 14$

We want two numbers whose product is 14 but whose sum is -9. The numbers are -7 and -2. So

$$x^2 - 9x + 14 = (x - 7)(x - 2) \quad \text{or} \quad (x - 2)(x - 7)$$

Practice Problem 5 Factor. $y^2 - 11y + 24$ ■

All the examples so far have had a positive last term. What happens when the last term is negative? If the last term is negative, we will need one positive number and one negative number. Why? The product of a positive and a negative number is negative.)

EXAMPLE 6 Factor. $x^2 + 2x - 8$

We want two numbers whose product is -8 but whose sum is $+2$.

$x^2 + 2x - 8 = (x + 4)(x - 2)$ *Think:* $(+4)(-2) = -8$ *and* $+4 + (-2) = +2$.

Practice Problem 6 Factor. $x^2 + 5x - 6$ ■

EXAMPLE 7 Factor. $x^2 - 3x - 10$

We want two numbers whose product is -10 but whose sum is -3. The two numbers are -5 and $+2$.

$$x^2 - 3x - 10 = (x - 5)(x + 2)$$

Practice Problem 7 Factor. $a^2 - 5a - 24$ ■

What if we made a sign error and *incorrectly* factored the trinomial $x^2 - 3x - 10$ as $(x + 5)(x - 2)$? We could detect the error immediately since the sum of $+5$ and $-2 = 3$. We need a sum of -3!

EXAMPLE 8 Factor. $x^2 + 10x - 24$ Check your answer.

The two numbers whose product is -24 but whose sum is $+10$ are the numbers $+12$ and -2.

$$x^2 + 10x - 24 = (x + 12)(x - 2)$$

WARNING It is very easy to make a sign error in these problems. Make sure that you mentally multiply your answer back to obtain the original expression. Check each sign carefully.

Check: $(x + 12)(x - 2) = x^2 - 2x + 12x - 24 = x^2 + 10x - 24$ ✓

Practice Problem 8 Factor. $x^2 + 17x - 60$ Multiply your answer to check. ■

EXAMPLE 9 Factor. $x^2 - 16x - 36$

We want two numbers whose product is -36 and whose sum is -16.

List all the possible factors of 36 (without regard to sign). Find the pair that has a difference of 16. We are looking for a difference because the signs of the factors are different.

Factors of 36	*The Difference between the Factors*
36 and 1	35
18 and 2	16 ⟵ This is the value we want.
12 and 3	9
9 and 4	5
6 and 6	0

Once we have picked the pair of numbers (18 and 2), it is easy to find the signs. For the coefficient of the middle term to be -16, we will have to add the numbers -18 and $+2$.

$$x^2 - 16x - 36 = (x - 18)(x + 2)$$

Practice Problem 9 Factor. $x^2 - 7x - 60$ You may find it helpful to list the pairs of numbers whose product is 60. ■

At this point you should work several problems to develop your factoring skill. This is one section where you really need to drill by doing many problems.

Feel a little confused about the signs? If you do, you may find these facts helpful.

Facts about Factoring Trinomials of the Form $x^2 + bx + c$

The *two numbers* at the end of the parentheses will have the *same sign* if the last term is *positive*.

$$x^2 + bx + c = (x \quad)(x \quad)$$

1. They will both be *positive* if the *coefficient* of the *second* term is *positive*.

$$x^2 + 5x + 6 = (x + 2)(x + 3)$$

2. They will both be *negative* if the *coefficient* of the *second* term is *negative*.

$$x^2 - 5x + 6 = (x - 2)(x - 3)$$

The two numbers at the end of the parentheses will have *opposite signs* if the last term is *negative*.

1. The *larger* of the two numbers *without regard to sign* will be given a plus sign if the coefficient of the *second term* is *positive*.

$$x^2 + 6x - 7 = (x + 7)(x - 1)$$

2. The larger of the two numbers *without regard to sign* will be given a negative sign if the coefficient of the *second term* is *negative*.

$$x^2 - 6x - 7 = (x - 7)(x + 1)$$

Do not memorize these facts; rather, try to understand the pattern.

Sometimes the exponent of the first term will be greater than 2. If the exponent is an even power, it is a square. For example, $x^4 = (x^2)(x^2)$. Likewise, $x^6 = (x^3)(x^3)$.

EXAMPLE 10 Factor. $y^4 - 2y^2 - 35$

Think: $y^4 = (y^2)(y^2)$

$(y^2 \quad)(y^2 \quad)$ *Place y^2 in the parentheses. Next determine*

$(y^2 + \quad)(y^2 - \quad)$ *the signs. The last term of the polynomial is negative. The signs will be different. Now think of factors of 35 whose difference is 2.*

$(y^2 - 5)(y^2 + 7)$ *Check by multiplying.*

Practice Problem 10 Factor. $a^4 + a^2 - 42$ ■

2 *Factoring Polynomials That Have a Common Factor*

Some factoring problems require two steps. Often we must first remove a common factor from each term of the polynomial. Once this is done, we may find that the other factor is a trinomial that can be factored using the methods previously discussed in this section.

EXAMPLE 11 Factor. $2x^2 + 36x + 160$

$2(x^2 + 18x + 80)$ *First remove the common factor of 2 from each term of the polynomial.*

$2(x + 8)(x + 10)$ *Then factor the remaining polynomial.*

The final answer is $2(x + 8)(x + 10)$. *Be sure to list all parts of the answer.*

Practice Problem 11 Factor. $3x^2 + 45x + 150$ ■

EXAMPLE 12 Factor. $3x^2 + 9x - 162$

$3(x^2 + 3x - 54)$ *First remove the common factor of 3 from each term of the polynomial.*

$3(x - 6)(x + 9)$ *Then factor the remaining polynomial.*

The final answer is $3(x - 6)(x + 9)$.

Practice Problem 12 Factor. $4x^2 - 8x - 140$ ■

5.3 Exercises

Verbal and Writing Skills

Fill in the blanks.

1. To factor $x^2 + 5x + 6$, find two numbers whose ______________ is 6 and whose ______________ is 5.

2. To factor $x^2 + 5x - 6$, find two numbers whose ______________ is -6 and whose ______________ is 5.

Factor.

3. $x^2 + 9x + 8$

4. $x^2 + 8x + 15$

5. $x^2 + 8x + 12$

6. $x^2 + 21x + 20$

7. $x^2 - 4x + 3$

8. $x^2 - 6x + 8$

9. $x^2 - 13x + 22$

10. $x^2 - 15x + 14$

11. $x^2 + x - 12$

12. $x^2 + 2x - 8$

13. $x^2 - 13x - 14$

14. $x^2 - 6x - 16$

15. $x^2 + x - 20$

16. $x^2 + 7x - 18$

17. $x^2 - 5x - 24$

18. $x^2 - 11x - 26$

Look over your answers to problems 3 through 18 carefully. Be sure that you are clear on your sign rules. Problems 19 through 50 contain a mixture of all the types of problems in this section. Make sure you can do them all. Check your answers by multiplication.

19. $x^2 - 7x + 10$

20. $x^2 + 3x - 18$

21. $x^2 - x - 2$

22. $x^2 + 10x + 21$

23. $x^2 + 5x - 14$

24. $x^2 - 2x - 15$

25. $x^2 - 9x + 20$

26. $x^2 - 8x + 12$

27. $x^2 + 17x + 30$

28. $x^2 - 3x - 28$

29. $y^2 - 4y - 5$

30. $y^2 - 8y + 7$

31. $a^2 + 6a - 16$

32. $a^2 - 13a + 30$

33. $x^2 - 12x + 32$

34. $x^2 - 6x - 27$

35. $x^2 + 4x - 21$

36. $x^2 - 11x + 18$

37. $x^2 + 13x + 40$

38. $x^2 + 15x + 50$

39. $x^2 - 21x - 22$

40. $x^2 + 12x - 28$

41. $x^2 + 9x - 36$

42. $x^2 - 13x + 36$

43. $x^2 - xy - 42y^2$

44. $x^2 - xy - 56y^2$

45. $x^2 - 16xy + 63y^2$

46. $x^2 + 19xy + 48y^2$

47. $x^4 - 3x^2 - 40$

48. $x^4 + 11x^2 + 28$

49. $x^4 + 32x^2 + 60$

50. $x^4 - 13x^2 - 30$

In Problems 51 through 62, first remove the greatest common factor from each term. Then factor the remaining polynomial. Refer to Examples 11 and 12.

51. $2x^2 - 12x + 16$

52. $2x^2 - 14x + 24$

53. $3x^2 - 6x - 72$

54. $3x^2 - 12x - 63$

55. $4x^2 + 24x + 20$

56. $4x^2 + 28x + 40$

57. $5x^2 + 10x - 15$

58. $5x^2 + 15x - 20$

59. $6x^2 + 18x + 12$

60. $6x^2 + 24x + 18$

61. $2x^2 - 32x + 126$

62. $2x^2 - 30x + 112$

Cumulative Review Problems

63. Solve for t. $A = P + Prt$

64. Solve. $2 - 3x \leq 7$ Give a graphical representation on the number line.

65. A car travels from Xanadu to Yreka in 2 hours, maintaining a constant speed. A train, traveling 20 mph faster, makes the trip in $1\frac{1}{2}$ hours. How far is it from Xanadu to Yreka? *Hint:* Let c = car's speed. First find the speed of the car and the speed of the train. Then be sure to answer the question.

66. Ada, a radio advertising sales representative, earns a flat \$400 (to cover expenses) plus 3% commission on her sales. She wants to earn \$4300 or more this month. At least how much must she generate in sales?

5.4 Factoring Trinomials of the Form $ax^2 + bx + c$

After studying this section, you will be able to:

1 *Factor a trinomial of the form $ax^2 + bx + c$ by the trial and error method.*
2 *Factor a trinomial of the form $ax^2 + bx + c$ by the grouping number method.*
3 *Factor a trinomial of the form $ax^2 + bx + c$ after a common factor has been removed for each term.*

1 *Using the Trial and Error Method to Factor*

When the coefficient of the x^2 term is not 1, the trinomial is more difficult to factor. Several possibilities must be considered.

EXAMPLE 1 Factor. $2x^2 + 5x + 3$

To get the coefficient of the first term to be 2, the factors would be 2 and 1. $2x^2 + 5x + 3 = (2x \quad)(x \quad)$.

To get the last term to be 3, the factors would be 3 and 1.

Since all signs are positive, we know that each set of parentheses will contain only positive signs. However, we still have two possibilities. They are

$$(2x + 3)(x + 1)$$

$$(2x + 1)(x + 3)$$

We check them by multiplying by the FOIL method:

$(2x + 1)(x + 3) = 2x^2 + 7x + 3$ *Wrong middle term.*

$(2x + 3)(x + 1) = 2x^2 + 5x + 3$ *Right middle term.*

Thus the correct answer is

$$(2x + 3)(x + 1) \quad \text{or} \quad (x + 1)(2x + 3)$$

Practice Problem 1 Factor. $2x^2 - 7x + 5$ ■

Some problems have many more possibilities.

EXAMPLE 2 Factor. $4x^2 - 13x + 3$

The Different Factors of 4 Are:	***The Factors of 3 Are:***
2 and 2	1 and 3
1 and 4	

Let us list the possible factoring combinations and compute the middle term by the FOIL method. Note that both signs will be negative. Why?

Possible Factors	***Middle Term***	***Correct?***
$(2x - 3)(2x - 1)$	$-8x$	No
$(4x - 3)(x - 1)$	$-7x$	No
$(4x - 1)(x - 3)$	$-13x$	Yes

The correct answer is $(4x - 1)(x - 3)$ or $(x - 3)(4x - 1)$

This method is called the **trial-and-error method.**

Practice Problem 2 Factor. $9x^2 - 64x + 7$ ■

EXAMPLE 3 Factor. $6x^2 + x - 5$

Let's consider the last number in each set of parentheses. Because -5 is negative, the two numbers will have opposite signs. Since the last number can be either positive or negative, we have more possibilities. In this case:

Factors of 6	*Factors of 5*
3 and 2	1 and 5
6 and 1	

Possible Factors	*Middle Term*	*Correct?*
$(3x + 1)(2x - 5)$	$-13x$	No
$(3x - 1)(2x + 5)$	$+13x$	No
$(3x + 5)(2x - 1)$	$+7x$	No
$(3x - 5)(2x + 1)$	$-7x$	No
$(6x + 1)(x - 5)$	$-29x$	No
$(6x - 1)(x + 5)$	$+29x$	No
$(6x + 5)(x - 1)$	$-x$	No
$(6x - 5)(x + 1)$	$+x$	Yes

The correct answer is

$$(6x - 5)(x + 1) \quad \text{or} \quad (x + 1)(6x - 5)$$

Practice Problem 3 Factor. $3x^2 + 4x - 4$ ■

Look back to Example 3. Let's see if we can shorten our work. Notice that there are four pairs of possible factors. The only difference between the two parentheses in the pair is that the last signs in each set of parentheses are reversed. This suggests that we list only half as many possibilities. Then, if the coefficient of the middle term is the opposite in sign of what we want, we can just reverse signs. Let's try this shorter method.

EXAMPLE 4 Factor. $3x^2 - 2x - 8$

Factors of 3	*Factors of 8*
3 and 1	8 and 1
	4 and 2

Let us list only one-half of the possibilities. The last term of that first set of parentheses will always be positive.

Possible Factors	*Middle Term*	*Correct Factors?*
$(x + 8)(3x - 1)$	$+23x$	No
$(x + 1)(3x - 8)$	$-5x$	No
$(x + 4)(3x - 2)$	$+10x$	No
$(x + 2)(3x - 4)$	$+2x$	No (but it is exactly opposite in sign from what is needed)

So we just *reverse* the signs of the last term in each set of parentheses.

	Middle Term	***Correct Factor?***
$(x - 2)(3x + 4)$	$-2x$	Yes

The correct answer is

$$(x - 2)(3x + 4) \quad \text{or} \quad (3x + 4)(x - 2)$$

Practice Problem 4 Factor. $3x^2 - x - 14$ ■

It takes a good deal of practice to readily factor problems of this type. The more problems you do, the more proficient you will become. The following method will help you factor more quickly.

2 *Using the Grouping Number Method for Factoring*

One way to factor a trinomial of the form $ax^2 + bx + c$ is to write it with four terms and factor by grouping, as we did in Section 5.2. For example, the trinomial $2x^2 + 13x + 20$ can be written as $2x^2 + 5x + 8x + 20$. Using the methods of Section 5.3, we factor

$$2x^2 + 5x + 8x + 20 = x(2x + 5) + 4(2x + 5)$$
$$= (2x + 5)(x + 4)$$

We can factor all factorable trinomials of the form $ax^2 + bx + c$ in this way. We will use the following procedure:

Grouping Number Method for Factoring Trinomials of the Form $ax^2 + bx + c = 0$

1. Obtain the grouping number ac.
2. Find the factors of the grouping number whose sum is b.
3. Use those factors to write bx as the sum of two terms.
4. Factor by grouping.
5. Multiply to check.

Let's try the same problem we did in Example 1.

EXAMPLE 5 Factor by grouping. $2x^2 + 5x + 3$

1. The grouping number is $(2)(3) = 6$.
2. The factors of 6 are $6 \cdot 1$ and $3 \cdot 2$. *We chose the factors whose sum is 5.*
3. We write $5x$ as the sum $3x + 2x$.
4. Factor by grouping.

$$2x^2 + 5x + 3 = 2x^2 + 2x + 3x + 3$$
$$= 2x(x + 1) + 3(x + 1)$$
$$= (x + 1)(2x + 3)$$

5. Multiply to check.

$$(x + 1)(2x + 3) = 2x^2 + 2x + 3x + 3$$
$$= 2x^2 + 5x + 3 \checkmark$$

Practice Problem 5 Factor by grouping. $2x^2 - 7x + 5$ ■

EXAMPLE 6 Factor by grouping. $4x^2 - 13x + 3$

1. The grouping number is $(4)(3) = 12$.
2. The factors of 12 are $(12)(1)$ or $(4)(3)$ or $(6)(2)$. We choose the factors whose sum is 13. Both factors receive the sign of the middle term in the trinomial.
3. We write $-13x$ as the sum $-12x + -1x$.
4. Factor by grouping.

$$4x^2 - 13x + 3 = 4x^2 - 12x - 1x + 3$$

$$= 4x(x - 3) - 1(x - 3)$$ *Remember to factor out a -1 from the last two terms so that both parentheses contain the same expression.*

$$= (x - 3)(4x - 1)$$

Practice Problem 6 Factor by grouping. $9x^2 - 64x + 7$ ■

EXAMPLE 7 Factor by grouping. $3x^2 - 2x - 8$

1. The grouping number is $(3)(-8) = -24$.
2. Since the last sign is negative we want two numbers whose product is 24 but whose difference is 2. They are 6, 4. The largest gets the middle sign; so we have -6 and $+4$.
3. We write $-2x$ as a sum $-6x + 4x$.
4. Factor by grouping.

$$3x^2 - 6x + 4x - 8 = 3x(x - 2) + 4(x - 2)$$

$$= (x - 2)(3x + 4)$$

Practice Problem 7 Factor by grouping. $3x^2 + 4x - 4$ ■

Use the method that works best for you, either trial and error or grouping.

3 *Factoring Polynomials That Have a Common Factor*

Some problems require first removing a common factor and then factoring the trinomial by one of the two methods of this section.

EXAMPLE 8 Factor. $9x^2 + 3x - 30$

$$9x^2 + 3x - 30 = 3(3x^2 + 1x - 10)$$ *We first remove the common factor of 3 from each term of the trinomial.*

$$= 3(3x - 5)(x + 2)$$ *We then factor the trinomial by the grouping method or by the trial and error method.*

Practice Problem 8 Factor. $8x^2 - 8x - 6$ ■

EXAMPLE 9 Factor. $32x^2 - 40x + 12$

$$32x^2 - 40x + 12 = 4(8x^2 - 10x + 3)$$ *We first remove the greatest common factor of 4 from each term of the trinomial.*

$$= 4(2x - 1)(4x - 3)$$ *We then factor the trinomial by the grouping method or by the trial and error method.*

Practice Problem 9 Factor. $24x^2 - 38x + 10$ ■

5.4 Exercises

Factor. Check your answers for Problems 1 through 20 using the FOIL method.

1. $3x^2 + 7x + 2$

2. $2x^2 + 7x + 3$

3. $2x^2 - 5x + 2$

4. $3x^2 - 8x + 4$

5. $2x^2 + 3x - 5$

6. $3x^2 + x - 2$

7. $2x^2 - 5x - 3$

8. $2x^2 - x - 6$

9. $5x^2 + 3x - 2$

10. $6x^2 + x - 2$

11. $6x^2 - 13x + 6$

12. $6x^2 - 7x + 2$

13. $2x^2 + 3x - 20$

14. $6x^2 + 11x - 10$

15. $9x^2 + 9x + 2$

16. $4x^2 + 11x + 6$

17. $6x^2 - 5x - 6$

18. $3x^2 - 13x - 10$

19. $8x^2 - 6x + 1$

20. $4x^2 - 8x + 3$

21. $4x^2 + 16x - 9$

22. $8x^2 + x - 9$

23. $9y^2 - 13y + 4$

24. $5y^2 - 11y + 2$

25. $5a^2 - 13a - 6$

26. $3a^2 - 10a - 8$

27. $12x^2 - 20x + 3$

28. $9x^2 + 5x - 4$

29. $15x^2 + 4x - 4$

30. $8x^2 - 11x + 3$

31. $10x^2 + 21x + 9$

32. $12x^2 + 11x + 2$

33. $16x^2 - 6x - 1$

34. $3x^2 - x - 14$

35. $2x^4 + 15x^2 - 8$

36. $4x^4 - 11x^2 - 3$

37. $4x^2 + 8xy - 5y^2$

38. $3x^2 + 8xy + 4y^2$

39. $5x^2 + 16xy - 16y^2$

40. $12x^2 + 11xy - 5y^2$

Factor each of the following by first removing the greatest common factor. See Examples 8 and 9.

41. $10x^2 + 22x + 12$

42. $4x^2 + 34x + 42$

43. $12x^2 - 24x + 9$

44. $8x^2 - 26x + 6$

45. $18x^2 + 39x - 15$

46. $12x^2 + 12x - 45$

47. $21x^2 + 60x - 9$

48. $18x^2 - 33x - 6$

Factor the following.

★**49.** $12x^2 + 16x - 35$

★**50.** $20x^2 - 53x + 18$

★**51.** $20x^2 - 27x + 9$

★**52.** $12x^2 - 29x + 15$

Cumulative Review Problems

53. Solve. $7x - 3(4 - 2x) = 2(x - 3) - (5 - x)$

54. Factor. $x^2 + 5x - 36$

55. A triangle has a second angle 5 degrees less than twice its first. The third angle measures as much as the first plus three times the second. Find the number of degrees in each angle.

56. A rectangle's length is 2 meters more than its width, w. A second rectangle is made by increasing the length of the first by 4 meters and decreasing its width by 2 meters. The resulting rectangle is equal to the first in area. Give the dimensions of the second rectangle.

5.5 Special Cases of Factoring

After studying this section, you will be able to:

1. *Recognize and factor problems of the type $a^2 - b^2$.*
2. *Recognize and factor problems of the type $a^2 + 2ab + b^2$.*
3. *Factor problems that require removing a common factor and then using the special case formula.*

As we proceed in this section you will be able to reduce the time it takes you to factor a problem by quickly recognizing and factoring two special types of factoring problems. By becoming very familiar with the appropriate formula, you will be able to factor the difference of two squares and perfect square trinomials.

1 *Factoring the Difference of Two Squares*

Recall Formula A from Section 4.5:

$$(a + b)(a - b) = a^2 - b^2$$

In reverse form we can use it for factoring.

Difference of Two Squares

$$a^2 - b^2 = (a + b)(a - b)$$

We can state it in words in this way: ''The difference of two squares can be factored into the sum and difference of those values that were squared.''

EXAMPLE 1 Factor. $9x^2 - 1$

We see that the problem is in the form of the difference of two squares. $9x^2$ is a square and 1 is a square. So using the formula we can write

$$9x^2 - 1 = (3x + 1)(3x - 1) \qquad \textit{Because } 9x^2 = (3x)^2 \textit{ and } 1 = (1)^2.$$

Practice Problem 1 Factor. $1 - 64x^2$ ■

EXAMPLE 2 Factor. $25x^2 - 16$

We see that by using the formula for the difference of squares, we can write

$$25x^2 - 16 = (5x + 4)(5x - 4) \qquad \textit{Because } 25x^2 = (5x)^2 \textit{ and } 16 = (4)^2.$$

Practice Problem 2 Factor. $36x^2 - 49$ ■

Sometimes the problem will contain two variables.

EXAMPLE 3 Factor. $4x^2 - 49y^2$

We see that

$$4x^2 - 49y^2 = (2x + 7y)(2x - 7y)$$

Practice Problem 3 Factor. $100x^2 - 81y^2$ ■

Some problems may involve more than one step.

EXAMPLE 4 Factor. $81x^4 - 1$

We see that

$81x^4 - 1 = (9x^2 + 1)(9x^2 - 1)$ *Because* $81x^4 = (9x)^2$ *and* $1 = (1)^2$.

Is the factoring complete? We can factor $9x^2 - 1$.

$81x^4 - 1 = (9x^2 + 1)(3x - 1)(3x + 1)$ *Because* $(9x^2 - 1) = (3x - 1)(3x + 1)$.

Practice Problem 4 Factor. $x^8 - 1$ ■

2 *Factoring Perfect Square Trinomials*

There is a formula that will help us to factor very quickly certain trinomials, called *perfect square trinomials.* Recall Formula B from Section 4.2:

$$(a + b)^2 = a^2 + 2ab + b^2$$

$$(a - b)^2 = a^2 - 2ab + b^2$$

In reverse form we can use these two equations for factoring.

Perfect Square Trinomials

$$a^2 + 2ab + b^2 = (a + b)^2$$

$$a^2 - 2ab + b^2 = (a - b)^2$$

A perfect square trinomial is a trinomial that is the result of squaring a binomial. How can we recognize a perfect square trinomial?

1. The first and last terms are *perfect squares.*
2. The middle term is twice the product of the values whose squares are the first and last terms.

EXAMPLE 5 Factor. $x^2 + 6x + 9$

This *is* a perfect square trinomial.

1. The first and last terms are perfect squares because $x^2 = (x)^2$ and $9 = (3)^2$.
2. The middle term is twice the product of x and 3.

Since $x^2 + 6x + 9$ is a perfect square trinomial, we can use the formula

$$a^2 + 2ab + b^2 = (a + b)^2$$

and we have

$$x^2 + 6x + 9 = (x + 3)^2$$

Practice Problem 5 Factor. $16x^2 + 8x + 1$ ■

EXAMPLE 6 Factor. $4x^2 - 20x + 25$

This *is* a perfect square trinomial. $20x = 2(2x \cdot 5)$ Note the negative sign. We have

$4x^2 - 20x + 25 = (2x - 5)^2$ *Since* $a^2 - 2ab + b^2 = (a - b)^2$.

Practice Problem 6 Factor. $25x^2 - 30x + 9$ ■

More than one variable may be involved and the exponents may be higher than 2. The same principles apply.

EXAMPLE 7 Factor. **(a)** $49x^2 + 42xy + 9y^2$ **(b)** $36x^4 - 12x^2 + 1$

(a) This is a perfect square trinomial. Why?

$$49x^2 + 42xy + 9y^2 = (7x + 3y)^2$$ *Because* $49x^2 = (7x)^2$ *and* $9y^2 = (3y)^2$. *Also,* $42xy = 2(7x \cdot 3y)$.

(b) This is a perfect square trinomial. Why?

$$36x^4 - 12x^2 + 1 = (6x^2 - 1)^2$$ *Because* $36x^4 = (6x^2)^2$ *and* $1 = (1)^2$. *Also,* $12x^2 = 2(6x^2 \cdot 1)$.

Practice Problem 7 Factor. **(a)** $25x^2 - 60xy + 36y^2$ **(b)** $64x^6 - 48x^3 + 9$ ■

Some problems appear to be perfect square trinomials but are not. They were factored as other trinomials in Section 5.4.

EXAMPLE 8 Factor. $49x^2 + 35x + 4$

This is *not* a perfect square trinomial! Although the first and last terms are perfect squares since $(7x)^2 = 49x^2$ and $(2)^2 = 4$, the middle term is not double the product of 2 and $7x$! $35x \neq 28x$! So we must factor by trial and error or by grouping to obtain

$$49x^2 + 35x + 4 = (7x + 4)(7x + 1)$$

Practice Problem 8 Factor. $9x^2 - 15x + 4$ ■

3 *Factoring Polynomials by First Removing a Common Factor and Then Using One of the Special Factoring Formulas*

In some cases, we will first remove a common factor. Then we will find an opportunity to use the difference of two squares formula or one of the perfect square trinomial formulas.

EXAMPLE 9 Factor. $12x^2 - 48$

$12x^2 - 48 = 12(x^2 - 4)$ *First we remove the greatest common factor* 12.

$= 12(x + 2)(x - 2)$ *Then we use the difference of two squares formula:* $a^2 - b^2 = (a + b)(a - b)$.

Practice Problem 9 Factor. $20x^2 - 45$ ■

EXAMPLE 10 Factor. $24x^2 + 72x + 54$

$24x^2 + 72x + 54 = 6(4x^2 + 12x + 9)$ *First we remove the greatest common factor* 6.

$= 6(2x + 3)^2$ *Then we use the perfect square trinomial formula:* $a^2 + 2ab + b^2 = (a + b)^2$.

Practice Problem 10 Factor. $75x^2 - 60x + 12$ ■

5.5 Exercises

Factor.

1. $9x^2 - 16$
2. $36x^2 - 1$
3. $1 - 100s^2$
4. $144x^2 - 25$
5. $25x^2 - 16$
6. $81x^2 - 1$
7. $4x^2 - 25$
8. $16x^2 - 25$
9. $36x^2 - 25$
10. $1 - 25x^2$
11. $1 - 49x^2$
12. $1 - 36x^2$
13. $81x^2 - 49y^2$
14. $16x^4 - y^4$
15. $25 - 121x^2$
16. $9x^2 - 49$
17. $81x^2 - 100y^2$
18. $25x^4 - 16y^4$
19. $25a^2 - 1$
20. $36a^2 - 49$
21. $49x^2 - 25y^2$
22. $4x^2 - 49y^2$
23. $49x^2 + 14x + 1$
24. $36x^2 + 12x + 1$
25. $y^2 - 6y + 9$
26. $y^2 - 8y + 16$
27. $9x^2 - 24x + 16$
28. $4x^2 + 20x + 25$
29. $49x^2 + 28x + 4$
30. $25x^2 + 30x + 9$
31. $x^2 + 10x + 25$
32. $x^2 + 12x + 36$
33. $9x^2 - 42x + 49$
34. $36x^2 - 60x + 25$
35. $25x^2 - 40x + 16$
36. $49x^2 - 42x + 9$
37. $81x^2 + 36xy + 4y^2$
38. $36x^2 + 60xy + 25y^2$
39. $9x^2 - 42xy + 49y^2$
40. $49x^2 - 70xy + 25y^2$
41. $16a^2 + 72ab + 81b^2$
42. $169a^2 + 26ab + b^2$
43. $9x^4 - 6x^2y + y^2$
44. $y^4 - 22y^2 + 121$
45. $49x^2 + 70x + 9$
46. $25x^2 - 50x + 16$
47. $x^6 - 25$
48. $x^8 - 1$
49. $x^{10} - 36y^{10}$
50. $x^4 - 49y^6$
51. $4x^8 + 12x^4 + 9$
52. $4x^6 + 4x^3 + 1$

To Think About

53. When you factor Example 4 as $81x^4 - 1 = (9x^2 + 1)(9x^2 - 1)$, we can then factor $9x^2 - 1 = (3x + 1)(3x - 1)$. Show why you cannot factor $9x^2 + 1$.

54. What two numbers could replace the b in $25x^2 + bx + 16$ so that the resulting trinomial would be a perfect square? (*Hint:* One number is negative.)

55. What value could you give to c so that $16y^2 - 56y + c$ would become a perfect square trinomial? Is there only one answer or more than one?

56. Jerome says that he can find two numbers that will fit into the blank that will result in a perfect square.

$$100x^2 + \_\_\_\_\_\_\_\_\_\_ x - 9$$

Kesha says there is only one that fits, and Larry says there are none. Who is correct, and why?

Factor by first removing the greatest common factor. See Examples 9 and 10.

57. $27x^2 - 3$

58. $32x^2 - 2$

59. $125x^2 - 5$

60. $28x^2 - 63$

61. $12x^2 - 36x + 27$

62. $125x^2 - 100x + 20$

63. $98x^2 + 84x + 18$

64. $128x^2 + 96x + 18$

Mixed Practice

Factor. Be sure to remove common factors first.

65. $x^2 - 9x + 14$

66. $x^2 - 9x - 36$

67. $2x^2 + 5x - 3$

68. $15x^2 - 11x + 2$

69. $16x^2 - 121$

70. $9x^2 - 100y^2$

71. $25x^2 - 20x + 4$

72. $49x^2 - 42x + 9$

73. $3x^2 + 6x - 45$

74. $4x^2 + 24x + 32$

75. $7x^2 - 63$

76. $11x^2 - 44$

77. $5x^2 + 20x + 20$

78. $8x^2 + 48x + 72$

79. $2x^2 - 32x + 126$

80. $2x^2 - 32x + 110$

Cumulative Review Problems

81. Multiply. $(x + 4)(2x + 3)(x - 2)$

82. Solve and graph. $7 - 11x < -(13 + 3x)$

83. Divide. $(x^3 + x^2 - 2x - 11) \div (x - 2)$

84. Divide. $(6x^3 + 11x^2 - 11x - 20) \div (3x + 4)$

5.6 A Brief Review of Factoring

After studying this section, you will be able to:

1 *Identify and factor any polynomial that can be factored.*
2 *Determine if a polynomial is prime.*

1 *Review of Factoring*

Often the various types of factoring problems are all mixed together. We need to be able to identify each type of factoring problem quickly. The following table may be helpful in summarizing your knowledge of factoring.

FACTORING ORGANIZER

Number of Terms	Identifying Name and/or Formula	Example
A. Any number of terms	**Common factor** Each term has a common factor consisting of a number, a variable, or both.	$2x^2 - 16x = 2x(x - 8)$ $3x^2 + 9y - 12 = 3(x^2 + 3y - 4)$ $4x^2y + 2xy^2 - wxy + xyz = xy(4x + 2y - w + z)$
B. Two terms	**Difference of two squares** First and last terms are perfect squares. $a^2 - b^2 = (a + b)(a - b)$	$16x^2 - 1 = (4x + 1)(4x - 1)$ $25y^2 - 9x^2 = (5y + 3x)(5y - 3x)$
C. Three terms	**Perfect square trinomial** First and last terms are perfect squares. $a^2 + 2ab + b^2 = (a + b)^2$ $a^2 - 2ab + b^2 = (a - b)^2$	$25x^2 - 10x + 1 = (5x - 1)^2$ $16x^2 + 24x + 9 = (4x + 3)^2$
D. Three terms	**Trinomial of the form $x^2 + bx + c$** It starts with x^2. The two numbers at the end of each parentheses are numbers whose product is c but whose sum is b.	$x^2 - 7x + 12 = (x - 3)(x - 4)$ $x^2 + 11x - 26 = (x + 13)(x - 2)$ $x^2 - 8x - 20 = (x - 10)(x + 2)$
E. Three terms	**Trinomial of the form $ax^2 + bx + c$** It starts with ax^2, where a is any number but 1.	Use trial and error or the grouping number method. Factor. $12x^2 - 5x - 2$ 1. The grouping number yields -24. 2. Find two factors whose sum is -5. -8 and 3 3. $12x^2 - 5x - 2 = 12x^2 + 3x - 8x - 2$ $= 3x(4x + 1) - 2(4x + 1)$ $= (4x + 1)(3x - 2)$
F. Four terms	**Factor by grouping** Rearrange the order if the first two terms do not have a common factor.	Factor. $wx - 6yz + 2wy - 3xz$ $= wx + 2wy - 3xz - 6yz$ $= w(x + 2y) - 3z(x + 2y)$ $= (x + 2y)(w - 3z)$

Many problems involve more than one step when factoring. When you are asked to factor a polynomial, it is expected that you will factor it completely. Usually, the first step of factoring is removing the common factor, then the next step will be more apparent.

EXAMPLE 1 Factor.

(a) $25x^3 - 10x^2 + x$ **(b)** $20x^2y^2 - 45y^2$

(c) $2ax + 4ay + 4x + 8y$ **(d)** $15x^2 - 3x^3 + 18x$

(a) $25x^3 - 10x^2 + x = x(25x^2 - 10x + 1)$ *Remove the common factor of x. The other factor is a perfect square trinomial.*

$= x(5x - 1)^2$

(b) $20x^2y^2 - 45y^2 = 5y^2(4x^2 - 9)$ *Remove the common factor of $5y^2$. The other factor is the difference of squares.*

$= 5y^2(2x - 3)(2x + 3)$

(c) $2ax + 4ay + 4x + 8y = 2[ax + 2ay + 2x + 4y]$ *Remove the common factor of 2.*

$= 2[a(x + 2y) + 2(x + 2y)]$ *Factor the terms inside the bracket by the grouping method.*

$= 2[(x + 2y)(a + 2)]$ *Remove the common factor of $(x + 2y)$.*

(d) $15x^2 - 3x^3 + 18x = -3x^3 + 15x^2 + 18x$ *Rearrange the terms in descending order of powers of x.*

$= -3x(x^2 - 5x - 6)$ *Remove the common factor of $-3x$.*

$= -3x(x - 6)(x + 1)$ *Factor the trinomial.*

Practice Problem 1 Factor. Be careful. These practice problems are mixed.

(a) $9x^4y^2 - 9y^2$ **(b)** $12x - 9 - 4x^2$ **(c)** $3x^2 - 36x + 108$

(d) $5x^3 - 15x^2y + 10x^2 - 30xy$ ■

2 *Determining If a Polynomial Is Prime*

Not all polynomials can be factored using the methods in this chapter. If we cannot factor a polynomial by elementary methods, we will identify it as a **prime** polynomial. When a polynomial cannot be factored by our previously learned techniques, you will be comfortable in saying ''It cannot be done so it is prime,'' rather than ''I can't do it—I give up!''

EXAMPLE 2 Factor, if possible. $x^2 + 6x + 12$

Since we are only using rational numbers, we find this polynomial is prime. The rational factors of 12 are

$$(1)(12) \quad \text{or} \quad (2)(6) \quad \text{or} \quad (3)(4)$$

None of these pairs will add up to 6, the coefficient of the middle term. Thus the problem cannot be factored by the methods of this chapter. It is **prime.**

Practice Problem 2 Factor. $x^2 - 9x - 8$ ■

EXAMPLE 3 Factor, if possible. $25x^2 + 4$

We have a formula to factor the difference of two squares. There is no way to factor the sum of two squares. That is, $a^2 + b^2$ cannot be factored. Thus

$$25x^2 + 4 \text{ is prime}$$

Practice Problem 3 Factor, if possible. $25x^2 + 82x + 4$ ■

5.6 Exercises

Review the six basic types of factoring in the Factoring Organizer on page 000. Each of the six types is included in the first 12 exercises. Be sure you can find two of each type.

Factor. Check your answer by multiplying.

1. $9x - 6y + 15$

2. $x^2 + 5xy - 7x$

3. $36x^2 - 9y^2$

4. $100x^2 - 1$

5. $9x^2 - 12xy + 4y^2$

6. $16x^2 + 24xy + 9y^2$

7. $x^2 + 3x - 28$

8. $x^2 + 15x + 54$

9. $5x^2 - 13x + 6$

10. $21x^2 + 25x - 4$

11. $ax - 3ay - 6by + 2bx$

12. $ax - 20y - 5x + 4ay$

Factor, if possible. Be sure to factor completely. Always remove the greatest common factor first if one exists.

13. $3x^4 - 12$

14. $y^2 + 16y + 64$

15. $2x^2 - 10x - 132$

16. $16x^2 - 81y^2$

17. $2x^2 - 11x + 12$

18. $2xy^2 - 50x$

19. $x^2 - 3xy - 70y^2$

20. $2x^3 - 7x^2 + 4x - 14$

21. $ax + 20 - 5a - 4x$

22. $3x^2 - 9x + 15$

23. $45x - 5x^3$

24. $18y^2 + 3y - 6$

25. $5x^3y^3 - 10x^2y^3 + 5xy^3$

26. $12x^2 - 36x + 27$

27. $27xyz^2 - 12xy$

28. $12x^2 - 2x - 18x^3$

29. $3x^2 + 6x - 105$

30. $4x^2 - 28x - 72$

31. $5x^2 - 30x + 40$

32. $7x^2 + 3x - 2$

33. $7x^2 - 2x^4 + 4$

34. $2x^4 - 9x^2 - 5$

35. $6x^2 - 3x + 2$

36. $4x^3 + 8x^2 - 60x$

37. $3x^2 - 15xy + 18y^2$

38. $3x^2 - 24xy + 30y^2$

39. $36x^2 - 6xy - 6y^2$

40. $56x^2 - 14xy - 7y^2$

41. $8x^2 + 28x - 16$

42. $12x^2 - 30x + 12$

43. $99x^4 - 11x^2$

44. $6x^4 - 150x^2$

45. $a^2x + 2a^2y - 3ax - 6ay$

46. $2ax^2 + 2axy - 5x^2 - 5xy$

To Think About

47. A polynomial that cannot be factored by the methods of this chapter is called ______________.

48. A binomial of the form $x^2 - d$ can be quickly factored or identified as prime. If it can be factored, what is true of the number d?

Cumulative Review Problems

49. Three times a number is twelve more than seven times the number. Find this number.

50. When Frank took a job instructing at North West Community College, he took a 13% pay cut. He now earns \$26,100 per year. What did he earn in his previous job?

5.7 Solving Quadratic Equations by Factoring

After studying this section, you will be able to:

1. *Solve a quadratic equation by factoring.*
2. *Solve a variety of applied problems by using a quadratic equation.*

1 *Solving a Quadratic Equation by Factoring*

In Chapter 2 we learned how to solve linear equations such as $3x + 5 = 0$ to find the root (or value of x) that satisfied the equation. Now we turn to the question of how to solve equations like $3x^2 + 5x + 2 = 0$. Such equations are called ***quadratic equations.*** A quadratic equation is a polynomial equation that contains the power 2 of a variable as the highest power in the equation.

> The *standard form* of a quadratic equation is $ax^2 + bx + c = 0$, where a, b, and c are real numbers and $a \neq 0$.

In this section, we will study quadratic equations in standard form, where a, b, and c are integers.

Usually we can find two real roots that satisfy a quadratic equation. But how can we find them? The most direct approach is the factoring method. This method depends on a very powerful property.

> **Zero Factor Property**
>
> If $a \cdot b = 0$, then $a = 0$ or $b = 0$.

When we make a statement in mathematics using the word **or,** we intend it to mean one or the other or both. Therefore, *both a* and *b* can equal zero if the product is zero. We can use this principle to solve a quadratic equation. Before you start, make sure that the equation is in standard form.

> 1. Factor, if possible, the quadratic expression that equals 0.
> 2. Set each factor equal to 0.
> 3. Solve the resulting equations to find each root.
> 4. Check each root.

EXAMPLE 1 Find the two roots. $3x^2 + 5x + 2 = 0$

$3x^2 + 5x + 2 = 0$ *The equation is in standard form.*

$(3x + 2)(x + 1) = 0$ *Factor the quadratic expression.*

$3x + 2 = 0 \qquad x + 1 = 0$ *Set each factor equal to 0.*

$3x = -2 \qquad x = -1$ *Solve each equation to find the two roots.*

$x = -\frac{2}{3}$

The two roots are $-\frac{2}{3}$ and -1.

Check. We can determine if the two roots $-\frac{2}{3}$ and -1 are each a solution to the equation. Substitute $-\frac{2}{3}$ for x in the *original equation.* If an identity results, $-\frac{2}{3}$ is a solution. Do the same for -1.

$$3x^2 + 5x + 2 = 0 \qquad\qquad 3x^2 + 5x + 2 = 0$$

$$3\left(-\frac{2}{3}\right)^2 + 5\left(-\frac{2}{3}\right) + 2 \stackrel{?}{=} 0 \qquad\qquad 3(-1)^2 + 5(-1) + 2 \stackrel{?}{=} 0$$

$$3\left(\frac{4}{9}\right) + 5\left(-\frac{2}{3}\right) + 2 \stackrel{?}{=} 0 \qquad\qquad 3(1) + 5(-1) + 2 \stackrel{?}{=} 0$$

$$\frac{4}{3} - \frac{10}{3} + 2 \stackrel{?}{=} 0 \qquad\qquad 3 - 5 + 2 \stackrel{?}{=} 0$$

$$\frac{4}{3} - \frac{10}{3} + \frac{6}{3} \stackrel{?}{=} 0 \qquad\qquad -2 + 2 \stackrel{?}{=} 0$$

$$0 = 0 \checkmark \qquad\qquad 0 = 0 \checkmark$$

Thus $x = -\frac{2}{3}$ and $x = -1$ are both roots for the equation $3x^2 + 5x + 2 = 0$.

Practice Problem 1 Find the roots by factoring and check. $10x^2 - x - 2 = 0$ ■

EXAMPLE 2 Find the two roots. $2x^2 + 13x - 7 = 0$

$2x^2 + 13x - 7 = 0$	*The equation is in standard form.*
$(2x - 1)(x + 7) = 0$	*Factor.*
$2x - 1 = 0 \qquad x + 7 = 0$	*Set each factor equal to 0.*
$2x = 1 \qquad x = -7$	*Solve each equation to find the two roots.*
$x = \frac{1}{2}$	

The two roots are $\frac{1}{2}$ and -7.
Check. If $x = \frac{1}{2}$, then

$$2\left(\frac{1}{2}\right)^2 + 13\left(\frac{1}{2}\right) - 7 = 2\left(\frac{1}{4}\right) + 13\left(\frac{1}{2}\right) - 7$$

$$= \frac{1}{2} + \frac{13}{2} - \frac{14}{2} = 0 \checkmark$$

If $x = -7$, then

$$2(-7)^2 + 13(-7) - 7 = 2(49) + 13(-7) - 7$$

$$= 98 - 91 - 7 = 0 \checkmark$$

Thus $x = \frac{1}{2}$ and $x = -7$ are both roots for the equation $2x^2 + 13x - 7 = 0$.

Practice Problem 2 Solve. $3x^2 - 5x + 2 = 0$ ■

If the quadratic equation $ax^2 + bx + c = 0$ has no visible constant term, then $c = 0$. All such quadratic equations can be factored by removing a common factor to obtain two solutions that are real numbers.

EXAMPLE 3 Find the two roots. $7x^2 - 3x = 0$

$7x^2 - 3x = 0$ *The equation is in standard form. Here $c = 0$.*

$x(7x - 3) = 0$ *Factor by removing the common factor.*

$x = 0 \quad 7x - 3 = 0$ *Set each factor equal to 0 by the zero factor property.*

$7x = 3$

$x = \frac{3}{7}$ *Solve each equation to find the two roots.*

The two roots are $x = 0$ and $x = \frac{3}{7}$.

Check. Verify that $x = 0$ and $x = \frac{3}{7}$ are the roots of $7x^2 - 3x = 0$.

Practice Problem 3 Solve. $7x^2 + 11x = 0$ ■

If the quadratic equation is not in standard form, we use the same basic algebraic methods we studied in Sections 2.1 to 2.3 to set the equation equal to zero so that we can use the zero factor property.

EXAMPLE 4 Find the two roots. $x^2 = 12 - x$

$x^2 = 12 - x$ *The equation is not in standard form.*

$x^2 + x - 12 = 0$ *Add x and -12 to both sides of the equation so that the left side is equal to zero; we can now factor.*

$(x - 3)(x + 4) = 0$ *Factor.*

$x - 3 = 0 \quad x + 4 = 0$ *Set each factor equal to 0 by the zero factor property.*

$x = 3 \quad x = -4$ *Solve each equation for x.*

Check. If $x = 3$: $(3)^2 \stackrel{?}{=} 12 - 3$ If $x = -4$: $(-4)^2 \stackrel{?}{=} 12 - (-4)$

$9 \stackrel{?}{=} 12 - 3$ $16 \stackrel{?}{=} 12 + 4$

$9 = 9$ ✓ $16 = 16$ ✓

Both roots check.

Practice Problem 4 Solve. $x^2 - 6x + 4 = -8 + x$ ■

EXAMPLE 5 Solve. $(x + 3)(x - 2) = 50$

Be careful here! You must have a zero on the right hand side to use the zero factor property.

$$(x + 3)(x - 2) = 50$$

$$x^2 + 3x - 2x - 6 = 50 \qquad \textit{Remove the parentheses.}$$

$$x^2 + x - 56 = 0 \qquad \textit{Place in standard form.}$$

$$(x + 8)(x - 7) = 0 \qquad \textit{Factor.}$$

$$x + 8 = 0 \qquad x - 7 = 0 \qquad \textit{Set each factor equal to zero.}$$

$$x = -8 \qquad x = 7 \qquad \textit{Solve each equation for x.}$$

The check is left to the student.

Practice Problem 5 Solve. $x(3x - 7) = 20$ ■

EXAMPLE 6 Solve. $\dfrac{x^2 - x}{2} = 6$

We must first clear the fractions from the equation.

$$2\left(\frac{x^2 - x}{2}\right) = 2(6) \qquad \textit{Multiply each side by 2.}$$

$$x^2 - x = 12 \qquad \textit{Simplify.}$$

$$x^2 - x - 12 = 0 \qquad \textit{Place in standard form.}$$

$$(x - 4)(x + 3) = 0 \qquad \textit{Factor.}$$

$$x - 4 = 0 \qquad x + 3 = 0 \qquad \textit{Set each factor equal to zero.}$$

$$x = 4 \qquad x = -3 \qquad \textit{Solve each equation for x.}$$

The check is left to the student.

Practice Problem 6 Solve. $\dfrac{2x^2 - 7x}{3} = 5$ ■

2 *Using a Quadratic Equation to Solve Applied Problems*

Certain types of word problems lead to quadratic equations. These problems may be number problems, problems about geometric figures, or applied problems. We'll show the solution to some model problems in this section.

It is particularly important to check the apparent solutions to the quadratic equation with conditions stated in the word problem. Often a particular solution to the quadratic equation will be eliminated by the conditions of the word problem.

EXAMPLE 7 The length of a rectangle is 3 meters longer than twice the width. The area of the rectangle is 44 square meters. Find the length and width of the rectangle.

1. *Understand the problem.*
   Draw a picture.

   Let w = the width in meters

   Then $2w + 3$ = the length in meters

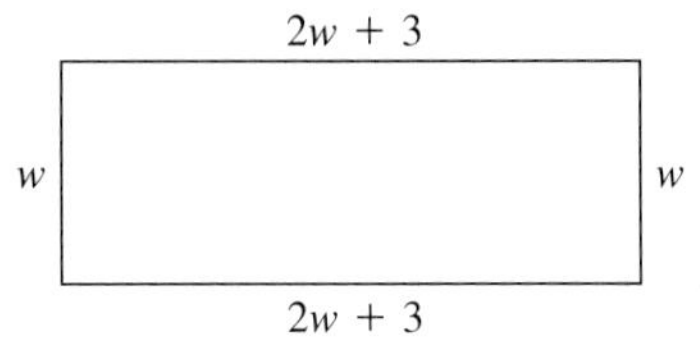

2. *Write an equation.*

$$\text{area} = (\text{width})(\text{length})$$

$$44 = w(2w + 3)$$

3. *Solve the equation and state the answer.*

$44 = w(2w + 3)$		
$44 = 2w^2 + 3w$		*Remove parentheses.*
$0 = 2w^2 + 3w - 44$		*Subtract 44 from both sides.*
$0 = (2w + 11)(w - 4)$		*Factor.*
$2w + 11 = 0$	$w - 4 = 0$	*Set each factor equal to* 0.
$2w = -11$	$w = 4$	*Simplify and solve.*
$w = -5\frac{1}{2}$		*Note that* $w = -5\frac{1}{2}$ *is not a valid solution. It would not make sense to have a rectangle with a negative number as a width.*

Since $w = 4$, the width of the rectangle is 4 meters. The length is $2w + 3$, so we have $2(4) + 3 = 8 + 3 = 11$. Thus the length of the rectangle is 11 meters.

4. *Check.* Is the length 3 meters more than twice the width?

$$11 \stackrel{?}{=} 3 + 2(4) \qquad 11 = 3 + 8 \ \checkmark$$

Is the area of the rectangle 44 square meters?

$$4 \times 11 \stackrel{?}{=} 44 \qquad 44 = 44 \ \checkmark$$

Practice Problem 7 The length of a rectangle is 2 meters longer than triple the width. The area of the rectangle is 85 square meters. Find the length and width of the rectangle. ■

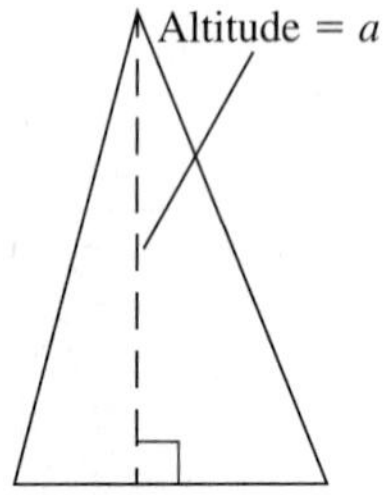

EXAMPLE 8 The area of a triangle is 49 square centimeters. The altitude of the triangle is 7 centimeters longer than the base. Find the altitude and the base of the triangle.

Let b = the length of the base in centimeters

$b + 7$ = the length of the altitude in centimeters

To find the area of a triangle, we use

$$\text{area} = \frac{1}{2}(\text{altitude})(\text{base}) = \frac{1}{2}ab = \frac{ab}{2}$$

$\frac{ab}{2} = 49$		*Write an equation.*
$\frac{(b + 7)(b)}{2} = 49$		*Substitute the expressions for altitude and base.*
$\frac{b^2 + 7b}{2} = 49$		*Simplify.*
$b^2 + 7b = 98$		*Multiply each side of the equation by 2.*
$b^2 + 7b - 98 = 0$		*Place the quadratic equation in standard form.*
$(b - 7)(b + 14) = 0$		*Factor.*
$b - 7 = 0$	$b + 14 = 0$	*Set each factor equal to zero.*
$b = 7$	$b = -14$	*Solve each equation for b.*

We cannot have a base of -14 centimeters, so we reject the negative answer. The only possible answer is $b = 7$. So the base is 7 centimeters. The altitude is $b + 7 = 7 + 7 = 14$. The altitude is 14 centimeters.

Check. When you do the check, answer the following two questions.

1. Is the altitude 7 centimeters longer than the base?

2. Is the area of a triangle with a base of 7 centimeters and an altitude of 14 centimeters actually 49 square centimeters?

Practice Problem 8 A triangle has an area of 35 square centimeters. The altitude of the triangle is 3 centimeters shorter than the base. Find the altitude and the base of the triangle. ■

Many examples in the sciences require the use of quadratic equations. You will study these in more detail if you take a course in physics or calculus in college. Often a quadratic equation is given as part of the problem.

When an object is thrown upward, its height (S) in meters is given, approximately, by the quadratic equation

$$S = -5t^2 + vt + h$$

The letter h represents the initial height in meters. The letter v represents the initial velocity of the object thrown. The letter t represents the time in seconds starting from the time the object is thrown.

EXAMPLE 9 Suppose that the initial height above the ground is 6 meters. A tennis ball is thrown upward with an initial velocity of 29 meters/second. At what time t will the ball hit the ground?

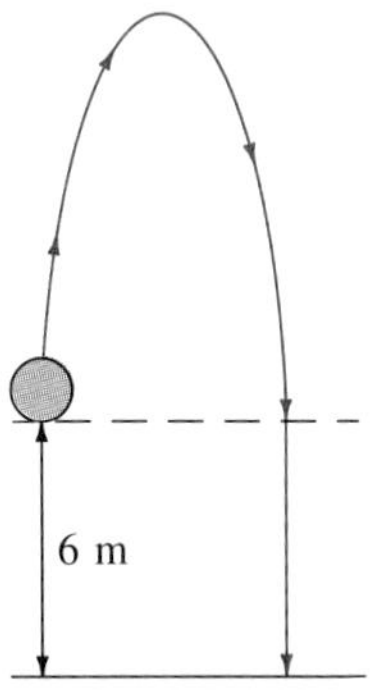

In this case $S = 0$ since the ball will hit the ground. The initial upward velocity is $v = 29$ meters/second. The initial height is 6 meters, so $h = 6$.

$S = -5t^2 + vt + h$		*Write an equation.*
$0 = -5t^2 + 29t + 6$		*Substitute all values into the equation.*
$5t^2 - 29t - 6 = 0$		*Transpose to left side. (Most students can factor more readily if the squared variable is positive.)*
$(5t + 1)(t - 6) = 0$		*Factor.*
$5t + 1 = 0$	$t - 6 = 0$	*Set each factor* = 0.
$5t = -1$	$t = 6$	*Solve for t.*
$t = -\frac{1}{5}$		

We want a positive time t in seconds; thus we do not use $t = -\frac{1}{5}$. Thus the ball will strike the ground 6 seconds after you throw it.

Check. Verify the solution.

Practice Problem 9 A diver does a handstand dive from the 45-meter platform. This constitutes free fall, so the initial $v = 0$, and there is no upward spring, so $h = 45$ meters. How long will it be until she breaks the water's surface? ■

5.7 Exercises

Using the factoring method, solve for the roots of each quadratic equation. Be sure to place the equation in standard form before factoring. Check your answers.

1. $x^2 - 3x - 10 = 0$

2. $x^2 + 2x - 24 = 0$

3. $x^2 + 13x + 40 = 0$

4. $x^2 + 16x + 63 = 0$

5. $2x^2 - 5x - 3 = 0$

6. $3x^2 - 5x - 2 = 0$

7. $4x^2 - 11x - 3 = 0$

8. $2x^2 + 15x - 8 = 0$

9. $x^2 - 6x = 16$

10. $x^2 - 21 = 4x$

11. $10 = 7x - x^2$

12. $6 = 7x - x^2$

13. $6x^2 - 13x = -6$

14. $10x^2 + 19x = 15$

15. $x^2 - 8x = 0$

16. $5x^2 - x = 0$

17. $3x^2 = 2x$

18. $8x = 6x^2$

19. $5x^2 + 3x = 8x$

20. $6x^2 - 4x = 3x$

21. $x(9 + x) = 4(2x + 5)$

22. $(x - 2)(x - 3) = 6(x - 3)$

23. $x^2 - 12x + 36 = 0$

24. $x^2 - 14x + 49 = 0$

25. $4x^2 - 3x + 1 = -7x$

26. $9x^2 - 2x + 4 = 10x$

27. $\dfrac{x^2}{2} - 8 + x = -8$

28. $4 + \dfrac{x^2}{3} = 2x + 4$

29. $\dfrac{x^2 - 3x}{2} = 27$

30. $\dfrac{x^2 - 9x}{2} = -9$

31. $\dfrac{2x^2 + x}{3} = 12$

32. $\dfrac{2x^2 - 2x}{2} = 6$

To Think About

33. Why can an equation in standard form with $c = 0$ (leaving $ax^2 + bx = 0$) always be solved?

34. Martha solved $(x + 3)(x - 2) = 14$ as follows:

$$x + 3 = 14 \quad \text{or} \quad x - 2 = 14$$
$$x = 11 \quad \text{or} \quad x = 16$$

Josette said this had to be wrong because these values did not check. Explain what is wrong with Martha's method.

Applications

35. The area of a rectangle is 140 square meters. The width is 3 meters longer than one-half of the length. Find the length and the width of the rectangle.

36. The area of a triangle is 33 square centimeters. The base of the triangle is 1 centimeter less than double the altitude. Find the altitude and the base of the triangle.

37. Find two positive numbers whose difference is 10 and whose product is 56.

38. The sum of two numbers is -17. The product of these two numbers is 66. Find the numbers.

Use the following information for Problems 39 and 40. When an object is thrown upward, its height (S), in meters, is given (approximately) by the quadratic equation

$$S = -5t^2 + vt + h$$

where $v =$ *the upward initial velocity in meters/second*

$t =$ *the time of flight in seconds*

$h =$ *the height above level ground from which the object is thrown*

39. You stand on a platform 6 meters high and you throw a ball straight up at a velocity of 13 meters/second. At what time t will the ball hit the ground? How far from the ground is the ball after 2 seconds have elapsed from the time of the throw? (Assume that the ball is 6 meters from the ground when it leaves your hand.)

40. You are standing on the edge of a sheer cliff overlooking the ocean. The place where you stand is 180 meters from the ocean. You drop a pebble into the water. (''Dropping'' the pebble implies that there is no initial velocity, so $v = 0$.) How many seconds will it take to hit the water? How far has the pebble dropped in 3 seconds?

41. Overtime production is expensive, but if sufficient extra goods are produced, it is worth putting a shift on overtime. In a certain company, the additional hourly cost in dollars for producing x additional stereo sets is given by the cost equation $C = 2x^2 - 7x$. If the additional hourly cost is \$15, how many additional stereo sets were produced?

42. A boat generator on a fishing boat is required to produce 64 watts of power. The amount of current I measured in amperes needed to produce the power for this generator is given by the equation $P = 40I - 4I^2$. What is the *minimum* number of amperes required to produce the necessary power?

Cumulative Review Problems

Simplify.

43. $(2x^2y^3)(-6xy^4)$

44. $(3a^4b^5)(4a^6b^8)$

45. $\dfrac{21a^5b^{10}}{-14ab^{12}}$

46. $\dfrac{18x^3y^6}{54x^8y^{10}}$

Putting Your Skills to Work

HOME IMPROVEMENT

Robert's parents are installing a swimming pool that is to be surrounded by a concrete deck. The deck is to have a uniform width of 5 feet. The deck will have 1150 square feet of area. The pool itself is to be $2\frac{1}{2}$ times longer than its width.

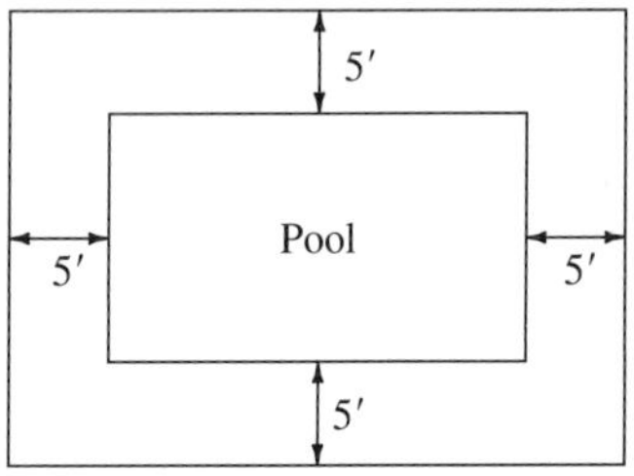

Solve each problem.

1. What is the length and width of the pool?

2. Robert's parents have changed their minds and have decided on a pool for which the length is 3 times longer than the width. The deck will now have 1300 square feet of area. What is the length and width of the new pool?

COOPERATIVE GROUP ACTIVITY

3. A contractor bids on the job of pouring the concrete deck in Problem 1. The concrete will be poured to a depth of 3 inches ($\frac{1}{4}$ foot). The contractor estimates such jobs at \$125 per cubic yard of concrete. What should be his estimate of the cost for the job? (*Hint:* You will need to find the number of cubic feet in one cubic yard.)

4. The same contractor bids on the job of pouring the concrete deck in Problem 2. The concrete will now be poured to a depth of 4 inches ($\frac{1}{3}$ foot). Because of the greater depth of the concrete, his estimate will be at \$120 per cubic yard of concrete. What should be his estimate of the cost of the job?

Chapter Organizer

Topic	Procedure	Examples
A. Common factor, p. 294.	Remove the largest common factor from each term.	$2x^2 - 2x = 2x(x - 1)$ $3a^2 + 3ab - 12a = 3a(a + b - 4)$ $8x^4y - 24x^3 = 8x^3(xy - 3)$
Special cases, B. Difference of squares, p. 314. C. Perfect square trinomials, p. 315.	If you recognize the special cases, you will be able to factor quickly. $a^2 - b^2 = (a + b)(a - b)$ $a^2 + 2ab + b^2 = (a + b)^2$ $a^2 - 2ab + b^2 = (a - b)^2$	$25x^2 - 36y^2 = (5x + 6y)(5x - 6y)$ $16x^4 - 1 = (4x^2 + 1)(4x^2 - 1)$ $25x^2 + 10x + 1 = (5x + 1)^2$ $49x^2 - 42xy + 9y^2 = (7x - 3y)^2$
D. Trinomial that starts with x^2, p. 305.	Factor trinomials of the form $x^2 + bx + c$ by asking what two numbers have a product of c and a sum of b. If each term of the trinomial has a common factor, remove the common factor as the first step.	$x^2 - 18x + 77 = (x - 7)(x - 11)$ $x^2 + 7x - 18 = (x + 9)(x - 2)$ $5x^2 - 10x - 40 = 5(x^2 - 2x - 8)$ $= 5(x - 4)(x + 2)$
E. Trinomial that starts with ax^2, where $a \neq 1$, p. 311.	Factor trinomials of the form $ax^2 + bx + c$ by the grouping number method or by trial and error.	$6x^2 + 11x - 10$ Grouping number $= -60$ Two numbers whose product is -60 and whose sum is $+11$ are $+15$ and -4 $6x^2 + 15x - 4x - 10 = 3x(2x + 5) - 2(2x + 5)$ $(2x + 5)(3x - 2)$
F. Four terms. Factor by grouping, p. 298.	Rearrange the terms if necessary so that the terms have a common factor. Then remove the common factors. $ax + ay - bx - by = a(x + y) - b(x + y)$ $= (a - b)(x + y)$	$2ax^2 + 21 + 3a + 14x^2$ $= 2ax^2 + 14x^2 + 3a + 21$ $= 2x^2(a + 7) + 3(a + 7)$ $= (a + 7)(2x^2 + 3)$
Prime polynomials, p. 320.	A polynomial that is not factorable is called *prime*.	$x^2 + y^2$ is *prime*. $x^2 + 5x + 7$ is *prime*.
Multistep factoring, p. 320.	Many problems require two or three steps of factoring. Always try to remove a common factor as the first step.	$3x^2 - 21x + 36 = 3(x^2 - 7x + 12)$ $= 3(x - 4)(x - 3)$ $2x^3 - x^2 - 6x = x(2x^2 - x - 6)$ $= x(2x + 3)(x - 2)$ $25x^3 - 49x = x(25x^2 - 49)$ $= x(5x + 7)(5x - 7)$ $8x^2 - 24x + 18 = 2(4x^2 - 12x + 9)$ $= 2(2x - 3)^2$
Solving quadratic equations by factoring, p. 322.	**1.** Write as $ax^2 + bx + c = 0$. **2.** Factor. **3.** Set each factor equal to 0. **4.** Solve the resulting equations.	Solve: $3x^2 + 5x = 2$ $3x^2 + 5x - 2 = 0$ $(3x - 1)(x + 2) = 0$ $3x - 1 = 0$ or $x + 2 = 0$ $x = \frac{1}{3}$ or $x = -2$
Using quadratic equations to solve applied problems, p. 325.	Word problems involving the product of two numbers, area, and formulas with a square variable can be solved using the factoring methods we have shown.	The length of a rectangle is 4 less than three times the width. Find the length and width if the area is 55 square inches. Let w = width $3w - 4$ = length $55 = w(3w - 4)$ $55 = 3w^2 - 4w$ $0 = 3w^2 - 4w - 55$ $0 = (3w + 11)(w - 5)$ $w = -\frac{11}{3}$ or $w = 5$ $-\frac{11}{3}$ is not a valid solution. width = 5 inches length = 11 inches

Chapter 5 Review Problems

5.1 *Factor out the greatest common factor from each term.*

1. $15x^3y - 9x^2y^2$

2. $21x^2 - 14x$

3. $7x^2y - 14xy^2 - 21x^3y^3$

4. $50a^4b^5 - 25a^4b^4 + 75a^5b^5$

5. $27x^3 - 9x^2$

6. $2x - 4y + 6z + 12$

7. $2a(a + 3b) - 5(a + 3b)$

8. $15x^3y + 6xy^2 + 3xy$

5.2 *Factor by grouping.*

9. $3ax - 7a - 6x + 14$

10. $a^2 + 5ab - 4a - 20b$

11. $x^2y + 3y - 2x^2 - 6$

12. $4ad - 25c - 5d + 20ac$

13. $15x^2 - 3x + 10x - 2$

14. $30w^2 - 18w + 5wz - 3z$

5.3 *Factor completely. Be sure to remove any common factors as your first step.*

15. $x^2 - 3x - 18$

16. $x^2 - 10x + 24$

17. $x^2 + 14x + 48$

18. $x^2 + 8xy + 15y^2$

19. $x^4 + 13x^2 + 42$

20. $x^2 - 10x - 39$

21. $5x^2 + 20x + 15$

22. $3x^2 + 39x + 36$

23. $2x^2 - 28x + 96$

24. $4x^2 - 44x + 120$

5.4 *Factor completely. Be sure to remove any common factors as your first step.*

25. $4x^2 + 7x - 15$

26. $12x^2 + 11x - 5$

27. $15x^2 + 7x - 4$

28. $6x^2 - 13x + 6$

29. $2x^2 - x - 3$

30. $3x^2 + 14x - 5$

31. $20x^2 + 48x - 5$

32. $20x^2 + 21x - 5$

33. $4a^2 - 11a - 3$

34. $4a^2 - a - 3$

35. $6x^2 + 4x - 10$

36. $6x^2 - 4x - 10$

37. $4x^2 - 26x + 30$

38. $4x^2 - 20x - 144$

39. $10x^2 - 22x + 12$

40. $2x^2 + 3x - 44$

41. $12x^2 + 5x - 3$

42. $16x^2 + 14x - 30$

43. $6x^2 - 19xy + 10y^2$

44. $6x^2 - 32xy + 10y^2$

45. $3x^4 - 5x^2 - 8$

46. $3x^4 + 10x^2 - 8$

5.5 *Factor these special cases. Be sure to remove any common factors.*

47. $49x^2 - y^2$

48. $16x^2 - 36y^2$

49. $9x^2 - 12x + 4$

50. $16x^2 - 1$

51. $25x^2 - 36$

52. $100x^2 - 9$

53. $1 - 49x^2$

54. $4 - 49x^2$

55. $36x^2 + 12x + 1$

56. $25x^2 - 20x + 4$

57. $16x^2 - 24xy + 9y^2$

58. $49x^2 - 28xy + 4y^2$

59. $2x^2 - 18$

60. $3x^2 - 75$

61. $8x^2 + 40x + 50$

62. $50x^2 + 120x + 72$

5.6 *If possible, factor each problem completely. If it cannot be factored, state "prime."*

63. $4x^2 - 9y^2$

64. $x^2 + 6x + 9$

65. $x^2 - 9x + 18$

66. $x^2 + 13x - 30$

67. $6x^2 + x - 7$

68. $10x^2 + x - 2$

69. $12x + 16$

70. $8x^2y^2 - 4xy$

71. $50x^3y^2 + 30x^2y^2 - 10x^2y^2$

72. $26a^3b - 13ab^3 + 52a^2b^4$

73. $x^3 - 16x^2 + 64x$

74. $2x^2 + 40x + 200$

75. $3x^2 - 18x + 27$

76. $25x^3 - 60x^2 + 36x$

77. $7x^2 + 3x - 10$

78. $5x^2 - 8x - 4$

79. $9x^3y - 4xy^3$

80. $3x^3a^3 - 11x^4a^2 - 20x^5a$

81. $12a^2 + 14ab - 10b^2$

82. $16a^2 - 40ab + 25b^2$

83. $7a - 7 - ab + b$

84. $3d - 4 - 3cd + 4c$

85. $2x - 1 + 2bx - b$

86. $5xb - 35x + 4by - 28y$

87. $2a^2x - 15ax + 7x$

88. $x^5 - 17x^3 + 16x$

89. $x^4 - 81y^{12}$

90. $6x^4 - x^2 - 15$

91. $28yz - 16xyz + x^2yz$

92. $12x^3 + 17x^2 + 6x$

93. $16w^2 - 2w - 5$

94. $12w^2 - 12w + 3$

95. $4y^3 + 10y^2 - 6y$

96. $10y^2 + 33y - 7$

97. $8y^{10} - 16y^8$

98. $49x^4 - 49$

99. $x^2 + 13x + 54$

100. $8x^2 - 19x - 6$

101. $8y^5 + 4y^3 - 60y$

102. $9xy^2 + 3xy - 42x$

103. $16x^4y^2 - 56x^2y + 49$

104. $128x^3y - 2xy$

105. $2ax + 5a - 10b - 4bx$

106. $2x^3 - 9 + x^2 - 18x$

5.7 *Solve these quadratic equations by factoring.*

107. $x^2 - 3x - 18 = 0$

108. $x^2 + 3x - 10 = 0$

109. $5x^2 = 2x - 7x^2$

110. $8x^2 + 5x = 2x^2 - 6x$

111. $2x^2 + 9x - 5 = 0$

112. $x^2 + 11x + 24 = 0$

113. $x^2 + 14x + 45 = 0$

114. $5x^2 = 7x + 6$

115. $3x^2 + 6x = 2x^2 - 9$

116. $4x^2 + 9x - 9 = 0$

117. $5x^2 - 11x + 2 = 0$

Solve the following applied problems.

118. The area of a triangle is 35 square centimeters. The base is 3 centimeters longer than the altitude of the triangle. Find the length of the base and the altitude.

119. The area of a rectangle is 105 square feet. The length of the rectangle is 1 foot longer than double the width. Find the length and width of the rectangle.

120. The height in feet that a model rocket attains is given by height $= -16t^2 + 80t + 96$, where t is the time measured in seconds. How many seconds will it take until the rocket finally reaches the ground? (*Hint:* At ground level the height = 0.)

121. An electronic technician is working with a 100-volt electric generator. The output power of the generator is given by the equation, power $= -5x^2 + 100x$, where x is the amount of current measured in amperes. The technician wants to find the value for x when the power is 480 watts. Can you find the two answers?

Putting Your Skills to Work

MAXIMUM PROFIT

Quadratic equations can be used to describe the profit a company makes. One such equation is

$$P = -\frac{7}{400}t^2 + 7t - 400$$

where t is the number of items sold.

If the company does not sell enough items, it will not be able to meet its expenses and it will be running at a loss. The company may also experience a loss if it sells too many items. Why? What companies strive to do is to maximize their profit, that is, sell just the number of items that will give them the greatest profit.

Use the equation $P = -\frac{7}{400}t^2 + 7t - 400$ to answer the following.

1. What is the profit if 100 items are sold? (*Hint:* Evaluate the equation for $t = 100$.)

2. How many items need to be sold for a profit of \$188. (*Hint:* Let $P = 188$ and solve for t.)

3. What is the ideal number of items that need to be sold to maximize the profit? (*Hint:* For any quadratic $ax^2 + bx + c$, the maximum value occurs when $x = -\frac{b}{2a}$.)

4. Find the profit for the value of t in Problem 3.

5. Show that the profit in Problem 4 is the maximum profit by testing a number that is less than the value of t in Problem 3 and a number that is greater. What should occur?

COOPERATIVE GROUP ACTIVITY

An automobile manufacturer has a profit equation of

$$\text{profit} = -n^2 + 3000n - 2{,}000{,}000$$

where n is the number of cars sold per month.

6. What is the profit when 1200 cars are sold per month?

7. What is the profit when 500 cars are sold per month?

8. How many cars should be sold per month to break even?

9. How many cars should be sold per month to achieve the maximum profit?

10. What is the maximum profit?

11. If zero cars are sold in a given month, what is the loss to the company?

Chapter 5 Test

If possible, factor each problem completely. If a problem cannot be factored, state "prime."

1. $x^2 + 12x - 28$

2. $9x^2 - 100y^2$

3. $10x^2 + 27x + 5$

4. $9a^2 - 30ab + 25b^2$

5. $7x - 9x^2 + 14xy$

6. $3x^2 - 4wy - 2wx + 6xy$

7. $6x^3 - 20x^2 + 16x$

8. $5a^2c - 11abc + 2b^2c$

9. $100x^4 - 16y^4$

10. $9x^2 - 15xy + 4y^2$

11. $7x^2 - 42x$

12. $a^2 + 1$

1. ______________________

2. ______________________

3. ______________________

4. ______________________

5. ______________________

6. ______________________

7. ______________________

8. ______________________

9. ______________________

10. ______________________

11. ______________________

12. ______________________

13. $3x^2 + 5x + 1$

14. $60xy^2 - 20x^2y - 45y^3$

15. $16x^2 - 1$

16. $x^{16} - 1$

17. $2ax + 6a - 5x - 15$

18. $aw^2 - 8b + 2bw^2 - 4a$

19. $3x^2 - 3x - 90$

20. $2x^3 - x^2 - 15x$

Solve the following quadratic equations.

21. $x^2 + 14x + 45 = 0$

22. $14 + 3x(x + 2) = -7x$

23. $2x^2 + x - 10 = 0$

Solve the following problem by using a quadratic equation.

24. The park service is studying a rectangular piece of land that contains 91 square miles. The length of this piece of land is 1 mile shorter than double the width. Find the length and width of this rectangular piece of land.

13. ____

14. ____

15. ____

16. ____

17. ____

18. ____

19. ____

20. ____

21. ____

22. ____

23. ____

24. ____

Cumulative Test for Chapters 0–5

Approximately one-half of this test covers the content of Chapters 0–4. The remainder covers the content of Chapter 5.

1. What % of 480 is 72?

2. What is 13% of 3.8?

Simplify.

3. $-3.2 - 6.4 + 0.24 - 1.8 + 0.8$

4. $(-2x^3y^4)(-4xy^6)$

5. -2^6

6. $(7x - 3)(2x + 1)$

7. $(2x^2 - 6x + 1)(x - 3)$

Solve.

8. $3x - 4 \geq 6x + 5$

9. $3x - (7 - 5x) = 3(4x - 5)$

10. $\frac{1}{2}x - 3 = \frac{1}{4}(3x + 3)$

11. Solve for t: $s = \frac{1}{2}(2a + 3t)$.

1. __________

2. __________

3. __________

4. __________

5. __________

6. __________

7. __________

8. __________

9. __________

10. __________

11. __________

12.

13.

14.

15.

16.

17.

18.

19.

20.

21.

22.

23.

24.

Factor each problem completely. If a problem cannot be factored, state "prime."

12. $6x^2 - 5x + 1$

13. $6x^2 - 5x - 1$

14. $9x^2 + 3x - 2$

15. $121x^2 - 64y^2$

16. $4x + 120 - 80x^2$

17. $x^2 + 1$

18. $16x^3 + 40x^2 + 25x$

19. $81x^4 - 16b^4$

20. $2ax - 4bx + 3a - 6b$

21. $x^4 + 8x^2 + 15$

Solve the following quadratic equations.

22. $x^2 + 5x - 24 = 0$

23. $3x^2 - 11x + 10 = 0$

Solve the following problem by using a quadratic equation.

24. The park service is studying a triangular piece of land that contains 57 square miles. The altitude of this triangle is 7 miles longer than double the base. Find the altitude and the base of this triangular piece of land.

6

Rational Expressions and Equations

Some applied problems like the following one seem ''obvious,'' and most students can guess an answer quickly—but usually their answer is wrong! Here is the problem: The larger pump at the town pumping station can fill the town water tank in 15 hours. The smaller pump can do the same job in 20 hours. How long will it take the two pumps working together?

Using rational equations can help us find the correct answer quite easily. See Exercise 6.6, Problem 35.

Pretest Chapter 6

1. ______
2. ______
3. ______
4. ______
5. ______
6. ______
7. ______
8. ______
9. ______
10. ______
11. ______
12. ______
13. ______
14. ______
15. ______
16. ______
17. ______
18. ______

If you are familiar with the topics in this chapter, take this test now. Check your answers with those in the back of the book. If an answer was wrong or you couldn't do a problem, study the appropriate section of the chapter.

If you are not familiar with the topics in this chapter, don't take this test now. Instead, study the examples, work the practice problems, and then take the test.

This test will help you to identify those concepts that you have mastered and those that need more study.

Simplify the algebraic expressions in Problems 1 through 14.

Section 6.1

1. $\dfrac{6a - 4b}{2b - 3a}$ **2.** $\dfrac{3x^2 - 14x + 8}{4x^2 - 13x - 12}$ **3.** $\dfrac{a^2b + 2ab^2}{2a^3 + 3a^2b - 2ab^2}$

Section 6.2

4. $\dfrac{4a^2 - b^2}{6a - 6b} \cdot \dfrac{3a - 3b}{6a + 3b}$ **5.** $\dfrac{x^2 - 6x + 9}{x^2 - x - 6} \div \dfrac{x^2 + 2x - 15}{x^2 + 2x}$

6. $\dfrac{x^2 + 5x + 6}{3x^2 + 8x + 4} \cdot \dfrac{1}{x + 3} \cdot \dfrac{6x^2 - 11x - 10}{2x^2 + x - 15}$ **7.** $\dfrac{xy + 3y}{x^2 - x} \div \dfrac{x + 3}{x}$

Section 6.3

8. $\dfrac{3y + 1}{y + 2} + \dfrac{5}{y + 2}$ **9.** $\dfrac{2y - 1}{2y^2 + y - 3} - \dfrac{2}{y - 1}$

10. $\dfrac{4x + 10}{x^2 - 25} - \dfrac{3}{x - 5}$ **11.** $\dfrac{x + 3}{x^2 - 6x + 9} + \dfrac{2x + 3}{3x^2 - 9x}$

Section 6.4

12. $\dfrac{\dfrac{2}{a} - \dfrac{3}{a^2}}{5 + \dfrac{1}{a}}$ **13.** $\dfrac{\dfrac{a}{a + 1} - \dfrac{2}{a}}{3a}$ **14.** $\dfrac{\dfrac{x - y}{x} - \dfrac{x + y}{y}}{\dfrac{x - y}{x} + \dfrac{x + y}{y}}$

Section 6.5

Solve for x. If the equation has no solution, so state.

15. $\dfrac{5}{2x} = 2 - \dfrac{2x}{x + 1}$ **16.** $\dfrac{14}{x^2 - 3x - 10} = \dfrac{2}{x - 5} + \dfrac{3}{x + 2}$

Section 6.6

17. Solve for x and round your answer to the nearest tenth. $\dfrac{5}{x} = \dfrac{7}{13}$

18. When Marcia went to England, she found that \$3.10 in U.S. money was needed at the current exchange rate to receive 2 British pounds. How much U.S. money was needed to receive 170 British pounds?

6.1 Simplify Rational Expressions

After studying this section, you will be able to:

1 *Simplify an algebraic fraction by factoring.*

Recall that a rational number is a number that can be written as one integer divided by another integer such as $3 \div 4$ or $\frac{3}{4}$. We usually use the word fraction to mean $\frac{3}{4}$. We can extend this idea to algebraic expressions. A rational expression is an algebraic expression divided by another algebraic expression such as

$$(3x + 2) \div (x + 4) \quad \text{or} \quad \frac{3x + 2}{x + 4}$$

The last fraction is sometimes called a **fractional algebraic expression.** There is a special restriction for all fractions, including fractional algebraic expressions. The denominator of the fraction cannot be 0. For example, in the expression

$$\frac{3x + 2}{x + 4}$$

the denominator cannot be 0. Therefore, the value of x cannot be -4. The following important restriction will apply throughout this chapter. We state it here to avoid having to mention it repeatedly throughout this chapter.

Restriction

The denominator of an algebraic fraction cannot be zero. Any value of the variable that would make the denominator zero is not allowed.

We have discovered that fractions can be simplified (or reduced) in the following way.

$$\frac{15}{25} = \frac{3 \cdot 5}{5 \cdot 5} = \frac{3}{5}$$

This is sometimes referred to as the *basic rule of fractions* and can be stated as follows:

Basic Rule of Fractions

For any polynomials a, b, c (where $b \neq 0$ and $c \neq 0$),

$$\frac{ac}{bc} = \frac{a}{b}$$

There is one essential property that is revealed by the basic rule of fractions. It is this: If the numerator and denominator of a given fraction are multiplied by the same quantity, an equivalent fraction is obtained. The rule can be used two ways. You can start with $\frac{ac}{bc}$ and end with the equivalent fraction $\frac{a}{b}$. You can start with $\frac{a}{b}$ and end with the equivalent fraction $\frac{ac}{bc}$.

EXAMPLE 1

(a) Write a fraction equivalent to $\frac{3}{5}$ with a denominator of 10.

(b) Reduce $\dfrac{21}{39}$.

(a) $\dfrac{3}{5} = \dfrac{3 \cdot 2}{5 \cdot 2} = \dfrac{6}{10}$ *Use the rule* $\frac{a}{b} = \frac{ac}{bc}$. *Let* $c = 2$ *since* $5 \cdot 2 = 10$.

(b) $\dfrac{21}{39} = \dfrac{7 \cdot 3}{13 \cdot 3} = \dfrac{7}{13}$ *Use the rule* $\frac{ac}{bc} = \frac{a}{b}$. *Let* $c = 3$ *because* 3 *is the greatest common factor of* 21 *and* 39.

Practice Problem 1

(a) Find a fraction equivalent to $\frac{7}{3}$ with a denominator of 18.

(b) Reduce $\frac{28}{63}$. ■

1 *Simplifying Algebraic Fractions*

The process of reducing the fraction shown above is sometimes called *dividing out* common factors. Remember, only factors of both the numerator and the denominator can be divided out. Now, to apply this rule, it is usually necessary that the numerator and denominator of the fraction be completely factored. You will need to use your factoring skills from Chapter 4 to accomplish this step. When you are applying this rule you are *simplifying the fraction.*

EXAMPLE 2 Simplify. $\dfrac{4x + 12}{5x + 15}$

$$\frac{4x + 12}{5x + 15} = \frac{4(x + 3)}{5(x + 3)}$$ *Factor* 4 *from the numerator.* *Factor* 5 *from the denominator.*

$$= \frac{4\,(x + 3)}{5\,(x + 3)}$$ *Apply the basic rule of fractions.*

$$= \frac{4}{5}$$

Practice Problem 2 Simplify. $\dfrac{12x - 6}{14x - 7}$ ■

EXAMPLE 3 Simplify. $\dfrac{x^2 + 9x + 14}{x^2 - 4}$

$= \dfrac{(x + 7)(x + 2)}{(x - 2)(x + 2)}$ *Factor the numerator.* *Factor the denominator.*

$= \dfrac{(x + 7)(x + 2)}{(x - 2)(x + 2)}$ *Apply the basic rule of fractions.*

$= \dfrac{x + 7}{x - 2}$

Practice Problem 3 Simplify. $\dfrac{4x - 6}{2x^2 - x - 3}$ ■

EXAMPLE 4 Simplify. $\dfrac{ab + ac}{ax - 3ay}$

$= \dfrac{a(b + c)}{a(x - 3y)}$ *Factor the numerator.* *Factor the denominator.*

$= \dfrac{a(b + c)}{a(x - 3y)}$ *Apply the basic rule of fractions.*

$= \dfrac{b + c}{x - 3y}$

Practice Problem 4 Simplify. $\dfrac{7x^2 - 14xy}{14x^2 - 28xy}$ ■

Some problems may involve more than one step of factoring. Always remember to remove any common factors as the first step, if it is possible to do so.

EXAMPLE 5 Simplify. $\dfrac{9x - x^3}{x^3 + x^2 - 6x}$

$= \dfrac{x(9 - x^2)}{x(x^2 + x - 6)}$ *Remove a common factor from the polynomial in the numerator and in the denominator.*

$= \dfrac{x(3 + x)(3 - x)}{x(x + 3)(x - 2)}$ *Factor each polynomial and apply the basic rule of fractions.* *Note that $(3 + x)$ is equivalent to $(x + 3)$ since addition is commutative.*

$= \dfrac{3 - x}{x - 2}$

Practice Problem 5 Simplify. $\dfrac{x^3 + 11x^2 + 30x}{3x^3 + 17x^2 - 6x}$ ■

Be on the lookout for a special simplifying situation. Watch for a situation where *each term in the denominator is opposite in sign* from each term in the numerator. In such cases you should factor -1 or another negative number from one polynomial so that the expression in each set of parentheses is equivalent. Look carefully at the following two examples.

EXAMPLE 6 Simplify. $\dfrac{5x - 15}{6 - 2x}$

Notice that the variable term in the numerator, $5x$, and the variable term in the denominator, $-2x$, *are opposite in sign.* Likewise, the numerical terms -15 and 6 *are opposite in sign.* Factor out a negative number from the denominator.

$$= \frac{5(x-3)}{-2(-3+x)}$$

Factor 5 from the numerator.
Factor -2 from the denominator.

Note that $(x - 3)$ and $(-3 + x)$ are equivalent (since $+x - 3 = -3 + x$!).

$$= \frac{5(x-3)}{-2(-3+x)}$$

Apply the basic rule of fractions.

$$= -\frac{5}{2}$$

Note that $\frac{5}{-2}$ is not considered to be in simple form. We usually avoid leaving a negative number in the denominator. Therefore, to simplify, give the result as $-\frac{5}{2}$ or $\frac{-5}{2}$.

Practice Problem 6 Simplify. $\dfrac{2x - 5}{5 - 2x}$ ■

EXAMPLE 7 Simplify. $\dfrac{2x^2 - 11x + 12}{16 - x^2}$

$$= \frac{(x-4)(2x-3)}{(4-x)(4+x)}$$

Factor numerator and denominator. Observe that $(x - 4)$ and $(4 - x)$ are opposite in sign.

$$= \frac{(x-4)(2x-3)}{-1(-4+x)(4+x)}$$

Factor -1 out of $(+4 - x)$ to obtain $-1(-4 + x)$.

$$= \frac{(x-4)(2x-3)}{-1(-4+x)(4+x)}$$

Apply the basic rule of fractions, since $(x - 4)$ and $(-4 + x)$ are equivalent.

$$= \frac{2x-3}{-1(4+x)}$$

$$= -\frac{2x-3}{4+x}$$

Practice Problem 7 Simplify. $\dfrac{4x^2 + 3x - 10}{25 - 16x^2}$ ■

After doing Examples 6 and 7, you will notice a pattern. Whenever the factor in the numerator and the factor in the denominator are exactly opposite in sign, the value -1 results. We could actually make this a definition of that property.

> For all monomials A and B where $A \neq B$, it is true that
>
> $$\frac{A - B}{B - A} = -1$$

You may use this definition in reducing fractions if it is helpful to you. Otherwise, you may use the factoring method discussed in Examples 6 and 7.

Some problems will involve two or more variables. In such cases, you will need to factor carefully and make sure that each set of parentheses contains the correct letters.

EXAMPLE 8 Simplify. $\dfrac{x^2 - 7xy + 12y^2}{2x^2 - 7xy - 4y^2}$

$= \dfrac{(x - 4y)(x - 3y)}{(2x + y)(x - 4y)}$ *Factor the numerator.* *Factor the denominator.*

$= \dfrac{(x - 4y)(x - 3y)}{(2x + y)(x - 4y)}$ *Apply the basic rule of fractions.*

$= \dfrac{x - 3y}{2x + y}$

Practice Problem 8 Simplify. $\dfrac{4x^2 - 9y^2}{4x^2 + 12xy + 9y^2}$ ■

EXAMPLE 9 Simplify. $\dfrac{6a^2 + ab - 7b^2}{36a^2 - 49b^2}$

$= \dfrac{(6a + 7b)(a - b)}{(6a + 7b)(6a - 7b)}$ *Factor the numerator.* *Factor the denominator.*

$= \dfrac{(6a + 7b)(a - b)}{(6a + 7b)(6a - 7b)}$ *Apply the basic rule of fractions.*

$= \dfrac{a - b}{6a - 7b}$

Practice Problem 9 Simplify. $\dfrac{12x^3 - 48x}{6x - 3x^2}$ ■

6.1 Exercises

Simplify.

1. $\dfrac{6x - 3y}{2x - y}$

2. $\dfrac{7a + 2b}{14a + 4b}$

3. $\dfrac{x^2}{3x^2 - x^3}$

4. $\dfrac{xy - 2x}{3y - 6}$

5. $\dfrac{2x - 8}{x^2 - 8x + 16}$

6. $\dfrac{4x^2 + 4x + 1}{1 - 4x^2}$

7. $\dfrac{x^2 - 9y^2}{x + 3y}$

8. $\dfrac{xy(x + y^2)}{x^2y^2}$

9. $\dfrac{6x^2}{2x(x - 3y)}$

10. $\dfrac{x^2 + 3x - 4}{x^2 - 1}$

11. $\dfrac{x^2 + x - 2}{x^2 - x}$

12. $\dfrac{x^2 - x - 6}{x^2 - 5x + 6}$

13. $\dfrac{x^2 + 2x - 3}{4x^2 - 5x + 1}$

14. $\dfrac{4x^2 - 10x + 6}{2x^2 + x - 3}$

15. $\dfrac{x^3 - 8x^2 + 16x}{x^2 + 2x - 24}$

16. $\dfrac{x^2 - 10x - 24}{x^3 + 9x^2 + 14x}$

17. $\dfrac{x^2 + 14x + 45}{2x^2 + 17x - 9}$

18. $\dfrac{2x^2 - 5x - 12}{x^2 - 12x + 32}$

19. $\dfrac{3x^2 - 8x + 5}{4x^2 - 5x + 1}$

20. $\dfrac{5y^2 + 13y - 6}{5y^2 + 18y - 8}$

21. $\dfrac{2x^2 - 5x - 12}{2x^2 - x - 6}$

22. $\dfrac{2x^2 + 2x - 12}{x^2 + 3x - 4}$

23. $\dfrac{2x^3 + 11x^2 + 5x}{2x^3 - 5x^2 - 3x}$

24. $\dfrac{6x^3 + 5x^2 - 6x}{6x^3 - x^2 - 2x}$

25. $\dfrac{6 - 3x}{2x - 4}$

26. $\dfrac{5 - ay}{ax^2y - 5x^2}$

27. $\dfrac{2x^2 - 7x - 15}{25 - x^2}$

28. $\dfrac{49 - x^2}{2x^2 - 9x - 35}$

29. $\dfrac{(4x + 5)^2}{8x^2 + 6x - 5}$

30. $\dfrac{6x^2 - 13x - 8}{(3x - 8)^2}$

31. $\dfrac{2y^2 - 5y - 12}{8 + 2y - y^2}$

32. $\dfrac{6 + 13y - 5y^2}{2y^2 - 5y - 3}$

33. $\dfrac{a^2 + 2ab - 3b^2}{2a^2 + 5ab - 3b^2}$

34. $\dfrac{a^2 + 3ab - 10b^2}{3a^2 - 7ab + 2b^2}$

35. $\dfrac{16x^2 - 25y^2}{4x^2 + 3xy - 10y^2}$

36. $\dfrac{6x^2 - 11xy + 4y^2}{9x^2 - 16y^2}$

★37. $\dfrac{6x^4 - 9x^3 - 6x^2}{12x^3 + 42x^2 + 18x}$

★38. $\dfrac{xa - yb - ya + xb}{xa - ya + 2bx - 2by}$

Cumulative Review Problems

39. Multiply. $(3x - 7)^2$

40. $(10x + 9y)(10x - 9y)$

41. $(2x + 3)(x - 4)(x - 2)$

42. $(x^2 - 3x - 5)(2x^2 + x + 3)$

6.2 Multiplication and Division of Rational Expressions

After studying this section, you will be able to:

1 *Multiply algebraic fractions and write the answer in simplest form.*
2 *Divide algebraic fractions and write the answer in simplest form.*

1 *Multiplying Algebraic Fractions*

To multiply two fractional expressions, we multiply the numerators and we multiply the denominators. As before, the denominators cannot equal zero.

> For any two fractions with polynomials a, b, c, d where $b \neq 0$, $d \neq 0$
>
> $$\frac{a}{b} \text{ and } \frac{c}{d}$$
>
> The definition of multiplication of these two fractions is $\frac{a}{b} \cdot \frac{c}{d} = \frac{ac}{bd}$.

Simplifying or reducing fractions *prior to multiplying them* usually makes the problem easier to do. To do the problem the long way would certainly increase the chance for error! This long approach should be avoided.

As an example, let's do the same problem two ways to see which way seems easier. First let's multiply by the "long way":

$$\frac{5}{7} \times \frac{49}{125}$$

$\frac{5}{7} \times \frac{49}{125} = \frac{245}{875}$ *Multiply the numerators and multiply the denominators.*

$= \frac{7}{25}$ *Reduce the fraction.* (Note: *It takes a bit of trial and error to discover how to reduce it.)*

Compare this with the method of reducing the fractions prior to multiplying them. Let's multiply by simplifying *before* multiplication:

$$\frac{5}{7} \times \frac{49}{125}$$

$\frac{5}{7} \times \frac{7 \cdot 7}{5 \cdot 5 \cdot 5}$ Step 1. *It is easier to factor first. We factor the numerator and denominator of the second fraction.*

$\frac{5 \cdot 7 \cdot 7}{7 \cdot 5 \cdot 5 \cdot 5}$ Step 2. *We express the product as one fraction (by the definition of multiplication of fractions).*

$\frac{5 \cdot 7 \cdot 7}{7 \cdot 5 \cdot 5 \cdot 5} = \frac{7}{25}$ Step 3. *Then we apply the basic rule of fractions to divide common factors of 5 and 7 that appear in the denominator and in the numerator.*

A similar approach can be used with the multiplication of rational algebraic expressions. We first factor the numerator and denominator of each fraction wherever possible. Then we cancel any factor that is common to a numerator and a denominator. Finally, we multiply the remaining numerators and the remaining denominators.

EXAMPLE 1 Multiply. $\dfrac{x^2 - x - 12}{x^2 - 16} \cdot \dfrac{2x^2 + 7x - 4}{x^2 - 4x - 21}$

$$\frac{(x-4)(x+3)}{(x-4)(x+4)} \cdot \frac{(x+4)(2x-1)}{(x+3)(x-7)}$$

Factor wherever possible is always the first step.

$$= \frac{(x-4)(x+3)}{(x-4)(x+4)} \cdot \frac{(x+4)(2x-1)}{(x+3)(x-7)}$$

Apply the basic rule of fractions. (Three pairs of factors divide out.)

$$= \frac{2x-1}{x-7}$$

The final answer.

Practice Problem 1 Multiply. $\dfrac{10x}{x^2 - 7x + 10} \cdot \dfrac{x^2 + 3x - 10}{25x}$ ■

In some cases, a given numerator can be factored more than once. You should always be sure that you remove the *common factor first* wherever possible.

EXAMPLE 2 Multiply. $\dfrac{x^4 - 16}{x^3 + 4x} \cdot \dfrac{2x^2 - 8x}{4x^2 + 2x - 12}$

$$\frac{(x^2+4)(x^2-4)}{x(x^2+4)} \cdot \frac{2x(x-4)}{2(2x^2+x-6)}$$

Factor each numerator and denominator. Factoring out the common factor first is very important.

$$= \frac{(x^2+4)(x+2)(x-2)}{x(x^2+4)} \cdot \frac{2x(x-4)}{2(x+2)(2x-3)}$$

Factor again where possible.

$$= \frac{(x^2+4)(x+2)(x-2)}{x(x^2+4)} \cdot \frac{2x(x-4)}{2(x+2)(2x-3)}$$

Remove factors that appear in both numerator and denominator. (There are four such pairs to be removed.)

$$= \frac{(x-2)(x-4)}{(2x-3)} \quad \text{or} \quad \frac{x^2-6x+8}{2x-3}$$

Write the answer as one fraction. (Usually, if there is more than one set of parentheses in a numerator, the answer is left in factored form.)

Practice Problem 2 Multiply. $\dfrac{2y^2 - 6y - 8}{y^2 - y - 2} \cdot \dfrac{y^2 - 5y + 6}{2y^2 - 4y - 6}$. ■

For any two fractions $\frac{a}{b}$ and $\frac{c}{d}$ the operation of division can be performed by inverting the second fraction and multiplying it by the first fraction.

> The definition for division of fractions is
>
> $$\frac{a}{b} \div \frac{c}{d} = \frac{a}{b} \cdot \frac{d}{c}$$

This property holds whether a, b, c, and d are polynomials or numerical values. (It is assumed, of course, that no denominator is zero.)

In the first step for dividing two algebraic fractions, you should invert the second fraction and write the problem as a multiplication. Then you follow the procedure for multiplying algebraic fractions.

EXAMPLE 3 Divide. $\dfrac{6x + 12y}{2x - 6y} \div \dfrac{9x^2 - 36y^2}{4x^2 - 36y^2}$

$$\frac{6x + 12y}{2x - 6y} \cdot \frac{4x^2 - 36y^2}{9x^2 - 36y^2}$$ *Invert the second fraction and write the problem as the product of two fractions.*

$$= \frac{6(x + 2y)}{2(x - 3y)} \cdot \frac{4(x^2 - 9y^2)}{9(x^2 - 4y^2)}$$ *Factor each numerator and denominator.*

$$= \frac{(3)(2)(x + 2y)}{2(x - 3y)} \cdot \frac{(2)(2)(x + 3y)(x - 3y)}{(3)(3)(x + 2y)(x - 2y)}$$ *Factor again where possible.*

$$= \frac{(3)(2)(x + 2y)}{2(x - 3y)} \cdot \frac{(2)(2)(x + 3y)(x - 3y)}{(3)(3)(x + 2y)(x - 2y)}$$ *Remove factors that appear in both numerator and denominator.*

$$= \frac{(2)(2)(x + 3y)}{3(x - 2y)}$$ *Write the result as one fraction.*

$$= \frac{4(x + 3y)}{3(x - 2y)}$$ *Simplify. (Usually, answers are left in this form.)*

Practice Problem 3 Divide. $\dfrac{x^2 + 5x + 6}{x^2 + 8x} \div \dfrac{2x^2 + 5x + 2}{2x^2 + x}$ ■

A polynomial that is not in fraction form can be written as a fraction if you write a denominator of 1.

EXAMPLE 4 Divide. $\dfrac{15 - 3x}{x + 6} \div (x^2 - 9x + 20)$

Note that $x^2 - 9x + 20$ can be written as $\dfrac{x^2 - 9x + 20}{1}$.

$\dfrac{15 - 3x}{x + 6} \cdot \dfrac{1}{x^2 - 9x + 20}$ *Invert and multiply.*

$= \dfrac{-3(-5 + x)}{x + 6} \cdot \dfrac{1}{(x - 5)(x - 4)}$ *Factor where possible. Note we had to factor -3 from the first numerator so that we would have a common factor with the second denominator.*

$= \dfrac{-3(-5 + x)}{x + 6} \cdot \dfrac{1}{(x - 5)(x - 4)}$ *Divide out the common factor. $(-5 + x)$ is equivalent to $(x - 5)$.*

$= \dfrac{-3}{(x + 6)(x - 4)}$ *The final answer. Note that the answer can be written in several equivalent forms.*

or $-\dfrac{3}{(x + 6)(x - 4)}$

or $\dfrac{3}{(x + 6)(4 - x)}$ *Take a minute to study each. Each is equivalent to the others.*

Practice Problem 4 Divide. $\dfrac{x + 3}{x - 3} \div (9 - x^2)$ ■

6.2 Exercises

Verbal and Writing Skills

1. Before we do any multiplication in the multiplication of algebraic fractions, we always first try to ____.

2. Division of two algebraic functions is done by keeping the first fraction unchanged and then ______________________________________.

Perform the operation indicated.

3. $\dfrac{x+3}{x+7} \cdot \dfrac{x^2+3x-10}{x^2+x-6}$

4. $\dfrac{5x^2-5}{3x+3} \cdot \dfrac{3x+6}{5x^2-15x+10}$

5. $\dfrac{x^2+2x}{6x} \cdot \dfrac{3x^2}{x^2-4}$

6. $\dfrac{3x+12}{8x^3} \cdot \dfrac{16x^2}{9x+36}$

$\dfrac{3(x+4)}{(8)(x)(x)(x)} \cdot \dfrac{(2)(8)(x)(x)}{(3)(3)(x+4)} = \dfrac{2}{3x}$

7. $\dfrac{x^2-x-6}{x^2-2x-8} \cdot \dfrac{x^2+7x+12}{x^2-9}$

8. $\dfrac{x^2-x-6}{x^2+x-12} \cdot \dfrac{x^2+3x-4}{x^2+2x-3}$

9. $\dfrac{x^2+7x-8}{2x^2-18} \cdot \dfrac{2x^2+20x+42}{7x^2-7x}$

10. $\dfrac{3x^2-3x-18}{x^2+2x-15} \cdot \dfrac{2x^2+6x-20}{2x^2-12x+16}$

11. $\dfrac{5x^2+6x+1}{x^2+5x+6} \div (5x+1)$

12. $(3x-2) \div \dfrac{9x^2-4}{6x+4}$

13. $\dfrac{x^2+2xy+y^2}{x^2-2xy+y^2} \div \dfrac{3x+3y}{4x-4y}$

14. $\dfrac{3x^2+8x-3}{15x^2} \div \dfrac{3x-1}{5x^3}$

15. $\frac{xy - y^2}{x^2 + 2x + 1} \div \frac{2x^2 + xy - 3y^2}{2x^2 + 5xy + 3y^2}$

16. $\frac{4a^2 - b^2}{a^2 + 4ab + 4b^2} \div \frac{4a - 2b}{3a + 6b}$

17. $\frac{x^2 + 5x - 14}{x - 5} \div \frac{x^2 + 12x + 35}{15 - 3x}$

18. $\frac{3x^2 + 13x + 4}{16 - x^2} \div \frac{3x^2 - 5x - 2}{3x - 12}$

19. $\frac{(x + 4)^2}{2x^2 - 7x - 15} \cdot \frac{(x - 5)^2}{x + 4}$

20. $\frac{3x^2 - 10x - 8}{(4x + 5)^2} \cdot \frac{4x + 5}{(x - 4)^2}$

21. $\frac{y^2 - y - 2}{y^2 + 4y + 3} \cdot \frac{y^2 + y - 6}{y^2 - 4y + 4}$

22. $\frac{y^2 - y - 6}{y^2 + 5y - 24} \cdot \frac{y^2 + 6y - 16}{y^2 - 4}$

23. $\frac{3x^2 - 7x + 2}{3x^2 + 2x - 1} \cdot \frac{2x^2 - 9x - 5}{x^2 + x - 6} \cdot \frac{4x^2 + 11x - 3}{x^2 - 11x + 30}$

24. $\frac{6a^2 - a - 1}{4a^2 - 1} \cdot \frac{10a^2 + a - 2}{3a^2 - 17a - 6} \cdot \frac{a^2 - 3a - 18}{5a^2 + 3a - 2}$

Cumulative Review Problems

25. Solve. $5x^2 - 7x + 11 = 5x^2 - x + 2$

26. Multiply. $(7x^2 - x - 1)(x - 3)$

6.3 Addition and Subtraction of Rational Expressions

After studying this section, you will be able to:

1. *Add and subtract algebraic fractions with the same denominator.*
2. *Determine the LCD for two or more algebraic fractions with different denominators.*
3. *Add or subtract algebraic fractions with different denominators.*

1 *Adding and Subtracting Algebraic Fractions with the Same Denominator*

If fractional algebraic expressions have the same denominator, they can be combined in a fashion similar to that used in arithmetic. The numerators are added or subtracted and the denominator remains the same.

Adding Fractional Algebraic Expressions

For any polynomials a, b, c,

$$\frac{a}{b} + \frac{c}{b} = \frac{a + c}{b} \quad \text{where } b \neq 0$$

EXAMPLE 1 Add. $\dfrac{5a}{a + 2b} + \dfrac{6a}{a + 2b}$

$$\frac{5a}{a + 2b} + \frac{6a}{a + 2b} = \frac{5a + 6a}{a + 2b} = \frac{11a}{a + 2b}$$

Note the denominators are the same. Add the numerators.

Practice Problem 1 Add. $\dfrac{2s + t}{2s - t} + \dfrac{s - t}{2s - t}$ ■

Subtracting Fractional Algebraic Expressions

For any polynomials a, b, c

$$\frac{a}{b} - \frac{c}{b} = \frac{a - c}{b} \quad \text{where } b \neq 0$$

EXAMPLE 2 Subtract. $\dfrac{3x}{(x + y)(x - 2y)} - \dfrac{8x}{(x + y)(x - 2y)}$

$$\frac{3x}{(x + y)(x - 2y)} - \frac{8x}{(x + y)(x - 2y)} = \frac{3x - 8x}{(x + y)(x - 2y)}$$ *Write as one fraction.*

$$= \frac{-5x}{(x + y)(x - 2y)}$$ *Simplify.*

Practice Problem 2 Subtract. $\dfrac{b}{(a - 2b)(a + b)} - \dfrac{2b}{(a - 2b)(a + b)}$ ■

2 Determining the LCD

How do we add or subtract fractional expressions when the denominators are not the same? First we must find the **least common denominator** (LCD). You need to be clear on how to find a least common denominator and how to add and subtract fractions from arithmetic before you attempt this section. Review Sections 0.1 and 0.2 if you have any weaknesses on this topic.

> **How to Find the LCD of Two or More Algebraic Fractions**
>
> 1. Factor each denominator completely.
> 2. The LCD is the product of each *different factor.*
> 3. If a factor occurs more than once in any one denominator, the LCD will contain that factor repeated the greatest number of times that it occurs in any one denominator.

EXAMPLE 3 Find the LCD of the following algebraic fractions. $\dfrac{5}{2x-4}, \dfrac{6}{3x-6}$

Factor each denominator.

$$2x - 4 = 2(x - 2)$$
$$3x - 6 = 3(x - 2)$$

The different factors are 2, 3, and $(x - 2)$. Since no factor is repeated more than once in any one denominator, the LCD is the product of these three factors.

$$\text{LCD} = (2)(3)(x - 2) = 6(x - 2)$$

Practice Problem 3 Find the LCD of the following. $\dfrac{7}{6x+21}, \dfrac{13}{10x+35}$ ■

EXAMPLE 4 Find the LCD. **(a)** $\dfrac{5}{12ab^2c}, \dfrac{13}{18a^3bc^4}$

(b) $\dfrac{7}{x^2-2x}, \dfrac{9}{x^2-4x+4}$ **(c)** $\dfrac{8}{x^2-5x+4}, \dfrac{12}{x^2+2x-3}$

If a factor occurs more than once in any one denominator, the LCD will contain that factor repeated the greatest number of times that it occurs in any one denominator.

(a)
$$\begin{array}{rcccccccccccccc}
12ab^2c = & 2\cdot & 2\cdot & 3\cdot & & a & & & \cdot b\cdot & b\cdot & c & & & \\
18a^3bc^4 = & & 2\cdot & 3\cdot & 3\cdot & a\cdot & a\cdot & a\cdot & b\cdot & & c\cdot & c\cdot & c\cdot & c \\
 & \downarrow & \downarrow & \downarrow & \downarrow & \downarrow & \downarrow & \downarrow & \downarrow & \downarrow & \downarrow & \downarrow & \downarrow & \downarrow \\
 & 2\cdot & 2\cdot & 3\cdot & 3\cdot & a\cdot & a\cdot & a\cdot & b\cdot & b\cdot & c\cdot & c\cdot & c\cdot & c
\end{array}$$

$$\text{LCD} = 2^2 \cdot 3^2 \cdot a^3 \cdot b^2 \cdot c^4$$
$$= 36a^3b^2c^4$$

(b)
$$x^2 - 2x = x(x - 2)$$
$$x^2 - 4x + 4 = (x - 2)(x - 2)$$
$$\downarrow \quad \downarrow \quad \downarrow$$
$$\text{LCD} = x(x - 2)(x - 2)$$

(c)
$$x^2 - 5x + 4 = (x - 4)(x - 1)$$
$$x^2 + 2x - 3 = (x - 1)(x + 3)$$
$$\downarrow \quad \downarrow \quad \downarrow$$
$$\text{LCD} = (x - 4)(x - 1)(x + 3)$$

Practice Problem 4 Find the LCD. **(a)** $\dfrac{3}{50xy^2z}, \dfrac{19}{40x^3yz}$

(b) $\dfrac{12}{x^2+5x}, \dfrac{16}{x^2+10x+25}$ **(c)** $\dfrac{2}{x^2+5x+6}, \dfrac{6}{3x^2+5x-2}$ ■

3 *Adding and Subtracting Fractions with Different Denominators*

If two fractions have different denominators, we first change them to equivalent fractions with the least common denominator. Then we add or subtract the numerators and keep the common denominator.

EXAMPLE 5 Add. $\frac{5}{xy} + \frac{2}{y}$

The denominators are different. Find the LCD. The two factors are x and y. We observe that the LCD is xy.

$$\frac{5}{xy} + \frac{2}{y} = \frac{5}{xy} + \frac{2}{y} \cdot \frac{x}{x}$$ *Multiply the second fraction by* $\frac{x}{x}$.

$$= \frac{5}{xy} + \frac{2x}{xy}$$ *Now each fraction has a common denominator of* xy.

$$= \frac{5 + 2x}{xy}$$ *Write the sum as one fraction.*

Practice Problem 5 Add. $\frac{7}{a} + \frac{3}{abc}$ ■

EXAMPLE 6 Add. $\frac{3x}{(x + y)(x - y)} + \frac{5}{x + y}$

The two factors are $(x + y)$ and $(x - y)$. We observe that the LCD = $(x + y)(x - y)$.

$$\frac{3x}{(x + y)(x - y)} + \frac{5}{(x + y)} \cdot \frac{x - y}{x - y}$$ *Multiply the second fraction by* $\frac{x - y}{x - y}$.

$$= \frac{3x}{(x + y)(x - y)} + \frac{5x - 5y}{(x + y)(x - y)}$$ *Now each fraction has a common denominator of* $(x + y)(x - y)$.

$$= \frac{3x + 5x - 5y}{(x + y)(x - y)}$$ *Write the sum of the numerators over one common denominator.*

$$= \frac{8x - 5y}{(x + y)(x - y)}$$ *Collect like terms.*

Practice Problem 6 Add. $\frac{2a - b}{(a + 2b)(a - 2b)} + \frac{2}{(a + 2b)}$ ■

It is important to remember that the LCD is the smallest algebraic expression into which each denominator can be divided. For algebraic expressions the LCD must contain *each factor* that appears in any denominator. If the factor is repeated, the LCD must contain each factor the greatest number of times that it appears in any one denominator.

EXAMPLE 7 Add. $\dfrac{5}{xy^2} + \dfrac{3}{x^2y} + \dfrac{2}{y^3}$

The x factor is squared in one fraction. The y factor is cubed in one fraction. Therefore, the LCD is x^2y^3.

$$\frac{5}{xy^2} \cdot \frac{xy}{xy} + \frac{3}{x^2y} \cdot \frac{y^2}{y^2} + \frac{2}{y^3} \cdot \frac{x^2}{x^2}$$

Multiply each fraction by the appropriate value to obtain x^2y^3 in each denominator.

(Note: Here we always multiply the numerator and denominator of any one fraction by the same value to obtain an equivalent fraction.

$$= \frac{5xy}{x^2y^3} + \frac{3y^2}{x^2y^3} + \frac{2x^2}{x^2y^3}$$

Now all fractions have a common denominator.

$$= \frac{5xy + 3y^2 + 2x^2}{x^2y^3}$$

Write the sum as one fraction.

Practice Problem 7 Add. $\dfrac{3}{x^2} + \dfrac{4x}{xy} + \dfrac{7}{y}$ ■

In many cases, the denominators in an addition or subtraction problem are not in factored form. You must factor each denominator in order to determine the LCD. Collect like terms in the numerator; then look to see if that final numerator can be factored. If so, the fraction can sometimes be simplified.

EXAMPLE 8 Add. $\dfrac{5}{x^2 - y^2} + \dfrac{3x}{x^3 + x^2y}$

$$\frac{5}{x^2 - y^2} + \frac{3x}{x^3 + x^2y} = \frac{5}{(x + y)(x - y)} + \frac{3x}{x^2(x + y)}$$

Factor the two denominators. Observe that the LCD is $x^2(x + y)(x - y)$.

$$= \frac{5}{(x + y)(x - y)} \cdot \frac{x^2}{x^2} + \frac{3x}{x^2(x + y)} \cdot \frac{x - y}{x - y}$$

Multiply each fraction by the appropriate value to obtain a common denominator of $x^2(x + y)(x - y)$.

$$= \frac{5x^2}{x^2(x + y)(x - y)} + \frac{3x^2 - 3xy}{x^2(x + y)(x - y)}$$

$$= \frac{5x^2 + 3x^2 - 3xy}{x^2(x + y)(x - y)}$$

Write the sum of the numerators over one common denominator.

$$= \frac{8x^2 - 3xy}{x^2(x + y)(x - y)}$$

Collect like terms.

$$= \frac{x(8x - 3y)}{x^2(x + y)(x - y)}$$

Remove the common factor x in the numerator and denominator and simplify.

$$= \frac{8x - 3y}{x(x + y)(x - y)}$$

Practice Problem 8 Add. $\dfrac{7a}{a^2 + 2ab + b^2} + \dfrac{4}{a^2 + ab}$ ■

It is very easy to make a mistake in sign when subtracting two fractions. You will find it helpful to place parentheses around the numerator of the second fraction so that you will not forget to subtract the entire numerator.

EXAMPLE 9 Subtract. $\dfrac{3x+4}{x-2} - \dfrac{x-3}{2x-4}$

Factor the second denominator.

$$\frac{3x+4}{x-2} - \frac{x-3}{2(x-2)}$$ *Observe that the LCD is $2(x-2)$.*

$$= \frac{2}{2} \cdot \frac{(3x+4)}{x-2} - \frac{x-3}{2(x-2)}$$ *Multiply the first fraction by $\frac{2}{2}$ so that the resulting fraction will have the common denominator.*

$$= \frac{2(3x+4)-(x-3)}{2(x-2)}$$ *Write the indicated subtraction as one fraction. Note the parentheses around $(x-3)$.*

$$= \frac{6x+8-x+3}{2(x-2)}$$ *Remove the parentheses in the numerator.*

$$= \frac{5x+11}{2(x-2)}$$ *Collect like terms.*

Practice Problem 9 Subtract. $\dfrac{x+7}{3x-9} - \dfrac{x-6}{x-3}$ ■

To avoid making errors when subtracting two fractions, some students find it helpful to change subtraction to addition of the opposite of the second fraction.

EXAMPLE 10 Subtract. $\dfrac{8x}{x^2-16} - \dfrac{4}{x-4}$

$$\frac{8x}{x^2-16} - \frac{4}{x-4} = \frac{8x}{(x+4)(x-4)} + \frac{-4}{x-4}$$ *Factor the first denominator. Use the property that $\dfrac{a}{b} - \dfrac{c}{b} = \dfrac{a}{b} + \dfrac{-c}{b}$.*

$$= \frac{8x}{(x+4)(x-4)} + \frac{-4}{x-4} \cdot \frac{x+4}{x+4}$$ *Multiply the second fraction by $\dfrac{x+4}{x+4}$.*

$$= \frac{8x+(-4)(x+4)}{(x+4)(x-4)}$$ *Write the addition of the numerators over one common denominator.*

$$= \frac{8x-4x-16}{(x+4)(x-4)}$$ *Remove parentheses.*

$$= \frac{4x-16}{(x+4)(x-4)}$$ *Collect like terms. Note that the numerator can be factored.*

$$= \frac{4(x-4)}{(x+4)(x-4)}$$ *Divide out. Since $(x-4)$ is a* factor *of the numerator* and *the denominator, we may divide out the common factor.*

$$= \frac{4}{x+4}$$

Practice Problem 10 Subtract and simplify your answer. $\dfrac{x-2}{x^2-4} - \dfrac{x+1}{2x^2+4x}$ ■

6.3 Exercises

Perform the operation indicated. Be sure to simplify your answer.

1. $\dfrac{7}{12x^3} + \dfrac{1}{12x^3}$

2. $\dfrac{9}{21ab^2} + \dfrac{5}{21ab^2}$

3. $\dfrac{5}{x+2} + \dfrac{x+3}{2+x}$

4. $\dfrac{4x+5}{x+6} + \dfrac{8}{6+x}$

5. $\dfrac{x}{x-4} - \dfrac{x+1}{x-4}$

6. $\dfrac{x+3}{x+5} - \dfrac{x-8}{x+5}$

7. $\dfrac{8x+3}{5x+7} - \dfrac{6x+10}{5x+7}$

8. $\dfrac{6x+5}{2x+11} - \dfrac{4x+16}{2x+11}$

9. $\dfrac{3x-2}{5x-6} - \dfrac{x+5}{5x-6}$

Find the LCD *of each of the following fractions. Find the* LCD *only. Do not combine fractions.*

10. $\dfrac{13}{3ab}, \dfrac{7}{a^2b^2}$

11. $\dfrac{12}{5a^2}, \dfrac{9}{a^3}$

12. $\dfrac{3}{16xy^4}, \dfrac{5}{24x^3y^2}$

13. $\dfrac{7}{15x^3y^5}, \dfrac{13}{40x^4y}$

14. $\dfrac{8}{x+3}, \dfrac{15}{x^2-9}$

15. $\dfrac{13}{x^2-16}, \dfrac{7}{x-4}$

16. $\dfrac{6}{2x^2+x-3}, \dfrac{5}{4x^2+12x+9}$

17. $\dfrac{3}{16x^2-8x+1}, \dfrac{6}{4x^2+7x-2}$

Add or subtract the following fractions.

18. $\dfrac{5}{y^2} + \dfrac{3}{y^3}$

19. $\dfrac{8}{xy} + \dfrac{2}{xy^2}$

20. $\dfrac{2}{x+2} + \dfrac{3}{x^2-4}$

21. $\dfrac{5}{x^2+5x+6} + \dfrac{2}{x+3}$

22. $\dfrac{3y}{y+2} + \dfrac{y}{y-2}$

23. $\dfrac{2}{y-1} + \dfrac{2}{y+1}$

24. $\dfrac{4}{a+3}+\dfrac{2}{3a}$

25. $\dfrac{5}{2ab}+\dfrac{1}{2a+b}$

26. $\dfrac{2}{3xy}+\dfrac{1}{6yz}$

27. $\dfrac{x-3}{4x}+\dfrac{6}{x^2}$

28. $\dfrac{1}{x^2+3x+2}+\dfrac{1}{x^2+4x+3}$

29. $\dfrac{5}{x^2-25}+\dfrac{1}{x^2+10x+25}$

30. $\dfrac{3x-8}{x^2-5x+6}+\dfrac{x+2}{x^2-6x+8}$

31. $\dfrac{3x+5}{x^2+4x+3}+\dfrac{-x+5}{x^2+2x-3}$

32. $\dfrac{a+b}{3}-\dfrac{a-2b}{4}$

33. $\dfrac{a+1}{2}-\dfrac{a-1}{3}$

34. $\dfrac{a-1}{4a}-\dfrac{2a+3}{a}$

35. $\dfrac{4x-9}{3x}-\dfrac{3x-8}{4x}$

36. $\dfrac{8}{2x-3}-\dfrac{6}{x+2}$

37. $\dfrac{6}{3x-4}-\dfrac{5}{4x-3}$

38. $\dfrac{x}{x^2 + 2x - 3} - \dfrac{x}{x^2 - 5x + 4}$

39. $\dfrac{1}{x^2 - 2x} - \dfrac{5}{x^2 - 4x + 4}$

40. $\dfrac{3y}{8y^2 + 2y - 1} - \dfrac{5y}{2y^2 - 9y - 5}$

41. $\dfrac{2x}{x^2 + 5x + 6} - \dfrac{x + 1}{x^2 + 2x - 3}$

42. $\dfrac{5}{x} + \dfrac{x}{x - 5} + \dfrac{3}{5x}$

43. $\dfrac{3}{ab} + \dfrac{2}{a - b} + \dfrac{a}{b}$

44. $\dfrac{y - 6}{y + 6} - \dfrac{y + 6}{y - 6}$

45. $\dfrac{x}{x - 5} - \dfrac{2x}{5 - x}$

46. $\dfrac{4y}{y^2 + 4y + 3} + \dfrac{2}{y + 1}$

47. $\dfrac{y - 23}{y^2 - y - 20} + \dfrac{2}{y - 5}$

48. $\dfrac{x}{x - 1} + \dfrac{x}{x + 1} + \dfrac{2}{x^2 - 1}$

49. $\dfrac{2}{x^2 - 9} + \dfrac{x}{x + 3} + \dfrac{2x}{x - 3}$

50. $\dfrac{3x}{x^2 + 3x - 10} + \dfrac{5}{4 - 2x}$

51. $\dfrac{2y}{3y^2 - 8y - 3} + \dfrac{1}{6y - 2y^2}$

52. $\dfrac{5}{3x-6} - \dfrac{2}{4x-8}$

53. $\dfrac{12}{15x-3} + \dfrac{1}{10x-2}$

54. $\dfrac{2x-1}{x^2+6x+9} - \dfrac{3}{x^2+4x+3}$

55. $\dfrac{2}{x^2+4x+4} - \dfrac{2x-1}{x^2+5x+6}$

56. $\dfrac{1}{x-3y} + \dfrac{1}{x-4y} + \dfrac{y}{x^2-7xy+12y^2}$

57. $\dfrac{42x}{y^2-49} + \dfrac{3x}{y-7} + \dfrac{3x}{y+7}$

58. $\dfrac{3x}{x^2-9x+14} - \dfrac{2x}{x^2+x-6} + \dfrac{x}{x^2-4x-21}$

59. $\dfrac{5x}{x^2+x-6} + \dfrac{2x+1}{x^2+2x-3} - \dfrac{1-2x}{x^2-3x+2}$

Cumulative Review Problems

60. Solve. $\dfrac{1}{3}(x-2) + \dfrac{1}{2}(x+3) = \dfrac{1}{4}(3x+1)$

61. Solve for y. $5ax = 2(ay - 3bc)$

62. Solve for x. $\dfrac{1}{2}x < \dfrac{1}{3}x + \dfrac{1}{4}$

63. Simplify. $(3x^3y^4)^4$

Putting Your Skills to Work

AVERAGE RATES OF SPEED

Suppose that you drive slowly up a mountain road at x miles per hour and return, driving the same distance down the mountain road, more rapidly at y miles per hour. What is the average speed of the trip in miles per hour?

It can be shown that the average speed of any round trip is given by the complex fraction

$$\frac{2}{\frac{1}{x}+\frac{1}{y}}$$

where x is the average speed on the way to your destination and y is the average speed on your return trip.

1. Find the average speed if you travel 30 miles per hour going up a mountain road but 50 miles per hour on the return trip down the mountain road.

2. Simplify the complex fraction by combining the two fractions in the denominator, then use the rules for dividing two fractions.

COOPERATIVE GROUP ACTIVITY

3. When trying to answer question 1, one student said that the average rate of speed on the round trip was $\frac{30+50}{2}=\frac{80}{2}=40$ miles per hour.

What is wrong with this student's reasoning?

4. If $x = 20$ miles per hour and $y = 80$ miles per hour, what is the average speed of the entire trip? If $x = 10$ miles per hour and $y = 100$ miles per hour, what is the average speed of the entire trip? As the difference between speed x and speed y becomes greater, what happens to the average rate of speed of the trip?

6.4 Simplify Complex Rational Expressions

After studying this section, you will be able to:

1 *Simplify complex fractions by adding or subtracting in the numerator and denominator.*
2 *Simplify complex fractions by multiplying by the LCD of all the denominators.*

1 *Simplifying Complex Fractions by Adding or Subtracting in the Numerator and the Denominator*

A complex fraction has a fraction in the numerator or in the denominator, or both.

$$\frac{3+\frac{2}{x}}{\frac{x}{7}+2} \qquad \frac{\frac{x}{y}+1}{2} \qquad \frac{\frac{a+b}{3}}{\frac{x-2y}{4}}$$

The bar in a complex fraction is both a grouping symbol and a symbol for division.

$$\frac{\frac{a+b}{3}}{\frac{x-2y}{4}} \quad \text{is equivalent to} \quad \left(\frac{a+b}{3}\right) \div \left(\frac{x-2y}{4}\right)$$

We need a procedure for simplifying complex fractions.

> **Procedure to Simplify a Complex Fraction: Adding and Subtracting**
>
> **1.** Add or subtract so that you have a single fraction in the numerator and in the denominator.
>
> **2.** Divide the fraction in the numerator by the fraction in the denominator. This is done by inverting the fraction in the denominator and multiplying it by the numerator.

EXAMPLE 1 Simplify. $\dfrac{\frac{1}{x}}{\frac{2}{y^2}+\frac{1}{y}}$

Step 1 Add the two fractions in the denominator.

$$\frac{\frac{1}{x}}{\frac{2}{y^2}+\frac{1}{y}\cdot\frac{y}{y}} = \frac{\frac{1}{x}}{\frac{2+y}{y^2}}$$

Step 2 Divide the fraction in the numerator by the fraction in the denominator.

$$\frac{1}{x} \div \frac{2+y}{y^2} = \frac{1}{x}\cdot\frac{y^2}{2+y} = \frac{y^2}{x(2+y)}$$

Practice Problem 1 Simplify. $\dfrac{\frac{1}{a}+\frac{1}{b}}{\frac{2}{ab^2}}$ ■

In some problems there may be two or more fractions in the numerator and the denominator.

EXAMPLE 2 Simplify. $\dfrac{\dfrac{1}{x}+\dfrac{1}{y}}{\dfrac{3}{a}-\dfrac{2}{b}}$

We observe that the LCD of the fractions in the numerator is xy. The LCD of the fractions in the denominator is ab.

$$\dfrac{\dfrac{1}{x}\cdot\dfrac{y}{y}+\dfrac{1}{y}\cdot\dfrac{x}{x}}{\dfrac{3}{a}\cdot\dfrac{b}{b}-\dfrac{2}{b}\cdot\dfrac{a}{a}}$$
Multiply each fraction by the appropriate value to obtain common denominators.

$$=\dfrac{\dfrac{y+x}{xy}}{\dfrac{3b-2a}{ab}}$$
Add the two fractions in the numerator.
Subtract the two fractions in the denominator.

$$=\dfrac{y+x}{xy}\cdot\dfrac{ab}{3b-2a}$$
Invert the fraction in the denominator and multiply it by the numerator.

$$=\dfrac{ab(y+x)}{xy(3b-2a)}$$
Write the answer as one fraction.

Practice Problem 2 Simplify. $\dfrac{\dfrac{1}{a}+\dfrac{1}{b}}{\dfrac{1}{a}-\dfrac{1}{b}}$ ■

In some problems, factoring may be necessary to determine the LCD and to combine fractions.

EXAMPLE 3 Simplify. $\dfrac{\dfrac{1}{x^2-1}+\dfrac{2}{x+1}}{x}$

We observe the need for factoring $x^2 - 1$.

$$=\dfrac{\dfrac{1}{(x+1)(x-1)}+\dfrac{2}{(x+1)}\cdot\dfrac{x-1}{x-1}}{x}$$
The LCD for the fractions in the numerator is $(x+1)(x-1)$.

$$=\dfrac{\dfrac{1+2x-2}{(x+1)(x-1)}}{x}$$
Add the two fractions in the numerator.

$$=\dfrac{2x-1}{(x+1)(x-1)}\cdot\dfrac{1}{x}$$
Simplify the numerator. Invert the fraction in the denominator and multiply.

$$=\dfrac{2x-1}{x(x+1)(x-1)}$$
Write the answer as one fraction.

Practice Problem 3 Simplify. $\dfrac{\dfrac{x}{x^2+4x+3}+\dfrac{2}{x+1}}{x+1}$ ■

In simplifying complex fractions, always be alert to see if the final fraction can be reduced or simplified.

EXAMPLE 4 Simplify. $\dfrac{\dfrac{3}{a+b}-\dfrac{3}{a-b}}{\dfrac{5}{a^2-b^2}}$

The LCD of the two fractions in the numerator is $(a+b)(a-b)$.

$$\frac{\dfrac{3}{a+b}\cdot\dfrac{a-b}{a-b}-\dfrac{3}{a-b}\cdot\dfrac{a+b}{a+b}}{\dfrac{5}{a^2-b^2}}=\frac{\dfrac{3a-3b}{(a+b)(a-b)}-\dfrac{3a+3b}{(a+b)(a-b)}}{\dfrac{5}{a^2-b^2}}$$

Study carefully how we combine the two fractions in the numerator. Do you see how we obtain $-6b$?

$$=\frac{\dfrac{-6b}{(a+b)(a-b)}}{\dfrac{5}{(a+b)(a-b)}} \qquad \textit{Factor } a^2-b^2 \textit{ as } (a+b)(a-b).$$

$$=\frac{-6b}{(a+b)(a-b)}\cdot\frac{(a+b)(a-b)}{5}$$

Note that since $(a+b)(a-b)$ are factors in both numerator and denominator they may be divided out.

$$=\frac{-6b}{5}$$

Practice Problem 4 Simplify. $\dfrac{\dfrac{6}{x^2-y^2}}{\dfrac{1}{x-y}+\dfrac{3}{x+y}}$ ■

2 *Simplifying Complex Fractions by Multiplying by the LCD of All the Denominators*

There is another way to simplify complex fractions: Multiply the numerator and denominator of the complex fraction by the least common denominator of all the denominators appearing in the complex fraction.

Procedure to Simplify a Complex Fraction: Multiplying by the LCD

1. Determine the LCD of all individual denominators occurring in the numerator and denominator of the complex fraction.
2. Multiply both the numerator and the denominator of the complex fraction by the LCD.
3. Simplify if possible.

EXAMPLE 5 Simplify. $\dfrac{\dfrac{5}{ab^2} - \dfrac{2}{ab}}{3 - \dfrac{5}{2a^2b}}$

The LCD of all the denominators in the complex fraction is $2a^2b^2$.

$$\frac{2a^2b^2\left(\frac{5}{ab^2} - \frac{2}{ab}\right)}{2a^2b^2\left(3 - \frac{5}{2a^2b}\right)} = \frac{2a^2b^2\left(\frac{5}{ab^2}\right) - 2a^2b^2\left(\frac{2}{ab}\right)}{2a^2b^2(3) - 2a^2b^2\left(\frac{5}{2a^2b}\right)} \quad \textit{Multiply each term by } 2a^2b^2.$$

$$= \frac{10a - 4ab}{6a^2b^2 - 5b} \quad \textit{Simplify.}$$

Practice Problem 5 Simplify by using the alternative method. $\dfrac{\dfrac{2}{3x^2} - \dfrac{3}{y}}{\dfrac{5}{xy} - 4}$ ■

So that you can compare the two methods, we will redo Example 4 by the alternative method.

EXAMPLE 6 Simplify. $\dfrac{\dfrac{3}{a+b} - \dfrac{3}{a-b}}{\dfrac{5}{a^2 - b^2}}$ *Use alternative method.*

The LCD of all individual fractions contained in the complex fraction is $(a + b)(a - b)$.

$$\frac{(a+b)(a-b)\left(\frac{3}{a+b}\right) - (a+b)(a-b)\left(\frac{3}{a-b}\right)}{(a+b)(a-b)\left(\frac{5}{(a+b)(a-b)}\right)} \quad \textit{Multiply each term by the LCD.}$$

$$= \frac{3(a-b) - 3(a+b)}{5} \quad \textit{Simplify.}$$

$$= \frac{3a - 3b - 3a - 3b}{5} \quad \textit{Remove parentheses.}$$

$$= -\frac{6b}{5} \quad \textit{Simplify.}$$

Practice Problem 6 Simplify by using the alternative method. $\dfrac{\dfrac{6}{x^2 - y^2}}{\dfrac{7}{x-y} + \dfrac{3}{x+y}}$ ■

6.4 Exercises

Simplify.

1. $\dfrac{\frac{3}{x}}{\frac{2}{y}}$

2. $\dfrac{\frac{3}{ab}}{\frac{b}{a}}$

3. $\dfrac{\frac{1}{x}+\frac{1}{y}}{\frac{1}{xy}}$

4. $\dfrac{\frac{1}{x}+1}{\frac{1}{x}}$

5. $\dfrac{\frac{1}{x}+\frac{1}{y}}{x^2-y^2}$

6. $\dfrac{\frac{1}{x}+\frac{1}{y}}{x+y}$

7. $\dfrac{1-\frac{9}{x^2}}{\frac{3}{x}+1}$

8. $\dfrac{\frac{4}{x}+1}{1-\frac{16}{x^2}}$

9. $\dfrac{\frac{2}{x+1}-2}{3}$

10. $\dfrac{1-\frac{3}{xy}}{x+2y}$

11. $\dfrac{a+\frac{3}{a}}{\frac{a^2+2}{3a}}$

12. $\dfrac{a+\frac{1}{a}}{\frac{3}{a}-a}$

13. $\dfrac{2+\frac{3y}{x-y}}{2-\frac{3y}{x+2y}}$

14. $\dfrac{y-\frac{4}{y}}{1+\frac{3}{2y+1}}$

15. $\dfrac{\frac{x}{4}-\frac{1}{2}}{\frac{x}{4}+\frac{1}{x}}$

16. $\dfrac{\frac{1}{x}-\frac{1}{2x}}{\frac{1}{2}+\frac{1}{2x}}$

17. $\dfrac{\frac{1}{x^2-9}+\frac{2}{x+3}}{\frac{3}{x-3}}$

18. $\dfrac{\frac{5}{x+4}}{\frac{1}{x-4}-\frac{2}{x^2-16}}$

19. $\dfrac{\frac{2}{y-1}+2}{\frac{2}{y+1}-2}$

20. $\dfrac{\frac{y}{y+1}+1}{\frac{2y+1}{y-1}}$

★ 21. $\dfrac{\frac{y}{y+1}-\frac{y-1}{y}}{\frac{y+1}{y}+\frac{y}{y-1}}$

★ 22. $\dfrac{y+2-\frac{6}{2y+3}}{\frac{8y}{2y-1}+y}$

Cumulative Review Problems

23. Solve for w. $P = 2(l + w)$

24. Solve and graph. $7 + x < 11 + 5x$

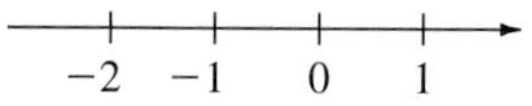

6.5 Equations Involving Rational Expressions

After studying this section, you will be able to:

1 *Solve equations involving algebraic fractions that have solutions.*
2 *Determine if an equation involving algebraic fractions has no solution.*

1 *Solving Equations Involving Algebraic Fractions*

In Section 2.4 we developed procedures to solve linear equations containing fractions whose denominators were numerical values. In this section we use a similar approach to solve equations containing fractions whose denominators are polynomials. It would be wise for you to review Section 2.4 briefly *before you begin this section.* It will be especially helpful to carefully study Examples 1 and 2.

To Solve a Rational Equation

1. Determine the LCD of all the denominators.
2. Multiply each term of the equation by the LCD.
3. Solve the resulting equation.
4. Check your solution. Exclude from your solution any value that would make the LCD equal to zero. If such a value is obtained, there is *no solution.*

EXAMPLE 1 Solve for x and check your solution. $\frac{5}{x} + \frac{2}{3} = 2 - \frac{2}{x} - \frac{1}{6}$

We observe that the LCD is $6x$.

$6x\left(\frac{5}{x}\right) + 6x\left(\frac{2}{3}\right) = 6x(2) - 6x\left(\frac{2}{x}\right) - 6x\left(\frac{1}{6}\right)$ *Multiply each term by $6x$.*

$30 + 4x = 12x - 12 - x$ *Simplify. Do you see how each term is obtained?*

$30 + 4x = 11x - 12$ *Collect like terms.*

$30 = 7x - 12$ *Subtract $4x$ from both sides.*

$42 = 7x$ *Add 12 to both sides.*

$6 = x$ *Divide both sides by 7.*

Check. $\frac{5}{6} + \frac{2}{3} \stackrel{?}{=} 2 - \frac{2}{6} - \frac{1}{6}$ *Replace each x by 6.*

$\frac{5}{6} + \frac{4}{6} \stackrel{?}{=} \frac{12}{6} - \frac{2}{6} - \frac{1}{6}$

$\frac{9}{6} = \frac{9}{6}$ ✓ *It checks.*

Practice Problem 1 Solve for x and check your solution.

$\frac{2}{x} + \frac{4}{x} = 3 - \frac{1}{x} - \frac{17}{8}$ ■

EXAMPLE 2 Solve and check. $\dfrac{6}{x+3} = \dfrac{3}{x}$

We observe that the LCD = $x(x + 3)$.

$$x(x+3)\left(\frac{6}{x+3}\right) = x(x+3)\left(\frac{3}{x}\right) \quad \textit{Multiply both sides by } x(x+3).$$

$$6x = 3(x+3) \quad \textit{Simplify. Do you see how this is done?}$$

$$6x = 3x + 9 \quad \textit{Remove parentheses.}$$

$$3x = 9 \quad \textit{Subtract } 3x \textit{ from both sides.}$$

$$x = 3 \quad \textit{Divide both sides by } 3.$$

Check. $\dfrac{6}{3+3} \stackrel{?}{=} \dfrac{3}{3}$ *Replace each x by 3.*

$\dfrac{6}{6} = \dfrac{3}{3}$ ✓ *It checks.*

Practice Problem 2 Solve and check. $\dfrac{4}{2x+1} = \dfrac{6}{2x-1}$ ■

It is sometimes necessary to factor denominators before the correct LCD can be determined.

EXAMPLE 3 Solve and check your solution. $\dfrac{3}{x+5} - 1 = \dfrac{4-x}{2x+10}$

We observe a need to factor the $2x + 10$.

$$\frac{3}{x+5} - 1 = \frac{4-x}{2(x+5)} \quad \textit{Factor. We determine that the LCD is } 2(x+5).$$

$$2(x+5)\left(\frac{3}{x+5}\right) - 2(x+5)(1) = 2(x+5)\left[\frac{4-x}{2(x+5)}\right] \quad \textit{Multiply each term by the LCD.}$$

$$2(3) - 2(x+5) = 4 - x \quad \textit{Simplify.}$$

$$6 - 2x - 10 = 4 - x \quad \textit{Remove parentheses.}$$

$$-2x - 4 = 4 - x \quad \textit{Collect like terms.}$$

$$-4 = 4 + x \quad \textit{Add } 2x \textit{ to both sides.}$$

$$-8 = x \quad \textit{Subtract 4 from both sides.}$$

Check. $\dfrac{3}{-8+5} - 1 \stackrel{?}{=} \dfrac{4-(-8)}{2(-8)+10}$ *Replace each x in the original equation by −8.*

$$\frac{3}{-3} - 1 \stackrel{?}{=} \frac{4+8}{-16+10}$$

$$-1 - 1 \stackrel{?}{=} \frac{12}{-6}$$

$-2 = -2$ ✓ *It checks. $x = -8$ is the solution.*

Practice Problem 3 Solve and check. $\dfrac{x-1}{x^2-4} = \dfrac{2}{x+2} + \dfrac{4}{x-2}$ ■

2 *Determining If an Equation Involving Algebraic Fractions Has No Solution*

Linear equations containing fractions that have variables in denominators sometimes appear to have solutions when in fact they do not. By this we mean that the "solutions" we get by using completely correct methods are, in actuality, not solutions.

In the case where a value makes a denominator in the equation equal to zero, we say there is **no solution** to the equation. Such a value is called an **extraneous solution.** An extraneous solution is an apparent solution that does *not* satisfy the original equation. It is important that you check the number in the original equation.

EXAMPLE 4 Solve and check. $\dfrac{y}{y-2} - 4 = \dfrac{2}{y-2}$

We observe that the LCD is $y - 2$.

$$(y-2)\left(\frac{y}{y-2}\right) - (y-2)(4) = (y-2)\left(\frac{2}{y-2}\right)$$ *Multiply each term by $(y - 2)$.*

$$y - 4(y-2) = 2$$ *Simplify. Do you see how this is done?*

$$y - 4y + 8 = 2$$ *Remove parentheses.*

$$-3y + 8 = 2$$ *Collect like terms.*

$$-3y = -6$$ *Subtract 8 from both sides.*

$$\frac{-3y}{-3} = \frac{-6}{-3}$$ *Divide both sides by negative 3.*

$$y = 2$$ *$y = 2$ is only an apparent solution.*

Why? We can see immediately that $y = 2$ is not a possible solution for the original equation. The use of the value $y = 2$ in a denominator would make the denominator equal to zero, and the expression is undefined.

Check. $\dfrac{y}{y-2} - 4 = \dfrac{2}{y-2}$ *Suppose that you try to check the apparent solution by substituting $y = 2$.*

$$\frac{2}{2-2} - 4 \stackrel{?}{=} \frac{2}{2-2}$$

$$\frac{2}{0} - 4 = \frac{2}{0}$$ *This does not check since you do not obtain a real number when you divide by zero.*

$\uparrow \qquad \uparrow$

These expressions are not defined.

There is no such number as $2 \div 0$. We see that $y = 2$ does *not* check. There is *no solution* to this problem.

Practice Problem 4 Solve and check. $\dfrac{2x}{x+1} = \dfrac{-2}{x+1} + 1$ ■

6.5 Exercises

Solve and check your solution.

1. $\dfrac{x+1}{2x} = \dfrac{2}{3}$

2. $\dfrac{9}{2x-1} = \dfrac{3}{x}$

3. $\dfrac{1}{x} = \dfrac{1}{4-x}$

4. $\dfrac{x-2}{4x} = \dfrac{1}{6}$

5. $\dfrac{y-3}{3y+2} = \dfrac{1}{5}$

6. $\dfrac{2}{y+3} = \dfrac{5}{y}$

7. $\dfrac{2}{x} + \dfrac{x}{x+1} = 1$

8. $\dfrac{5}{2} = 3 + \dfrac{2x+7}{x+6}$

9. $\dfrac{x+1}{x} = 1 + \dfrac{x-2}{2x}$

10. $\dfrac{7x-4}{5x} = \dfrac{9}{5} - \dfrac{4}{x}$

11. $\dfrac{x+1}{x} = \dfrac{1}{2} - \dfrac{4}{3x}$

12. $\dfrac{2x-1}{3x} + \dfrac{1}{9} = \dfrac{x+2}{x} + \dfrac{1}{9x}$

13. $\dfrac{x}{x-2} - 2 = \dfrac{2}{x-2}$

14. $4 - \dfrac{8}{x+1} = \dfrac{8x}{x+1}$

15. $\dfrac{2}{x+1} - \dfrac{1}{x-1} = \dfrac{2x}{x^2-1}$

16. $\dfrac{8x}{4x^2-1} = \dfrac{3}{2x+1} + \dfrac{3}{2x-1}$

17. $\dfrac{y+1}{y^2+2y-3} = \dfrac{1}{y+3} - \dfrac{1}{y-1}$

18. $\dfrac{2y}{y+1} + \dfrac{1}{3y-2} = 2$

19. $\dfrac{79-x}{x} = 5 + \dfrac{7}{x}$

20. $\dfrac{2x-14}{x^2+3x-28} + \dfrac{x-2}{x-4} = \dfrac{x+3}{x+7}$

21. $\dfrac{3}{x+3} + \dfrac{1}{x-1} = \dfrac{8}{x^2+2x-3}$

22. $\dfrac{2}{3x+6} + \dfrac{1}{2x+4} = \dfrac{1}{6}$

23. $\dfrac{x+11}{x^2-5x+4} + \dfrac{3}{x-1} = \dfrac{5}{x-4}$

24. $\dfrac{52-x}{x} = 9 + \dfrac{2}{x}$

25. $\dfrac{2x}{x+4} - \dfrac{8}{x-4} = \dfrac{2x^2+32}{x^2-16}$

26. $\dfrac{4x}{x+3} - \dfrac{12}{x-3} = \dfrac{4x^2+36}{x^2-9}$

27. $\frac{23}{5} - \frac{3}{5x - 10} = \frac{4}{x - 2}$

28. $\frac{x - 3}{x - 2} + \frac{x + 1}{x + 3} = \frac{2x^2 - 15x}{x^2 + x - 6}$

29. $\frac{2}{x + 7} + \frac{4x - 1}{x^2 + 5x - 14} = \frac{1}{x - 2}$

30. $\frac{2}{x - 2} + \frac{3}{x} = \frac{4}{-2x + x^2}$

★31. $\frac{10}{x - 2} - \frac{40}{x^2 + x - 6} = \frac{12}{x + 3}$

★32. $\frac{10 - 2y^2}{y^2 + 3y - 10} + \frac{4y + 2}{y - 2} = \frac{2y + 2}{y + 5}$

To Think About

In each of the following equations, determine the extraneous solutions. Do not solve the equation. (Hint: *The denominator can never be zero.)*

33. $\frac{3x}{x - 2} - \frac{4x}{x - 4} = \frac{3}{x^2 - 6x + 8}$

34. $\frac{2x}{x + 2} - \frac{x}{x + 3} = \frac{-3}{x^2 + 5x + 6}$

Cumulative Review Problems

35. Factor. $6x^2 - x - 12$

36. Solve. $4 - (3 - x) = 7(x + 3)$

37. The perimeter of a rectangle is 54 meters. The length is 1 meter less than triple the width. Find the dimensions of the rectangle.

38. A number plus its half is equal to the number tripled plus a half. Find the number.

6.6 Ratio, Proportion, and Other Applied Problems

After studying this section, you will be able to:

1. *Solve problems involving ratio and proportion.*
2. *Solve problems involving similar triangles.*
3. *Solve certain distance problems.*
4. *Solve work problems.*

1 *Solving Problems Involving Ratio and Proportion*

A **ratio** is a comparison of two quantities. You may be familiar with ratios that compare miles to hours or miles to gallons. A ratio is often written as a quotient in the form of a fraction. For example, the ratio of 7 to 9 can be written as $\frac{7}{9}$.

A **proportion** is an equation that states that two ratios are equal. For example,

$$\frac{7}{9} = \frac{21}{27}, \quad \frac{2}{3} = \frac{10}{15}, \quad \frac{a}{b} = \frac{c}{d}$$

Let's take a closer look at the last proportion. We can see that the LCD of the fractional equation is bd.

$(bd)\frac{a}{b} = (bd)\frac{c}{d}$ *Multiply each side by the LCD.*

$da = bc$

$ad = bc$ *Since multiplication is commutative, $da = ad$.*

Thus we have proved the following:

The Proportion Equation

If

$$\frac{a}{b} = \frac{c}{d} \quad \text{then} \quad ad = bc$$

for all real numbers a, b, c, and d, where $b \neq 0$, $d \neq 0$.

This is sometimes called **cross-multiplying.** It can be applied only if you have *one* fraction and nothing else on each side of the equation.

We can use a proportion and the technique of cross-multiplying to solve a variety of applied problems involving two ratios.

EXAMPLE 1 Michael took 5 hours to drive 245 miles on the turnpike. If he continues at the same rate, how many hours will it take him to drive a distance of 392 miles?

1. *Understand the problem.*
   Let x = the number of hours it will take to drive 392 miles. If 5 hours are needed to drive 245 miles, then x hours is needed to drive 392.
2. *Write an equation.*
   We can write this as a proportion. Compare time to distance in each ratio.

$$\begin{matrix}\text{Time} \longrightarrow \\ \text{Distance} \longrightarrow\end{matrix} \frac{5 \text{ hours}}{245 \text{ miles}} = \frac{x \text{ hours}}{392 \text{ miles}} \begin{matrix}\longleftarrow \text{Time} \\ \longleftarrow \text{Distance}\end{matrix}$$

3. *Solve the equation and state the answer.*

$$5(392) = 245x \qquad \textit{Cross-multiply.}$$

$$\frac{1960}{245} = x \qquad \textit{Divide both sides by 245.}$$

$$8 = x$$

It would take Michael 8 hours to drive 392 miles.

4. *Check.* Is $\frac{5}{245} = \frac{8}{392}$? Do the computation and see.

Practice Problem 1 It took Brenda 8 hours to drive 420 miles. At the same rate, how long would it take her to drive 315 miles? ■

EXAMPLE 2 If $\frac{3}{4}$ inch on a map represents an actual distance of 20 miles, how long is the distance represented by $4\frac{1}{8}$ inches on the same map?

Let x = the distance represented by $4\frac{1}{8}$ inches.

Initial measurement on map ⟶ $\frac{3}{4}$ ... $4\frac{1}{8}$ ⟵ Second measurement on the map

$$\frac{\frac{3}{4}}{20} = \frac{4\frac{1}{8}}{x}$$

Initial distance ⟶ 20 ... x ⟵ Second distance

$$\left(\frac{3}{4}\right)(x) = (20)\left(4\frac{1}{8}\right) \qquad \textit{Cross-multiply.}$$

$$\left(\frac{3}{4}\right)(x) = (\overset{5}{20})\left(\frac{33}{\underset{2}{8}}\right) \qquad \textit{Write } 4\tfrac{1}{8} \textit{ as } \tfrac{33}{8} \textit{ and simplify.}$$

$$\frac{3x}{4} = \frac{165}{2} \qquad \textit{Multiplication of fractions.}$$

$$4\left(\frac{3x}{4}\right) = \overset{2}{4}\left(\frac{165}{2}\right) \qquad \textit{Multiply each side by 4.}$$

$$3x = 330 \qquad \textit{Simplify.}$$

$$x = 110 \qquad \textit{Divide both sides by 3.}$$

$4\frac{1}{8}$ inches on the map represents an actual distance of 110 miles.

Practice Problem 2 If $\frac{5}{8}$ inch on a map represents 30 miles, how long is the distance represented by $2\frac{1}{2}$ inches on the same map? ■

2 *Solving Problems Involving Similar Triangles*

Similar triangles are triangles that have the same shape, but may be a different size. For example, if you draw a triangle on a sheet of paper and place the paper in a photocopy machine and make a copy that is reduced by 25%, you would create a triangle that was similar to the original triangle. The two triangles would have the same shape. The corresponding sides of the triangles would be proportional.

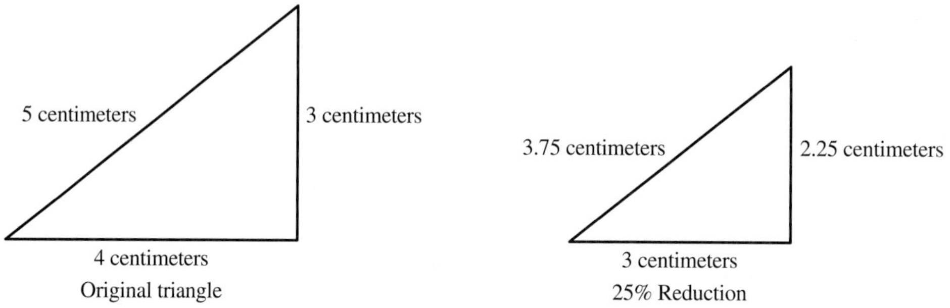

You can use the proportion equation to show that the corresponding sides of the above triangles are proportional. In fact, you can use the proportion equation to find the length of a side of a triangle if it is not known.

EXAMPLE 3 Find the length of side x in the two similar triangles pictured below.

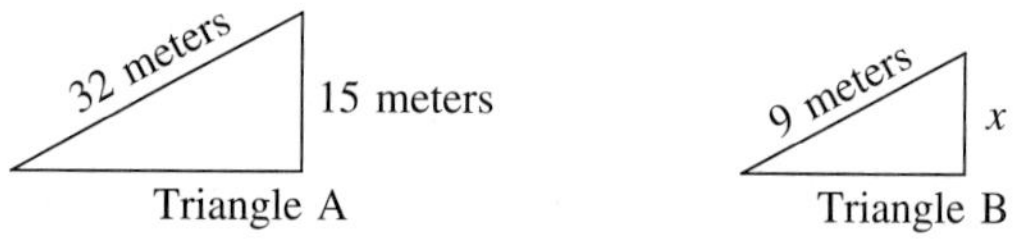

$$\begin{array}{rcl} \text{Triangle A, longest side} \longrightarrow & \dfrac{32}{9} = \dfrac{15}{x} & \longleftarrow \text{Shortest side, triangle A} \\ \text{Triangle B, longest side} \longrightarrow & & \longleftarrow \text{Shortest side, triangle B} \end{array}$$

$$32x = (9)(15) \qquad \textit{Cross-multiply.}$$

$$32x = 135 \qquad \textit{Multiply.}$$

$$x = \frac{135}{32} \qquad \textit{Divide both sides by 32.}$$

$$\text{or} \quad x = 4\frac{7}{32} \text{ meters}$$

Practice Problem 3 Triangle C is similar to triangle D. Find the length of side x. Leave your answer as a fraction.

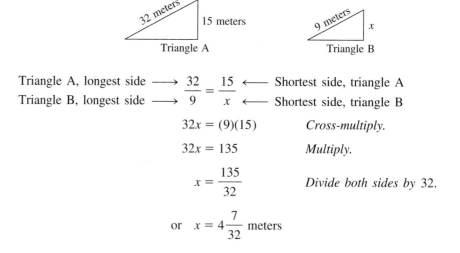

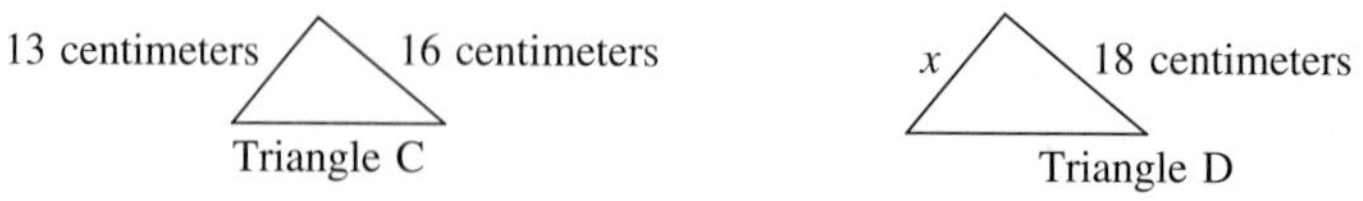

We can also use similar triangles for indirect measurement, for instance, to find the measure of an object that is too tall to measure using standard measuring devices. When the sun shines on two vertical objects at the same time, the shadows and the objects form similar triangles.

EXAMPLE 4 A woman who is 5 feet tall casts a shadow that is 8 feet long. At the same time of day, a building casts a shadow that is 72 feet long. How tall is the building?

1. *Understand the problem.*
First we draw a sketch.

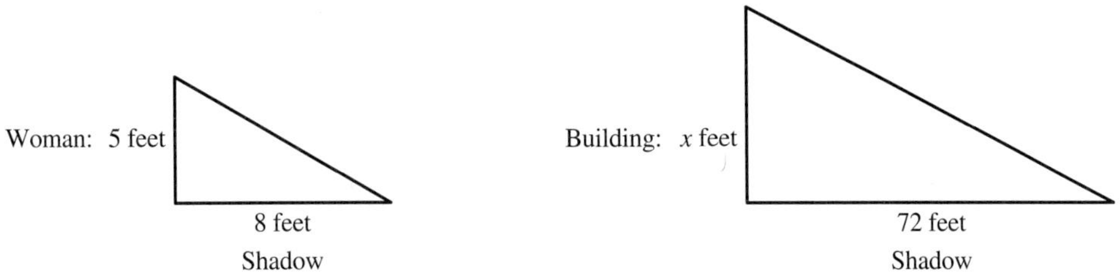

We do not know the height of the building, so we call it x.

2. *Write an equation and solve.*

Height of woman ⟶ $\frac{5}{8} = \frac{x}{72}$ ⟵ Height of building
Length of woman's shadow ⟶ (denominators) ⟵ Length of building's shadow

$$\frac{5}{8} = \frac{x}{72}$$

$$(5)(72) = 8x \qquad \textit{Cross-multiply.}$$

$$360 = 8x$$

$$45 = x$$

The height of the building is 45 feet.

Practice Problem 4 A man who is 6 feet tall casts a shadow that is 7 feet long. At the same time of day, a large flagpole casts a shadow that is 38.5 feet long. How tall is the flagpole? ■

3 *Solving Distance Problems Involving Algebraic Fractions*

Some distance problems are solved using rational equations. We will need the formula Distance = Rate × Time, $D = RT$, which we can write in the form $T = \frac{D}{R}$.

EXAMPLE 5 Plane A flies at a speed that is 50 kilometers per hour faster than plane B. Plane A flies 500 kilometers in the amount of time that plane B flies 400 kilometers. Find the speed of each plane.

1. *Understand the problem.*
   Let s = the speed of plane B in kilometers per hour.
   Let $s + 50$ = the speed of plane A in kilometers per hour.
   Make a simple table for D, R, and T.

	D	R	$T = \frac{D}{R}$
Plane A	500	$s + 50$	?
Plane B	400	s	?

   Since $T = \frac{D}{R}$, we divide the expression for D by the expressions for R and write it in the table in the column for time.

	D	R	$T = \frac{D}{R}$
Plane A	500	$s + 50$	$\frac{500}{s + 50}$
Plane B	400	s	$\frac{400}{s}$

2. *Write an equation and solve.*
   Each plane flies for the same amount of time. That is, the time for plane A equals the time for plane B.

$$\frac{500}{s + 50} = \frac{400}{s}$$

   You can solve this equation using the methods in Section 6.5 or you may cross-multiply. Here we will cross-mltiply.

$500s = (s + 50)(400)$ *Cross-multiply.*

$500s = 400s + 20{,}000$ *Remove parentheses.*

$100s = 20{,}000$ *Add 400s to each side.*

$s = 200$ *Divide each side by* 100.

Plane B travels 200 kilometers per hour. Since

$s + 50 = 200 + 50 = 250$

plane A travels 250 kilometers per hour.

Practice Problem 5 Two European freight trains traveled toward Paris for the same amount of time. Train A traveled 180 kilometers, while train B traveled 150 kilometers. Train A travels 10 kilometers per hour faster than train B. What is the speed of each train? ■

4 *Solving Work Problems*

Some applied problems involve the length of time needed to do a job. These problems are often referred to as work problems.

EXAMPLE 6 Reynaldo can sort a huge stack of mail on an old sorting machine in 9 hours. His brother Carlos can sort the same amount of mail using a newer sorting machine in 8 hours. How long would it take them to do the job working together? Express your answer in hours and minutes. Round to the nearest minute.

1. *Understand the problem.*
Let's do a little reasoning.

If Reynaldo can do the job in 9 hours, then in *1 hour* he could do $\frac{1}{9}$ of the job.
If Carlos can do the job in 8 hours, then in *1 hour* he could do $\frac{1}{8}$ of the job.
Let x = the number of hours it takes Reynaldo and Carlos to do the job together.

In *1 hour* together they could do $\frac{1}{x}$ of the job.

2. *Write an equation and solve.*
The amount of work Reynaldo can do in 1 hour plus the amount of work Carlos can do in 1 hour must be equal to the amount of work they could do together in 1 hour.

Amount of work done by Reynaldo	+	Amount of work done by Carlos	=	Amount of work done together
$\frac{1}{9}$	+	$\frac{1}{8}$	=	$\frac{1}{x}$

We observe the LCD is $72x$.

$$72x\left(\frac{1}{9}\right) + 72x\left(\frac{1}{8}\right) = 72x\left(\frac{1}{x}\right) \quad \textit{Multiply each term by the LCD.}$$

$$8x + 9x = 72 \quad \textit{Simplify.}$$

$$17x = 72 \quad \textit{Collect like terms.}$$

$$x = \frac{72}{17} \quad \textit{Divide each side by 17 hours.}$$

$$x = 4\frac{4}{17}$$

To change $\frac{4}{17}$ of an hour to minutes we multiply.

$$\frac{4}{17}\cancel{\text{hour}} \times \frac{60 \text{ minutes}}{1 \cancel{\text{hour}}} = \frac{240}{17} \text{ minutes, which is approximately 14.118 minutes}$$

To the nearest minute this is 14 minutes. Thus doing the job together will take 4 hours and 14 minutes.

Practice Problem 6 John and Dave obtained night custodian jobs at a local factory while going to college part time. Using the buffer machine, John can buff all the floors in the building in 6 hours. Dave takes a little longer and can do all the floors in the building in 7 hours. Their supervisor bought another buffer machine. How long will it take John and Dave to do all the floors in the building working together, each with his own machine? Express your answer in hours and minutes. Round your answer to the nearest minute. ■

6.6 Exercises

Solve each proportion.

1. $\frac{7}{5} = \frac{21}{x}$

2. $\frac{3}{8} = \frac{x}{14}$

3. $\frac{x}{17} = \frac{12}{5}$

4. $\frac{16}{x} = \frac{3}{4}$

5. $\frac{5}{3} = \frac{x}{8}$

6. $\frac{120}{84} = \frac{6}{x}$

7. $\frac{5}{x} = \frac{150}{75}$

8. $\frac{x}{18} = \frac{13}{2}$

Applications

Use a proportion to answer each question.

9. If 190 cups of coffee can be made from 5 pounds of coffee, how many cups can be made from 3 pounds of coffee?

10. It took 6 hours to register 380 students. How long will it take (at the same rate) for 500 students to be registered? Round your answer to nearest minute.

11. Ben's engine is using a mixture of 4 pints of oil for every 7 gallons of gas. He has 11 pints of oil on hand to prepare a mixture. How many gallons of gas should he use?

12. A family punch uses 5 quarts of ginger ale for every 8 quarts of fruit drink. How many quarts of ginger ale should be mixed with 31 quarts of fruit drink?

13. If a 25-yard length of copper tubing weighs 210 pounds, how much does 60 yards of this copper tubing weigh?

14. A 10-pound bag of fertilizer will cover 980 square feet of lawn. How many bags are needed to cover 1470 square feet of lawn?

15. Alfonse and Melinda took a drive in Mexico. They know that if you drive 100 kilometers per hour that this is about 62 miles per hour. They are now driving on a Mexican road that has a speed limit of 90 kilometers per hour. How many miles per hour is the speed limit? Round your answer to the nearest mile per hour.

16. Dick and Anne took a trip to France. They know that 50 kilograms is equivalent to 110 pounds. Their suitcases were weighed at the airport and the weight recorded was 39 kilograms. How many pounds did their suitcases weigh? Round your answer to the nearest pound.

17. On the highway, Marcia and Melissa found that their car uses 8 gallons of gas to travel 170 miles. They need to travel on the highway from Boston to Chicago, a distance of 970 miles. Approximately how many gallons of gas will they use? Round your answer to the nearest gallon.

18. Hector uses 11 ounces of enchilada sauce for his famous beef taco recipe. The recipe makes 14 tacos. He wants to make 30 tacos. How many ounces of enchilada sauce will he need? Round your answer to the nearest tenth of an ounce.

19. On a map the distance between two mountains is $3\frac{1}{2}$ inches. The actual distance between the mountains is 136 miles. Russ is camped at a location that on the map is $\frac{3}{4}$ inch from the base of the mountain. How many miles is he from the base of the mountain? Round your answer to the nearest mile.

20. Maria is adding a porch to her house that is 18 feet long. On the drawing done by the carpenter the length is shown as 11 inches. The drawing shows that the width of the porch is 8 inches. How many feet wide will the porch be? Round your answer to the nearest foot.

Triangles A and B are similar. Use this in answering questions 21–24. Leave your answers as fractions.

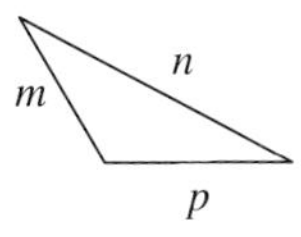

Triangle A

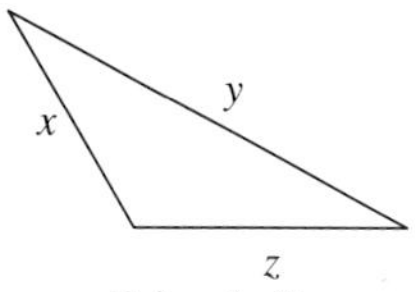

Triangle B

21. If $x = 20$ in., $y = 29$ in., and $m = 13$ in., find the length of side n.

22. If $p = 14$ in., $m = 17$ in. and $z = 23$ in., find the length of side x.

23. If $n = 40$ meters, $m = 35$ meters, and $x = 100$ meters, find the length of side y.

24. If $z = 18$ cm, $y = 25$ cm, and $n = 9$ cm, find the length of side p.

Use a proportion to solve each problem.

25. A 6-foot-tall man casts a shadow of 5 feet. At the same time, a building casts a shadow of 47 feet. How tall is the building? Round your answer to the nearest foot.

26. A 5-foot-tall woman casts a shadow of 4 feet. At the same time, a tree casts a shadow of 31 feet. How tall is the tree? Round your answer to the nearest foot.

27. A kite is held out on a line that is almost perfectly straight. When the kite is held out on 7 meters of line, it is 5 meters high off the ground. Assuming the same relationship, how high would the kite be if it is held out on 120 meters of line? Round your answer to the nearest meter.

28. A wire line helps to secure a radio transmission tower. The wire measures 23 meters from the tower to the ground anchor pin. The wire is secured 14 meters up on the tower. Assuming the same relationship, if a wire is secured 130 meters up on the tower, how long would it need to be to reach from the tower to an anchor pin on the ground? Round your answer to the nearest meter.

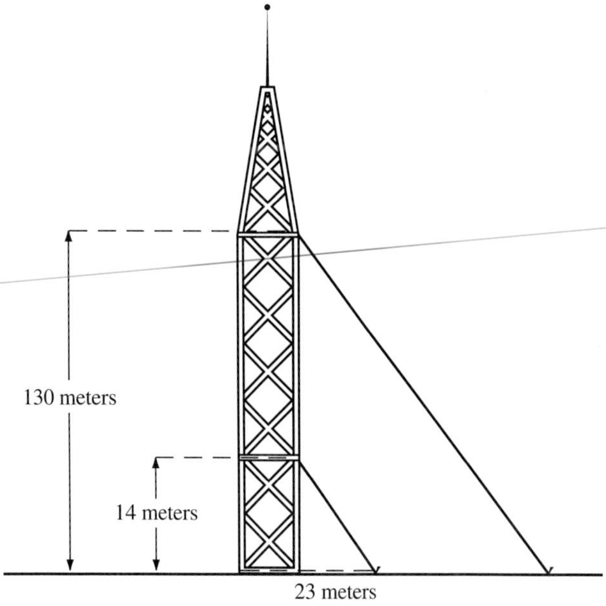

29. Speedboat A travels 10 kilometers per hour faster than speedboat B. Speedboat A has traveled 240 kilometers, while speedboat B has only traveled 210 kilometers in the same time. How fast did each boat travel?

30. Car A travels 20 kilometers per hour faster than car B. Car A has traveled 360 kilometers, while car B has only traveled 300 kilometers. How fast did each car travel?

31. Helicopter B at maximum speed travels 40 kilometers per hour faster than helicopter A. Helicopter B travels 1250 kilometers, while helicopter A only travels 1050 kilometers. What is the maximum speed of each helicopter?

32. Melissa drove to Dallas, while Marcia drove to Houston. Melissa drove 360 kilometers, while Marcia drove 280 kilometers. Melissa averaged 20 kilometers per hour faster than Marcia on her trip. What is the average speed in kilometers per hour for each woman?

33. It takes a person using a large rotary mower 4 hours to mow all the laws at the town park. It takes a person using a small rotary mower 5 hours to mow these same lawns. If both mowers are used, how long should it take two people to mow these lawns? Round your answer to the nearest minute.

34. It takes 7 hours to write the addresses on envelopes for a business mailing list with a secretary doing the work by typewriter. It takes 3 hours to have the computer print out the addresses and have the secretary put the computer labels on the envelopes. The company boss is in a big rush and wants one secretary to do part of the job with the computer and another secretary to do the rest of the list with a typewriter. If they work efficiently, how long should it take with the two secretaries working together? Round your answer to the nearest minute.

35. The larger pump at the town pumping station can fill the town water tank in 15 hours. The smaller pump can do the same job in 20 hours. How long will it take the pumps working together? Round your answer to the nearest minute.

36. It took 400 hours for highway team A to clean all the streets of the city. It took 300 hours for highway team B to do the same job. The mayor of the city is in a big hurry to get the streets clean for visiting dignitaries. If team A and team B both work together, how many hours should it take to clean all the streets of the city? Round your answer to the nearest hour.

Cumulative Review Problems

37. Write in scientific notation. 0.0000006316

38. Write in decimal notation. 5.82×10^8

39. Write with positive exponents. $\dfrac{x^{-3}y^{-2}}{z^4w^{-8}}$

40. Evaluate. $\left(\dfrac{2}{3}\right)^{-3}$

Putting Your Skills to Work

ELECTRICAL RESISTANCE

When electricity passes through an appliance, such as a television, the appliance resists some of the electrical current passing through it. If all the appliances in a house were electrically connected with one straight wire (wired in series), then when you use one appliance it would decrease the amount of electricity available for the next appliance. By the time the last appliance was turned on, there would not be enough electricity to run the appliance.

To avoid this problem, when electricians wire a house they use many parallel circuits. The electricity comes into the house at one point but follows separate paths to different appliances.

Suppose a house is wired with two parallel circuits. The total resistance R of the circuits in the house is given by the equation

$$\frac{1}{R} = \frac{1}{x} + \frac{1}{y}$$

where x is the amount of resistance in the first circuit and y is the amount of resistance in the second circuit. Electrical resistance in commonly measured in *ohms.*

1. If the resistance in one circuit is 10 ohms and the resistance in the second circuit is 6 ohms, what is the total resistance in the two circuits?

2. If the resistance in one circuit is 12 ohms and the resistance in the second circuit is 4 ohms, what is the total resistance in the two circuits?

COOPERATIVE GROUP ACTIVITY

3. Suppose a house is wired with four parallel circuits. The total resistance R of the circuits in the house is given by the equation

$$\frac{1}{R} = \frac{1}{x} + \frac{1}{y} + \frac{1}{w} + \frac{1}{z}$$

where x is the amount of resistance in the first circuit, y the amount in the second, w the amount in the third, and z the amount in the fourth. If the resistance in each of the four circuits is 8 ohms, 12 ohms, 6 ohms, and 3 ohms, respectively, what is the total resistance in the four circuits? Round your answer to the nearest tenth of an ohm.

4. In a house with two circuits with resistances x and y, respectively, what would happen to the total resistance R if the value of x was doubled but the value of y was divided by 2? Would the resistance R remain the same or change? Why? Test your results with a couple of specific cases. What pattern do you see?

Chapter Organizer

Topic	Procedure	Examples
Simplifying rational expressions, p. 342.	**1.** Factor numerator and denominator. **2.** Divide out any factor common to both numerator and denominator.	$\dfrac{36x^2 - 16y^2}{18x^2 + 24xy + 8y^2} = \dfrac{\overset{2}{4}(3x - 2y)(3x + 2y)}{2(3x + 2y)(3x + 2y)}$ $= \dfrac{2(3x - 2y)}{3x + 2y}$
Multiplying rational expressions, p. 348.	**1.** Factor all numerators and denominators, writing the problem as one fraction. **2.** Proceed as in simplifying, above.	$\dfrac{x^2 - y^2}{x^2 + 2xy + y^2} \cdot \dfrac{x^2 + 4xy + 3y^2}{x^2 - 4xy + 3y^2}$ $= \dfrac{(x + y)(x - y)(x + y)(x + 3y)}{(x + y)(x + y)(x - y)(x - 3y)}$ $= \dfrac{x + 3y}{x - 3y}$
Dividing rational expressions, p. 350.	**1.** Invert the second fraction and change the problem to multiplication. **2.** Proceed as for multiplication.	$\dfrac{14x^2 + 17x - 6}{x^2 - 25} \div \dfrac{4x^2 - 8x - 21}{x^2 + 10x + 25}$ $= \dfrac{(2x + 3)(7x - 2)}{(x + 5)(x - 5)} \cdot \dfrac{(x + 5)(x + 5)}{(2x - 7)(2x + 3)}$ $= \dfrac{(7x - 2)(x + 5)}{(x - 5)(2x - 7)}$
Adding rational expressions, p. 354.	**1.** If the denominators differ, factor them and determine the least common denominator (LCD). **2.** Change, by multiplication, each fraction into an equivalent one with the LCD. **3.** Add (add numerators; put answer over the LCD). **4.** Simplify as needed.	$\dfrac{x - 1}{x^2 - 4} + \dfrac{x - 1}{3x + 6}$ $(x + 2)(x - 2) \quad 3(x + 2)$ LCD $= 3(x + 2)(x - 2)$ $\dfrac{x - 1}{(x + 2)(x - 2)} = \dfrac{?}{3(x + 2)(x - 2)}$ Need to multiply by $\dfrac{3}{3}$. $\dfrac{(x - 1)}{3(x + 2)} = \dfrac{?}{3(x + 2)(x - 2)}$ Need to multiply by $\dfrac{x - 2}{x - 2}$. $= \dfrac{(x - 1) \cdot 3}{3(x + 2)(x - 2)} + \dfrac{(x - 1)(x - 2)}{3(x + 2)(x - 2)}$ $= \dfrac{3x - 3 + x^2 - 3x + 2}{3(x + 2)(x - 2)}$ $= \dfrac{x^2 - 1}{3(x + 2)(x - 2)}$
Subtracting rational expressions, p. 354.	Move a subtraction sign to the numerator of the following fraction. Add. Simplify if possible. $\dfrac{a}{b} - \dfrac{c}{b} = \dfrac{a}{b} + \dfrac{-c}{b}$	$\dfrac{5x}{x - 2} - \dfrac{3x + 4}{x - 2}$ $= \dfrac{5x}{x - 2} + \dfrac{-(3x + 4)}{x - 2}$ $= \dfrac{5x - 3x - 4}{x - 2} = \dfrac{2x - 4}{x - 2}$ $= \dfrac{2(x - 2)}{x - 2} = 2$

Chapter Organizer

Topic	Procedure	Examples
Simplifying complex fractions, p. 364.	1. Add together the two fractions in the numerator. 2. Add together the two fractions in the denominator. 3. Divide the fraction in the numerator by the fraction in the denominator. This is done by inverting the fraction in the denominator and multiplying by the numerator. 4. Simplify.	$\dfrac{\dfrac{x}{x^2-4}+\dfrac{1}{x+2}}{\dfrac{3}{x+2}-\dfrac{4}{x-2}}$ $=\dfrac{\dfrac{x}{(x+2)(x-2)}+\dfrac{1(x-2)}{(x+2)(x-2)}}{\dfrac{3(x-2)}{(x+2)(x-2)}+\dfrac{-4(x+2)}{(x+2)(x-2)}}$ $=\dfrac{\dfrac{x+x-2}{(x+2)(x-2)}}{\dfrac{3x-6-4x-8}{(x+2)(x-2)}}$ $=\dfrac{2x-2}{(x+2)(x-2)}\div\dfrac{-x-14}{(x+2)(x-2)}$ $=\dfrac{2(x-1)}{(x+2)(x-2)}\cdot\dfrac{(x+2)(x-2)}{-x-14}$ $=\dfrac{2x-2}{-x-14}$ or $\dfrac{-2x+2}{x+14}$
Solving equations involving algebraic fractions, p. 369.	1. Determine the LCD of all denominators. 2. Note what values will make the LCD equal to 0. These are excluded from your solutions. 3. Multiply each side by the LCD, distributing as needed. 4. Solve the resulting polynomial equation. 5. Check. Be sure to exclude those values found in step 2.	$\dfrac{3}{x-2}=\dfrac{4}{x+2}$ LCD = $(x-2)(x+2)$. Since LCD $\neq 0$, $x \neq 2, -2$. $(x-2)(x+2)\dfrac{3}{x-2}=\dfrac{4}{x+2}(x-2)(x+2)$ $3(x+2)=4(x-2)$ $3x+6=4x-8$ $-x=-14$ $x=14$ (Since $x \neq 2, -2$ this should check unless an error has been made.) *Check:* $\dfrac{3}{14-2} \stackrel{?}{=} \dfrac{4}{14+2}$ $\dfrac{3}{12} \stackrel{?}{=} \dfrac{4}{16}$ $\dfrac{1}{4}=\dfrac{1}{4}$ ✓
Solving applied problems with proportions, p. 374.	1. Organize the data. 2. Write a proportion equating the respective parts, letting x represent the value that is not known. 3. Solve the proportion.	Renee can make 5 cherry pies with 3 cups of flour. How many cups of flour does she need to make 8 cherry pies? $\dfrac{5 \text{ cherry pies}}{3 \text{ cups flour}}=\dfrac{8 \text{ cherry pies}}{x \text{ cups flour}}$ $\dfrac{5}{3}=\dfrac{8}{x}$ $5x=24$ $x=\dfrac{24}{5}$ $x=4\dfrac{4}{5}$ $4\dfrac{4}{5}$ cups of flour are needed for 8 cherry pies.

Chapter 6 Review Problems

6.1 *Simplify.*

1. $\dfrac{4x - 4y}{5y - 5x}$

2. $\dfrac{bx}{bx - by}$

3. $\dfrac{2x^2 + 5x - 3}{2x^2 - 9x + 4}$

4. $\dfrac{6x^2 + 7x + 2}{3x^2 + 5x + 2}$

5. $\dfrac{x^2 - 9}{x^2 - 10x + 21}$

6. $\dfrac{2x^2 + 18x + 40}{3x + 15}$

7. $\dfrac{4x^2 + 4x - 3}{4x^2 - 2x}$

8. $\dfrac{x^3 + 3x^2}{x^3 - 2x^2 - 15x}$

9. $\dfrac{2x^2 - 2xy - 24y^2}{2x^2 + 5xy - 3y^2}$

10. $\dfrac{4 - y^2}{3y^2 + 5y - 2}$

11. $\dfrac{5x^3 - 10x^2}{25x^4 + 5x^3 - 30x^2}$

12. $\dfrac{16x^2 - 4y^2}{4x - 2y}$

6.2 *Multiply or divide.*

13. $\dfrac{3x^2 - 13x - 10}{3x^2 + 2x} \cdot \dfrac{x^2 - 25x}{x^2 - 25}$

14. $\dfrac{x^2 + 7x + 12}{4x + 12} \cdot \dfrac{x^2 - 16}{x^2 + 8x + 16}$

15. $\dfrac{2x^2 - 18}{3y^2 + 3y} \div \dfrac{y^2 + 6y + 9}{y^2 + 4y + 3}$

16. $\dfrac{2y^2 + 3y - 2}{2y^2 + y - 1} \div \dfrac{2y^2 + y - 1}{2y^2 - 3y - 2}$

17. $\dfrac{x^3 - 36x}{12x^2 + 2x} \cdot \dfrac{36x + 6}{x^3 - 6x^2} \cdot \dfrac{x^2 + 2x}{x^2 + 11x + 30}$

18. $\dfrac{10x^2}{x^2 - 7x + 10} \cdot \dfrac{x^2 + 3x - 10}{25x^2} \cdot \dfrac{2x - 8}{x^2 + x - 20}$

19. $\dfrac{6y^2 + 13y - 5}{9y^2 + 3y} \div \dfrac{4y^2 + 20y + 25}{12y^2}$

20. $\dfrac{3xy^2 + 12y^2}{2x^2 - 11x + 5} \div \dfrac{2xy + 8y}{8x^2 + 2x - 3}$

21. $\dfrac{11}{x - 2} \cdot \dfrac{2x^2 - 8}{44}$

22. $\dfrac{2x^2 + 10x + 2}{8x - 8} \cdot \dfrac{3x - 3}{4x^2 + 20x + 4}$

23. $\dfrac{5x^7}{16x^2 - 9} \cdot \dfrac{8x + 6}{10x^8}$

24. $\dfrac{x^2 - 5xy - 24y^2}{2x^2 - 2xy - 24y^2} \cdot \dfrac{4x^2 + 4xy - 24y^2}{x^2 - 10xy + 16y^2}$

6.3 *Add or subtract.*

25. $\dfrac{7}{x + 1} + \dfrac{4}{2x}$

26. $5 + \dfrac{1}{x} + \dfrac{1}{x + 1}$

27. $\dfrac{2}{x^2 - 9} + \dfrac{x}{x + 3}$

28. $\dfrac{7}{x + 2} + \dfrac{3}{x - 4}$

29. $\dfrac{x}{y} + \dfrac{3}{2y} + \dfrac{1}{y + 2}$

30. $\dfrac{4}{a} + \dfrac{2}{b} + \dfrac{3}{a + b}$

31. $\dfrac{3x + 1}{3x} - \dfrac{1}{x}$

32. $\dfrac{x + 4}{x + 2} - \dfrac{1}{2x}$

33. $\dfrac{1}{x^2 + 7x + 10} - \dfrac{x}{x + 5}$

34. $\dfrac{7}{x^2 - 9x + 20} - \dfrac{3x}{x - 4}$

35. $\dfrac{2x + 1}{x^2 + 5x + 6} + \dfrac{4x}{x^2 - 9}$

36. $\dfrac{3x}{x^2 + x - 2} - \dfrac{2x + 1}{x^2 + 6x + 8}$

6.4 *Simplify.*

37. $\dfrac{\dfrac{3}{2y} - \dfrac{1}{y}}{\dfrac{4}{y} + \dfrac{3}{2y}}$

38. $\dfrac{\dfrac{2}{x} + \dfrac{1}{2x}}{x + \dfrac{x}{2}}$

39. $\dfrac{w - \dfrac{4}{w}}{1 + \dfrac{2}{w}}$

40. $\dfrac{1 - \dfrac{w}{w - 1}}{1 + \dfrac{w}{1 - w}}$

41. $\dfrac{1 + \dfrac{1}{y^2 - 1}}{\dfrac{1}{y + 1} - \dfrac{1}{y - 1}}$

42. $\dfrac{\dfrac{1}{y} + \dfrac{1}{x + y}}{1 + \dfrac{2}{x + y}}$

43. $\dfrac{\dfrac{1}{a + b} - \dfrac{1}{a}}{b}$

44. $\dfrac{\dfrac{2}{a + b} - \dfrac{3}{b}}{\dfrac{1}{a + b}}$

45. $\left(\dfrac{1}{x + 2y} - \dfrac{1}{x - y}\right) \div \dfrac{2x - 4y}{x^2 - 3xy + 2y^2}$

46. $\dfrac{x + 5y}{x - 6y} \div \left(\dfrac{1}{5y} - \dfrac{1}{x + 5y}\right)$

6.5 *Solve for the variable. If no solution is possible, say so.*

47. $\dfrac{8}{a - 3} = \dfrac{12}{a + 3}$

48. $\dfrac{8a - 1}{6a + 8} = \dfrac{3}{4}$

49. $\dfrac{2x - 1}{x} - \dfrac{1}{2} = -2$

50. $\dfrac{5 - x}{x} - \dfrac{7}{x} = -\dfrac{3}{4}$

51. $\dfrac{5}{2} - \dfrac{2y + 7}{y + 6} = 3$

52. $\dfrac{5}{4} - \dfrac{1}{2x} = \dfrac{1}{x} + 2$

53. $\dfrac{1}{3x} + 2 = \dfrac{5}{6x} - \dfrac{1}{2}$

54. $\dfrac{7}{8x} - \dfrac{3}{4} = \dfrac{1}{4x} + \dfrac{1}{2}$

55. $\dfrac{x - 8}{x - 2} = \dfrac{2x}{x + 2} - 2$

56. $\dfrac{3}{y - 3} = \dfrac{3}{2} + \dfrac{y}{y - 3}$

57. $\dfrac{3y - 1}{3y} - \dfrac{6}{5y} = \dfrac{1}{y} - \dfrac{4}{15}$

58. $\dfrac{9}{2} - \dfrac{7y - 4}{y + 2} = -\dfrac{1}{4}$

59. $\dfrac{4}{x^2 - 1} = \dfrac{2}{x - 1} + \dfrac{2}{x + 1}$

60. $\dfrac{y + 18}{y^2 - 16} = \dfrac{y}{y + 4} - \dfrac{y}{y - 4}$

61. $\dfrac{9y - 3}{y^2 + 2y} - \dfrac{5}{y + 2} = \dfrac{3}{y}$

62. $\dfrac{2}{3 - 3y} + \dfrac{2}{2y - 1} = \dfrac{4}{3y - 3}$

6.6 *Solve each proportion. Round your answer to the nearest tenth.*

63. $\frac{8}{5} = \frac{2}{x}$

64. $\frac{x}{4} = \frac{12}{17}$

65. $\frac{33}{10} = \frac{x}{8}$

66. $\frac{5}{x} = \frac{22}{9}$

Use a proportion to answer each question.

67. It took 9 hours to register 450 students. At the same rate, how long will it take to register 550 students? Round your answer to the nearest minute.

68. Fred is driving in Canada. He knows that 100 kilometers per hour is about the same as 62 miles per hour. The speed limit on this back road is 65 kilometers per hour. How many miles per hour is the speed limit? Round your answer to the nearest mile per hour.

69. A 5-gallon can of paint will cover 240 square feet. How many gallons of paint will be needed to cover 400 square feet? Round your answer to the nearest tenth of a gallon.

70. Aunt Lexie uses 3 pounds of sugar to make 100 cookies. How many cookies can she make with 5 pounds of sugar? Round to nearest whole number.

71. Ron found that his car used 7 gallons of gas to travel 200 miles. He plans to drive 1300 miles from his home to Denver, Colorado. How many gallons of gas will he use if his car continues to use gas at the same rate? Round your answer to the nearest whole number.

72. The distance on a map between two cities is 4 inches. The actual distance between these cities is 122 miles. Two lakes are 3 inches apart on the same map. How many miles is the actual distance between these two lakes?

73. A train travels 180 miles in the same time that a car travels 120 miles. The speed of the train is 20 miles per hour faster than the speed of the car. Find the speed of the train and the speed of the car.

74. A professional painter can paint the interior of the Jackson's house in 5 hours. John Jackson can do the same job in 8 hours. How long will it take the two people working together on the painting job? Round your answer to the nearest minute.

75. A flagpole that is 8 feet tall casts a shadow of 3 feet. At the same time of day, a tall office building in the city casts a shadow of 450 feet. How tall is the office building? Round your answer to the nearest foot.

76. Mary took a walk across a canyon in New Mexico. She stands 5.75 feet tall and her shadow is 3 feet long. At the same time the shadow from the peak of the canyon wall casts a shadow that is 95 feet long. How tall is the peak of the canyon? Round your answer to the nearest foot.

Chapter 6 Test

Perform the operation indicated. Simplify your answers.

1. $\dfrac{2ac + 2ad}{3a^2c + 3a^2d}$

2. $\dfrac{8x^2 - 2x^2y^2}{y^2 + 4y + 4}$

3. $\dfrac{x^2 + 2x}{2x - 1} \cdot \dfrac{10x^2 - 5x}{12x^3 + 24x^2}$

4. $\dfrac{a^2 + ab}{b} \cdot \dfrac{a^2 - 2ab + b^2}{a^4 - b^4} \cdot \dfrac{a^2 + b^2}{3a^2b - 3ab^2}$

5. $\dfrac{2a^2 - 3a - 2}{4a^2 + a - 14} \div \dfrac{2a^2 + 5a + 2}{16a^2 - 49}$

6. $\dfrac{9}{x - 2} + \dfrac{2}{x - 3} + \dfrac{1}{2x - 6}$

7. $\dfrac{x - y}{xy} - \dfrac{a - y}{ay}$

8. $\dfrac{3x}{x^2 - 3x - 18} - \dfrac{x - 4}{x - 6}$

9. $\dfrac{\dfrac{x}{3y} - \dfrac{1}{2}}{\dfrac{4}{3y} - \dfrac{2}{x}}$

10. $\dfrac{\dfrac{2}{x + 3} + \dfrac{1}{x}}{3x + 9}$

11. $\dfrac{2x^2 + 3xy - 9y^2}{4x^2 + 13xy + 3y^2}$

12. $\dfrac{1}{x + 4} - \dfrac{2}{x^2 + 6x + 8}$

1. ______________

2. ______________

3. ______________

4. ______________

5. ______________

6. ______________

7. ______________

8. ______________

9. ______________

10. ______________

11. ______________

12. ______________

13.
14.
15.
16.
17.
18.
19.
20.

Solve for x. Check your answers.

13. $\dfrac{15}{x} + \dfrac{9x - 7}{x + 2} = 9$

14. $\dfrac{x - 3}{x - 2} = \dfrac{2x^2 - 15}{x^2 + x - 6} - \dfrac{x + 1}{x + 3}$

15. $3 - \dfrac{7}{x + 3} = \dfrac{x - 4}{x + 3}$

16. $\dfrac{3}{3x - 5} = \dfrac{7}{5x + 4}$

17. $\dfrac{9}{x} = \dfrac{13}{5}$

18. Katie is typing a term paper for her English class. She typed 3 pages of the paper in 55 minutes. If she continues at the same rate, how long will it take her to type the entire 21-page paper? Express your answer in hours and minutes.

19. In northern Michigan the Gunderson family heats their home with firewood. They used \$100 worth of wood in 25 days. Mr. Gunderson estimates that he needs to burn wood at that rate for about 92 days during the winter. If that is so, how much will the 92-day supply of wood cost?

20. A hiking club is trying to construct a rope bridge across a canyon. A 6-foot construction pole held upright casts a 7-foot shadow. At the same time of day, a tree at the edge of the canyon casts a shadow that exactly covers the distance that is needed for the rope bridge. The tree is exactly 87 feet tall. How long should the rope bridge be? Round your answer to the nearest foot.

Cumulative Test for Chapters 0–6

Approximately one-half of this test covers the content of Chapters 0–5.
The remainder covers the content of Chapter 6.

1. To the nearest thousandth, how much error can be tolerated in the length of a wire that is supposed to be 2.57 centimeters long? Specifications allow an error of no more than 0.25%.

2. Fifteen percent of what number is \$37.50?

3. After the quake of 1989, California raised its sales tax by $\frac{1}{4}$% (one-quarter of 1%) to aid earthquake relief. How much extra sales tax will be paid on a new \$19,990 car?

4. Solve.

$5(x - 3) - 2(4 - 2x) = 7(x - 1) - (x - 2)$

5. Solve for h.

$A = \pi r^2 h$

6. Solve and graph on a number line. $4(2 - x) < 3$

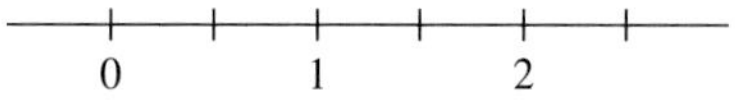

7. Solve for x.

$$\frac{1}{4}x + \frac{3}{4} < \frac{2}{3}x - \frac{4}{3}$$

8. Factor. $3ax + 3bx - 2ay - 2by$

9. Factor. $8a^3 - 38a^2b - 10ab^2$

10. Simplify. $(-3x^2y)(4x^3y^6)$

11. Simplify. $\dfrac{4x^2 - 25}{2x^2 + 9x - 35}$

1. ________
2. ________
3. ________
4. ________
5. ________
6. ________
7. ________
8. ________
9. ________
10. ________
11. ________

12.

13.

14.

15.

16.

17.

18.

19.

20.

21.

Perform the indicated operations.

12. $\dfrac{x^2 - 4}{x^2 - 25} \cdot \dfrac{3x^2 - 14x - 5}{3x^2 + 6x} \div \dfrac{x - 2}{3x^2 + 15x}$

13. $\dfrac{5}{2x + 4} + \dfrac{3}{x - 3}$

14. $\dfrac{2x + 1}{x^2 + x - 12} - \dfrac{x - 1}{x^2 - x - 6}$

Solve for x.

15. $\dfrac{3x - 2}{3x + 2} = 2$

16. $\dfrac{x - 3}{x} = \dfrac{x + 2}{x + 3}$

Simplify.

17. $\dfrac{\dfrac{1}{x - 3} + \dfrac{5}{x^2 - 9}}{\dfrac{6x}{x + 3}}$

18. $\dfrac{\dfrac{3}{a} + \dfrac{2}{b}}{\dfrac{5}{a^2} - \dfrac{2}{b^2}}$

19. Solve for x. $\dfrac{5}{7} = \dfrac{14}{x}$

20. Jane is looking at a road map. The distance between two cities is 130 miles. This distance is represented by $2\frac{1}{2}$ inches on the map. She sees that the distance she has to drive today is a totally straight interstate highway represented by 4 inches on the map. How many miles does she have to drive? Round your answer to the nearest mile.

21. Roberto is working as a telemarketing salesperson for a large corporation. In the last 22 telephone calls he has made, he was able to make a sale 5 times. His goal is to make 110 sales this month. How many phone calls must he make if his rate for making a sale continues as in the past?

7
Graphing and Functions

The pressure of water on a diver increases for every additional foot the diver goes down. The equation that describes this and the graph of this function relationship yield very useful information. See page 426.

Pretest Chapter 7

1. ______

2. ______

3. (a) ______

(b) ______

4. ______

5. ______

6. ______

7. ______

8. (a) ______

(b) ______

If you are familiar with the topics in this chapter, take this test now. Check your answers with those in the back of the book. If an answer was wrong or you couldn't do a problem, study the appropriate section of the chapter.

If you are not familiar with the topics in this chapter, don't take this test now. Instead, study the examples, work the Practice Problems, and then take the test.

This test will help you to identify those concepts that you have mastered and those that need more study.

Section 7.1

1. Graph the points.
$(-3, 1), (4, -4), (-2, -6)$

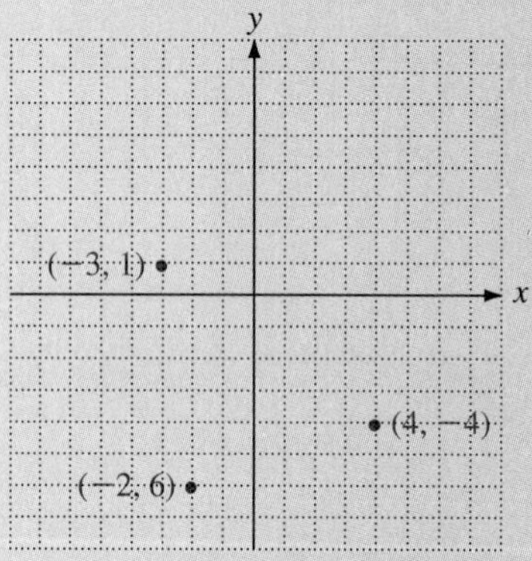

2. Write the coordinates of each point below.

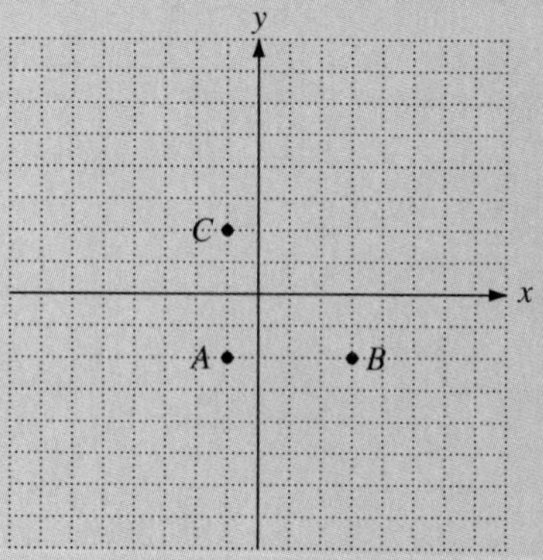

3. Complete each ordered pair so that it is a solution to the equation $y = -3x + 7$.
(a) $(-1, \quad)$ **(b)** $(\quad, 1)$

Section 7.2

4. Graph $5x - 2y = -10$.

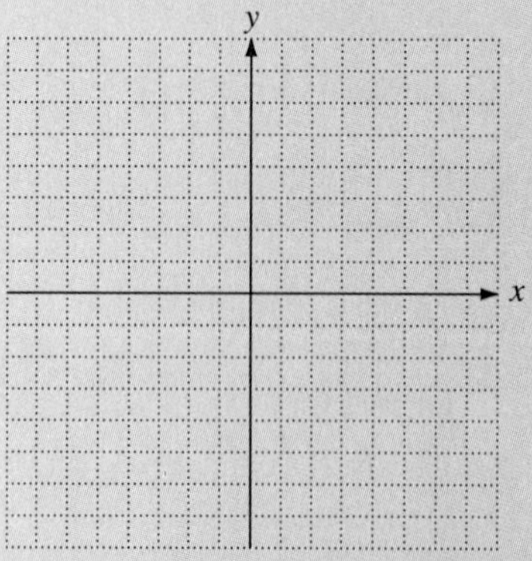

5. Graph $y = 4$.

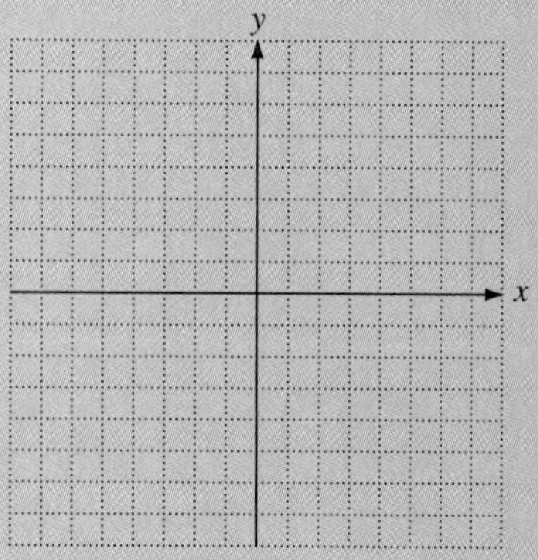

Section 7.3

6. What is the slope of the line passing through $(4, -2)$ and $(-3, -4)$?

7. Find the slope of the line $4x - 3y - 7 = 0$.

8. Find the equation of the line with slope 4 and y-intercept -5.
(a) Write the equation in slope–intercept form.
(b) Write the equation in $Ax + By = C$ form.

Section 7.4

9. Graph the equation $y = 2x + 1$.

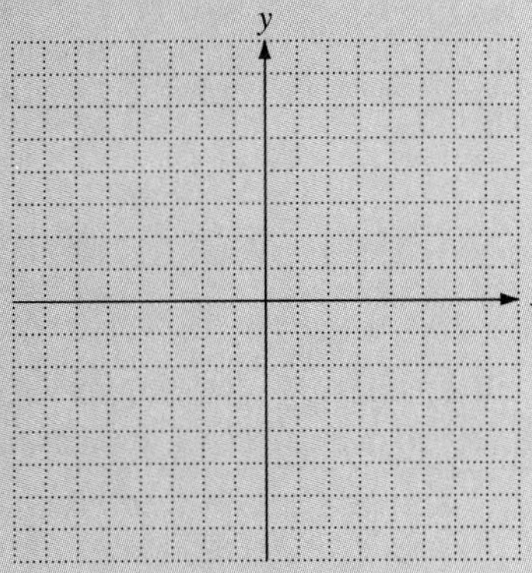

10. Write the equation of the graph.

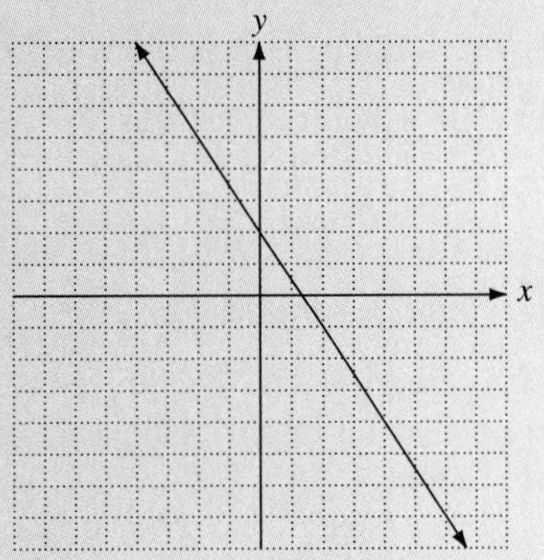

11. Line p has a slope of $-\frac{1}{4}$.
(a) What is the slope of a line parallel to line p?
(b) What is the slope of a line perpendicular to line p?

12. Find the equation of a line with slope $\frac{2}{3}$ that passes through the point $(-3, -3)$.

13. Find the equation of a line that passes through $(-1, 6)$ and $(2, 3)$.

Section 7.5

14. Graph $y \geq 3x - 2$.

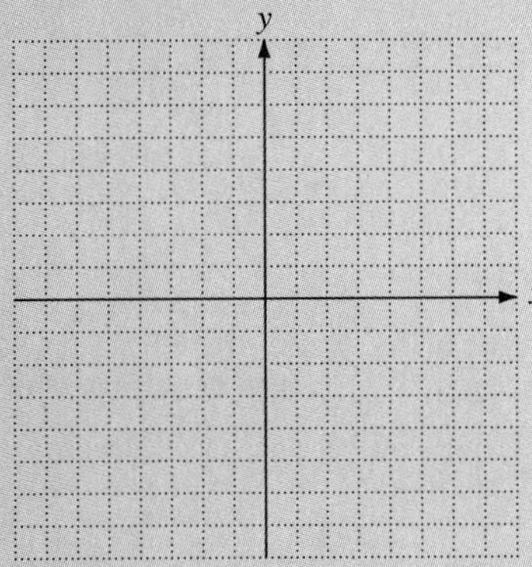

15. Graph $4x + 2y < -12$.

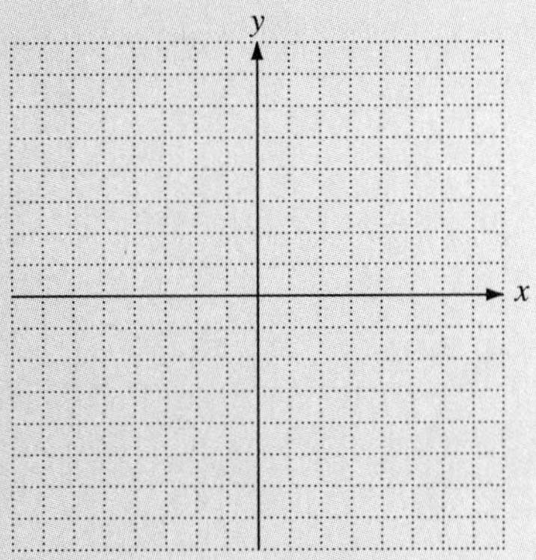

Section 7.6

Determine whether each of the following is a function.

16. Circle

r	1	5	10	15
C	6.28	3.14	62.8	94.2

17.

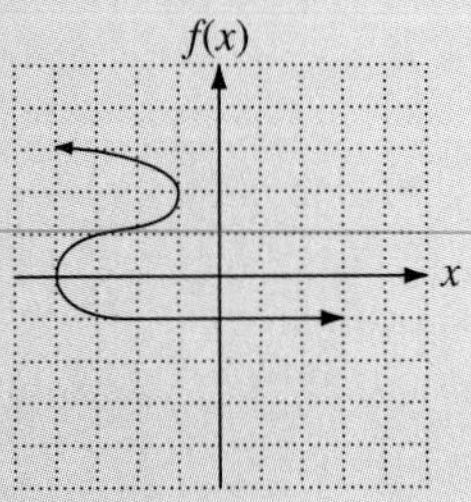

Find the indicated values for each function.

18. $f(x) = 3x - 7$
(a) $f(-2)$ **(b)** $f(4)$

19. $g(x) = 2x^2$
(a) $g(-6)$ **(b)** $g(\frac{1}{2})$

9. ____________

10. ____________

11. (a) ____________

(b) ____________

12. ____________

13. ____________

14. ____________

15. ____________

16. ____________

17. ____________

18. (a) ____________

(b) ____________

19. (a) ____________

(b) ____________

7.1 Rectangular Coordinate System

After studying this section, you will be able to:

1 *Plot a point, given the coordinates.*
2 *Name the coordinates of a plotted point.*
3 *Find ordered pairs for a given linear equation.*

1 *Plotting a Point*

Many things in everyday life are clearer if we see a picture. Similarly, in mathematics, we can picture algebraic relationships by drawing a graph. To draw a graph, we need a frame of reference.

In Chapter 1 we showed that any real number can be represented on a number line. Look at the number line below. The arrow indicates the positive direction.

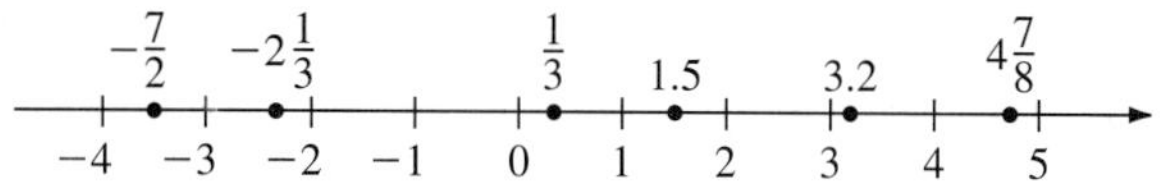

To form a rectangular coordinate system, we draw a second number line vertically. We construct it so that the 0 point on each number line is exactly at the same place. We refer to this location as the **origin.** The horizontal number line is often called the x-**axis.** The vertical number line is often called the y-**axis.** Arrows show the positive direction for each axis.

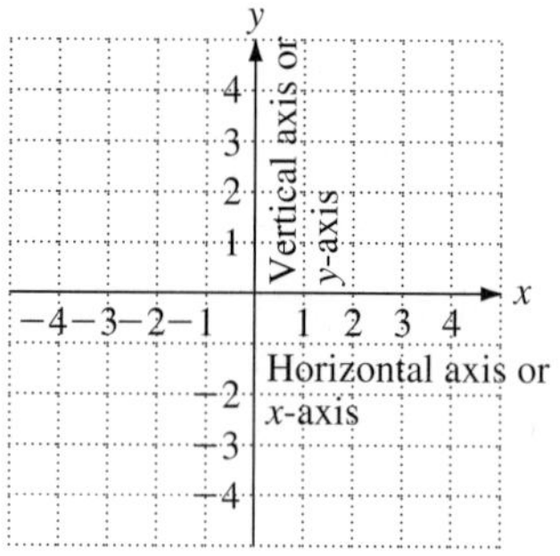

We can represent a point in this rectangular coordinate system by using an **ordered pair** of numbers. The first number of the pair represents the distance from the origin measure along the horizontal or x-axis. The second number of the pair represents the distance measured on the y-axis or on a line parallel to the y-axis. For example (5, 2) is an ordered pair that represents a point in the rectangular coordinate system.

The ordered pair of numbers that represent a point are often referred to as the **coordinates** of a point. The first value is called the x-coordinate. The second value is called the y-coordinate. If the x-coordinate is positive, we count the proper number of squares to the right in the positive direction. If the x-coordinate is negative, we count to the left. If the y-coordinate is positive, we count the proper number of squares upward in the positive direction. If the y-coordinate is negative, we count downward.

(5, 2)

x-coordinate y-coordinate

EXAMPLE 1 Plot the point (5, 2) on a rectangular coordinate system. Label this as point A.

Since the x-coordinate is 5, we first count 5 units to the right on the x-axis. Then, because the y-coordinate is 2, we count 2 units up from the point where we stopped on the x-axis. This locates the point corresponding to (5, 2). We label this point A.

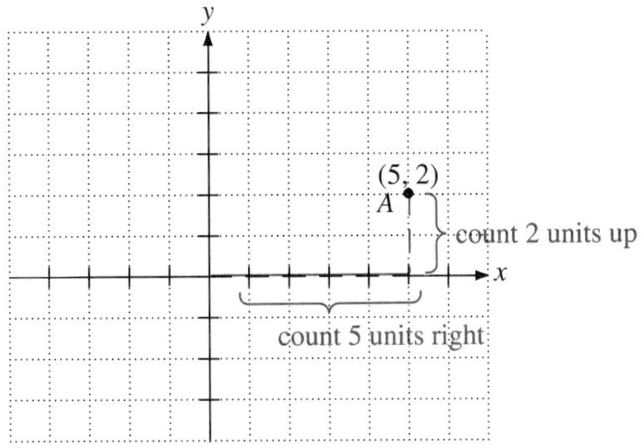

Practice Problem 1 Plot the point (3, 4) on the coordinate system above. Label this as point B. ■

It is important to remember that the first number in an ordered pair is the x-coordinate, and the second number is the y-coordinate. The points represented by (5, 2) and (2, 5) are totally different, as we shall see in the next example.

EXAMPLE 2 Plot the point (2, 5) on a rectangular coordinate system.

We count 2 units to the right for the x value of the point. Then we count 5 units upward for the y value of the point. We draw a dot and label it C at the location of the ordered pair (2, 5).

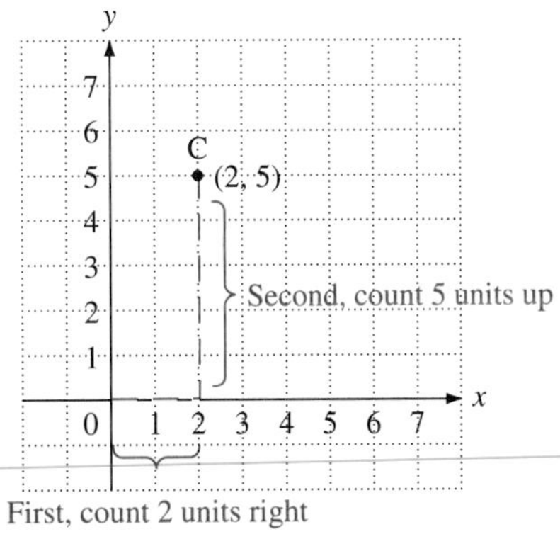

We observe that point C (2, 5) is in a very different location from point A (5, 2) in Example 1.

Practice Problem 2 Plot the point (1, 4) and the point (4, 1) on the coordinate system in the margin. Label the points D and E, respectively. ■

To Think About Describe how the location of points D and E differ.

PRACTICE PROBLEM 2

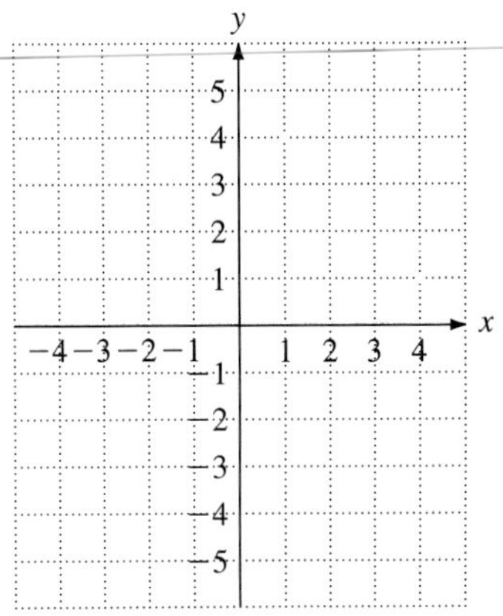

EXAMPLE 3 Plot each of the following points on a rectangular coordinate system. Label each point *F*, *G*, and *H*, respectively.

(a) $(-5, 3)$ **(b)** $(2, -6)$ **(c)** $(-4, -5)$

(a) Notice that the *x*-coordinate, -5, is negative. On the coordinate grid the *x*-coordinate is negative to the left of the origin. Thus, we will begin by counting 5 squares to the left starting at the origin. Since the *y*-coordinate, 3, is positive, we will count 3 units up from the point where you stopped on the *x*-axis.

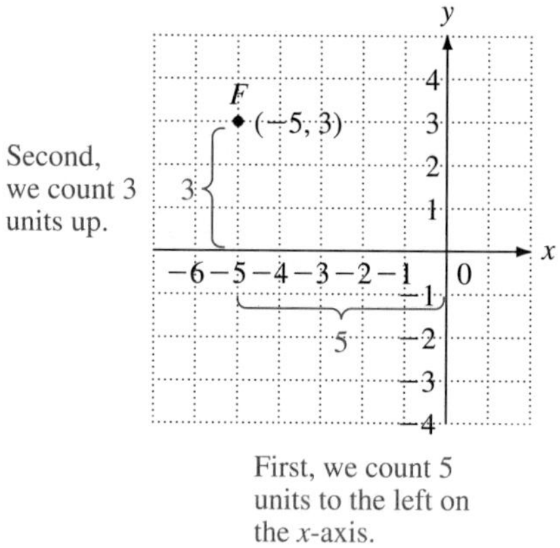

(b) The *x*-coordinate is positive. Begin by counting squares to the right of the origin. Then count down because the *y*-coordinate is negative.

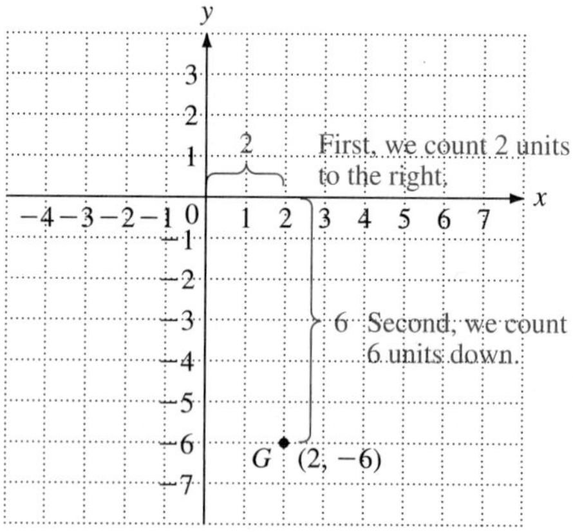

(c) The *x*-coordinate is negative. Begin by counting squares to the left of the origin. Then count down because the *y*-coordinate is negative.

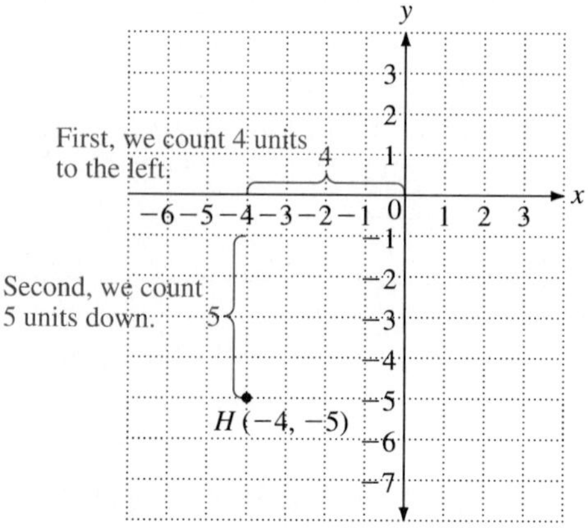

PRACTICE PROBLEM 3

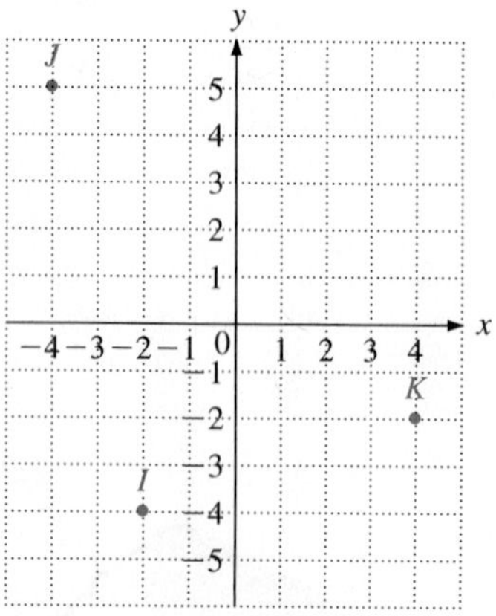

Practice Problem 3 Use the rectangular coordinate system in the margin to plot the following points. Label the points *I*, *J*, and *K*, respectively.

(a) $(-2, -4)$ **(b)** $(-4, 5)$ **(c)** $(4, -2)$ ■

EXAMPLE 4 Plot the following points.

F: (0, 5) G: $(3, \frac{3}{2})$ H: (−6, 4) I: (−3, −4)

J: (−4, 0) K: (2, −3) L: (6.5, −7.2)

These points are plotted in the figure.

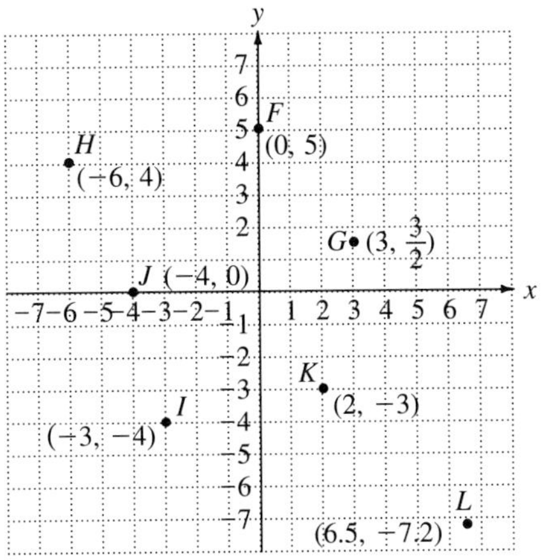

Note: When you are plotting decimal values like (6.5, −7.2), put the location of the plotted point halfway between 6 and 7 for the 6.5 and at your best approximation in the y-direction for the −7.2.

PRACTICE PROBLEM 4

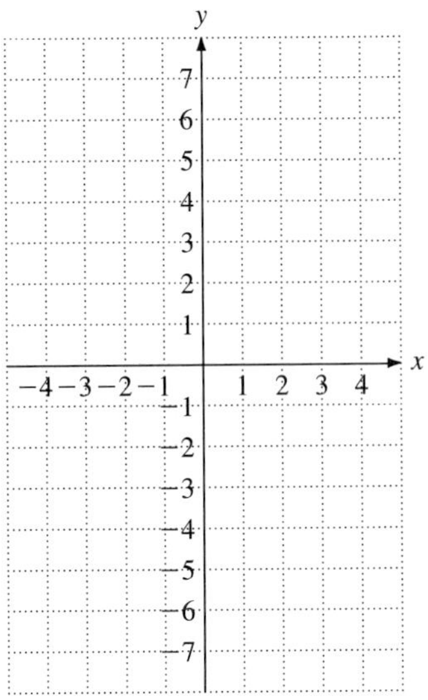

Practice Problem 4 Plot the following points. Label each point with both the letter and the ordered pair. Use the coordinate system provided in the margin.

A: (3, 7) B: (0, −6) C: (3, −4.5) D: $(-\frac{7}{2}, 2)$ ■

2 *Determining the Coordinates of a Plotted Point*

Sometimes, we need to find the coordinates of a point that has been plotted. First, we count the units we need on the x-axis to get as close as possible to that point. Next we count the units up or down we need to go from the x-axis to finally reach that point.

EXAMPLE 5 What ordered pair of numbers represents the point A in the figure on the right?

We see that the point A is represented by the ordered pair (5, 4). We first counted 5 units to the right of the origin on the x-axis. Thus we obtain 5 as the first number of the ordered pair. Then we counted 4 units upward on a line parallel to the y-axis. Thus we obtain 4 as the second number on the ordered pair.

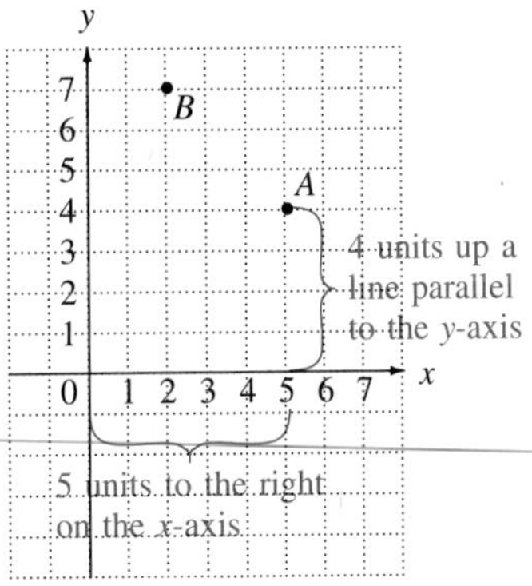

Practice Problem 5 Give the ordered pair of numbers that represents point B in the graph on the right. ■

EXAMPLE 6 Write the coordinates of each point plotted below.

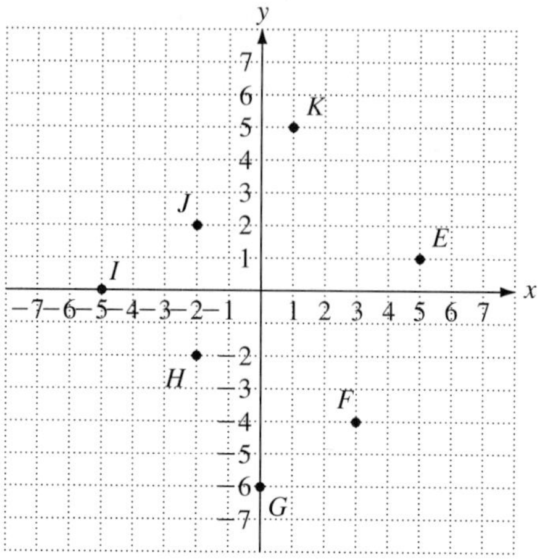

The coordinates of each point are:

$E = (5, 1)$ $\quad I = (-5, 0)$

$F = (3, -4)$ $\quad J = (-2, 2)$

$G = (0, -6)$ $\quad K = (1, 5)$

$H = (-2, -2)$

Be very careful that you put the x-coordinate first and the y-coordinate second. Be careful that each sign is correct.

PRACTICE PROBLEM 6

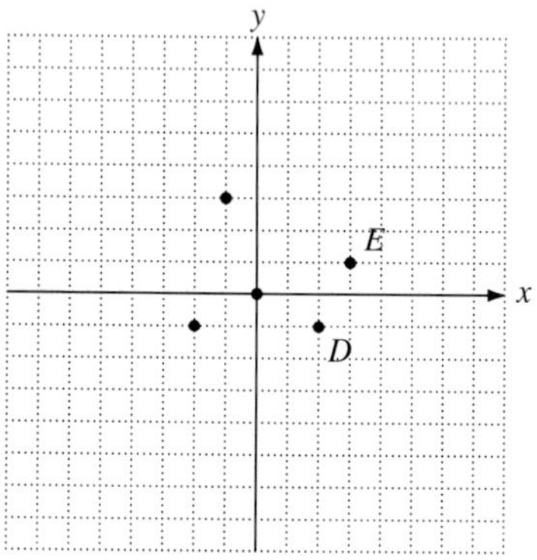

Practice Problem 6 Give the coordinate of each point in the graph in the margin. ■

3 *Finding Ordered Pairs for a Given Linear Equation*

Equations such as $3x + 2y = 5$ and $6x + y = 3$ are called linear equations in two variables. A **linear equation in two variables** is an equation that can be written in the form $Ax + By = C$ where A, B, and C are real numbers but A and B cannot *both* be zero. Replacement values for x and y that make *true mathematical statements* of the equation are called *truth values* and an ordered pair of these truth values is called a **solution.**

Consider the equation $3x + 2y = 5$. The ordered pair $(1, 1)$ is a solution to the equation. When we replace x by 1 and y by 1 we obtain a true statement.

$$3(1) + 2(1) = 5 \text{ or } 3 + 2 = 5$$

There are an infinite number of solutions for any given linear equation in two variables.

We obtain ordered pairs from examining such a linear equation. If one value of an ordered pair is known, the other can be quickly obtained. To do so, we replace one variable in the linear equation by the known value. Then using the methods learned in Chapter 2 we solve the resulting equation for the other variable.

EXAMPLE 7 Find the missing coordinate to complete the following ordered pairs for the equation $2x + 3y = 15$.

(a) $(0, ?)$ **(b)** $(?, 1)$

(a) In the ordered pair $(0, ?)$ we know that $x = 0$. Replace x by 0 in the equation.

$$2x + 3y = 15$$
$$2(0) + 3y = 15$$
$$0 + 3y = 15$$
$$y = 5$$

Thus we have the ordered pair $(0, 5)$.

(b) In the ordered pair (, 1), we *do not know* the value of x. However, we do know that $y = 1$. So we start by replacing the variable y by 1. We will end up with an equation with one variable x. We can then solve for x.

$$2x + 3y = 15$$
$$2x + 3(1) = 15$$
$$2x + 3 = 15$$
$$2x = 12$$
$$x = 6$$

Thus we have the ordered pair (6, 1).

Practice Problem 7 Find the missing coordinate to complete the following ordered pairs for the equation $3x - 4y = 12$.

(a) (0,) **(b)** (, 3) **(c)** (, −6) ■

The linear equations that we work with are not always written in the form $Ax + By = C$ but are sometimes solved for y as in $y = -6x + 3$. Consider the equation $y = -6x + 3$. The ordered pair $(2, -9)$ is a solution to the equation. When we replace x by 2 and y by -9 we obtain a true mathematical statement:

$$(-9) = -6(2) + 3 \text{ or } -9 = -12 + 3.$$

EXAMPLE 8 Find the missing coordinate to complete the following ordered pairs for the equation $y = -3x + 4$

(a) (2, ?) **(b)** (−3, ?)

(a) For the ordered pair (2, ?) we know that x is 2, so we replace x by 2 in the equation and solve for y as in Chapter 2.

$y = -3x + 4$	Check your answers.
$y = -3(2) + 4$	$y = -3x + 4$
$y = -6 + 4$	$(-2) = -3(2) + 4$
$y = -2$	$-2 = -6 + 4$ ✓

Thus, the solution is the complete ordered pair $(2, -2)$.

(b) For the ordered pair (−3, ?) we replace x by -3 in the equation.

$y = -3x + 4$	Check. $y = -3x + 4$
$y = -3(-3) + 4$	$13 = -3(-3) + 4$
$y = 9 + 4$	$13 = 9 + 4$ ✓
$y = 13$	

Thus, the solution is the ordered pair $(-3, 13)$.

Practice Problem 8 Find the missing coordinate to complete the following ordered pairs for the equation $y = -2x - 5$.

(a) (3, ?) **(b)** (−4, ?) ■

7.1 Exercises

1. Plot the following points.
   A: $(-2, -7)$ B: $(3, -4)$
   C: $(0, 8)$ D: $(4, -8)$

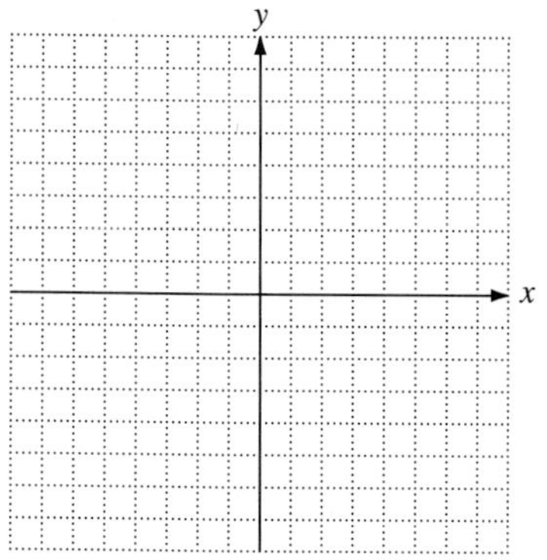

2. Plot the following points.
   E: $(-6, 2)$ F: $(3, 5)$
   G: $(-6, -1)$ H: $(0, -5)$

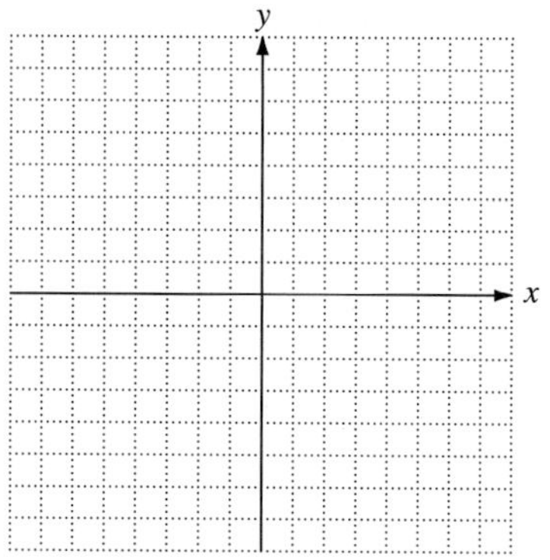

3. Plot the following points.
   J: $(5, 0)$ K: $(-5, 0)$
   L: $\left(-3\frac{1}{2}, 7\right)$ M: $(-2, -4.5)$

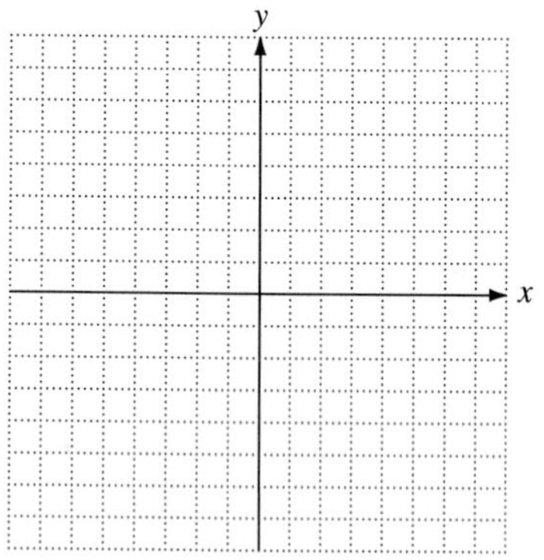

4. Plot the following points.
   N: $(0, 4)$ O: $(-4, 0)$
   P: $\left(4\frac{1}{2}, -3\right)$ Q: $(-2, 3.5)$

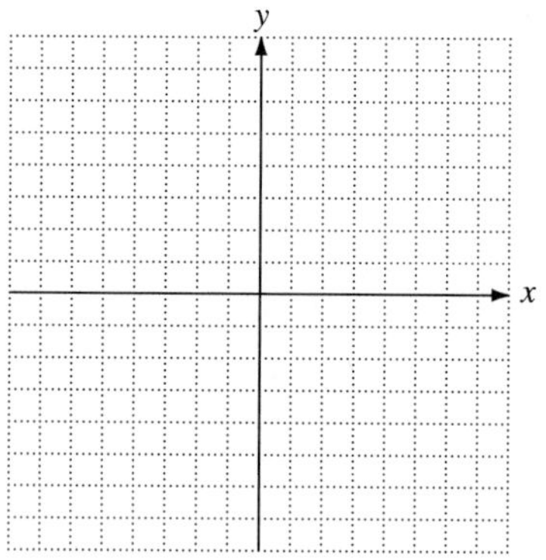

Consider the points plotted in the graph on the right.

5. Give the coordinates for the points R, S, X, and Y.

6. Give the coordinates for points T, V, W, and Z.

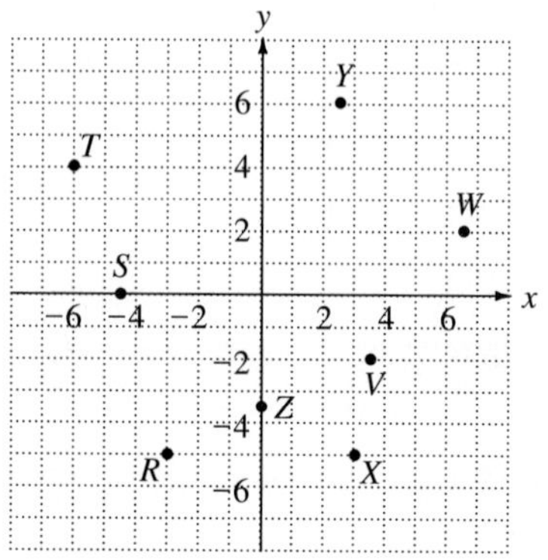

Verbal and Writing Skills

7. What is the x-coordinate of the origin?

8. What is the y-coordinate of the origin?

9. Explain why (5, 1) is referred to as an ordered pair of numbers.

10. Ten points are plotted in the figure shown. List all the ordered pairs needed to represent these points.

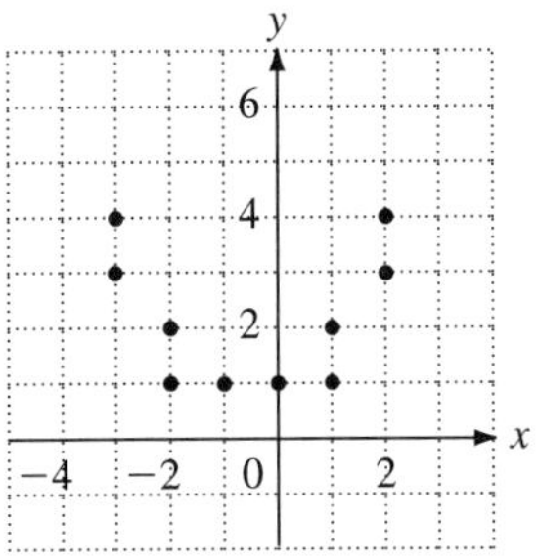

11. In the figure below, 10 points are plotted. List all the ordered pairs needed to represent these points.

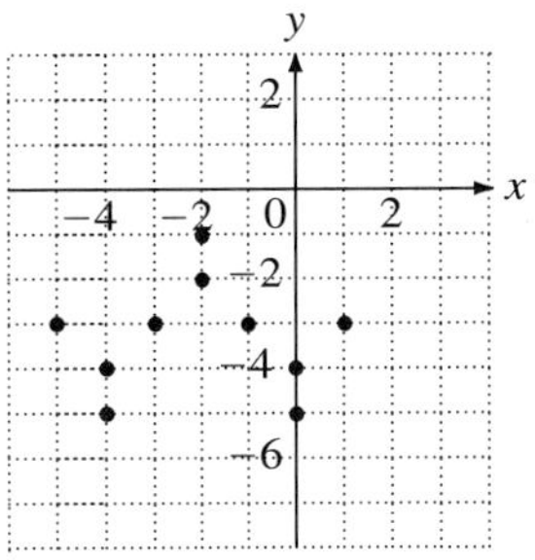

Complete the ordered pairs so that each is a solution to the given linear equation.

12. $y = 2x + 5$
(a) (0,) **(b)** (3,)

13. $y = 3x + 8$
(a) (0,) **(b)** (4,)

14. $y = -4x + 2$
(a) (−5,) **(b)** (4,)

15. $y = -2x + 3$
(a) (−6,) **(b)** (3,)

16. $6x - 2y = 24$
(a) (1,) **(b)** (, −3)

17. $2x - 8y = 16$
(a) (8,) **(b)** (, −5)

18. $3x + 4y = -18$
(a) (2,) **(b)** (, −3)

19. $5x + 2y = -20$
(a) (2,) **(b)** (, −5)

20. $-10x + 11y = 30$
(a) $\left(-\frac{4}{5}, \quad\right)$ **(b)** (, 3)

Cumulative Review Problems

21. Simplify. $\dfrac{8x}{x + y} + \dfrac{8y}{x + y}$

22. Simplify. $\dfrac{x^2 - y^2}{x^2 + 4xy + 3y^2} \cdot \dfrac{x^2 - 9y^2}{x^2 - 6xy + 9y^2}$

23. Find the area of a circular flower garden whose radius is 20 yards. (Use $\pi = 3.14$.)

24. The price of a new two-door sports coupe from Sushi Motors is \$26,640, up 11% over last year's model. What was the price of the older model?

7.2 Graphing Linear Equations

After studying this section, you will be able to:

1. *Graph a straight line by finding three ordered pairs that are solutions to the linear equation.*
2. *Graph a straight line by finding its* x- *and* y-*intercepts.*
3. *Graph horizontal and vertical lines.*

1 *Graphing a Linear Equation by Plotting Three Ordered Pairs*

We have seen that the graph of an ordered pair is a point. An ordered pair can also be a solution to a linear equation in two variables. Since there are an infinite number of solutions, there are an infinite number of ordered pairs and an infinite number of points. If we can plot these points, we will be graphing the equation. What will this graph look like? Let's look at the equation $y = -3x + 4$. To look for a solution to the equation, we can choose any value for x. For convenience we will choose $x = 0$. That is, the first coordinate of the ordered pair will be 0. To complete the ordered pair (0,), we substitute 0 for x in the equation:

$$\begin{aligned} y &= -3x + 4 \\ &= -3(0) + 4 \\ &= 0 + 4 \\ &= 4 \end{aligned}$$

Thus the ordered pair is (0, 4).

To find another ordered pair that is a solution to the equation $y = -3x + 4$, let $x = 1$.

$$\begin{aligned} y &= -3x + 4 \\ &= -3(1) + 4 \\ &= -3 + 4 \\ &= 1 \end{aligned}$$

Thus the ordered pair (1, 1) is another solution to the equation.

The graph of these two ordered pairs or solutions is shown at the right. From geometry we know that two points determine a line. Thus we can say that the line that contains the two points is the graph of the equation $y = -3x + 4$. In fact, *the graph of any linear equation in two variables is a straight line.*

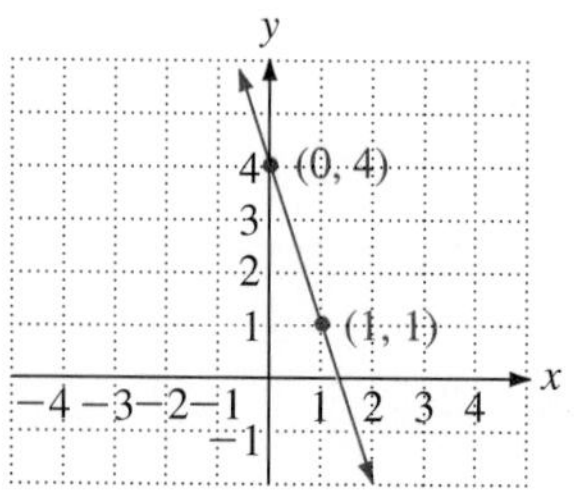

While you only need two points to determine a line, we recommend that you use three points to graph an equation. Two points to determine the line and a third point to verify. For ease in plotting the points it is better that the ordered pairs contain integers.

To Graph a Linear Equation

1. Look for three ordered pairs that are solutions to the equation.
2. Plot the points.
3. Draw a line through the points.

EXAMPLE 1 Find three ordered pairs that satisfy $2x + y = 4$. Then graph the resulting straight line.

Since we can choose any value for x, choose those numbers that are convenient. To organize the results, we will make a table of values. We will let $x = 0$, $x = 1$, and $x = 3$, respectively. We write these numbers under x in our table of values. For each of these x values, we find the corresponding y value in the equation $2x + y = 4$.

TABLE OF VALUES

x	y
0	4
1	2
3	−2

$$\begin{array}{ccc} 2x + y = 4 & 2x + y = 4 & 2x + y = 4 \\ 2(0) + y = 4 & 2(1) + y = 4 & 2(3) + y = 4 \\ 0 + y = 4 & 2 + y = 4 & 6 + y = 4 \\ y = 4 & y = 2 & y = -2 \end{array}$$

We record these results by placing each y value in the table next to its corresponding x value. Keep in mind that these values represent ordered pairs each of which is a solution to the equation. Since there are an infinite number of ordered pairs, we will prefer those with integers whenever possible. If we plot these ordered pairs and connect the three points, we get a straight line that is the graph of the equation $2x + y = 4$. The graph of the equation is shown in the figure on the right.

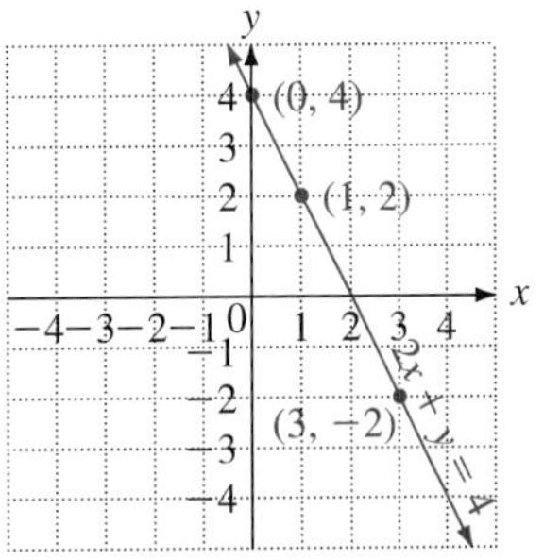

Practice Problem 1 Graph $x + y = 10$ on the coordinate system below. ■

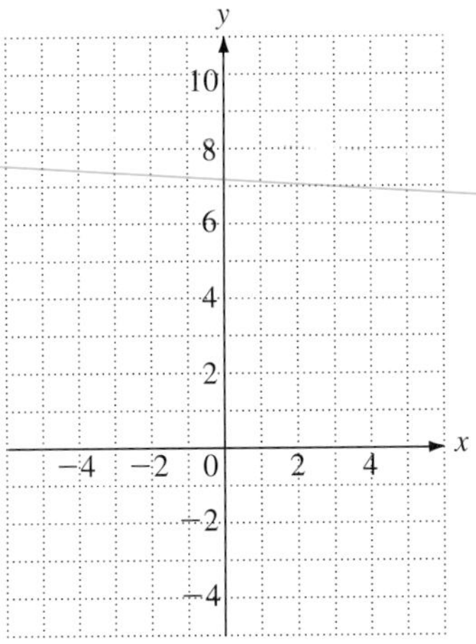

EXAMPLE 2 Graph $5x - 4y + 2 = 2$.

First, we simplify the equation by adding -2 to each side.

$$5x - 4y + 2 = 2$$
$$5x + 4y + 2 - 2 = 2 - 2$$
$$5x - 4y = 0$$

Since we are free to choose any value of x, $x = 0$ is a natural choice. Calculate the value of y when $x = 0$.

$$5x - 4y = 0$$
$$5(0) - 4y = 0$$
$$-4y = 0$$
$$y = 0$$ *Remember: Any number times 0 is 0. Since $-4y = 0$, y must equal 0.*

Now let's see what happens when $x = 1$.

$$5x - 4y = 0$$
$$5(1) - 4y = 0$$
$$5 - 4y = 0$$
$$-4y = -5$$
$$y = \frac{-5}{-4} \text{ or } \frac{5}{4}$$ *This is not an easy number to graph.*

A better choice for a replacement of x is a number that is divisible by 4. Let's see why. Let $x = 4$ and let $x = -4$.

$5x - 4y = 0$	$5x - 4y = 0$
$5(4) - 4y = 0$	$5(-4) - 4y = 0$
$20 - 4y = 0$	$-20 - 4y = 0$
$-4y = -20$	$-4y = 20$
$y = \frac{-20}{-4}$ or 5	$y = \frac{20}{-4}$ or -5

PRACTICE PROBLEM 2

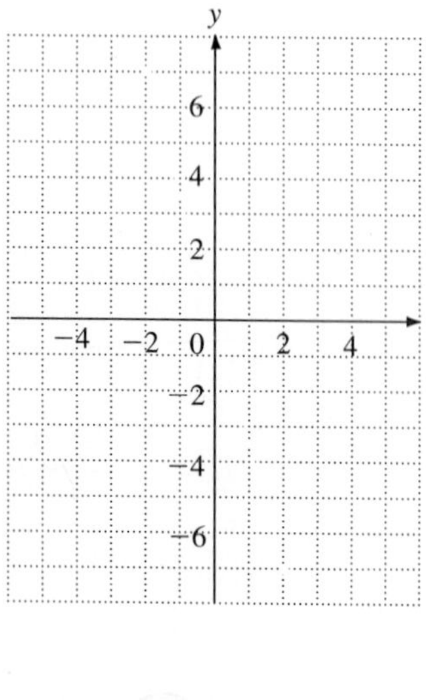

Now we can put these numbers in our table of values and graph.

TABLE OF VALUES	
x	y
0	0
4	5
−4	−5

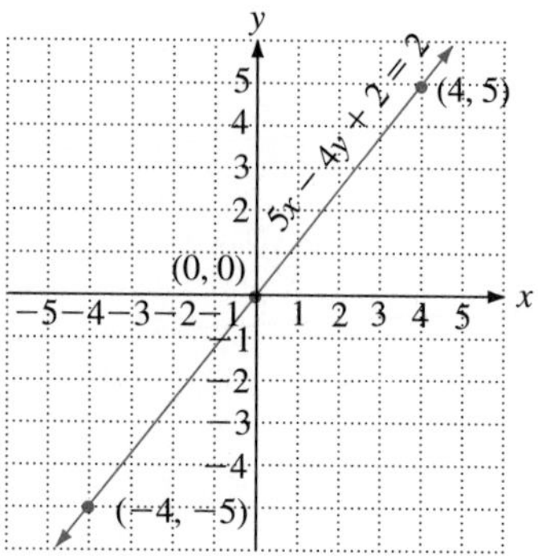

Practice Problem 2 Graph $7x + 3 = -2y + 3$ on the coordinate system in the margin. ■

To Think About In Example 2, we picked values of x and found the corresponding value for y. An alternative approach is to *first* solve the equation for the variable y. Thus

$$5x - 4y = 0$$

$$-4y = -5x \qquad \textit{Add } -5x \textit{ to each side.}$$

$$\frac{-4y}{-4} = \frac{-5x}{-4} \qquad \textit{Divide each side by } -4.$$

$$y = \frac{5}{4}x$$

Now let $x = -4$, $x = 0$, and $x = 4$, and find the corresponding values of y. Explain why you would choose multiples of 4 as replacements of x in this equation. Graph the equation and compare it to the graph in Example 2.

In the previous two examples we began by picking values for x. We could just as easily have chosen values for y.

EXAMPLE 3 Graph $3x - 4y = 12$.

We first find three ordered pairs by arbitrarily picking three values for y and in each case solving for x. We will choose $y = 0$, $y = 3$, and $y = -3$. Can you see why?

$3x - 4y = 12$	$3x - 4y = 12$	$3x - 4y = 12$
$3x - 4(0) = 12$	$3x - 4(3) = 12$	$3x - 4(-3) = 12$
$3x - 0 = 12$	$3x - 12 = 12$	$3x + 12 = 12$
$3x = 12$	$3x = 24$	$3x = 0$
$x = 4$	$x = 8$	$x = 0$

Our table of values is:

x	y
4	0
8	3
0	−3

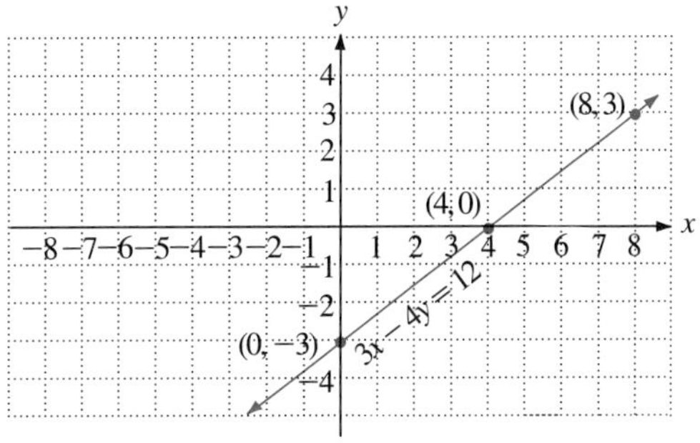

Practice Problem 3 Graph $3x - 2y = 6$ by arbitrarily picking three values for y and in each case solving for x. Use the coordinate system below. ■

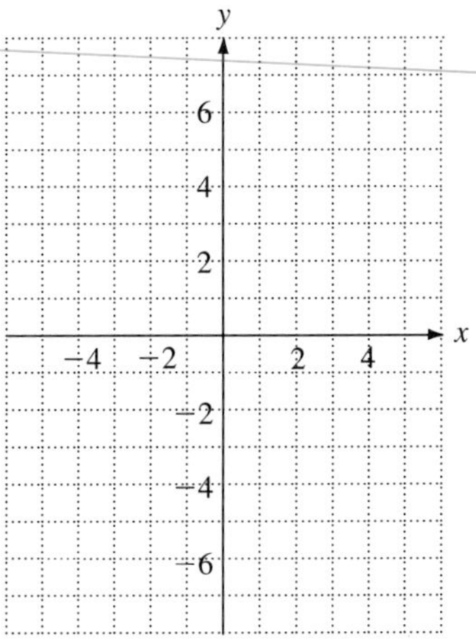

2 *Graphing a Straight Line by Plotting Its Intercepts*

What values should we pick for x and y? Which points should we use for plotting? For many straight lines it is easiest to pick the two *intercepts*. A few lines have only one intercept. We will discuss these separately.

> The number a is the *x-intercept* of a line if the line crosses the x-axis at $(a, 0)$. The number b is the *y-intercept* of a line if the line crosses the y-axis at the point $(0, b)$.

Intercept Method of Graphing

To graph an equation by using intercepts, we:

1. Find the x-intercept by letting $y = 0$ and solve for x.
2. Find the y-intercept by letting $x = 0$ and solve for y.
3. Find one additional ordered pair so that we have three points with which to plot the line.

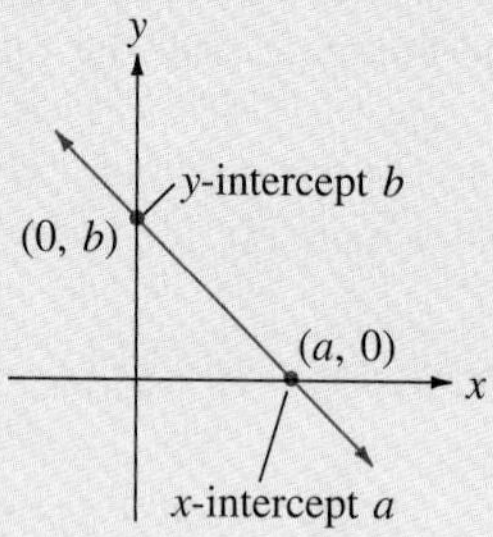

EXAMPLE 4 Graph $5y - 3x = 15$ by the intercept method.

Let $y = 0$. $\quad 5(0) - 3x = 15$ *Replace y by* 0.

$-3x = 15$ *Divide both sides by* -3.

$x = -5$ *The x-intercept is* -5.

The ordered pair is $(-5, 0)$ is the x-intercept point.

$5y - 3x = 15$

Let $x = 0$. $\quad 5y - 3(0) = 15$ *Replace x by* 0.

$5y = 15$ *Divide both sides by* 5.

$y = 3$ *The y-intercept is* 3.

The ordered pair is $(0, 3)$ is the y-intercept point.

We find another pair to have a third point.

Let $y = 6$. $\quad 5(6) - 3x = 15$ *Replace y by* 6.

$30 - 3x = 15$ *Simplify.*

$-3x = -15$ *Subtract* 30 *from both sides.*

$$x = \frac{-15}{-3} \text{ or } 5$$

The ordered pair is $(5, 6)$.

Our table of values is

x	y
0	3
-5	0
5	6

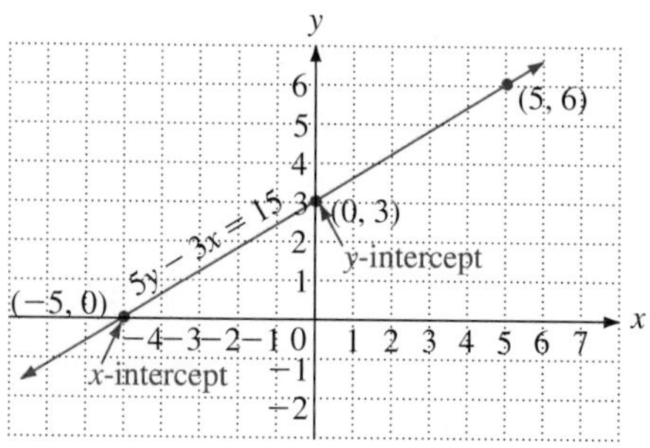

PRACTICE PROBLEM 4

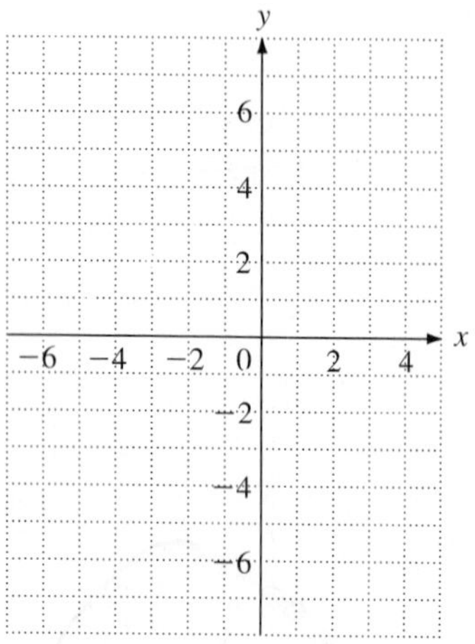

Practice Problem 4 Use the intercept method to graph $2y - x = 6$. Use the coordinate system in the margin. ■

SIDELIGHT Can you draw all straight lines by the intercept method? Not really. Some straight lines may go through the origin and have only one intercept. If a line goes through the origin, it will have an equation of the form $Ax + By = 0$, where $A \neq 0$ or $B \neq 0$ or both. In such cases you should plot two additional points besides the origin. Be sure to simplify each equation before attempting to graph it.

3 *Graphing Horizontal and Vertical Lines*

You will notice that the x-axis is a horizontal line. It is the line $y = 0$, since for any x, y is 0. Try a few points. (1, 0), (3, 0), (−2, 0) all lie on the x-axis. Any horizontal line will be parallel to the x-axis. Lines such as $y = 5$ and $y = -2$ are horizontal lines. What does $y = 5$ mean? It means that, for any x, y is 5. Likewise $y = -2$ means that, for any x, $y = -2$.

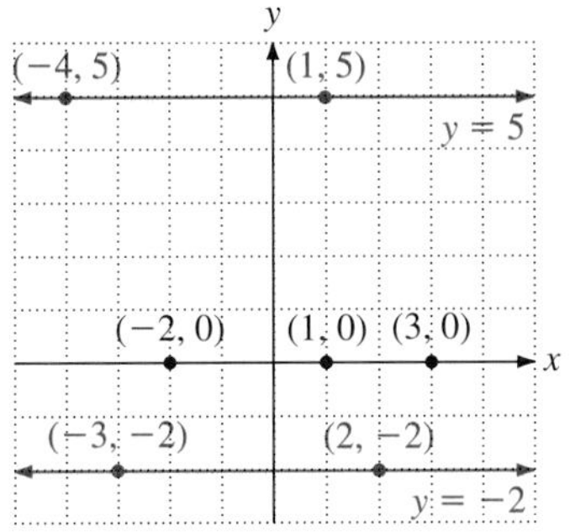

How can we recognize the equation of a line that is horizontal, that is, parallel to the x-axis.

> If an equation is a straight line that is parallel to the x-axis (that is, a horizontal line), it will be of the form $y = b$, where b is some real number.

EXAMPLE 5 Graph $y = -3$.

You could write the equation as $0x + y = -3$. Then it is clear that for any value of x that you substitute you will always obtain $y = -3$. Thus, as shown in the figure, (4, −3), (0, −3), and (−3, −3) are all ordered pairs that satisfy the equation $y = -3$. Since the y-coordinate of every point on this line is −3, it is easy to see that the horizontal line will be 3 units below the x-axis.

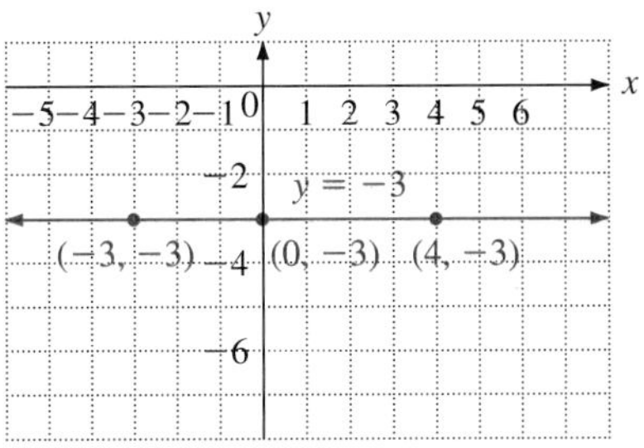

Practice Problem 5 Graph $2y - 3 = 0$ on the coordinate system in the margin. ■

PRACTICE PROBLEM 5

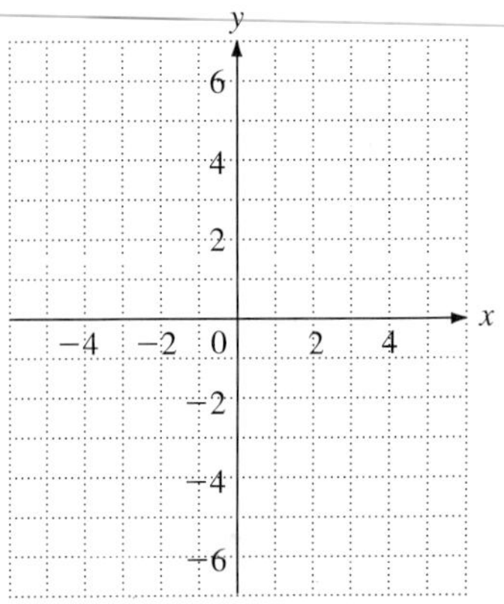

Notice that the y-axis is a vertical line. This is the line $x = 0$, since, for any y, x is 0. Try a few points. $(0, 2)$, $(0, -3)$, $(0, \frac{1}{2})$ all lie on the y-axis. Any vertical line will be parallel to the y-axis. Lines such as $x = 2$ and $x = -3$ are vertical lines. Think of what $x = 2$ means. It means that, for any value of y, x is 2. $x = 2$ is a vertical line two units to the right of the y-axis.

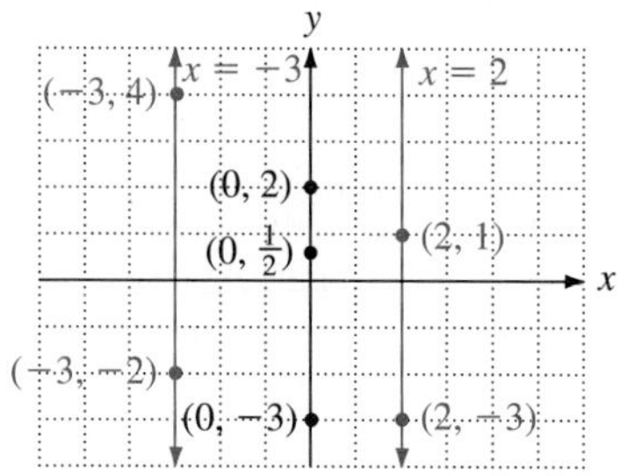

How can we recognize the equation of a line that is vertical, that is, parallel to the y-axis?

> If the graph of an equation is a straight line that is parallel to the y-axis, that is, a vertical line, its equation will be of the form $x = a$, where a is some real number.

EXAMPLE 6 Graph $x = 5$.

This can be done immediately by drawing a vertical line 5 units to the right of the origin. The x-coordinate of every point on this line is 5.

The equation $x - 5 = 0$ can be rewritten as $x = 5$ and graphed as shown.

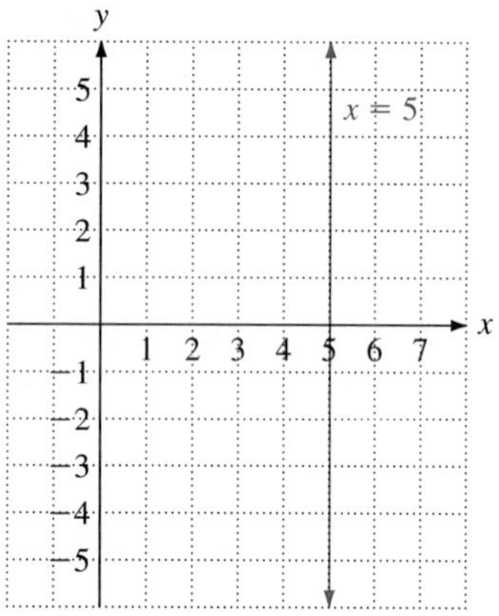

Practice Problem 6 Graph $x + 3 = 0$ on the coordinate system below. ■

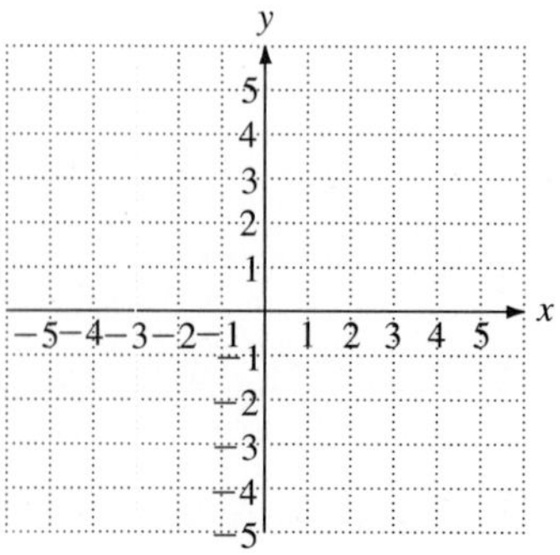

7.2 Exercises

Verbal and Writing Skills

1. Is the point $(-2, 5)$ a solution to the equation $2x + 5y = 0$? Why or why not?

2. The graph of a linear equation in two variables is a ______________

______________.

3. The x-intercept of a line is the point where the line crosses the ______________.

4. The graph of the equation $y = b$ is a ______________ line.

Complete the ordered pairs so that each is a solution of the given linear equation. Then plot each solution and graph the equation by connecting the points by a straight line.

5. $y = -2x - 3$
$(-2, \quad)$
$(-1, \quad)$
$(0, \quad)$

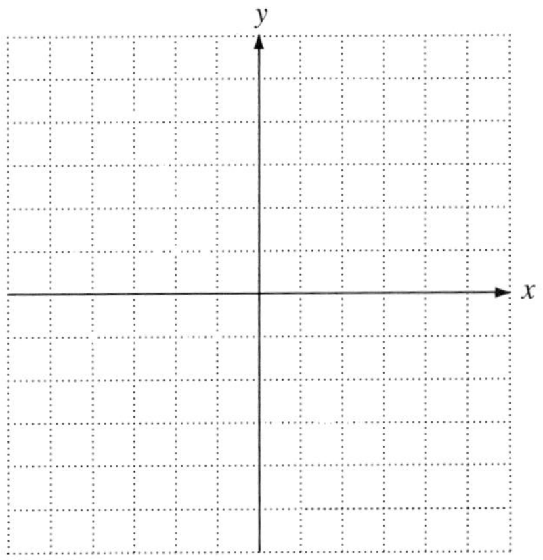

6. $y = -3x - 4$
$(-2, \quad)$
$(-1, \quad)$
$(0, \quad)$

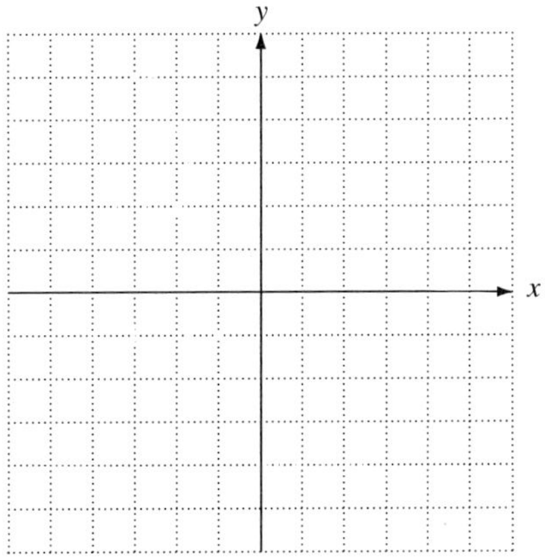

7. $y = x - 4$
$(0, \quad)$
$(2, \quad)$
$(4, \quad)$

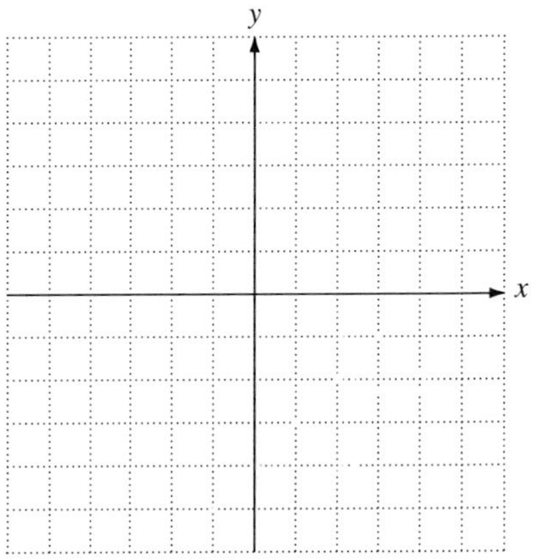

Graph each equation by plotting three points and connecting them.

8. $y = 2x - 3$

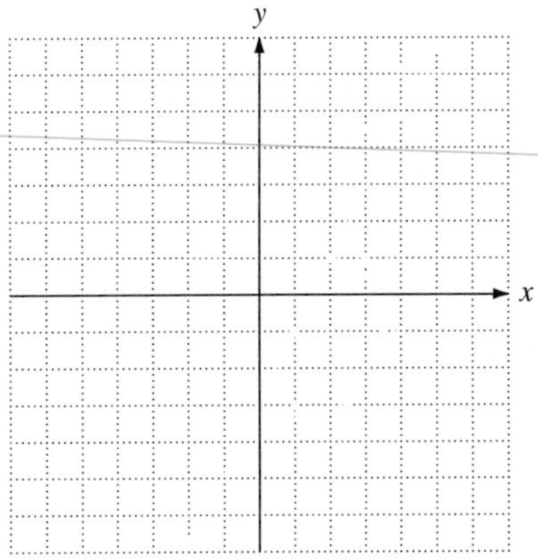

9. $y = 3x - 1$

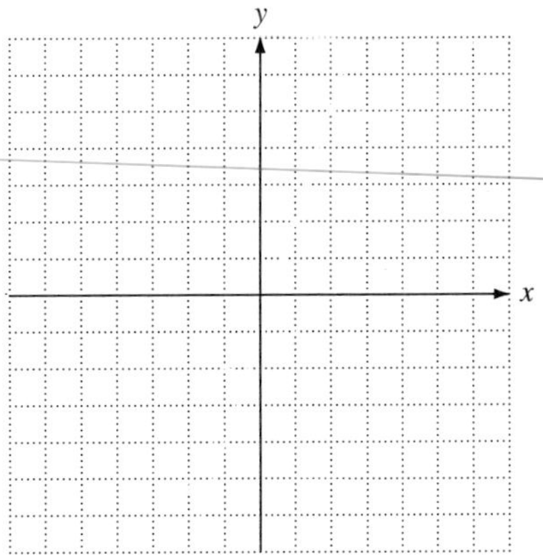

10. $y = 3x + 2$

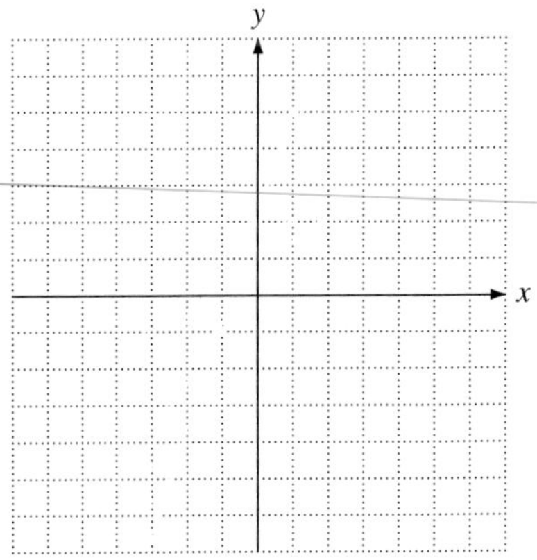

11. $y = 2x + 1$

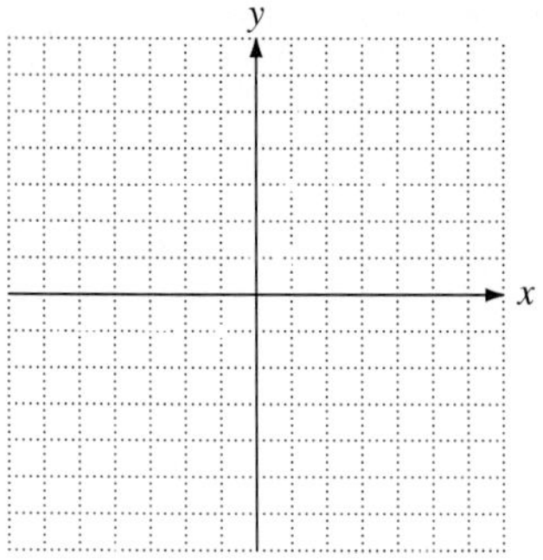

12. $3x - 5y = 0$

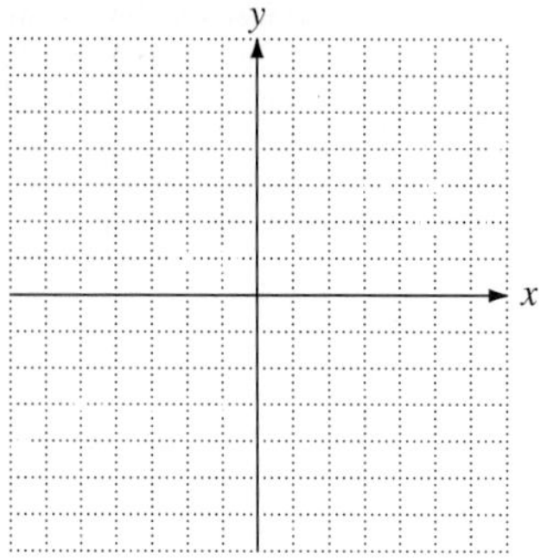

13. $2y - 4x = 0$

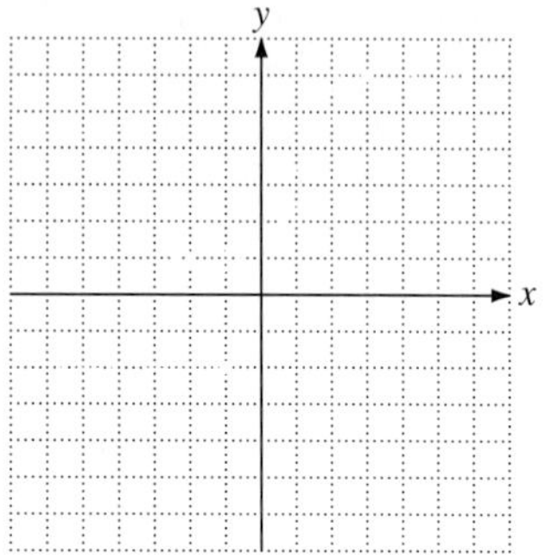

Graph each equation by plotting the intercepts and one other point.

14. $y = 6 - 2x$

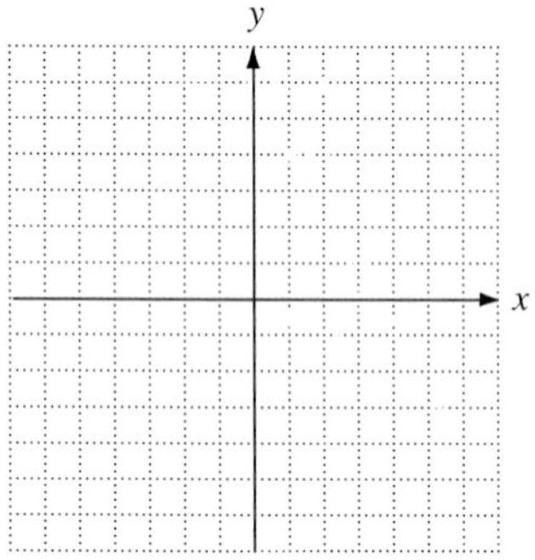

15. $y = 4 - 2x$

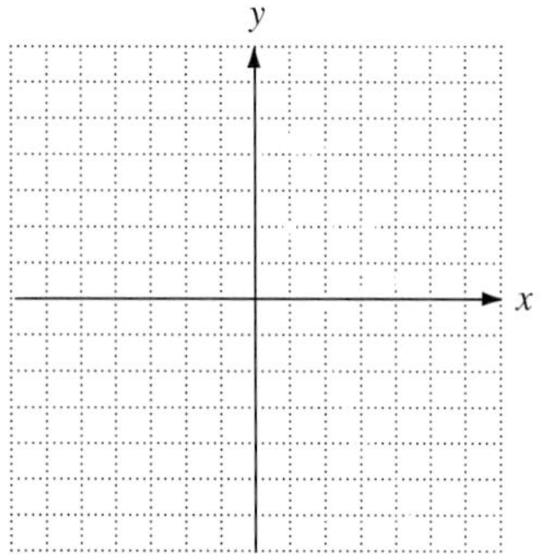

16. $4x + 3y = 12$

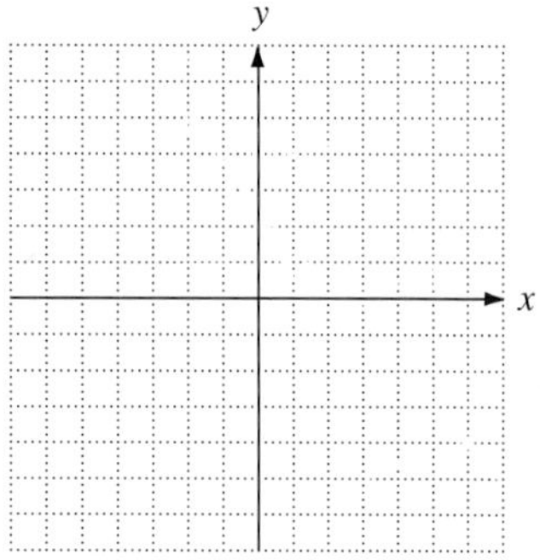

17. $3x + 2y = 6$

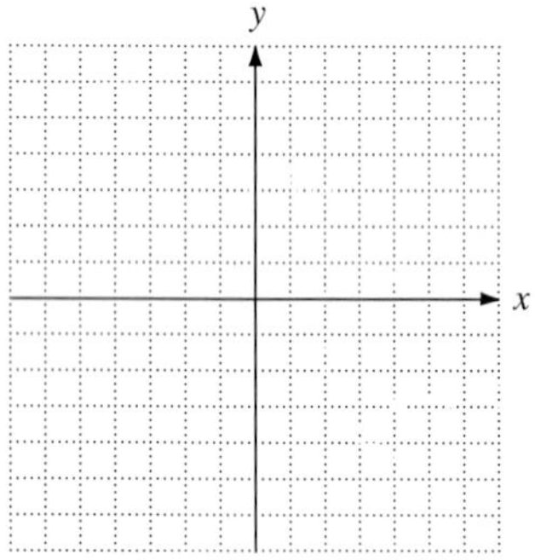

18. $x - 4 = 2y$

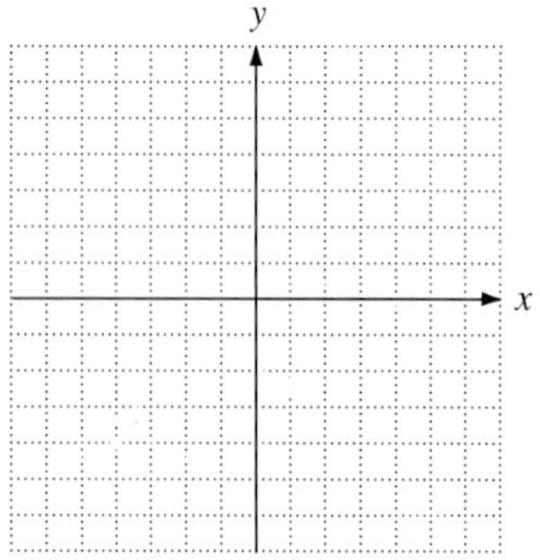

19. $x - 2 = 4y$

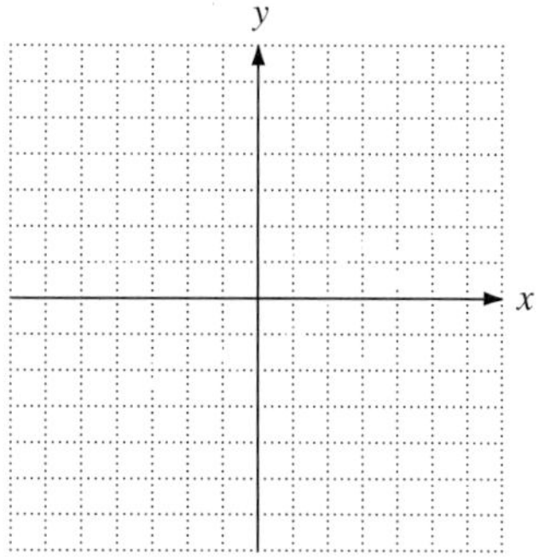

Graph each equation. Be sure to simplify your equation before graphing it.

20. $4x = 8$

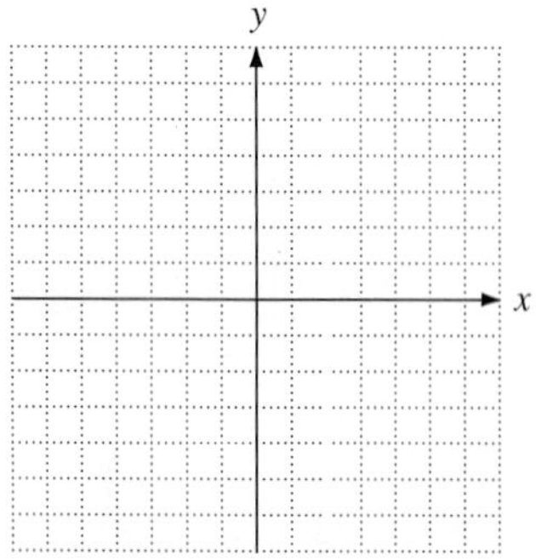

21. $y - 2 = 3y$

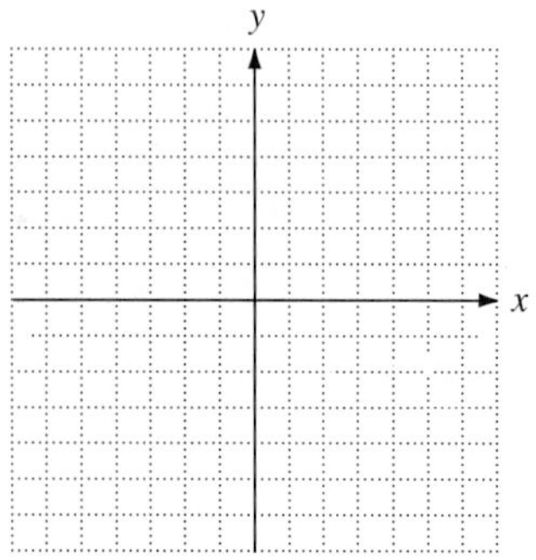

22. $2x + 5y - 2 = -12$

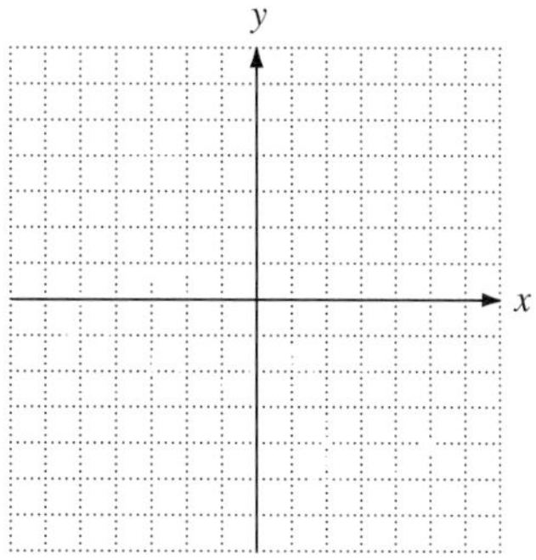

Cumulative Review Problems

23. Solve. $\frac{x}{x+4} + \frac{x-1}{x-4} = 2$

24. Solve and graph on a number line. $4 - 3x \leq 18$

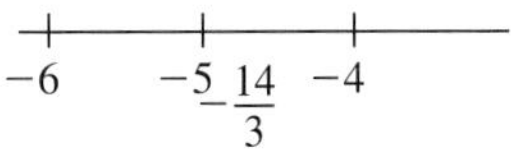

25. A rectangle's width is 1 meter more than half of its length. Find the dimensions if the perimeter measures 53 meters.

26. There are 29 students for every 2 faculty members at Skyline University. Last semester the school had 3074 students. If this ratio is true, how many faculty members are there?

7.3 Slope of a Line

After studying this section, you will be able to:

1. *Find the slope of a straight line given two points on the line.*
2. *Find the slope and y-intercept of a straight line, given the equation of the line.*
3. *Write the equation of a line, given the slope and the y-intercept.*
4. *Graph a line using the slope and y-intercept.*
5. *Find the slopes of lines that are parallel or perpendicular.*

1 *Find the Slope of a Straight Line, Given Two Points on the Line*

We often use the word slope to describe the incline (the steepness) of a hill. A carpenter or a builder will refer to the *pitch* or *slope* of a roof. The slope is the change in the vertical distance compared to the change in the horizontal distance as you go from one point to another point along the roof. If the change in the vertical distance is greater than the change in the horizontal distance, the slope will be steep. If the change in the horizontal distance is greater than the change in the vertical distance, the slope will be gentle.

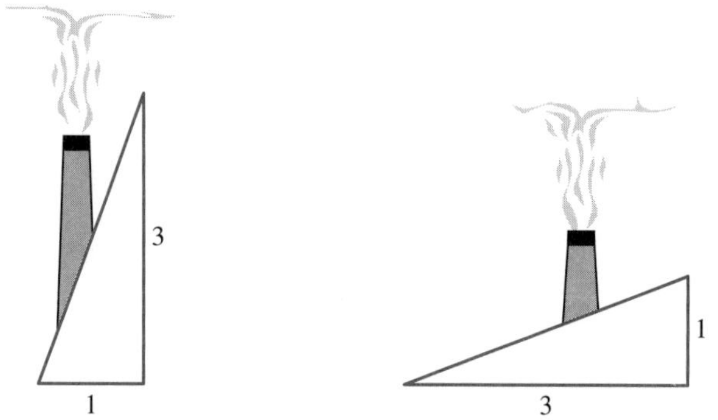

In a coordinate plane, the **slope** of a straight line is defined by the change in y divided by the change in x.

$$\text{Slope} = \frac{\text{change in } y}{\text{change in } x}$$

Consider the line drawn through points A and B in the figure. If we measure the change from point A to point B in the x-direction and the y-direction, we will have an idea of the steepness (or the slope) of the line. From point A to point B the change in y values is from 2 to 4, a *change of* 2. From point A to point B the change in x values is from 1 to 5, a *change of* 4. Thus

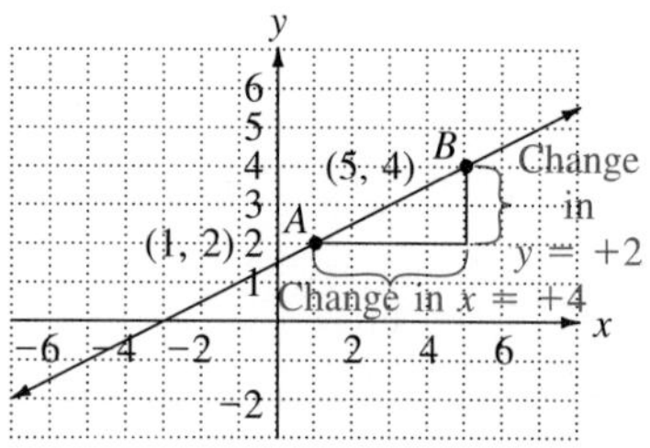

$$\text{slope} = \frac{\text{change in } y}{\text{change in } x} = \frac{2}{4} = \frac{1}{2}$$

Informally, we can describe this move as the rise over and the run: **slope** $= \dfrac{\text{rise}}{\text{run}}$.

We now state a more formal (and more frequently used) definition.

Definition of Slope of a Line

The *slope of any nonvertical straight line* that contains the points with coordinates (x_1, y_1) and (x_2, y_2) has a slope defined by the difference ratio

$$\text{slope} = m = \frac{y_2 - y_1}{x_2 - x_1} \qquad \text{where } x_2 \neq x_1$$

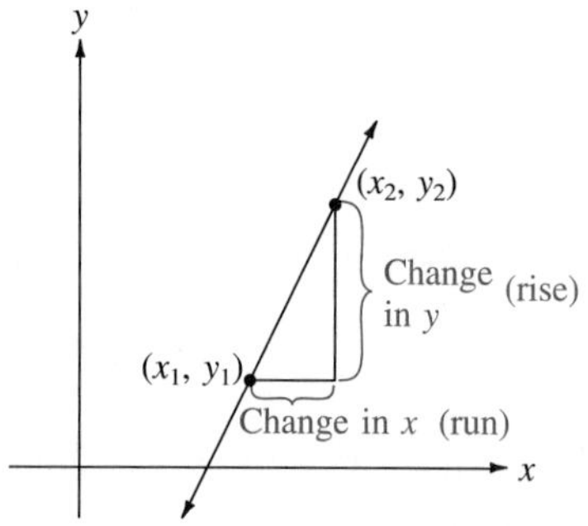

The use of subscripted terms such as x_1, x_2, and so on, is just a way of indicating that the first x value is x_1 and the second x value is x_2. Thus (x_1, y_1) are the coordinates of the first point and (x_2, y_2) are the coordinates of the second point. The letter m is commonly used for the slope.

EXAMPLE 1 Find the slope of the line that passes through (2, 0) and (4, 2).

Let (2, 0) be the first point (x_1, y_1)

(4, 2) be the second point (x_2, y_2)

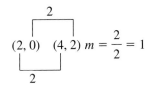

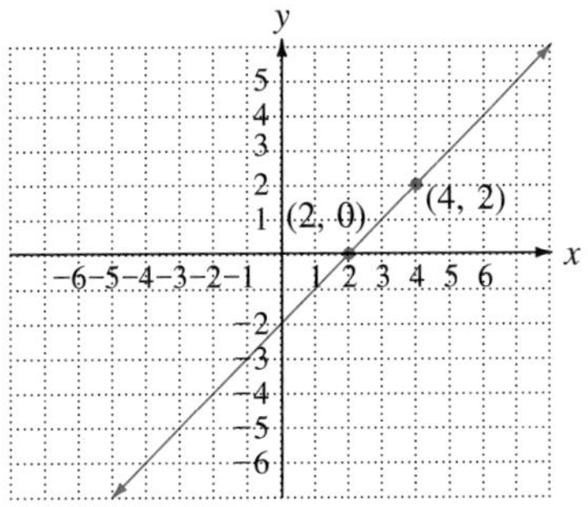

2

(2, 0) (4, 2) $m = \dfrac{2}{2} = 1$

2

By the formula

$$\text{slope} = m = \frac{y_2 - y_1}{x_2 - x_1} = \frac{2 - 0}{4 - 2} = \frac{2}{2} = 1$$

The sketch of the line is shown in the figure on the right.

Note that the slope of the line will be the same if we let (4, 2) be the first point (x_1, y_1) and (2, 0) be the second point (x_2, y_2).

−2

(4, 2) (2, 0)

−2

$$m = \frac{y_2 - y_1}{x_2 - x_1} = \frac{0 - 2}{2 - 4} = \frac{-2}{-2} = 1$$

Thus, given two points it does not matter which you call (x_1, y_1) and which you call (x_2, y_2).

WARNING Be careful, however, not to put the x's in one order and the y's in another order when finding the slope, given two points on a line.

Practice Problem 1 Find the slope of the line through (6, 1) and (−4, −1). ■

EXAMPLE 2 Find the slope of the line that passes through (2, 0) and (4, 4).

Let (2, 0) be the first point (x_1, y_1)
(4, 4) be the second point (x_2, y_2)

$$m = \frac{y_2 - y_1}{x_2 - x_1} = \frac{4 - 0}{4 - 2} = \frac{4}{2} = 2$$

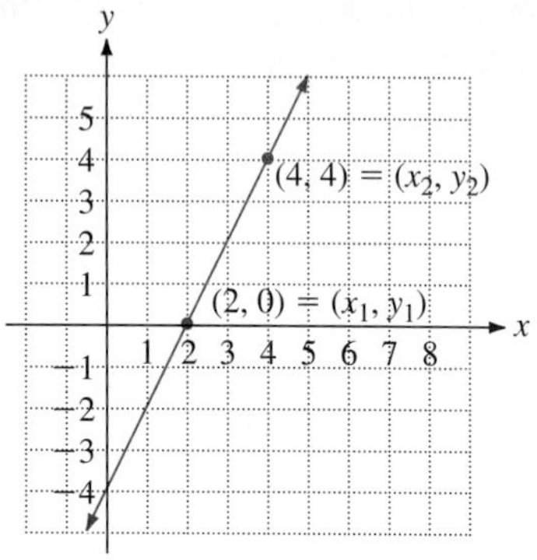

The slope of the line is 2. Notice that the slope (rate of change) of this line is twice that of the slope of the line in Example 1.

Practice Problem 2 Find the slope of the line through (−1, −4) and (1, 6). ■

To Think About Compare the slope of the line in Practice Problem 1 with the slope of the line in Practice Problem 2. Which line is steeper? Why? Draw the lines to verify.

It is important to see how positive and negative slopes affect the graphs of the line.

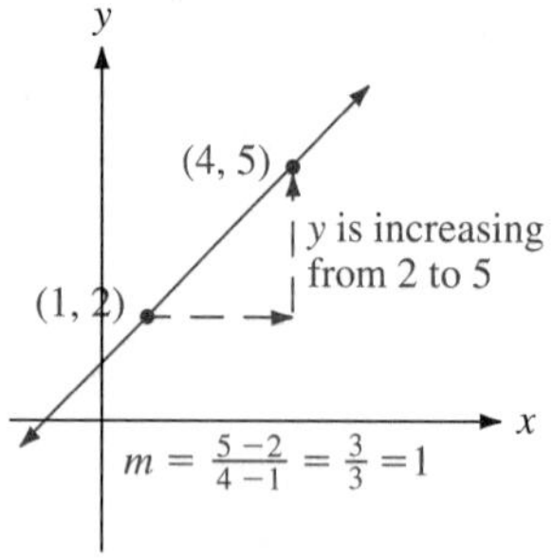

1. Line with positive slope

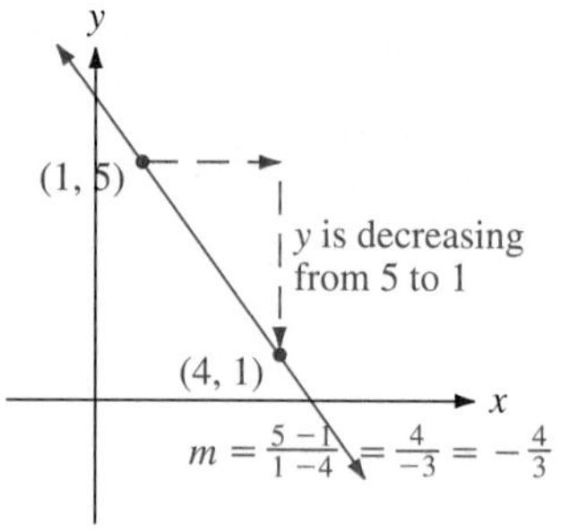

2. Line with negative slope

1. If the y values increase as you go from left to right, the slope of the line is positive.
2. If the y values decrease as you go from left to right, the slope of the line is negative.

EXAMPLE 3 Find the slope of the line through (−3, 2) and (2, −4).

Let (−3, 2) be (x_1, y_1)
(2, −4) be (x_2, y_2)

$$m = \frac{y_2 - y_1}{x_2 - x_1} = \frac{-4 - 2}{2 - (-3)} = \frac{-4 - 2}{2 + 3} = \frac{-6}{5} = -\frac{6}{5}$$

The slope of this line is negative. We would expect this since the y value decreased from 2 to −4. What does the graph of this line look like? Plot the points and draw the line to verify.

Practice Problem 3 Find the slope of the line through (2, 0) and (−1, 1). ■

To Think About Describe the line in Practice Problem 3 by looking at its slope. Then verify by drawing the graph.

EXAMPLE 4 Find the slope of the line that passes through the given points.

(a) (0, 2) and (5, 2) **(b)** (−4, 0) and (−4, −4)

(a) Take a moment to look at the y values. What do you notice? What does this tell you about the line? Now calculate the slope.

$$m = \frac{2 - 2}{5 - 0} = \frac{0}{5} = 0$$

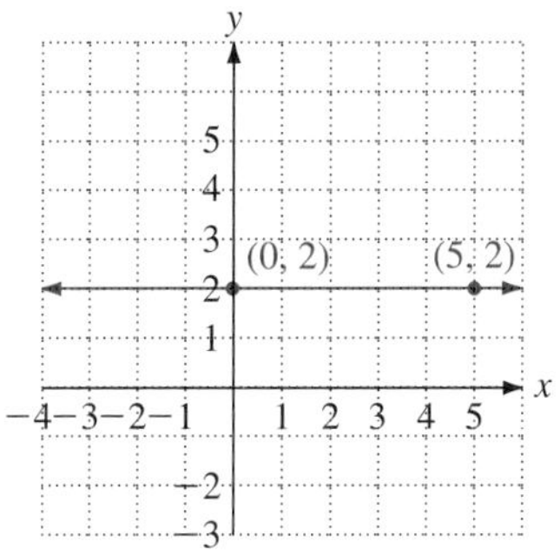

A horizontal line has a slope of 0.

(b) Take a moment to look at the x values. What do you notice? What does this tell you about the line? Now calculate the slope.

$$m = \frac{4 - 0}{-4 - (-4)} = \frac{4}{0}$$

Recall that division by 0 is undefined.

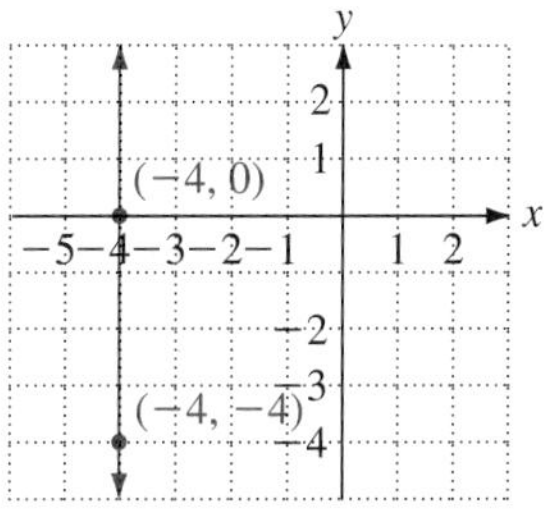

The slope of a vertical line is undefined.

Notice in our definition of slope that $x_2 \neq x_1$. Thus it is not appropriate to use the formula for slope for the points in (b). We do so to illustrate what would happen if $x_2 = x_1$. We get an impossible situation, $\frac{y_2 - y_1}{0}$. Now you can see why we include the restriction $x_2 \neq x_1$ in our definition.

Practice Problem 4 Find the slope of the line that passes through the given points.

(a) (−4, 0) and (−4, 4) **(b)** (−7, −11) and (3, −11) ■

Slope of a Straight Line

Positive slope
Line goes upward to the right

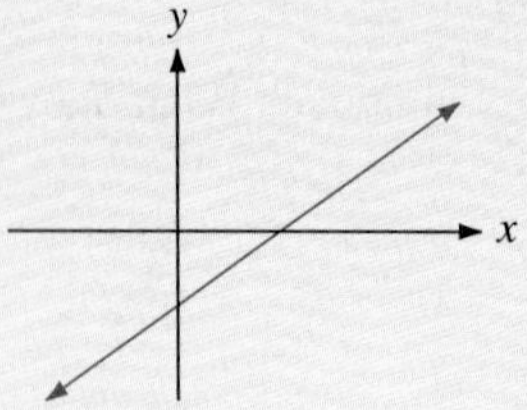

1. *Lines with positive slopes go upward as x increases* (that is, as you go from left to right).

Negative slope
Line goes downward to the right

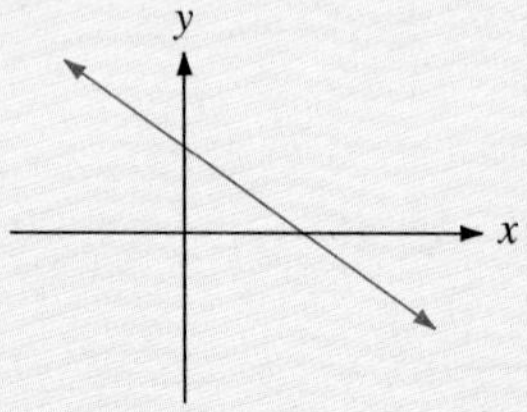

2. *Lines with negative slopes go downward as x increases* (that is, as you go from left to right).

Zero slope
Horizontal line

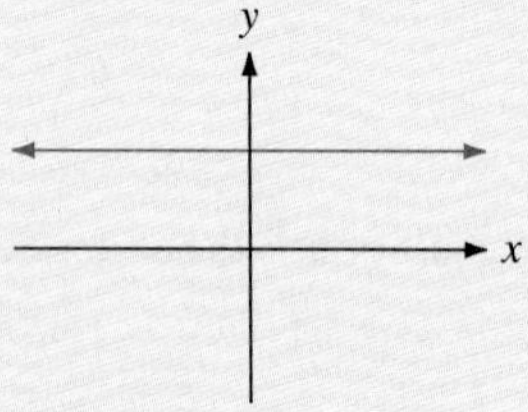

3. Horizontal lines have a slope of 0.

Undefined slope
Vertical line

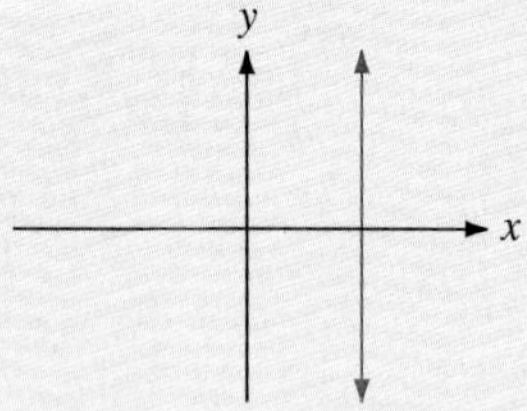

4. A vertical line is said to have undefined slope. The slope of a vertical line is not defined.

2 *Finding the Slope and y-Intercept of a Line, Given the Equation of a Line*

Recall that the equation of a a line is a linear equation in two variables. This equation can be written in several different ways. A very useful form of the equation of a straight line is the slope–intercept form. The form can be derived in the following way. Suppose that a straight line, with slope m, crosses the y-axis at a point $(0, b)$. Consider any other point on the line and label the point (x, y).

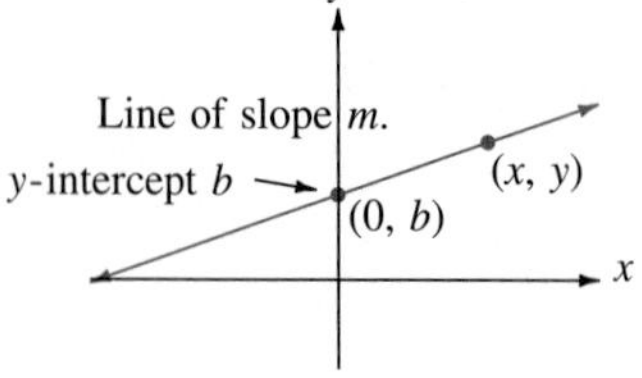

$\dfrac{y_2 - y_1}{x_2 - x_1} = m$	*Definition of slope.*
$\dfrac{y - b}{x - 0} = m$	Substitute $(0, b)$ *for* (x_1, y_1) *and* (x, y) *for* (x_2, y_2).
$\dfrac{y - b}{x} = m$	*Simplify.*
$y - b = mx$	*Multiply both sides by x.*
$y = mx + b$	*Add b to both sides.*

This form of a linear equation immediately reveals the slope of the line, m, and where the line intercepts (crosses) the y-axis, b.

Slope–Intercept Form of a Line

The slope–intercept form of the equation of a line that has slope m and the y-intercept b at point $(0, b)$ is given by

$$y = mx + b$$

EXAMPLE 5 What is the slope and the y-intercept of the line whose equation is $y = 4x - 5$?

The equation is in the form $y = mx + b$.

$$y = 4x + (-5)$$

$$\text{slope} = 4 \quad y\text{-intercept} = -5$$

The slope is 4 and the y-intercept is -5.

Practice Problem 5 What is the slope and the y-intercept of $y = -2x + 6$? ■

By using algebraic operations, we can write any linear equation in slope–intercept form and use this form to identify the slope and the y-intercept of the line.

EXAMPLE 6 What is the slope and the y-intercept of the line $5x + 3y = 2$?

We want to solve for y and get the equation in the form $y = mx + b$. We need to isolate the y variable.

$5x + 3y = 2$

$3y = -5x + 2$ *Subtract $5x$ from both sides.*

$y = \dfrac{-5x + 2}{3}$ *Divide both sides by 3.*

$y = -\dfrac{5}{3}x + \dfrac{2}{3}$ *Write the right-hand side as two fractions, using the property* $\dfrac{a + b}{c} = \dfrac{a}{c} + \dfrac{b}{c}$.

The *slope* is $-\dfrac{5}{3}$. The *y-intercept* is $\dfrac{2}{3}$.

Practice Problem 6 Find the slope and y-intercept of the line $4x - 2y = -5$. ■

3 Writing the Equation of a Line, Given the Slope and the y-Intercept

If we know the slope of a line m, and the y-intercept, b, we can write the equation of the line, $y = mx + b$.

EXAMPLE 7 Find the equation of a line with slope $\frac{2}{5}$ and y-intercept -3.

(a) Write the equation in slope–intercept form, $y = mx + b$.

(b) Write the equation in $Ax + By = C$ form.

(a) We are given that $m = \frac{2}{5}$ and $b = -3$. Since

$$\begin{aligned} y &= mx + b \\ &= \frac{2}{5}x + (-3) \\ &= \frac{2}{5}x - 3 \end{aligned}$$

(b) Recall, for the form $Ax + By = C$, that A, B, and C are integers. We first clear the equation of fractions. Then we move the x term to the left side.

$5y = 5\frac{2x}{5} - 5(3)$ *Multiply each term by 5.*

$5y = 2x - 15$ *Simplify.*

$-2x + 5y = -15$ *Subtract 2x from each side.*

$2x - 5y = 15$ *Multiply each term by −1. The form Ax + By = C dictates that A be a positive integer.*

Practice Problem 7 Find the equation of the line with slope $-\frac{3}{7}$ and y-intercept $\frac{2}{7}$.

(a) Write the equation in slope–intercept form.

(b) Write the equation in the form $Ax + By = C$. ■

4 Graphing a Line Using the Slope and the y-Intercept

If we know the slope of a line and the y-intercept, we can draw the graph of the line.

EXAMPLE 8 Graph a line with slope $m = \frac{2}{3}$ and y-intercept of -3. Use the coordinate system in the margin.

Recall that the y-intercept is the point where the line crosses the y-axis. The x-coordinate of this point is 0. Thus the coordinates of the y-intercept for this line are $(0, -3)$. We plot the point.

Recall that slope $= \frac{\text{rise}}{\text{run}}$. Since the slope for this line is $\frac{2}{3}$, we will go up (rise) 2 units and go over (run) to the right 3 units from the point $(0, -3)$. Look at the figure on the right. This is the point $(3, -1)$. Plot the point. Draw a line that connects the two points $(0, -3)$ and $(3, -1)$.

This is the graph of the line with slope $\frac{2}{3}$ and y-intercept of -3.

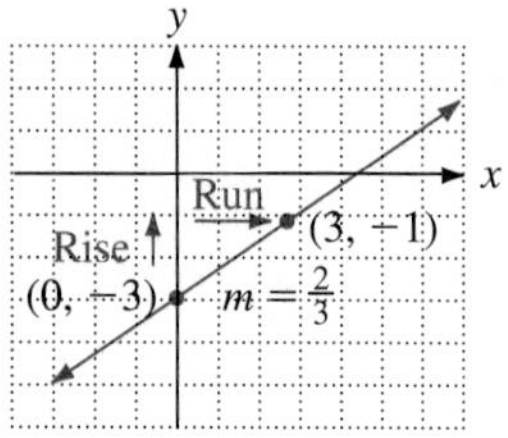

PRACTICE PROBLEM 8

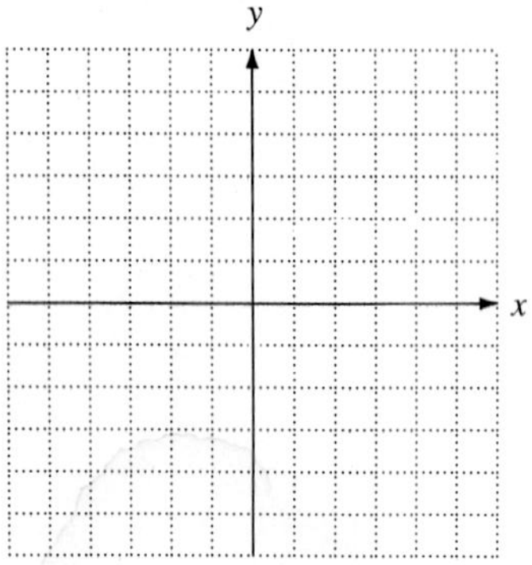

Practice Problem 8 Graph a line with slope $= \frac{3}{4}$ and a y-intercept of -1. Use the coordinate system in the margin. ■

EXAMPLE 9 Graph the equation $y = -\frac{1}{2}x + 4$. Use the coordinate system in the margin.

Begin with the y-intercept. Since the y-intercept is 4, plot the point (0, 4). Now look at the slope $-\frac{1}{2}$. This can be written as $-\frac{1}{2}$. Begin at (0, 4) and go *down* 1 unit and to the right 2 units. This is the point (2, 3). Plot the point. Draw the line that connects the points (0, 4) and (2, 3).

This is the graph of the equation $y = -\frac{1}{2}x + 4$.

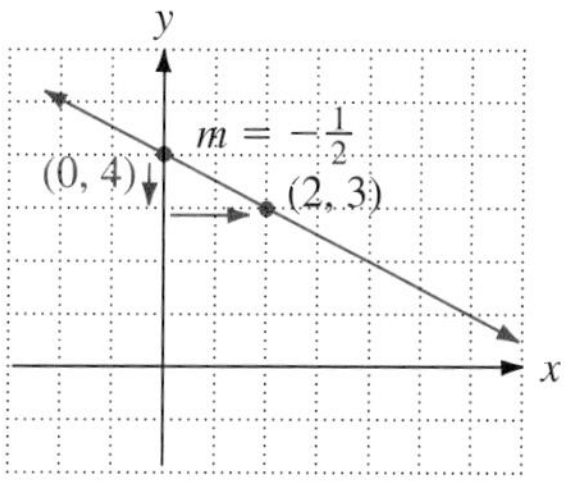

PRACTICE PROBLEM 9

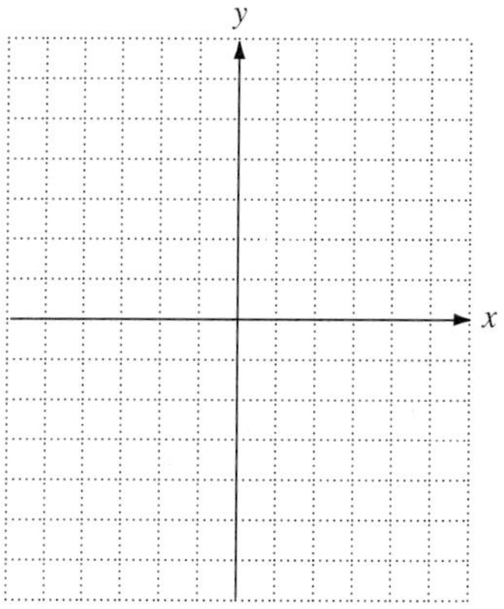

Practice Problem 9 Graph the equation $y = -\frac{2}{3}x + 5$ on the coordinate system in the margin. ■

To Think About Explain why you count down 1 unit and move to the right 2 units to represent the slope $-\frac{1}{2}$. Could you have done this in another way? Try it. Verify that this is the same line.

5 *Finding the Slopes of Lines That are Parallel or Perpendicular*

Parallel lines are two straight lines that never touch. Look at the parallel lines in the figure at the right. Notice that the slope of line a is -3 and the slope of line b is also -3. Why do you think the slopes must be equal? What would happen if the slope of line b was -1. Try it.

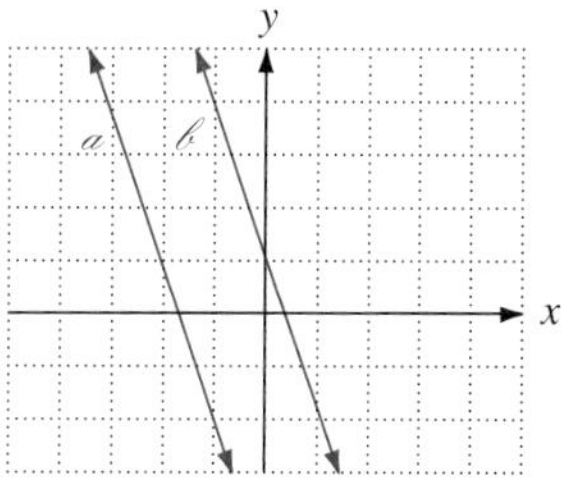

Parallel Lines

Parallel lines are two straight lines that never touch.

Parallel lines have the same slope.

$$m_1 = m_2$$

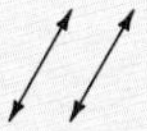

Parallel lines

Perpendicular lines are two lines that meet at a 90° angle. Look at the perpendicular lines in the figure on the right. The slope of line c is -3. The slope of line d is $\frac{1}{3}$. Notice that

$$(-3)\left(\frac{1}{3}\right) = \left(-\frac{3}{1}\right)\left(\frac{1}{3}\right) = -1$$

You may wish to draw several pairs of perpendicular lines to determine if this will always happen.

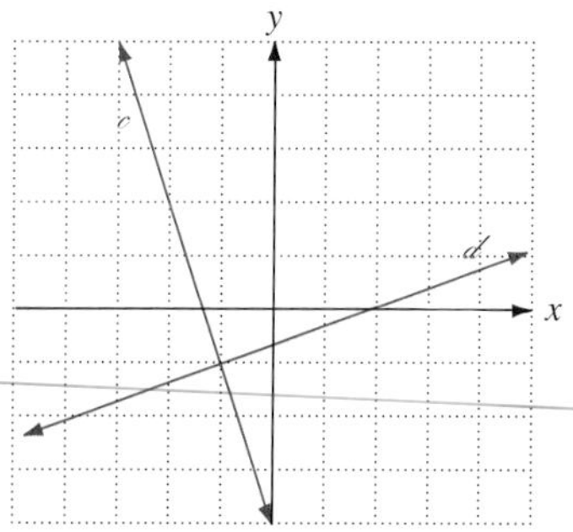

Perpendicular Lines

Perpendicular lines are two lines that meet at a 90° angle.

Perpendicular lines have slopes whose product is -1. If m_1 and m_2 are slopes of perpendicular lines, then

$$m_1 m_2 = -1 \quad \text{or} \quad m_1 = -\frac{1}{m_2}$$

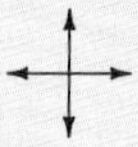

Perpendicular lines

EXAMPLE 10 Line $\mathfrak{e}$ has a slope of $-\frac{2}{3}$.

(a) If line $\mathfrak{f}$ is parallel to line $\mathfrak{e}$, what is its slope?

(b) If line $\mathfrak{g}$ is perpendicular to line $\mathfrak{e}$, what is its slope?

(a) Parallel lines have the same slope.
Line $\mathfrak{f}$ will have a slope of $-\frac{2}{3}$.

(b) Perpendicular lines have slopes whose product is -1.

$$m_1m_2 = -1$$

$$-\frac{2}{3}m_2 = -1 \qquad \textit{Substitute } -\tfrac{2}{3} \textit{ for } m_1.$$

$$\left(-\frac{3}{2}\right)\left(-\frac{2}{3}\right)m_2 = -1\left(-\frac{3}{2}\right) \qquad \textit{Multiply both sides by } -\tfrac{3}{2}.$$

$$m_2 = \frac{3}{2} \qquad \textit{Check.}$$

Practice Problem 10 Line $\mathfrak{h}$ has a slope of $\frac{1}{4}$.

(a) If line $\mathfrak{j}$ is parallel to line $\mathfrak{h}$, what is its slope?

(b) If line $\mathfrak{k}$ is perperpendicular to line $\mathfrak{h}$, what is its slope? ■

EXAMPLE 11 The equation of line ℓ is $y = -2x + 3$.

(a) What is the slope of a line that parallel to line ℓ?

(b) What is the slope of a line that is perpendicular to line ℓ?

(a) Looking at the equation, we can see that the slope of line ℓ is -2.

$$y = -2x + 3$$

$$m_1 = -2$$

The slope of a line that is parallel to line ℓ is -2.

(b) Perpendicular lines have slopes whose product is -1.

$$m_1m_2 = -1$$

$$(-2)m_2 = -1 \qquad \textit{Substitute } -2 \textit{ for } m_1.$$

$$m_2 = \frac{1}{2} \qquad \textit{Because } (-2)\left(\frac{1}{2}\right) = -1.$$

The slope of a line that is perpendicular to line ℓ is $\frac{1}{2}$.

Practice Problem 11 The equation of line $\mathfrak{n}$ is $y = \frac{1}{4}x - 1$.

(a) What is the slope of a line that is parallel to line $\mathfrak{n}$?

(b) What is the slope of a line that is perpendicular to line $\mathfrak{n}$? ■

7.3 Exercises

Find the slope of a straight line that passes through the given pair of points.

1. (7, 3) and (6, 9)

2. (5, 2) and (8, 5)

3. (−2, 1) and (3, 4)

4. (5, 6) and (−3, 1)

5. (−7, −6) and (−3, 2)

6. (2, −4) and (−6, −3)

7. (3, −4) and (−6, −4)

8. (7, −2) and (−12, −2)

9. (−3,0) and (0, −4)

10. (0, 5) and (5, 3)

11. $(\frac{1}{3}, -2)$ and $(\frac{2}{3}, -5)$

12. $(4, -3)$ and $(\frac{5}{2}, 5)$

Verbal and Writing Skills

13. Can you find the slope of the line passing through (5, −12) and (5, −6)? Why? Why not?

14. Can you find the slope of the line passing through (6, −2) and (−8, −2)? Why? Why not?

Find the slope and the y-intercept of each line.

15. $y = 8x + 9$

16. $y = 2x + 10$

17. $y = 3x - 7$

18. $y = 5x - 8$

19. $y = \frac{5}{6}x - \frac{2}{9}$

20. $y = -\frac{3}{4}x + \frac{5}{6}$

21. $y = -6x$

22. $y = -2$

23. $4x + y = -\frac{1}{3}$

24. $3x + y = \frac{2}{5}$

25. $3x - y = -2$

26. $4x - y = 5$

27. $5x + 2y = 3$

28. $7x + 3y = 4$

29. $4x - 3y = 7$

30. $6x - 5y = 8$

Write the equation of the line (a) in the slope–intercept form and (b) in the form $Ax + By = C$.

31. $m = \frac{3}{4}, b = 2$

32. $m = \frac{4}{5}, b = 3$

33. $m = 6, b = -3$

34. $m = 5, b = -6$

35. $m = \dfrac{-5}{2}, b = 6$

36. $m = \dfrac{-2}{3}, b = 7$

37. $m = -2, b = \frac{1}{2}$

38. $m = -3, b = \frac{1}{4}$

Graph the line.

39. $m = \frac{1}{2}, b = -3$

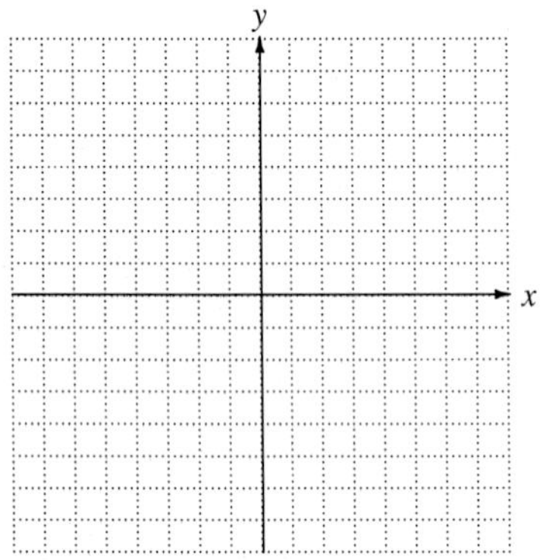

40. $m = \frac{2}{3}, b = -4$

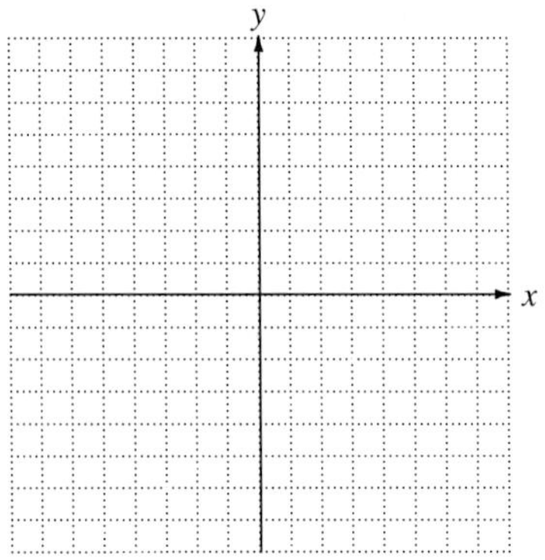

41. $m = -\frac{2}{5}, b = 3$

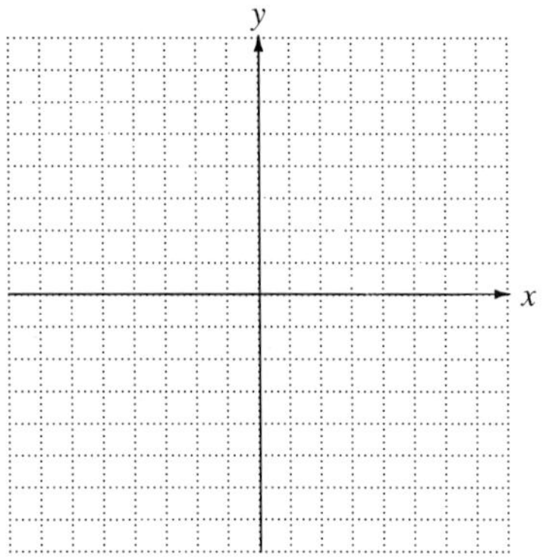

42. $m = -\frac{1}{4}, b = 2$

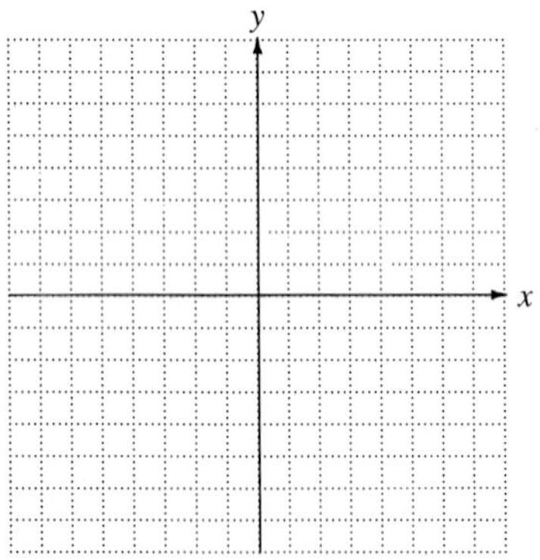

43. $y = \frac{3}{4}x - 5$

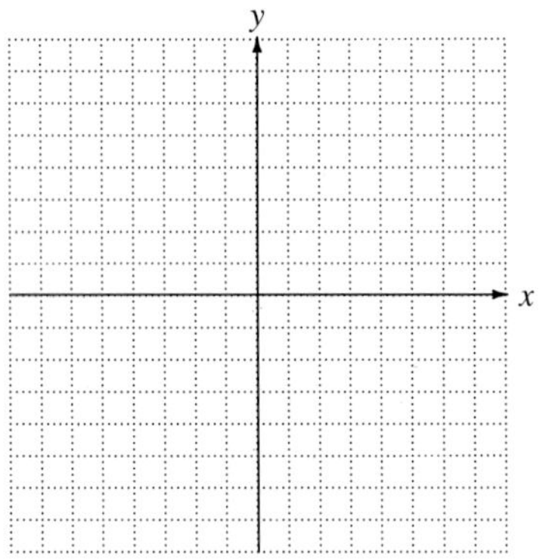

44. $y = \frac{2}{3}x - 6$

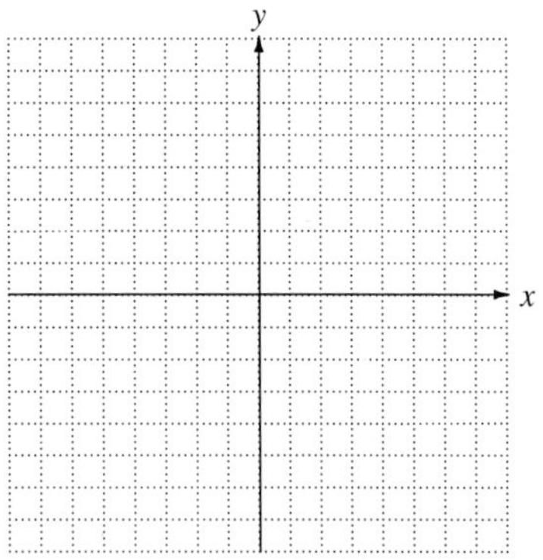

45. $y + 3x = 2$

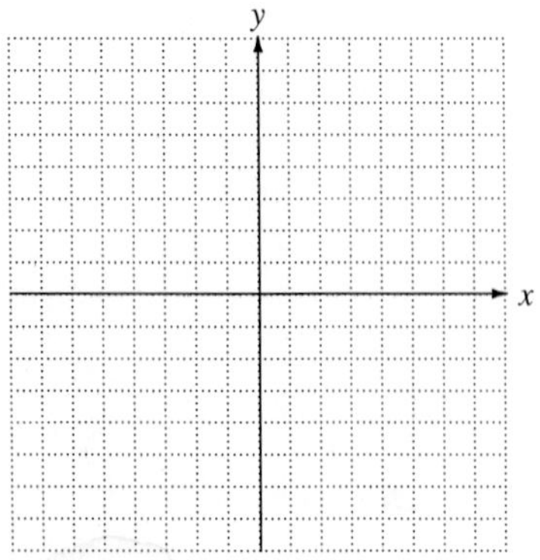

46. $y + 2x = 4$

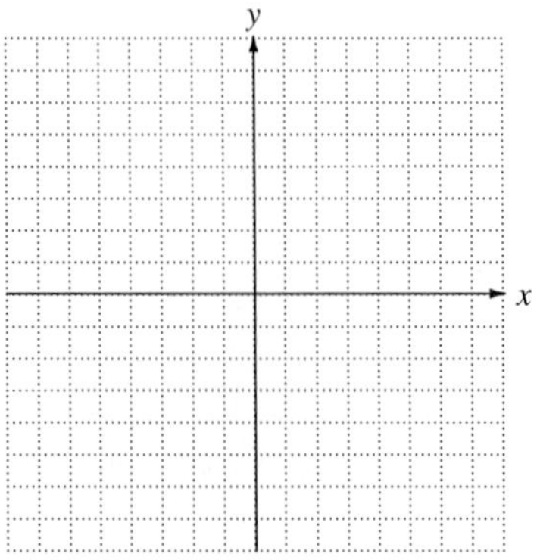

47. $y = 2x$

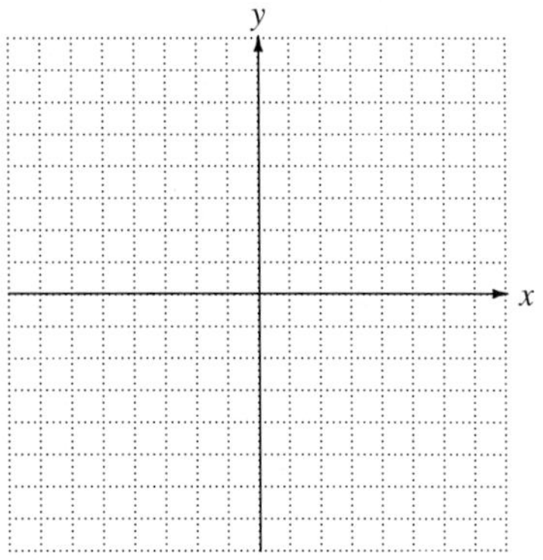

48. A line has a slope of $\frac{2}{3}$.

(a) What is the slope of a line parallel to it?

(b) What is the slope of a line perpendicular to it?

49. A line has a slope of $\frac{13}{5}$.

(a) What is the slope of a line parallel to it?

(b) What is the slope of a line perpendicular to it?

50. A line has a slope of -4.

(a) What is the slope of a line parallel to it?

(b) What is the slope of a line perpendicular to it?

51. A line has a slope of $\frac{1}{8}$.

(a) What is the slope of a line parallel to it?

(b) What is the slope of a line perpendicular to it?

52. The equation of a line is $y = 2x - 5$.

(a) What is the slope of a line parallel to it?

(b) What is the slope of a line perpendicular to it?

53. The equation of a line is $y = -5x + 1$.

(a) What is the slope of a line parallel to it?

(b) What is the slope of a line perpendicular to it?

54. The equation of a line is $y = \frac{1}{3}x + 2$.

(a) What is the slope of a line parallel to it?

(b) What is the slope of a line perpendicular to it?

55. The equation of a line is $y = -\frac{1}{2}x - 3$.

(a) What is the slope of a line parallel to it?

(b) What is the slope of a line perpendicular to it?

Cumulative Review Problems

Solve for x and graph your solution.

56. $3x + 8 > 2x + 12$

57. $\dfrac{1}{4}x + 3 > \dfrac{2}{3}x + 2$

58. $\dfrac{1}{2}(x + 2) \leq \dfrac{1}{3}x + 5$

59. $7x - 2(x + 3) \leq 4(x - 7)$

Putting Your Skills to Work

UNDERWATER PRESSURE

The underwater pressure in pounds per square inch in the ocean is given by $p = \frac{5}{11}d + 15$, where d is the depth in feet below the surface.

1. Find the underwater pressure at a depth of 22 feet.

2. The underwater pressure was found to be 35 pounds per square inch. What was the depth at which this measurement was taken?

3. Using the coordinates of the two points obtained in problems 1 and 2, calculate the slope of the line.

4. From the linear equation, determine the coordinates of a third point. Plot the point and the points obtained in problems 1 and 2. Draw a line.

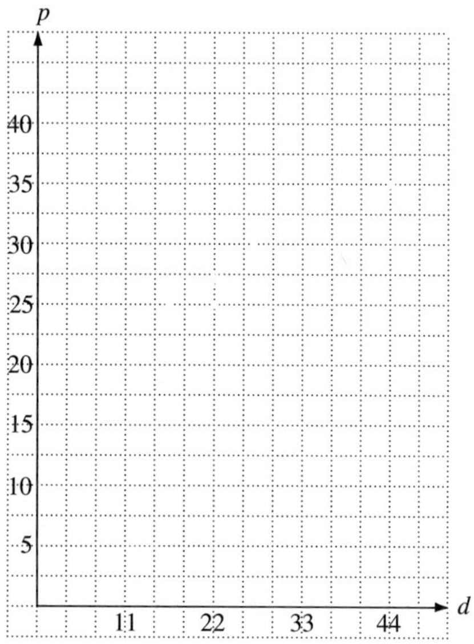

5. From the graph, determine the slope of the line. Does this agree with the slope in problem 3? Does this agree with the equation? Why? Why not?

6. Complete the following table of values.

d	0	11	22	33	44
p					

LOOK FOR A PATTERN

(a) What is the difference between successive p values?

(b) What is the difference between successive d values?

(c) How is this related to the slope? Why?

COOPERATIVE GROUP ACTIVITY

Scientists are working to find a pressure equation similar to $p = \frac{5}{11}d + 15$ for a liquid that has different properties than water. They have compiled the following table of values in which p is the pressure in pounds per square inch and d is the depth in feet below the surface of this liquid:

d	0	5	12
p	18	22	27.6

7. Using as coordinates for (d, p) the values $(0, 18)$ and $(5, 22)$, find the slope of the line.

8. Using as coordinates for (d, p) the values $(5, 22)$ and $(12, 27.6)$, find the slope of the line. How does this compare with your answer from problem 7? What does this imply?

9. Write the equation of the pressure for this new liquid in the form $p = md + b$ by determining the slope m and the p-intercept b.

10. What would the pressure be if an object is submerged in this liquid at a depth of 44 feet? How does this compare with the pressure of an object submerged in water at a depth of 44 feet?

11. Is this new liquid more dense than water or less dense than water? What evidence do you have to support your conclusions?

7.4 Obtaining the Equation of a Line

After studying this section, you will be able to:

1. *Write the equation of a line, given a point and a slope.*
2. *Write the equation of a line, given two points.*
3. *Write the equation of a line, given a graph of the line.*

1 *Writing the Equation of a Line, Given a Point and a Slope*

If we know the slope of a line and the y-intercept, we can write the equation of the line in slope–intercept form. Sometimes we are given the slope and a point on the line. We use the information to find the y-intercept. Then we can write the equation of the line.

EXAMPLE 1 Find the equation of a line with a slope of 4 that passes through the point (3, 1).

We are given the following values: $m = 4$, $x = 3$, $y = 1$.

$y = mx + b$

$y = 4x + b$ *We are given that the slope of the line is* 4.

$1 = 4(3) + b$ *Since* (3, 1) *is a point on the line, it satisfies the equation. Substitute* $x = 3$ *and* $y = 1$ *into the equation.*

$1 = 12 + b$ *Solve for* b.

$-11 = b$

Thus the y-intercept is -11. We can now write the equation of the line.

$$y = 4x - 11$$

Practice Problem 1 Find the equation of the line through $(-2, 4)$ with slope $\frac{1}{2}$. ■

It may be helpful to summarize our approach.

> **To Find the Equation of a Line Given a Point and a Slope**
>
> **1.** Substitute the given values of x, y, and m into the equation $y = mx + b$.
>
> **2.** Solve for b.
>
> **3.** Use the values of b and m to write the equation in the form $y = mx + b$.

EXAMPLE 2 Find the equation of a line with slope $-\frac{2}{3}$ that passes through the point $(-3, 6)$.

We write out the values $m = -\frac{2}{3}$, $x = -3$, $y = 6$.

$y = mx + b$

$6 = \left(-\frac{2}{3}\right)(-3) + b$ *Substitute known values.*

$6 = 2 + b$

$4 = b$

The equation of the line is $y = -\frac{2}{3}x + 4$.

Practice Problem 2 Find the equation of a line with slope $-\frac{3}{4}$ that passes through $(-8, 12)$. ■

2 *Finding the Equation of a Line, Given Two Points*

Our procedure can be extended to the case for which two points are given.

EXAMPLE 3 Find the equation of a line that passes through (2, 5) and (6, 3).

We first find the slope of the line. Then we proceed as in Example 2.

$$m = \frac{y_2 - y_1}{x_2 - x_1}$$

$$m = \frac{3 - 5}{6 - 2}$$ *Substitute* $(x_1, y_1) = (2, 5)$ *and* $(x_2, y_2) = (6, 3)$ *into the formula.*

$$m = \frac{-2}{4} = -\frac{1}{2}$$

Choose either point, say (2, 5) and proceed as in Example 2.

$$y = mx + b$$

$$5 = -\frac{1}{2}(2) + b$$

$$5 = -1 + \text{b}$$

$$6 = b$$

The equation of the line is $y = -\frac{1}{2}x + 6$.

Note: After finding the slope $m = -\frac{1}{2}$, we could have used the other point (6, 3) and would have arrived at the same answer. Try it.

Practice Problem 3 Find the equation of the line through (3, 5) and (−1, 1). ■

3 *Finding the Equation of a Line from the Graph*

EXAMPLE 4 What is the equation of the line in the figure on the right?

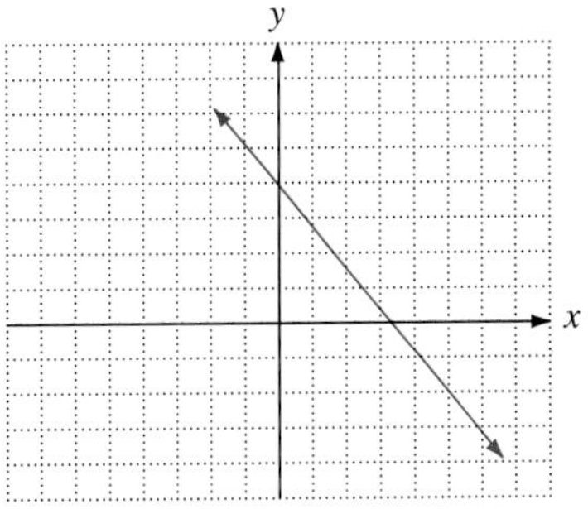

First, look for the y-intercept. The line crosses the y-axis at (0, 4). Thus $b = 4$.

Second, find the slope.

$$m = \frac{\text{change in } y}{\text{change in } x}$$

Look for another point on the line. We chose (5, −2). Count the number of vertical units from 4 to −2 (rise). Count the number of horizontal units from 0 to 5 (run).

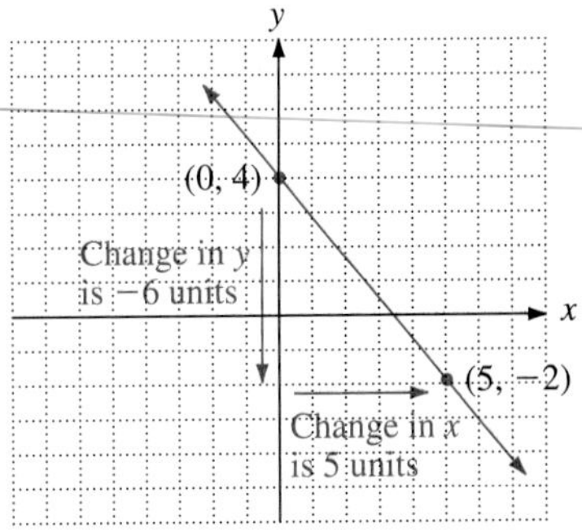

$$m = \frac{-6}{5}$$

Now we can write the equation of the line, $m = -\frac{6}{5}$ and $b = 4$.

$$y = mx + b$$

$$= -\frac{6}{5}x + 4$$

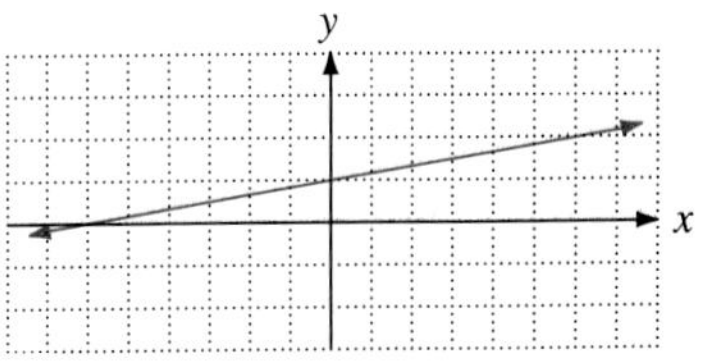

Practice Problem 4 What is the equation of the line in the figure on the right? ■

7.4 Exercises

Find the equation of the line that has the given slope and passes through the given point.

1. $m = 6, (-3, 2)$ **2.** $m = 5, (4, -1)$ **3.** $m = -2, (4, 3)$ **4.** $m = -4, (5, 7)$

5. $m = \frac{2}{3}, (-3, -2)$ **6.** $m = \frac{2}{5}, (-4, -2)$ **7.** $m = -3, \left(\frac{1}{2}, 2\right)$ **8.** $m = -2, \left(3, \frac{1}{3}\right)$

9. $m = \frac{3}{4}, (4, -2)$ **10.** $m = \frac{2}{3}, (3, -2)$ **11.** $m = 0, (7, -4)$ **12.** $m = 0, (-3, 2)$

Write the equation of a line passing through the two points given.

13. $(1, 5)$ and $(4, 11)$ **14.** $(-1, -5)$ and $(2, 7)$ **15.** $(1, -2)$ and $(3, -8)$ **16.** $(2, -7)$ and $(-1, 8)$

17. $(0, -4)$ and $(1, -7)$ **18.** $(1, -8)$ and $(2, -14)$ **19.** $(3, 5)$ and $(-1, -15)$ **20.** $(-1, -19)$ and $(2, 2)$

21. $(1, -3)$ and $(2, 6)$ **22.** $(0, 0)$ and $(1, \frac{1}{2})$ ★ **23.** $(1, \frac{5}{6})$ and $(3, \frac{3}{2})$ ★ **24.** $(2, 0)$ and $(\frac{3}{2}, \frac{1}{2})$

Write the equation of each line.

25.

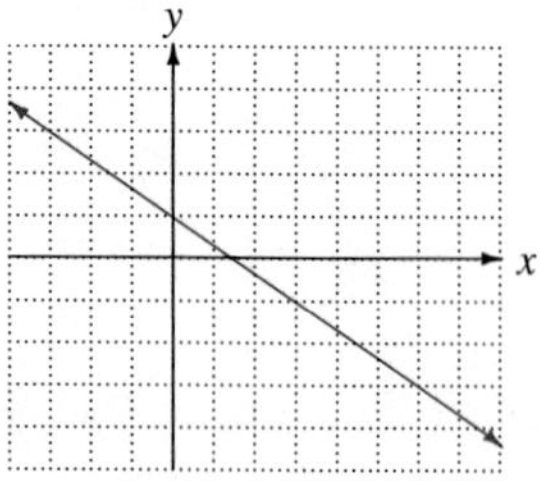

26.

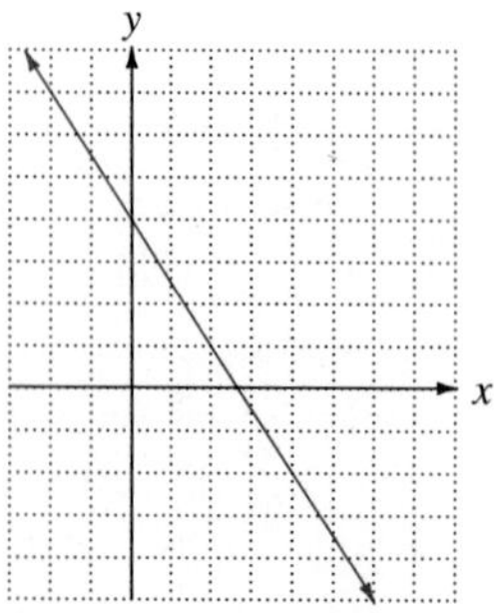

27.

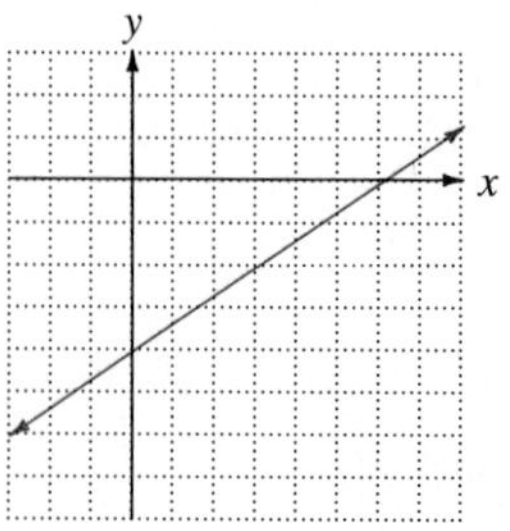

28.

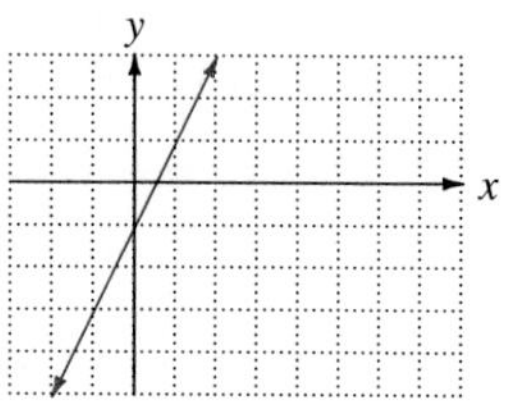

29.

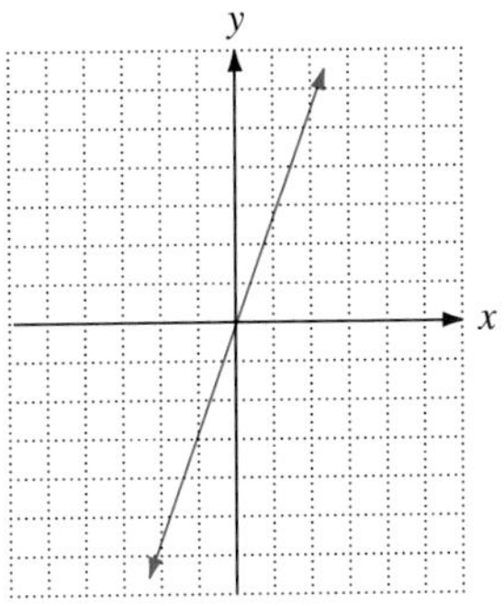

30.

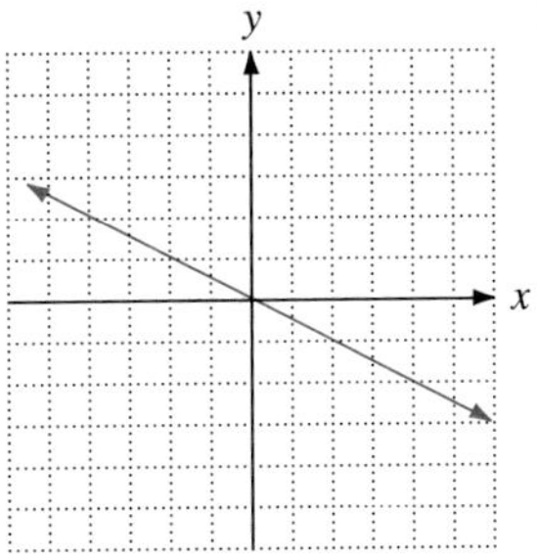

31.

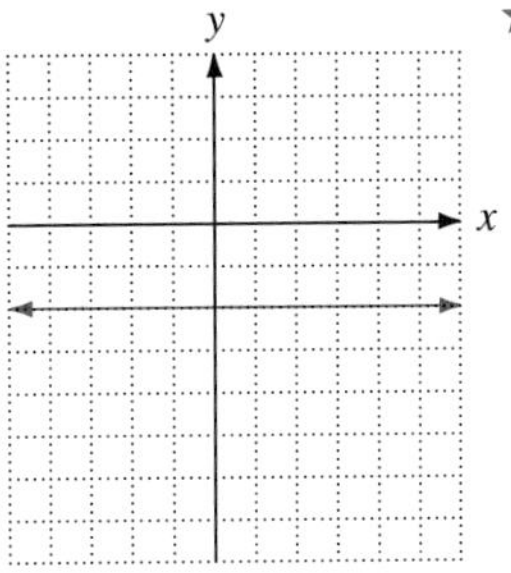

★ 32.

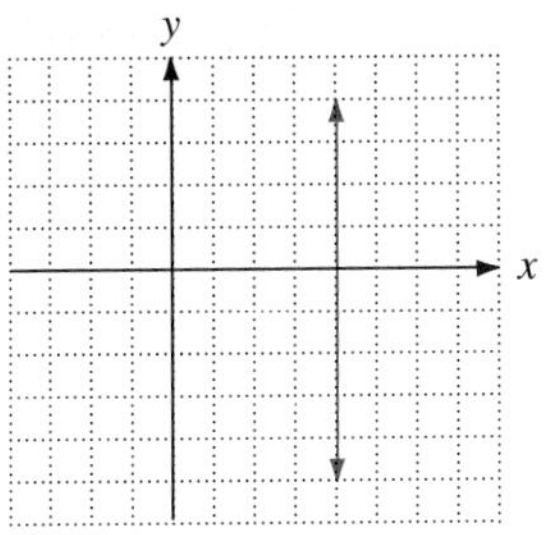

To Think About

Find the equation of the line passing through the following.

33. $(6, -1)$ with undefined slope.

34. $(2, 5)$ with slope 0.

35. $(-5, 7)$ that is parallel to the y-axis.

36. $(-3, -3)$ that is parallel to the x-axis.

37. $(0, 2)$ that is perpendicular to the y-axis.

38. $(-1, 4)$ that is perpendicular to the x-axis.

39. $(0, 3)$ that is parallel to $y = 2x + 1$.

40. $(0, 3)$ that is perpendicular to $y = 2x + 1$.

41. $(1, 5)$ that is parallel to $y = 3x - 1$.

42. $(3, 0)$ that is perpendicular to $y = 3x - 1$.

Cumulative Review Problems

43. Solve. $\frac{1}{8} + \frac{1}{t} = \frac{1}{3}$

44. Simplify. $\frac{4x^2 - 25}{4x^2 - 20x + 25} \cdot \frac{x^2 - 9}{2x^2 + x - 15} \div (2x^2 - x - 15)$.

45. A recent shipment of microchips from plant 17 of Electra-Tech Corporation was found to contain 231 defective chips. If 44 cases were shipped, each containing 512 microchips, what percent of the shipment was defective? Round your answer to the nearest hundredth of a percent.

46. Total interest of \$750 was earned in one year by investing part of \$11,000 in an account earning 6% simple interest. The balance was invested in a somewhat more risky venture, which earned 9%. How much was invested in the risky venture?

7.5 Graphing Linear Inequalities

After studying this section, you will be able to:

1 *Graph linear inequalities in two variables.*

In Section 2.7 we discussed inequalities in one variable. Look at the inequality $x < -2$ (x is less than -2). Some of the solutions to the inequality are -3, -5, and $-5\frac{1}{2}$. In fact all numbers to the left of -2 on the number line are solutions. The graph of the inequality is given below. Notice that the open circle at 2 indicates that 2 is *not* a solution.

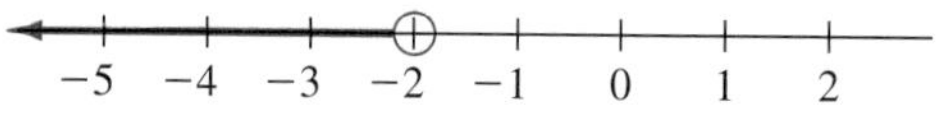

$x < -2$

1 Graphing Linear Inequalities in Two Variables

Consider the inequality $y \geq x$. The solution of the inequality is the set of all possible ordered pairs that when substituted into the inequality will yield a true statement. Which ordered pairs will make the statement $y \geq x$ true? Let's try some.

$(0, 6)$	$(-2, 1)$	$(1, -2)$	$(3, 5)$	$(4, 4)$
$6 \geq 0$, true	$1 \geq -2$, true	$-2 \geq 1$, false	$5 \geq 3$, true	$4 \geq 4$, true

$(0, 6)$, $(-2, 1)$, $(3, 5)$ and $(4, 4)$ are solutions to the inequality $y \geq x$. In fact, every point at which the y-coordinate is greater than or equal to the x-coordinate is a solution to the inequality. This is shown by the shaded region in the graph at the right. Notice that the solution set includes the points on the line $y = x$.

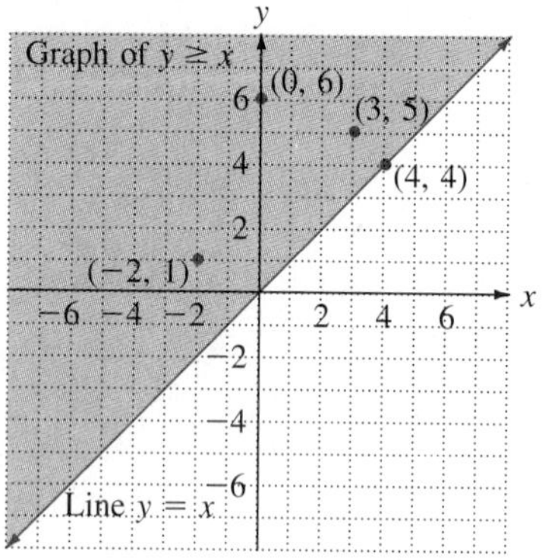

Did we need to test so many points? The solution set for an inequality in two variables will be the region above the line or the region below the line. It is sufficient to test one point. If the point is a solution to the inequality, shade the region that includes the point. If the point is not a solution, shade the region on the other side of the line.

EXAMPLE 1 Graph $5x + 3y > 5$. Use the coordinate system in the margin.

We begin by graphing the line $5x + 3y = 15$. You may use any method discussed previously to graph the line. Since there is no equal sign in the inequality, we will draw a dashed line to indicate that the line is *not* part of the solution set.

Look for a test point. The easiest point to test is $(0, 0)$. Substitute $(0, 0)$ for (x, y) in the inequality.

$$5x + 3y > 15$$
$$5(0) + 3(0) > 15$$
$$0 > 15 \quad \text{false}$$

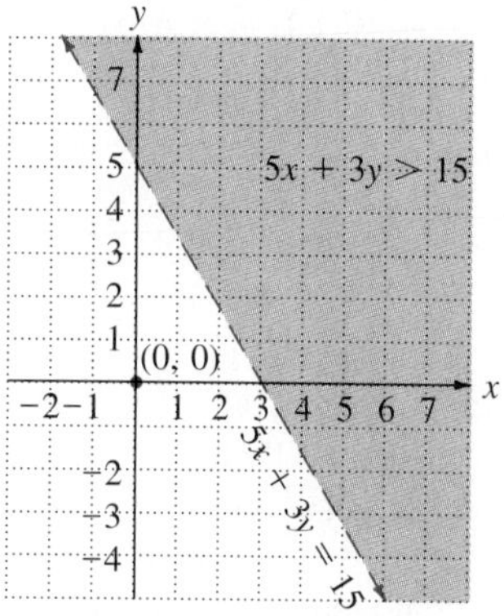

$(0, 0)$ is not a solution. Shade the side of the line that does *not* include $(0, 0)$.

PRACTICE PROBLEM 1

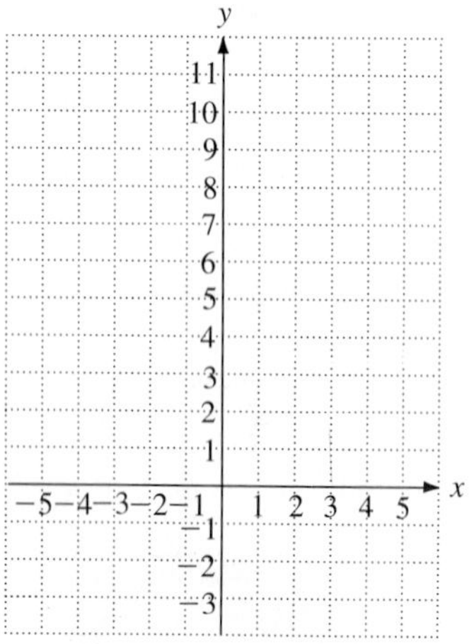

Practice Problem 1 Graph $x - y \geq -10$ on the coordinate system in the margin.

■

Graphing Linear Inequalities

1. Replace the inequality symbol by an equality symbol. Graph the line.
   (a) The line will be solid if the inequality is $\geq$ or $\leq$.
   (b) The line will be dashed if the inequality is $>$ or $<$.
2. Test the point (0, 0) in the inequality.
   (a) If the inequality is true, shade the side of the line that includes (0, 0).
   (b) If the inequality is false, shade the side of the line that does not include (0, 0).
3. If the point (0, 0) is a point on the line, choose another test point and proceed accordingly.

EXAMPLE 2 Graph $2y \leq -3x$.

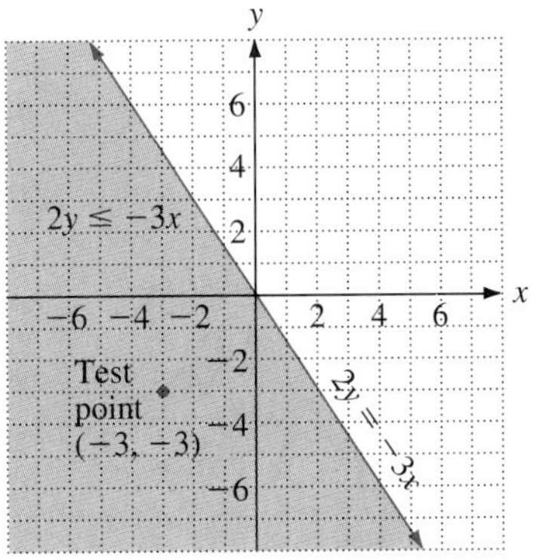

Step 1 Graph $2y = -3x$. Since $\leq$ is used, the line should be a solid line.
Step 2 We see that the line passes through (0, 0).
Step 3 Choose another test point. We will choose $(-3, -3)$.

$$2y \leq -3x$$
$$2(-3) \leq -3(-3)$$
$$-6 \leq 9 \quad \text{true}$$

Shade the region that includes $(-3, -3)$.
Shade the region below the line.

PRACTICE PROBLEM 2

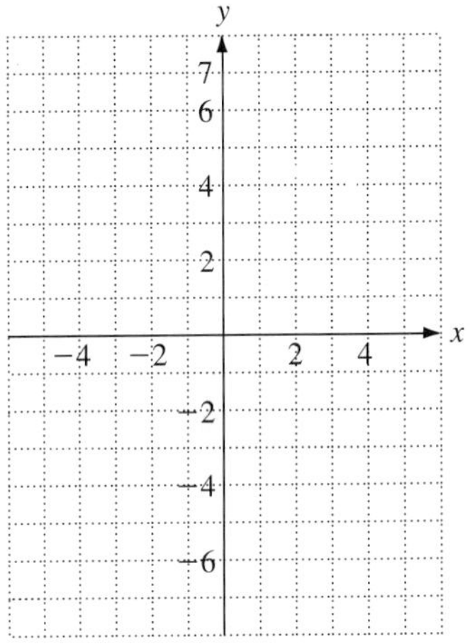

Practice Problem 2 Graph $y > \frac{1}{2}x$ on the coordinate system in the margin. ■

If we are graphing the inequality $x < -2$ on the coordinate plane, the solution will be a region. Notice that this is very different from the solution $x < -2$ on the number line discussed earlier.

EXAMPLE 3 Graph $x < -2$ on the coordinate plane.

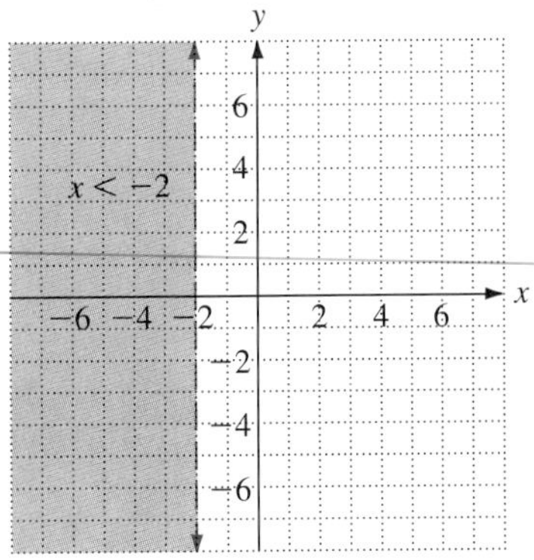

Step 1 Graph $x = -2$. Since $<$ is used, the line should be dashed.
Step 2 Test (0, 0) in the inequality.

$$x < -2$$
$$0 < -2 \quad \text{false}$$

Shade the region that does not include (0, 0).
Shade the region to the left of the line $x = -2$.

PRACTICE PROBLEM 3

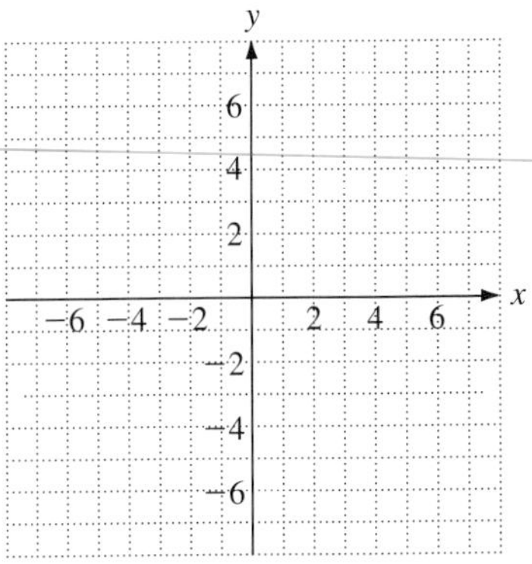

Practice Problem 3 Graph $y \geq -3$ on the coordinate system in the margin. ■

7.5 Exercises

Verbal and Writing Skills

1. Does it matter what point you use as your test point? Justify your response.

2. Explain when to use a solid line or a dashed line when graphing linear inequalities in two variables.

Graph the region described by the inequality.

3. $y > 2x - 2$

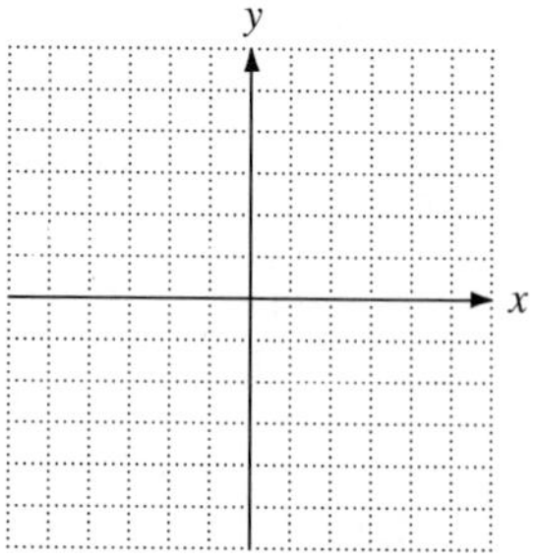

4. $y > 3x - 1$

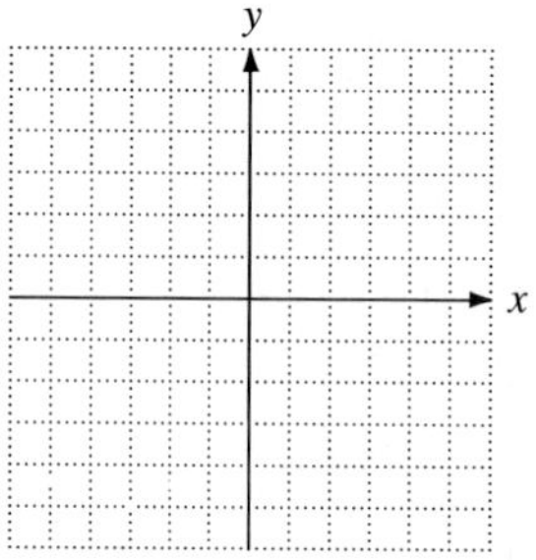

5. $2x - 3y < 6$

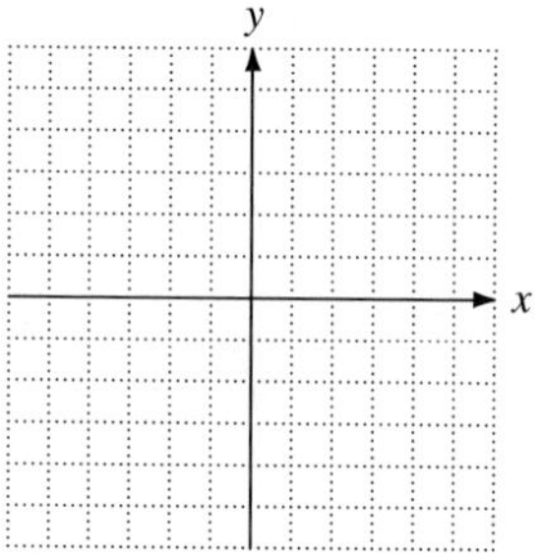

6. $3x + 2y < -6$

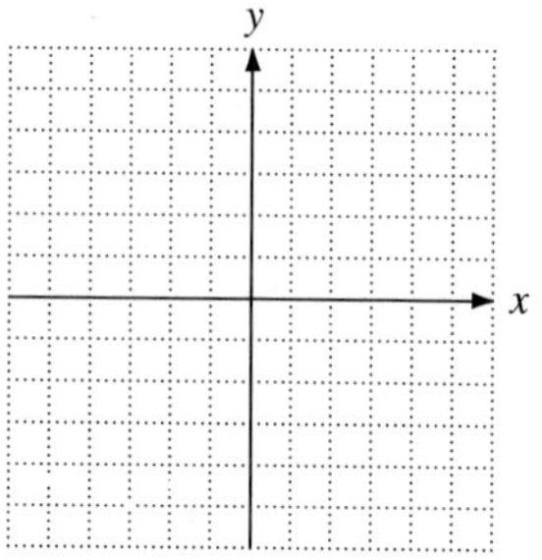

7. $x - y \geq 2$

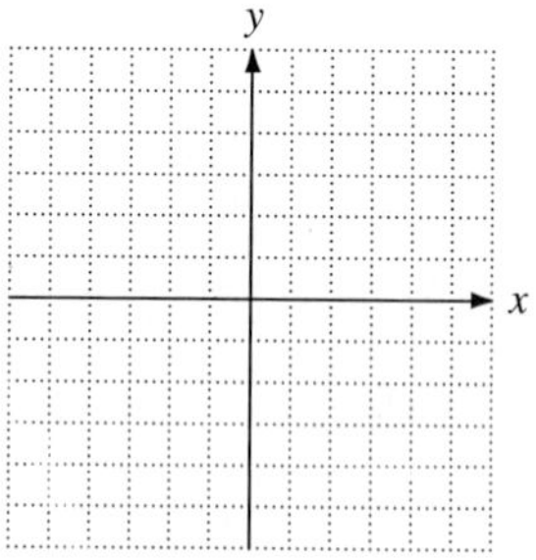

8. $-y + x \geq 3$

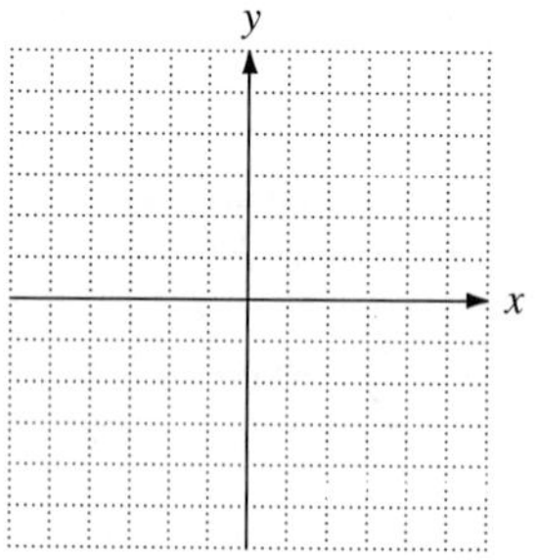

9. $y \geq 3x$

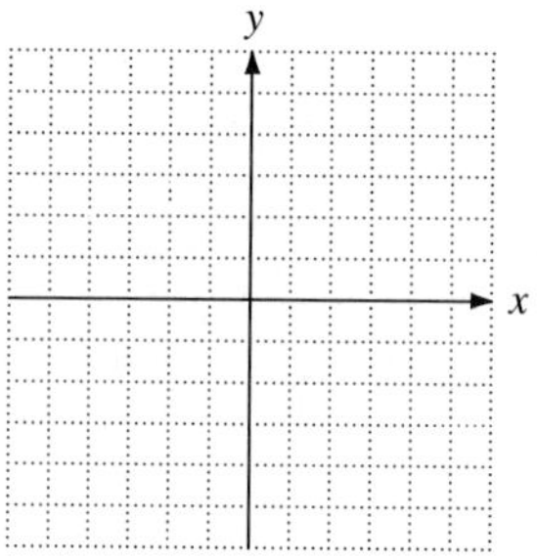

10. $y \geq -4x$

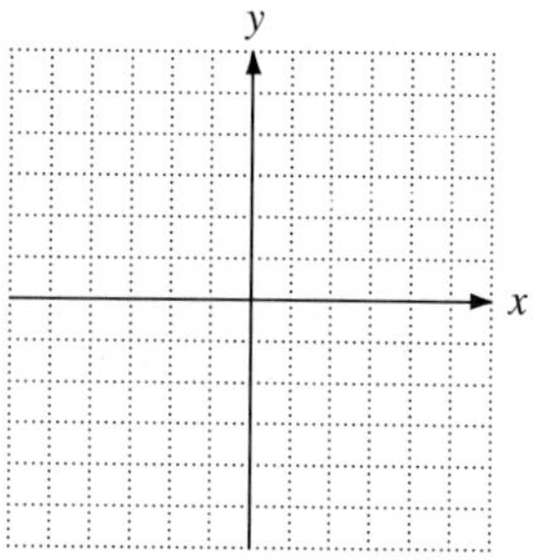

11. $y < -\frac{1}{2}x$

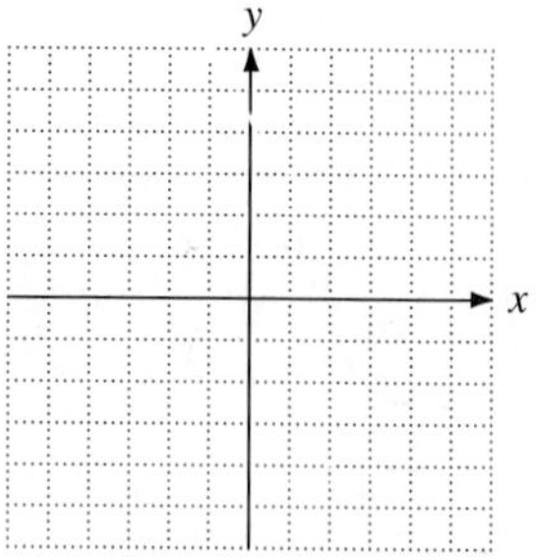

12. $y > \frac{1}{5}x$

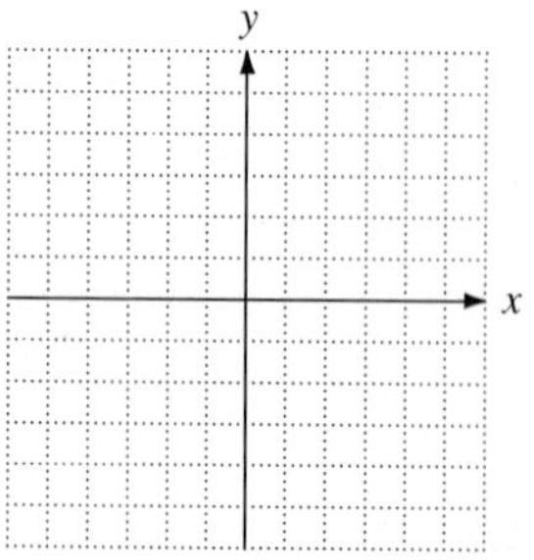

13. $x \leq -3$

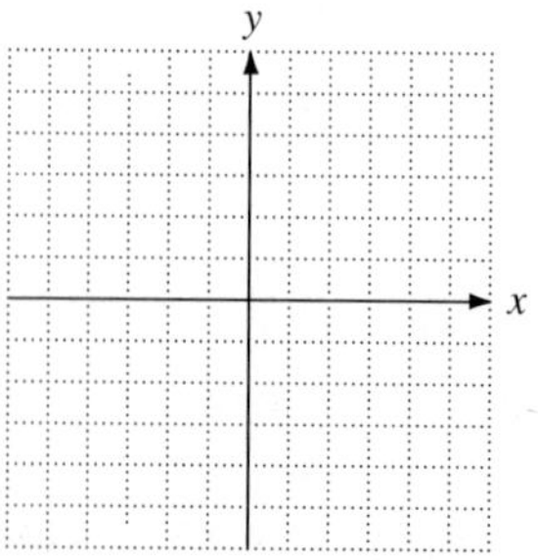

14. $y \geq -4$

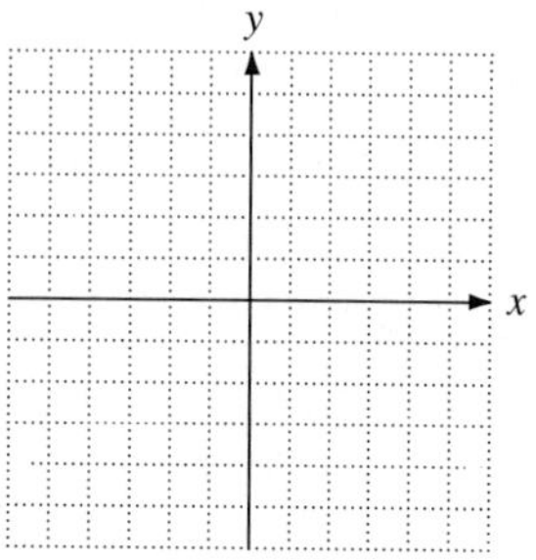

15. $5x - y - 5 \leq 0$

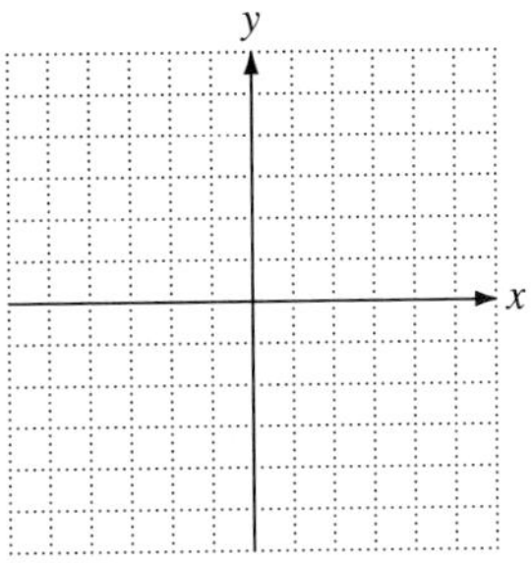

16. $4x + y - 4 \geq 0$

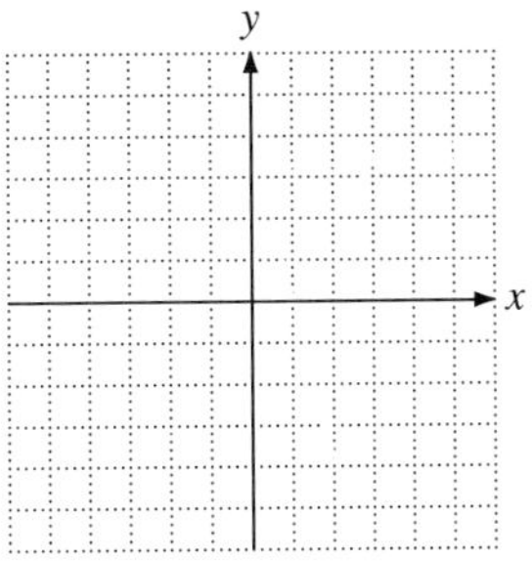

17. $3y \geq -2x$

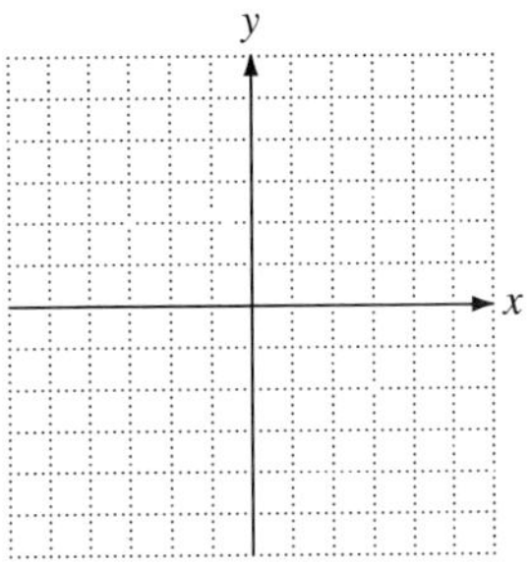

18. $2y < 5x$

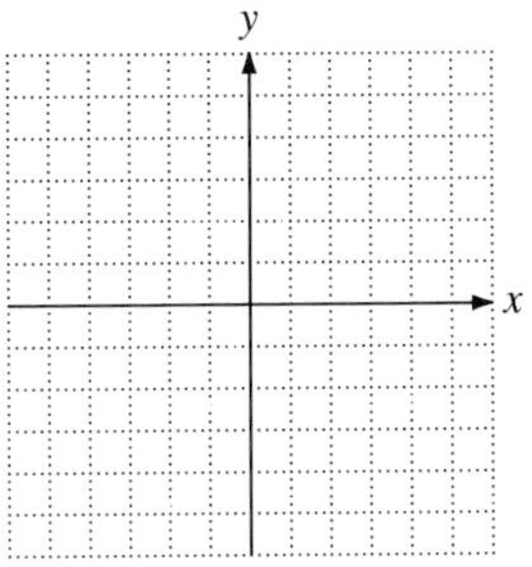

★**19.** $2x > 3 - y$

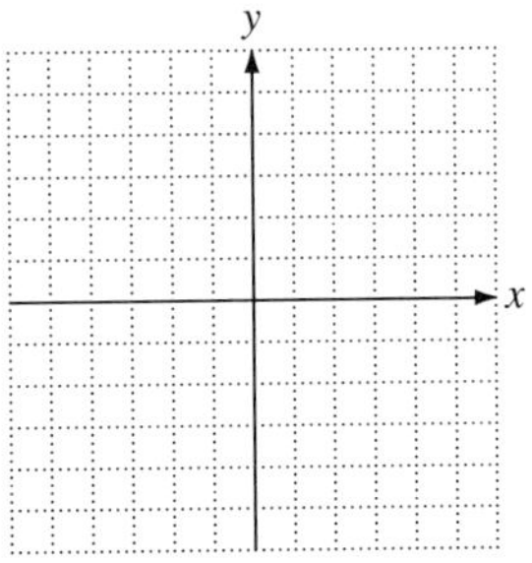

★**20.** $3x > 1 + y$

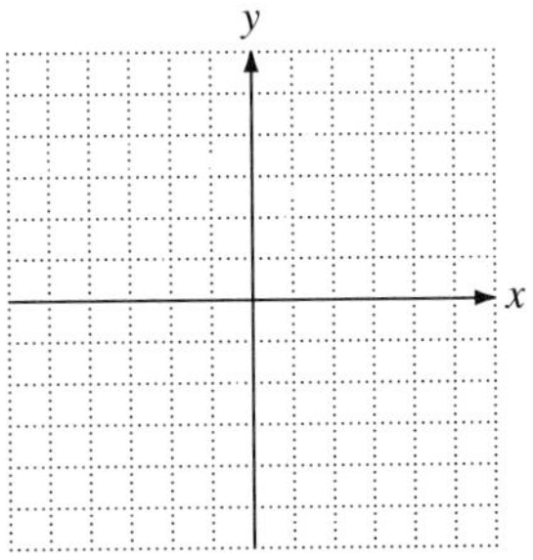

★**21.** $x > -2y$

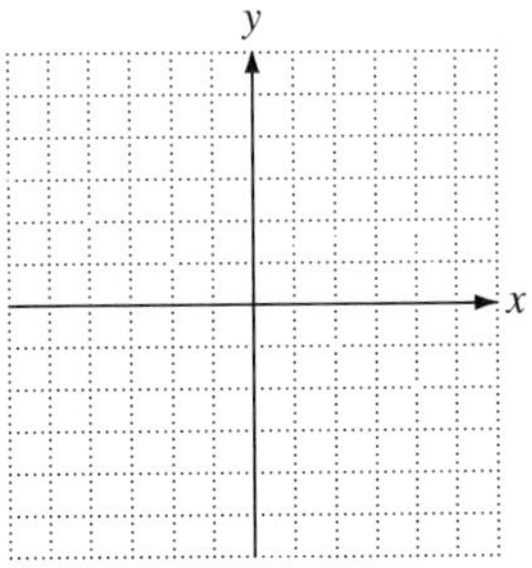

★**22.** $x < -3y$

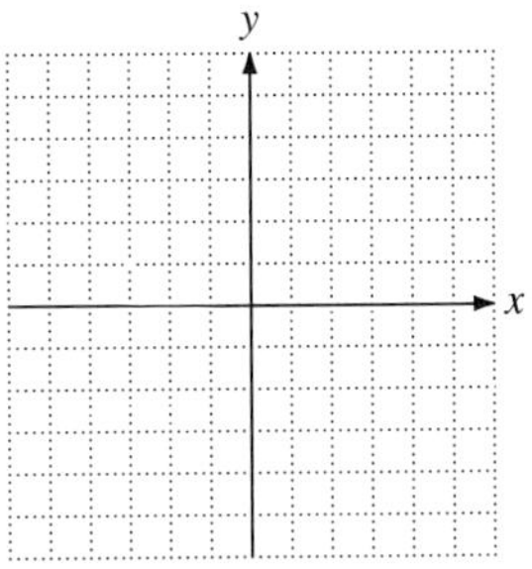

To Think About

23. Graph the inequality $3x + 3 > 5x - 3$ by graphing the lines $y = 3x + 3$ and $y = 5x - 3$ on the same coordinate plane.

(a) For which values of x does the graph of $y = 3x + 3$ lie above the graph of $y = 5x - 3$?

(b) For which values of x is $3x + 3 > 5x - 3$ true?

(c) Explain how the answer to (a) can help you to find the answer to (b).

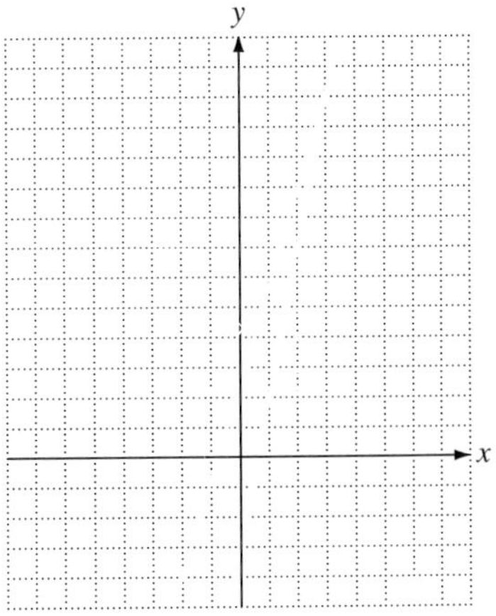

Cumulative Review Problems

24. What is the slope–intercept form of the equation for the line through (3, 0) and (6, −7)?

25. Solve for x. $2x^2 + 7x + 6 = 0$

26. With three games left to play, Lefty Louie has a batting average of 0.298 in 500 at-bats (batting average = number of hits ÷ number of at-bats). If he gets 10 more at-bats in the remaining games, how many hits will he need to finish the year with a 0.300 batting average?

27. Three consecutive odd integers have a sum of 243. What is the largest of these three numbers?

7.6 Functions

After studying this section, you will be able to:

1. *Understand and use the definition of a relation and a function.*
2. *Graph simple nonlinear equations.*
3. *Determine if a graph represents a function.*
4. *Use function notation.*

1 *Understanding and Using the Definition of Relation and Function*

Thus far you have studied linear equations in two variables. You have seen that such an equation can be represented by a table of values, by the algebraic equation itself, and by a graph.

x	0	1	3
y	4	1	-5

$y = -3x + 4$

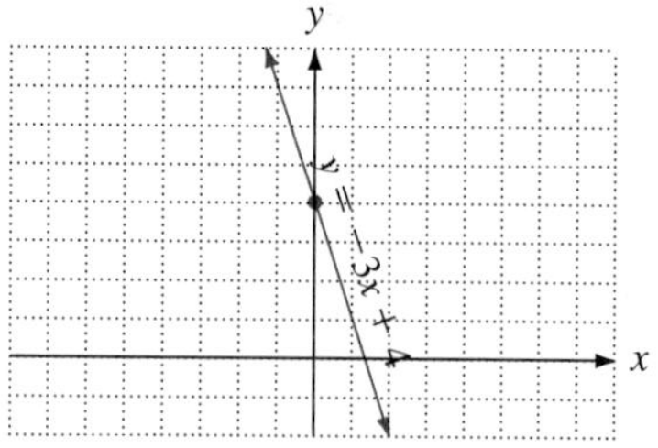

The solutions to the linear equation are all the ordered pairs that satisfy the equation (make the equation true). They are all the points that lie on the graph of the line. These ordered pairs can be represented in a table of values. Notice the relationship between the ordered pairs. We can choose any value for x. But once we have chosen a value for x, the value of y is determined. For the above equation, if x is 0, then y must be 4. We say that x is the *independent variable* and that y is the *dependent variable.*

Mathematicians call such a pairing of two values a *relation.*

Definition of a Relation

A **relation** is any set of ordered pairs.

All the first coordinates in each ordered pair make up the **domain** of the relation. All the second coordinates in each ordered pair make up the **range** of the relation. Notice that the definition of a relation is very broad. Some relations cannot be described by an algebraic expression. These relations may simply be a set of discrete ordered pairs.

EXAMPLE 1 State the domain and range of the following relation.

$$\{(5, 7), (9, 11), (10, 7), (12, 14)\}$$

The domain consists of all the first coordinates in the ordered pairs.

Domain

$$\{(5, 7), (9, 11), (10, 7), (12, 14)\}$$

Range

The range consists of all of the second coordinates in the ordered pairs. Thus

The domain is $\{5, 9, 10, 12\}$

The range is $\{7, 11, 14\}$ *We only list 7 once.*

Practice Problem 1 State the domain and range of the following relation.

$$\{(-3, -5), (3, 5), (0, -5), (20, 5)\}$$ ■

Some relations have a special property and are called *functions*. The relation $y = -3x + 4$ is a function. If you look at its table of values, no two different ordered pairs have the same first coordinate. If you look at its graph, you will notice that any vertical line intersects the graph only once. Sometimes we get a better idea if we look at some relations that are *not* functions. Look at $y^2 = x$.

x	1	4	9
y	± 1	± 2	± 3

$y^2 = x$

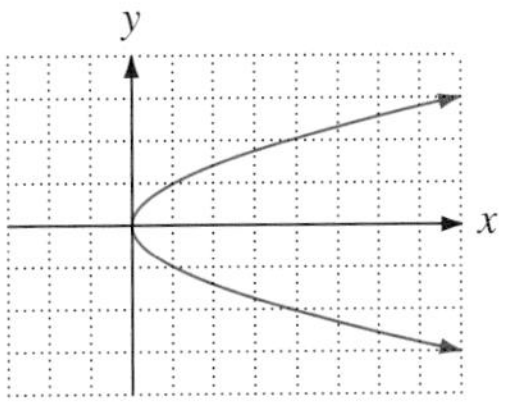

$\{(1, -1), (1, 1), (4, -2), (4, 2), (9, -3), (9, 3)\}$

Notice that, if x is 1, y could be 1 since $1^2 = 1$ or y could be -1 since $(-1)^2 = 1$. That is, the ordered pairs $(1, -1)$ and $(1, 1)$ both have the same first coordinate. $y^2 = x$ is not a function. What do you notice about the graph?

Definition of a Function

A **function** is a relation in which no two different ordered pairs have the same first coordinates.

EXAMPLE 2 Determine if the relation is a function or not a function.

(a) $\{(3, 9), (4, 16), (5, 9), (6, 36)\}$

(b) $\{(7, 8), (9, 10), (12, 13), (7, 14)\}$

(a) Look at the ordered pairs. No two ordered pairs have the same first coordinate. Thus (a) is a function. Note that the ordered pairs (3, 9) and (5, 9) have the same second coordinate, but this does not keep the relation from being a function.

(b) Look at the ordered pairs. Two different ordered pairs, (7, 8) and (7, 14), have the same first coordinate. Thus this relation is *not* a function.

Practice Problem 2 Determine if the relation is a function or not a function.

(a) $\{(-5, -6), (9, 30), (-3, -3), (8, 30)\}$

(b) $\{(60, 30), (40, 20), (20, 10), (60, 120)\}$ ■

Functions are what we hope to find when we analyze two sets of data. Look at the table of values that compares Celsius temperature with Fahrenheit temperature. Is there a relationship between degrees Fahrenheit and degrees Celsius? Is the relation a function?

Temperature

°F	23	32	41	50
°C	−5	0	5	10

Since every Fahrenheit temperature produces a unique Celsius temperature, we would expect this to be a function. We can verify our assumption by looking at the formula $C = \frac{5}{9}(F - 32)$ and its graph. The formula is a linear equation, and its graph is a line with a slope $\frac{5}{9}$ and y-intercept at about -17.8. The relation is a function. Notice that the *dependent variable* is C, since the value of C *depends* on the value of F. We say that F is the *independent variable*. The *domain* can be described as the set of possible values of the independent variable. The *range* is the set of corresponding values of the dependent variable. Thus,

Domain = {all possible Fahrenheit temperatures from absolute zero to infinity}

Range = {all corresponding Celsius temperatures from −273°C to infinity}

EXAMPLE 3 Determine if the relation is a function or not a function. If it is a function, identify the domain and range.

(a) Circle

Radius	1	2	3	4	5
Area	3.14	12.56	28.26	50.24	78.5

(b) $4000 loan at 8%

Time (yr)	1	2	3	4	5
Interest	$320	$665.60	$1038.85	$1441.96	$1877.32

(a) Looking at the table, we see that no two different ordered pairs have the same first coordinate. The area of a circle is a function of the length of the radius.

Next we need to identify the independent variable in order to determine the domain. Sometimes it is easier to identify the dependent variable. Here we notice that the area of the circle depends on the length of the radius. Thus radius is the independent variable. Since a negative length does not make sense, the radius cannot be a negative number.

Domain = {all positive real numbers}

Range = {all positive real numbers}

(b) No two different ordered pairs have the same first coordinate. Interest is a function of time.

Since the amount of interest paid on a loan depends on the number of years (term of the loan), interest is the dependent variable and time is the independent variable. Negative numbers do not apply in this situation.

Domain = {all positive real numbers greater than 1}

Range = {all positive real numbers greater than $320}

Practice Problem 3 Determine if the relation is a function or not a function. If it is a function, identify the domain and the range.

(a) 28 mpg at $1.20 per gallon

Distance	0	28	42	56	70
Cost	0	1.20	1.80	2.40	3.00

(b) Store's inventory of shirts

Number of shirts	5	10	5	2	8
Price of shirt	$20	$25	$30	$45	$50

■

To Think About Look at the following bus schedule. Determine if the relation is a function. Which is the independent variable? Explain your choice.

Bus Schedule

Bus Stop	Main St.	8th Ave.	42nd St.	Sunset Blvd.	Cedar Lane
Time	7:00	7:10	7:15	7:30	7:39

2 *Graphing Simple Nonlinear Equations*

Thus far in this chapter we have graphed linear equations in two variables. We now turn to graphing a few nonlinear equations. We will need to plot more than three points to get a good idea of what the graph will look like.

EXAMPLE 4 Graph $y = x^2$.

Begin by constructing a table of values. We select values for x and then determine by the equation the corresponding values of y. We will include negative values for x as well as positive values. We then plot the ordered pairs and connect the points with a smooth curve.

x	$y = x^2$	y
-2	$y = (-2)^2 = 4$	4
-1	$y = (-1)^2 = 1$	1
0	$y = (0)^2 = 0$	0
1	$y = (1)^2 = 1$	1
2	$y = (2)^2 = 4$	4

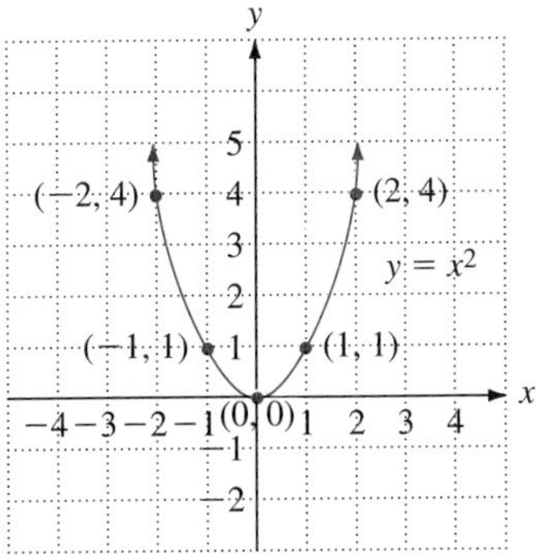

PRACTICE PROBLEM 4

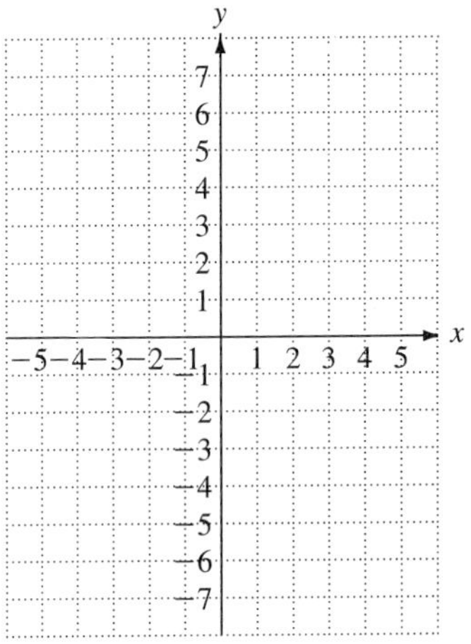

This type of curve is called a *parabola*. We will study the graph of these types of curves more extensively in Chapter 10.

Practice Problem 4 Graph $y = x^2 - 2$ on the coordinate system in the margin. ■

To Think About Compare the graph for Example 4 with the graph for Practice Problem 4. How are the graphs the same? How are the graphs different? Now compare the equations. How does the -2 in the equation $y = x^2 - 2$ affect the graph? What do you think the graph of $y = x^2 + 3$ will look like? Verify by drawing the graph.

Some equations are solved for x. Usually, in those cases we pick values of y and then obtain the corresponding values of x from the equation.

EXAMPLE 5 Graph $x = y^2 + 2$.

We will select a value of y and then substitute it into the equation to obtain x. For convenience in graphing, we will repeat the y column at the end so that it is easy to write the ordered pairs (x, y).

y	$x = y^2 + 2$	x	y
-2	$x = (-2)^2 + 2 = 4 + 2 = 6$	6	-2
-1	$x = (-1)^2 + 2 = 1 + 2 = 3$	3	-1
0	$x = (0)^2 + 2 = 0 + 2 = 2$	2	0
1	$x = (1)^2 + 2 = 1 + 2 = 3$	3	1
2	$x = (2)^2 + 2 = 4 + 2 = 6$	6	2

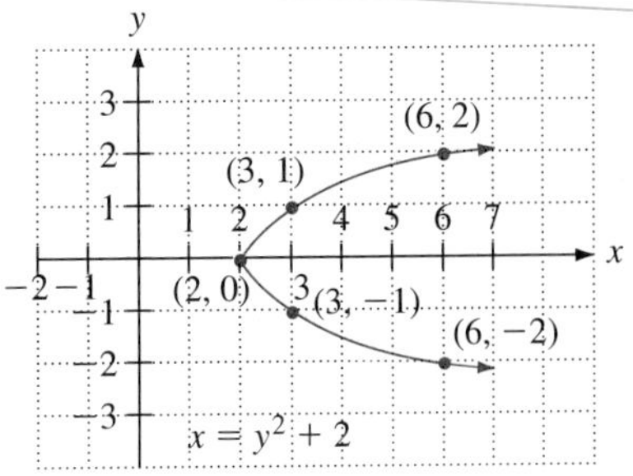

PRACTICE PROBLEM 5

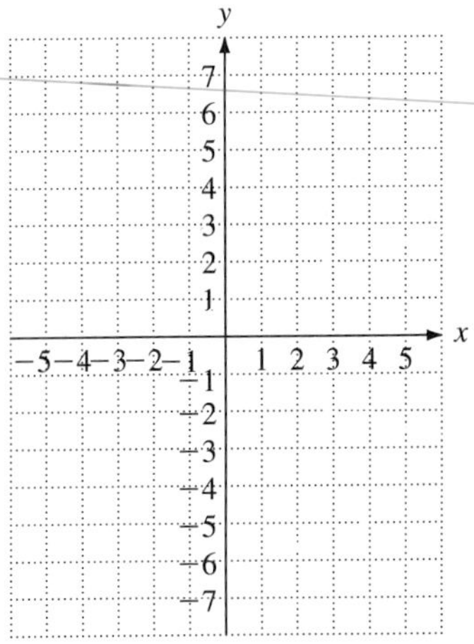

Practice Problem 5 Graph $x = y^2 - 1$ on the coordinate system in the margin. ■

If the equation involves fractions with variables in the denominator, we must use extra caution. You may never divide by zero.

EXAMPLE 6 Graph $y = \frac{4}{x}$.

It is important to note that x cannot be zero because division by zero is not defined. $y = \frac{4}{0}$ is not allowed! Observe that when we draw the graph we get two separate branches of the curve that do not touch.

x	$y = \frac{4}{x}$	y
-4	$y = \frac{4}{-4} = -1$	-1
-2	$y = \frac{4}{-2} = -2$	-2
-1	$y = \frac{1}{-1} = -4$	-4
0	We cannot divide by zero.	There is no value.
1	$y = \frac{4}{1} = 4$	4
2	$y = \frac{4}{2} = 2$	2
4	$y = \frac{4}{4} = 1$	1

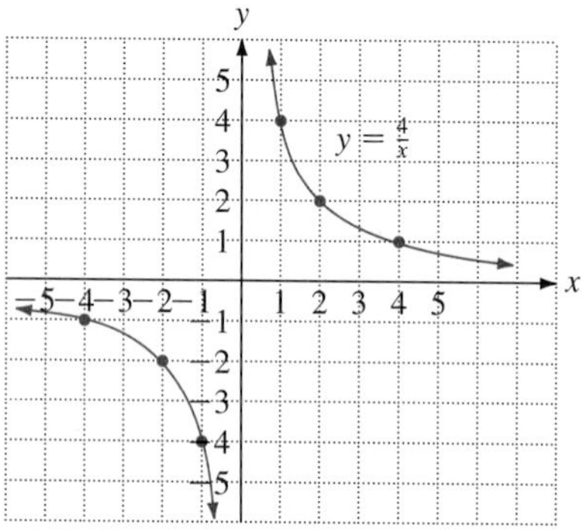

PRACTICE PROBLEM 1

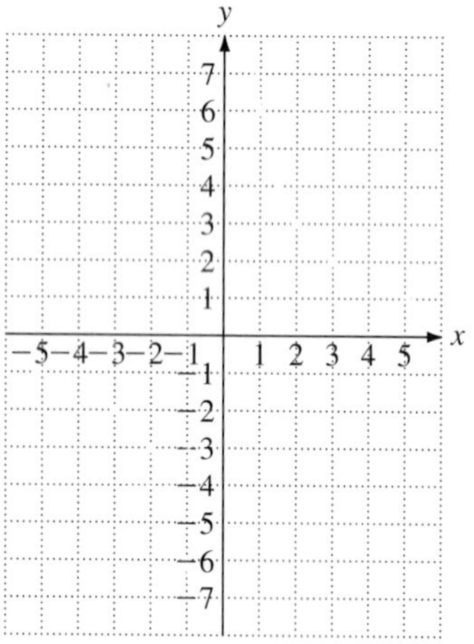

Practice Problem 6 Graph $y = \frac{6}{x}$ on the coordinate system in the margin. ■

3 *Determining If a Graph Represents a Function*

Can we tell from a graph if it is the graph of a function? Recall that a function must have the property that no two different ordered pairs have the same first coordinates. That is, each value of x must have a separate unique value of y. Look at the graph of $y = x^2$ in Example 4. Each x value has a unique y value. Look at the graph of $x = y^2 + 2$ in Example 5. At $x = 3$ there are two y values, 1 and -1. In fact, for every x value greater than 2 there are two y values. $x = y^2 + 2$ is not a function.

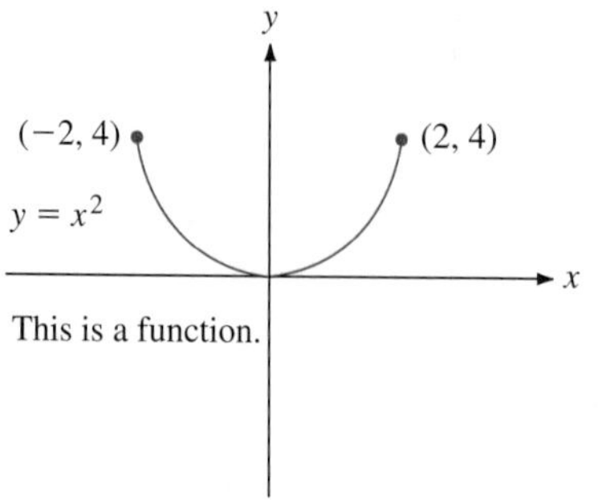

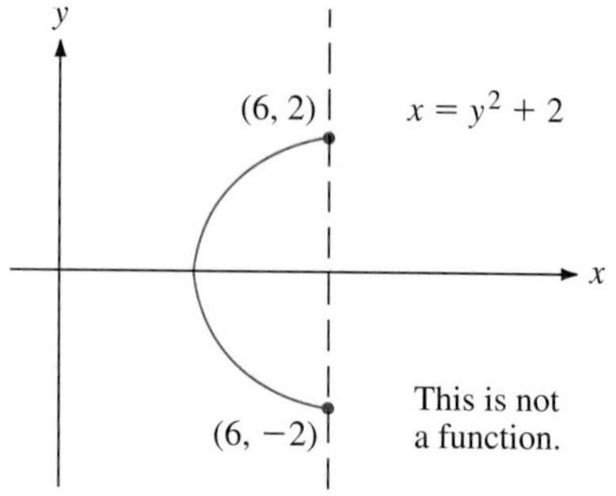

Observe that we can draw a vertical line through (6, 2) and (6, -2). Any graph that is not a function will have at least one region in which a vertical line will cross the curve more than once.

Vertical Line Test

If a vertical line can intersect the graph of a relation more than once, the relation is not a function.

EXAMPLE 7 Determine if each of the following is the graph of a function.

(a)

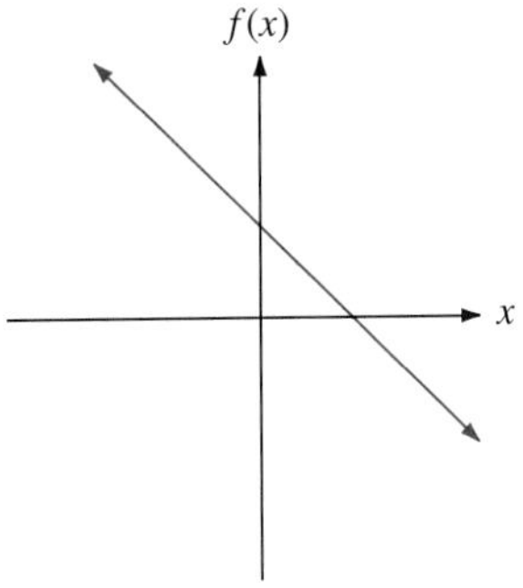

(b)

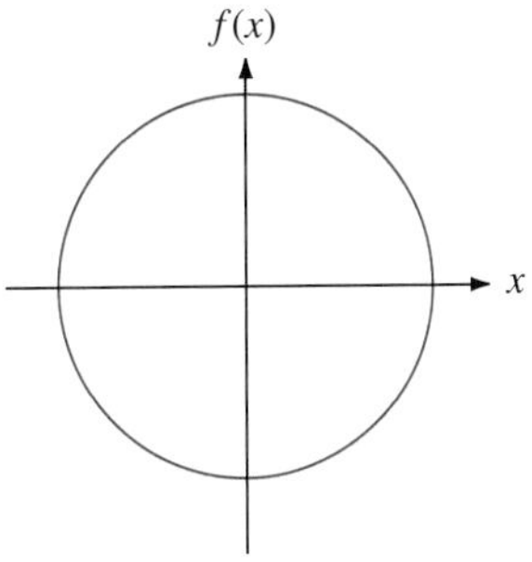

(c)

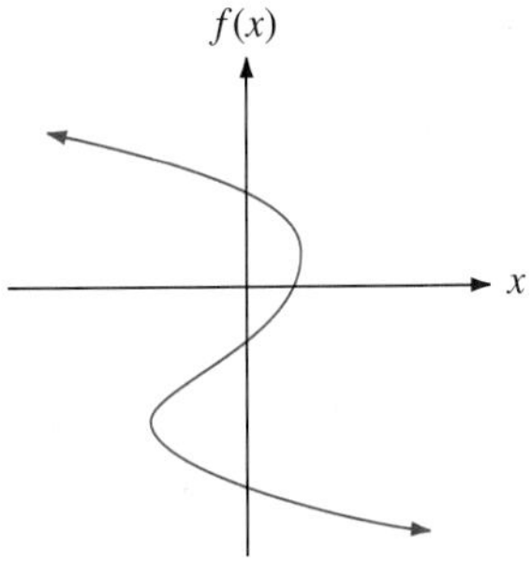

(a) The graph of the straight line is a function. A vertical line can only cross this straight line in one location.

(b) and **(c)** Each of these graphs is not the graph of a function. In each case a vertical line can cross the curve in more than one place.

(a)

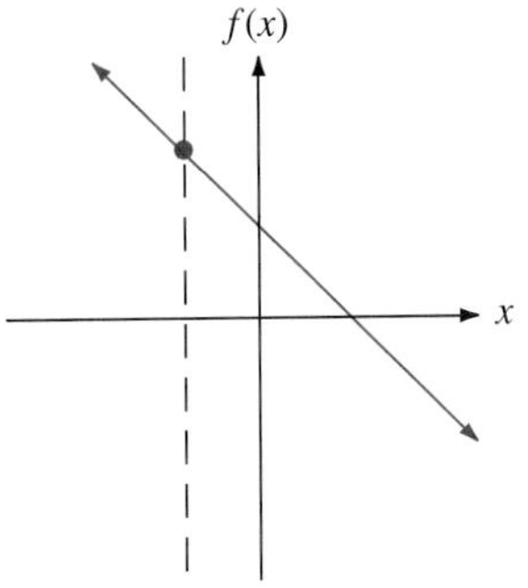

A function

(b)

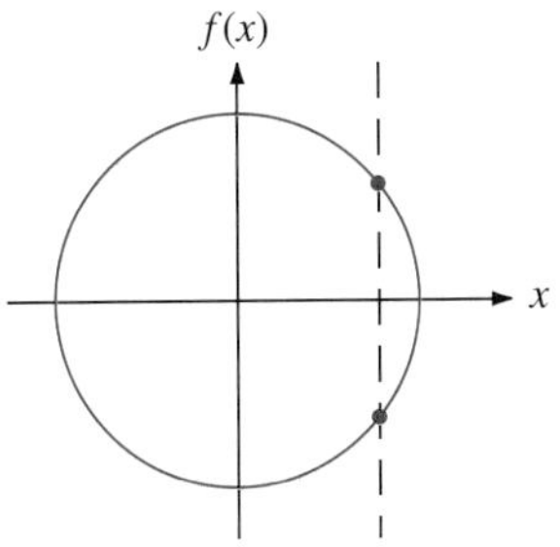

Not a function

(c)

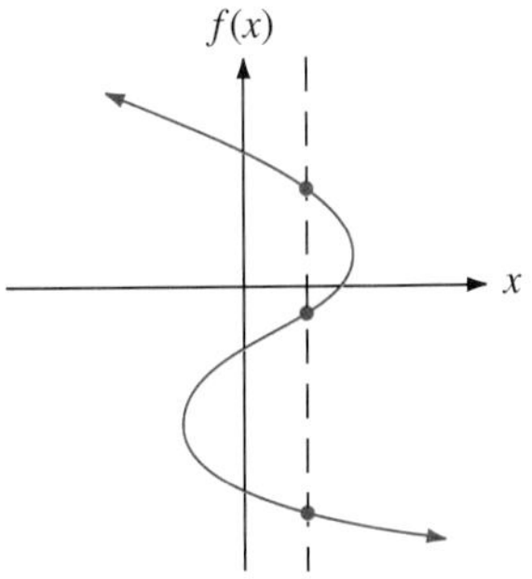

Not a function

Practice Problem 7 Determine if each of the following is the graph of a function.

(a)

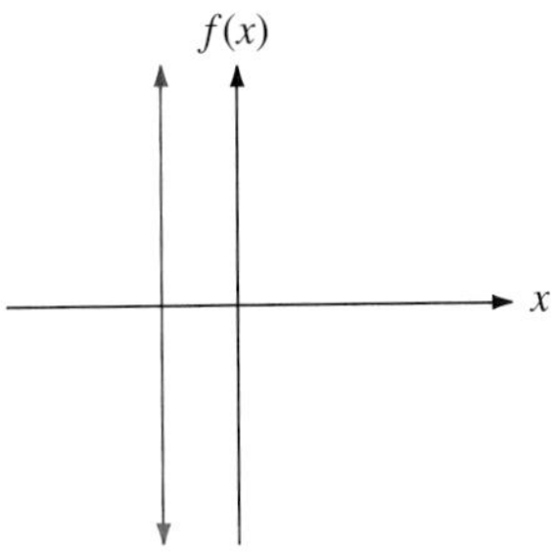

(b)

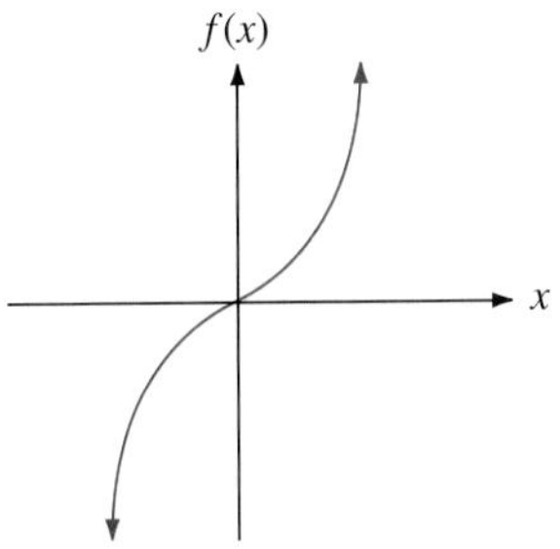

(c)

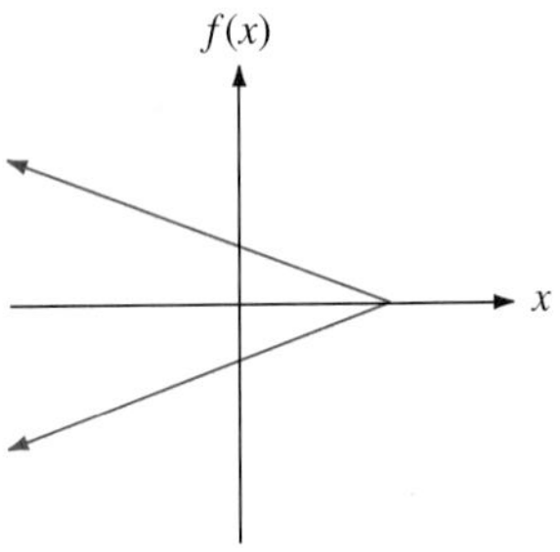

■

4 *Using Function Notation*

We have seen that an equation like $y = 2x + 7$ is a function. For each value of x, the equation assigns a unique value to y. We could say "y is a function of x." This statement can be symbolized by using the function notation $y = f(x)$. Thus you can avoid the y variable completely and describe an expression like $2x + 7$ as a function of x by writing $f(x) = 2x + 7$.

WARNING Be careful. The notation $f(x)$ does not mean f multiplied by x.

EXAMPLE 8 If $f(x) = 2x + 7$, then find $f(3)$.

$f(3)$ means that we want to find the value of the function $f(x)$ when $x = 3$. In this example $f(x)$ is $2x + 7$. Thus we substitute 3 for x wherever x occurs in the equation and evaluate.

$$f(x) = 2x + 7$$

$$f(3) = 2(3) + 7 = 6 + 7 = 13$$

$$= 13$$

Practice Problem 8 If $f(x) = -3x + 4$, then find $f(5)$. ■

EXAMPLE 9 If $f(x) = 3x^2 - 4x + 5$, find:

(a) $f(2)$ **(b)** $f(-2)$ **(c)** $f(4)$ **(d)** $f(0)$

(a) $f(2) = 3(2)^2 - 4(2) + 5 = 3(4) - 4(2) + 5 = 12 - 8 + 5 = 9$

(b) $f(-2) = 3(-2)^2 - 4(-2) + 5 = 3(4) - 4(-2) + 5 = 12 + 8 + 5 = 25$

(c) $f(4) = 3(4)^2 - 4(4) + 5 = 3(16) - 4(4) + 5 = 48 - 16 + 5 = 37$

(d) $f(0) = 3(0)^2 - 4(0) + 5 = 3(0) - 4(0) + 5 = 0 - 0 + 5 = 5$

Practice Problem 9 If $f(x) = -2x^2 + 3x - 8$, find:

(a) $f(-1)$ **(b)** $f(2)$ **(c)** $f(-3)$ **(d)** $f(0)$ ■

7.6 Exercises

Verbal and Writing Skills

1. What are the three ways you can use to describe a function?

2. What is the difference between a function and a relation?

3. The domain of a function is the set of ____________ ____________ of the ____________ variable.

4. The range of a function is the set of ____________ ____________ of the ____________ variable.

5. How can we tell if a graph is the graph of a function?

(a) Find the domain and range of each relation. (b) Determine if each relation is a function.

6. $\{(9, 3), (4, 5), (7, 8), (1, 3)\}$

7. $\{(2, 6), (6, 8), (9, 9), (6, 2)\}$

8. $\{(\frac{1}{2}, \frac{1}{2}), (-10, \frac{1}{2}), (7, \frac{1}{4}), (\frac{1}{2}, \frac{1}{4})\}$

9. $\{(\frac{1}{2}, 5), (\frac{1}{4}, 10), (\frac{3}{4}, 6), (\frac{1}{2}, 6)\}$

10. $\{(8.6, 3), (9.2, 6), (8.6, -3)\}$

11. $\{(10, 1), (1, 10), (8, 3), (3, 8)\}$

12. $\{(12, 1), (14, 3), (1, 12), (9, 12)\}$

13. $\{(5.6, 8), (5.8, 6), (6, 5.8), (5, 6)\}$

14. $\{(1, 20), (2, 20), (3, 20), (1, 40)\}$

15. $\{(40, 2), (50, 30), (30, 60), (40, 80)\}$

Graph the equation. Compare the graph with its algebraic expression. Is the graph what you might expect? If not, check your work.

16. $y = x^2 + 3$

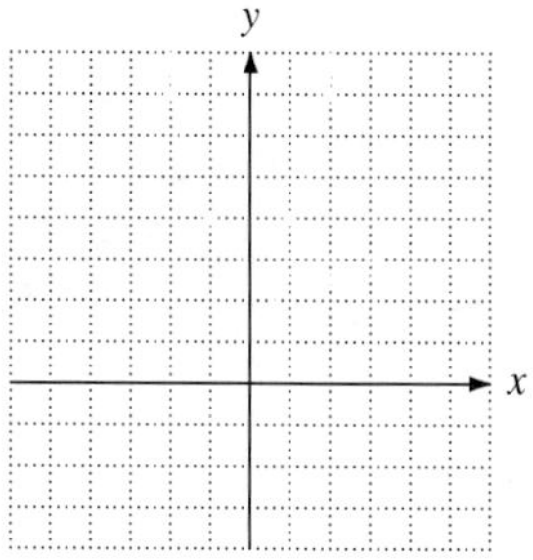

17. $y = x^2 - 1$

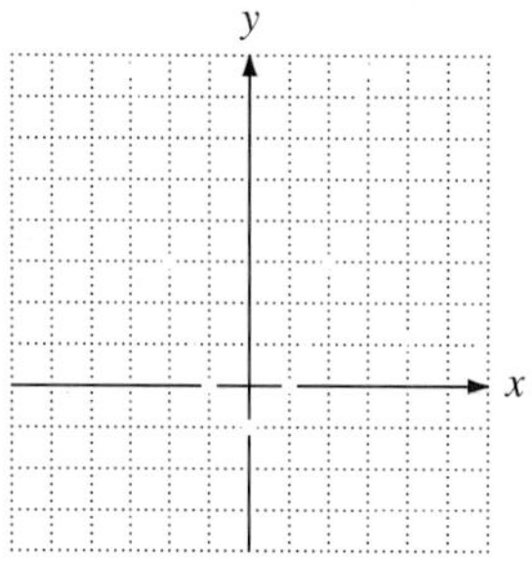

18. $(2, 8)(-2, 8)$

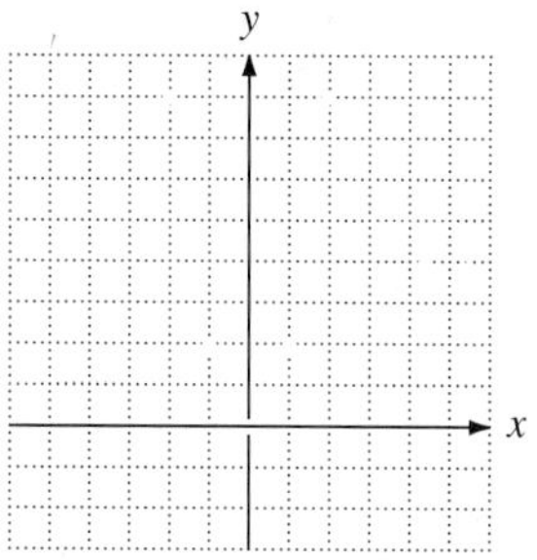

19. $y = \frac{1}{2}x^2$

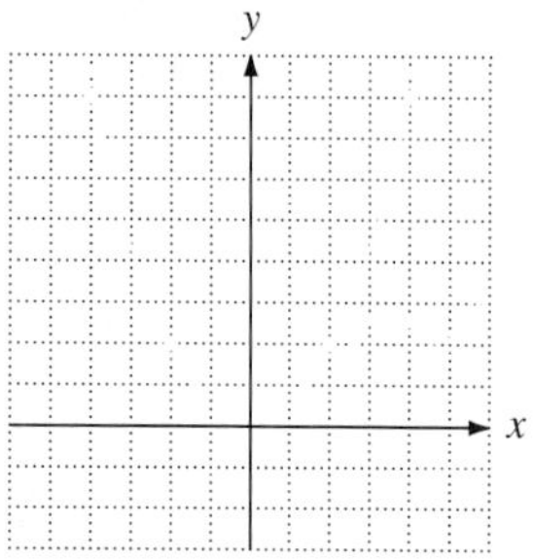

20. $x = y^2$

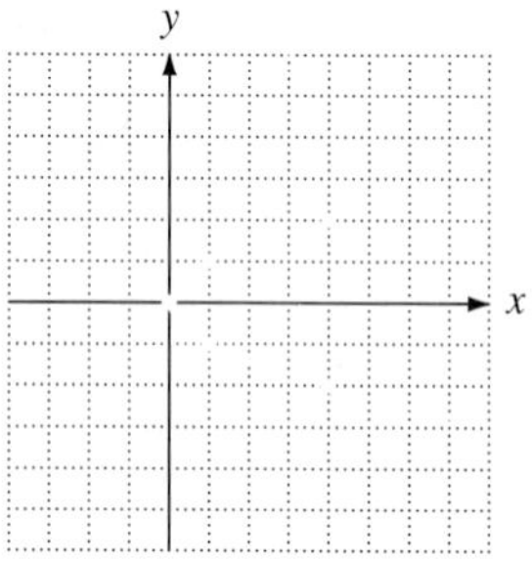

21. $x = y^2 - 3$

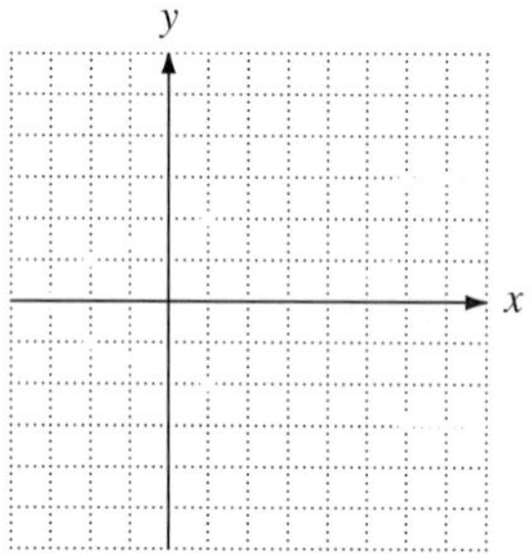

22. $x = y^2 + 1$

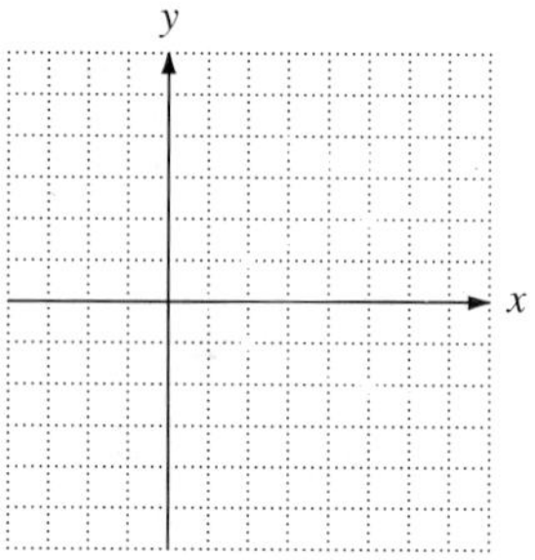

23. $x = 2y^2$

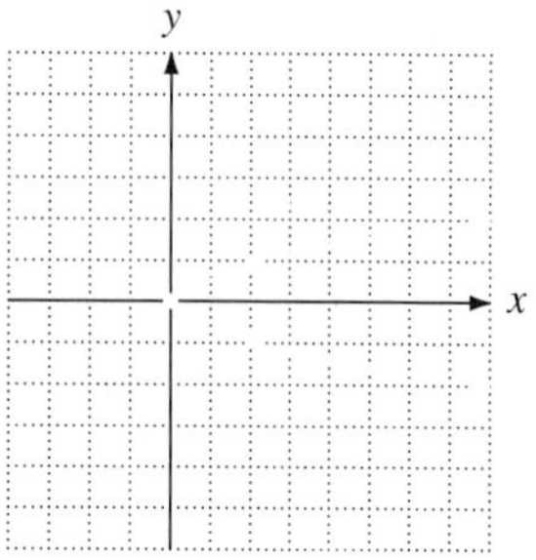

24. $y = \frac{2}{x}$

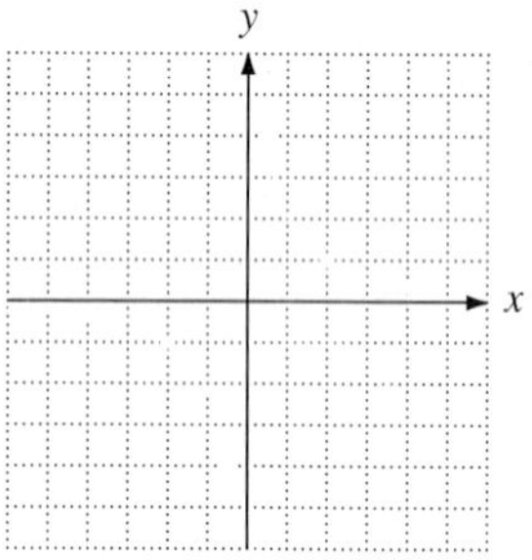

25. $y = -\frac{2}{x}$

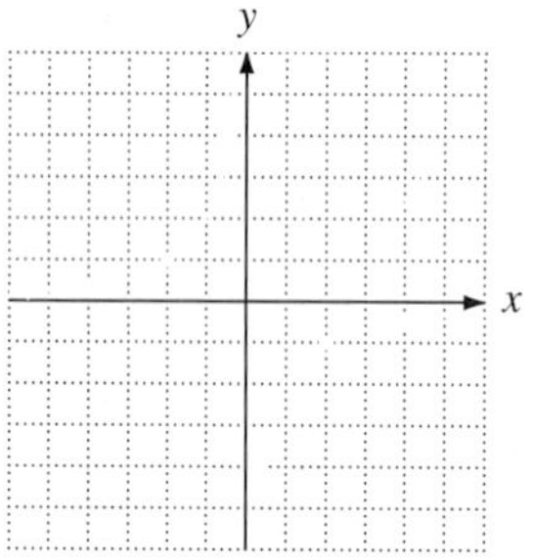

26. $y = \frac{1}{x}$

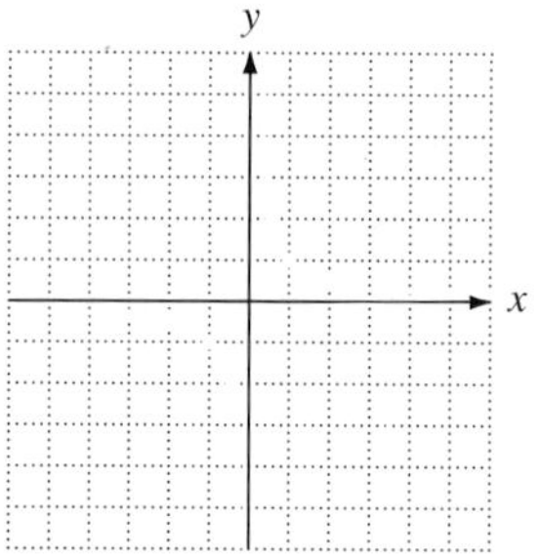

27. $y = -\frac{1}{x}$

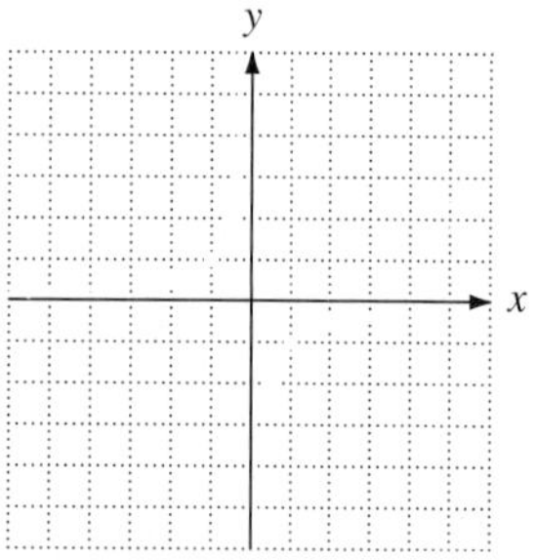

★ **28.** $y = (x + 1)^2$

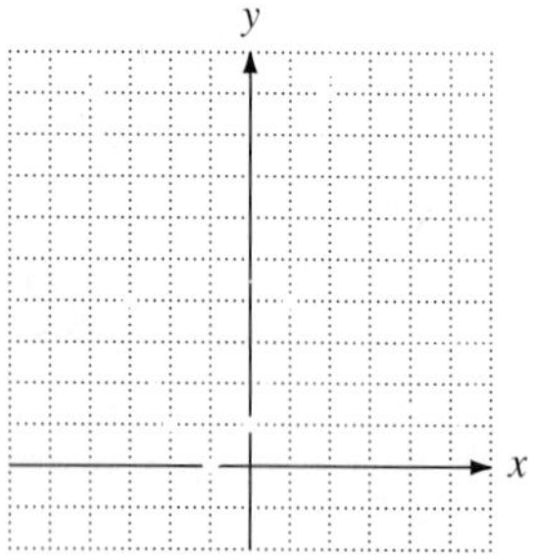

★ **29.** $x = (y - 2)^2$

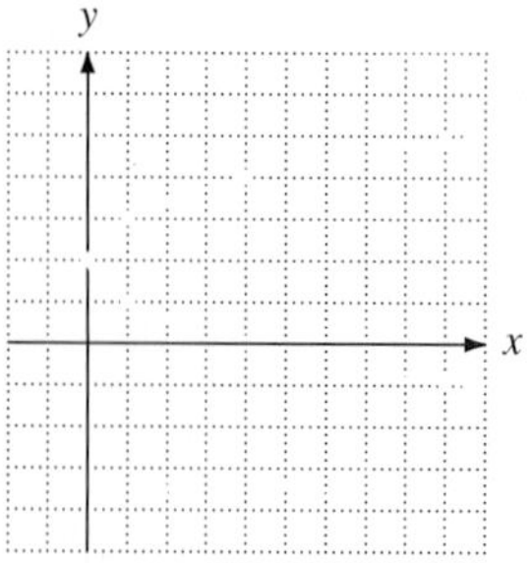

★ **30.** $y = \frac{4}{x - 2}$

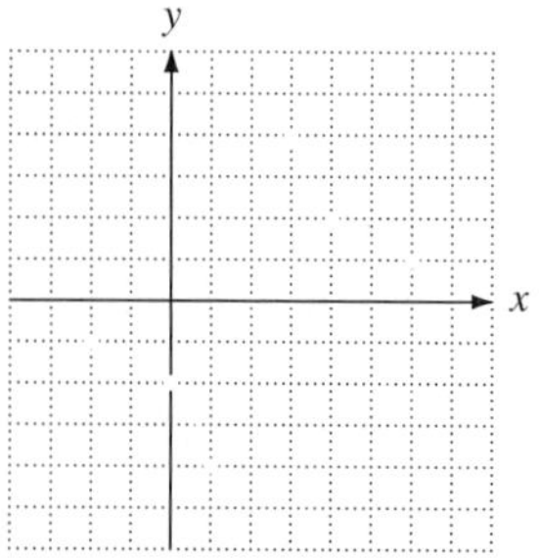

★ **31.** $x = \frac{2}{y + 1}$

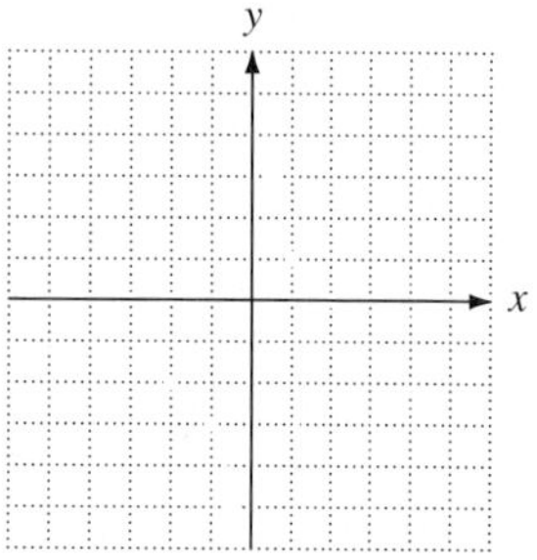

Determine if each relation is a function.

32.

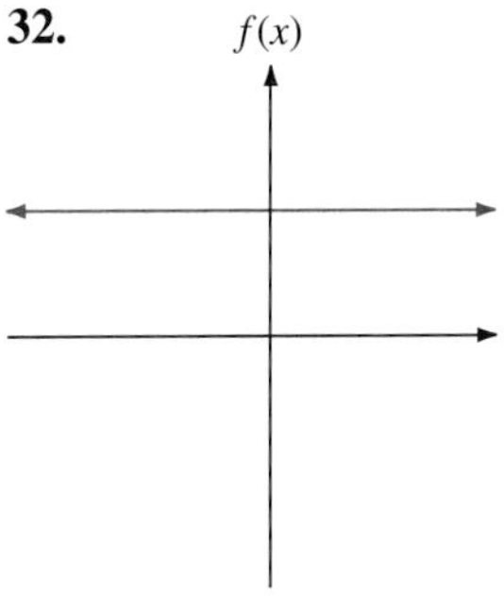

33.

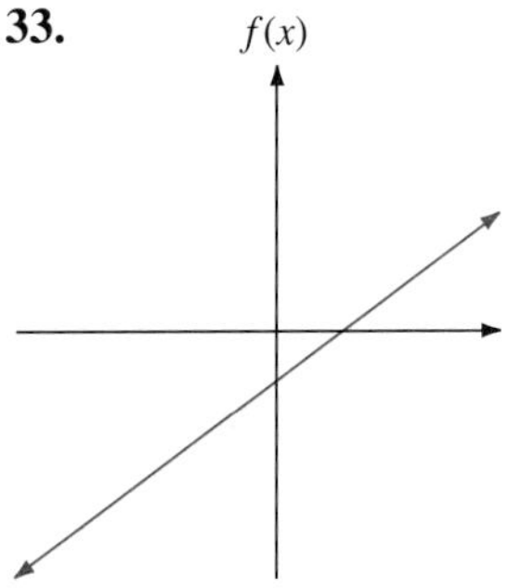

34.

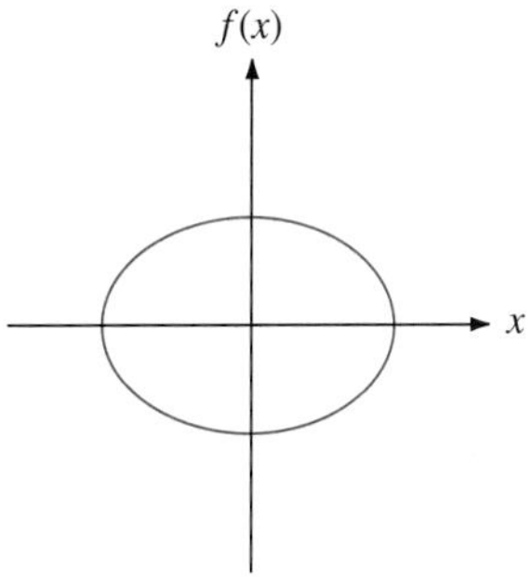

35.

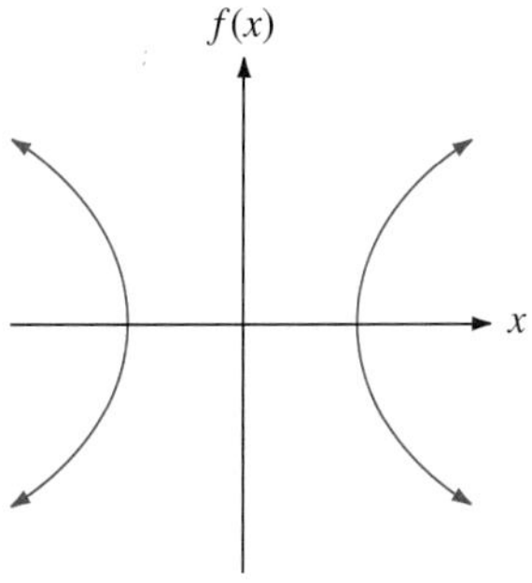

36.

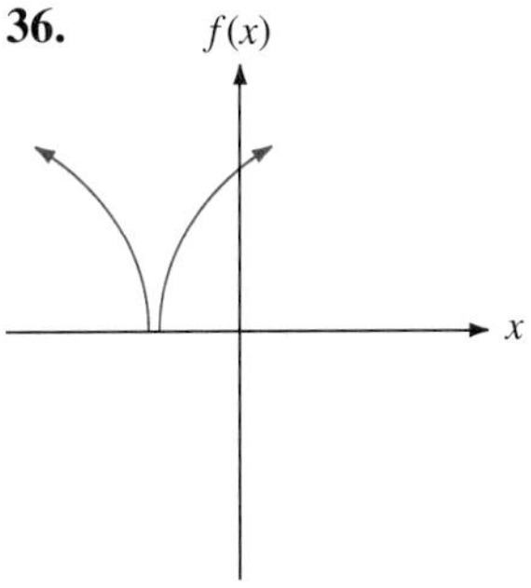

37.

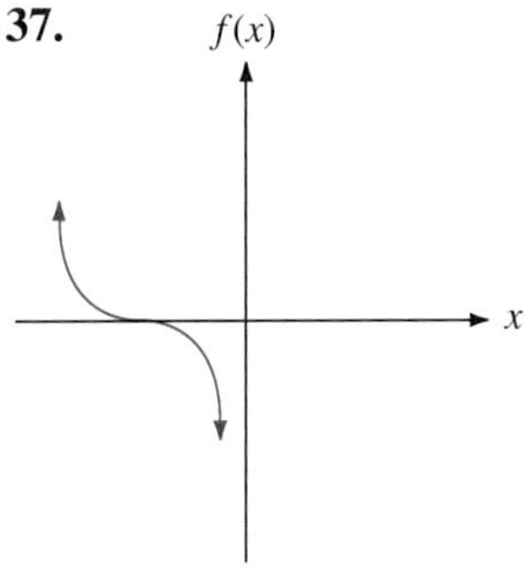

38.

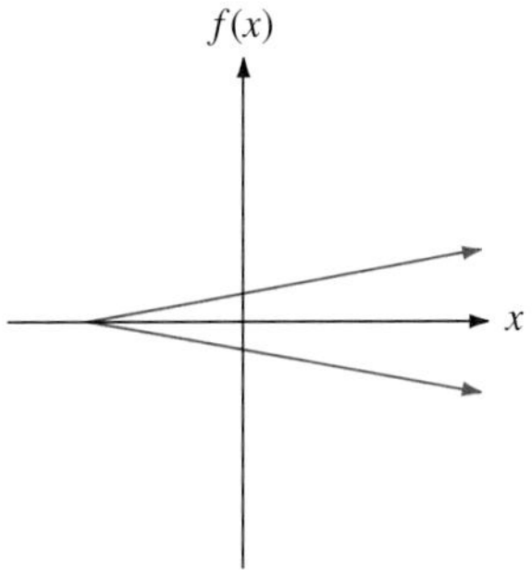

39.

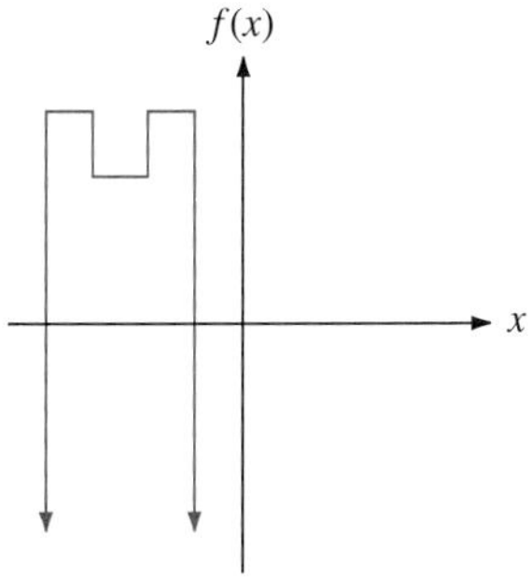

Given the following functions, find the indicated values.

40. $f(x) = 2x - 6$ (a) $f(-1)$ (b) $f(2)$ (c) $f(-3)$

41. $f(x) = 3x - 7$ (a) $f(1)$ (b) $f(-2)$ (c) $f(3)$

42. $f(x) = 4 - 8x$ (a) $f(2)$ (b) $f(0.5)$ (c) $f(-4)$

43. $f(x) = 1 - 10x$ (a) $f(0.5)$ (b) $f(-0.1)$ (c) $f(3)$

44. $f(x) = x^2 + 3x + 4$ (a) $f(2)$ (b) $f(0)$ (c) $f(-3)$

45. $f(x) = x^2 + 2x - 3$ (a) $f(-1)$ (b) $f(0)$ (c) $f(3)$

46. $g(x) = 2 - 6x^2$ (a) $g(0)$ (b) $g(4)$ (c) $g(-2)$

47. $g(x) = 3 - 4x^2$ (a) $g(0)$ (b) $g(2)$ (c) $g(-3)$

48. $f(x) = \frac{4}{x} + 1$ (a) $f(2)$ (b) $f(-2)$ (c) $f(-8)$

49. $f(x) = \frac{6}{x} - 2$ (a) $f(6)$ (b) $f(-3)$ (c) $f(-12)$

Cumulative Review Problems

Simplify each of the following

50. $(5x^2 - 6x + 2) + (-7x^2 - 8x + 4)$

51. $(-2x^2 - 3x + 8) - (x^2 + 4x - 3)$

52. $(x^2 + 2x + 1) + (x^2 - 3x + 2)$

53. $(6x^2 + 10x - 5) \div (3x - 1)$

Putting Your Skills to Work

COMPANY INCOME

Felicia works as a sales representative for a local cable television company. The company has tried to increase the number of subscribers in the local community. At the present time, 10,000 families subscribe to the basic cable package for \$15.00 per month. The company has tried lowering the package cost by \$0.25 per month. Each time the price is lowered by \$0.25, an additional 200 families subscribe to the basic monthly service.

1. Complete the adjacent table of values to obtain a pattern to describe the income. Does this table represent a function? Why?

Number of Subscribers	Cost per Month	Monthly Income for the Cable Company
10,000	\$15.00	\$150,000
10,200	\$14.75	\$150,450
10,400	\$14.50	\$150,800
10,600		
10,800		
11,000		
11,200		

2. Based on the values in your table, what price should be charged per month to obtain the maximum monthly income for the cable company?

3. Choose an appropriate scale for number of subscribers (N) as the horizontal axis and income (I) for the vertical axis, and graph the seven values in your table as ordered pairs (N, I). Connect the points by a smooth curve. What do you notice about the shape of the curve? How does this correspond to the maximum monthly income for the cable company?

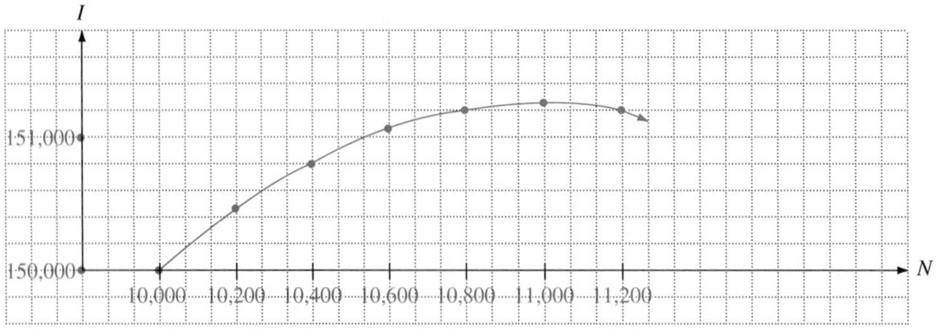

COOPERATIVE GROUP ACTIVITY

Let us represent the income function discussed above as a function in the form $I(X) = (10{,}000 + 200x)(15 - 0.25x)$.

4. Using the above equation, find the values of the income function when $x = 1$ and $x = 2$. When $x = 1$, what is $I(1)$? When $x = 2$, what is $I(2)$?

5. When $x = 1$, what is the number of subscribers? When $x = 2$, what is the number of subscribers?

6. What is the relationship between x and the number of subscribers?

7. For what value of x is the maximum income obtained based on the information that you studied in this exercise?

Chapter Organizer

Topic	Procedure	Examples
Plotting points, p. 396.	To plot (x, y): **1.** Begin at the origin. **2.** If x is positive, move to the right along the x-axis. If x is negative, move to the left along the x-axis. **3.** If y is positive, move up. If y is negative, move down.	To plot $(-2, 3)$: 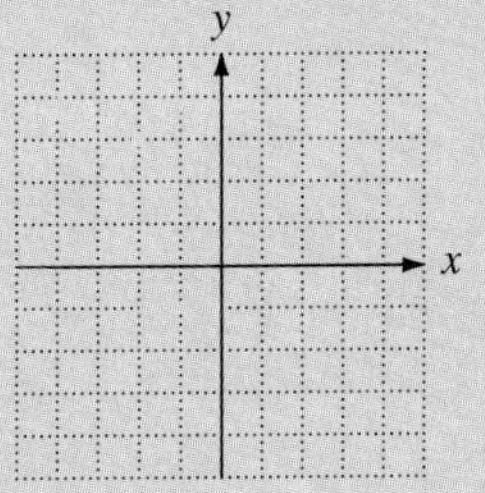
Graphing straight lines, p. 404.	An equation of the form $Ax + By = C$ has a graph that is a straight line. To graph such an equation, plot any three points; two give the line and the third checks it. (Where possible, use the x- and y-intercepts.)	Graph $3x + 2y = 6$. x | y 0 | 3 4 | −3 2 | 0 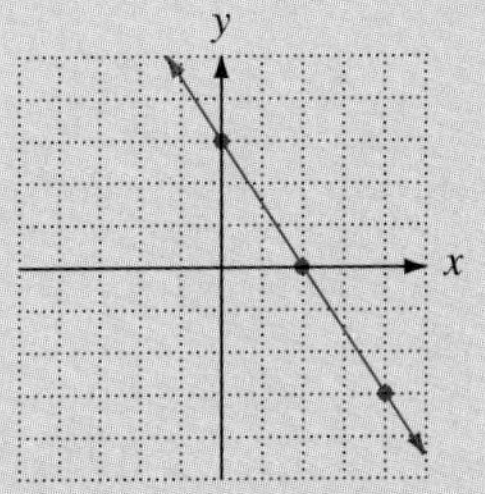
Finding the slope given two points, p. 414.	Nonvertical lines passing through distinct points (x_1, y_1) and (x_2, y_2) have slope $m = \frac{y_2 - y_1}{x_2 - x_1}$ The slope of a horizontal line is 0. The slope of a vertical line is undefined.	What is the slope of the line through (2, 8) and (5, 1)? $m = \frac{1 - 8}{5 - 2} = -\frac{7}{3}$
Finding the slope and y-intercept of a line given the equation, p. 418.	**1.** Place the equation in the form $y = mx + b$. **2.** The slope is m. **3.** The y-intercept is b.	Find the slope and y-intercept of $3x - 4y = 8$ $-4y = -3x + 8$ $y = \frac{3}{4}x - 2$ The slope is $\frac{3}{4}$. The y-intercept is -2.
Finding the equation of a line given the slope and y-intercept, p. 420.	The slope-intercept form of the equation of a line is $y = mx + b$ The slope is m and the y-intercept is at $(0, b)$.	Find the equation of the line through (0, 7) with slope $m = 3$. $y = 3x + 7$
Graphing a line using slope-intercept, p. 420.	**1.** Plot the y-intercept. **2.** Starting from $(0, b)$, plot a second point using the slope. slope $= \frac{\text{rise}}{\text{run}}$ **3.** Connect the two points.	Graph $y = -4x + 1$. Slope $= -4$ or $\frac{-4}{1}$ 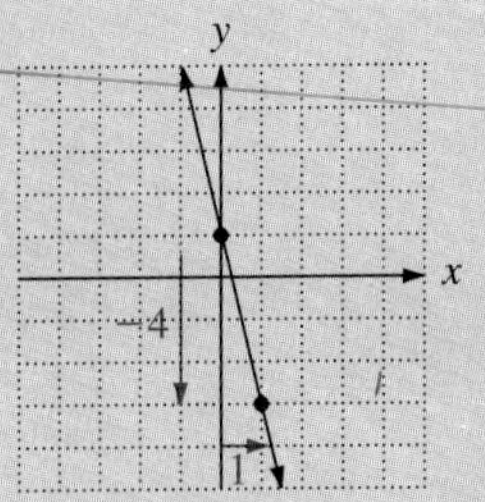

Chapter Organizer

Topic	Procedure	Examples
Finding the slope of parallel and perpendicular lines, p. 421.	Parallel lines have the same slope. Perpendicular lines have slopes whose product is -1.	Line q has a slope of 2. The slope of a line parallel to q is 2. The slope of a line perpendicular to q is $-\frac{1}{2}$.
Finding the equation of a line through a point with a given slope, p. 428.	**1.** Substitute the known values in the equation $y = mx \pm b$. **2.** Solve for b. **3.** Use the values of m and b to write the general equation.	Find the equation of the line through (3, 2) with slope $m = \frac{4}{5}$. $y = mx + b$ $2 = \frac{4}{5}(3) + b$ $2 = \frac{12}{5} + b$ $-\frac{2}{5} = b$ The equation is $y = \frac{4}{5}x - \frac{2}{5}$.
Finding the equation of a line through two points, p. 429.	**1.** Find the slope. **2.** Use the procedure when given a point and the slope.	Find the equation of the line through (3, 2) and (13, 10). $m = \frac{y_2 - y_1}{x_2 - x_1} = \frac{10 - 2}{13 - 3} = \frac{8}{10} = \frac{4}{5}$ We choose to use the point (3, 2). $y = mx + b$ $2 = \frac{4}{5}(3) + b$ $2 = \frac{12}{5} + b$ $-\frac{2}{5} = b$ The equation is $y = \frac{4}{5}x - \frac{2}{5}$.
Writing the equation of a line given the graph, p. 429.	**1.** Identify the y-intercept. **2.** Find the slope. $\text{slope} = \frac{\text{change in } y}{\text{change in } x}$	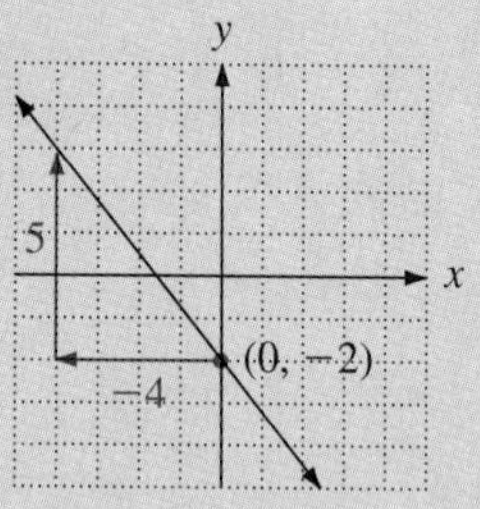 The equation is $y = -\frac{5}{4}x - 2$
Graphing linear inequalities, p. 432.	**1.** Graph as if it were an equation. If the inequality symbol is > or <, use a dashed line. If the inequality symbol is ≥ or ≤, use a solid line. **2.** Look for a test point. The easiest test point is (0, 0). **3.** Substitute the coordinates of the test point into the inequality. **4.** If it is a true statement, shade the side of the line containing the test point. If it is a false statement, shade the side of the line that does *not* contain the test point.	Graph $y \geq 3x + 2$. Graph the line $y = 3x + 2$. Use a solid line. Test (0, 0). $0 \geq 3(0) + 2$ $0 \geq 2$ false Shade the side of the line that does not contain (0, 0). 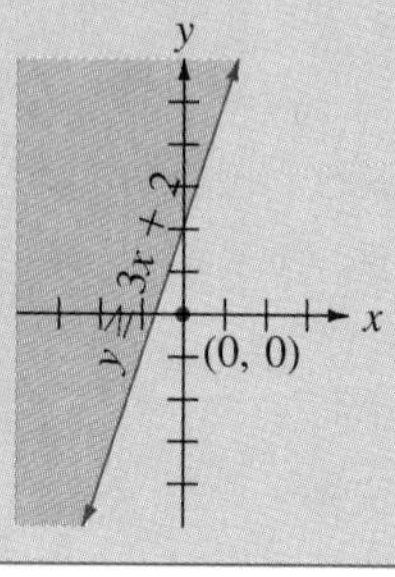

Chapter Organizer

Topic	Procedure	Examples
Determining if a relation is a function, p. 436.	A function is a relation in which no two different ordered pairs have the same first coordinates.	Is this relation a function? $\{(5, 7), (3, 8), (5, 10)\}$ It is *not* a function since in (5, 7) and (5, 10) we have two different ordered pairs with an x-coordinate of 5.
Determining if a graph is a function, p. 440.	If a vertical line can intersect the graph of a relation more than once, the relation is not a function.	Is this graph a function? 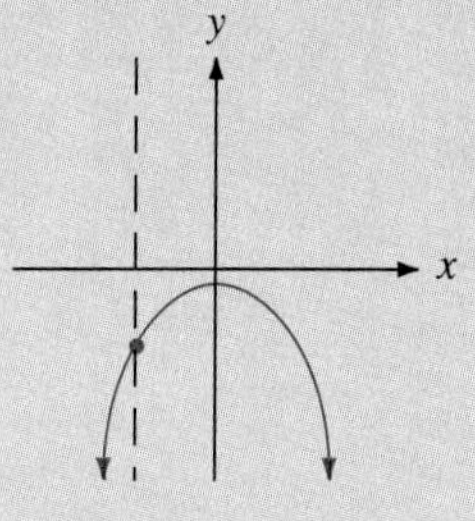 Yes. A vertical line can only intersect it once.

Chapter 7 Review Problems

7.1

1. Plot the following points.
A: (2, −3) B: (−1, 0) C: (3, 2) D: (−2, −3)

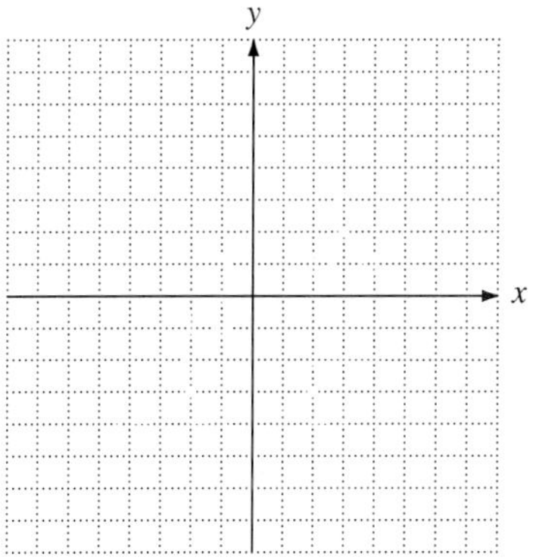

2. Give the coordinates of each point.

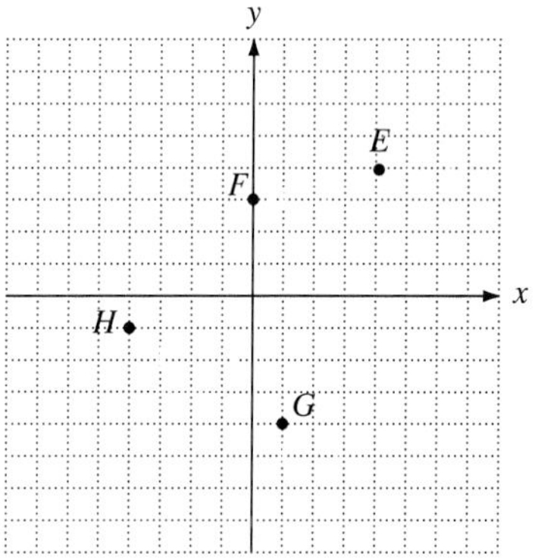

Complete the ordered pairs so that each is a solution to the given equation.

3. $y = 3x - 5$
(a) (0,   ) **(b)** (3,   )

4. $2x + 5y = 12$
(a) (1,   ) **(b)** (, 4)

5. $x = 6$
(a) (, −1) **(b)** (, 3)

7.2

6. Graph $3y = 2x + 6$.

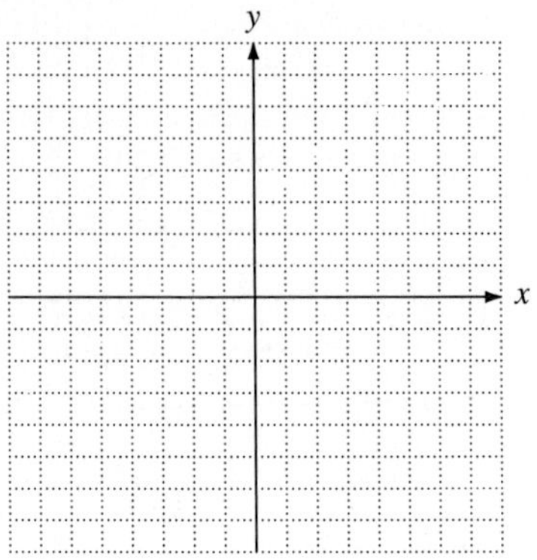

7. Graph $5y + x = -15$.

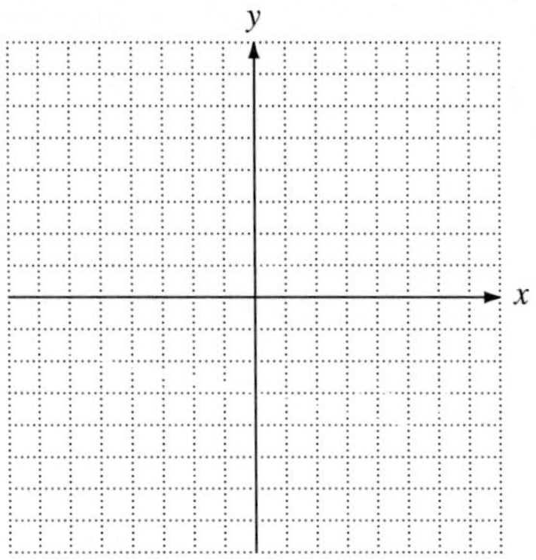

8. Graph $3x + 5y = 15 + 3x$.

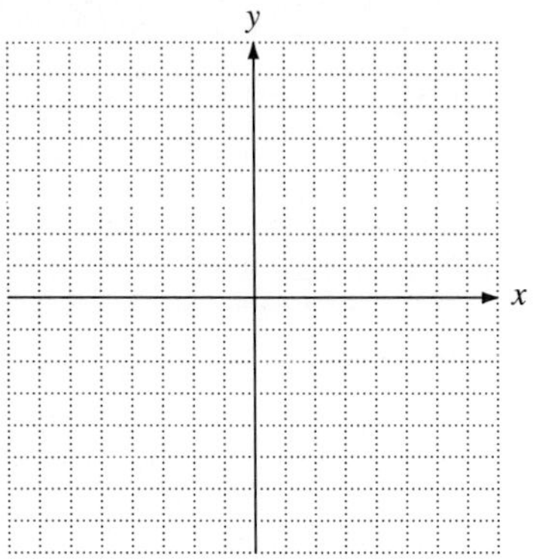

7.3

9. Find the slope of a line passing through $(5, -3)$ and $(2, -\frac{1}{2})$.

10. Find the slope and y-intercept of the line $5x - 8y + 12 = 0$.

11. Write the equation of a line with slope $-\frac{1}{2}$ and y-intercept of 3.

12. Write the equation of a line with slope 0 and y-intercept of 7.

13. A line has a slope of $-\frac{2}{3}$. What is the slope of a line perpendicular to that line?

14. A line has a slope of $\frac{1}{7}$. What is the slope of a line parallel to that line?

15. Graph $y = -\frac{1}{2}x + 3$.

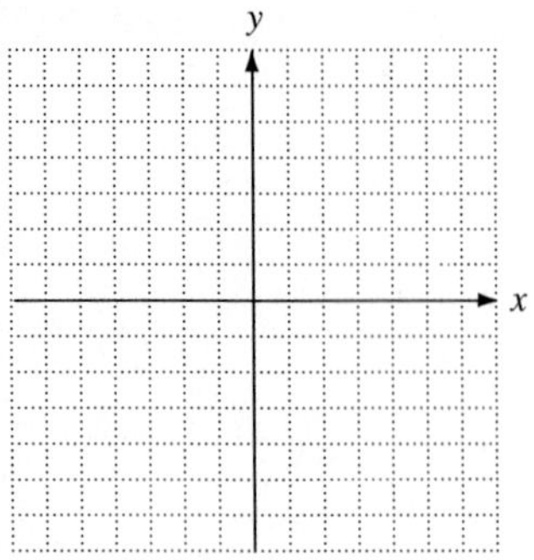

16. Graph $2x - 3y = -12$.

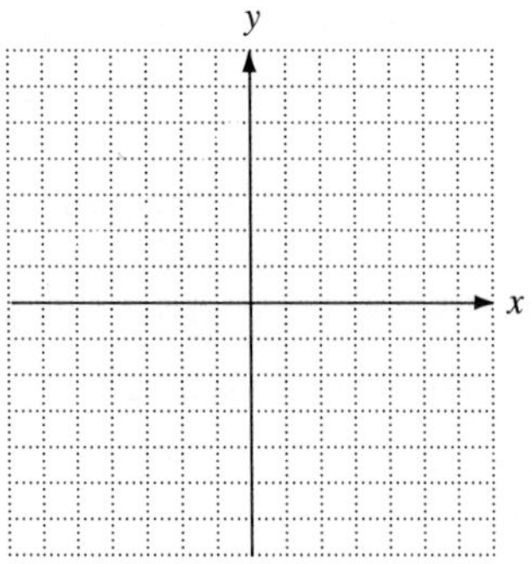

17. Graph $5x + 2y = 20 + 2y$.

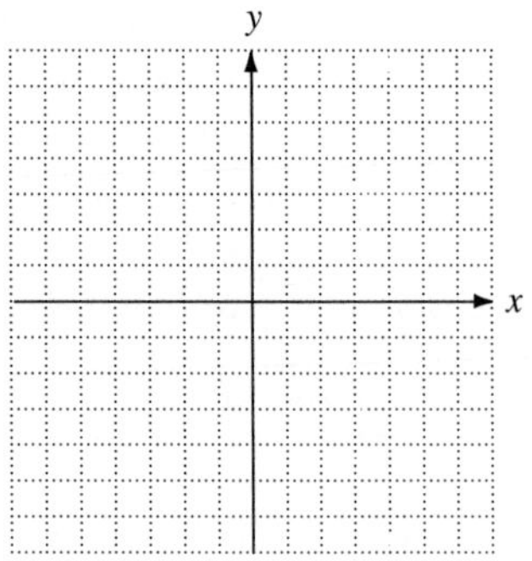

7.4

18. Write the equation of a line passing through $(5, 6)$ having a slope of 2.

19. Write the equation of a line passing through $(3, -4)$ having a slope of -6.

20. Write the equation of a line passing through $(6, -3)$ having a slope of $\frac{1}{3}$.

21. Write the equation of a line passing through (5, 2) and (−3, 6).

22. Write the equation of a line passing through (6, 1) and (2, −3).

23. Write the equation of a line passing through (0, 5) and (−1, 5).

Write the equation of each graph.

24.

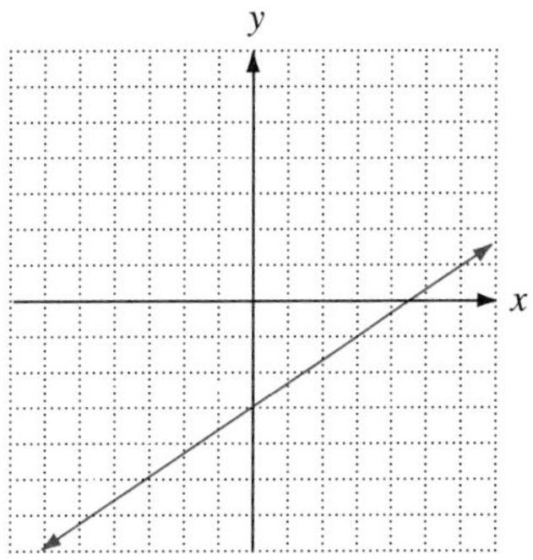

25.

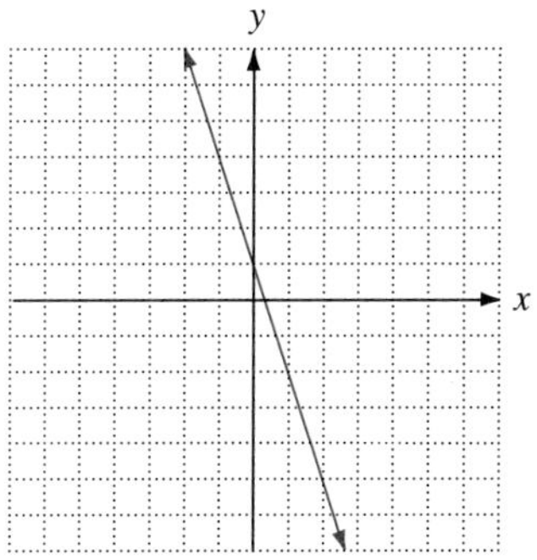

26.

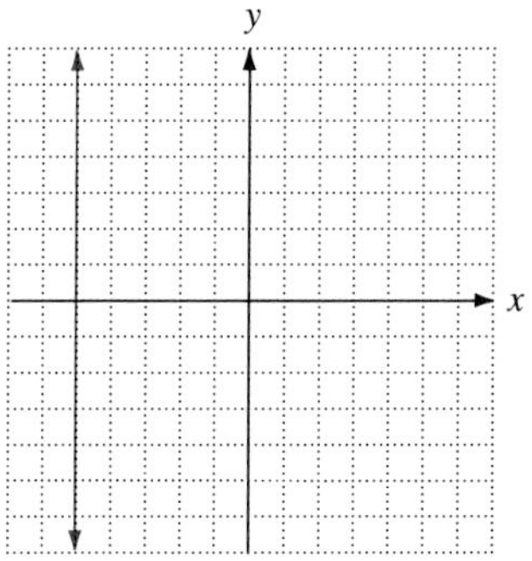

7.5

27. Graph $y < \frac{1}{3}x + 2$.

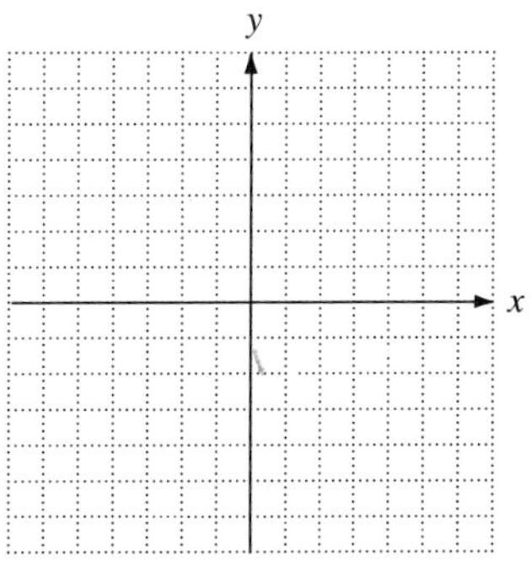

28. Graph $3y + 2x \geq 12$.

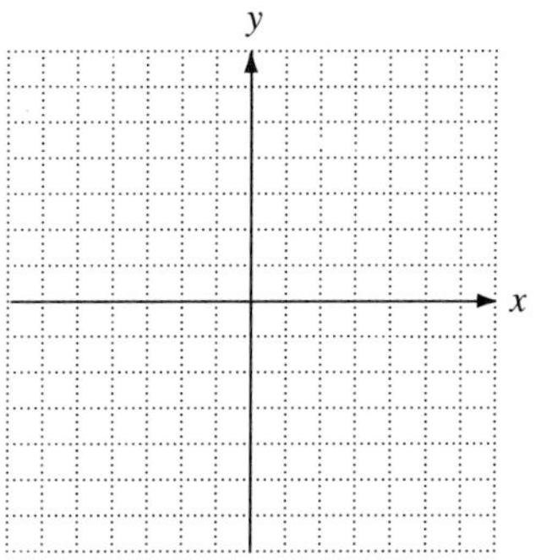

29. Graph $y \geq -2$.

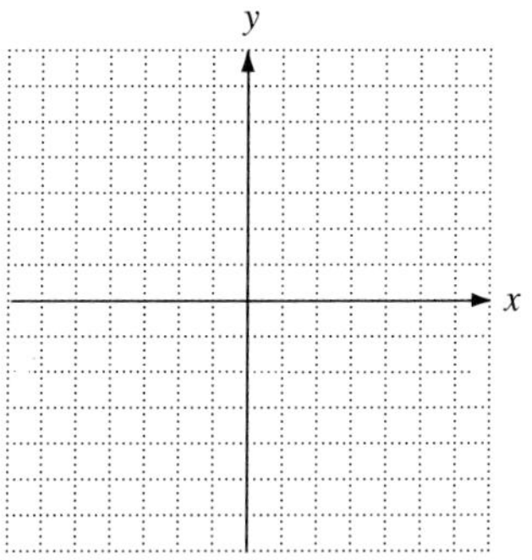

7.6

Determine the domain and range of each relation. Determine if each relation is a function.

30. {(−3, −2), (0, 4), (1, 7), (6, −2)}

31. {(0, 6), (1, 4), (2, 8), (0, 8)}

32. {(5, −6), (−6, 5), (−5, 5), (−6, −6)}

33. {(3, −7), (−7, 3), (−3, 7), (7, −3)}

Which of the following graphs represent a function?

34.

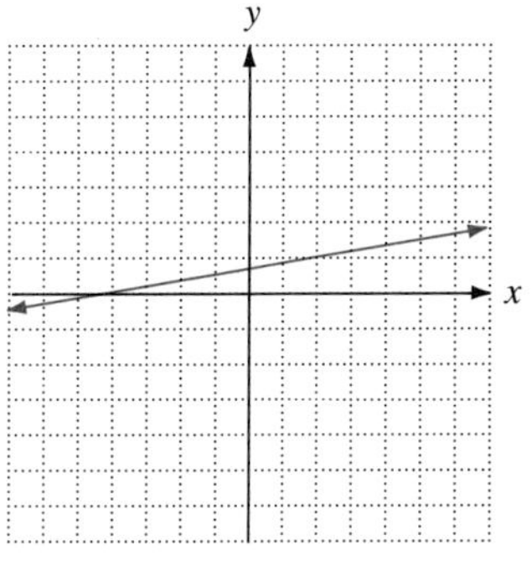

35.

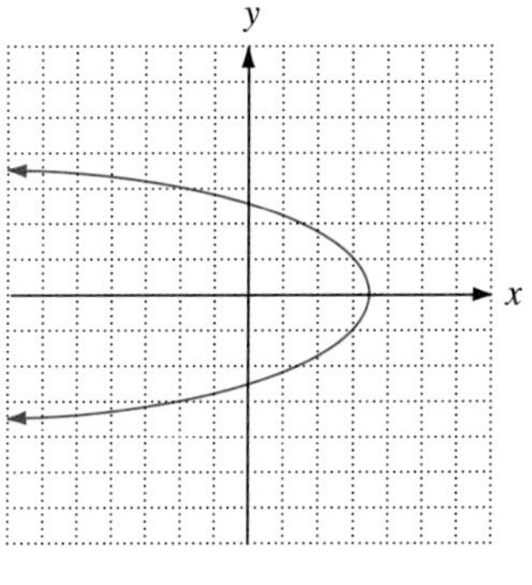

36.

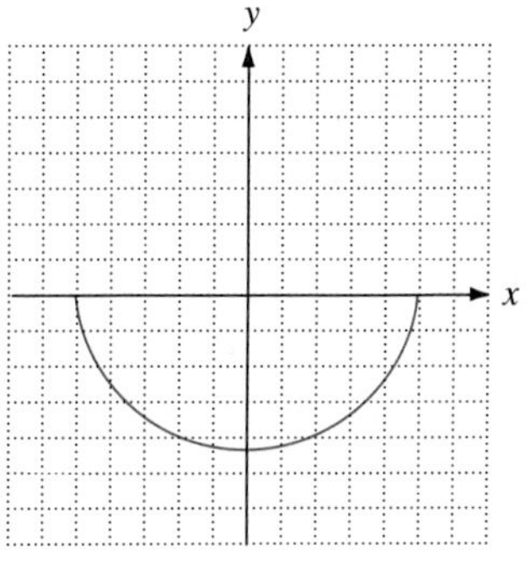

37. Graph $y = x^2 - 5$.

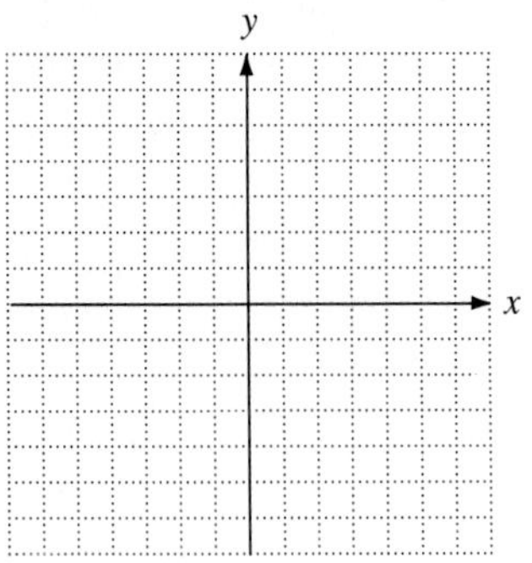

38. Graph $x = y^2 + 3$.

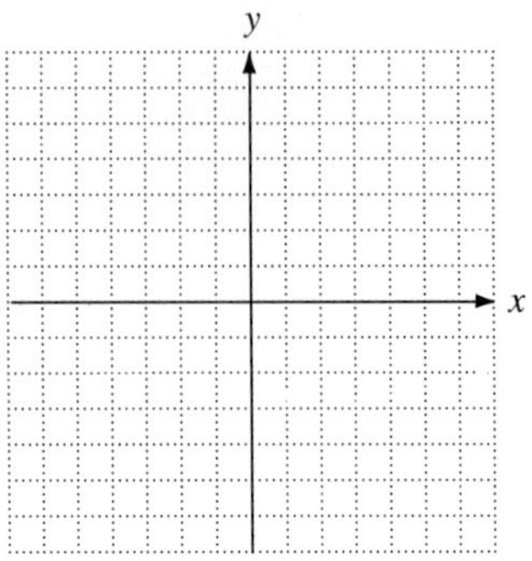

39. Graph $y = (x - 3)^2$.

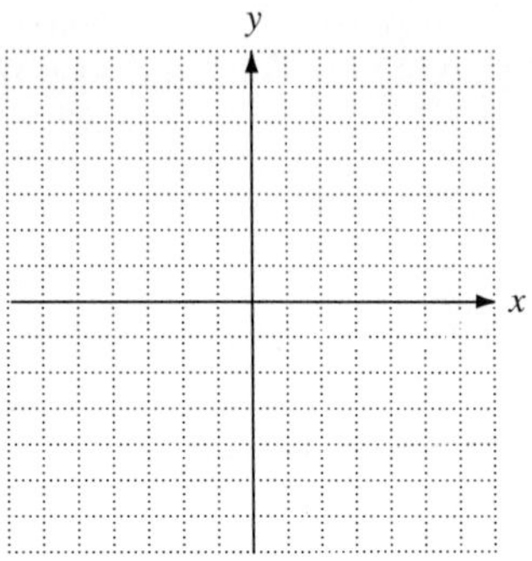

Given the following functions, find the indicated values.

40. $f(x) = 7 - 6x$ **(a)** $f(0)$ **(b)** $f(-2)$

41. $f(x) = 2x^2 - x + 3$ **(a)** $f(2)$ **(b)** $f(-3)$

42. $g(x) = -2x^2 + 3x + 4$ **(a)** $g(-1)$ **(b)** $g(2)$

43. $h(x) = -5 + 4x^2$ **(a)** $h(-3)$ **(b)** $h(4)$

44. $p(x) = \dfrac{-3}{x}$ **(a)** $p(-3)$ **(b)** $p(0.5)$

45. $f(x) = \dfrac{2}{x + 4}$ **(a)** $f(-2)$ **(b)** $f(6)$

Chapter 7 Test

1. Plot and label the following points:
C: $(-4, -3)$ B: $(6, 1)$
D: $(-3, 0)$ E: $(5, -2)$

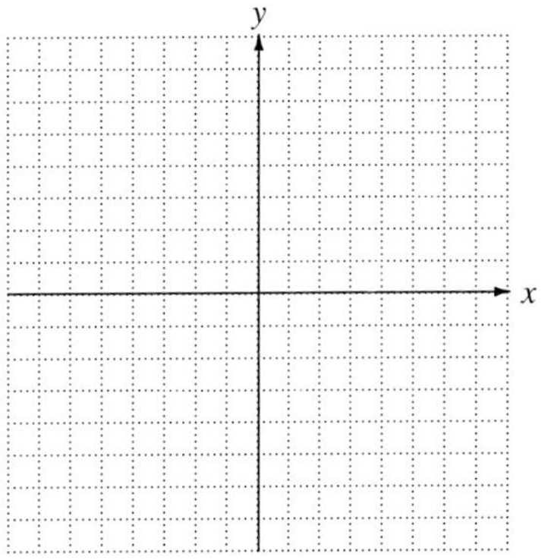

2. Graph the line $6x - 3 = 5x - 2y$.

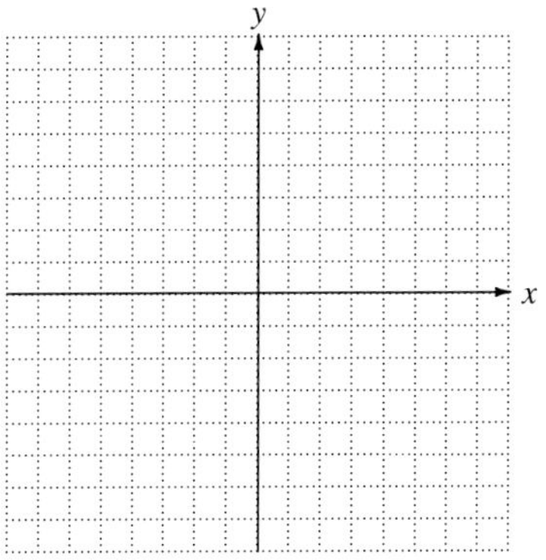

3. Graph the line $12x - 3y = 6$.

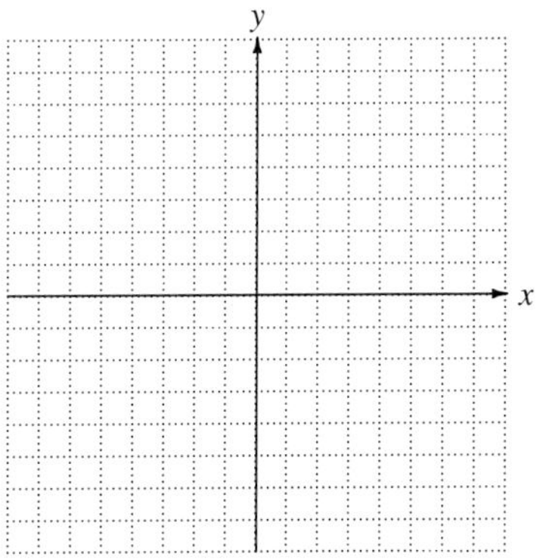

4. Graph $y = \frac{2}{3}x - 4$.

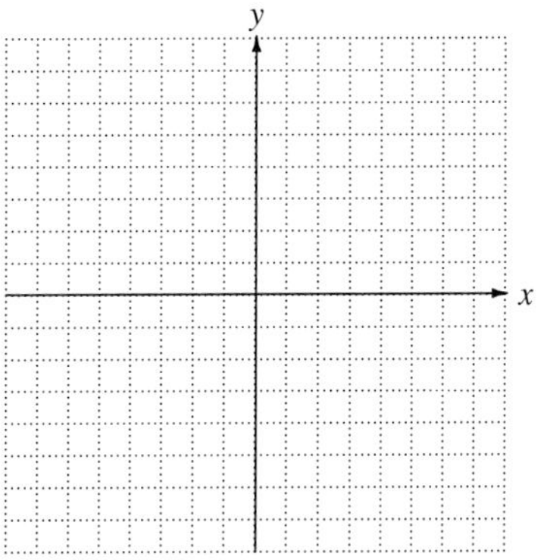

5. What is the slope and the y-intercept of the line $3x + 2y - 5 = 0$?

6. Find the slope of a line that passes through $(5, -2)$ and $(-3, -6)$.

7. Write the equation of the line that passes through $(4, -2)$ and has a slope of $\frac{1}{2}$.

8. Find the slope of the line through $(-3, 11)$ and $(6, 11)$.

9. Find the equation of a line passing through $(2, 5)$ and $(8, 3)$.

10. Write an equation for the line through $(2, 7)$ and $(2, -2)$. What is the slope of this line?

1. ______

2. ______

3. ______

4. ______

5. ______

6. ______

7. ______

8. ______

9. ______

10. ______

11.

12.

13.

14.

15.

16. (a)

(b)

17. (a)

(b)

11. Graph the region described by $4y \leq 3x$.

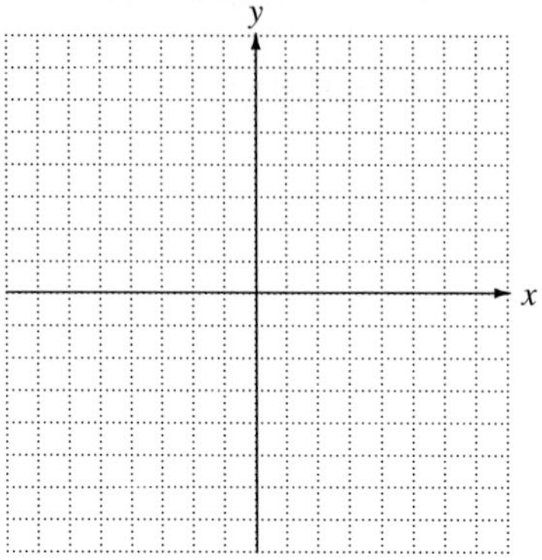

12. Graph the region described by $-3x - 2y > 10$.

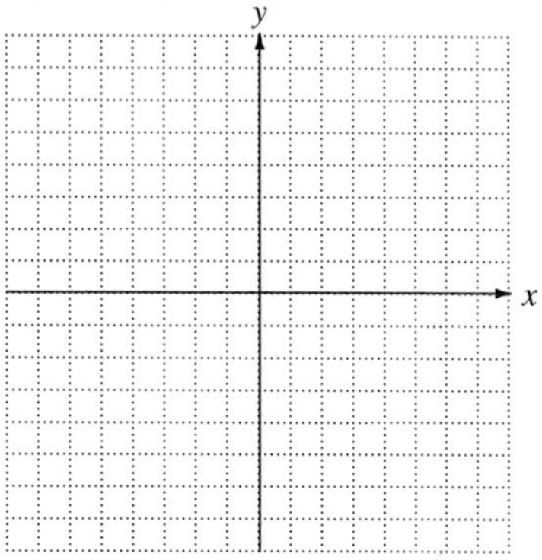

13. Is this relation a function? $\{(-8, 2), (3, 2), (5, -2)\}$ Why?

14. Is this relation a function? Why?

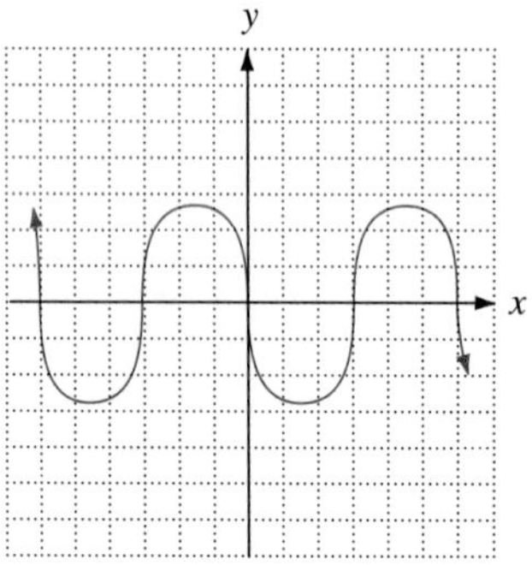

15. Graph $y = 2x^2 - 3$.

x	y
-2	
-1	
0	
1	
2	

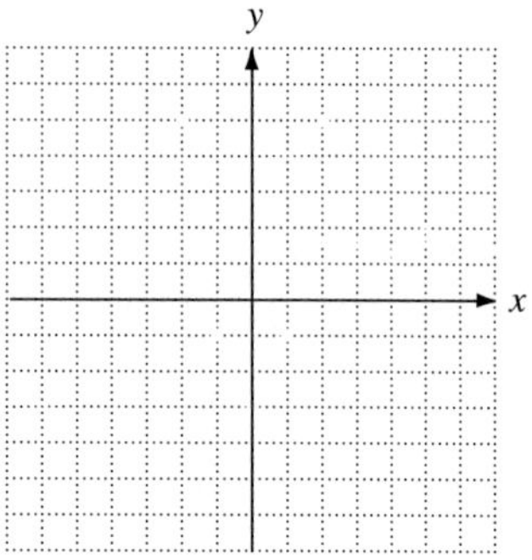

16. For $f(x) = 2x^2 + 3x - 4$:
(a) Find $f(0)$.
(b) Find $f(-2)$.

17. For $g(x) = \dfrac{3}{x - 4}$:
(a) Find $g(3)$.
(b) Find $g(0)$.

Cumulative Test for Chapters 0–7

Approximately one-half of this test covers the content of Chapters 0–6. The remainder covers the content of Chapter 7.

1. Simplify. $(-3x^3y^4z)^3$

2. Simplify. $-\dfrac{18a^3b^4}{24ab^6}$

Solve.

3. $2 - 3(4 - x) = x - (3 - x)$

4. $\dfrac{600}{R} = \dfrac{600}{R + 50} + \dfrac{500(R + 20)}{R^2 + 50R}$

5. $x = 2^3 - 6 \div 3 - 1$

6. $A = lw + lh$, for w

7. Factor. $50a^2 - 98b^2$

8. Factor. $2x^2 + 13x + 20$

9. Simplify. $\dfrac{x - \dfrac{1}{x}}{\dfrac{1}{2} + \dfrac{1}{2x}}$

10. Find the equation of the line through (6, 8) and (7, 11).

1. ______
2. ______
3. ______
4. ______
5. ______
6. ______
7. ______
8. ______
9. ______
10. ______

11.

12.

13.

14.

15.

16.

17.

18.

19.

20. (a)

(b)

Give the equations of the lines.

11. Through (7, −4) and vertical.

12. Through (−2, 3) with slope of $\frac{1}{3}$.

13. Find the slope of a line through (−8, −3) and (11, −3).

14. What is the slope of the line $3x - 7y = -2$?

15. Graph $y = \frac{2}{3}x - 4$.

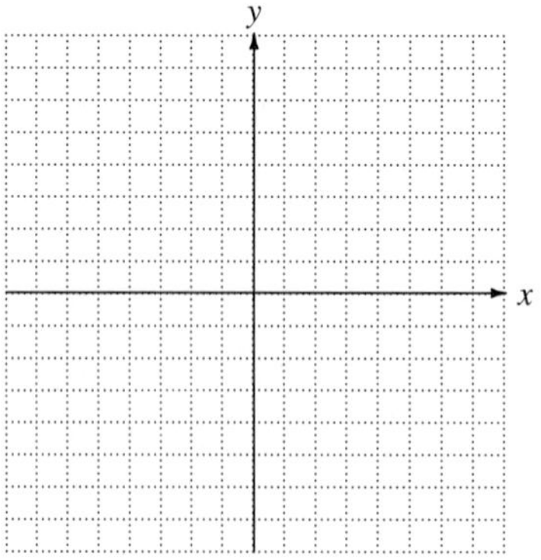

16. Graph $3x + 8 = 5x$.

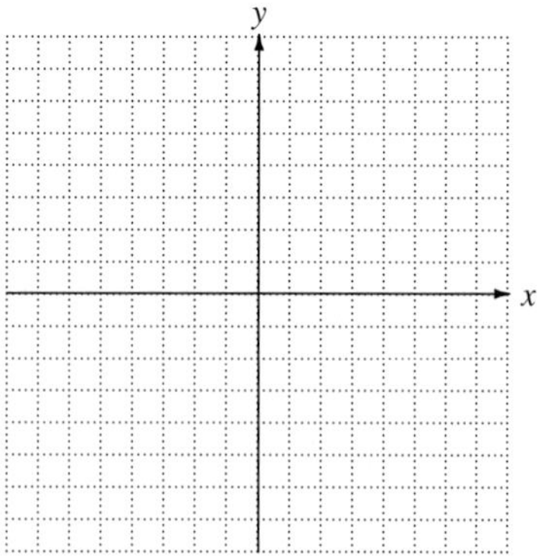

17. Graph the region $2x + 5y \leq -10$.

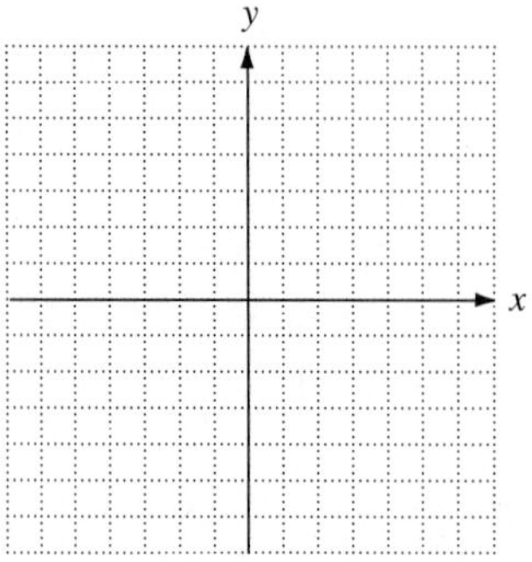

18. Graph $x = y^2 + 3$.
Is this a function?

x	y
	−2
	−1
	0
	1
	2

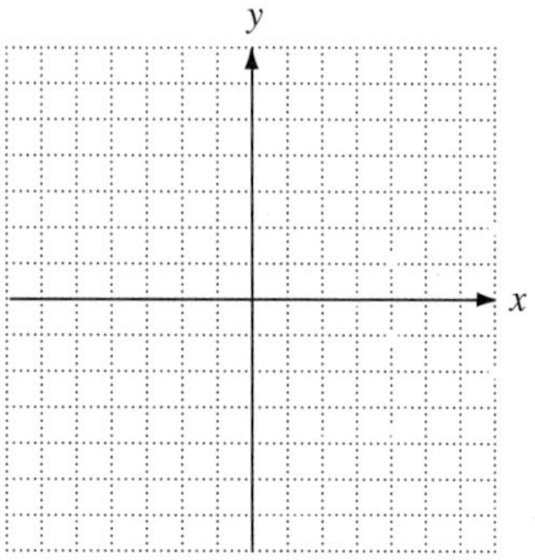

19. Is this relation a function?
{(3, 10), (10, −3), (0, −3), (5, 10)}

20. For $f(x) = -2x^2 - 3x + 5$:
(a) Find $f(0)$.
(b) Find $f(-2)$.

8
Systems of Equations

Systems of equations are used to solve a variety of problems in business, from small family-owned businesses to giant corporations. See Putting Your Skills to Work, pages 492 and 498.

Pretest Chapter 8

1. ______
2. ______
3. ______
4. ______
5. ______
6. ______
7. ______
8. ______
9. ______
10. ______
11. ______
12. ______
13. ______
14. ______

If you are familiar with the topics in this chapter, take this test now. Check your answers with those in the back of the book. If an answer was wrong or you couldn't do a problem, study the appropriate section of the chapter.

If you are not familiar with the topics in this chapter, don't take this test now. Instead, study the examples, work the practice problems, and then take the test.

This test will help you identify those concepts that you have mastered and those that need more study.

Section 8.1

Solve each system by the graphing method.

1. $4x + 2y = -8$
$-2x + 3y = 12$

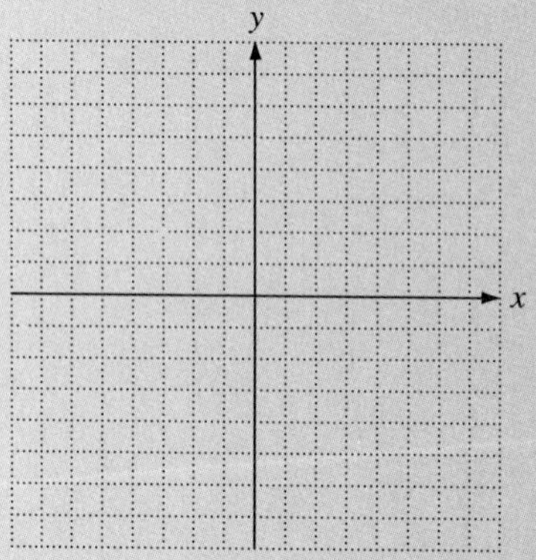

2. $x - 2y = -2$
$3x - 6y = 12$

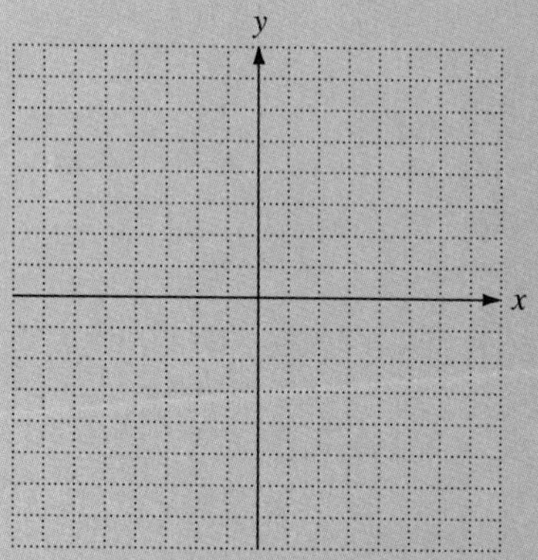

Section 8.2

Solve each system by the substitution method.

3. $3x + y = -11$
$4x - 3y = -6$

4. $\frac{2}{3}x + y = 1$
$3x + 2y = 3$

Section 8.3

Solve each system by the addition method.

5. $5x - 3y = 10$
$4x + 7y = 8$

6. $\frac{1}{2}x - \frac{3}{4}y = \frac{3}{2}$
$x + y = 8$

7. $0.8x - 0.3y = 0.7$
$1.2x + 0.6y = 4.2$

8. If a system has one solution, the graphs of the equations will be lines that ______.

9. If a system has no solutions, the graphs of the equations will be lines that ______.

10. If a system has an infinite number of solutions, the graphs of the equations will be lines that ______.

Section 8.4

Solve by any method.

11. $x + 8y = 7$
$5x - 2y = 14$

12. $3x - y = 5$
$-5x + 2y = -10$

13. $2x - 3y = 7$
$3x + 5y = 1$

Section 8.5

14. James and Michael work sorting letters for the post office. Yesterday James sorted letters for 3 hours and Michael for 2 hours. Together they sorted 2250 letters. Today James sorted letters for 2 hours and Michael for 3 hours. Together they sorted 2650 letters. How many letters per hour can be sorted by James? by Michael?

8.1 Solve a System of Equations by Graphing

After studying this section, you will be able to:

1 *Solve a system of linear equations that has a unique solution by graphing.*
2 *Graph a linear system of equations to determine the type of solution.*

In Chapter 7 we examined linear equations and inequalities in two variables. In this chapter we will examine systems of linear equations and inequalities. Recall that the graph of a linear equation is a straight line. If we have two linear equations on one coordinate plane, what are the possibilities with regard to their graphs?

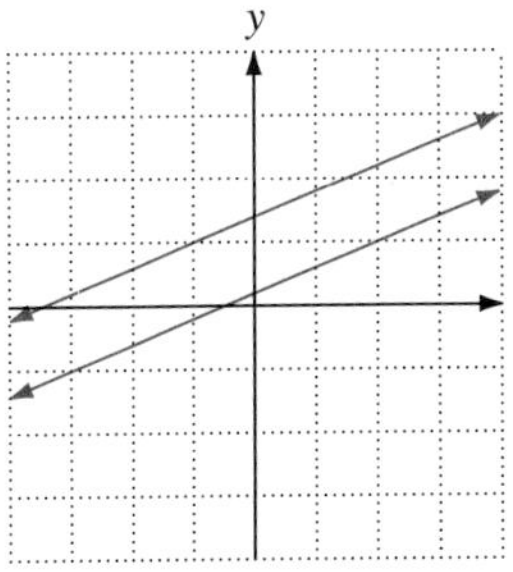

The lines may be parallel.

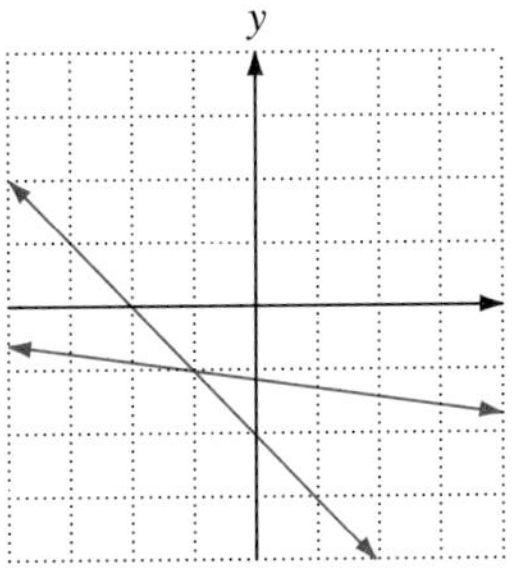

The lines may intersect.

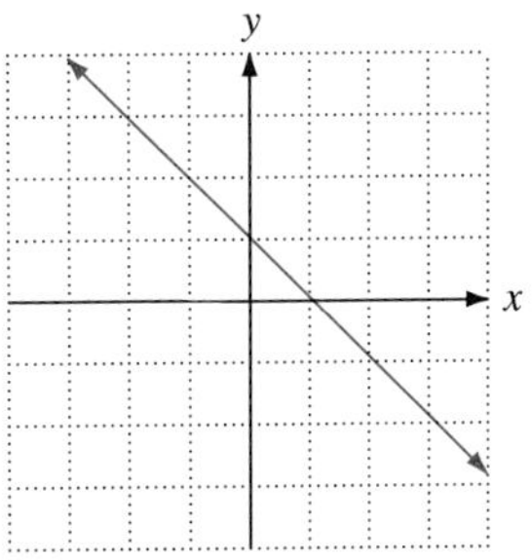

The lines may coincide.

The solution to the system of equations is those points that both lines have in common. Note that parallel lines have no points in common. The system has no solution. Intersecting lines intersect at one point. The system has one solution. Lines that coincide have an infinite number of points in common. The system has an infinite number of solutions. You can determine the solution to a system of equations by graphing.

1 *Solving a System of Linear Equations That Has a Unique Solution by Graphing*

To solve a system of equations by graphing, graph each equation on the same coordinate system. If the lines intersect, the system has a unique solution. The coordinates of the point of intersection is the solution to the system of equations.

EXAMPLE 1 Solve the following system of equations by graphing.

$$-2x + 3y = 6$$
$$2x + 3y = 18$$

Graph each equation on the same coordinate plane and determine the point of intersection. The graph of the above system is shown in the figure on the right. The lines intersect at the point (3, 4). Thus the unique solution to the system of equations is $x = 3$ and $y = 4$. Always check.

$-2(3) + 3(4) \stackrel{?}{=} 6$ $\qquad$ $2(3) + 3(4) \stackrel{?}{=} 18$

$6 = 6$ $\qquad$ $18 = 18$ ✓

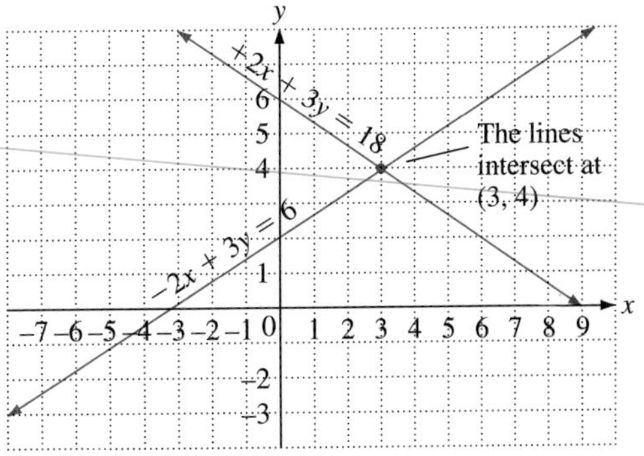

A linear system of equations that has one solution is said to be *consistent.*

Practice Problem 1 Solve by graphing. $x + y = 12$

$$-x + y = 4$$ ■

PRACTICE PROBLEM 1

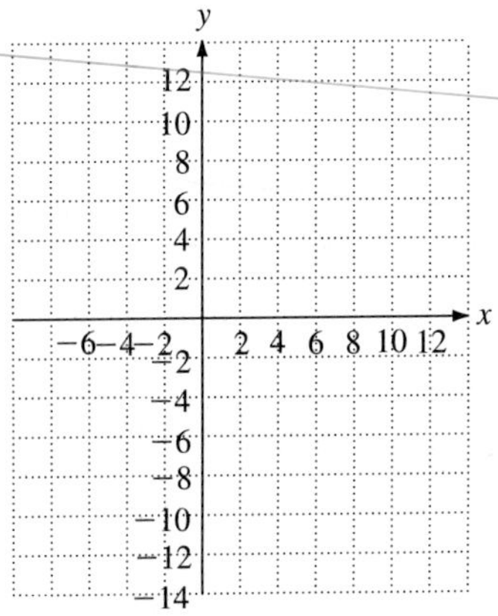

2 *Graphing a Linear System of Equations to Determine the Type of Solution*

Two lines in a plane may intersect, may never intersect (be parallel lines), or may be the same line. The corresponding system of equations will have one solution, no solution, or an infinite number of solutions. Thus far we have focused on systems of linear equations that have one unique solution. We now draw your attention to the other two cases.

EXAMPLE 2 Solve the following system of equations by graphing.

$$3x - y = -1$$
$$3x - y = -7$$

We graph each equation on the same coordinate plane. The graph is on the right. Notice that the lines are parallel. They do not intersect. Hence there is no solution to this system of equations.

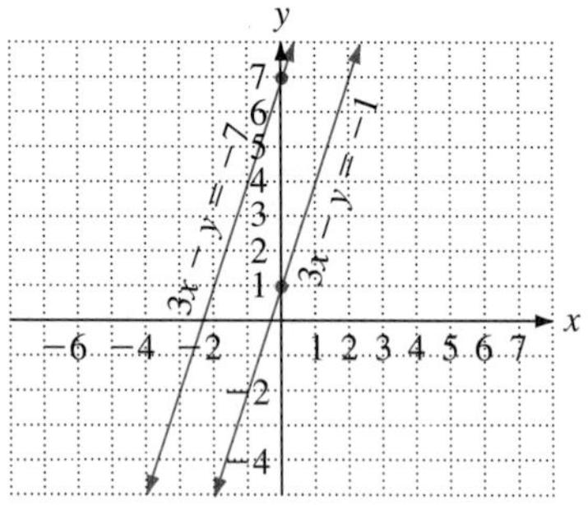

A linear system of equations that has no solution is called *inconsistent.*

PRACTICE PROBLEM 2

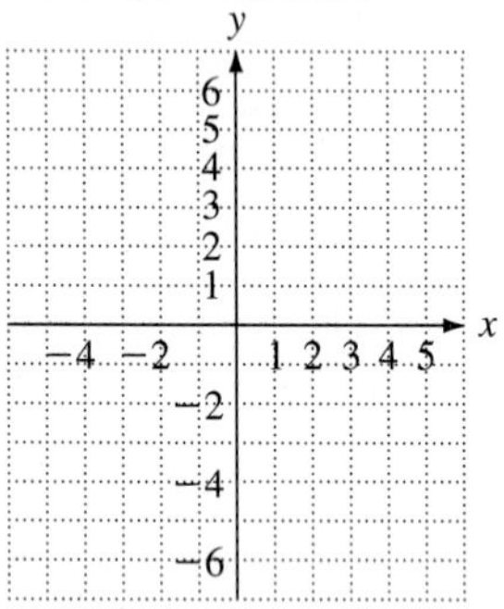

Practice Problem 2 Solve the following system of equations by graphing.

$$4x + 2y = 8$$
$$-6x - 3y = 6 \quad \blacksquare$$

To Think About Without graphing, how could you tell that the system in Example 2 represented parallel lines? Without graphing, determine which of the following systems is inconsistent (has no solution) or is consistent (has a solution).

(a) $2x - 3y = 6$
$6y = 4x + 1$

(b) $3x + y = 2$
$2x + y = 3$

(c) $x - 2y = 1$
$3x - 6y = 3$

EXAMPLE 3 Solve the following system of equations by graphing.

$$x + y = 4$$
$$3x + 3y = 12$$

We graph each equation on the same coordinate plane on the right. Notice that both equations represent the same line (coincide). The coordinates of every point on the line will satisfy both equations. That is, every point on the line represents a solution to $x + y = 4$ and to $3x + 3y = 12$. Thus there is *an infinite number of solutions* to this system.

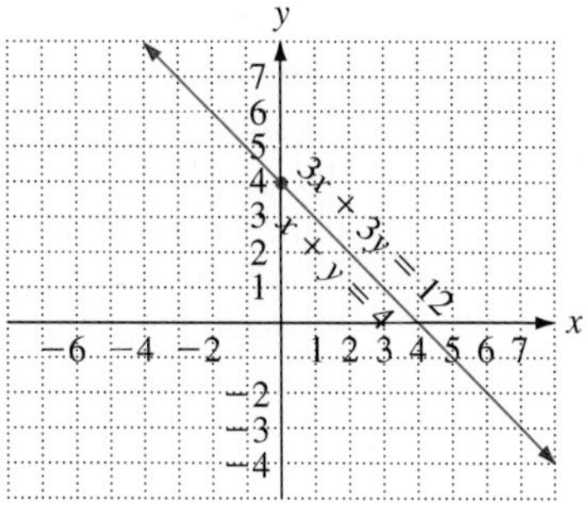

These equations are said to be *dependent.*

PRACTICE PROBLEM 3

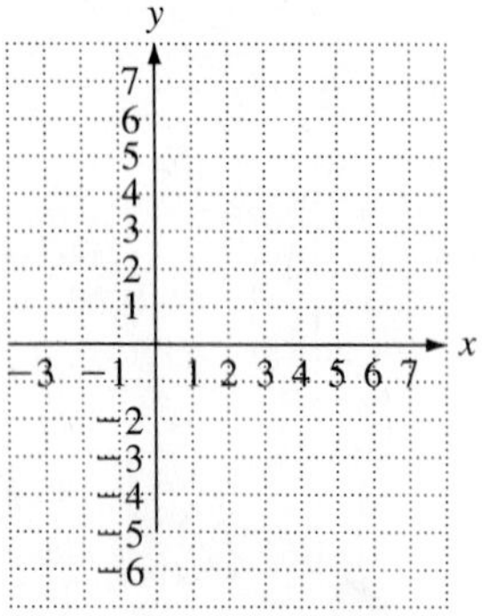

Practice Problem 3 Solve the following system of equations by graphing.

$$3x - 9y = 18$$
$$-4x + 12y = -24 \quad \blacksquare$$

8.1 Exercises

Solve each system by graphing. If there is not one solution to a system, state the reason.

1. $x + y = 8$
$x - y = 2$

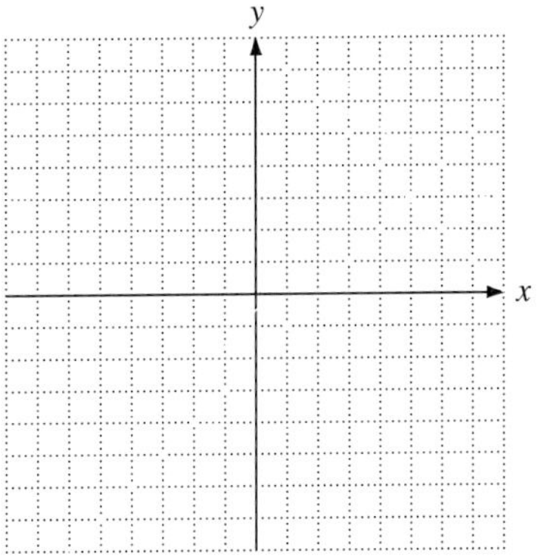

2. $x + y = 6$
$-x + y = -2$

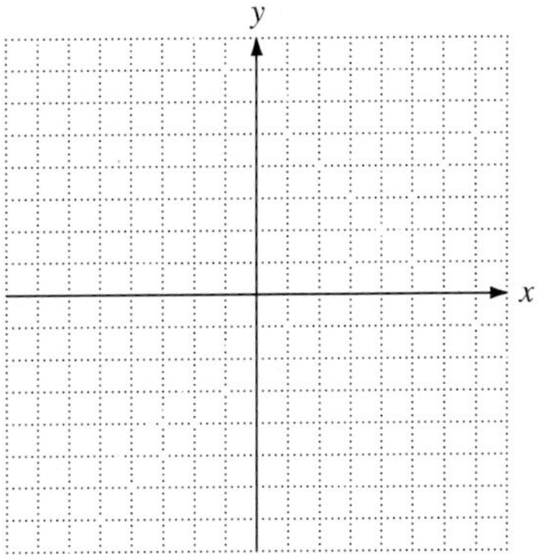

3. $2x + y = 6$
$x - 2y = 8$

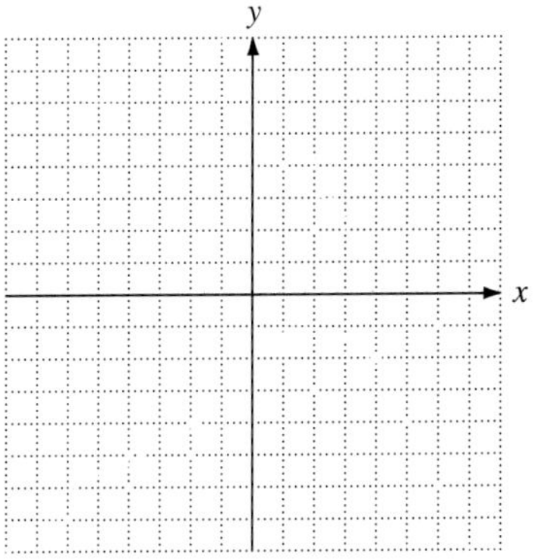

4. $2x + y = 0$
$-3x + y = 5$

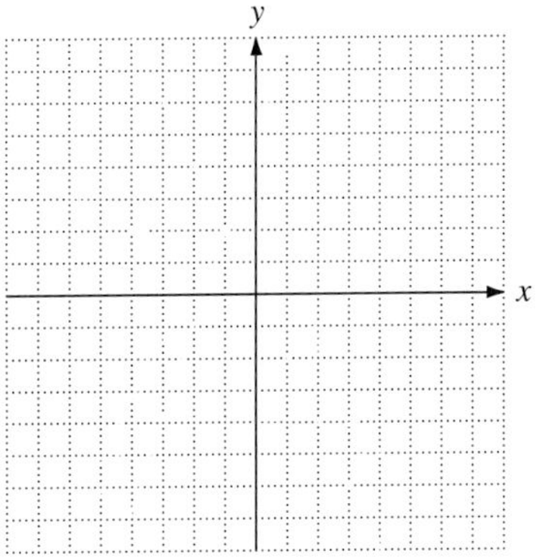

5. $2x + y = 6$
$-2x + y = 2$

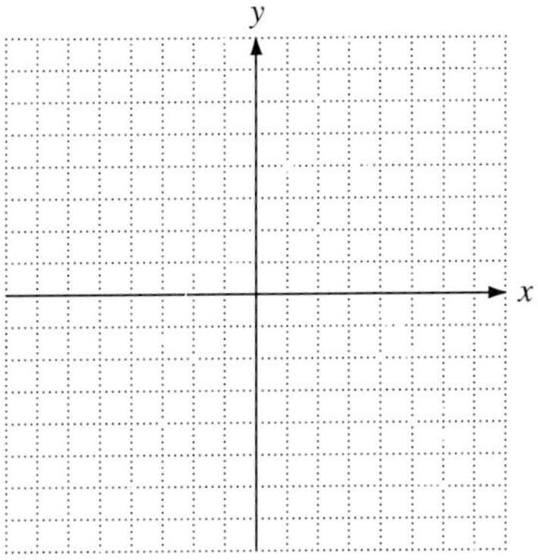

6. $x - 3y = 6$
$4x + 3y = 9$

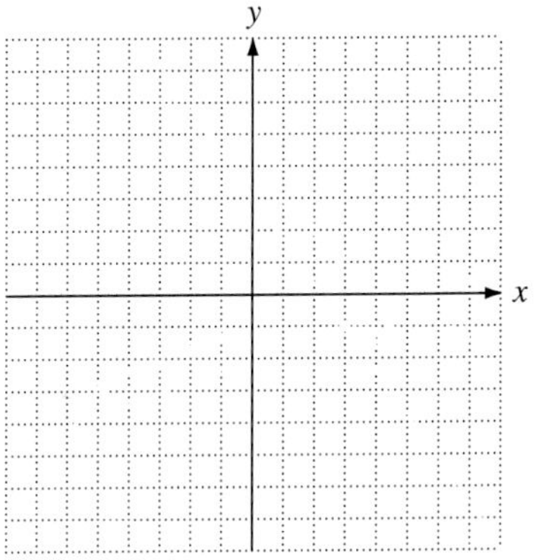

7. $-2x + y - 3 = 0$
$4x + y + 3 = 0$

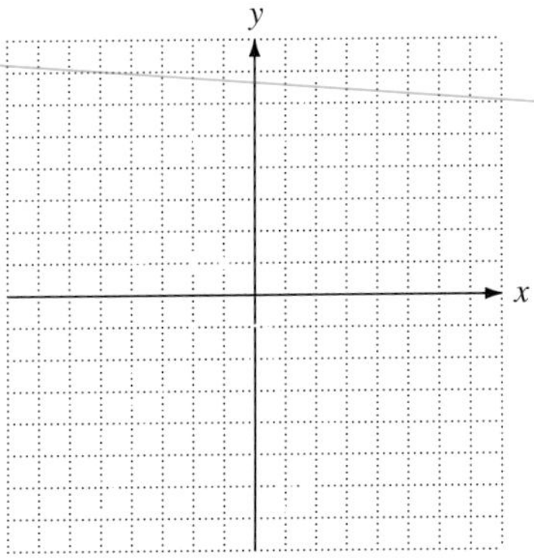

8. $x - 2y - 10 = 0$
$2x + 3y - 6 = 0$

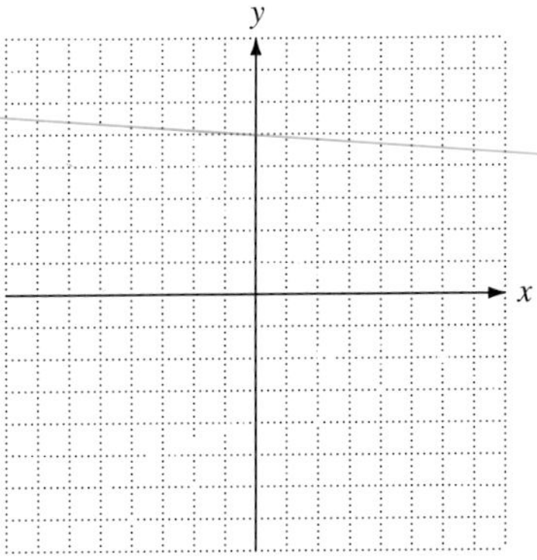

9. $3x - 2y = -18$
$2x + 3y = 14$

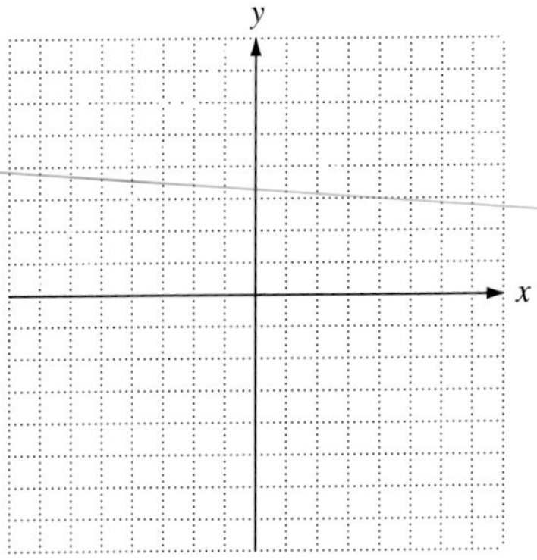

10. $3x + 2y = -10$

$-2x + 3y = 24$

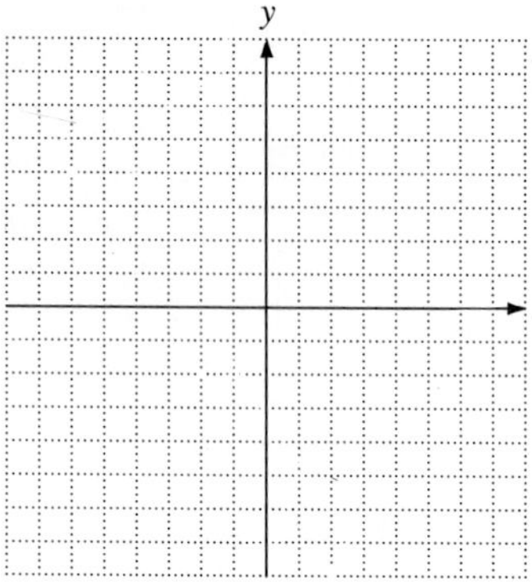

11. $y = \frac{2}{3}x - 3$

$y = -2x + 5$

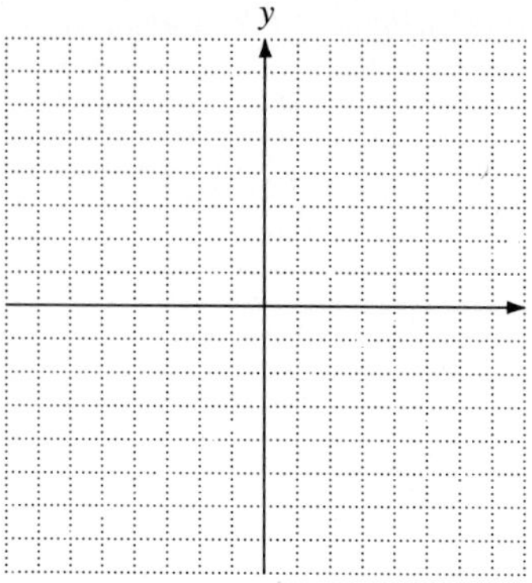

12. $y = \frac{3}{4}x + 2$

$y = x + 3$

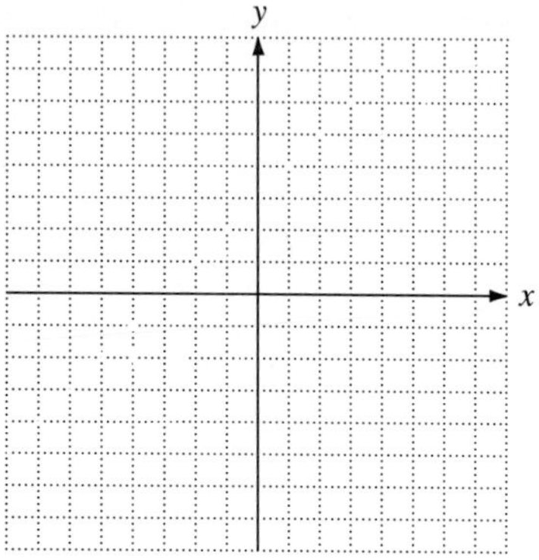

13. $4x - 6y = 8$

$-2x + 3y = -4$

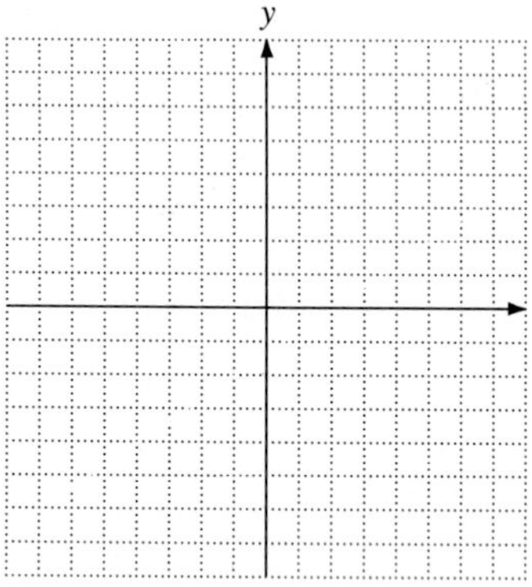

14. $3x - 2y = -4$

$-9x + 6y = -8$

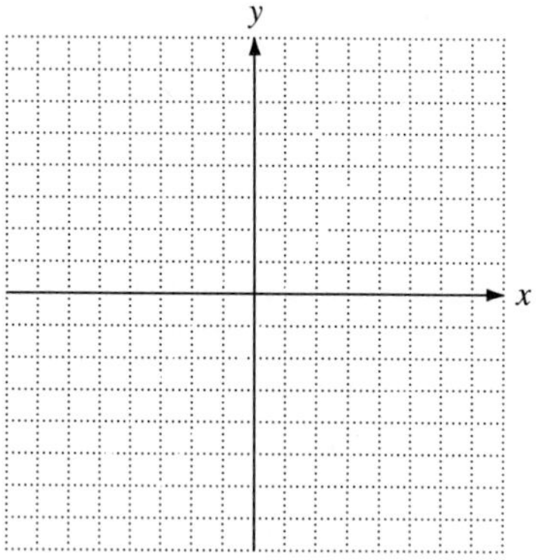

15. $2y + x - 6 = 0$

$y + \frac{1}{2}x = 4$

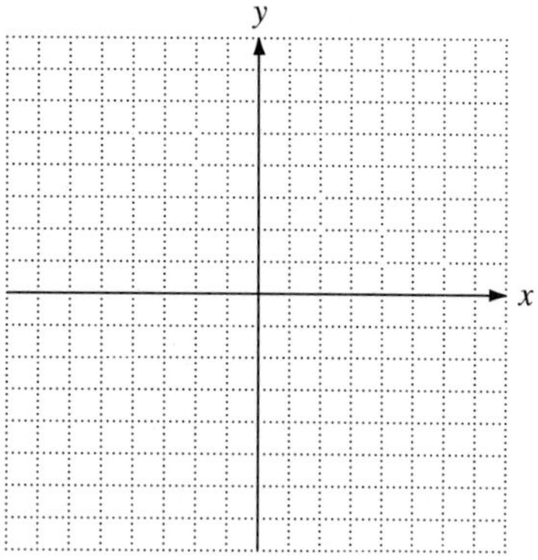

16. $y - 2x - 6 = 0$

$\frac{1}{2}y - 3 = x$

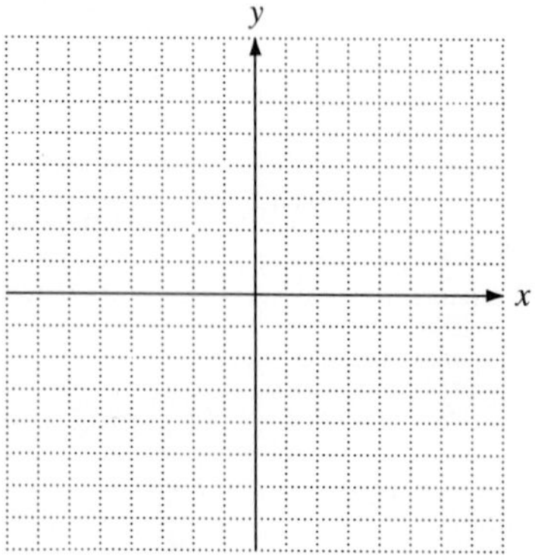

★ **17.** $\frac{1}{7}x - \frac{3}{7}y = 1$

$\frac{1}{2}x - \frac{1}{4}y = -1$

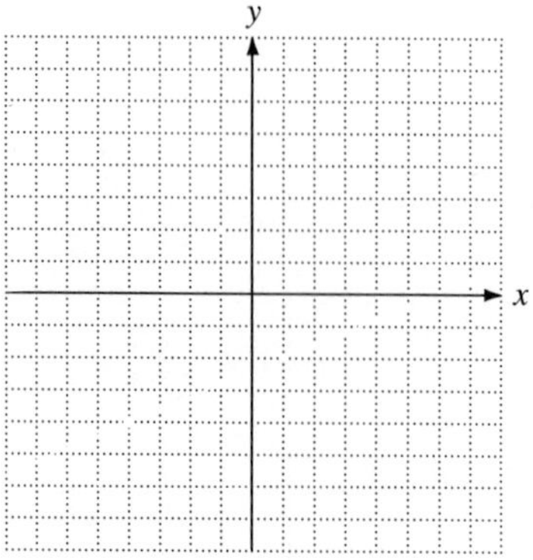

★ **18.** $\frac{1}{4}x - \frac{5}{12}y = -\frac{5}{4}$

$-\frac{1}{2}x + \frac{5}{6}y = \frac{5}{2}$

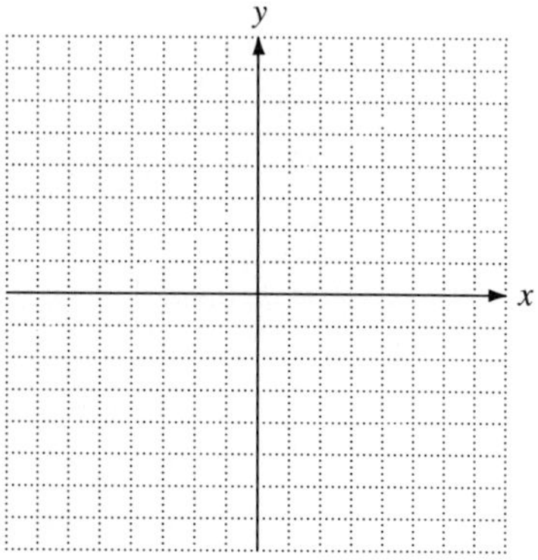

Verbal and Writing Skills

19. In the system $y = 2x - 5$ and $y = 2x + 6$, the lines have the same slope but different y-intercepts. What does this tell you about the graph of this system?

20. In the system $y = -3x + 4$ and $y = -3x + 4$, the lines have the same slope and the same y-intercepts. What does this tell you about the graph of the system?

21. Before you graph a system of equations, if you notice that the lines have different slopes, you can conclude they ______________ and have ____________________ solution(s).

22. If lines have the same slope, you cannot conclude that the lines must be parallel. Why? How else might the lines look when graphed?

23. Two lines have different slopes but the same y-intercept. What does this tell you about the graphs of the lines and about the solution of the system?

Cumulative Review Problems

24. Solve. $7 - 3(4 - x) \leq 11x - 6$

25. Solve for d. $3dx - 5y = 2(dx + 8)$

26. Multiply. $(3x - 7)^2$

27. Factor. $25x^2 - 60x + 36$

8.2 Solve a System of Equations by the Substitution Method

After studying this section, you will be able to:

1. *Solve a system of two linear equations with integer coefficients by the substitution method.*
2. *Solve a system of two linear equations with fractional coefficients by the substitution method.*

1 *Solving a System of Two Linear Equations with Integer Coefficients by the Substitution Method*

Since the solution to a system of equations may be a point where two lines intersect, we would expect the solution to be an ordered pair. Since the point must lie on both lines, the ordered pair (x, y) must satisfy both equations. For example, the solution to the system $3x + 2y = 6$ and $x + y = 3$ is $(0, 3)$. Let's check.

$$3x + 2y = 6 \qquad\qquad x + y = 3$$

$$3(0) + 2(3) \stackrel{?}{=} 6 \qquad\qquad 0 + 3 \stackrel{?}{=} 3$$

$$6 = 6 \;\checkmark \qquad\qquad 3 = 3 \;\checkmark$$

Given a system of linear equations, we can develop an algebraic method to find the solution. We know how to solve an equation with one variable. The following methods reduce the system to one equation in one variable. Once we have solved for the variable, we can substitute this known value into any one of the original equations and solve for the second variable.

EXAMPLE 1 Find the solution to the following system of equations.

$$y = 3x - 1$$

$$3x - 2y = -4$$

We observe that the first equation is solved for the variable y in terms of x. This means that y is equivalent to the expression $3x - 1$. Let us substitute $3x - 1$ for y in the second equation. The second equation will then have only one variable, and we will be able to solve for that variable.

$3x - 2y = -4$ *This is the original second equation.*

$3x - 2(3x - 1) = -4$ *Replace the variable y by $3x - 1$. Note that we need parentheses here.*

$3x - 6x + 2 = -4$ *Remove the parentheses.*

$-3x + 2 = -4$ *Collect like terms.*

$-3x = -6$ *Subtract 2 from both sides.*

$x = 2$ *Divide both sides by -3.*

We have now found the value for x. Now we substitute this value for x into one of the original equations and solve for y.

$y = 3x - 1$ *This is the original first equation.*

$= 3(2) - 1$ *Replace the variable x by 2.*

$= 6 - 1$ *Simplify.*

$= 5$

Thus we see that the solution to the system is $x = 2$, $y = 5$.

Check. $y = 3x - 1$ $\qquad$ $3x - 2y = -4$

$5 \stackrel{?}{=} 3(2) - 1$ $\qquad$ $3(2) - 2(5) \stackrel{?}{=} -4$

$5 \stackrel{?}{=} 6 - 1$ $\qquad$ $6 - 10 \stackrel{?}{=} -4$

$5 = 5$ ✓ $\qquad$ $-4 = -4$ ✓

Practice Problem 1 Find the solution to the following system of equations.

$$y = 3x + 10$$
$$2x - 3y = -16$$ ■

Sometimes neither equation is solved for a variable. In this case we need to begin by taking one equation (the choice is up to you) and solve for one variable. We have broken the procedure to follow into five steps.

EXAMPLE 2 Find the solution to the following system of equations.

$$x - 2y = 7$$
$$-5x + 4y = -5$$

Step 1 Solve for one variable.

$x - 2y = 7$ — *The first equation is the easiest one in which to isolate a variable.*

$x = 7 + 2y$ — *Add 2y to both sides to solve for x.*

Step 2 Substitute the expression into the other equation.

$-5x + 4y = -5$ — *This is the original second equation.*

$-5(7 + 2y) + 4y = -5$ — *Substitute the value 7 + 2y for x in this equation.*

Step 3 Solve this equation.

$-35 - 10y + 4y = -5$ — *Remove the parentheses.*

$-35 - 6y = -5$ — *Collect like terms.*

$-6y = 30$ — *Add 35 to both sides.*

$y = -5$ — *Divide both sides by −6.*

Step 4 Obtain the value of the second variable. We will now use this value $y = -5$ to find the value for x in one of the equations that contains both variables.

$x = 7 + 2y$ — *The easiest equation to use is the one that is already solved for x.*

$= 7 + 2(-5)$ — *Replace y by −5.*

$= 7 - 10$

$= -3$

The solution is $x = -3$ and $y = -5$.

Step 5 Check. To be sure that we have the correct solution, we will need to check that the obtained values of x and y can be substituted into *both original* equations to obtain true mathematical statements.

$x - 2y = 7$ $\qquad$ $-5x + 4y = -5$

$-3 - 2(-5) \stackrel{?}{=} 7$ $\qquad$ $-5(-3) + 4(-5) \stackrel{?}{=} -5$

$-3 + 10 \stackrel{?}{=} 7$ $\qquad$ $15 - 20 \stackrel{?}{=} -5$

$7 = 7$ ✓ $\qquad$ $-5 = -5$ ✓

Procedure for Solving a System of Equations by the Substitution Method

1. Solve for one variable in one of the two equations.
2. Substitute the expression you obtain in step 1 for this variable in the *other* equation.
3. You now have one equation with one variable. Solve this equation to find the value for that one variable.
4. Substitute this value for the variable in one of the original equations to obtain a value for the second variable.
5. Check each solution against each equation to verify your results.

Practice Problem 2 Solve the system.

$$5x + 3y = 19$$
$$2x - y = 12 \quad \blacksquare$$

2 *Solving a System of Two Linear Equations with Fractional Coefficients by the Substitution Method*

If a system of equations contains fractions, clear the equations of fractions *before* doing any other steps.

EXAMPLE 3 Find the solution to the following system of equations.

$$\frac{3}{2}x + y = \frac{5}{2}$$
$$-y + 2x = -1$$

We want to clear the first equation of the fractions. We observe that the LCD of the fractions is 2.

$$2\left(\frac{3}{2}x\right) + 2(y) = 2\left(\frac{5}{2}\right) \qquad \textit{Multiply each term of the equation by the LCD.}$$

$$3x + 2y = 5 \qquad \textit{This is equivalent to } \frac{3}{2}x + y = \frac{5}{2}$$

Now follow the five-step procedure.

Step 1 Solve for one variable. We will use the second equation.

$$-y + 2x = -1$$
$$-y = -1 - 2x \qquad \textit{Add } -2x \textit{ to both sides.}$$
$$y = 1 + 2x \qquad \textit{Multiply each term by } -1.$$

Step 2 Substitute the expression into the other equation.

$$3x + 2(1 + 2x) = 5$$

$$3x + 2 + 4x = 5 \qquad \textit{Remove parentheses.}$$

$$7x + 2 = 5$$

Step 3 Solve this equation for the variable.

$$7x + 2 = 5$$

$$7x = 3$$

$$x = \frac{3}{7}$$

Step 4 Find the value of the second variable. Always use one of the original equations. We will use the second equation.

$$-y + 2x = -1$$

$$-y + 2\left(\frac{3}{7}\right) = -1$$

$$-y + \frac{6}{7} = -1$$

$$y - \frac{6}{7} = 1$$

$$y = 1 + \frac{6}{7} = \frac{13}{7}$$

The solution to the system is $x = \frac{3}{7}$ and $y = \frac{13}{7}$.

Step 5 The check is up to you. Remember you must check the solution in both original equations.

Practice Problem 3 Solve the system of equations. Be sure to check your answer.

$$\frac{x}{3} - \frac{y}{2} = 1$$

$$x + 4y = -8 \quad \blacksquare$$

8.2 Exercises

Find the solution for each system of equations by the substitution method. Check your answers.

1. $y = 4x - 3$
$3x - y = 2$

2. $x = 6 + 2y$
$3x - 2y = 12$

3. $4x + 3y = 9$
$x - 3y = 6$

4. $x + 4y = 4$
$-x + 2y = 2$

5. $2x + y = 4$
$2x - y = 0$

6. $7x - 3y = -10$
$x + 3y = 2$

7. $3x - 4y = -8$
$y = 2x$

8. $2x - 7y = -3$
$x = 3y$

9. $x + 3y = 2$
$2x + 3y = 7$

10. $4x + 2y = 4$
$3x + y = 4$

11. $-x + 2y = -3$
$-2x + 3y = -6$

12. $6x - y = 6$
$-x + 3y = -1$

13. $3a - 5b = 2$
$3a + b = 32$

14. $a - 5b = 21$
$3a + 4b = -13$

15. $3x - 4y = 6$
$-2x - y = -4$

16. $-2x - 3y = -1$
$3x - y = 7$

17. $5c + 8d + 1 = 0$
$3c + d - 7 = 0$

18. $4s - 3t - 9 = 0$
$-2s + t + 3 = 0$

19. $3x - y - 9 = 0$
$8x + 5y - 1 = 0$

20. $8x + 2y - 7 = 0$
$-2x - y + 2 = 0$

21. $\frac{5}{3}x + \frac{1}{3}y = -3$
$-2x + 3y = 24$

22. $-x + y = -4$
$\frac{3x}{7} + \frac{2y}{3} = 5$

23. $3x - 2y = -11$
$x - \frac{1}{3}y = \frac{-10}{3}$

24. $\frac{1}{6}x - \frac{1}{9}y = -1$
$x + 4y = 8$

25. $\frac{3}{5}x - \frac{2}{5}y = 5$
$x + 4y = -1$

26. $-\frac{1}{3}x + \frac{1}{2}y = 1$
$2x - y = 2$

27. $-\frac{1}{2}x + \frac{1}{6}y = 2$
$3x - 2y = -15$

28. $\frac{1}{4}x + \frac{1}{8}y = 1$
$3x - 2y = 5$

★ **29.** $3(x + y) = 10 - y$
$x - 2 = 3(y - 1)$

★ **30.** $2(x - 8) = 1 - y$
$2(y - 2x) = -6$

★ **31.** $\frac{1}{4}x - 3y = 0$
$\frac{3}{4}x + 6y = \frac{15}{4}$

★ **32.** $\frac{1}{8}x + \frac{1}{6}y = 4$
$-\frac{1}{4}x + \frac{1}{2}y = 2$

Solve for x and y by the substitution method. Round your answer to four decimal places.

33. $2.9864x + y = 9.8632$
$1.8926x + 4.7618y = 10.7891$

34. $2.0153x + 3.1087y = -13.1286$
$x - 2.2154y = 4.1387$

Verbal and Writing Skills

35. How many equations do you think you would need in order to solve a system with three unknowns? With seven unknowns? Explain your reasoning.

36. The point(s) where the graphs of the linear equations intersect is the ______________ to the system.

37. The solution to the system must satisfy ______________ equations.

38. How many solutions will a system of equations have if the graphs of the lines intersect? Justify your answer.

39. How many solution(s) will a system of equations have if the graphs are parallel lines? Why?

Cumulative Review Problems

40. Find the slope and y-intercept of $7x + 11y = 19$.

41. Graph $3x - y < 2$.

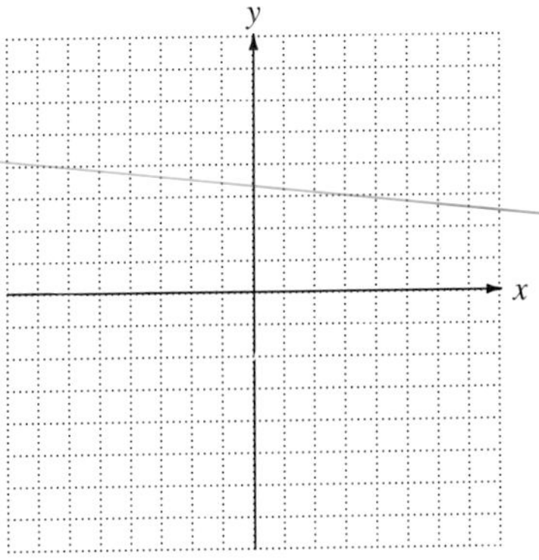

42. The sum of two numbers is 50. When three times the smaller is subtracted from the larger, the result is 10. Find the numbers.

43. Josh's coin box contains only quarters and dimes. It has 154 coins worth $34. How many of each coin does Josh have?

8.3 Solve a System of Equations by the Addition Method

After studying this section, you will be able to:

1. *Use the addition method to solve a system of two linear equations with integer coefficients.*
2. *Use the addition method to solve a system of two linear equations with fractional coefficients.*
3. *Use the addition method to solve a system of two linear equations with decimal coefficients.*

1 *Using the Addition Method to Solve a System of Two Linear Equations with Integer Coefficients*

The substitution method is useful when the coefficient of the variable is 1. In the following system, the variable x in the first equation has a coefficient of 1, and we can easily solve for x.

$$x - 2y = 7 \Rightarrow x = 7 + 2y$$
$$-5x + 4y = -5$$

This makes the substitution method for solving a system a natural choice. We may not always have 1 as a coefficient of any variable. We need a method of solving systems of equations that will work for integer coefficients, fractional coefficients, and decimal coefficients. One such method is the addition method.

EXAMPLE 1 Find the solution to the following system of equations by the *addition method.*

$2x + 5y = -2$ $\underline{-7x - 5y = 32}$ $-5x = 30$	*Add the two equations. Since $5y - 5y = 0$, we have eliminated the y variable.*
$\dfrac{-5x}{-5} = \dfrac{30}{-5}$	*Divide both sides by -5.*
$x = -6$	*Simplify.*
$2(-6) + 5y = -2$	*Substitute $x = -6$ into one of the original equations.*
$-12 + 5y = -2$	*Remove parentheses.*
$5y = 10$	*Add 12 to both sides.*
$y = 2$	*Divide both sides by 5.*

The solution is $x = -6$ and $y = 2$.

Notice that when we add the two equations together, one variable is eliminated. The addition method is therefore often called the *elimination* method.

Practice Problem 1 Find the solution to this system of equations by the addition method.

$$3x + 5y = 1$$
$$y - 3x = 11$$

(Be careful to note the position of the variables in this problem, you may change the position of the variables if needed.) ■

Most systems of equations cannot be solved this easily by the addition method. If you look back at Example 1, you will notice that there was a $5y$ in the first equation and a $-5y$ in the second equation. When these equations were added, the y terms became 0 and were eliminated. Our goal in the following examples will be to eliminate one of the variables.

EXAMPLE 2 Use the addition method to solve this system.

$$5x + 2y = 7 \quad \textbf{(1)}$$
$$3x - y = 13 \quad \textbf{(2)}$$

We would like the coefficients of the y terms to be equal in value and opposite in sign. One should be $+2y$ and the other $-2y$. Therefore, we multiply each term of equation **(2)** by 2.

$2(3x) - 2(y) = 2(13)$ *Multiply each term of equation* **(2)** *by* 2.

$5x + 2y = 7$ *We now have an equivalent system of equations.*

$6x - 2y = 26$ *Add the two equations. This will eliminate the*
$11x \quad = 33$ *y variable.*

$x = 3$ *Divide both sides by* 11.

$5(3) + 2y = 7$ *Substitute* $x = 3$ *into one of the original equations. Arbitrarily we pick equation* **(1)**.

$15 + 2y = 7$ *Remove parentheses.*

$2y = -8$ *Subtract* 15 *from both sides.*

$y = -4$ *Divide both sides by* 2.

The solution is $x = 3$ and $y = -4$.

Check: Replace x by 3 and y by -4 in *both* original equations.

$$5x + 2y = 7 \qquad 3x - y = 13$$
$$5(3) + 2(-4) \stackrel{?}{=} 7 \qquad 3(3) - (-4) \stackrel{?}{=} 13$$
$$15 - 8 \stackrel{?}{=} 7 \qquad 9 + 4 \stackrel{?}{=} 13$$
$$7 = 7 \checkmark \qquad 13 = 13 \checkmark$$

Practice Problem 2 Solve by addition.

$$3x + y = 7$$
$$5x - 2y = 8 \blacksquare$$

For convenience, we will make a list of the steps we are using to solve a system of equations by the addition method.

Procedure for Solving a System of Equations by the Addition Method

1. Multiply each term of one or both equations by some nonzero integer so that the coefficients of one of the variables are opposites.
2. Add the equations of this new system.
3. Solve the resulting equation for the remaining variable.
4. Substitute the value found in step 3 into either of the original equations to find the value of the second variable.
5. Check your solution in both of the original equations.

EXAMPLE 3 Use the addition method to solve.

$$3x + 4y = 7 \quad \textbf{(1)}$$

$$2x + 7y = 9 \quad \textbf{(2)}$$

Suppose we want the x terms to have opposite coefficients. One equation could have $+6x$ and the other $-6x$. This would happen if we multiply equation **(1)** by $+2$ and equation **(2)** by -3. We will do so.

$(2)(3x) + (2)(4y) = (2)(7)$ *Multiply each term of equation* **(1)** *by 2.*

$(-3)(2x) + (-3)(7y) = (-3)(9)$ *Multiply each term of equation* **(2)** *by* -3.

(Often this step of multiplying each equation by 2 or -3 can be done mentally.)

$6x + 8y = 14$ *We now have an equivalent system of equations.*

$-6x - 21y = -27$ *The coefficients of x terms are opposite.*

$-13y = -13$ *Add the two equations to eliminate the variable x.*

$y = 1$ *Divide both sides by* -13.

Substitute $y = 1$ into one of the original equations.

$$3x + 4(1) = 7$$

$3x + 4 = 7$ *Solve for x.*

$$3x = 3$$

$$x = 1$$

The solution is $x = 1$ and $y = 1$. Check the solution in both original equations.

Alternative solution: You could also have eliminated the y variable. For example, if you multiply the first equation by 7 and the second equation by -4, you would obtain the equivalent system.

$$21x + 28y = 49$$

$$-8x - 28y = -36$$

You can add these equations to eliminate the y variable. Since the numbers involved in this approach are somewhat larger, it is probably wiser to eliminate the x variable in this example.

Practice Problem 3 Solve by addition.

$$4x + 5y = 7$$

$$3x + 7y = 2 \quad ■$$

2 *Using the Addition Method to Solve a System of Two Linear Equations with Fractional Coefficients*

If the system of equations has fractional coefficients, you should first clear each equation of the fractions. To do so, you will need to multiply each term in the equation by the LCD of the fractions.

EXAMPLE 4 Solve.

$$x - \frac{5}{2}y = \frac{5}{2} \quad \textbf{(1)}$$

$$\frac{4}{3}x + \quad y = \frac{23}{3} \quad \textbf{(2)}$$

$2(x) - 2\left(\frac{5}{2}y\right) = 2\left(\frac{5}{2}\right)$ *Multiply each term of equation* **(1)** *by* 2.

$2x - 5y = 5$

$3\left(\frac{4}{3}x\right) + 3(y) = 3\left(\frac{23}{3}\right)$ *Multiply each term of equation* **(2)** *by* 3.

$4x + 3y = 23$

We now have an equivalent system of equations that does not contain fractions.

$$2x - 5y = 5 \quad \textbf{(3)}$$

$$4x + 3y = 23 \quad \textbf{(4)}$$

Let us eliminate the x variable. We want the coefficients of x to be opposites.

$(-2)(2x) - (-2)5y = (-2)(5)$ *Multiply each term of equation* **(3)** *by* -2.

$-4x + 10y = -10$ *We now have an equivalent system of equations.*

$4x + 3y = 23$ *The coefficients of the x terms are opposites.*

$13y = 13$ *Add the two equations to eliminate the variable x.*

$y = 1$ *Divide both sides by* 13.

$4x + 3(1) = 23$ *Substitute* $y = 1$ *into equation* **(4)**.

$4x + 3 = 23$

$4x = 20$

$x = 5$

The solution is $x = 5$ and $y = 1$.

Check: Check the solution in both of the *original* equations.

Practice Problem 4 Solve.

$$\frac{2}{3}x - \frac{3}{4}y = 3$$

$$-2x + \quad y = 6 \; ■$$

3 *Using the Addition Method to Solve a System of Equations with Decimal Coefficients*

Some linear equations will have decimal coefficients. It will be easier to work with the equations if we change the decimal coefficients to integer coefficients. To do so, we will multiply each term of the equation by a power of 10.

EXAMPLE 5 Solve.

$$0.12x + 0.05y = -0.02 \quad \textbf{(1)}$$

$$0.08x - 0.03y = -0.14 \quad \textbf{(2)}$$

Since the decimals are hundredths, we will multiply each term of both equations by 100.

$$100(0.12x) + 100(0.05y) = 100(-0.02)$$

$$100(0.08x) + 100(-0.03y) = 100(-0.14)$$

$$12x + 5y = -2 \quad \textbf{(3)}$$

$$8x - 3y = -14 \quad \textbf{(4)}$$

We now have an equivalent system of equations that has integer coefficients.

We will eliminate the variable y. We want the coefficients of the y terms to be opposites.

$$36x + 15y = -6$$
$$40x - 15y = -70$$

Multiply equation **(3)** *by* 3 *and equation* **(4)** *by* 5.

$$76x \qquad = -76$$

Add the equations.

$$x = -1$$

Solve for x.

$$12(-1) + 5y = -2$$

Replace x by -1 *in equation* **(3)**, *and solve for y.*

$$-12 + 5y = -2$$

$$5y = -2 + 12$$

$$5y = 10$$

$$y = 2$$

The solution is $x = -1$ and $y = 2$.

Check: We substitute $x = -1$ and $y = 2$ into the original equations. I'm sure most students would rather not use equation **(1)** and equation **(2)**! However, *we must check the solutions in the original equations* if we want to be sure of our answers. It is possible to make an error going from equations **(1)** and **(2)** to equations **(3)** and **(4)**. The solutions would satisfy equations **(3)** and **(4)**, but would not satisfy equations **(1)** and **(2)**.

$$0.12x + 0.05y = -0.02 \qquad\qquad 0.08x - 0.03y = -0.14$$

$$0.12(-1) + 0.05(2) \stackrel{?}{=} -0.02 \qquad\qquad 0.08(-1) - 0.03(2) \stackrel{?}{=} -0.14$$

$$-0.12 + 0.10 \stackrel{?}{=} -0.02 \qquad\qquad -0.08 - 0.06 \stackrel{?}{=} -0.14$$

$$-0.02 = -0.02 \ \checkmark \qquad\qquad -0.14 = -0.14 \ \checkmark$$

Practice Problem 5 Solve. $0.2x + 0.3y = -0.1$

$0.5x - 0.1y = -1.1$ ■

Warning

A common error is to find the value for x and then stop. You have not solved a system in x and y until you find both the x and the y values that make both equations true.

8.3 Exercises

Verbal and Writing Skills

1. Look at the following systems of equations. Decide which variable to eliminate in each system, and explain how you would eliminate that variable.

(a) $3x + 2y = 5$
$5x - y = 3$

(b) $2x - 9y = 1$
$2x + 3y = 2$

(c) $4x - 3y = 10$
$5x + 4y = 0$

Find the solution to each system of equations by the addition method. *Check your answers.*

2. $2x + y = -1$
$3x - y = -9$

3. $2x + y = 2$
$4x - y = -8$

4. $2x + 3y = 1$
$x - 2y = 4$

5. $2x + y = 4$
$3x - 2y = -1$

6. $x + 5y = 2$
$2x + 3y = -3$

7. $x + y = 5$
$2x - y = -5$

8. $2x + 5y = 1$
$x - 2y = -4$

9. $2x + 3y = 6$
$-3x + y = 13$

10. $2x + 3y = 1$
$3x + 2y = -6$

11. $2x + 5y = 11$
$3x + 8y = 16$

12. $a - 2b = 0$
$6a - 13b = 5$

13. $a + 3b = 0$
$-3a - 10b = -2$

14. $8x + 6y = -2$
$10x - 9y = -8$

15. $4x + 9y = 0$
$8x - 5y = -23$

16. $2x + 3y = -8$
$5x + 4y = -34$

17. $9x - 2y = 14$
$4x + 3y = 14$

18. $5x - 2y = 6$
$6x + 7y = 26$

19. $-5x + 4y = -13$
$11x + 6y = -1$

20. $3x + 7y = 15$
$-5x + 2y = 16$

21. $5x - 2y = 24$
$4x - 5y = 9$

Verbal and Writing Skills

22. Before using the addition method to solve each of the following systems, you will need to simplify the equation(s). Explain how you would change the fractional coefficients to integer coefficients in each system.

(a) $\frac{1}{4}x - \frac{3}{4}y = 2$
$2x + 3y = 1$

(b) $5x + 2y = 9$
$\frac{4}{9}x - \frac{2}{3}y = 4$

(c) $\frac{1}{2}x + \frac{2}{3}y = \frac{1}{3}$
$\frac{3}{4}x - \frac{4}{5}y = 2$

Find the solution to each system of equations. Check your answers.

23. $x + \frac{5}{4}y = \frac{9}{4}$
$\frac{2}{5}x - y = \frac{3}{5}$

24. $\frac{2}{3}x + y = 2$
$x + \frac{1}{2}y = 7$

25. $\frac{4}{3}x + \frac{4}{5}y = 0$
$-2x + 6y = 36$

26. $\frac{5}{6}x - 5y = \frac{15}{2}$
$6x - 2y = 20$

27. $\frac{5}{6}x + y = -\frac{1}{3}$
$-8x + 9y = 28$

28. $\frac{2}{9}x - \frac{1}{3}y = 1$
$4x + 9y = -2$

29. $\frac{1}{5}x - \frac{1}{5}y = 1$
$\frac{1}{6}x + \frac{1}{3}y = -\frac{1}{6}$

30. $\frac{1}{2}x + \frac{5}{6}y = \frac{7}{3}$
$\frac{1}{3}x - \frac{5}{3}y = 6$

Verbal and Writing Skills

31. Explain how you would change each of the following systems to an equivalent system with integer coefficients. Write the equivalent system.

(a) $0.5x - 0.3y = 0.1$
$5x + 3y = 6$

(b) $0.08x + y = 0.05$
$2x - 0.1y = 3$

(c) $4x + 0.5y = 9$
$0.2x - 0.05y = 1$

Solve for x and y using the addition method.

32. $0.2x - 0.3y = 0.4$
$0.3x - 0.4y = 0.9$

33. $0.5x - 0.2y = 0.5$
$0.4x + 0.7y = 0.4$

34. $0.02x - 0.04y = 0.26$
$0.07x - 0.09y = 0.66$

35. $0.04x - 0.03y = 0.05$
$0.05x + 0.08y = -0.76$

★ **36.** $0.4x - 5y = -1.2$
$-0.03x + 0.5y = 0.14$

★ **37.** $-0.6x - 0.08y = -4$
$3x + 2y = 4$

38. $2x + 3y = 13$
$351x + 274y = 2226$
(Round answer to four decimal places)

39. $3x + 5y = 11$
$291x + 243y = 1551$

To Think About

Find the solution to each system of equations.

40. $y + 1 = 2(x - 6)$
$4(-x - 2) = 9y - 1$

41. $5(x - y) = -7 - 3y$
$3(2 + x) = y + 1$

42. $2(x + y) = 11 + y$
$3(y - x) = 2(9 - 2x)$

43. $4(x - y) = 8 - y$
$y = 2(7 - x)$

Let a and b represent constants. Find the solution to each system of equations by the addition method.

44. $ax + by = b^2 + a^2$
$-bx + ay = b^2 + a^2$

45. $-\frac{x}{a} + \frac{y}{b} = -2$
$\frac{x}{a} + \frac{y}{b} = 6$

Cumulative Review Problems

46. Solve. $\frac{8}{x - 3} = \frac{13x - 1}{x^2 - 9}$

47. Two numbers have a difference of eleven. Three times the lesser plus twice the greater gives 7. What are the numbers?

48. Factor $75x^2 - 3$ completely.

49. Factor. $6x^2 - 25x + 25$

Putting Your Skills to Work

FINDING MISSING INFORMATION FROM RECEIPTS

Larry and Nella Coates purchased 1 quart of oil and 5 gallons of gas for their car on May 1. On June 2 they purchased 2 quarts of oil and 9 gallons of gas for their car at the same service station. When their monthly statement came from the oil company, it recorded that the purchase on May 1 was for \$6.50 and the purchase on June 2 was for \$11.90. The monthly statement did not list the amount charged for a quart of oil nor did it list the amount charged for a gallon of gasoline.

1. Let $x =$ the cost of 1 quart of oil and $y =$ the cost of 1 gallon of gasoline. Write a system of equations that describes the two purchases.

2. Use the addition method to solve the system of equations, and find the cost of 1 quart of oil and the cost of 1 gallon of gasoline.

3. On July 1, the oil company increased the prices of motor oil and gasoline. On July 20, Larry and Nella purchased 3 quarts of oil and 10 gallons of gas. On July 30, they purchased 2 quarts of oil and 11 gallons of gas. The monthly statement recorded the July 20 purchase for \$20.30, whereas the July 30 purchase was for \$19.60. How much did 1 quart of oil now cost? How much did 1 gallon of gas now cost?

COOPERATIVE GROUP ACTIVITY

4. Ken and Carolyn Umenhoefer went Christmas shopping and purchased 5 blouses and 4 skirts for relatives. The total cost was \$326.00. Each blouse cost the same amount. The skirts were a different price, but all 4 were purchased at the same price. The prices were not recorded on the sales slip; it only read ''5 blouses and 4 skirts . . . total \$326.00'' Some blouses and skirts did not fit, so Ken and Carolyn returned 2 blouses and 3 skirts for a credit of \$192.00. How much did they pay for each blouse? How much did they pay for each skirt?

5. Dave and Elsie Bagley had an unpleasant experience at Hersh Jordan Department store. They purchased 4 women's sweaters and 3 men's sweaters for Christmas shopping. All the women's sweaters cost one price. The men's sweaters were a different price, but each cost the same. The purchase price was \$340. However, when they returned 1 women's sweater and 2 men's sweaters, the store was having a 20% off sale. They could not find their original sales receipt. Thus, for each sweater they returned, they only received 80% of the original purchase price in return. Their total refund was \$108.00. How much did they originally pay for each men's sweater? How much did they originally pay for each women's sweater?

8.4 Review of Methods for Solving Systems of Equations

After studying this section, you will be able to:

1 *Choose an appropriate method to solve a system of equations algebraically.*
2 *Identify inconsistent and dependent systems by algebraic methods.*

1 *Chose an Appropriate Method to Solve a System of Equations Algebraically*

At this point we will review the algebraic methods for solving systems of linear equations and discuss the advantages and disadvantages of each method.

Method	Advantage	Disadvantage
Substitution	Works well if one or more variables have a coefficient of 1 or -1.	Often becomes difficult to use if no variable has a coefficient of 1 or -1.
Addition	Works well if equations have fractional or decimal coefficients. Works well if no variable has a coefficient of 1 or -1.	None

EXAMPLE 1 Select a method and solve each system of equations.

(a) $x + y = 3080$
$2x + 3y = 8740$

(b) $5x - 2y = 19$
$-3x + 7y = 35$

(a) Since both x and y have coefficients of 1, we will select the substitution method.

$y = 3080 - x$ *Solve the first equation for y.*

$2x + 3(3080 - x) = 8740$ *Substitute the expression into the second equation.*

$2x + 9240 - 3x = 8740$ *Remove parentheses.*

$-1x = -500$ *Simplify.*

$x = 500$ *Divide each side by* -1.

$y = 3080 - x$

$= 3080 - 500$ *Substitute* 500 *for x.*

$= 2580$ *Simplify.*

The solution is $x = 500$ and $y = 2580$.

(b) Because neither x nor y has a coefficient of 1 or -1, we select the addition method. We choose to eliminate the y variable. Thus, we would like the coefficients of y to be -14 and 14.

$7(5x) - 7(2y) = 7(19)$ *Multiply each term of the first equation by* 7.

$2(-3x) + 2(7y) = 2(35)$ *Multiply each term of the second equation by* 2.

$35x - 14y = 133$ *We now have an equivalent system of equations.*
$-6x + 14y = 70$

$29x = 203$ *Add the two equations.*

$x = 7$ *Divide each side by* 29.

Substitute $x = 7$ into one of the original equations.

$$5(7) - 2y = 19$$

$$35 - 2y = 19 \qquad \textit{Solve for } y.$$

$$-2y = -16$$

$$y = 8$$

The solution is $x = 7$ and $y = 8$.

Practice Problem 1 Select a method and solve each system of equations.

(a) $3x + 5y = 1485$
$x + 2y = 564$

(b) $7x + 6y = 45$
$6x - 5y = -2$ ■

2 *Identify Inconsistent and Dependent Systems by Algebraic Methods*

Recall that an inconsistent system of linear equations is a system of parallel lines. Since parallel lines never intersect, the system has no solution. Could we determine this algebraically?

EXAMPLE 2 Solve the following system algebraically.

$$3x - y = -1$$

$$3x - y = -7$$

Clearly, the addition method would be very convenient in this case.

$$3x - y = -1 \qquad \textit{Keep the first equation unchanged.}$$
$$-3x + y = +7 \qquad \textit{Multiply each term in the second equation by } -1.$$
$$0 = 6 \qquad \textit{Add the two equations.}$$

Notice we have $0 = 6$, which we know is not true. The statement $0 = 6$ is inconsistent with known mathematical facts. No possible x and y values can make that equation true. There is no solution to this system of equations.

This example shows us that, if we obtain a mathematical statement that is not true (inconsistent with known facts), we can identify the entire system as inconsistent. If graphed, these lines would be parallel.

Practice Problem 2 Solve this system algebraically.

$$4x + 2y = 2$$

$$-6x - 3y = 6 \text{ ■}$$

What happens if we try to solve a dependent system of equations algebraically? Recall that a dependent system consists of two lines that coincide (are the same line).

EXAMPLE 3 Solve the following system algebraically.

$$x + y = 4$$
$$3x + 3y = 12$$

Let us use the substitution method.

$y = 4 - x$	*Solve the first equation for y.*
$3x + 3(4 - x) = 12$	*Substitute $4 - x$ for y in the second equation.*
$3x + 12 - 3x = 12$	*Remove parentheses.*
$12 = 12$	

Notice we have $12 = 12$, which is always true. Thus we obtain an equation that is true for any value of x. An equation that is always true is called an **identity.** All the points that lie on the line will satisfy both equations. There are an infinite number of solutions.

This example shows us that, if we obtain a mathematical statement that is always true (an identity), we can identify the equations as dependent. There are an unlimited number of solutions to the system that has dependent equations.

Practice Problem 3 Solve this system algebraically.

$$3x - 9y = 18$$
$$-4x + 12y = -24$$ ■

TWO LINEAR EQUATIONS WITH TWO VARIABLES		
Graph	Number of Solutions	Algebraic Interpretation
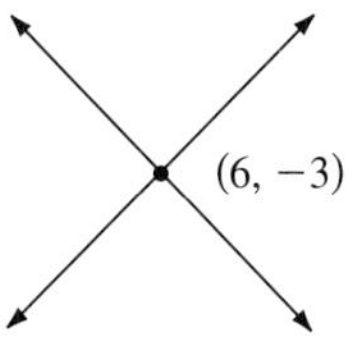 Two lines intersect at one point	One unique solution	You obtain one value for x and one value for y. For example, $x = 6, y = -3$
 Parallel lines	No solution	You obtain an equation that is inconsistent with known facts. For example, $0 = 6$
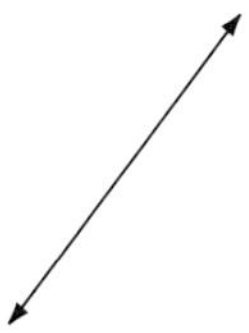 Lines coincide	Infinite number of solutions	You obtain an equation that is always true. For example, $8 = 8$

8.4 Exercises

Verbal and Writing Skills

1. If there is no solution to a system of linear equations, the graphs of the equations are ______________. Solving the system algebraically, you will obtain an equation that is ______________ with known facts.

2. If an algebraic attempt at solving a system of linear equations results in an identity, the system is said to be ______________. There are an ______________ number of solutions, and the graphs of the lines ______________.

3. If there is exactly one solution, the graphs of the equations ______________. This system is said to be ______________ and ______________.

4. A student solved a system of equations, but obtained the equation $-36x = 0$. He was unsure of what to do next. What should he do next? What should he conclude about the system of equations?

If possible, solve each system by an algebraic method (substitution or addition method, without the use of graphing). Otherwise, state that the problem has no solution or an infinite number of solutions.

5. $x - 4y = 6$
$-x + 2y = 4$

6. $-2x + 2y = 7$
$4x - 2y = 1$

7. $4x - 6y = 10$
$-10x + 15y = -25$

8. $3x + 2y = -2$
$-4x + 5y = 18$

9. $-2x - 3y = 15$
$5x + 2y = 1$

10. $2x - 4y = 5$
$-4x + 8y = 9$

11. $5x + 10y = 15$
$2x + 4y = -1$

12. $3x - y = 4$
$-9x + 3y = -12$

13. $5x - 3y = 9$
$7x + 2y = -6$

14. $4x + 5y = 16$
$6x - 7y = 24$

15. $4x - 3y = 3$
$6x + 5y = 33$

16. $x + y = 100$
$0.3x + 0.8y = 50$

17. $0.3x - 0.5y = 0.4$
$0.6x + 1.0y = 2.8$

18. $-0.05x + 0.02y = -0.01$
$0.10x + 0.04y = -0.22$

19. $4x = 3(5 - y)$
$2(x - 2y) = 1 + y$

20. $6(x + y) = 7 + y$
$7(2 + y) = 3x + 1$

21. $3(b - 1) = 2(a + b)$
$4(a - b) = -b + 1$

22. $a + 2(b + 1) = 1$
$4(a - b) = 22 + b$

Solve for x and y. Round your answer to three decimal places.

23. $26x - 39y = 208$
$51x + 68y = 119$

24. $0.625x + 0.375y = 0.875$
$0.118x + 0.294y = 0.058$

To Think About

25. Let c and d represent constants in the system

$$cx + dy = -3$$

$$dx - cy = -3$$

Find the values for c and d so that the solution to the system is $x = 1$ and $y = 2$.

Cumulative Review Problems

Solve for x in each of the following equations.

26. $\frac{1}{6}x - \frac{3}{2}x = -\frac{2}{3}$

27. $4(x + 3) + 3(x - 4) = -3 + 2(x - 1)$

28. $0.3(150) + x = 0.4(x + 150)$

29. $20x + 12(\frac{9}{2} - x) = 70$

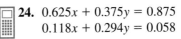

8.5 Solve Word Problems Using a System of Equations

After studying this section, you will be able to:

1 *Solve a variety of applied word problems by using a system of two linear equations containing two variables.*

1 *Using a System of Equations to Solve Word Problems*

Word problems can be solved in a variety of ways. In Chapter 3 and throughout the text, you have used one equation with one unknown to describe situations found in real-life applications. Some of these same problems can be solved using two equations in two unknowns. The method you use will depend on what you find works best for you. Become familiar with both methods before you decide. Generally, you will want to use an equation to represent each fact provided in the problem situation.

EXAMPLE 1 The student concert committee on campus sold tickets for a concert given by a popular music group. Tickets purchased in advance were \$4. Tickets purchased at the door were \$5. A total of 480 students bought tickets and attended the concert. The revenue from the ticket sales was \$2100. How many students bought tickets in advance and how many bought tickets at the door?

1. *Understand the problem.*

We are provided with two basic pieces of information

(1) The number of tickets bought

(2) The total revenue

This will be the basis for our two equations.

There are two different prices for the tickets. These will represent the two unknowns.

Let a = the number of tickets purchased in advance

d = the number of tickets purchased at the door

2. *Write an equation(s).*

Pick out the information you need to represent (1) total bought and (2) total revenue.

$$\underbrace{a}_{\text{number bought in advance}} + \underbrace{d}_{\text{number bought at door}} = \underbrace{480}_{\text{total bought}} \quad \textbf{(1)}$$

$$\underbrace{4a}_{\substack{\text{revenue from \$4 tickets} \\ \text{\$4 (number bought)}}} + \underbrace{5d}_{\substack{\text{revenue from \$5 tickets} \\ \text{\$5(number bought)}}} = \underbrace{2100}_{\text{total revenue}} \quad \textbf{(2)}$$

3. *Solve the system of equations and state the answer.*

We will solve the system by the substitution method.

$$a + d = 480 \quad \textbf{(1)}$$

$$4a + 5d = 2100 \quad \textbf{(2)}$$

$$d = 480 - a \qquad \textit{Solve equation } \textbf{(1)} \textit{ for variable } d.$$

$$4a + 5(480 - a) = 2100 \qquad \textit{Use equation } \textbf{(2)} \textit{ and replace } d \textit{ by } 480 - a.$$

$$4a + 2400 - 5a = 2100 \qquad \textit{Solve for } a.$$

$$-1a + 2400 = 2100$$

$$-1a = -300$$

$$a = 300$$

$$d = 480 - a \qquad \textit{Substitute } 300 \textit{ for } a \textit{ to find } d.$$

$$d = 480 - 300$$

$$d = 180$$

Thus 300 tickets were sold in advance and 180 tickets were sold at the door.

4. *Check.*
Do these answers seem reasonable? *Yes.*
Does the number of tickets bought in advance and at the door total 480?

$$300 + 180 = 480$$

Does revenue from advance sales and from sales at the door total $2100?

$$4(300) + 5(180) \stackrel{?}{=} 2100$$

$$1200 + 900 = 2100$$

Practice Problem 1 During the first week of college, Alicia took math and verbal placement tests. She just received her test scores. The combined score of the math test and the verbal test was 1153 points. She shared that figure with her friends, but she did not reveal the score for the separate tests. Later, she reluctantly admitted that the difference of the two scores was 409. Can you determine her actual score on each test if you know her math test score was the higher of the two? ■

EXAMPLE 2 A technician in a data processing office is trying to verify the rate at which two electronic card-sorting machines operate. Yesterday the first machine sorted for 3 minutes and the second machine sorted for 4 minutes. The total work load both machines processed during that time period was 10,300 cards. Two days ago the first machine sorted for 2 minutes and the second machine for 3 minutes. The total work load both machines processed during that time period was 7400 cards. Can you determine the number of cards per minute sorted by each machine?

1. *Understand the problem.*
The number of cards processed by two different machines on two different days provides the basis for two linear equations with two unknowns.

The two equations will represent what occurred on two different days.

(1) yesterday

(2) two days ago

The number of cards sorted by each machine is what we need to find.

Let $x =$ the number of cards per minute sorted by the first machine

$y =$ the number of cards per minute sorted by the second machine

2. *Write the equation(s).*
Yesterday the first machine sorted for 3 minutes and the second machine sorted for 4 minutes, processing 10,300 cards.

first machine		second machine		total work	
3(no. of cards per minute)		4(no. of cards per minute)			
$3x$	$+$	$4y$	$=$	10,300	**(1)**

Two days ago the first machine sorted for 2 minutes and the second machine sorted for 3 minutes, processing 7400 cards.

first machine		second machine		total work	
2(no. of cards per minute)		3(no. of cards per minute)			
$2x$	$+$	$3y$	$=$	7400	**(2)**

3. *Solve the system of equations and state the answer.*
We will use the addition method to solve this system.

$$3x + 4y = 10{,}300 \quad \textbf{(1)}$$
$$2x + 3y = 7{,}400 \quad \textbf{(2)}$$
$$6x + 8y = 20{,}600$$ *Multiply equation* **(1)** *by* 2.
$$-6x - 9y = -22{,}200$$ *Multiply equation* **(2)** *by* -3.
$$-1y = -1600$$ *Add the two equations to eliminate* x *and solve for* y.
$$y = 1600$$
$$2x + 3(1600) = 7400$$ *Substitute this value of* y *into equation* **(2)**.
$$2x + 4800 = 7400$$ *Simplify and solve for* x.
$$2x = 2600$$
$$x = 1300$$

Thus, the first machine sorts at the rate of 1300 cards per minute and the second machine at the rate of 1600 cards per minute.

4. *Check:* The numbers appear to be reasonable. Just by guessing, we would probably estimate that each machine would sort somewhere between 1000 and 2000 cards per minute. We can verify each statement in the problem:

Yesterday: Were 10,300 cards sorted?

$$3(1300) + 4(1600) \stackrel{?}{=} 10{,}300$$
$$3900 + 6400 \stackrel{?}{=} 10{,}300$$
$$10{,}300 = 10{,}300 \quad \checkmark$$

Two days ago: Were 7400 cards sorted?

$$2(1300) + 3(1600) \stackrel{?}{=} 7400$$
$$2600 + 4800 \stackrel{?}{=} 7400$$
$$7400 = 7400 \quad \checkmark$$

Practice Problem 2 After a recent disaster, the Red Cross brought in a pump to evacuate water from a basement. After 3 hours a larger pump was brought in and the two ran for an additional 5 hours, removing 49,000 gallons of water. The next day, after running both pumps for 3 hours, the larger one was taken to another site. The smaller pump finished the job after another 2 hours. If on the second day 30,000 gallons were pumped, how much water does each pump remove per hour? ■

EXAMPLE 3 Fred is considering working for a company that sells encyclopedias. During his job interview the company was not willing to release the specific salary structure for beginning sales representatives. Fred found out that all starting representatives receive the same annual base salary and a standard commission of a certain percentage of the sales they make during the first year. He has been told that one representative sold \$50,000 worth of encyclopedias her first year and that she earned \$14,000. He was able to find out that another representative sold \$80,000 worth of encyclopedias and that he earned \$17,600. Determine the base salary and the commission rate of a beginning sales representative.

The earnings of the two different sales representatives will represent the two equations. Earnings is based on

base salary + commission = total earnings

What is unknown? The base salary and the commission rate are what we need to find.

Let b = the amount of base salary

c = the commission rate

$b + 50{,}000c = 14{,}000$ (1) earnings for the first sales representative

$b + 80{,}000c = 17{,}600$ (2) earnings for the second sales representative

We will use the addition method to solve these equations.

$-b - 50{,}000c = -14{,}000$ *Multiply equation* **(1)** *by* -1.

$b + 80{,}000c = 17{,}600$ *Leave equation* **(2)** *unchanged.*

$30{,}000c = 3600$ *Add the two equations and solve for c.*

$$c = \frac{3600}{30{,}000}$$

$c = 0.12$ *The commission rate is* 12% *or* 0.12.

$b + (50{,}000)(0.12) = 14{,}000$ *Substitute this value into equation* **(1)**.

$b + 6000 = 14{,}000$ *Simplify and solve for b.*

$b = 8000$ *The base salary is* \$8000 *per year.*

Thus the base salary is \$8000 per year, and the commission rate is 12%.

Check: You should verify that these values are reasonable and that they satisfy the statements in the original problem.

Practice Problem 3 Last week Rick bought 7 gallons of unleaded premium and 8 gallons of unleaded regular gasoline and paid \$27.22. This week he purchased 8 gallons of unleaded premium and 4 gallons of regular and paid \$22.16. He forgot to record how much each type of fuel cost per gallon. Can you determine these values? ■

The use of two variables in two equations is very helpful in solving a class of problems often referred to as mixture problems.

EXAMPLE 4 A lab technician is required to prepare 200 liters of a solution. The prepared solution must contain 42% fungicide. The technician wishes to use a solution that contains 30% fungicide with a solution that contains 50% fungicide. How much of each solution should he use?

Understand the problem. The technician needs to make one solution from two different solutions. We need to find the amount (number of liters) of each solution that will give us 200 liters of a 42% solution. The two unknowns are easy to identify.

Let x = the number of liters of the 30% solution needed

y = the number of liters of the 50% solution needed

Now, what are the two equations? We know that the 200 liters of the final solution will be made by combining x and y.

$$x + y = 200 \qquad \textbf{(1)}$$

The second piece of information is about the percent of fungicide in each solution. We will draw a picture.

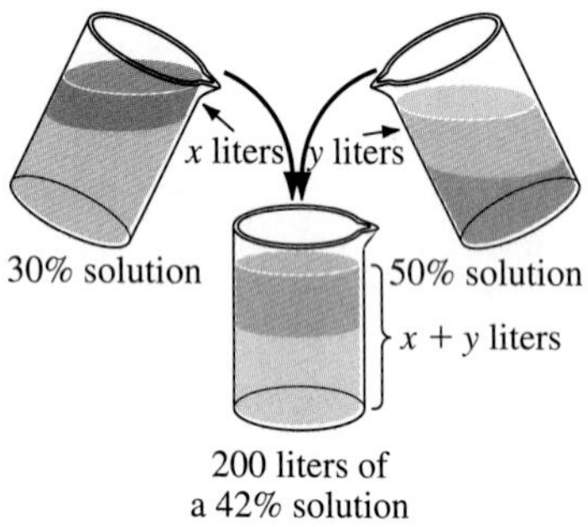

$$0.3x + 0.5y = 0.42(200) \quad \textbf{(2)}$$

Thus our system of equations is

$$x + y = 200 \qquad \textbf{(1)}$$
$$0.3x + 0.5y = 84 \qquad \textbf{(2)}$$

We will solve this system by the addition method.

$$1.0x + 1.0y = 200$$ *Rewrite equation* **(1)** *in an equivalent form, but write the coefficients as decimals.*

$$-0.6x - 1.0y = -168$$ *Multiply equation* **(2)** *by* -2.

$$0.4x = 32$$ *Solve for x.*

$$\frac{0.4x}{0.4} = \frac{32}{0.4}$$

$$x = 80$$

$$x + y = 200$$ *Substitute the value* $x = 80$ *into one of the original equations and solve for y.*

$$80 + y = 200$$

$$y = 120$$

The technician will use 80 liters of the 30% fungicide and 120 liters of the 50% fungicide to obtain the required solution.

Check. Do the amounts total 200 liters?

$$80 + 120 = 200 \ \checkmark$$

Do the amounts yield a 42% strength mixture?

$$0.30x + 0.50y \stackrel{?}{=} 0.42(200)$$

$$0.30(80) + 0.50(120) \stackrel{?}{=} 84$$

$$24 + 60 = 84 \ \checkmark$$

Practice Problem 4 Dwayne Laboratories needs to ship 4000 liters of H_2SO_4 (sulfuric acid). Dwayne keeps in stock solutions that are 20% H_2SO_4 and 80% H_2SO_4. If the shipment is to be 65% strength, how should Dwayne's stock be mixed? ■

EXAMPLE 5 Mike has a boat on Lazy River. He would like to find the speed of his boat and the speed of the current on the river. He took a 48-mile trip up the river traveling against the current in exactly 3 hours. He refueled and made the return trip in exactly 2 hours. What is the speed of his boat in still water and the speed of the current on the river?

$$\text{Let } x = \text{the speed of the boat in mph}$$

$$y = \text{the speed of the river current in mph}$$

You may want to draw a picture to show the boat traveling in each direction.

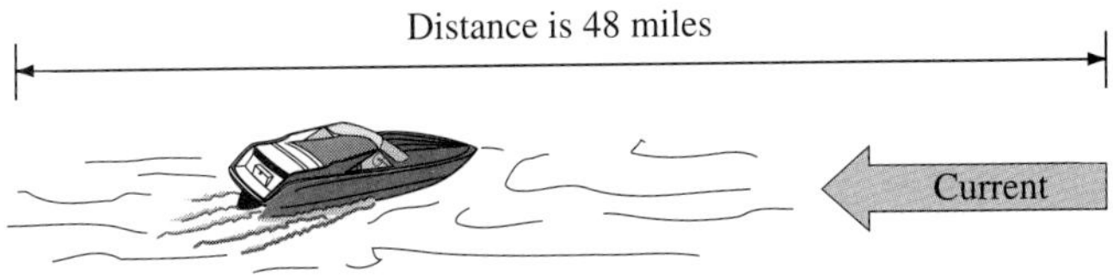

Against the current boat: takes 3 hours

With the current boat: takes 2 hours

Traveling *against* the current it took 3 hours to travel 48 miles. We can calculate the rate (speed) the boat travels upstream. Use $d = r \cdot t$.

$$48 = 3 \cdot r$$

$$r = 16 \text{ mph}$$

The boat is going *against* the current.

$$x - y = 16 \qquad \textbf{(1)}$$

Traveling with the current it took 2 hours to travel 48 miles. We can calculate the rate (speed) the boat travels downstream. Use $d = r \cdot t$.

$$48 = 2 \cdot r$$

$$r = 24 \text{ mph}$$

The boat is going *with* the current.

$$x + y = 24 \qquad \textbf{(2)}$$

This system is solved most easily by the addition method.

$$x - y = 16 \qquad \textbf{(1)}$$

$$\underline{x + y = 24} \qquad \textbf{(2)}$$

$$2x \quad = 40 \qquad \textit{Add the two equations.}$$

$$x = 20 \qquad \textit{Solve for } x \textit{ by dividing both sides by 2.}$$

$$x + y = 24 \qquad \textbf{(2)} \qquad \textit{Substitute the value of } x = 20 \textit{ into one of the original equations and solve for } y.$$

$$20 + y = 24 \qquad \textit{Substitute } x = 20 \textit{ into equation } \textbf{(2)}.$$

$$y = 4$$

Thus the speed of Mike's boat is 20 mph and the current speed of Lazy River is 4 mph.

Check. Can you verify these answers?

Practice Problem 5 Mr. Caminetti traveled downstream in his boat a distance of 72 miles in a time of 3 hours. His return trip up the river against the current took 4 hours. Find the speed of the boat in still water. Find the speed of the current in the river. ■

8.5 Exercises

Applications

Solve each problem by using two equations with two variables.

1. The sum of two numbers is -4. The difference between the two numbers is 30. Find the two numbers.

2. The sum of two numbers is 24. The difference between the two numbers is 14. Find the two numbers.

3. A new compact car has an EPA estimated highway mileage that is 12 miles per gallon more than the estimated city mileage. If you average the city and highway mileage rating for the car, you obtain 36 miles per gallon. Find the city and the highway mileage ratings.

4. In all right triangles the sum of the two smaller interior angles is 90°. In a certain right triangle the difference between those two angles is 28°. Find the number of degrees in each of these two angles.

5. There are 46 managers in a large company. The majority of the managers have attended college. The difference between the number of managers who have attended college and those who have not is 18. How many managers attended college? How many managers did not attend college?

6. A large department store starts its salespersons at a salary of \$4.50 per hour. The starting assistant managers earn \$6.50 per hour. Two roommates work at the same store. The person who was an assistant manager worked 4 more hours than the salesperson. The first seven days of the job they earned a total of \$532.00. How many hours did each work?

7. John's father is a theater manager. A recent matinee performance of a new musical was attended by 420 paying customers. In the theater mezzanine, tickets cost \$8 and orchestra tickets are \$13 for matinee showings. The sales receipts for this performance were \$4610. How many tickets of each type were sold?

8. A temporary office employment agency pays secretaries \$65 per day and receptionists \$50 per day. Yesterday, 46 people were employed by the agency. Yesterday's payroll was \$2780. How many secretaries and how many receptionists were employed yesterday?

9. Sioko planted a rooftop garden for her apartment in the city. Last year she needed 48 feet of fencing to enclose the rectangular garden. This year her neighbor gave her permission to use her adjoining roof space. This allowed Sioko to triple the width of her garden and to double the length. Now she needs 118 feet of fencing. What was the length and the width of her original garden last year?

10. Abdul is selling cars and trucks at a local dealership. He is considered the top salesperson. Last week he sold 14 vehicles. The manager said the rest of the entire dealership sales staff (eight other salespersons) sold four times as many cars as Abdul and three times as many trucks as Abdul. This amounted to a sale of 51 vehicles. How many cars and how many trucks did Abdul sell?

11. Bob's leaking radiator holds 16 quarts of water. When the radiator was partially full, Bob patched the leak temporarily. He added a mixture from his workbench of 80% antifreeze. The radiator is now full and Bob tests it and finds it is 65% antifreeze. Before he added anything, Bob had tested the radiator and found that it was 50% antifreeze. How many quarts of solution were in the radiator just before Bob added antifreeze? How many quarts of 80% solution antifreeze did Bob add?

12. A manufacturing company wants to utilize 20 kilograms of a special 45% aluminum alloy. To obtain this, a 50% aluminum alloy is combined with a 30% aluminum alloy in a special heating process. How many kilograms of the 50% alloy and how many kilograms of the 30% alloy are used to make the required 20 kilograms?

13. The hiking club wants to sell 50 pounds of an energy snack food that will cost them only \$1.80 per pound. The mixture will be made up of nuts that cost \$2.00 per pound and raisins that cost \$1.50 per pound. How much of each should they buy to obtain 50 pounds of the desired mixture?

14. How much flour that costs \$0.85 per pound should be mixed with an unrefined flour that costs \$0.90 per pound to obtain 10 pounds of a mixture that will cost \$0.88 per pound?

15. An airplane travels between two cities that are 630 miles apart. The trip with a tailwind was completed in 3 hours. The return trip traveling against the wind took 3.5 hours. Find the speed of the wind. Find the speed of the airplane in still air.

16. An airplane travels between two cities that are 3000 kilometers apart. The trip against the wind took 6 hours. The return trip with the benefit of the wind was 5 hours. What is the wind speed in kilometers per hour? What is the speed of the plane in still air?

17. Alicia took her boat on a 45-mile trip. The trip against the current took 4 hours. Her return trip with the current took only 3 hours. Find the speed of the boat in still water. Find the speed of the current.

18. Marcia can row a boat with the current at 5.5 miles per hour. She can row that boat at 4.2 miles per hour against the current. How fast is the current? How fast can she row the boat in still water?

★ **19.** Margaret's commission is a given percent of the total sales cost of the books she sells. Margaret received an 8% commission for selling encyclopedia sets. She received a 6% commission for selling collegiate dictionaries. Last week she earned \$350 in commissions. This week she sold the same quantity of encyclopedias and the same quantity of dictionaries. However, her company gave her a raise and she now earns 10% commission for selling encyclopedia sets and an 8% commission for selling dictionaries. This week she earned \$440. What is the value of the encyclopedia sets she sold last week? What is the value of the collegiate dictionaries she sold last week?

Cumulative Review Problems

20. Add. $\dfrac{1}{x-3} - \dfrac{1}{x+3} + 2$

21. Evaluate when $x = -2$ and $y = 3$.
$2x^2 - 3y + 4y^2$

22. Simplify. $(-4x^3y^2)^3$

Putting Your Skills to Work

CHANGING TWO RATES IN BUSINESS SITUATIONS

Often businesses will price one item at a loss and one item at a large profit in order to attract new customers. Ben Portwood recently found this out at his summer job as a desk clerk at MidWestern Hotel. His hotel caters to travelers. All the rooms are the same size, but the hotel charges a different rate for two people (a double) than for one person (a single). During the summer, the hotel charges \$60 a night for a double but only \$52 a night for a single. During off season, the price is \$40 for a double and \$35 for a single.

You may want to use a calculator to solve these problems.

1. On August 1, the hotel took in \$9480 in receipts. The hotel manager said that if they increased the cost of a double to \$64, but decreased the cost of a single to \$49, the total of receipts for the exact same number of people would be \$9530. How many double rooms were rented that night? How many single rooms were rented that night?

2. On August 15, the hotel took in \$8520 at the regular price rates. On September 7, the hotel went to off-season rates. The manager said that on September 15 exactly the same number of double rooms and single rooms as on August 15 were rented, but the hotel only took in \$5700 in receipts. How many double rooms were rented on those days? How many single rooms were rented on those days?

COOPERATIVE GROUP ACTIVITY

3. The manager stated that, if they increased the daily cost of a double in the summer to \$70, that would decrease the number of people renting a double room by 10%. If they decreased the cost of a single room in the summer to \$40, that would increase the number of people renting a single room by 50%. What would happen to the total hotel room rental receipts for one day if the number of people who came in on August 1 reacted exactly as the manager predicted?

4. Would all the increases in revenue obtained in Problem 3 be direct profit for the hotel? List some reasons why you feel this would or would not be the case.

Chapter Organizer

Topic	Procedure	Examples
Solving a system of equations: substitution method, p. 464.	1. Solve for one variable in terms of the other variable in one of the two equations. 2. Substitute the expression you obtain for this variable in the other equation. 3. Solve this equation to find the value of the variable. 4. Substitute this value for the variable in one of the equations with two variables. 5. Check each solution in both equations.	Solve: $2x + 3y = -6$ $3x - \ \ y = 13$ Solve second equation for y: $-y = -3x + 13$ $y = \ \ 3x - 13$ Substitute this expression into the first equation: $2x + 3(3x - 13) = -6$ $2x + 9x - 39 = -6$ $11x = -6 + 39$ $x = 3$ Substitute $x = 3$ into $y = 3x - 13$: $y = 3(3) - 13 = 9 - 13 = -4$ *Check:* $2(3) + 3(-4) \stackrel{?}{=} -6$ $3(3) - (-4) \stackrel{?}{=} 13$ $6 - 12 \stackrel{?}{=} -6$ $9 - (-4) \stackrel{?}{=} 13$ $-6 = -6$ ✓ $9 + 4 = 13$ ✓ The solution is $(3, -4)$.
Solving a system of equations: addition method, p. 470.	1. Multiply each term of one or both equations by some nonzero integer so that the coefficients of one of the variables are opposites. 2. Add the equations of this new system. 3. Solve the resulting equation for the variable. 4. Substitute the value found into either of the original equations to find the value of the second variable. 5. Check your solution in both of the original equations.	Solve: $3x - 2y = 0$ $-4x + 3y = 1$ Eliminate the x's: a common multiple of 3 and 4 is 12. Multiply the first equation by 4 and the second by 3, then add. $12x - 8y = 0$ $-12x + 9y = 3$ $y = 3$ Substitute $y = 3$ in either equation and solve for x. Let us pick $3x - 2y = 0$: $3x - 2(3) = 0$ $3x - 6 = 0,$ $x = 2$ *Check:* $3(2) - 2(3) \stackrel{?}{=} 0$ $-4(2) + 3(3) \stackrel{?}{=} 1$ $6 - 6 \stackrel{?}{=} 0$ $-8 + 9 \stackrel{?}{=} 1$ $0 = 0$ ✓ $1 = 1$ ✓ The solution is $(2, 3)$.
Solving a system of equations by graphing, p. 459.	Graph each equation on the same coordinate system. One of the following will be true. 1. The lines intersect. The point of intersection is the solution to the system. Such a system is consistent. 2. The lines are parallel. There is no point of intersection. The system has no solution. Such a system is inconsistent. 3. The lines coincide. Every point on the line represents a solution. There are an infinite number of solutions. Such a system is dependent.	1. 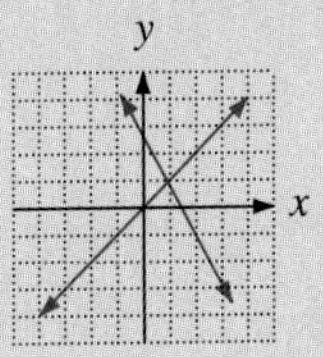 2. 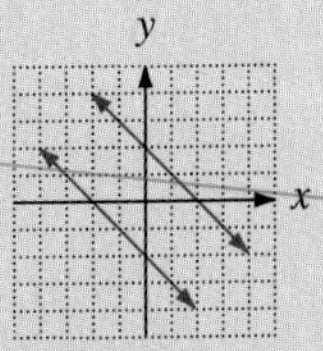 3. 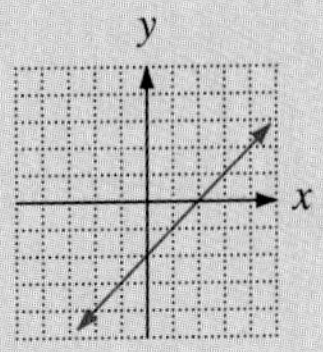

Chapter Organizer

<table>
<tr><th>Topic</th><th>Procedure</th><th>Examples</th></tr>
<tr><td>Choosing a method to solve a system of equations, p. 479.</td><td>1. Substitution works well if one or more variables have a coefficient of 1 or −1.
2. Addition works well for integer, fractional, or decimal coefficients.
3. Graphing gives you a picture of the system. It works well if the system has integer solutions or if the system is inconsistent or dependent.</td><td>Use substitution.
$x + y = 8$
$2x - 3y = 9$
Use addition.
$\frac{3}{5}x - \frac{1}{4}y = 4$
$\frac{1}{5}x + \frac{3}{4}y = 8$
Graph to get a picture of the system. If the system is consistent and if it is difficult to determine the solutions because they are fractions, use an algebraic method.</td></tr>
<tr><td>To identify inconsistent and dependent systems algebraically, p. 480.</td><td>1. If you obtain an equation that is inconsistent with known facts, such as 0 = 2, the system is inconsistent. There is no solution.
2. If you obtain an equation that is always true, such as 0 = 0, the system is dependent. There are an infinite number of solutions.</td><td>Solve: $x + 2y = 1$
$3x + 6y = 12$
Multiply the first equation by −3:
$-3x - 6y = -3$
$3x + 6y = 12$
$0 = 9$
0 is not equal to 9, so there is no solution. The system is inconsistent.
Solve: $2x + y = 1$
$6x + 3y = 3$
Multiply the first equation by −3:
$-6x - 3y = -3$
$6x + 3y = 3$
$0 = 0$
Since 0 is always equal to 0, the system is dependent. There are an infinite number of solutions.</td></tr>
<tr><td>Solving word problems with a system of equations, p. 484.</td><td>1. Understand the problem. Choose a variable to represent each unknown quantity.
2. Write a system of equations in two variables.
3. Solve the system of equations and state the answer.
4. Check the answer.</td><td>Apples sell for $0.35 per pound. Oranges sell for $0.40 per pound. Nancy bought 12 pounds of apples and oranges for $4.45. How many pounds of each fruit did she buy?
Let x = number of pounds of apples
y = number of pounds of oranges
12 pounds in all were purchased. $x + y = 12$
The purchase cost $4.45. $0.35x + 0.40y = 4.45$
Multiply the second equation by 100 and we have the system
$x + y = 12$
$35x + 40y = 445$
Multiply the top equation by −35 and add the equations.
$-35x - 35y = -420$
$35x + 40y = 445$
$5y = 25$
$y = 5$
Substituting into $x + y = 12$:
$x + 5 = 12$
$x = 7$
Nancy purchased 7 pounds of apples and 5 pounds of oranges.
Check: Are there 12 pounds of apples and oranges?
$7 + 5 = 12$ ✓
Would this purchase cost $4.45?
$0.35(7) + 0.40(5) = 4.45$
$2.45 + 2.00 = 4.45$ ✓</td></tr>
</table>

Chapter 8 Review Problems

8.1 *Solve each system of equations by graphing each line and finding the point of intersection.*

1. $3x + y = 4$
$3x - 2y = 10$

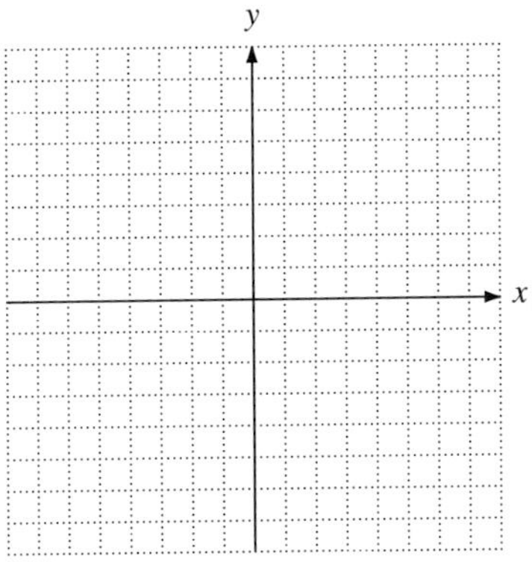

2. $-3x + y = -2$
$-2x - y = -8$

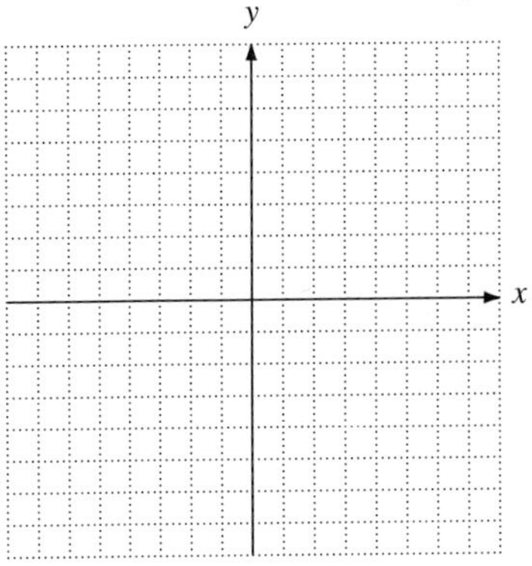

3. $-4x + 3y = 15$
$2x + y = -5$

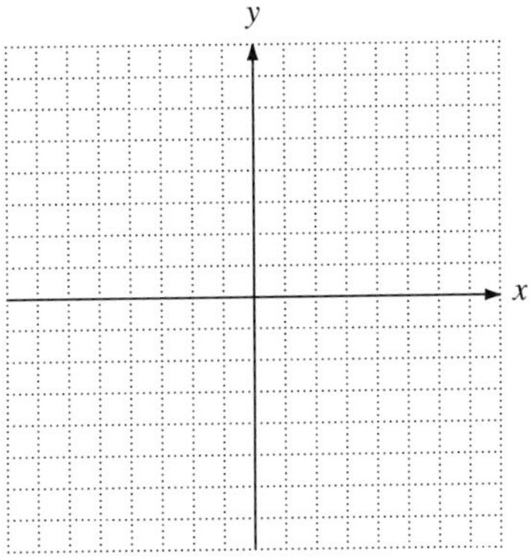

4. $2x - y = 1$
$3x + y = -6$

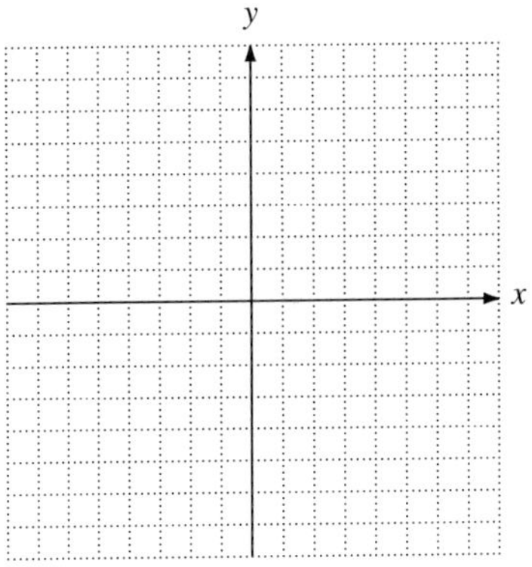

8.2 *Solve by the substitution method. Check your solution.*

5. $x + 3y = 18$
$2x + y = 11$

6. $x + y = 6$
$-2x + y = -3$

7. $3x + 2y = 0$
$2x - y = 7$

8. $5x - y = 4$
$2x - \frac{1}{3}y = 6$

8.3 *Solve by the addition method. Check your solution.*

9. $6x - 2y = 10$
$2x + 3y = 7$

10. $4x - 5y = -4$
$-3x + 2y = 3$

11. $-5x + 8y = 4$
$2x - 7y = 6$

12. $2x + 4y = 10$
$4x - 3y = 9$

8.4 *If possible, solve for both variables by any appropriate method to obtain one solution. If it is not possible to obtain one solution, state the reason.*

13. $3x - 2y = 17$
$x + y = 9$

14. $x - 2y = 2$
$2x + 3y = 11$

15. $11x + 3y = 12$
$4x + y = 5$

16. $7x - 5y = 8$
$-x + y = -2$

17. $5x - 3y = 32$
$5x + 7y = -8$

18. $4x + 3y = 46$
$-2x + 3y = -14$

19. $7x + 3y = 2$
$-8x - 7y = 2$

20. $x + 5y = 7$
$-3x - 4y = 1$

21. $5x - 11y = -4$
$6x - 8y = -10$

22. $3x + 5y = -9$
$4x - 3y = 17$

23. $2x - 4y = 6$
$-3x + 6y = 7$

24. $4x - 7y = 8$
$5x + 9y = 81$

25. $2x - 9y = 0$
$3x + 5 = 6y$

26. $2x + 10y = 1$
$-4x - 20y = -2$

27. $5x - 8y = 3x + 12$
$7x + y = 6y - 4$

28. $1 + x - y = y + 4$
$4(x - y) = 3 - x$

29. $3x + y = 9$
$x - 2y = 10$

30. $x + 12y = 0$
$3x - 5y = -2$

31. $2(x + 3) = y + 4$
$4x - 2y = -4$

32. $x + y = 3000$
$x - 2y = -120$

33. $4x - 3y + 1 = 6$
$5x + 8y + 2 = -74$

34. $5x + 4y + 3 = 23$
$8x - 3y - 4 = 75$

35. $\frac{2x}{3} - \frac{3y}{4} = \frac{7}{12}$
$8x + 5y = 9$

36. $\frac{x}{2} + \frac{y}{5} = 4$
$\frac{x}{3} + \frac{y}{5} = \frac{10}{3}$

37. $\frac{1}{5}a + \frac{1}{2}b = 6$
$\frac{3}{5}a - \frac{1}{2}b = 2$

38. $\frac{2}{3}a + \frac{3}{5}b = -17$
$\frac{1}{2}a - \frac{1}{3}b = -1$

39. $0.2s - 0.3t = 0.3$
$0.4s + 0.6t = -0.2$

40. $0.3s + 0.2t = 0.9$
$0.2s - 0.3t = -0.6$

41. $3m + 2n = 5$
$4n = 10 - 6m$

42. $12n - 8m = 6$
$4m + 3 = 6n$

43. $3(x + 2) = -2 - (x + 3y)$
$3(x + y) = 3 - 2(y - 1)$

44. $13 - x = 3(x + y) + 1$
$14 + 2x = 5(x + y) + 3x$

45. $0.2b = 1.4 - 0.3a$
$0.1b + 0.6 = 0.5a$

46. $0.3a = 1.1 - 0.2b$
$0.3b = 0.4a - 0.9$

47. $\frac{b}{5} = \frac{2}{5} - \frac{a - 3}{2}$
$4(a - b) = 3b - 2(a - 2)$

48. $9(b + 4) = 2(2a + 5b)$
$\frac{b}{5} + \frac{a}{2} = \frac{18}{5}$

Solve for x and y. Assume that a and b are constants.

★ **49.** $ax - by = ab + b^2$
$ax + by = 2a^2 + ab - b^2$

★ **50.** $5 + ax + by = 2a + 5$
$2b + by = ax$

8.5 *Solve each word problem.*

51. During a concert, \$3280 was received. A total of 760 students attended the concert. Reserved seats were \$6 and general admission seats were \$4. How many of each kind of ticket was sold?

52. In a local fruit market, apples are on sale for \$0.25 per pound. Oranges are on sale for \$0.39 per pound. Coach Smith purchased 12 pounds of a mixture of apples and oranges before the basketball team left on a bus trip to play at Center City. If the coach paid \$3.70, how many pounds of each fruit did he get?

53. A plane travels 1500 miles in 5 hours with the benefit of a tailwind. On the return trip it requires 6 hours to fly against the wind. Can you find the speed of the plane in still air and the wind speed?

54. A chemist needs to combine two solutions of a certain acid. How many liters of a 30% acid solution should be added to 40 liters of a 12% acid solution to obtain a final mixture that is 20% acid?

55. The girls and boys sponsoring the Scouts Car Wash had 86 customers on Saturday. They charged \$4 to wash regular-sized cars and \$3.50 to wash compact cars. The gross receipts for the day were \$323. How many cars of each type were washed?

56. Carl manages a highway department garage in Maine. A mixture of salt and sand is stored in a garage for use during the winter. Carl needs 24 tons of a salt/sand mixture that is 25% salt. He will combine shipments of 15% salt and shipments of 30% salt to achieve the desired 24 tons. How much of each type should he require?

57. A water ski boat traveled 23 kilometers per hour going with the current. It returned and went back in the opposite direction against the current and traveled only 15 kilometers per hour. What was the speed of the boat? What was the speed of the current?

58. Fred is analyzing the cost to produce two different items at an electronics company. An electrical sensing device requires 5 grams of copper and requires 3 hours to assemble. A smaller sensing device made by the same company requires 4 grams of copper but requires 5 hours to assemble. The first device has a production cost of \$27. The second device has a production cost of \$32. How much does it cost the company for a gram of this type of copper? What is the hourly labor cost at this company? (Assume that production cost is obtained by adding copper cost and labor cost.)

Putting Your Skills to Work

BREAK EVEN POINT

Systems of equations are used in the business world to determine the **break-even point.** Below the break-even point, a business is operating at a loss. Above the break-even point, a business is operating at a profit. We look at Dave's Candy Company. We know that:

1. The company has fixed costs from salaries, rent, light, heat, and telephone of \$300 per day.
2. Each pound of candy produced costs \$2 per pound for ingredients.
3. Each pound of candy sold yields \$3 in revenue.

To analyze this information, we begin by writing a system of equations.

Let x = the number of pounds of candy
C = cost per day
R = revenue per day

$$C = \underbrace{\$300}_{\text{fixed expenses}} + \underbrace{2x}_{\text{\$2 per pound}}$$

$$R = \underbrace{3x}_{\text{\$3 per pound}}$$

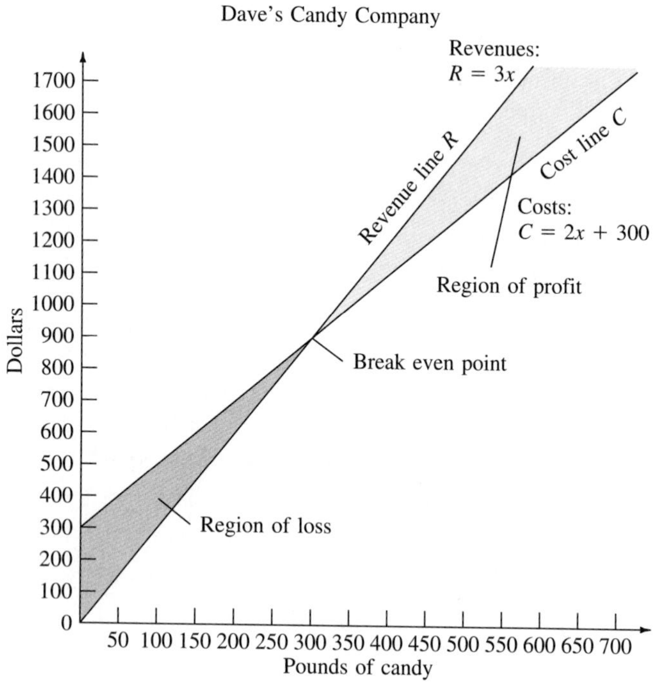

The graph of the system is on the right. The break-even point is where cost equals revenue, the intersection of the two lines. Algebraically,

$$R = C$$
$$3x = 2x + 300$$
$$x = 300$$

1. 200 pounds of candy are made and sold in one day.
   (a) Find the cost.
   (b) Find the revenue.
   (c) Is this a loss or a profit? Why?

2. If you manufacture and sell daily 625 pounds of candy, how much is the profit or loss?

3. If you manufacture and sell daily 275 pounds of candy, how much is the profit or loss?

COOPERATIVE GROUP ACTIVITY

Campus T-Shirts, Inc., is a local student-run enterprise. It has daily fixed costs of \$400. Every T-shirt costs the company \$5 to make. Every T-shirt sold yields \$7 in revenue.

4. Write a system of equations to describe the cost and revenue for Campus T-shirts, Inc.

5. Draw a graph to illustrate profit and loss for the company.

6. Determine the break-even point by solving the system of equations.

7. What is the profit or loss if 320 shirts are made and sold each day?

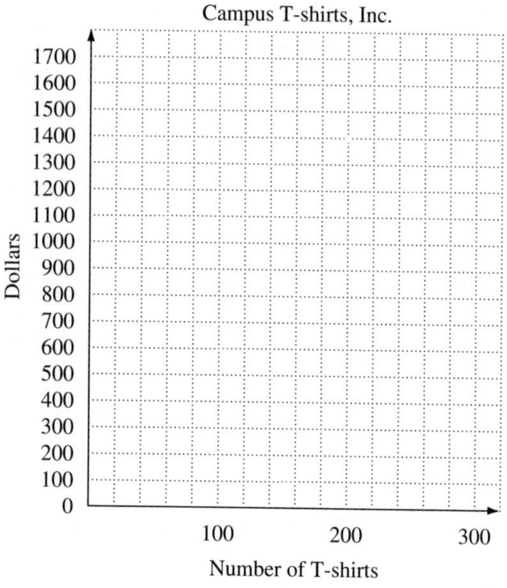

Chapter 8 Test

Solve by the method specified.

1. Substitution method.

$$3x - y = -5$$
$$-2x + 5y = -14$$

2. Addition method.

$$3x + 4y = 7$$
$$2x + 3y = 6$$

3. Graphing method.

$$2x - y = 4$$
$$4x + y = 2$$

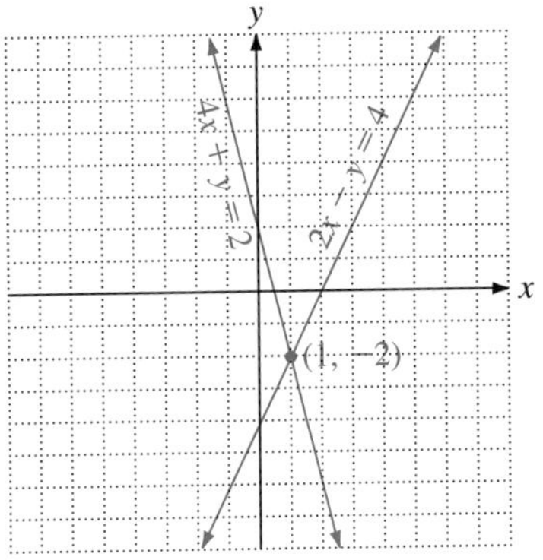

4. Any method.

$$x - 3y = 0$$
$$2x - 3y = -6$$

Solve by any method to find one solution if this is possible. If there is not one solution to a system, state why.

5. $2x - y = 5$
$-x + 3y = 5$

6. $2x + y = 4$
$3x + 7y = 17$

7. $\frac{2}{3}x - \frac{1}{5}y = 2$
$\frac{4}{3}x + 4y = 4$

8. $-2x + 4y = 14$
$3x - 6y = 5$

9. $5x - 2 = y$
$10x = 4 + 2y$

10. $0.3x + 0.2y = 0$
$1.0x + 0.5y = -0.5$

11. $2(x + y) = 2(1 - y)$
$5(-x + y) = 2(23 - y)$

12. $3(x - y) = 11 - y$
$8(y - 1) = 12x - 5$

1. ______
2. ______
3. ______
4. ______
5. ______
6. ______
7. ______
8. ______
9. ______
10. ______
11. ______
12. ______

13.

14.

15.

16.

17.

Solve each word problem.

13. Twice one number plus three times a second is one. Twice the second plus three times the first is nine. Find the numbers.

14. Both $8 and $12 tickets were sold for the game. In all, 30,790 paid for admission, resulting in a ''gate'' of $297,904. How many of each ticket was sold?

15. Five shirts and three pairs of slacks cost $172. Three of the same shirts and four pairs of the same slacks cost $156. How much does each shirt cost? How much is a pair of slacks?

16. Two machines place address labels on business envelopes. The first machine operated for 8 hours and the second machine for 3 hours. That day 5200 envelopes were affixed with address labels. The next day 4400 envelopes were addressed when the first machine was operated for 4 hours and the second machine for 6 hours. How many envelopes are addressed per hour by each machine?

17. A jet plane flew 2000 kilometers against the wind in a time of 5 hours. It refueled and flew back with the wind in 4 hours. Find the wind speed in kilometers per hour. Find the speed of the jet in still air in kilometers per hour.

Cumulative Test for Chapters 0–8

Approximately one-half of this test covers the content of Chapters 0–7. The remainder covers the content of Chapter 8.

Simplify.

1. 11% of \$37.20 (Round your answer to the nearest cent)

2. $\dfrac{-2^4 \cdot (2^2)}{2^3}$

3. $(3x^2 - 2x + 1)(5x - 3)$

4. $\dfrac{x^2 - 7x + 12}{x^2 - 9} \div (x^2 - 8x + 16)$

Solve.

5. $5(3 - x) \geq 6$

6. $x^2 - 4x - 45 = 0$

7. 12% of what number is 360?

8. $\dfrac{3x - 7}{5x^2 - 7x - 1} = 0$

9. Factor. $5a^3 - 5a^2 - 60a$

10. Solve for x. $4x = 2(12 - 2x)$

1. __________

2. __________

3. __________

4. __________

5. __________

6. __________

7. __________

8. __________

9. __________

10. __________

Solve for x and y, if possible.

11.

12.

13.

14.

15.

16.

17.

18.

11. $2x + y = 8$
$3x + 4y = -8$

12. $\frac{1}{3}x - \frac{1}{2}y = 4$
$\frac{3}{8}x - \frac{1}{4}y = 7$

13. $1.3x - 0.7y = 0.4$
$-3.9x + 2.1y = -1.2$

14. $x + 2y = 3$
$5x - 6y = -1$

Solve each word problem.

15. The equations in a system are graphed and parallel lines are obtained. What is this system called? What does that mean?

16. A mixture of 8% and 20% solutions yields 12,000 liters of a 16% solution. What quantities of the original solutions were used?

17. A boat trip 6 miles upstream takes 3 hours. The return takes 90 minutes. How fast is the stream?

18. Two printers were used for a 5-hour job of printing 15,000 labels. The next day a second run of the same labels took 7 hours because one printer broke down after 2 hours. How many labels can each printer process per hour?

9
Radicals

An accident has occurred but there are no eye witnesses. How fast was the car traveling? One of the powers of mathematics, is the use of formulas and equations to determine a value indirectly when a direct measurement is impossible. Once you have mastered the contents of this chapter, you will be given a chance to see if you can decide how fast the car was traveling, see page 552.

Pretest Chapter 9

1. ____	18. ____
2. ____	19. ____
3. ____	20. ____
4. ____	21. ____
5. ____	22. ____
6. ____	23. ____
7. ____	24. ____
8. ____	25. ____
9. ____	26. ____
10. ____	27. ____
11. ____	28. ____
12. ____	29. ____
13. ____	30. ____
14. ____	31. ____
15. ____	32. ____
16. ____	33. ____
17. ____	

If you are familiar with the topics in this chapter, take this test now. Check your answers with those in the back of the book. If an answer was wrong or you couldn't do a problem, study the appropriate section of the chapter.

If you are not familiar with the topics in this chapter, don't take this test now. Instead, study the examples, work the practice problems, and then take the test.

This test will help you to identify those concepts that you have mastered and those that need more study. Assume all variables in the radicand represent positive real numbers.

Section 9.1

Find the square roots, if possible. (If it is impossible to obtain a real number for an answer, say so.) Approximate to the nearest thousandth.

1. $\sqrt{121}$ **2.** $\sqrt{\frac{25}{64}}$ **3.** $\sqrt{169}$

4. $-\sqrt{100}$ **5.** $\sqrt{5}$ **6.** $-\sqrt{-1}$

Section 9.2

Simplify.

7. $\sqrt{x^4}$ **8.** $\sqrt{25x^4y^6}$ **9.** $\sqrt{98}$

10. $\sqrt{x^5}$ **11.** $\sqrt{36x^3}$ **12.** $\sqrt{8a^4b^3}$

Section 9.3

Combine, if possible.

13. $5\sqrt{2} - \sqrt{2}$ **14.** $\sqrt{2} + \sqrt{3}$

15. $3\sqrt{2} - \sqrt{8} + \sqrt{18}$ **16.** $2\sqrt{28} + 3\sqrt{63} - \sqrt{49}$

17. $3\sqrt{8} + \sqrt{12} + \sqrt{50} - 4\sqrt{75}$

Section 9.4

Multiply. Simplify your answer.

18. $(2\sqrt{6})(3\sqrt{3})$ **19.** $\sqrt{2}(\sqrt{8} + 2\sqrt{3})$

20. $(\sqrt{7} + 4)^2$ **21.** $(\sqrt{11} - \sqrt{10})(\sqrt{11} + \sqrt{10})$

22. $(\sqrt{m} + \sqrt{m+1})(\sqrt{m} - \sqrt{m+1})$ **23.** $(2\sqrt{3} - \sqrt{5})^2$

Section 9.5

Simplify. Rationalize all denominators.

24. $\sqrt{\frac{36}{x^2}}$ **25.** $\frac{5}{\sqrt{3}}$ **26.** $\frac{3}{\sqrt{5}+2}$ **27.** $\frac{\sqrt{7}-\sqrt{6}}{\sqrt{7}+\sqrt{6}}$

Section 9.6

Use the Pythagorean theorem *to find the length of the third side of the right triangle labeled x.*

28.

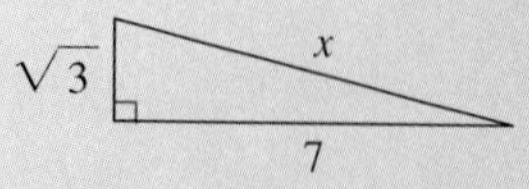

29.

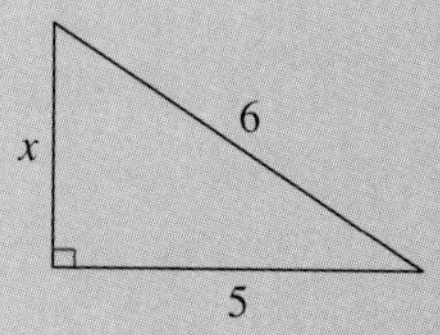

Solve each radical equation and verify your answer.

30. $2 + \sqrt{3x+4} = x$ **31.** $\sqrt{3x+1} = 6$ **32.** $\sqrt{11x-5} = \sqrt{4-7x}$

Section 9.7

33. If y varies inversely as x and $y = \frac{2}{3}$ when $x = 21$, find the value of y if $x = 2$.

9.1 Square Roots

After studying this section, you will be able to:

1. *Evaluate a square root of a perfect square.*
2. *Approximate a square root to the nearest thousandth.*

1 *Evaluating the Square Root of a Perfect Square*

How long is the side of a square whose area is 4? Recall that the formula for the area of a square is

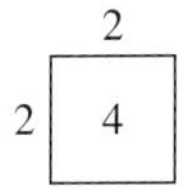

area of a square $= s^2$	*Out question then becomes, what number times itself is* 4?
$s^2 = 4$	
$s = 2$	*Because* $(2)(2) = 4$.
$s = -2$	*Because* $(-2)(-2) = 4$.

We say that 2 is the square root of 4 because $(2)(2) = 4$. We can also say that -2 is the square root of 4 because $(-2)(-2) = 4$. Note that 4 is a **perfect square.** The square root of a perfect square is an integer.

The symbol $\sqrt{}$ is used in mathematics for the **square root** of a number. The symbol, itself $\sqrt{}$ is called the *radical sign*. When we use this symbol, it is understood to be the nonnegative square root, or principal square root. If we want to find the negative square root of a number, we use the symbol $-\sqrt{}$.

> **Definition of Square Root**
>
> For all nonnegative numbers N, the square root of N (written $\sqrt{N}$) is defined to be the nonnegative number a if and only if $a^2 = N$.

Notice in the definition that we did not use the words "for all *positive* numbers N" because we also want to include the square root of 0. $\sqrt{0} = 0$ since $0^2 = 0$. So, for positive numbers *and* for zero, the definition of a square root is $\sqrt{N} = a$ if and only if $a^2 = N$.

Let us now examine the use of this definition with a few examples.

EXAMPLE 1 Find. **(a)** $\sqrt{144}$ **(b)** $-\sqrt{9}$

(a) $\sqrt{144} = 12$ since $(12)^2 = (12)(12) = 144$.

(b) The symbol $-\sqrt{9}$ is read "the negative square root of 9."

$$-\sqrt{9} = -3 \text{ since } (-3)^2 = (-3)(-3) = 9.$$

Practice Problem 1 Find. **(a)** $\sqrt{64}$ **(b)** $-\sqrt{121}$ ■

The number beneath the radical sign is called the **radicand.** The radicand will not always be an integer.

EXAMPLE 2 Find. **(a)** $\sqrt{\frac{1}{4}}$ **(b)** $-\sqrt{\frac{4}{9}}$

(a) $\sqrt{\frac{1}{4}} = \frac{1}{2}$ since $\left(\frac{1}{2}\right)^2 = \left(\frac{1}{2}\right)\left(\frac{1}{2}\right) = \frac{1}{4}$.

(b) $-\sqrt{\frac{4}{9}} = -\frac{2}{3}$ since $\left(-\frac{2}{3}\right)\left(-\frac{2}{3}\right) = \frac{4}{9}$.

Practice Problem 2 Find. **(a)** $\sqrt{\frac{9}{25}}$ **(b)** $-\sqrt{\frac{121}{169}}$ ■

EXAMPLE 3 Find. **(a)** $\sqrt{0.09}$ **(b)** $\sqrt{1600}$

(a) $\sqrt{0.09} = 0.3$ since $(0.3)(0.3) = 0.09$.
Notice that $\sqrt{0.09}$ is *not* 0.03 because $(0.03)(0.03) = 0.0003$. Remember to count the decimal places when you multiply decimals. When finding the square root of a decimal, you should do the multiplication as a check.

(b) $\sqrt{1600} = 40$ since $(40)(40) = 1600$.

Practice Problem 3 Find. **(a)** $-\sqrt{0.0036}$ **(b)** $\sqrt{2500}$ ■

2 *Approximating the Square Root*

Not all numbers are perfect squares. How can we find the square root of 2? That is, what number times itself is 2? The $\sqrt{2}$ is an irrational number. It cannot be written as $\frac{p}{q}$, where p, q are integers, $q \neq 0$. The $\sqrt{2}$ is a nonterminating, nonrepeating decimal and the best we can do is to approximate its value. To do so, you can use the square root key on your calculator or you can use the table of square roots in Appendix A.

To use a calculator, enter 2 and push the $\boxed{\sqrt{x}}$ or $\boxed{\sqrt{\ }}$ key.

$\boxed{2}\ \boxed{\sqrt{x}}\ \boxed{=}\ \boxed{\qquad\qquad}$

Remember, this is only an approximation. The calculator is limited to an eight-digit display.

To use the table, look under the x column until you find the 2 row. Then look under the $\sqrt{x}$ column in that row. The value is rounded to the nearest thousandth.

$$\sqrt{2} \approx 1.414$$

x	$\sqrt{x}$
1	1.000
2	1.414
3	1.732
4	2.000
5	2.236
6	2.449
7	2.646

EXAMPLE 4 Use a calculator or a table to approximate each square root. Round to the nearest thousandth.

(a) $\sqrt{28}$ ***(b)*** $\sqrt{191}$

Using a calculator, we have

(a) 28 $\boxed{\sqrt{x}}$ 5.291502622, thus $\sqrt{28} \approx 5.292$

(b) 191 $\boxed{\sqrt{x}}$ 13.82027496, thus $\sqrt{191} \approx 13.820$

Practice Problem 4 Approximate each square root.

(a) $\sqrt{3}$ **(b)** $\sqrt{35}$ **(c)** $\sqrt{127}$ ■

9.1 Exercises

Verbal and Writing Skills

1. Define a square root in your own words.

2. Is $\sqrt{0.04} = 0.02$? Why or why not?

3. Is $\sqrt{900} = 30$? Why or why not?

4. How would you find $\sqrt{90}$?

Find the two square roots of each number.

5. 9 **6.** 4 **7.** 64 **8.** 25

9. 81 **10.** 100 **11.** 121 **12.** 144

Find the square root. Do not *use a calculator or a table of square roots.*

13. $\sqrt{25}$ **14.** $\sqrt{36}$ **15.** $\sqrt{0}$ **16.** $\sqrt{1}$

17. $\sqrt{49}$ **18.** $\sqrt{16}$ **19.** $\sqrt{100}$ **20.** $\sqrt{9}$

21. $-\sqrt{36}$ **22.** $-\sqrt{25}$ **23.** $\sqrt{0.81}$ **24.** $\sqrt{0.64}$

25. $\sqrt{0.01}$ **26.** $\sqrt{0.16}$ **27.** $\sqrt{\frac{16}{25}}$ **28.** $\sqrt{\frac{9}{64}}$

29. $\sqrt{\frac{49}{64}}$ **30.** $\sqrt{\frac{49}{100}}$ **31.** $\sqrt{400}$ **32.** $\sqrt{3600}$

33. $-\sqrt{10{,}000}$ **34.** $\sqrt{40{,}000}$ **35.** $\sqrt{169}$ **36.** $\sqrt{225}$

37. $-\sqrt{\frac{1}{64}}$ **38.** $\sqrt{\frac{81}{100}}$ **39.** $\sqrt{\frac{25}{49}}$ **40.** $-\sqrt{\frac{9}{64}}$

41. $\sqrt{0.36}$ **42.** $\sqrt{0.25}$ **43.** $\sqrt{14{,}400}$ **44.** $-\sqrt{19{,}600}$

★ **45.** $\sqrt{441}$ ★ **46.** $\sqrt{961}$ **47.** $\sqrt{0.0064}$ **48.** $\sqrt{0.000081}$

Use a calculator or the square root table to approximate to the nearest thousandth each square root.

49. $\sqrt{42}$

50. $\sqrt{46}$

51. $\sqrt{59}$

52. $\sqrt{52}$

53. $-\sqrt{133}$

54. $-\sqrt{146}$

55. $-\sqrt{195}$

56. $-\sqrt{182}$

First estimate the answer. Then use a calculator to find each answer. Round your results to two decimal places.

57. $-\sqrt{360}$

58. $\sqrt{12.5}$

59. $\sqrt{45.5}$

60. $-\sqrt{800}$

Applications

The time in seconds it takes for an object to fall s feet is given by $t = \frac{1}{4}\sqrt{s}$. *Thus an object falling 100 feet would take* $t = \frac{1}{4}\sqrt{100} = \frac{1}{4}(10) = 2.5$ *seconds.*

61. How long would it take an object dropped out of a building to fall 900 feet?

62. How long would it take an object dropped out of a tower to fall 1600 feet?

63. The Sears Tower in Chicago is the world's tallest building. It is 1454 feet tall. If a bolt fell out of the building a distance of 1444 feet, how long would it take the bolt to hit the ground?

64. The CN Tower in Toronto, Ontario, is the world's tallest self-supporting structure. It is 1821 feet tall. If a bolt fell out of the tower a distance of 1764 feet, how long would it take the bold to hit the ground?

To Think About

Without the aid of a calculator, see if you can evaluate the following. (You may need to make several educated guesses!)

65. $\sqrt{\sqrt{81}}$

66. $\sqrt{\sqrt{\sqrt{256}}}$

Cumulative Review Problems

67. Solve. $3x + 2y = 8$
$7x - 3y = 11$

68. Solve if possible. $2x - 3y = 1$
$-8x + 12y = 4$

69. A taxi company bought 3 tires and 2 car batteries last month for $360. This month they bought 4 tires and 3 car batteries for $500. How much did each tire cost? How much did each battery cost?

70. With a tail wind an overseas plane flight from Los Angeles covered 7150 miles in 11 hours. The return flight took 13 hours. If the airspeed of the plane remained constant, and if the return flight encountered a constant head wind equal to the earlier tail wind, what was the plane's airspeed?

9.2 Simplify Radicals

After studying this section, you will be able to:

1 *Simplify radical expressions with a radicand that is a perfect square.*
2 *Simplify radical expressions with a radicand that is not a perfect square.*

1 *Simplify a Radical Expression with a Radicand That Is a Perfect Square*

We know that $\sqrt{25} = \sqrt{5^2} = 5$. $\sqrt{9} = \sqrt{3^2} = 3$ and $\sqrt{36} = \sqrt{6^2} = 6$. If we write the radicand as a square, the square root is the base. That is $\sqrt{x^2} = x$. We can use this idea to simplify radicals.

EXAMPLE 1 Find. **(a)** $\sqrt{9^4}$ **(b)** $\sqrt{126^2}$ **(c)** $\sqrt{17^6}$

(a) Using the law of exponents, we rewrite 9^4 as a square.

$\sqrt{9^4} = \sqrt{(9^2)^2}$ *Raise a power to a power* $(9^2)^2 = 9^4$.

$= 9^2$ *To check, multiply* $(9^2)(9^2) = 9^4$.

(b) $\sqrt{126^2} = 126$

(c) $\sqrt{17^6} = \sqrt{(17^3)^2}$ *Raise a power to a power* $(17^3)^2 = 17^6$.

$= 17^3$ *To check, multiply* $(17^3)(17^3) = 17^6$.

Practice Problem 1 Find. **(a)** $\sqrt{6^2}$ **(b)** $\sqrt{13^4}$ **(c)** $\sqrt{18^{12}}$ ■

This same concept can be used with variable expressions. We will restrict the values of each variable to be nonnegative if the variable is under the radical sign. Thus *all variable radicands in this chapter* are assumed to *represent positive numbers or zero.*

EXAMPLE 2 Find. **(a)** $\sqrt{x^6}$ **(b)** $\sqrt{x^{16}}$ **(c)** $\sqrt{y^{24}}$

(a) $\sqrt{x^6} = \sqrt{(x^3)^2}$ *By the law of exponents,* $(x^3)^2 = x^6$.

$= x^3$

(b) $\sqrt{x^{16}} = x^8$ **(c)** $\sqrt{y^{24}} = y^{12}$

Practice Problem 2 Find. **(a)** $\sqrt{x^4}$ **(b)** $\sqrt{y^{18}}$ **(c)** $\sqrt{x^{30}}$ ■

In each of these examples we found the square root exactly. It is as if we "removed the radical sign" when we found the square root. When you find a square root exactly, do *not* leave a radical sign in your answer.

Often the coefficient of the variable will not be written in exponent form. You will want to be able to recognize the squares of numbers from 1 to 15, 20, and 25. If you do, you will be able to find the square root faster than by using the calculator or a table of square roots.

Before you begin, you need to know the multiplication rule for square roots.

> **Multiplication Rule for Square Roots**
> For all nonnegative numbers a, b,
> $$\sqrt{a} \cdot \sqrt{b} = \sqrt{ab} \quad \text{and} \quad \sqrt{ab} = \sqrt{a} \cdot \sqrt{b}$$

EXAMPLE 3 Find. (a) $\sqrt{225x^2}$ (b) $\sqrt{x^6 y^{14}}$ (c) $\sqrt{169 x^8 y^{10}}$

(a) $\sqrt{225x^2} = 15x$ Using the multiplication rule for square roots,
$\sqrt{225x^2} = \sqrt{225}\sqrt{x^2} = 15x$.

(b) $\sqrt{x^6 y^{14}} = x^3 y^7$ (c) $\sqrt{169 x^8 y^{10}} = 13 x^4 y^5$

To check, square each answer. The square of the answer will be the expression under the radical sign.

Practice Problem 3 Find.
(a) $\sqrt{625 y^4}$ (b) $\sqrt{x^{16} y^{22}}$ (c) $\sqrt{121 x^{12} y^6}$ ∎

2 Simplifying a Radical Expression with a Radicand That Is Not a Perfect Square

Most of the time when we encounter square roots the radicand is not a perfect square. Thus, we cannot remove the radical sign and find the square root. It may, however, be possible to simplify the expression. We will want to do so, so that we can combine these expressions, as we will be doing in Section 9.3.

EXAMPLE 4 Simplify. (a) $\sqrt{20}$ (b) $\sqrt{50}$ (c) $\sqrt{48}$

To begin, when you look at a radicand, look for perfect squares.

(a) $\sqrt{20} = \sqrt{4 \cdot 5}$ Since 4 is a perfect square, we can write 20 as $4 \cdot 5$.
$\phantom{\sqrt{20}} = \sqrt{4}\sqrt{5}$ Recall $\sqrt{4} = 2$
$\phantom{\sqrt{20}} = 2\sqrt{5}$

(b) $\sqrt{50} = \sqrt{25 \cdot 2}$ (c) $\sqrt{48} = \sqrt{16 \cdot 3}$
$\phantom{\sqrt{50}} = \sqrt{25}\sqrt{2}$ $\phantom{\sqrt{48}} = \sqrt{16}\sqrt{3}$
$\phantom{\sqrt{50}} = 5\sqrt{2}$ $\phantom{\sqrt{48}} = 4\sqrt{3}$

Practice Problem 4 Simplify. (a) $\sqrt{98}$ (b) $\sqrt{12}$ (c) $\sqrt{75}$ ∎

The same procedure can be used with expressions containing variables. The key is to think squares. That is, think, "How can I rewrite the expression so that it contains a perfect square."

EXAMPLE 5 Simplify. **(a)** $\sqrt{x^3}$ **(b)** $\sqrt{y^5}$ **(c)** $\sqrt{x^7y^9}$

(a) Recall that the law of exponents tells us to add the exponents when multiplying two numbers that have the same base. Think of rewriting 3 as a sum that contains an even number. $3 = 2 + 1$

$\sqrt{x^3} = \sqrt{x^2}\sqrt{x}$ *Because* $(x^2)(x) = x^3$ *by the law of exponents.*

$= x\sqrt{x}$

(b) $\sqrt{y^5} = \sqrt{y^4}\sqrt{y}$ *Notice that* y^4 *is the largest possible even exponent.*

$= y^2\sqrt{y}$

In general, to simplify a square root that has a variable with an exponent in the radicand, first write it as a product of the form $\sqrt{x^n}\sqrt{x}$, where n is the largest possible even exponent.

(c) $\sqrt{x^7y^9} = \sqrt{x^6}\sqrt{x}\sqrt{y^8}\sqrt{y}$

$= x^3\sqrt{x}y^4\sqrt{y}$

$= x^3y^4\sqrt{xy}$

In the final form, we place the factors with exponents first and the radical factor second.

Practice Problem 5 Simplify. **(a)** $\sqrt{x^{11}}$ **(b)** $\sqrt{x^5y^3}$ ■

Putting this all together, to simplify a radical, you factor each part of the expression under the radical sign and take out the perfect squares.

EXAMPLE 6 Simplify. **(a)** $\sqrt{12y^5}$ **(b)** $\sqrt{27x^5y^6}$ **(c)** $\sqrt{18x^3y^7w^{10}}$

(a) $\sqrt{12y^5} = \sqrt{4 \cdot 3 \cdot y^4 \cdot y}$

$= 2y^2\sqrt{3y}$ *Remember,* $y^4 = (y^2)^2$ *and* $\sqrt{(y^2)^2} = y^2$.

(b) $\sqrt{27x^5y^6} = \sqrt{9 \cdot 3 \cdot x^4 \cdot x \cdot y^6}$

$= 3x^2y^3\sqrt{3x}$

Notice that since the exponent of y is even y^6 is a perfect square. y will not appear under the radical sign in the answer.

(c) $\sqrt{18x^3y^7w^{10}} = \sqrt{9 \cdot 2 \cdot x^2 \cdot x \cdot y^6 \cdot y \cdot w^{10}}$

$= 3xy^3w^5\sqrt{2xy}$

Practice Problem 6 Simplify.

(a) $\sqrt{48x^{11}}$ **(b)** $\sqrt{72x^8y^9}$ **(c)** $\sqrt{121x^6y^7z^8}$ ■

9.2 Exercises

Simplify. Leave the answer in exponent form.

1. $\sqrt{7^2}$
2. $\sqrt{6^2}$
3. $\sqrt{10^4}$
4. $\sqrt{8^4}$
5. $\sqrt{5^{10}}$
6. $\sqrt{7^6}$
7. $\sqrt{56^6}$
8. $\sqrt{39^8}$
9. $\sqrt{3^{150}}$
10. $\sqrt{2^{180}}$

Simplify. Assume that all variables represent positive numbers.

11. $\sqrt{w^{12}}$
12. $\sqrt{y^6}$
13. $\sqrt{x^{24}}$
14. $\sqrt{x^{30}}$
15. $\sqrt{y^{26}}$
16. $\sqrt{w^{14}}$
17. $\sqrt{49x^6}$
18. $\sqrt{81y^4}$
19. $\sqrt{144x^2}$
20. $\sqrt{196y^{10}}$
21. $\sqrt{400x^8}$
22. $\sqrt{2500y^{12}}$
23. $\sqrt{x^2y^6}$
24. $\sqrt{x^4y^{12}}$
25. $\sqrt{25x^8y^4}$
26. $\sqrt{16x^2y^{20}}$
27. $\sqrt{64x^2y^4}$
28. $\sqrt{100x^{12}y^8}$
29. $\sqrt{121x^6y^{14}}$
30. $\sqrt{225x^4y^{16}}$

Simplify. Assume that all variables represent positive numbers.

31. $\sqrt{20}$
32. $\sqrt{12}$
33. $\sqrt{24}$
34. $\sqrt{27}$
35. $\sqrt{18}$
36. $\sqrt{32}$
37. $\sqrt{50}$
38. $\sqrt{48}$
39. $\sqrt{72}$
40. $\sqrt{60}$
41. $\sqrt{90}$
42. $\sqrt{125}$
43. $\sqrt{98}$
44. $\sqrt{40}$
45. $\sqrt{150}$

46. $\sqrt{8x^3}$ **47.** $\sqrt{12y^5}$ **48.** $\sqrt{27w^5}$ **49.** $\sqrt{18y^{13}}$ **50.** $\sqrt{16x^7}$

51. $\sqrt{49x^5}$ **52.** $\sqrt{75x^3y^5}$ **53.** $\sqrt{50x^7y}$ **54.** $\sqrt{48y^3w}$ **55.** $\sqrt{12x^2y^3}$

56. $\sqrt{75x^2y^3}$ **57.** $\sqrt{27a^8y^7}$ **58.** $\sqrt{135x^5y^7}$ **59.** $\sqrt{140x^7y^{11}}$ **60.** $\sqrt{150a^3b^{13}}$

61. $\sqrt{45ab^3c^4}$ **62.** $\sqrt{32a^3bc^4}$ **63.** $\sqrt{63a^4b^6c^3}$ **64.** $\sqrt{28a^2b^5c^8}$ **65.** $\sqrt{81x^{12}y^{11}w^5}$

To Think About

Simplify. Assume that all variables represent positive numbers.

66. $\sqrt{(x+3)^2}$ **67.** $\sqrt{x^2 + 12x + 36}$ **68.** $\sqrt{4x^2 + 12x + 9}$

69. $\sqrt{9x^2 + 12x + 4}$ **70.** $\sqrt{x^2y^2 + 14xy^2 + 49y^2}$ **71.** $\sqrt{16x^2 + 16x + 4}$

Cumulative Review Problems

72. Solve. $3x - 2y = 14$
$y = x - 5$

73. Solve. $\frac{1}{3}(2x - 4) = \frac{1}{4}x - 3$

74. Three more than seven times a number is a dozen less than twice the number. Find the number.

75. The sum of two numbers is twenty. Twice their difference is eleven. Find these two numbers.

9.3 Addition and Subtraction of Radicals

After studying this section, you will be able to:

1 *Add or subtract radical expressions with common radicals.*
2 *Add or subtract radical expressions that must be first simplified.*

1 *Adding or Subtracting Radicals with Common Radicals*

Recall that when you simplify an algebraic expression you combine like terms. $8b + 5b = 13b$ because both terms contain the same variable b, and you add the coefficients. This same idea is used when combining radicals. To add or subtract radicals, the number or the algebraic expression under the radical sign must be the same.

EXAMPLE 1 Combine. **(a)** $5\sqrt{2} - 8\sqrt{2}$
(b) $7\sqrt{a} + 3\sqrt{a} - 5\sqrt{a}$

First check to see if the radicands are the same. If they are, you can combine the radicals.

(a) $5\sqrt{2} - 8\sqrt{2} = -3\sqrt{2}$ **(b)** $7\sqrt{a} + 3\sqrt{a} - 5\sqrt{a} = 5\sqrt{a}$

Practice Problem 1 Combine. **(a)** $7\sqrt{11} + 4\sqrt{11}$
(b) $4\sqrt{t} - 7\sqrt{t} + 6\sqrt{t} - 2\sqrt{t}$ ■

Be careful. Not all the terms in an expression will have common radicals.

EXAMPLE 2 Combine. $5\sqrt{2a} + 3\sqrt{a} - 7\sqrt{2} + 3\sqrt{2a}$

The only terms that have the same radicand are $5\sqrt{2a}$ and $3\sqrt{2a}$. These terms may be combined. All other terms stay the same.

$$5\sqrt{2a} + 3\sqrt{a} - 7\sqrt{2} + 3\sqrt{2a} = 8\sqrt{2a} + 3\sqrt{a} - 7\sqrt{2}$$

Practice Problem 2 Combine. $3\sqrt{x} - 2\sqrt{xy} - 5\sqrt{y} + 7\sqrt{xy}$ ■

2 *Adding and Subtracting Radical Expressions That Must First Be Simplified*

Sometimes it is necessary to simplify one or more radicals before their terms can be combined. Be alert to combine only like radicals.

EXAMPLE 3 Combine. **(a)** $2\sqrt{3} + \sqrt{12}$ **(b)** $\sqrt{12} - \sqrt{27} + \sqrt{50}$

(a) $2\sqrt{3} + \sqrt{12} = 2\sqrt{3} + \sqrt{4 \cdot 3}$ *Look for perfect square factors.*

$= 2\sqrt{3} + 2\sqrt{3}$ *Simplify and combine like radicals.*

$= 4\sqrt{3}$

(b) $\sqrt{12} - \sqrt{27} + \sqrt{50} = \sqrt{4 \cdot 3} - \sqrt{9 \cdot 3} + \sqrt{25 \cdot 2}$

$= 2\sqrt{3} - 3\sqrt{3} + 5\sqrt{2}$ *Only $2\sqrt{3}$ and $-3\sqrt{3}$ can be combined.*

$= -\sqrt{3} + 5\sqrt{2}$ or $5\sqrt{2} - \sqrt{3}$

Practice Problem 3 Combine. **(a)** $\sqrt{50} - \sqrt{18} + \sqrt{98}$
(b) $\sqrt{12} + \sqrt{18} - \sqrt{50} + \sqrt{27}$ ■

EXAMPLE 4 Combine. $\sqrt{2a} + \sqrt{8a} + \sqrt{27a}$

$$\sqrt{2a} + \sqrt{8a} + \sqrt{27a} = \sqrt{2a} + \sqrt{4 \cdot 2a} + \sqrt{9 \cdot 3a}$$

Look for perfect square factors.

$$= \sqrt{2a} + 2\sqrt{2a} + 3\sqrt{3a}$$

Be careful. $\sqrt{2a}$ and $\sqrt{3a}$ are ***not*** *like radicals. The expression under the radical sign must be the same.*

$$= 3\sqrt{2a} + 3\sqrt{3a}$$

Practice Problem 4 Combine. $\sqrt{9x} + \sqrt{8x} - \sqrt{4x} + \sqrt{50x}$ ■

Special care should be taken if the original radical has a numerical coefficient. If the radical can be simplified, the two resulting numerical coefficients should be multiplied.

EXAMPLE 5 Combine. **(a)** $3\sqrt{18} - 5\sqrt{2}$ **(b)** $2\sqrt{20} + 3\sqrt{45} - 4\sqrt{80}$

(a) $3\sqrt{18} - 5\sqrt{2} = 3 \cdot \sqrt{9} \cdot \sqrt{2} - 5\sqrt{2}$

$$= 3 \cdot 3\sqrt{2} - 5\sqrt{2}$$

Remember to multiply $3 \cdot 3$ here.

$$= 9\sqrt{2} - 5\sqrt{2}$$

$$= 4\sqrt{2}$$

(b) $2\sqrt{20} + 3\sqrt{45} - 4\sqrt{80} = 2 \cdot \sqrt{4} \cdot \sqrt{5} + 3 \cdot \sqrt{9} \cdot \sqrt{5} - 4 \cdot \sqrt{16} \cdot \sqrt{5}$

$$= 2 \cdot 2 \cdot \sqrt{5} + 3 \cdot 3 \cdot \sqrt{5} - 4 \cdot 4 \cdot \sqrt{5}$$

$$= 4\sqrt{5} + 9\sqrt{5} - 16\sqrt{5}$$

$$= -3\sqrt{5}$$

Practice Problem 5 Combine. **(a)** $2\sqrt{18} + 3\sqrt{8}$ **(b)** $\sqrt{27} - 4\sqrt{3} + 2\sqrt{75}$ ■

EXAMPLE 6 Combine. $3a\sqrt{8a} + 2\sqrt{50a^3}$

$3a\sqrt{8a} + 2\sqrt{50a^3}$ *Simplify each radical.*

$$= 3a\sqrt{4}\sqrt{2a} + 2\sqrt{25a^2}\sqrt{2a}$$

$$= 3a \cdot 2 \cdot \sqrt{2a} + 2 \cdot 5a \cdot \sqrt{2a}$$

$$= 6a\sqrt{2a} + 10a\sqrt{2a}$$

$$= 16a\sqrt{2a}$$

(*Note*: If you are unsure of the last step, show the use of the distributive property in performing the addition.)

$$6a\sqrt{2a} + 10a\sqrt{2a} = (6 + 10)a\sqrt{2a}$$

$$= 16a\sqrt{2a}$$

Practice Problem 6 Combine. $2\sqrt{12x} - 3\sqrt{45x} - 3\sqrt{27x} + \sqrt{20}$ ■

9.3 Exercises

Verbal and Writing Skills

1. List the steps involved in simplifying an expression such as $\sqrt{75x} + \sqrt{48x} + \sqrt{16x^2}$.

2. Julio said that, since $\sqrt{8}$ and $\sqrt{2}$ are not similar, $\sqrt{8} + \sqrt{2}$ could not be simplified. What do you think?

Combine each of the following, if possible. Do not use a calculator or a table of square roots.

3. $\sqrt{3} - 5\sqrt{3} + 2\sqrt{3}$

4. $\sqrt{6} + 4\sqrt{6} - 7\sqrt{6}$

5. $\sqrt{2} + 8\sqrt{3} - 5\sqrt{3} + 4\sqrt{2}$

6. $\sqrt{5} - \sqrt{6} + 3\sqrt{5} - 2\sqrt{6}$

7. $3\sqrt{2x} + 5\sqrt{x} - 7\sqrt{2x}$

8. $\sqrt{ab} - 2\sqrt{b} + 3\sqrt{ab}$

9. $\sqrt{3} - \sqrt{12}$

10. $\sqrt{5} - \sqrt{20}$

11. $\sqrt{50} + 3\sqrt{32}$

12. $2\sqrt{12} + \sqrt{48}$

13. $2\sqrt{8} - 3\sqrt{2}$

14. $2\sqrt{27} - 4\sqrt{3}$

15. $\sqrt{75} + \sqrt{3} - 2\sqrt{27}$

16. $\sqrt{28} - 3\sqrt{7} + 2\sqrt{63}$

17. $2\sqrt{12} + \sqrt{20} + \sqrt{36}$

18. $\sqrt{24} - 2\sqrt{54} + 2\sqrt{18}$

19. $\sqrt{20} - \sqrt{80} + 3\sqrt{48}$

20. $\sqrt{50} + \sqrt{28} - \sqrt{9}$

21. $3\sqrt{3x} + \sqrt{12x}$

22. $5\sqrt{2y} - \sqrt{18y}$

23. $3\sqrt{8xy} - \sqrt{50xy}$

24. $2\sqrt{45ab} - \sqrt{20ab}$

25. $1.2\sqrt{3x} - 0.5\sqrt{12x}$

26. $-1.5\sqrt{2a} + 0.2\sqrt{8a}$

27. $\sqrt{20y} + 2\sqrt{45y} - \sqrt{5y}$

28. $\sqrt{72w} - 3\sqrt{2w} - \sqrt{50w}$

29. $3\sqrt{y^3} + 2y\sqrt{y}$

30. $4x\sqrt{x} - \sqrt{16x^3}$

31. $2x\sqrt{8x} + 5\sqrt{18x}$

32. $-3a\sqrt{12a} + \sqrt{75a}$

33. $5\sqrt{8x^3} - 3x\sqrt{50x}$

34. $-2\sqrt{27y^3} + y\sqrt{12y}$

35. $3\sqrt{27x^2} - 2\sqrt{48x^2}$

36. $-5\sqrt{8y^2} + 2\sqrt{32y^2}$

37. $2\sqrt{6y^3} - 2y\sqrt{54}$

38. $3\sqrt{24x^4} - 2\sqrt{54x^4}$

39. $-5\sqrt{72x^6} - 2\sqrt{50x^6}$

40. $3\sqrt{20x^3} - 4x\sqrt{5x^2}$

41. $5x\sqrt{8x} - 24\sqrt{50x^3}$

Applications

To estimate the cost of erecting a radio tower, we need to know how many yards of cable are in the tower. Each level of the tower consists of two long panels and two short panels made up of cable. Since the tower narrows at the top, each level is a different size.

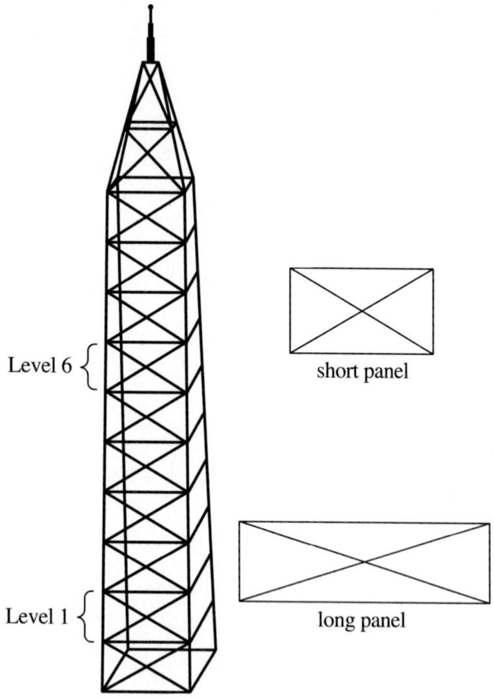

42. The size of the panels at level 1 is illustrated in the scale drawing below.

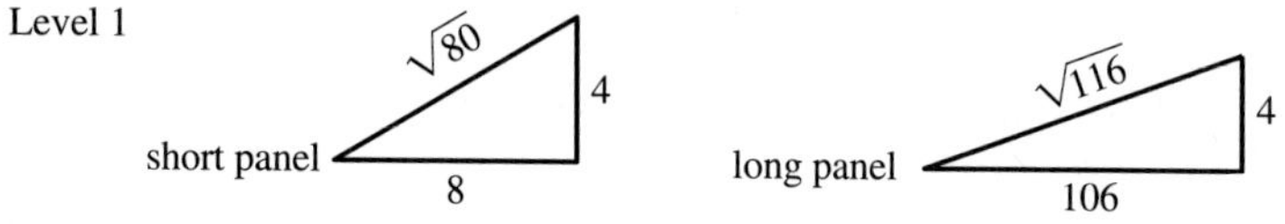

How many yards of cable are needed for level 1? (Simplify the radicals and combine where possible.)

43. The size of the panels at level 6 is illustrated in the scale drawing below.

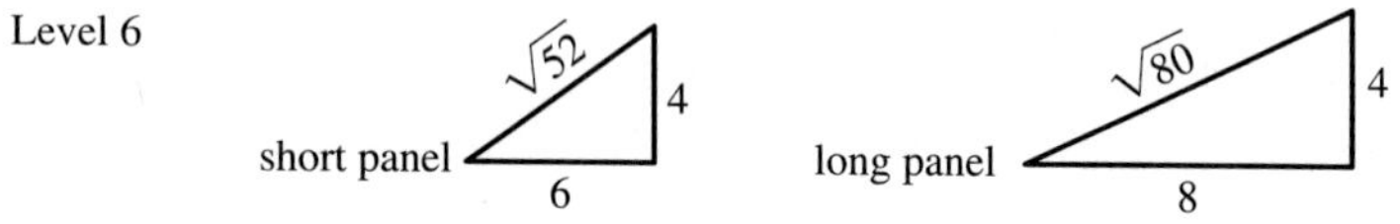

How many yards of cable are needed for level 6? (Simplify the radicals and combine where possible.)

To Think About

Simplify, if possible, by combining like radicals.

44. $\sqrt{4x^2 + 4} + \sqrt{9x^2 + 9}$

45. $\sqrt{x^5 - 2x^2} + \sqrt{4x^5 - 8x^2}$

46. $\sqrt{x^2 + 2x + 1} + \sqrt{x^2 - 2x + 1}$

47. $\sqrt{x^2 + 2x + 1} - \sqrt{x^2 - 2x + 1}$

48. $\sqrt{3} + \sqrt{\frac{1}{3}}$

49. $\sqrt{2} + \sqrt{\frac{1}{2}}$

50. $\sqrt{4} + \sqrt{\frac{1}{4}}$

Cumulative Review Problems

51. Graph on a coordinate system. $3y - 2x \leq 6$

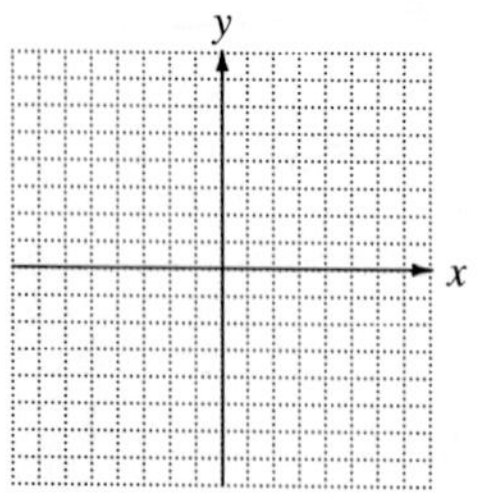

52. Solve and graph on a number line. $7 - 3x \leq 11$

53. Martin went to the post office and purchased 132 stamps for \$27.50. Some were 15¢ postal card stamps, the rest cost 25¢ each. How many of each did Martin purchase?

54. Maria invested \$8000 in two places. She invested part of the money at 5% interest and the rest of the money at 8% interest. At the end of one year she earned \$490. How much did she invest at each rate?

9.4 Multiplication of Radicals

After studying this section, you will be able to:

1. *Multiply monomial radical expressions.*
2. *Multiply a monomial radical expression by a binomial.*
3. *Multiply two polynomial radical expressions.*

1 *Multiplying Monomial Radical Expressions*

Recall the basic rule for multiplication of radicals

$$\sqrt{a}\sqrt{b} = \sqrt{ab}$$

We will use this concept to multiply radical expressions. Note that the direction ''multiply'' means express the product in simplest form.

EXAMPLE 1 Multiply. $\sqrt{7}\sqrt{14x}$

$$\sqrt{7}\sqrt{14x} = \sqrt{98x}$$

We do *not* stop here, because the radical $\sqrt{98x}$ can be simplified.

$\sqrt{98x} = \sqrt{49 \cdot 2x} = \sqrt{49}\sqrt{2x} = 7\sqrt{2x}$

Practice Problem 1 Multiply. $\sqrt{3a}\sqrt{6a}$ ■

EXAMPLE 2 Multiply.

(a) $(2\sqrt{3})(5\sqrt{7})$ **(b)** $(x\sqrt{2})(4\sqrt{10})$ **(c)** $(2a\sqrt{3a})(3\sqrt{6a})$

(a) $(2\sqrt{3})(5\sqrt{7}) = 10\sqrt{21}$ *Multiply the coefficients* $(2)(5) = 10$.
Multiply the radicals $\sqrt{3}\sqrt{7} = \sqrt{21}$.

(b) $(x\sqrt{2})(4\sqrt{10}) = 4x\sqrt{20}$ *Simplify* $\sqrt{20}$.

$= 4x\sqrt{4}\sqrt{5}$

$= 8x\sqrt{5}$

(c) $(2a\sqrt{3a})(3\sqrt{6a}) = 6a\sqrt{18a^2}$ *Simplify* $\sqrt{18a^2}$.

$= 6a\sqrt{9}\sqrt{2}\sqrt{a^2}$ *Multiply* $(6a)(3)(a)$.

$= 18a^2\sqrt{2}$

Practice Problem 2 Multiply.

(a) $(2\sqrt{3})(5\sqrt{5})$ **(b)** $(2x\sqrt{6})(3\sqrt{6})$ **(c)** $(4\sqrt{3x})(2x\sqrt{6x})$ ■

2 *Multiplying a Monomial Radical Expression by a Binomial*

Recall that a binomial consists of two terms. $\sqrt{2} + \sqrt{5}$ is a binomial. We use the distributive property when multiplying a binomial.

EXAMPLE 3 Multiply. $\sqrt{5}(\sqrt{2} + 3\sqrt{7})$

$$= \sqrt{5}\sqrt{2} + 3\sqrt{5}\sqrt{7}$$
$$= \sqrt{10} + 3\sqrt{35}$$

Practice Problem 3 Simplify. $2\sqrt{3}(\sqrt{3} + \sqrt{5} - \sqrt{12})$ ■

Note that $\sqrt{2} \cdot \sqrt{2} = 2$. This is a specific case of the multiplication rule for radicals.

> For any nonnegative real number a,
> $$\sqrt{a} \cdot \sqrt{a} = a$$

EXAMPLE 4 Multiply. $\sqrt{a}(3\sqrt{a} - 2\sqrt{5})$

$= 3\sqrt{a}\sqrt{a} - 2\sqrt{5}\sqrt{a}$ *Use the distributive property.*
$= 3a - 2\sqrt{5a}$

Practice Problem 4 Multiply. $2\sqrt{x}(4\sqrt{x} - x\sqrt{2})$ ■

Be sure to simplify all radicals after multiplying.

EXAMPLE 5 Multiply. $2\sqrt{2x}(3\sqrt{10x} - 2\sqrt{6})$

$= 6\sqrt{20x^2} - 4\sqrt{12x}$ *Use the distributive property.*
$= 6\sqrt{4x^2}\sqrt{5} - 4\sqrt{4}\sqrt{3x}$ *Simplify $\sqrt{20x^2}$ and $\sqrt{12x}$.*
$= 6(2x)\sqrt{5} - 4(2)\sqrt{3x}$
$= 12x\sqrt{5} - 8\sqrt{3x}$

Practice Problem 5 Multiply. $\sqrt{6x}(\sqrt{15x} + \sqrt{10x^2})$ ■

3 *Multiplying Two Polynomial Radical Expressions*

Recall the FOIL method used to multiply two binomials for algebraic expressions. The same method can be used to multiply two binomials as for radical expressions.

Algebraic Expressions

$$(2x + y)(x - 2y) = 2x^2 - 4xy + xy - 2y^2$$
$$= 2x^2 - 3xy - 2y^2$$

Radical Expressions

$$(2\sqrt{3} + \sqrt{5})(\sqrt{3} - 2\sqrt{5}) = 2\sqrt{9} - 4\sqrt{15} + \sqrt{15} - 2\sqrt{25}$$
$$= 2(3) - 3\sqrt{15} - 2(5)$$
$$= 6 - 3\sqrt{15} - 10$$
$$= -4 - 3\sqrt{15}$$

Let's look at this procedure more closely.

EXAMPLE 6 Multiply. $(\sqrt{2} + 5)(\sqrt{2} - 3)$

$(\sqrt{2} + 5)(\sqrt{2} - 3)$

$= \sqrt{4} - 3\sqrt{2} + 5\sqrt{2} - 15$ *Multiply to obtain the four products.*

$= 2 + 2\sqrt{2} - 15$ *Combine the middle terms.*

$= -13 + 2\sqrt{2}$

Practice Problem 6 Multiply. $(\sqrt{2} + \sqrt{6})(2\sqrt{2} - \sqrt{6})$ ■

EXAMPLE 7 Multiply. $(\sqrt{2} - 3\sqrt{6})(\sqrt{2} + \sqrt{6})$

$(\sqrt{2} - 3\sqrt{6})(\sqrt{2} + \sqrt{6})$

$= \sqrt{4} + \sqrt{12} - 3\sqrt{12} - 3\sqrt{36}$ *Multiply to obtain the four products.*

$= 2 - 2\sqrt{12} - 18$ *Combine the middle terms.*

$= -16 - 4\sqrt{3}$ *Simplify* $-2\sqrt{12}$.

Practice Problem 7 Multiply. $(\sqrt{6} + \sqrt{5})(\sqrt{2} + 2\sqrt{5})$ ■

If an expression with radicals is squared, write the expression as the product of two binomials and multiply.

EXAMPLE 8 Multiply. $(2\sqrt{3} - \sqrt{6})^2$

$$= (2\sqrt{3} - \sqrt{6})(2\sqrt{3} - \sqrt{6})$$
$$= 4\sqrt{9} - 2\sqrt{18} - 2\sqrt{18} + \sqrt{36}$$
$$= 12 - 4\sqrt{18} + 6$$
$$= 18 - 12\sqrt{2}$$

Practice Problem 8 Multiply. $(3\sqrt{5} - \sqrt{10})^2$ ■

9.4 Exercises

Multiply. Be sure to simplify any radicals in your answer. Do not use a calculator or a table of square roots.

1. $\sqrt{7}\sqrt{5}$

2. $\sqrt{3}\sqrt{11}$

3. $\sqrt{2}\sqrt{22}$

4. $\sqrt{5}\sqrt{15}$

5. $\sqrt{3a}\sqrt{5a}$

6. $\sqrt{6x}\sqrt{3}$

7. $\sqrt{5x}\sqrt{10}$

8. $\sqrt{12b}\sqrt{2b}$

9. $(3\sqrt{5})(2\sqrt{6})$

10. $(-3a\sqrt{2})(5a\sqrt{3a})$

11. $(2x\sqrt{x})(3x\sqrt{5x})$

12. $(4\sqrt{12})(2\sqrt{6})$

13. $(-3\sqrt{ab})(2\sqrt{b})$

14. $(2\sqrt{x})(5\sqrt{xy})$

15. $\sqrt{3}(\sqrt{2} + 2\sqrt{5})$

16. $\sqrt{2}(\sqrt{5} - 2\sqrt{2})$

17. $\sqrt{2}(\sqrt{10} - 3\sqrt{6})$

18. $\sqrt{5}(\sqrt{15} + 3\sqrt{5})$

19. $2\sqrt{x}(\sqrt{x} - 8\sqrt{5})$

20. $-3\sqrt{b}(2\sqrt{a} + 3\sqrt{b})$

21. $\sqrt{6}(\sqrt{2} - 3\sqrt{6} + 2\sqrt{10})$

22. $\sqrt{10}(\sqrt{5} - 3\sqrt{10} + 5\sqrt{2})$

23. $2\sqrt{a}(3\sqrt{b} + \sqrt{ab} - 2\sqrt{a})$

24. $3\sqrt{x}(\sqrt{y} + 2\sqrt{xy} - 4\sqrt{x})$

25. $(\sqrt{7} + \sqrt{2})(2\sqrt{7} + \sqrt{2})$

26. $(3\sqrt{5} + \sqrt{3})(\sqrt{5} + \sqrt{3})$

27. $(8 + \sqrt{3})(2 + 2\sqrt{3})$

28. $(5 + \sqrt{2})(6 + \sqrt{2})$

29. $(2\sqrt{7} - 3\sqrt{3})(\sqrt{7} + \sqrt{3})$

30. $(5\sqrt{6} - \sqrt{2})(\sqrt{6} + 3\sqrt{2})$

31. $(\sqrt{3} + 2\sqrt{6})(2\sqrt{3} - \sqrt{6})$

32. $(\sqrt{2} + 3\sqrt{10})(2\sqrt{2} - \sqrt{10})$

33. $(3\sqrt{7} - \sqrt{8})(\sqrt{8} + 2\sqrt{7})$

34. $(\sqrt{12} - \sqrt{5})(\sqrt{5} + 2\sqrt{12})$

35. $(2\sqrt{5} - 3)^2$

36. $(3\sqrt{2} + 5)^2$

37. $(\sqrt{3} + 5\sqrt{2})^2$

38. $(\sqrt{5} - 2\sqrt{6})^2$

39. $(\sqrt{6} - 2\sqrt{3})^2$

40. $(2\sqrt{3} + 5\sqrt{5})^2$

41. $(\sqrt{a} - \sqrt{ab})^2$

42. $(\sqrt{xy} - 2\sqrt{x})^2$

43. $(2a\sqrt{3} - \sqrt{a})(\sqrt{3} + \sqrt{a})$

44. $(5x\sqrt{y} + \sqrt{2})(\sqrt{y} - 2\sqrt{2})$

45. $(5x\sqrt{x} - 2\sqrt{5x})(\sqrt{5} - \sqrt{x})$

46. $(4a\sqrt{a} - 2\sqrt{3a})(\sqrt{a} - 6\sqrt{3})$

To Think About

47. (a) If a can be any number, can you think of an example of when it is not true that $\sqrt{a} \cdot \sqrt{a} = a$?
(b) Why is it necessary to state the restriction "for any nonnegative real number a"?

48. Hank was trying to evaluate $(\sqrt{a} + \sqrt{b})^2$. He wrote $(\sqrt{a} + \sqrt{b})^2 = (\sqrt{a})^2 + (\sqrt{b})^2 = a + b$. This was a serious error! What is wrong with his work? How should he have done the problem?

Cumulative Review Problems

49. Factor. $36x^2 - 49y^2$.

50. Factor. $x^8 - 1$

51. Maria invested $40,000 into two accounts, one paying 6%, the other 11%. She earned $3775 interest for the year. How much was invested in each of these two accounts?

52. The cost of a pair of shoes, including an 8% sales tax, is $59.40. What was the marked price before adding the sales tax?

9.5 Division of Radicals

After studying this section, you will be able to:

1. *Simplify a fraction involving radicals by using the quotient rule for square roots.*
2. *Rationalize the denominator of a fraction with a square root in the denominator.*
3. *Rationalize the denominator of a fraction with a binomial in the denominator containing at least one square root.*

1 *Simplifying a Fraction Involving Radicals by Using the Quotient Rule*

Just as there is a multiplication rule for square roots, there is a quotient rule for square roots. The rules are similar.

Quotient Rule for Square Roots

For all positive numbers a and b,

$$\frac{\sqrt{a}}{\sqrt{b}} = \sqrt{\frac{a}{b}} \quad \text{and} \quad \sqrt{\frac{a}{b}} = \frac{\sqrt{a}}{\sqrt{b}}$$

This can be used to divide or to simplify radical expressions involving division. We will be using both parts of the quotient rule interchangeably.

EXAMPLE 1 Simplify. **(a)** $\dfrac{\sqrt{75}}{\sqrt{3}}$ **(b)** $\sqrt{\dfrac{25}{36}}$

(a) We notice that 3 is a factor of 75. We use the quotient rule to rewrite the expression.

$$\frac{\sqrt{75}}{\sqrt{3}} = \sqrt{\frac{75}{3}} \qquad \textit{Divide.}$$

$$= \sqrt{25} \qquad \textit{Simplify.}$$

$$= 5$$

(b) Since both 25 and 36 are perfect squares, we will rewrite this as the quotient of square roots.

$$\sqrt{\frac{25}{36}} = \frac{\sqrt{25}}{\sqrt{36}}$$

$$= \frac{5}{6}$$

Practice Problem 1 Simplify. **(a)** $\sqrt{\dfrac{x^3}{x}}$ **(b)** $\sqrt{\dfrac{49}{x^2}}$ ■

EXAMPLE 2 Simplify. $\sqrt{\dfrac{20}{x^6}}$

$$\sqrt{\frac{20}{x^6}} = \frac{\sqrt{20}}{\sqrt{x^6}} = \frac{\sqrt{4}\sqrt{5}}{x^3} = \frac{2\sqrt{5}}{x^3} \qquad \textit{Don't forget to simplify } \sqrt{20} \textit{ to } 2\sqrt{5}.$$

Practice Problem 2 Simplify. $\sqrt{\dfrac{50}{a^4}}$ ■

2 *Rationalizing the Denominator of a Fraction with a Square Root in the Denominator*

Sometimes, when calculating with fractions that contain radicals, it is advantageous to have an integer in the denominator. If a fraction has a radical in the denominator, we rationalize the denominator. That is, we change the radical in the denominator into an integer. Remember, we do not want to change the value of the fraction. Thus we will multiply the numerator and the denominator by the same number.

EXAMPLE 3 Rationalize the denominator. **(a)** $\frac{3}{\sqrt{2}}$ **(b)** $\frac{a}{\sqrt{5}}$

(a) $\frac{3}{\sqrt{2}} \times 1 = \frac{3}{\sqrt{2}} \times \frac{\sqrt{2}}{\sqrt{2}} = \frac{3\sqrt{2}}{\sqrt{4}} = \frac{3\sqrt{2}}{2}$

(b) $\frac{a}{\sqrt{5}} = \frac{a}{\sqrt{5}} \times 1 = \frac{a}{\sqrt{5}} \times \frac{\sqrt{5}}{\sqrt{5}} = \frac{a\sqrt{5}}{5}$

Practice Problem 3 Simplify. **(a)** $\frac{9}{\sqrt{7}}$ **(b)** $\frac{\sqrt{5}}{\sqrt{6}}$

(Note that a fraction is not simplified unless the denominator is rationalized.) ■

We will want to use the smallest possible radical that will yield the square root of a perfect square. Very often we will not use the radical that is in the denominator.

EXAMPLE 4 Simplify. **(a)** $\frac{\sqrt{7}}{\sqrt{8}}$ **(b)** $\frac{3}{\sqrt{x^3}}$

(a) Think, "What times 8 will make a perfect square?"

$$\frac{\sqrt{7}}{\sqrt{8}} = \frac{\sqrt{7}}{\sqrt{8}} \times \frac{\sqrt{2}}{\sqrt{2}} = \frac{\sqrt{14}}{\sqrt{16}} = \frac{\sqrt{14}}{4}$$

(b) Think, "What times x^3 will give an even exponent?"

$$\frac{3}{\sqrt{x^3}} \times \frac{\sqrt{x}}{\sqrt{x}} = \frac{3\sqrt{x}}{\sqrt{x^4}} = \frac{3\sqrt{x}}{x^2}$$

You may want to simplify $\sqrt{x^3}$ before you rationalize the denominator.

$$\frac{3}{\sqrt{x^3}} = \frac{3}{x\sqrt{x}} \times \frac{\sqrt{x}}{\sqrt{x}} = \frac{3\sqrt{x}}{x\sqrt{x^2}} = \frac{3\sqrt{x}}{(x)(x)} = \frac{3\sqrt{x}}{x^2}$$

Practice Problem 4 Simplify. **(a)** $\frac{\sqrt{2}}{\sqrt{12}}$ **(b)** $\frac{6a}{\sqrt{a^7}}$ ■

EXAMPLE 5 Simplify. $\frac{\sqrt{2}}{\sqrt{27x}}$

Since "what multiplied by 27 will be a perfect square" is not apparent, we will begin by simplifying the denominator.

$$\frac{\sqrt{2}}{\sqrt{27x}} = \frac{\sqrt{2}}{3\sqrt{3x}}$$

Now it is easy to see that we multiply by $\frac{\sqrt{3x}}{\sqrt{3x}}$ *to rationalize the denominator.*

$$= \frac{\sqrt{2}}{3\sqrt{3x}} \times \frac{\sqrt{3x}}{\sqrt{3x}}$$

$$= \frac{\sqrt{6x}}{3\sqrt{9x^2}}$$

$$= \frac{\sqrt{6x}}{9x}$$

Practice Problem 5 Simplify. $\frac{\sqrt{2x}}{\sqrt{8x}}$ ■

3 *Rationalizing the Denominator of a Fraction with a Binomial Denominator*

Sometimes the denominator of a fraction is a binomial with a radical term. $\frac{1}{5 - 3\sqrt{2}}$ is just such a fraction. How can we eliminate the radical term in the denominator? Recall that when we multiply $(x + 3)(x - 3)$ the product does not have a middle term. That is, $(x + 3)(x - 3) = x^2 - 9$. Both terms are perfect squares. We can use this idea to eliminate the radical in $5 - 3\sqrt{2}$.

$$(5 - 3\sqrt{2})(5 + 3\sqrt{2}) = 5^2 - (3\sqrt{2})^2 = 25 - 36 = -11$$

$(5 - 3\sqrt{2})$ and $(5 + 3\sqrt{2})$ are called *conjugates.*

EXAMPLE 6 Rationalize the denominator. **(a)** $\frac{2}{\sqrt{3} - 4}$ **(b)** $\frac{\sqrt{x}}{\sqrt{5} + \sqrt{3}}$

(a) The conjugate of $\sqrt{3} - 4$ is $\sqrt{3} + 4$.

$$\frac{2}{(\sqrt{3} - 4)} \cdot \frac{(\sqrt{3} + 4)}{(\sqrt{3} + 4)} = \frac{2\sqrt{3} + 8}{\sqrt{9} + 4\sqrt{3} - 4\sqrt{3} - 16}$$

$$= \frac{2\sqrt{3} + 8}{3 - 16}$$

$$= \frac{2\sqrt{3} + 8}{-13}$$

$$= -\frac{2\sqrt{3} + 8}{13}$$

(b) The conjugate of $\sqrt{5} + \sqrt{3}$ is $\sqrt{5} - \sqrt{3}$.

$$\frac{\sqrt{x}}{(\sqrt{5} + \sqrt{3})} \cdot \frac{(\sqrt{5} - \sqrt{3})}{(\sqrt{5} - \sqrt{3})} = \frac{\sqrt{5x} - \sqrt{3x}}{\sqrt{25} - \sqrt{15} + \sqrt{15} - \sqrt{9}}$$

$$= \frac{\sqrt{5x} - \sqrt{3x}}{5 - 3}$$

$$= \frac{\sqrt{5x} - \sqrt{3x}}{2}$$

Be careful not to combine $\sqrt{5x}$ and $\sqrt{3x}$ in the numerator. They are not like square roots.

Practice Problem 6 Simplify. **(a)** $\dfrac{4}{\sqrt{3} + \sqrt{5}}$ **(b)** $\dfrac{\sqrt{a}}{\sqrt{10} - 3}$ ■

EXAMPLE 7 Rationalize the denominator. $\dfrac{\sqrt{3} + \sqrt{2}}{\sqrt{3} - \sqrt{2}}$

The conjugate of $\sqrt{3} - \sqrt{2}$ is $\sqrt{3} + \sqrt{2}$.

$$\frac{(\sqrt{3} + \sqrt{2})}{(\sqrt{3} - \sqrt{2})} \cdot \frac{(\sqrt{3} + \sqrt{2})}{(\sqrt{3} + \sqrt{2})}$$

$$= \frac{\sqrt{9} + \sqrt{6} + \sqrt{6} + \sqrt{4}}{(\sqrt{3})^2 - (\sqrt{2})^2} \qquad \textit{Multiply.}$$

$$= \frac{3 + 2\sqrt{6} + 2}{3 - 2} \qquad \textit{Simplify and combine like terms.}$$

$$= \frac{5 + 2\sqrt{6}}{1}$$

$$= 5 + 2\sqrt{6}$$

Practice Problem 7 Rationalize the denominator. $\dfrac{\sqrt{7} - \sqrt{x}}{\sqrt{7} + \sqrt{x}}$ ■

9.5 Exercises

Simplify. Be sure to rationalize all denominators. Do not use a calculator or a table of square roots.

1. $\frac{\sqrt{32}}{\sqrt{2}}$ **2.** $\frac{\sqrt{7}}{\sqrt{28}}$ **3.** $\frac{\sqrt{5}}{\sqrt{20}}$ **4.** $\frac{\sqrt{24}}{\sqrt{6}}$ **5.** $\frac{\sqrt{27}}{\sqrt{3}}$ **6.** $\frac{\sqrt{15}}{\sqrt{5}}$

7. $\frac{\sqrt{6}}{\sqrt{x^4}}$ **8.** $\frac{\sqrt{12}}{\sqrt{x^2}}$ **9.** $\frac{\sqrt{75}}{\sqrt{3}}$ **10.** $\frac{\sqrt{3}}{\sqrt{48}}$ **11.** $\frac{\sqrt{8}}{\sqrt{x^6}}$ **12.** $\frac{\sqrt{20}}{\sqrt{x^8}}$

13. $\frac{3}{\sqrt{7}}$ **14.** $\frac{2}{\sqrt{10}}$ **15.** $\frac{5}{\sqrt{6}}$ **16.** $\frac{3}{\sqrt{5}}$ **17.** $\frac{x}{\sqrt{3}}$ **18.** $\frac{\sqrt{y}}{\sqrt{6}}$

19. $\frac{\sqrt{8}}{\sqrt{x}}$ **20.** $\frac{\sqrt{18}}{\sqrt{y}}$ **21.** $\frac{3}{\sqrt{12}}$ **22.** $\frac{4}{\sqrt{20}}$ **23.** $\frac{7}{\sqrt{a^3}}$ **24.** $\frac{5}{\sqrt{b^5}}$

25. $\frac{x}{\sqrt{2x^5}}$ **26.** $\frac{y}{\sqrt{5x^3}}$ **27.** $\frac{\sqrt{3}}{\sqrt{12x^7}}$ **28.** $\frac{\sqrt{5}}{\sqrt{18x^5}}$ **29.** $\sqrt{\frac{2}{7}}$ **30.** $\sqrt{\frac{3}{10}}$

31. $\sqrt{\frac{3}{ab^2}}$ **32.** $\sqrt{\frac{5}{2a^2b}}$ **33.** $\frac{9}{\sqrt{32x}}$ **34.** $\frac{3}{\sqrt{50x}}$ **35.** $\frac{3}{\sqrt{2}-1}$ **36.** $\frac{2}{\sqrt{3}+1}$

37. $\frac{3}{\sqrt{2}+\sqrt{5}}$ **38.** $\frac{4}{\sqrt{7}-\sqrt{2}}$ **39.** $\frac{\sqrt{6}}{\sqrt{6}-\sqrt{3}}$ **40.** $\frac{\sqrt{3}}{\sqrt{5}+\sqrt{3}}$ **41.** $\frac{x}{\sqrt{7}+2\sqrt{2}}$ **42.** $\frac{x}{\sqrt{6}+3\sqrt{5}}$

43. $\frac{x}{\sqrt{3}-2x}$ **44.** $\frac{y}{\sqrt{5}-2y}$ **45.** $\frac{\sqrt{7}}{\sqrt{8}+\sqrt{7}}$ **46.** $\frac{\sqrt{5}}{\sqrt{5}+\sqrt{6}}$ **47.** $\frac{\sqrt{5}-\sqrt{2}}{\sqrt{5}+\sqrt{2}}$ **48.** $\frac{\sqrt{7}+\sqrt{3}}{\sqrt{7}-\sqrt{3}}$

49. $\frac{2\sqrt{3}+\sqrt{6}}{\sqrt{6}-\sqrt{3}}$

50. $\frac{2\sqrt{5}+3}{\sqrt{5}+\sqrt{3}}$

★ 51. $\frac{3\sqrt{5}+6}{\sqrt{7}-\sqrt{8}}$

★ 52. $\frac{\sqrt{6}+2\sqrt{2}}{\sqrt{7}-\sqrt{6}}$

53. $\frac{x-4}{\sqrt{x}+2}$

54. $\frac{x-9}{\sqrt{x}-3}$

★ 55. $\frac{\sqrt{x}+2}{\sqrt{x}-\sqrt{2}}$

★ 56. $\frac{\sqrt{x}-4}{\sqrt{x}+\sqrt{3}}$

To Think About

57. **(a)** Multiply $(x + \sqrt{2})$ by its conjugate.

(b) Use the results of part (a) to show that, while $x^2 - 2$ cannot be factored by earlier methods, it can be factored using radicals.

(c) Factor using radicals. $x^2 - 12$

58. $\frac{3}{\sqrt{11}}$ when rationalized is $\frac{3\sqrt{11}}{11}$. Evaluate each expression with a calculator. What do you notice? Why?

59. $\frac{7}{\sqrt{35}}$ when rationalized is $\frac{\sqrt{35}}{5}$. Evaluate each expression with a calculator. What do you notice? Why?

Cumulative Review Problems

60. Solve for x (round to two decimal places). $1.06(0.80x) = 32.64$

61. Evaluate $3x^2 - x \div y + y^2 - y$ for $x = 12$ and $y = -3$.

62. The perimeter of an isosceles triangle is 34 centimeters. Find the short side if each of the equal sides is 2 centimeters more than twice the short side.

63. How many points must Marcus get on his 200-point final to maintain an average of 80 if on five tests, each worth 100 points, he has scored 70, 70, 60, 90, and 80?

9.6 The Pythagorean Theorem and Radical Equations

After studying this section, you will be able to:

1 *Use the Pythagorean theorem.*
2 *Solve radical equations.*

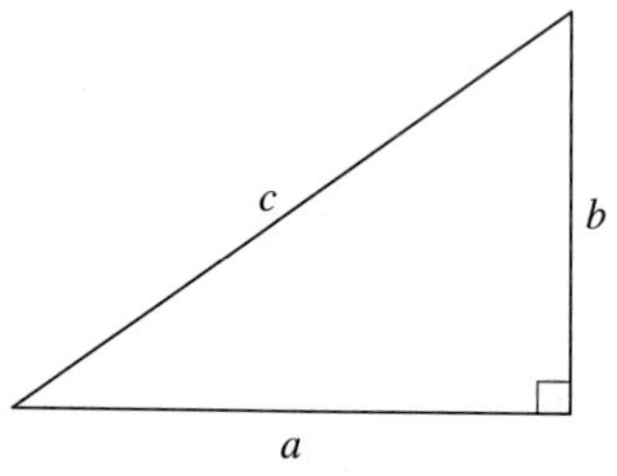

1 *Using the Pythagorean Theorem*

In ancient Greece, mathematicians studied a number of properties of a right triangle. They were able to prove that the square of the longest side of a right triangle is equal to the sum of the squares of the other two sides. This property is called the Pythagorean theorem, in honor of the Greek mathematician Pythagoras (ca. 590 B.C.). The shorter sides, a, b, are referred to as the **legs** of a right triangle. The longest side c is called the **hypotenuse** of a right triangle. We state the theorem thus:

> **Pythagorean Theorem**
>
> In any right triangle, if c is the length of the hypotenuse and a and b are the lengths of the two legs, then
>
> $$c^2 = a^2 + b^2$$

If we know any two sides of a right triangle, we can find the third side using this theorem.

Recall that in Section 9.1 we showed that if $s^2 = 4$ then $s = 2$ or -2. In general, if $x^2 = a$, then $x = \pm\sqrt{a}$. This is sometimes called "taking the square root of each side of an equation." The abbreviation $\pm\sqrt{a}$ means $+\sqrt{a}$ or $-\sqrt{a}$. It is read "plus or minus $\sqrt{a}$."

EXAMPLE 1 Find the length of the hypotenuse of a right triangle whose legs are 5 meters and 12 meters.

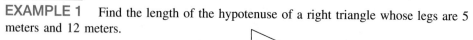

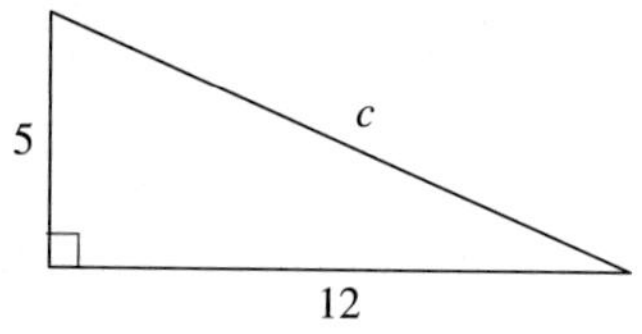

1. *Understand the problem.*
   Draw and label a diagram.
2. *Write an equation.*
   Write the Pythagorean theorem. Substitute the known values into the equation.

$$c^2 = a^2 + b^2$$
$$c^2 = 5^2 + 12^2$$

3. *Solve and state the answer.*

$$c^2 = 25 + 144$$
$$c^2 = 169 \qquad \textit{Take the square root of each side of the equation.}$$
$$c = \pm\sqrt{169}$$
$$= \pm 13$$

   The hypotenuse is 13 meters. We do not use -13 because length is not negative.

4. *Check.*
   Substitute the values for a, b, and c in the Pythagorean theorem and evaluate.
   Does $13^2 = 5^2 + 12^2$? Yes. The solution checks. ✓

Practice Problem 1 Find the length of the hypotenuse of a right triangle with legs of 9 centimeters and 12 centimeters. ■

Sometimes you will need to find one of the legs of a right triangle given the hypotenuse and the other leg.

EXAMPLE 2 Find the leg of a right triangle that has a hypotenuse of 10 yards and one leg of $\sqrt{19}$ yards.

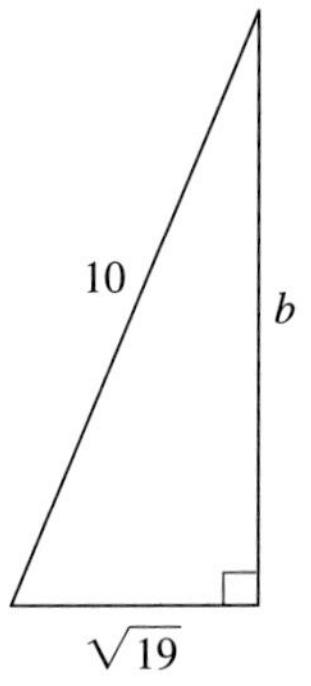

Draw a diagram. The diagram is in the margin on the right,

$c^2 = a^2 + b^2$	*Write the Pythagorean theorem.*
$10^2 = (\sqrt{19})^2 + b^2$	*Substitute the known values into the equation.*
$100 = 19 + b^2$	
$81 = b^2$	
$\pm 9 = b$	*Take the square root of each side of the equation.*

The leg is 9 yards.

Practice Problem 2 The hypotenuse of a right triangle is $\sqrt{17}$ meters. One leg is 1 meter long. Find the length of the other leg. ■

Sometimes the answer will be an irrational number such as $\sqrt{33}$. It is best to leave the answer in radical form unless you are asked for an approximate answer. To find the approximate value, use a calculator or the square root table. $\sqrt{33}$ is an exact answer.

EXAMPLE 3 A 25-foot ladder is placed against a building. The foot of the ladder is 8 feet from the wall. Approximately how high on the building will the ladder reach? Round your answer to the nearest tenth.

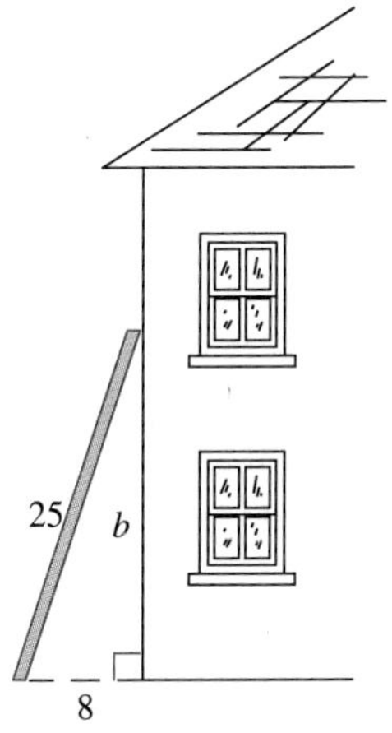

$$c^2 = a^2 + b^2$$
$$25^2 = 8^2 + b^2$$
$$625 = 64 + b^2$$
$$561 = b^2$$
$$\pm\sqrt{561} = b$$

We want only positive value for the distance, so $b = \sqrt{561}$. Using a calculator, we have

561 [√] 23.68543856 Rounding, we obtain $b \approx 23.7$

If you do not have a calculator, you will need to use the square root table in Appendix A. However, this square root table only goes to number 200! To approximate from the square root table, you will have to write the radical in a different fashion.

$b = \sqrt{3}\sqrt{187}$	*Since* $\sqrt{ab} = \sqrt{a}\sqrt{b}$ *and* $3 \cdot 187 = 561$.
$b \approx (1.732)(13.675)$	*Replace each radical by a decimal approximation from the table.*
$b \approx 23.7$	*Multiply and round to the nearest tenth.*

The ladder reaches up approximately 23.7 feet on the building.

Practice Problem 3 A support line is placed 3 meters away from the base of an 8-meter pole. If the support line is attached to the top of the pole and pulled tight (assume that it is a straight line), how long is the support line from the ground to the pole? Approximate your answer to the nearest tenth. ■

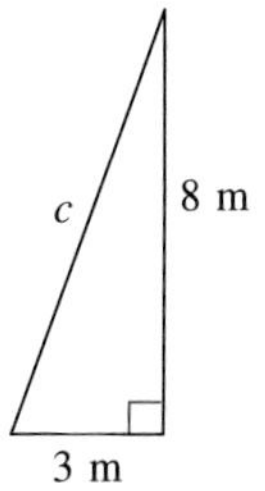

2 *Solving Radical Equations*

A **radical equation** is an equation with an unknown letter in one or more of the radicands. This type of equation can be simplified by squaring each side of the equation. The apparent solution to a radical equation *must* be checked by substitution into the *original* equation.

EXAMPLE 4 Solve and check your answer. $\sqrt{x+4} = 11$

$$\sqrt{x+4} = 11$$

$$(\sqrt{x+4})^2 = (11)^2 \qquad \textit{Square each side of the equation.}$$

$$x + 4 = 121 \qquad \textit{Simplify and solve for } x.$$

$$x = 117$$

Check: $\sqrt{117+4} \stackrel{?}{=} 11$

$$\sqrt{121} \stackrel{?}{=} 11$$

$$11 = 11 \quad ✓$$

Thus $x = 117$ is the solution.

Practice Problem 4 Solve and check. $\sqrt{2x-5} = 6$ ■

For some equations, you will need to *isolate the radical before* you square each side of the equation.

EXAMPLE 5 Solve and check your answer. $1 + \sqrt{5x-4} = 5$

$$1 + \sqrt{5x-4} = 5 \qquad \textit{We want to isolate the radical first.}$$

$$\sqrt{5x-4} = 4 \qquad \textit{Subtract 1 from each side to isolate the radical.}$$

$$(\sqrt{5x-4})^2 = (4)^2 \qquad \textit{Square each side.}$$

$$5x - 4 = 16 \qquad \textit{Simplify and solve for } x.$$

$$5x = 20$$

$$x = 4$$

Check: $1 + \sqrt{5(4)-4} \stackrel{?}{=} 5$

$$1 + \sqrt{20-4} \stackrel{?}{=} 5$$

$$1 + \sqrt{16} \stackrel{?}{=} 5$$

$$1 + 4 = 5 \quad ✓$$

Thus $x = 4$ is the solution.

Practice Problem 5 Solve and check. $\sqrt{3x-2} - 7 = 0$ ■

Some problems will have an apparent solution, but that solution does not always check. An obtained solution that does not satisfy the original equation is called an **extraneous root.**

EXAMPLE 6 Solve and check. $\sqrt{x+3} = -7$

$$\sqrt{x+3} = -7$$

$$(\sqrt{x+3})^2 = (-7)^2 \quad \textit{Square each side.}$$

$$x + 3 = 49 \quad \textit{Simplify and solve for } x.$$

$$x = 46$$

Check: $\sqrt{46+3} \stackrel{?}{=} -7$

$$\sqrt{49} \stackrel{?}{=} -7$$

$$7 \neq -7! \quad \textit{Does not check! The apparent solution cannot be verified.}$$

There is no solution to Example 6.

Practice Problem 6 Solve and check. $\sqrt{5x+4} + 2 = 0$ ■

Wait a minute! Is that original equation possible in Example 6? No. $\sqrt{x+3} = -7$ is an impossible statement. By our definition of the $\sqrt{\ }$ symbol, we defined it to mean the positive square root of the radicand. If you noticed this in the first step, you could write down "no solution" immediately.

When you square both sides of the equation you may obtain a quadratic equation. In such cases, you must check both apparent solutions.

EXAMPLE 7 Solve and check. $\sqrt{3x+1} = x + 1$.

$$\sqrt{3x+1} = x + 1$$

$$(\sqrt{3x+1})^2 = (x+1)^2 \quad \textit{Square each side.}$$

$$3x + 1 = x^2 + 2x + 1 \quad \textit{Simplify and set the equation equal to 0.}$$

$$0 = x^2 - x$$

$$0 = x(x-1) \quad \textit{Factor the quadratic equation.}$$

$$x = 0 \qquad x - 1 = 0 \quad \textit{Set each factor equal to 0 and solve.}$$

$$x = 1$$

Check: If $x = 0$: $\sqrt{3(0)+1} \stackrel{?}{=} 0 + 1$ If $x = 1$: $\sqrt{3(1)+1} \stackrel{?}{=} 1 + 1$

$$\sqrt{0+1} \stackrel{?}{=} 0 + 1 \qquad \sqrt{3+1} \stackrel{?}{=} 2$$

$$\sqrt{1} \stackrel{?}{=} 1 \qquad \sqrt{4} \stackrel{?}{=} 2$$

$$1 = 1 \ \checkmark \qquad 2 = 2 \ \checkmark$$

Thus $x = 0$ and $x = 1$ are both solutions.

It is possible in these cases for both apparent solutions to check, for both to be extraneous, or for one to check and the other to be extraneous.

Practice Problem 7 Solve and check. $-2 + \sqrt{6x-1} = 3x - 2$ ■

EXAMPLE 8 Solve and check. $\sqrt{2x-1} = x - 2$

$$\sqrt{2x-1} = x - 2$$

$$(\sqrt{2x-1})^2 = (x-2)^2 \qquad \textit{Square each side.}$$

$$2x - 1 = x^2 - 4x + 4 \qquad \textit{Simplify and set the equation to 0.}$$

$$0 = x^2 - 6x + 5 \qquad \textit{Solve for } x.$$

$$0 = (x-5)(x-1)$$

$$x - 5 = 0 \qquad x - 1 = 0$$

$$x = 5 \qquad x = 1$$

Check: If $x = 5$: $\sqrt{2(5) - 1} \stackrel{?}{=} 5 - 2$

$$\sqrt{10 - 1} \stackrel{?}{=} 3$$

$$\sqrt{9} \stackrel{?}{=} 3$$

$$3 = 3 \ \checkmark$$

If $x = 1$: $\sqrt{2(1) - 1} \stackrel{?}{=} 1 - 2$

$$\sqrt{2 - 1} \stackrel{?}{=} -1$$

$$\sqrt{1} \stackrel{?}{=} -1$$

$$1 \neq -1$$

It does not check. In this case $x = 1$ *is called an* extraneous root.

Thus only $x = 5$ is a solution to this equation.

Practice Problem 8 Solve and check. $2 - x + \sqrt{x+4} = 0$ ■

EXAMPLE 9 Solve and check. $\sqrt{3x+3} = \sqrt{5x-1}$

Here there are two radicals. Each radical is already isolated.

$$(\sqrt{3x+3})^2 = (\sqrt{5x-1})^2 \qquad \textit{Square each side.}$$

$$3x + 3 = 5x - 1 \qquad \textit{Simplify and solve for } x.$$

$$3 = 2x - 1$$

$$4 = 2x$$

$$2 = x$$

Check: $\sqrt{3(2) + 3} \stackrel{?}{=} \sqrt{5(2) - 1}$

$$\sqrt{6 + 3} \stackrel{?}{=} \sqrt{10 - 1}$$

$$\sqrt{9} = \sqrt{9} \ \checkmark$$

Thus $x = 2$ is a solution.

Practice Problem 9 Solve and check. $\sqrt{2x+1} = \sqrt{x-10}$ ■

9.6 Exercises

Use the Pythagorean theorem to find the length of the third side of each right triangle. Leave any irrational answers in radical form.

1.

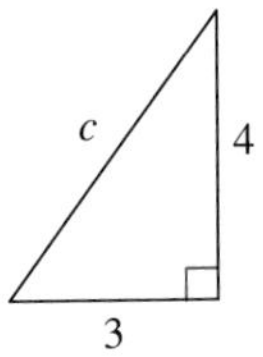

2.

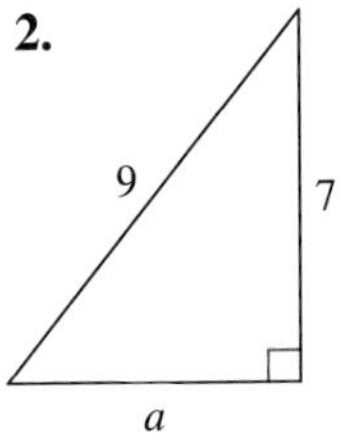

3.

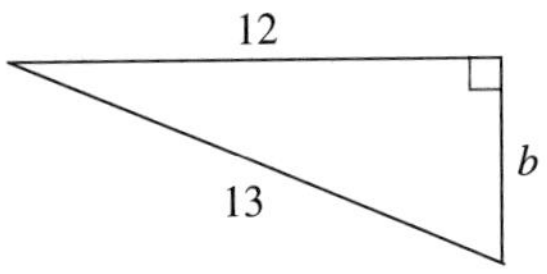

4.

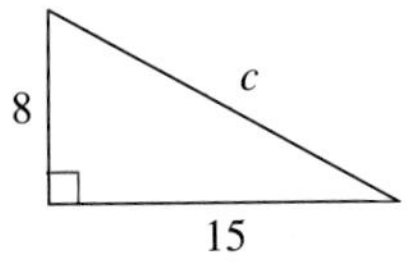

Problems 5 through 14 refer to a right triangle. Find the exact length of the missing side.

5. $a = 5, b = 7$
Find c.

6. $a = 6, b = 9$
Find c.

7. $a = \sqrt{18}, b = 4$
Find c.

8. $a = \sqrt{19}, b = 7$
Find c.

9. $c = 20, b = 18$
Find a.

10. $c = 17, b = 16$
Find a.

11. $c = \sqrt{23}, a = 4$
Find b.

12. $c = \sqrt{27}, a = 5$
Find b.

13. $c = 12.96$, $b = 8.35$ Find a to the nearest tenth.

14. $a = 7.61$, $b = 5.38$ Find c to the nearest tenth.

Applications

Draw a diagram and use the Pythagorean theorem to solve each word problem. Approximate the answer to the nearest tenth.

15. A ladder is 18 feet long. It reaches 15 feet up on a wall. How far is the base of the ladder from the wall?

16. A kite is flying on a string 100 feet long. Assume that the string is a straight line. How high is the kite if it is flying above a point 20 feet away along the ground?

17. A baseball diamond is really constructed from a square. Each side of the square is 90 feet long. How far is it from home plate to second base? *Hint:* Draw the diagonal.

18. A boat must be moored 20 feet away from a dock. The boat will be 10 feet below the level of the dock at low tide. What is the minimum length of rope needed to go from the boat to the dock at low tide?

19. A man is standing 6 miles from the center of a mountain. He is at the same level as the base of the mountain. The mountain is 3 miles high. How far is the man from the top of the mountain?

20. A plane flies 7 miles north of the airport. A helicopter flies 5 miles east of the airport. How far apart are the plane and helicopter?

Solve for the variable. Check your solutions.

21. $\sqrt{x} = 11$

22. $\sqrt{x} = 3.5$

23. $\sqrt{x + 8} = 5$

24. $\sqrt{x + 5} = 7$

25. $\sqrt{3x + 6} = 2$

26. $\sqrt{3x - 8} = 4$

27. $\sqrt{2x + 2} = \sqrt{3x - 5}$

28. $\sqrt{5x - 5} = \sqrt{4x + 1}$

29. $\sqrt{2x} - 5 = 4$

30. $\sqrt{3x} + 5 = 8$

31. $\sqrt{3x + 10} = x$

32. $\sqrt{5x - 6} = x$

33. $\sqrt{5y + 1} = y + 1$

34. $\sqrt{2y + 9} = y + 3$

35. $x - 3 = \sqrt{2x - 3}$

36. $x - 2 = \sqrt{2x - 1}$

37. $\sqrt{3y + 1} - y = 1$

38. $\sqrt{3y - 8} + 2 = y$

39. $\sqrt{5 + 2x} - 1 - x = 0$

40. $\sqrt{4x + 13} + 1 - 2x = 0$

41. $y + \sqrt{3y - 14} = 6$

42. $3y - \sqrt{6y - 1} = 0$

To Think About

Solve for x.

43. $\sqrt{2x + 5} = 2\sqrt{2x} + 1$

44. $\sqrt{x - 2} + 3 = \sqrt{4x + 1}$

Cumulative Review Problems

45. Solve if possible. $\dfrac{2x}{x - 7} + 3 = \dfrac{14}{x - 7}$

46. Multiply mentally. $(3x - 1)^2$

47. $f(x) = 2x^2 - 3x + 6$
Find $f(-2)$.

48. $x^2 + y^2 = 9$ Is this a relation or a function?

9.7 Word Problems Involving Radicals: Direct and Inverse Variation

After studying this section, you will be able to:

1 *Solve problems involving direct variation.*
2 *Solve problems involving inverse variation.*

1 *Solving Problems Involving Direct Variation*

Often in daily life there is a relationship between two measureable quantities. For example, we turn up the thermostat and the heating bill increases. We say that the heating bill varies directly as the temperature on the thermostat. If y varies directly as x, then $y = kx$, where k is a constant. The constant is often called the *constant of variation.* Consider the following example.

EXAMPLE 1 Cliff works part-time in a local supermarket while going to college. His salary varies directly as the number of hours worked. Last week he earned \$33.60 for working 7 hours. This week he earned \$52.80. How many hours did he work?

Let S = his salary
h = the number of hours he worked
k = the constant of variation

Since his salary varies directly as the number of hours worked, we write

$$S = k \cdot h$$

We can find the constant k by substituting the known values of $S = 33.60$ when $h = 7$.

$$33.60 = k \cdot 7 = 7k$$

$$\frac{33.60}{7} = \frac{7k}{7} \qquad \textit{Solve for k.}$$

$$4.80 = k \qquad \textit{The constant of variation is } 4.80.$$

$$S = 4.80h \qquad \textit{Replace k in the variation equation by } 4.80.$$

How many hours did he work to earn \$52.80?

$$S = 4.80h \qquad \textit{The direct variation equation with } k = 4.80.$$

$$52.80 = 4.80h \qquad \textit{Substitute } 52.80 \textit{ for S and solve for h.}$$

$$\frac{52.80}{4.80} = \frac{4.80h}{4.80}$$

$$11 = h$$

Cliff worked 11 hours.

Practice Problem 1 The *change* in temperature measured in Celsius varies directly as the change measured in Fahrenheit. The change from freezing water to boiling water is 100° in Celsius and 180° in Fahrenheit. If the Fahrenheit temperature drops 20°, what will be the change in temperature on the Celsius scale? ■

Solving a Direct Variation Problem

1. Write the direct variation equation.
2. Solve for the constant k by substituting in known values.
3. In the direct variation equation replace k by the value obtained in step 2.
4. Solve for the desired value.

EXAMPLE 2 If y varies directly as x and $y = 20$ when $x = 3$, find the value of y when $x = 21$.

Step 1 $y = kx$

Step 2 To find k, we substitute in $y = 20$ and $x = 3$.

$$20 = k(3)$$

$$20 = 3k$$

$$\frac{20}{3} = k$$

Step 3 We now write the variation equation with k replaced by $\frac{20}{3}$.

$$y = \frac{20}{3}x$$

Step 4 We replace x by 21. Now we find y.

$$y = \left(\frac{20}{\cancel{3}}\right)(\overset{7}{\cancel{21}})$$

$$= (20)(7) = 140$$

Thus $y = 140$ when $x = 21$.

Practice Problem 2 If y varies directly as x and $y = 18$ when $x = 5$, find the value of y when $x = \frac{20}{23}$. ■

Direct variation equations apply to a number of cases. The following table shows some of the more common forms of direct variation. In each case, k = the constant of variation.

Sample Direct Variation Situations

VERBAL DESCRIPTION	VARIATION EQUATION
y varies directly as x	$y = kx$
b varies directly as the square of c	$b = kc^2$
l varies directly as the cube of m	$l = km^3$
d varies directly as the square root of h	$d = k\sqrt{h}$

EXAMPLE 3 In a certain class of racing cars the maximum speed varies directly as the square root of the horsepower of the engine. If a car with 225 horsepower is able to achieve a maximum speed of 120 mph, what speed could it achieve if the engine developed 256 horsepower?

Let V = the maximum speed
h = the horsepower of the engine
k = the constant of variation

Step 1 Since the maximum speed (V) varies directly as the square root of the horsepower of the engine,

$$V = k\sqrt{h}$$

Step 2 $120 = k\sqrt{225}$ *Substitute known values of V and h.*
$120 = k \cdot 15$ *Solve for k.*
$8 = k$

Step 3 Now we can write the direct variation equation with a known value for k.

$$V = 8\sqrt{h}$$

Step 4 $V = 8\sqrt{256}$ *Substitute the value of h = 256.*
$= (8)(16)$ *Solve for V.*
$= 128$

Thus a maximum speed of 128 mph could be achieved if the engine developed 256 horsepower.

Practice Problem 3 For a certain class of automobiles, the distance to stop the car varies directly as the square of its speed. On ice-covered roads a car that is traveling at a speed of 20 mph can stop in 60 feet. If the car is traveling at 40 mph, how many feet will be necessary for it to come to a stop? ■

2 *Solving Problems Involving Inverse Variation*

If one variable is a constant multiple of the reciprocal of the other, the two variables are said to *vary inversely*. If y varies inversely as x, we express this by the equation $y = \frac{k}{x}$, where k is the constant of variation. Inverse variation problems can be solved by a four-step procedure similar to the direct variation problems.

Solving an Inverse Variation Problem

1. Write the inverse variation equation.
2. Solve for the constant k by substituting in known values.
3. In the inverse variation equation, replace k by the value obtained in step 2.
4. Solve for the desired value.

EXAMPLE 4 If y varies inversely as x and $y = 12$ when $x = 7$, find the value of y when $x = \frac{2}{3}$.

Step 1 $y = \frac{k}{x}$

Step 2 $12 = \frac{k}{7}$ *Substitute known values of x and y to find k.*

$84 = k$

Step 3 $y = \frac{84}{x}$

Step 4 To find y when $x = \frac{2}{3}$, we substitute: $y = \frac{84}{\frac{2}{3}}$

On the right side we have $\frac{84}{1} \div \frac{2}{3} = \frac{\overset{42}{\cancel{84}}}{1} \cdot \frac{3}{\underset{1}{\cancel{2}}} = 42 \cdot 3 = 126$

Thus $y = 126$ when $x = \frac{2}{3}$.

Practice Problem 4 If y varies inversely as x and $y = 8$ when $x = 15$, find the value of y when $x = \frac{3}{5}$. ■

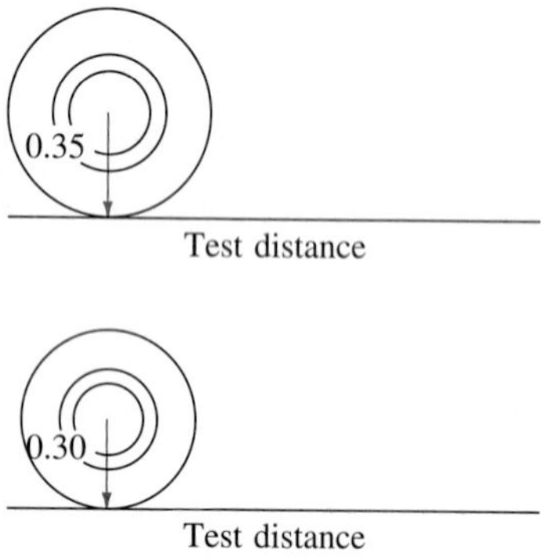

EXAMPLE 5 A car manufacturer is thinking of reducing the size of the wheel used in a subcompact car. The number of times a car wheel must turn to cover a given distance varies inversely as the radius of the wheel. (Notice that this says that the smaller the wheel, the more times it must turn to cover a given distance.) A wheel with a radius of 0.35 meter must turn 400 times to cover a specified distance on a test track. How many times would it have to turn if the radius is reduced to 0.30 meter (see the sketch)?

Let n = the number of times the car wheel will turn
r = the radius of the wheel
k = the constant of variation

Step 1 Since the number of turns varies inversely as the radius, we can write

$n = \frac{k}{r}$ *Write the variation equation.*

Step 2 $400 = \frac{k}{0.35}$ *Substitute known values of n and r to find k.*

$140 = k$

Step 3 $n = \frac{140}{r}$ *Use the variation equation where k is known.*

How many times will the wheel turn if the radius is 0.30 meter?

Step 4 $n = \frac{140}{0.30}$ *Substitute* 0.30 *for r.*

$n = 466\frac{2}{3}$

The wheel would have to turn $466\frac{2}{3}$ times to cover the same distance if the radius were only 0.30 meter.

Practice Problem 5 Over the last three years the market research division of a calculator company found the volume of sales of scientific calculators varies inversely as the price of the calculator. One year 120,000 calculators were sold at $30 each. How many calculators were sold the next year when the price was $24 for each calculator? ■

The following table demonstrates some situations involving inverse variation. In each case $k =$ the constant of variation.

Sample Inverse Variation Situations

VERBAL DESCRIPTION	VARIATION EQUATION
y varies inversely as x	$y = \frac{k}{x}$
b varies inversely as the square of c	$b = \frac{k}{c^2}$
l varies inversely as the cube of m	$l = \frac{k}{m^3}$
d varies inversely as the square root of t	$d = \frac{k}{\sqrt{t}}$

EXAMPLE 6 The illumination of a light source varies inversely as the square of the distance from the source. The illumination measures 25 candlepower when a certain light is 4 meters away. Find the illumination when the light is 8 meters away (see figure).

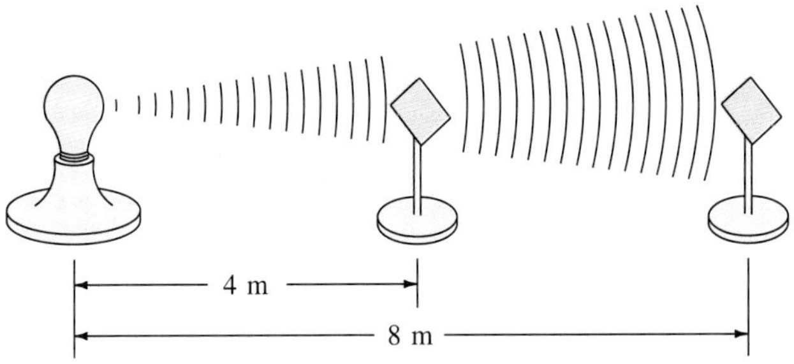

Let $I =$ the measurement of illuminance
$d =$ the distance from the light source
$k =$ the constant of variation

Step 1 Since the illumination varies inversely as the square of the distance,

$$I = \frac{k}{d^2}$$

Step 2 We evaluate the constant by substituting the given values.

$25 = \frac{k}{4^2}$ *Substitute* $I = 25$ *and* $d = 4$.

$25 = \frac{k}{16}$ *Simplify and solve for k.*

$400 = k$

Step 3 We may now write the variation equation with the constant evaluated.

$$I = \frac{400}{d^2}$$

Step 4 $I = \dfrac{400}{8^2}$ *Substitute a distance of* 8 *meters.*

$= \dfrac{400}{64}$ *Square* 8.

$= \dfrac{25}{4}$ or $6\dfrac{1}{4}$

The illumination is $6\frac{1}{4}$ candlepower when the light source is 8 meters away.

Practice Problem 6 If the amount of power in an electrical circuit is held constant, the resistance in the circuit varies inversely as the square of the amount of current. If the amount of current is 0.01 ampere, the resistance is 800 ohms. What is the resistance if the amount of current is 0.02 ampere? ■

Graphs of Variation Equations

The graphs of direct variation equations are shown below. Notice that as x increases y also increases.

Variation Statement	Equation	Graph
1. y varies directly as x	$y = kx$	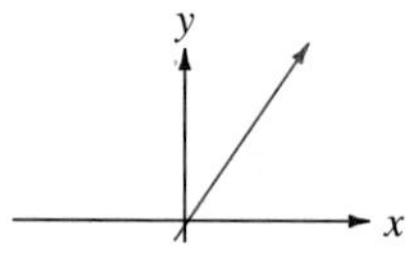
2. y varies directly as x^2	$y = kx^2$	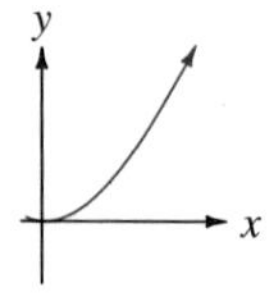
3. y varies directly as x^3	$y = kx^3$	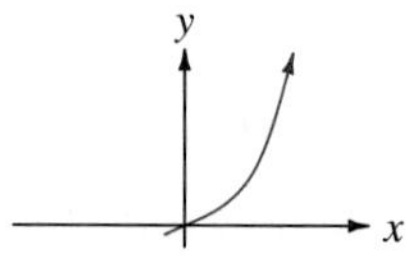
4. y varies directly as the square root of x	$y = k\sqrt{x}$,	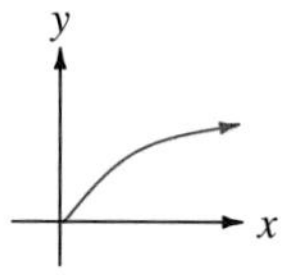

The graphs of indirect variation equations are shown below. Notice that as x increases y decreases.

Variation Statement	Equation	Graph
5. y varies inversely as x	$y = \frac{k}{x}$	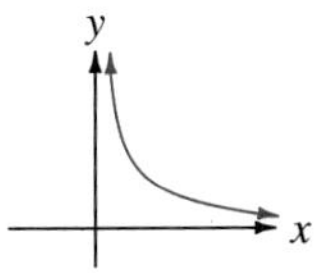
6. y varies inversely as x^2	$y = \frac{k}{x^2}$	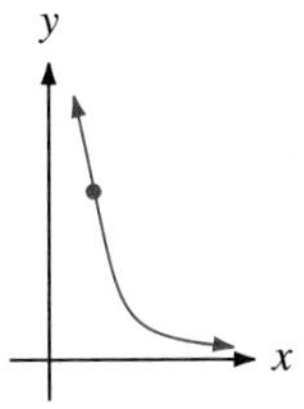

9.7 Exercises

1. If y varies directly as x, and $y = 7$ when $x = 4$, find y when $x = 9$.

2. If y varies directly as x, and $y = 9$ when $x = 12$, find y when $x = 5$.

3. If y varies directly as x, and $y = 350$ when $x = 100$, find y when $x = 20$.

4. If y varies directly as x and $y = 900$ when $x = 200$, find y when $x = 30$.

5. If y varies directly as the cube of x, and $y = 12$ when $x = 2$, find y when $x = 7$.

6. If y varies directly as the square root of x, and $y = 7$ when $x = 4$, find y when $x = 25$.

7. If y varies directly as the square of x, and $y = 800$ when $x = 40$, find y when $x = 25$.

8. If y varies directly as the square of x, and $y = 15{,}000$ when $x = 100$, find y when $x = 8$.

9. If y varies directly as x^2, and $y = 1146.88$ when $x = 12.80$, find y when $x = 7.30$.

10. If y varies directly as x^3, and $y = 214.37$ when $x = 3.5$, find y when $x = 8.2$.

Applications

11. The pressure of water on an object submerged beneath the surface varies directly as the distance beneath the surface. A submarine experiences a pressure of 26 pounds per square inch at 60 feet below the surface. How much pressure will the submarine experience at 390 feet below the surface?

12. The weight of an object on the surface of the moon varies directly as the weight of an object on the surface of Earth. An astronaut with his protective suit weighs 80 kilograms on Earth's surface; on the moon and wearing the same suit, he will weigh 12.8 kilograms. If his moon rover vehicle weighs 900 kilograms on Earth, how much will it weigh on the moon?

13. The time it takes to fill a storage cube with sand varies directly as the cube of the side of the box. A storage cube that is 2.0 meters on each side can be filled in 7 minutes by a sand loader. How long will it take to fill a storage cube that is 4.0 meters on each side?

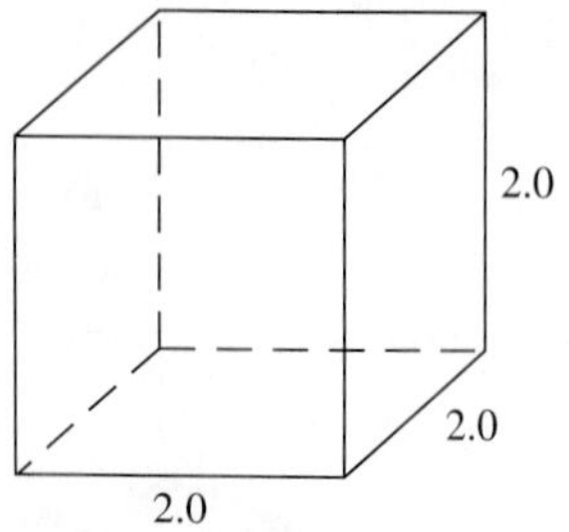

14. The distance that an object falls under the influence of gravity varies directly as the square of the time of fall. An object falls 64 feet in 2 seconds. How far will it fall in 7 seconds?

15. If y varies inversely as x, and $y = 12$ when $x = 4$, find y when $x = 7$.

16. If y varies inversely as x, and $y = 39$ when $x = 3$, find y when $x = 11$.

17. If y varies inversely as x, and $y = \frac{1}{4}$ when $x = 8$, find y when $x = 1$.

18. If y varies inversely as x, and $y = \frac{1}{5}$ when $x = 20$, find y when $x = 2$.

19. If y varies inversely as x^2, and $y = 100$ when $x = 5$, find y when $x = 3$.

20. If y varies inversely as the cube of x, and $y = 12$ when $x = 3$, find y when $x = 6$.

21. If y varies inversely as the square root of x, and $y = 2$ when $x = 4$, find y when $x = 81$.

22. If y varies inversely as the square root of x and $y = 2$ when $x = 25$, find y when $x = 121$.

23. If y varies inversely as $\sqrt{x}$, and $y = 5.00$ when $x = 73.96$, find y when $x = 18.49$.

24. If y varies inversely as $\sqrt{x}$, and $y = 10.0$ when $x = 47.61$, find y when $x = 5.29$.

25. The amount of current in an electric circuit varies inversely as the amount of resistance. In a certain circuit when the resistance is 5 ohms the amount of current is 42 amperes. If the resistance is 4 ohms, what is the amount of current?

26. The thickness of fiber-glass insulation is an indicator of its thermal effectiveness. The heat loss through a certain type of fiber-glass insulation varies inversely as the thickness of the fiber glass. The heat loss through 6 inches of this type of fiber glass is 3000 Btu/hour. Find the heat loss through 14 inches of fiber glass. (Round to the nearest tenth.)

27. The weight of an object on Earth's surface varies inversely as the square of its distance from the center of Earth. An object weighs 1000 pounds at Earth's surface. This is approximately 4000 miles from the center of Earth. How much will the object weigh 6000 miles from the center of Earth?

28. The pressure exerted by a given amount of gas on a cubical storage container varies inversely as the cube of the side of the container. A certain amount of gas exerts a pressure of 60 pounds per square inch when placed in a cubical storage container that measures 0.2 meter in each dimension. If the gas is stored in a box 0.4 meter in each dimension, what will be the pressure?

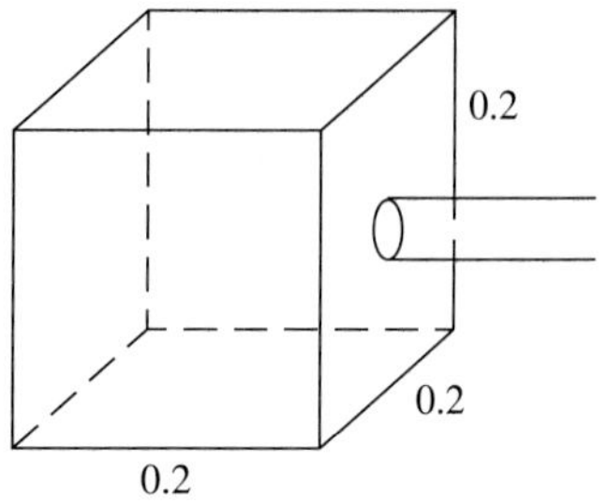

Verbal and Writing Skills

29. Variable y varies directly as variable x. Thus $y = kx$. Describe the graphs of all such direct variations.

30. Variable y varies inversely as variable x. Thus $y = \frac{k}{x}$. Describe the graphs of all such inverse variations for positive values of k and positive values of x.

To Think About

Write an equation to describe the variation and solve each word problem.

31. The amount of heat that passes through a wall can be related to the factors of the temperature difference on the two sides of the wall, the area of the wall, and the thickness of the wall. The heat transfer varies directly as the product of the area of the wall and the temperature difference between the two sides of the wall and inversely as the thickness of the wall. In a certain house 6000 Btu/hour was transferred by a wall measuring 80 square feet. The wall is 0.5 foot thick. The outside temperature was 40°F, the inside temperature was 70°F. Suppose that the wall was exactly 1.0 foot thick and the outside temperature was 20°F. Assuming the same wall area and inside temperature, how much heat would be transferred?

32. An important element of sailboat design is a calculation of the force of wind on the sail. The force varies directly with the square of the wind speed and directly with the area of the sail. On a certain class of racing boat the force on 100 square feet of sail is 280 pounds when the wind speed is 6 mph. What would be the force on the sail if it is rigged for a storm (the amount of sail exposed to the wind is reduced by lashing part of the sail to the boom) so that only 60 square feet of sail is exposed but the windspeed is 40 mph?

Cumulative Review Problems

33. Solve this equation for a. $\frac{80{,}000}{320} = \frac{120{,}000}{320 + a}$

34. Graph the equation $S = kh$, where $k = 4.80$. Use h for the horizontal axis, S for the vertical.

35. Factor. $12x^2 - 20x - 48$

36. Solve. $2x - (5 - 3x) = 6(x - 2) - 3(2 - x)$

Chapter Organizer

Topic	Procedure	Examples
Square roots, p. 505.	**1.** The square root of perfect squares up to $\sqrt{225} = 15$ (or so) should be memorized. Approximate values of others can be found in tables or by calculator. **2.** Negative numbers do not have real number square roots.	$\sqrt{196} = 14$ $\quad \sqrt{169} = 13$ $\sqrt{200} \approx 14.142$ $\quad \sqrt{13} \approx 3.606$ $\sqrt{-5}$ does not exist.
Simplifying radicals, p. 510.	$\sqrt{ab} = \sqrt{a}\sqrt{b}$ Radicals are simplified by taking the square root of all perfect squares and leaving other factors under the radical sign.	$\sqrt{48} = \sqrt{16}\sqrt{3} = 4\sqrt{3}$ $\sqrt{12x^2y^3z} = \sqrt{2^2x^2y^2}\sqrt{3yz}$ $= 2xy\sqrt{3yz}$ $\sqrt{a^{11}} = \sqrt{a^{10} \cdot a} = \sqrt{a^{10}}\sqrt{a} = a^5\sqrt{a}$
Adding and subtracting radicals, p. 515.	Simplify all radicals and then add or subtract like radicals.	$\sqrt{12} + \sqrt{18} + \sqrt{27} + \sqrt{50}$ $= \sqrt{4}\sqrt{3} + \sqrt{9}\sqrt{2} + \sqrt{9}\sqrt{3} + \sqrt{25}\sqrt{2}$ $= 2\sqrt{3} + 3\sqrt{2} + 3\sqrt{3} + 5\sqrt{2}$ $= 5\sqrt{3} + 8\sqrt{2}$
Multiplying radicals, p. 519.	Radicals are multiplied in a way analogous to the multiplication of polynomials. Apply the rule $\sqrt{a}\sqrt{b} = \sqrt{ab}$ and simplify all results.	$(2\sqrt{6} + \sqrt{3})(\sqrt{6} - 3\sqrt{3})$ $= 2\sqrt{36} - 6\sqrt{18} + \sqrt{18} - 3\sqrt{9}$ $= 2(6) - 5\sqrt{18} - 3(3)$ $= 12 - 5\sqrt{9}\sqrt{2} - 9$ $= 12 - 5(3)\sqrt{2} - 9$ $= 3 - 15\sqrt{2}$
Dividing radicals, p. 524.	Division of radical expressions is accomplished by writing the division as a fraction and rationalizing the denominator.	Divide. $\sqrt{x} \div \sqrt{3}$ $\frac{\sqrt{x}}{\sqrt{3}} \cdot \frac{\sqrt{3}}{\sqrt{3}} = \frac{\sqrt{3x}}{\sqrt{9}} = \frac{\sqrt{3x}}{3}$
Conjugate pairs, p. 526.	Note that the sum and difference of two radical expressions multiply to the difference of their squares—eliminating the radicals. These are called conjugate pairs.	$(3\sqrt{6} + \sqrt{10})$ and $(3\sqrt{6} - \sqrt{10})$ are conjugate pairs. $(3\sqrt{6} + \sqrt{10})(3\sqrt{6} - \sqrt{10})$ $= 9\sqrt{6^2} - \sqrt{10^2}$ $= 9 \cdot 6 - 10 = 54 - 10 = 44$
Rationalizing denominators, p. 525.	To simplify a fraction with a radical in the denominator, the radical (in the denominator) must be removed. **1.** Multiply the denominator by itself if it is a monomial. **2.** Multiply the denominator by its conjugate if it is a binomial. In each case the numerator is multiplied by the same quantity as the denominator.	$\frac{\sqrt{10}}{\sqrt{6}} = \frac{\sqrt{10} \cdot \sqrt{6}}{\sqrt{6} \cdot \sqrt{6}} = \frac{2\sqrt{15}}{6} = \frac{\sqrt{15}}{3}$ $\frac{\sqrt{3}}{\sqrt{3} - \sqrt{2}} = \frac{\sqrt{3}(\sqrt{3} + \sqrt{2})}{(\sqrt{3} - \sqrt{2})(\sqrt{3} + \sqrt{2})}$ $= \frac{\sqrt{9} + \sqrt{6}}{3 - 2} = 3 + \sqrt{6}$
Radicals with fractions, p. 525.	Note that $\sqrt{\frac{5}{3}}$ is the same as $\frac{\sqrt{5}}{\sqrt{3}}$ and so must be rationalized.	$\sqrt{\frac{5}{3}} = \frac{\sqrt{5}}{\sqrt{3}} \cdot \frac{\sqrt{3}}{\sqrt{3}} = \frac{\sqrt{15}}{3}$
Pythagorean theorem, p. 530.	In any right triangle with hypotenuse of length c and sides of lengths a and b, $c^2 = a^2 + b^2$ 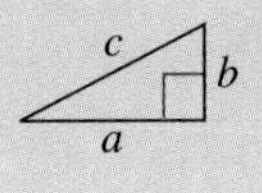	Find c to the nearest tenth if $a = 7$ and $b = 9$. $c^2 = 7^2 + 9^2 = 49 + 81 = 130$ $c = \sqrt{130} \approx 11.4$

Chapter Organizer

Topic	Procedure	Examples
Radical equations, p. 532.	To solve an equation containing a square root radical: **1.** Isolate one radical by itself on one side of the equation. **2.** Square both sides. **3.** Solve the resulting equation. **4.** Check all apparent solutions. Extraneous roots may have been introduced in step 2.	Solve. $\sqrt{3x+2} - 4 = 8$ $\sqrt{3x+2} = 12$ $(\sqrt{3x+2})^2 = 12^2$ $3x + 2 = 144$ $3x = 142$ $x = \frac{142}{3}$ $\sqrt{3 \cdot \frac{142}{3} + 2} \stackrel{?}{=} 12$ $\sqrt{144} = 12$ ✓ So $x = \frac{142}{3}$.
Variation: direct, p. 537.	If y varies directly with x, there is a constant of variation, k, such that $y = kx$. Once k is determined, other values of y or x can easily be computed.	y varies directly with x. When $x = 2$, $y = 7$. $y = kx$ $7 = k(2)$ Substituting. $k = \frac{7}{2}$ Solving. $y = \frac{7}{2}x$ What is y when $x = 8$? $y = \frac{7}{2}x = \frac{7}{2} \cdot 8 = 28$
Variation: inverse, p. 539.	If y varies inversely with x, the constant k is such that $y = \frac{k}{x}$	y varies inversely with x. When x is 5, y is 12. What is y when x is 15? $y = \frac{k}{x}$ $12 = \frac{k}{5}$ Substituting. $k = 60$ Solving. $y = \frac{60}{x}$ Substituting. $x = 15 \quad y = \frac{60}{15} = 4$

Chapter 9 Review Problems

9.1 *Simplify or evaluate, if possible.*

1. $\sqrt{9}$ **2.** $\sqrt{25}$ **3.** $\sqrt{36}$ **4.** $\sqrt{49}$ **5.** $\sqrt{100}$

6. $\sqrt{4}$ **7.** $\sqrt{-81}$ **8.** $\sqrt{-64}$ **9.** $\sqrt{16}$ **10.** $\sqrt{121}$

11. $\sqrt{225}$ **12.** $\sqrt{169}$ **13.** $\sqrt{0.81}$ **14.** $\sqrt{0.36}$ **15.** $\sqrt{\frac{1}{25}}$

16. $\sqrt{\frac{36}{49}}$ **17.** $-\sqrt{0.0004}$ **18.** $\sqrt{0.0009}$ **19.** $\sqrt{-0.0016}$ **20.** $-\sqrt{0.0081}$

Approximate, using the square root table in Appendix A or a calculator. Round to nearest thousandth, if necessary.

21. $\sqrt{105}$ **22.** $\sqrt{198}$ **23.** $\sqrt{77}$ **24.** $\sqrt{88}$

9.2

25. $\sqrt{50}$ 26. $\sqrt{48}$ 27. $\sqrt{98}$ 28. $\sqrt{72}$ 29. $\sqrt{40}$ 30. $\sqrt{80}$

31. $\sqrt{x^8}$ 32. $\sqrt{y^{10}}$ 33. $\sqrt{x^5y^6}$ 34. $\sqrt{a^3b^4}$ 35. $\sqrt{16x^3y^5}$ 36. $\sqrt{98x^4y^6}$

37. $\sqrt{12x^5}$ 38. $\sqrt{27x^7}$ 39. $\sqrt{75x^{10}}$ 40. $\sqrt{125x^{12}}$

41. $\sqrt{120a^3b^4c^5}$ 42. $\sqrt{121a^6b^4c}$ 43. $\sqrt{56x^7y^9}$ 44. $\sqrt{99x^{13}y^7}$

9.3

45. $\sqrt{2} - \sqrt{8} + \sqrt{32}$

46. $3\sqrt{5} - 2\sqrt{20} - \sqrt{5}$

47. $x\sqrt{3} + 3x\sqrt{3} + \sqrt{27x^2}$

48. $a\sqrt{2} + \sqrt{12a^2} + a\sqrt{98}$

49. $5\sqrt{5} - 6\sqrt{20} + 2\sqrt{10}$

50. $3\sqrt{6} - 5\sqrt{18} + 3\sqrt{24}$

51. $2\sqrt{28} - 3\sqrt{63} + 2x\sqrt{7}$

52. $5a\sqrt{3} - 12\sqrt{27} + 2\sqrt{75}$

9.4

53. $(2\sqrt{x})(3\sqrt{x^3})$

54. $(-5\sqrt{a})(2\sqrt{ab})$

55. $(\sqrt{2a^3})(\sqrt{8b^2})$

56. $(5x\sqrt{x})(-3x^2\sqrt{x})$

57. $\sqrt{5}(3\sqrt{5} - \sqrt{20})$

58. $2\sqrt{3}(\sqrt{27} - 6\sqrt{3})$

59. $\sqrt{2}(\sqrt{5} - \sqrt{3} - 2\sqrt{2})$

60. $\sqrt{5}(\sqrt{6} - 2\sqrt{5} + \sqrt{10})$

61. $(\sqrt{11} + 2)(2\sqrt{11} - 1)$

62. $(\sqrt{10} + 3)(3\sqrt{10} - 1)$

63. $(2 + 3\sqrt{6})(4 - 2\sqrt{3})$

64. $(5 - \sqrt{2})(3 - \sqrt{12})$

65. $(2\sqrt{3} + 3\sqrt{6})^2$

66. $(5\sqrt{2} - 2\sqrt{6})^2$

67. $(a\sqrt{b} - 2\sqrt{a})(\sqrt{b} - 3\sqrt{a})$

68. $(x\sqrt{y} - 2\sqrt{x})(3\sqrt{y} + \sqrt{x})$

9.5

69. $\dfrac{1}{\sqrt{3x}}$ **70.** $\dfrac{2y}{\sqrt{5}}$ **71.** $\dfrac{x^2y}{\sqrt{8}}$ **72.** $\dfrac{3ab}{\sqrt{2b}}$

73. $\sqrt{\dfrac{3}{7}}$ **74.** $\sqrt{\dfrac{2}{9}}$ **75.** $\dfrac{15\sqrt{60}}{5\sqrt{15}}$ **76.** $\dfrac{\sqrt{120}}{2\sqrt{5}}$

77. $\dfrac{\sqrt{a^5}}{\sqrt{2a}}$ **78.** $\dfrac{\sqrt{x^3}}{\sqrt{3x}}$ **79.** $\dfrac{a\sqrt{a^2b}}{ab\sqrt{ab^2}}$ **80.** $\dfrac{2xy\sqrt{x^3}}{2y\sqrt{y^3}}$

81. $\dfrac{3}{\sqrt{5}+\sqrt{2}}$ **82.** $\dfrac{2}{\sqrt{6}-\sqrt{3}}$ **83.** $\dfrac{1-\sqrt{5}}{2+\sqrt{5}}$ **84.** $\dfrac{1-\sqrt{3}}{3+\sqrt{3}}$

9.6

Use the Pythagorean theorem to find the length of the missing side of each right triangle.

85. $a = 5,\ b = 8$ **86.** $c = \sqrt{11},\ b = 3$ **87.** $c = 5,\ a = 3.5$

88. Find c and round your answer to the nearest thousandth, if $a = 2.400$ and $b = 2.000$.

89. A flagpole is 24 meters tall. A man stands 18 meters from the base of the pole. How far is it from the feet of the man to the top of the pole?

90. A city is 20.0 miles east of a major airport. A small town is directly south of the city. The town is 50.0 miles from the airport. How far is it from the town to the city? (Round your answer to the nearest tenth of a mile.)

91. The diagonal measurement of a small color television screen is 10 inches. The screen is square. How long is each side of the screen? (Round your answer to the nearest tenth of an inch.)

Solve each radical equation. Be sure to verify your answers.

92. $\sqrt{x+3} = 4$

93. $\sqrt{2x-3} = 9$

94. $\sqrt{1-3x} = \sqrt{5+x}$

95. $\sqrt{-5+2x} = \sqrt{1+x}$

96. $\sqrt{x} - 6 = x$

97. $\sqrt{2x-5} = 10 - x$

98. $6 - \sqrt{5x-1} = x + 1$

99. $4x + \sqrt{x+2} = 5x - 4$

9.7

100. If y varies directly with the square root of x, and $y =$ 35 when $x = 25$, find the value of y when $x = 121$.

101. If y varies inversely with the square of x, and $y = \frac{6}{5}$ when $x = 5$, find the value of y with $x = 15$.

102. If y varies inversely with the cube of x, and $y = 4$ when $x = 2$, find the value of y when $x = 4$.

103. The insect population in a potato field varies inversely with the amount of pesticide used. When 40 pounds of pesticide was used, the insect population of the field was estimated to be 1000 bugs. How many pounds of pesticide should be required to reduce the number of bugs in the field to 100?

104. The length of skid marks on the road when a car slams on the brakes varies directly with the square of the speed of the car. At 30 mph a certain car had skid marks 40 feet long. How long will the skid marks be if the car is traveling at 55 mph?

105. The horsepower that is needed to drive a racing boat through water varies directly with the cube of the speed of the boat. What will happen to the horsepower requirement if someone wants to double the maximum speed of a given boat?

Putting Your Skills to Work

Police use the formula

$$r = 2\sqrt{5L}$$

to estimate the speed a car has been traveling by measuring the skid marks. In the formula, r is the speed in miles per hour and L is the length of the skid mark in feet.

Estimate the speed of a car that left the following skid marks.

1. 5 feet long

2. 45 feet long

3. 125 feet long

4. 245 feet long

How long will the skid marks be at the following speeds.

5. 20 miles per hour

6. 40 miles per hour

7. 80 miles per hour

8. 100 miles per hour

COOPERATIVE GROUP ACTIVITY

9. Make a table of values and graph the equation $r = 2\sqrt{5L}$.

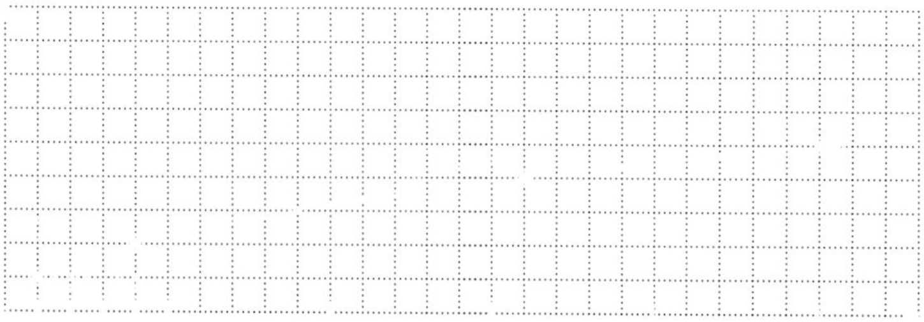

10. Describe what happens when the speed of the car increases from 20 miles per hour to 40 miles per hour. How much longer are the skid marks?

11. Describe what happens when the speed of the car increases from 80 miles per hour to 100 miles per hour. How much longer are the skid marks?

12. At higher rates of speed, when the speed is increased, what can be said about its effect on the increase in the length of the skid marks?

13. If the police were uncertain about the length of the skid marks (because the car went off the road), what would that do to the projected rate of speed? What would that mean in a court of law?

Evaluate.

1. $\sqrt{81}$

2. $\sqrt{\dfrac{9}{100}}$

Simplify.

3. $\sqrt{48x^2y^7}$

4. $\sqrt{100x^3yz^4}$

Combine and simplify.

5. $\sqrt{5} + 3\sqrt{20} - 2\sqrt{45}$

6. $\sqrt{4a} + \sqrt{8a} + \sqrt{36a} + \sqrt{18a}$

Multiply.

7. $(2\sqrt{a})(3\sqrt{b})(2\sqrt{ab})$

8. $\sqrt{3}(\sqrt{6} - \sqrt{2} + 5\sqrt{27})$

9. $(2 - \sqrt{6})^2$

10. $(4\sqrt{2} - \sqrt{5})(3\sqrt{2} + \sqrt{5})$

Simplify.

11. $\sqrt{\dfrac{x}{5}}$

12. $\dfrac{3}{\sqrt{12}}$

1. ______

2. ______

3. ______

4. ______

5. ______

6. ______

7. ______

8. ______

9. ______

10. ______

11. ______

12. ______

Simplify.

13. $\dfrac{3 - \sqrt{2}}{\sqrt{2} - 6}$

14. $\dfrac{3a}{\sqrt{5} + \sqrt{2}}$

15. Use the table of square roots or a calculator to approximate $\sqrt{156}$. Round your answer to nearest hundredth.

Find the missing side by using the Pythagorean theorem.

16.

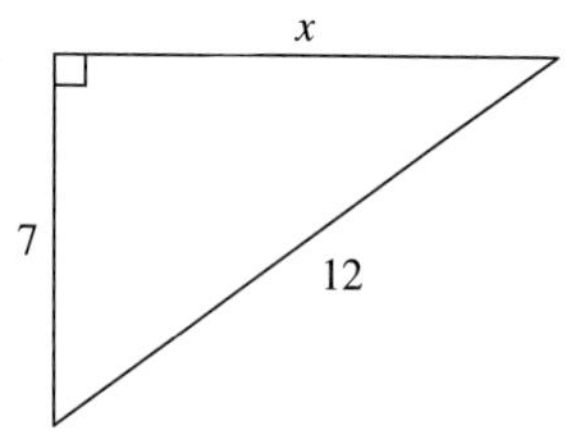

17.

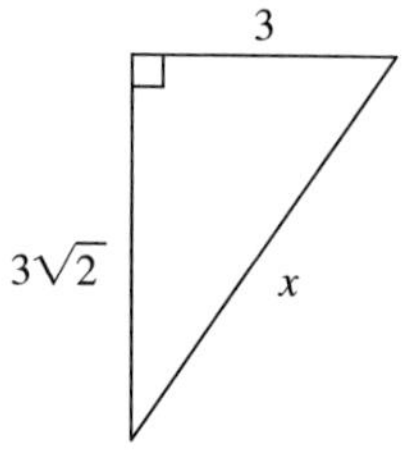

Solve each radical equation. Verify your solutions.

18. $6 - \sqrt{2x + 1} = 0$

19. $x = 5 + \sqrt{x + 7}$

Solve each word problem.

20. The intensity of illumination from a light source varies inversely with the square of the distance from the light source. A photoelectric cell is placed 8 inches from a light source. Then it is moved so that it only receives one-fourth as much illumination. How far is it from the light source?

21. Sales tax varies directly with the cost of the sale. If \$26.10 is the tax on a purchase of \$360, what will the tax be on a purchase of \$12,580?

22. The area of an equilateral triangle varies directly as the square of its perimeter. If the perimeter is 12 centimeters, the area is 6.93 square centimeters. What is the area of an equilateral triangle whose perimeter is 21 centimeters? Round to the nearest tenth.

13. ____________

14. ____________

15. ____________

16. ____________

17. ____________

18. ____________

19. ____________

20. ____________

21. ____________

22. ____________

Cumulative Test for Chapters 0–9

Approximately one-half of this test covers the content of Chapters 0–8. The remainder covers the content of Chapter 9.

Simplify.

1. $\dfrac{12}{15}$

2. $3\dfrac{1}{4} + 5\dfrac{2}{3}$

3. $6\dfrac{1}{4} \div 6\dfrac{2}{3}$

4. Write 0.07% as a decimal.

Simplify.

5. $-11 - 16 + 8 + 4 - 13 + 31$

6. $(-3)^2 \cdot 4 - 8 \div 2 - (3 - 2)^3$

7. $3x^3yz^2 + 4xyz - 5x^3yz^2$

8. $(3x - 2)^3$

9. $(2x - 3)^2$

10. $(3x - 11)(3x + 11)$

11. $\dfrac{4x}{x + 2} + \dfrac{4}{x + 2}$

12. $\dfrac{x^2 - 5x + 6}{x^2 - 5x - 6} \div \dfrac{x^2 - 4}{x^2 + 2x + 1}$

1. __________

2. __________

3. __________

4. __________

5. __________

6. __________

7. __________

8. __________

9. __________

10. __________

11. __________

12. __________

13.	
14.	
15.	
16.	
17.	
18.	
19.	
20.	
21.	
22.	
23.	
24.	
25.	
26.	

Simplify, if possible.

13. $\sqrt{98x^5y^6}$

14. $\sqrt{12} - \sqrt{27} + 3\sqrt{75}$

15. $(\sqrt{6} - \sqrt{3})^2$

16. $\dfrac{\sqrt{3} + \sqrt{2}}{\sqrt{3} - \sqrt{2}}$

17. $(3\sqrt{6} - \sqrt{2})(\sqrt{6} + 4\sqrt{2})$

18. $\dfrac{1}{3 - \sqrt{2}}$

19. $-\sqrt{16}$

20. $\sqrt{-16}$

Use the Pythagorean theorem to find the missing side of each triangle. Be sure to simplify your answer.

21. $a = 4$, $b = \sqrt{7}$

22. $a = 19$, $c = 21$

Solve for the variable. Be sure to verify your answer(s).

23. $\sqrt{4x + 5} = x$

24. $\sqrt{3y - 2} + 2 = y$

25. If y varies inversely as x, and $y = 4$ when $x = 5$, find the value of y when $x = 2$.

26. The surface area of a balloon varies directly with the square of the cross-sectional radius. When the radius is 2 meters, the surface area is 1256 square meters. Find the surface area if the radius is 0.5 meter.

10
Quadratic Equations

Classical art, ancient architecture, and modern building designs often reflect the Golden Ratio. This ratio, which ‘‘seems obvious,’’ actually comes from the solution of a quadratic equation. See page 599.

Pretest Chapter 10

1. ______
2. ______
3. ______
4. ______
5. ______
6. ______
7. ______
8. ______
9. ______
10. ______
11. ______
12. ______
13. ______
14. ______
15. ______

If you are familiar with the topics in this chapter, take this test now. Check your answers with those in the back of the book. If an answer was wrong or you couldn't do a problem, study the appropriate section of the chapter.

If you are not familiar with the topics in this chapter, don't take this test now. Instead, study the examples, work the practice problems, and then take the test.

This test will help you to identify those concepts that you have mastered and those that need more study.

Section 10.1

Solve each quadratic equation by the factoring method.

1. $x^2 - 12x - 28 = 0$ **2.** $3x^2 + 2x = 6x$

3. $5x^2 = 22x - 8$ **4.** $2x + 28 = 6x^2$

Section 10.2

Solve each quadratic equation by taking the square root of each side of the equation.

5. $x^2 - 12 = 0$ **6.** $3x^2 + 1 = 76$

Solve each quadratic equation by the completing the square method.

7. $x^2 + 8x + 5 = 0$ **8.** $2x^2 + 3x - 7 = 0$

Section 10.3

Solve each quadratic equation if possible by using the quadratic formula.

9. $2x^2 - 6x + 3 = 0$ **10.** $4x^2 = -7x + 2$

11. $3x^2 + 8x + 1 = 0$ **12.** $5x^2 + 3 = 4x$

Section 10.4

Graph the quadratic equations. Locate the vertex.

13. $y = 5 - 3x^2$

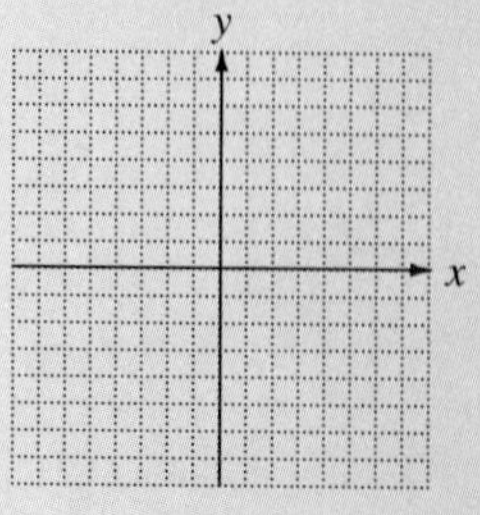

14. $y = 2x^2 - 4x - 1$

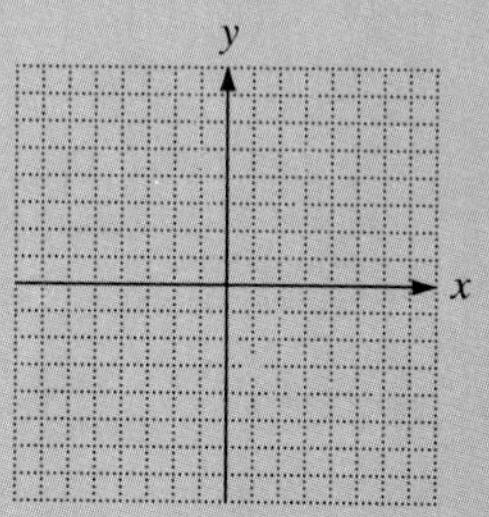

Section 10.5

15. A triangle has an area of 39 square centimeters. The altitude of the triangle is 1 centimeter longer than double the length of the base. Find the dimensions of the triangle.

10.1 Introduction to Quadratic Equations

After studying this section, you will be able to:

1 *Place a quadratic equation in the standard form $ax^2 + bx + c = 0$, where $a > 0$, and determine the value of the coefficients a, b, c.*
2 *Solve quadratic equations of the form $ax^2 + bx = 0$ by factoring.*
3 *Solve quadratic equations of the form $ax^2 + bx + c = 0$ by factoring.*
4 *Solve applied problems that require the use of quadratic equations.*

1 *Placing a Quadratic Equation in Standard Form*

In Section 5.7 we introduced quadratic equations. All quadratic equations contain polynomials of the second degree. For example,

$$3x^2 + 5x - 7 = 0, \qquad \tfrac{1}{2}x^2 - 5x = 0, \qquad 2x^2 = 5x + 9, \qquad 3x^2 = 8$$

are all quadratic equations. It is particularly useful to place quadratic equations in standard form.

> The **standard form of a quadratic equation** is $ax^2 + bx + c = 0$, where a, b, and c are real numbers, and a is a positive real number.

You will need to be able to recognize the real numbers that represent a, b, and c in specific equations. It will be easier to do so if these equations are placed in standard form.

EXAMPLE 1 Place each of the following quadratic equations in standard form and identify the real numbers a, b, and c.

(a) $5x^2 - 6x + 3 = 0$ — *This equation is in standard form.*
$ax^2 + bx + c = 0$ — *Match each term to the standard form.*
$a = 5, \quad b = -6, \quad c = 3$

(b) $2x^2 + 5x = 4$ — *The right-hand side is not zero. It is not in standard form.*
$2x^2 + 5x - 4 = 0$ — *Add −4 to each side of the equation.*
$ax^2 + bx + c = 0$ — *Match each term to the standard form.*
$a = 2, \quad b = 5, \quad c = -4$

(c) $-2x^2 + 15x + 4 = 0$ — *This is not in standard form since the coefficient of x^2 is negative.*
$2x^2 - 15x - 4 = 0$ — *Multiply each term on both sides of the equation by −1.*
$ax^2 + bx + c = 0$
$a = 2, \quad b = -15, \quad c = -4$

(d) $7x^2 - 9x = 0$ — *This equation is in standard form.*
$ax^2 + bx + c = 0$ — *Note that the constant term is missing, so we know $c = 0$.*
$a = 7, \quad b = -9, \quad c = 0$

(e) $x^2 + 3x + 5 = 10x - 6$ — *This equation is not in standard form.*
$x^2 + 3x - 10x + 5 + 6 = 0$ — *Add $-10x + 6$ to each side.*
$x^2 - 7x + 11 = 0$ — *Simplify.*
$ax^2 + bx + c = 0$ — *Observe that the coefficient of x^2 is 1.*
$a = 1, \quad b = -7, \quad c = 11$

Practice Problem 1 Place each quadratic equation in standard form and identify the real numbers a, b, and c.

(a) $2x^2 + 12x - 9 = 0$ **(b)** $7x^2 = 6x - 8$ **(c)** $-x^2 - 6x + 3 = 0$

(d) $10x^2 - 12x = 0$ **(e)** $5x^2 - 4x + 20 = 4x^2 - 3x - 10$ ■

2 *Solving Quadratic Equations of the Form $ax^2 + bx = 0$ by Factoring*

Notice that the terms in quadratic equations of the form $ax^2 + bx = 0$ both have x as a factor. To solve such equations, begin by factoring out the x.

EXAMPLE 2 Solve. **(a)** $25x^2 - 30x = 0$ **(b)** $7x^2 + 9x - 2 = -8x - 2$

(a) $25x^2 - 30x = 0$

$5x(5x - 6) = 0$ *Factor.*

$5x = 0 \qquad 5x - 6 = 0$ *Set each factor equal to zero.*

$\dfrac{5x}{5} = \dfrac{0}{5} \qquad 5x = 6$

$\dfrac{5x}{5} = \dfrac{6}{5}$

$x = 0$

$x = \dfrac{6}{5}$

(b) $7x^2 + 9x - 2 = -8x - 2$ *The equation is not in standard form.*

$7x^2 + 9x - 2 + 8x + 2 = 0$ *Add $8x + 2$ to each side.*

$7x^2 + 17x = 0$ *Collect like terms.*

$x(7x + 17) = 0$ *Factor.*

$x = 0 \qquad 7x + 17 = 0$ *Set each factor equal to zero.*

$7x = -17$

$x = -\dfrac{17}{7}$

One of the solutions to each of the above equations is $x = 0$. This will always be true of equations of this form. One root will be zero. The other root will be a nonzero real number.

Practice Problem 2 Solve. **(a)** $3x^2 + 18x = 0$ **(b)** $2x^2 - 7x - 6 = 4x - 6$ ■

If the equation has fractions, clear the fractions by multiplying each term by the least common denominator of all the fractions in the equation. Sometimes the original equation does not look like a quadratic equation. However, it takes the quadratic form once the fractions have been cleared. A note of caution. Always check a possible solution in the original equation. Any solution that would make the denominator of any fraction in the original equation 0 is not a valid solution.

EXAMPLE 3 Solve and check. $8x - 6 + \frac{1}{x} = 0$

$8x - 6 + \frac{1}{x} = 0$	*The equation has a fractional term.*
$x(8x) - (x)(6) + x\left(\frac{1}{x}\right) = x(0)$	*Multiply each term by the LCD = x.*
$8x^2 - 6x + 1 = 0$	*Simplify. The equation is now in standard form.*
$(4x - 1)(2x - 1) = 0$	*Factor.*
$4x - 1 = 0 \quad 2x - 1 = 0$	*Set each factor equal to 0.*
$4x = 1 \quad 2x = 1$	*Solve each equation for x.*
$x = \frac{1}{4} \quad x = \frac{1}{2}$	

Check. Checking fractional roots is more difficult, but you should be able to do it if you work carefully.

$$\text{If } x = \frac{1}{4}\text{:} \quad 8\left(\frac{1}{4}\right) - 6 + \frac{1}{\frac{1}{4}} = 2 - 6 + 4 = 0 \quad \left(\text{since } 1 \div \frac{1}{4} = 1 \cdot \frac{4}{1} = 4\right)$$

$$0 = 0 \checkmark$$

$$\text{If } x = \frac{1}{2}\text{:} \quad 8\left(\frac{1}{2}\right) - 6 + \frac{1}{\frac{1}{2}} = 4 - 6 + 2 = 0 \quad \left(\text{since } 1 \div \frac{1}{2} = 1 \cdot \frac{2}{1} = 2\right)$$

$$0 = 0 \checkmark$$

Both roots check, so $x = \frac{1}{2}$ and $x = \frac{1}{4}$ are the two roots that satisfy the original equation.

$$8x - 6 + \frac{1}{x} = 0$$

Practice Problem 3 Solve and check. $3x + 10 - \frac{8}{x} = 0$ ■

3 *Solving Quadratic Equation of the Form $ax^2 + bx + c = 0$ by Factoring*

If it is possible to factor a quadratic equation, the zero property allows us to find a solution. Recall that if $a \cdot b = 0$, then either $a = 0$ or $b = 0$. Thus we can set each factor of the quadratic equation equal to 0 and solve for x.

EXAMPLE 4 Solve and check. $6x^2 + 7x - 10 = 0$

$(6x - 5)(x + 2) = 0$ *Factor the quadratic equation.*

$6x - 5 = 0 \quad x + 2 = 0$ *Set each factor equal to zero and solve for x.*

$6x = 5 \quad x = -2$

$$x = \frac{5}{6}$$

Thus the solutions to the equation are $x = \frac{5}{6}$ and $x = -2$.

Check each solution in the original equation.

$$6\left(\frac{5}{6}\right)^2 + 7\left(\frac{5}{6}\right) - 10 \stackrel{?}{=} 0 \qquad 6(-2)^2 + 7(-2) - 10 \stackrel{?}{=} 0$$

$$6\left(\frac{25}{36}\right) + 7\left(\frac{5}{6}\right) - 10 \stackrel{?}{=} 0 \qquad 6(4) + 7(-2) - 10 \stackrel{?}{=} 0$$

$$\frac{25}{6} + \frac{35}{6} - 10 \stackrel{?}{=} 0 \qquad 24 + (-14) - 10 \stackrel{?}{=} 0$$

$$\frac{60}{6} - 10 \stackrel{?}{=} 0 \qquad 10 - 10 = 0 \checkmark$$

$$10 - 10 = 0 \checkmark$$

Practice Problem 4 Solve and check. $2x^2 + 9x - 18 = 0$ ■

It is important to place the quadratic equation in standard form before factoring.

EXAMPLE 5 Solve. $x + (x - 6)(x - 2) = 2(x - 1)$

$x + x^2 - 6x - 2x + 12 = 2x - 2$ *Remove parentheses.*

$x^2 - 7x + 12 = 2x - 2$ *Collect like terms.*

$x^2 - 9x + 14 = 0$ *Add $-2x + 2$ to each side.*

$(x - 7)(x - 2) = 0$ *Factor.*

$x = 7, \quad x = 2$ *Use the zero property.*

The solutions to the equation are $x = 7$ and $x = 2$.

Practice Problem 5 Solve. $-4x + (x - 5)(x + 1) = 5(1 - x)$ ■

EXAMPLE 6 Solve. $\dfrac{8}{x+1} - \dfrac{x}{x-1} = \dfrac{2}{x^2-1}$

The LCD $= (x + 1)(x - 1)$. We multiply each term by the LCD.

$$(x+1)(x-1)\left(\frac{8}{x+1}\right) - (x+1)(x-1)\left(\frac{x}{x-1}\right) = (x+1)(x-1)\left[\frac{2}{(x+1)(x-1)}\right]$$

$8(x-1) - x(x+1) = 2$ *Simplify.*

$8x - 8 - x^2 - x = 2$ *Remove parentheses.*

$-x^2 + 7x - 8 = 2$ *Collect like terms.*

$-x^2 + 7x - 8 - 2 = 0$ *Add -2 to each side.*

$-x^2 + 7x - 10 = 0$ *Simplify.*

$x^2 - 7x + 10 = 0$ *Multiply each term by -1.*

$(x-5)(x-2) = 0$ *Factor and solve for x.*

$x = 5, \quad x = 2$ *Use the zero property.*

Check. The check is left to the student.

Practice Problem 6 Solve. $\dfrac{10x+18}{x^2+x-2} = \dfrac{3x}{x+2} + \dfrac{2}{x-1}$ ■

4 *Solving Applied Problems That Require the Use of Quadratic Equations*

A manager of a truck delivery service is determining the possible routes he may have to send trucks on during a given day. His delivery service must have the capability to send trucks between any two of the following Texas cities: Austin, Dallas, Tyler, Lufkin, and Houston. He made a rough map of the cities.

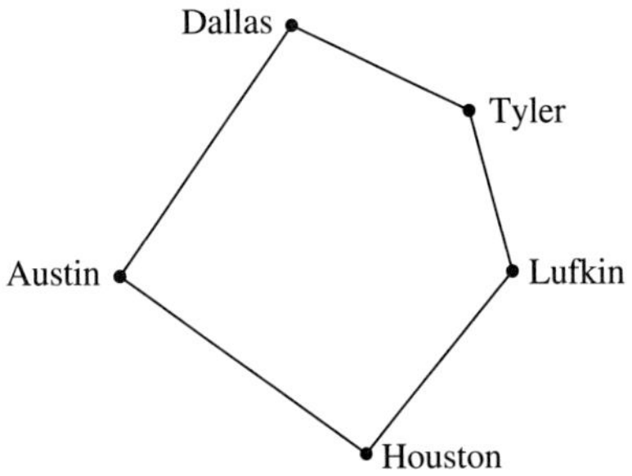

How many possible trips may he have to send trucks on in a given day, if he must be able to send a truck between any two cities? Of course, one possible way is to draw a straight line between every two cities. Let us assume that no three cities lie on a straight line. Therefore, there are always separate lines to connect any two cities. If we draw all these lines, we see there are exactly 10 lines, so the answer is 10 possible truck routes. What if he has to service 18 cities? How many truck routes will there be then?

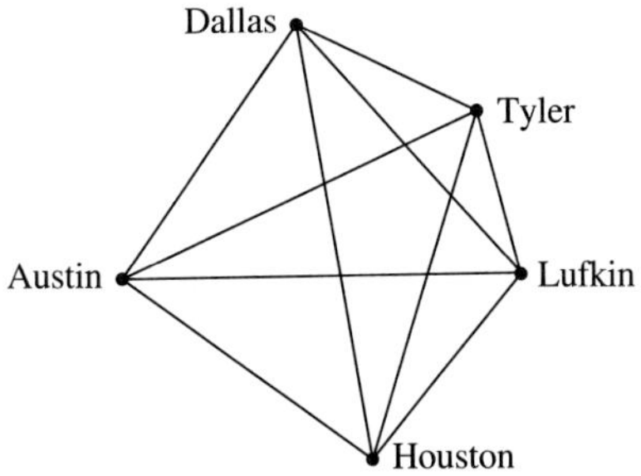

It gets quite difficult (and time consuming) to determine the number of lines using a drawing. In a more advanced math course, it can be proved that the maximum number of truck routes (t) can be found by using the number of cities (n) in the following equation.

$$t = \frac{n^2 - n}{2}$$

Thus, to find the number of truck routes for 18 cities, we merely substitute.

$$\begin{aligned} t &= \frac{(18)^2 - (18)}{2} \\ &= \frac{324 - 18}{2} \\ &= \frac{306}{2} \\ &= 153 \end{aligned}$$

There are a possible 153 truck routes to service 18 cities if no three of those cities lie on one straight line.

EXAMPLE 7 A truck delivery company can handle a maximum of 36 truck routes in one day. How many separate cities can the truck company service?

$$t = \frac{n^2 - n}{2}$$

$$36 = \frac{n^2 - n}{2}$$ *Substitute 36 for the number of truck routes.*

$72 = n^2 - n$ *Multiply both sides of the equation by 2.*

$0 = n^2 - n - 72$ *Add −72 to each side to obtain standard form.*

$0 = (n - 9)(n + 8)$ *Factor.*

$n-9 = 0 \qquad n + 8 = 0$ *Set each factor equal to zero.*

$n = 9 \qquad n = -8$ *Solve for n.*

We reject $n = -8$ as meaningless for this problem. We cannot have a negative number of cities. Thus our answer is that the company can service 9 cities.

Check. If we service $n = 9$ cities, does that really result in a maximum number of $t = 36$ truck routes?

$$t = \frac{n^2 - n}{2}$$ *The equation in which t = the number of truck routes and n = the number of cities to be connected.*

$$36 \stackrel{?}{=} \frac{(9)^2 - (9)}{2}$$ *Substitute $t = 36$ and $n = 9$.*

$$36 \stackrel{?}{=} \frac{81 - 9}{2}$$

$$36 \stackrel{?}{=} \frac{72}{2}$$

$36 = 36$ ✓

Practice Problem 7 How many cities can they service with 28 truck routes? ■

10.1 Exercises

Write each quadratic equation in standard form. Determine the value of a, b, and c.

1. $x^2 + 5x + 6 = 0$

2. $x^2 - 7x + 2 = 0$

3. $7x^2 - 11x - 2 = 0$

4. $6x^2 + 3x - 5 = 0$

5. $2 = -12x^2 + 3x$

6. $8 = -14x^2 - 9x$

7. $3x^2 - 8x = 0$

8. $5x^2 + 15x = 0$

9. $x^2 + 15x - 7 = 12x + 8$

10. $3x^2 + 10x + 8 = -17x - 2$

11. $2x(x - 3) = (x + 5)(x - 2)$

12. $6x(x + 2) = (x + 7)(x - 3)$

Using the factoring method, solve for the roots of each quadratic equation. Be sure to place your equation in standard form before factoring.

13. $18x^2 + 9x = 0$

14. $8x^2 - 4x = 0$

15. $11x^2 = 14x$

16. $8x^2 = 20x$

17. $7x^2 - 3x = 5x$

18. $9x^2 + 2x = -3x$

19. $15x^2 + 16x = 3x^2 - 20x$

20. $41x^2 - 26x = 3x^2 + 50x$

21. $x^2 + 5x - 24 = 0$

22. $x^2 - 14x + 40 = 0$

23. $2x^2 - 15x - 8 = 0$

24. $2x^2 + 13x - 7 = 0$

25. $3x^2 - 13x + 4 = 0$

26. $5x^2 - 16x + 3 = 0$

27. $x^2 = 9x - 14$

28. $x^2 = 12x - 27$

29. $0 = x^2 + 15x + 50$

30. $0 = x^2 + 14x + 48$

31. $0 = 6x^2 + 19x + 10$

32. $0 = 8x^2 + 18x + 9$

33. $y^2 + 8y + 13 = y + 1$

34. $y^2 + 10y + 17 = 2y + 2$

35. $6y^2 = 11y - 3$

36. $8y^2 = 14y - 3$

37. $9y^2 - 24y + 16 = 0$

38. $4y^2 - 12y + 9 = 0$

39. $n^2 - 10n + 25 = 0$

40. $n^2 - 12n + 36 = 0$

41. $n^2 + 5n - 17 = 6n + 3$

42. $n^2 + 2n - 11 = 5n + 7$

43. $3x^2 + 8x - 10 = -6x - 10$

44. $5x^2 - 7x + 12 = -8x + 12$

45. $2x(x - 4) = 4 - x$

46. $2x(x - 3) = 12 - x$

47. $(x - 6)(x + 1) = 8$

48. $(x - 3)(x + 1) = 12$

49. $x(x + 8) = 16(x - 1)$

50. $x(x + 9) = 4(x + 6)$

★**51.** $2x(x + 3) = (3x + 1)(x + 1)$

★**52.** $2(2x + 1)(x + 1) = x(7x + 1)$

★**53.** $-3x + (x + 6)(x + 2) = 12 - 3x$

★**54.** $-4x + (x - 6)(x + 2) = 20 - 4x$

Solve each equation. Check your solutions.

55. $\frac{4}{x} + \frac{3}{x+5} = 2$

56. $\frac{7}{x} + \frac{6}{x-1} = 2$

57. $\frac{12}{x-2} = \frac{2x-3}{3}$

58. $\frac{5}{x+2} = \frac{2x-1}{5}$

59. $\frac{x}{2} + \frac{5}{2} = -\frac{3}{x}$

60. $\frac{x}{2} + \frac{4}{x} = -3$

61. $x - 8 = -\frac{15}{x}$

62. $x + \frac{12}{x} = 8$

63. $\frac{24}{x^2-4} = 1 + \frac{2x-6}{x-2}$

64. $\frac{5x+7}{x^2-9} = 1 + \frac{2x-8}{x-3}$

To Think About

65. Why can an equation in standard form with $c = 0$ always be solved?

66. Martha solved $(x + 3)(x - 2) = 14$ as follows:

$$x + 3 = 14 \qquad x - 2 = 14$$
$$x = 11 \qquad x = 16$$

Josette said this had to be wrong because these values did not check. Explain what is wrong with Martha's method.

67. The equation $ax^2 - 7x + c = 0$ has the roots $x = -\frac{3}{2}$ and $x = 6$. Find the value of a and c.

Applications

The maximum number of truck routes, t, that is needed to provide direct service to n cities, where no three cities lie in a straight line, is given by the equation

$$t = \frac{n^2 - n}{2}$$

Use this equation in finding the solution to Problems 68 through 73.

68. How many truck routes will be needed to run service between 20 cities?

69. How many truck routes will be needed to run service between 22 cities?

70. A company can handle a maximum of 21 truck routes in one day. How many separate cities can be serviced?

71. A company can handle a maximum of 15 truck routes in one day. How many separate cities can be serviced?

72. A company can handle a maximum of 45 truck routes in one day. How many separate cities can be serviced?

73. A company can handle a maximum of 55 truck routes in one day. How many separate cities can be serviced?

Cumulative Review Problems

Simplify the following complex fractions.

74. $\dfrac{\dfrac{1}{x} + \dfrac{3}{x - 2}}{\dfrac{4}{x^2 - 4}}$

75. $\dfrac{\dfrac{1}{2x} + \dfrac{3}{4x}}{\dfrac{7}{x} - \dfrac{3}{2x}}$

Simplify.

76. $\dfrac{2x - 8}{x^3 - 16x}$

Add the fractions.

77. $\dfrac{2x}{3x - 5} + \dfrac{2}{3x^2 - 11x + 10}$

10.2 Solutions by the Square Root Property and by Completing the Square

After studying this section, you will be able to:

1. *Solve quadratic equations by the square root property.*
2. *Solve quadratic equations by completing the square.*

1 *Solving Quadratic Equations by the Square Root Property*

Recall that the quadratic equation $x^2 = a$ has two possible solutions, $x = \sqrt{a}$ or $x = -\sqrt{a}$, where a is a nonnegative real number. That is, $x = \pm\sqrt{a}$. This basic idea is called the *square root property*.

> **Square Root Property**
>
> If $x^2 = a$, then $x = \sqrt{a}$ or $x = -\sqrt{a}$, for all nonnegative real numbers a.

EXAMPLE 1 Solve. **(a)** $x^2 = 49$ **(b)** $x^2 = 20$ **(c)** $5x^2 = 125$

(a) $x^2 = 49$

$x = \pm\sqrt{49}$

$x = \pm 7$

(b) $x^2 = 20$

$x = \pm\sqrt{20}$

$x = \pm 2\sqrt{5}$

(c) $5x^2 = 125$

$x^2 = 25$ *Divide both sides by 5 before taking the square root.*

$x = \pm\sqrt{25}$

$x = \pm 5$

Practice Problem 1 Solve. **(a)** $x^2 = 1$ **(b)** $x^2 - 50 = 0$ **(c)** $3x^2 = 81$ ■

EXAMPLE 2 Solve. $3x^2 + 5x = 18 + 5x + x^2$

Simplify the equation by placing all the variable terms on the left and the numerical terms on the right.

$$2x^2 = 18$$

$$x^2 = 9$$

$$x = \pm 3$$

Practice Problem 2 Solve. $4x^2 - 5 = 319$ ■

This technique can also be used if the term that is squared is a binomial.

EXAMPLE 3 Solve. $(x - 2)^2 = 36$

$(x - 2)^2 = 36$

$x - 2 = \pm\sqrt{36}$ *Take the square root of both sides.*

$x - 2 = \pm 6$

Note that we now have two equations since $\pm$ indicates plus or minus. We will solve each equation.

$$x - 2 = +6 \qquad x - 2 = -6$$

$$x = 8 \qquad x = -4$$

The two roots of the equation are $x = 8$ and $x = -4$.

Practice Problem 3 Solve. $(2x + 3)^2 = 25$ ■

Sometimes when we use this method we obtain an irrational number. This will occur if the number on the right side of the equation is not a perfect square.

EXAMPLE 4 Solve. $(3x + 1)^2 = 8$

$$(3x + 1)^2 = 8$$

$$3x + 1 = \pm\sqrt{8} \qquad \textit{Take the square root of both sides.}$$

$$3x + 1 = \pm 2\sqrt{2} \qquad \textit{Simplify as much as possible.}$$

Now we must solve the two equations expressed by the plus or minus statement.

$$3x + 1 = +2\sqrt{2} \qquad 3x + 1 = -2\sqrt{2}$$

$$3x = -1 + 2\sqrt{2} \qquad 3x = -1 - 2\sqrt{2}$$

$$x = \frac{-1 + 2\sqrt{2}}{3} \qquad x = \frac{-1 - 2\sqrt{2}}{3}$$

The roots of this quadratic equation are irrational numbers. They are

$$x = \frac{-1 + 2\sqrt{2}}{3} \quad \text{and} \quad x = \frac{-1 - 2\sqrt{2}}{3}$$

We cannot simplify these roots further, so we have them in this form.

Practice Problem 4 Solve. $(2x - 3)^2 = 12$ ■

2 *Solving Quadratic Equations by Completing the Square*

If a quadratic equation is not in a form where we can use the square root property, we can rewrite the equation by a method called *completing the square*. That is, we try to rewrite the equation with a perfect square on the left side so that we have $(ax + b)^2 = c$.

EXAMPLE 5 Solve. **(a)** $x^2 - 2x = 8$ **(b)** $x^2 + 12x = 4$

(a) Think, what can we add to $x^2 - 2x$ to make a perfect square. Let's draw a picture of $x^2 - 2x$. Begin with $(x \quad)(x \quad)$. To have a $-2x$ as a middle term, the binomial must be $(x - 1)(x - 1)$. Now the picture is complete and we can complete the square.

$$x^2 - 2x = 8$$

$$(x - 1)(x - 1) \qquad \textit{We need a } +1 \textit{ to complete the square.}$$

$$x^2 - 2x + 1 = 8 + 1 \qquad \textit{Be sure to add } +1 \textit{ to both sides of the equation.}$$

$$(x - 1)^2 = 9 \qquad \textit{Now we have a perfect square on the left.}$$

$$x - 1 = \pm 3$$

$$x - 1 = 3 \qquad x - 1 = -3$$

$$x = 4 \qquad x = -2$$

The two roots are $x = 4$ and $x = -2$.

(b) Draw a picture of $x^2 + 12x$. Fill in the pieces you know.

$$x^2 + 12x \qquad = 4$$

$$(x + 6)(x + 6) = 4 \qquad \textit{We need a } +36 \textit{ to complete the square.}$$

$$x^2 + 12x + 36 = 4 + 36 \qquad \textit{Add } +36 \textit{ to both sides of the equation.}$$

$$(x + 6)^2 = 40 \qquad \textit{Write the left side as a perfect square.}$$

$$x + 6 = \pm\sqrt{40} \qquad \textit{Solve for } x.$$

$$x + 6 = +2\sqrt{10} \qquad\qquad x + 6 = -2\sqrt{10}$$

$$x = -6 + 2\sqrt{10} \qquad\qquad x = -6 - 2\sqrt{10}$$

The two roots are $x = -6 + 2\sqrt{10}$ and $x = -6 - 2\sqrt{10}$.

Practice Problem 5 Solve. **(a)** $x^2 - 8x = 20$ **(b)** $x^2 + 10x = 3$ ■

Completing the square is a little more difficult when the coefficient of x is an odd number. Let's see what happens.

EXAMPLE 6 Solve by completing the square. $x^2 - 3x - 1 = 0$

Write the equation so that all the x terms are on the left and the numbers are on the right.

$$x^2 - 3x \qquad = 1$$

$$(x \qquad)(x \qquad)$$

The missing number when added to itself must be -3.

Since $-\frac{3}{2} + (-\frac{3}{2}) = -\frac{6}{2} = -3$, the missing number is $-\frac{3}{2}$.

Complete the picture.

$$\left(x - \frac{3}{2}\right)\left(x - \frac{3}{2}\right)$$

$$x^2 - 3x + \frac{9}{4} = 1 + \frac{9}{4} \qquad \textit{Complete the square } (-\tfrac{3}{2})(-\tfrac{3}{2}) = \tfrac{9}{4}.$$

$$\left(x - \frac{3}{2}\right)^2 = \frac{13}{4} \qquad \textit{Solve for } x.$$

$$x - \frac{3}{2} = \pm\sqrt{\frac{13}{4}}$$

$$x - \frac{3}{2} = \pm\frac{\sqrt{13}}{2}$$

$$x - \frac{3}{2} = +\frac{\sqrt{13}}{2} \qquad\qquad x - \frac{3}{2} = -\frac{\sqrt{13}}{2}$$

$$= \frac{3}{2} + \frac{\sqrt{13}}{2} \qquad\qquad = \frac{3}{2} - \frac{\sqrt{13}}{2}$$

$$= \frac{3 + \sqrt{13}}{2} \qquad\qquad = \frac{3 - \sqrt{13}}{2}$$

The two roots are $x = \dfrac{3 \pm \sqrt{13}}{2}$.

Practice Problem 6 Solve by completing the square. $x^2 - 5x = 7$ ■

If the coefficient a is not 1, we divide all terms of the equation by a so that the coefficient of the squared term will be 1.

EXAMPLE 7 Solve by completing the square. $4y^2 + 4y - 3 = 0$

$$4y^2 + 4y = 3 \qquad \textit{Place the numerical term on the right.}$$

$$y^2 + y = \frac{3}{4} \qquad \textit{Divide all terms by 4.}$$

$$\left(y + \frac{1}{2}\right)\left(y + \frac{1}{2}\right) \qquad \textit{Think } \tfrac{1}{2} + \tfrac{1}{2} = 1. \textit{ The middle term is } 1y.$$

$$y^2 + y + \frac{1}{4} = \frac{3}{4} + \frac{1}{4} \qquad \textit{Complete the square } (\tfrac{1}{2})^2 = \tfrac{1}{4}.$$

$$\left(y + \frac{1}{2}\right)^2 = 1 \qquad \textit{Solve for } y.$$

$$y + \frac{1}{2} = \pm\sqrt{1}$$

$$= \pm 1$$

$$y + \frac{1}{2} = 1 \qquad\qquad y + \frac{1}{2} = -1$$

$$= \frac{2}{2} - \frac{1}{2} \qquad\qquad = -\frac{2}{2} - \frac{1}{2}$$

$$= \frac{1}{2} \qquad\qquad = -\frac{3}{2}$$

Thus the two roots are $y = \frac{1}{2}$ and $y = -\frac{3}{2}$.

This method of completing the square will enable us to solve any quadratic equation that has real roots. It is usually faster, however, to factor the quadratic equation if the polynomial is factorable.

Practice Problem 7 Solve by factoring. $4y^2 + 4y - 3 = 0$ ■

10.2 Exercises

Solve by using the square root property.

1. $x^2 = 49$

2. $x^2 = 100$

3. $x^2 = 98$

4. $x^2 = 48$

5. $x^2 - 75 = 0$

6. $x^2 - 40 = 0$

7. $3x^2 = 75$

8. $4x^2 = 64$

9. $6x^2 = 120$

10. $3x^2 = 150$

11. $4x^2 - 12 = 0$

12. $5x^2 - 15 = 0$

13. $2x^2 - 196 = 0$

14. $5x^2 - 140 = 0$

15. $2x^2 - 10 = 62$

16. $3x^2 - 18 = 57$

17. $5x^2 + 13 = 73$

18. $6x^2 + 5 = 53$

19. $13x^2 + 17 = 82$

20. $11x^2 + 14 = 47$

21. $(x - 3)^2 = 5$

22. $(x + 2)^2 = 3$

23. $(x + 6)^2 = 4$

24. $(x - 5)^2 = 36$

25. $(2x + 5)^2 = 2$

26. $(3x - 4)^2 = 6$

27. $(3x - 1)^2 = 7$

28. $(2x + 3)^2 = 10$

29. $(7x + 2)^2 = 12$

30. $(8x + 5)^2 = 20$

31. $(6x - 5)^2 = 18$

32. $(9x - 4)^2 = 24$

Solve for x. Round your answer to the nearest thousandth.

33. $x^2 - 19.263 = 0$

34. $x^2 - 22.986 = 0$

35. $3.120x^2 - 7.986 = 0$

36. $5.961x^2 - 12.382 = 0$

Solve each quadratic equation by completing the square.

37. $x^2 + 14x = 15$

38. $x^2 - 2x = 5$

39. $x^2 - 4x = 1$

40. $x^2 + 4x = 12$

41. $x^2 + 6x + 7 = 0$

42. $x^2 - 12x - 4 = 0$

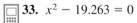

43. $x^2 - 10x - 4 = 0$

44. $x^2 + 20x + 72 = 0$

45. $x^2 + 3x = 0$

46. $x^2 + 5x = 3$

47. $x^2 - x - 6 = 0$

48. $x^2 + x - 20 = 0$

49. $4x^2 - 8x + 3 = 0$

50. $2x^2 - 3x = 9$

51. $2x^2 + 10x + 11 = 0$

To Think About

52. If an equation in standard form has $b = 0$, it reduces to $ax^2 - c = 0$. It is easy to find the solutions, if they exist. Explain how.

53. Let b represent a constant value. Solve for x by completing the square: $x^2 + bx - 7 = 0$.

54. Solve for x in terms of b and c. $x^2 + bx + c = 0$

Cumulative Review Problems

55. Solve. $3a - 5b = 8$

$5a - 7b = 8$

56. Solve. $\dfrac{x-4}{x-2} - \dfrac{1}{x} = \dfrac{x-3}{x^2 - 2x}$

57. Evaluate. $\left(\dfrac{3}{4}\right)^{-2}$

58. Find a number that equals ten more than eleven times itself.

10.3 Solutions by the Quadratic Formula

After studying this section, you will be able to:

1 *Solve a quadratic equation with real roots using the quadratic formula.*
2 *Find a decimal approximation to the real roots of a quadratic equation.*
3 *Determine if a given quadratic equation has no real solution.*

1 *Solving a Quadratic Equation by the Quadratic Formula*

To find solutions of a quadratic trinomial $ax^2 + bx + c = 0$, you can factor or you can complete the square when the trinomial is not factorable. The method of completing the square can be used on the general equation, and we can derive a general formula for the solution of any quadratic trinomial. This is called the *quadratic formula.*

> **Quadratic Formula**
>
> The roots of any quadratic equation in the form $ax^2 + bx + c = 0$, where a, b, and c are real numbers and $a \neq 0$, are
>
> $$x = \frac{-b \pm \sqrt{b^2 - 4ac}}{2a}$$

All you will need to know are the values of a, b, and c to solve a quadratic trinomial using the quadratic formula.

EXAMPLE 1 Solve by the quadratic formula. $3x^2 + 10x + 7 = 0$

In our given equation, $3x^2 + 10x + 7 = 0$, we have

$a = 3$ (the coefficient of x^2) $b = 10$ (the coefficient of x)

$c = 7$ (the constant term)

$$x = \frac{-b \pm \sqrt{b^2 - 4ac}}{2a} = \frac{-10 \pm \sqrt{(10)^2 - 4(3)(7)}}{2(3)}$$ *Write the quadratic formula and substitute values for a, b, and c.*

$$= \frac{-10 \pm \sqrt{100 - 84}}{6}$$ *Simplify.*

$$= \frac{-10 \pm \sqrt{16}}{6} = \frac{-10 \pm 4}{6}$$

$$x = \frac{-10 + 4}{6} = \frac{-6}{6} = -1$$ *Using the positive sign.*

$$x = \frac{-10 - 4}{6} = \frac{-14}{6} = -\frac{7}{3}$$ *Using the negative sign.*

Thus the two solutions are $x = -1$ and $x = -\frac{7}{3}$.

[*Note:* Here we obtain rational roots. We would obtain the same answer by factoring $3x^2 + 10x + 7 = 0$ into $(3x + 7)(x + 1) = 0$ and setting each factor $= 0$.]

Practice Problem 1 Solve by using the quadratic formula. $x^2 - 7x + 6 = 0$ ■

Often the roots will be irrational numbers. Sometimes these roots can be simplified. You should always leave your answer in simplest form.

EXAMPLE 2 Solve by the quadratic formula. $2x^2 - 4x - 9 = 0$

$$a = 2, \quad b = -4, \quad c = -9$$

$$x = \frac{-b \pm \sqrt{b^2 - 4ac}}{2a} = \frac{-(-4) \pm \sqrt{(-4)^2 - 4(2)(-9)}}{2(2)}$$

$$= \frac{4 \pm \sqrt{16 + 72}}{4} = \frac{4 \pm \sqrt{88}}{4}$$

Notice that we can simplify $\sqrt{88}$, so

$$= \frac{4 \pm 2\sqrt{22}}{4} = \frac{2(2 \pm \sqrt{22})}{4} = \frac{\overset{1}{\cancel{2}}(2 \pm \sqrt{22})}{\underset{2}{\cancel{4}}}$$

Be careful. Here we were able to divide out by 2 because 2 was a factor of every term.

$$x = \frac{2 \pm \sqrt{22}}{2}$$

Practice Problem 2 Solve. $3x^2 - 8x + 3 = 0$ ■

The quadratic equation *must* be written in the general form $ax^2 + bx + c = 0$ *before* the quadratic equation can be used. Several algebraic steps may be needed to accomplish that objective.

EXAMPLE 3 Solve by the quadratic formula. $2x^2 = 1 - \frac{7}{3}x$

$$2x^2 = 1 - \frac{7}{3}x$$ *The equation is not in general form.*

$$3(2x^2) = 3(1) - 3\left(\frac{7}{3}x\right)$$ *Multiply each term by the LCD of 3.*

$$6x^2 = 3 - 7x$$ *Simplify.*

$$6x^2 + 7x - 3 = 0$$ *Add 7x and −3 to both sides.*

$$a = 6, \quad b = 7, \quad c = -3$$

$$x = \frac{-7 \pm \sqrt{(7)^2 - 4(6)(-3)}}{2(6)}$$ *Substitute $a = 6$, $b = 7$, $c = -3$ into the quadratic formula.*

$$= \frac{-7 \pm \sqrt{49 + 72}}{12}$$

$$= \frac{-7 \pm \sqrt{121}}{12} = \frac{-7 \pm 11}{12}$$

Thus we have $x = -\frac{18}{12} = -\frac{3}{2}$ or $x = \frac{4}{12} = \frac{1}{3}$.

Practice Problem 3 Solve. $2x^2 = 3x + 1$ ■

2 *Finding a Decimal Approximation to the Real Roots of a Quadratic Equation*

EXAMPLE 4 Find the roots of $3x^2 - 5x = 7$. Approximate your answer to the nearest thousandth.

$$3x^2 - 5x - 7 = 0 \qquad \textit{Place the equation in standard form.}$$

$$a = 3, \qquad b = -5, \qquad c = -7$$

$$x = \frac{-(-5) \pm \sqrt{(-5)^2 - 4(3)(-7)}}{2(3)} = \frac{5 \pm \sqrt{25 + 84}}{6}$$

$$= \frac{5 \pm \sqrt{109}}{6}$$

Using most scientific calculators, we can find $\frac{5 + \sqrt{109}}{6}$ with these keystrokes.

5 [+] 109 [√] [=] [÷] 6 [=] 2.5733844

Rounding to the nearest thousandth, we have $x \approx 2.573$.

To find $\frac{5 - \sqrt{109}}{6}$ we have

5 [−] 109 [√] [=] [÷] 6 [=] -0.9067178

Rounding to the nearest thousandth, we have $x \approx -0.907$.
If you do not have a calculator with a square root key, look up $\sqrt{109}$ in the square root table. $\sqrt{109} \approx 10.440$. Using this result, we have

$$x \approx \frac{5 + 10.440}{6} = \frac{15.440}{6} = 2.573 \quad \text{(rounded to the nearest thousandth)}$$

and

$$x \approx \frac{5 - 10.440}{6} = \frac{-5.440}{6} = -0.907 \quad \text{(rounded to nearest thousandth)}$$

Practice Problem 4 Find the roots of $2x^2 = 13x + 5$. Approximate your answer to the nearest thousandth. ■

3 *Quadratic Equations with No Real Solutions*

EXAMPLE 5 Solve the quadratic formula. $2x^2 + 5 = -3x$

The equation is not in proper form.

$2x^2 + 5 = -3x$ *Add 3x to both sides.*

$2x^2 + 3x + 5 = 0$ *We can now find a, b, and c.*

$a = 2, \quad b = 3, \quad c = 5$ *Substitute these values into the quadratic formula.*

$$x = \frac{-3 \pm \sqrt{(3)^2 - 4(2)(5)}}{2(2)} = \frac{-3 \pm \sqrt{9 - 40}}{4} = \frac{-3 \pm \sqrt{-31}}{4}$$

There is no real number that is $\sqrt{-31}$. Since we are using only real numbers in this book, we say there is no solution to the problem. (*Note:* In more advanced math courses, these types of numbers, called complex numbers, will be studied.)

Practice Problem 5 Solve. $5x^2 + 2x = -3$ ■

In any given quadratic equation, we can tell if the roots are not real numbers. Look at the quadratic formula.

$$\frac{-b \pm \sqrt{b^2 - 4ac}}{2a}$$

The expression under the radical sign is called the *discriminant.*

$b^2 - 4ac$ is the discriminant.

If the discriminant is a negative number, the roots are not real numbers. There is no real number solution to the equation.

EXAMPLE 6 Determine whether or not $3x^2 = 5x - 4$ has real number solution(s).

First, we place the equation in standard form. Then we need only check the discriminant.

$3x^2 = 5x - 4$

$3x^2 - 5x + 4 = 0$

$a = 3, \quad b = -5, \quad c = 4$ *Substitute these values for a, b, and c into the discriminant and evaluate.*

$$b^2 - 4ac = (-5)^2 - 4(3)(4) = 25 - 48 = -23$$

The discriminant is negative. Thus $3x^2 = 5x - 4$ has no real number solution(s).

Practice Problem 6 Determine whether or not each equation has real number solution(s).

(a) $5x^2 = 3x + 2$ **(b)** $2x^2 - 5 = 4x$ ■

10.3 Exercises

Solve by the quadratic formula. If no real roots are possible, say so.

1. $x^2 + 4x + 1 = 0$

2. $x^2 - 2x - 4 = 0$

3. $x^2 - 5x - 2 = 0$

4. $x^2 + 5x + 3 = 0$

5. $2x^2 - 7x - 9 = 0$

6. $2x^2 - 6x + 1 = 0$

7. $5x^2 + x - 1 = 0$

8. $2x^2 + 12x - 5 = 0$

9. $5x - 1 = 2x^2$

10. $5x^2 = -3x + 2$

11. $6x^2 - 3x = 1$

12. $3 = 7x - 4x^2$

13. $x - 1 = -\frac{2}{5}x^2$

14. $2x^2 + 3x = \frac{3}{4}$

15. $9x^2 - 6x = 2$

16. $6x^2 = 13x + 5$

17. $4x^2 + 3x + 2 = 0$

18. $\frac{x}{2} + \frac{2}{x} = \frac{5}{2}$

19. $\frac{x}{3} + \frac{2}{x} = \frac{7}{3}$

20. $3x^2 = 2x - 5$

21. $5y^2 = 3 - 7y$

22. $5y - 6 = -2y^2$

23. $\frac{1}{3}m^2 = \frac{1}{2}m + \frac{3}{2}$

24. $\frac{3}{5}m^2 - m = \frac{2}{5}$

Solve for x by using the quadratic formula. Round your answer to the nearest thousandth.

25. $1.04x^2 - 17.982x + 74.108 = 0$

26. $0.986x^2 + 9.685x - 4.231 = 0$

Use the quadratic formula to find the roots of the following equations. Find a decimal approximation to the nearest thousandth.

27. $x^2 + 3x - 7 = 0$

28. $x^2 + 5x - 2 = 0$

29. $4x^2 - 4x - 1 = 0$

30. $3x^2 - 2x - 2 = 0$

31. $5x^2 + 10x + 1 = 0$

32. $2x^2 + 9x + 2 = 0$

Mixed Practice

Solve for x. Choose the method you feel works best.

33. $t^2 + 1 = \frac{13}{6}t$

34. $2t^2 - 1 = \frac{7}{3}t$

35. $3(s^2 + 1) = 10s$

36. $3 + p(p + 2) = 18$

37. $(p - 2)(p + 1) = 3$

38. $(s - 8)(s + 7) = 5$

39. $\frac{7}{3}y^2 - \frac{8}{3} = y$

40. $y = \frac{1}{3} - \frac{1}{9}y^2$

41. $3x^2 - 7 = 0$

42. $7x^2 + 5x = 0$

43. $x(x - 2) = 7$

44. $(3x + 1)^2 = 12$

Cumulative Review Problems

45. Simplify. $\frac{3x}{x - 2} + \frac{4}{x + 2} - \frac{x + 22}{x^2 - 4}$

46. Give equations in **(a)** slope-intercept form and **(b)** standard form with integer coefficients for the line through $(-8, 2)$ and $(5, -3)$.

47. Divide.
$(x^3 + 8x^2 + 17x + 6) \div (x + 3)$

48. Solve for g.
$2p = 3gf - 7p + 20$

10.4 Graphing Quadratic Equations

After studying this section, you will be able to:

1 *Graph equations of the form $y = ax^2 + bx + c$ by graphing ordered pairs of numbers.*

2 *Graph equations of the form $y = ax^2 + bx + c$ by using the vertex.*

1 *Graphing $y = ax^2 + bx + c$ Using Ordered Pairs*

Recall that the graph of a linear equation is always a straight line. What is the graph of a quadratic equation? What happens to the graph if one of the variable terms is squared? We will use point plotting to look for a pattern. We begin with a table of values.

EXAMPLE 1 Graph. **(a)** $y = x^2$ **(b)** $y = -x^2$

(a) $y = x^2$

x	y
−3	9
−2	4
−1	1
0	0
1	1
2	4
3	9

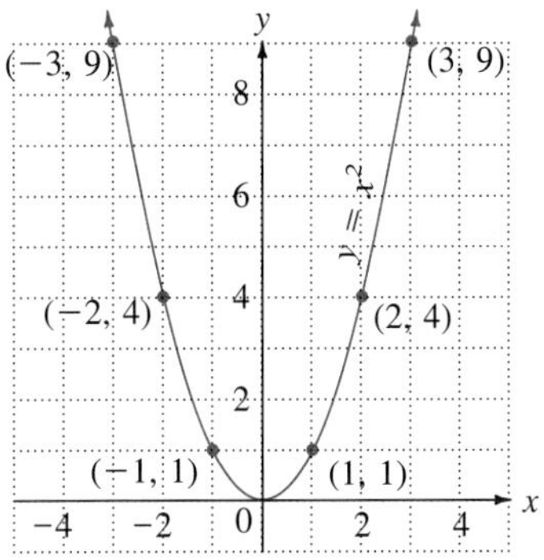

(b) $y = -x^2$

x	y
−3	−9
−2	−4
−1	−1
0	0
1	−1
2	−4
3	−9

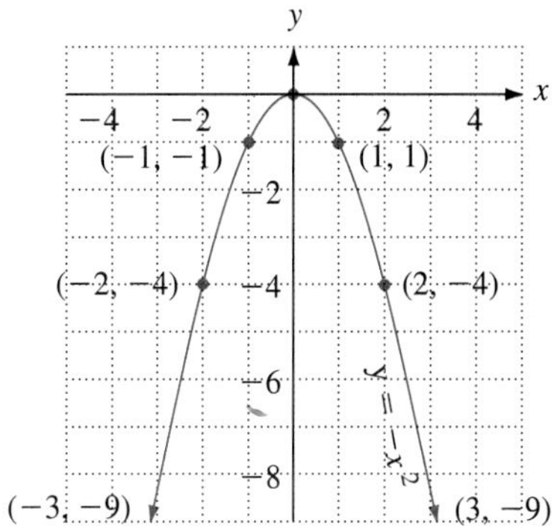

Practice Problem 1 Graph. **(a)** $y = 2x^2$ **(b)** $y = -2x^2$

$y = 2x^2$

x	y
−2	
−1	
0	
1	
2	

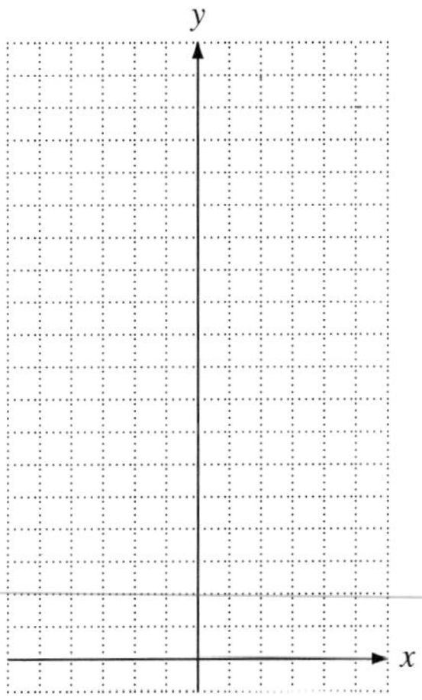

$y = -2x^2$

x	y
−2	
−1	
0	
1	
2	

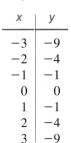

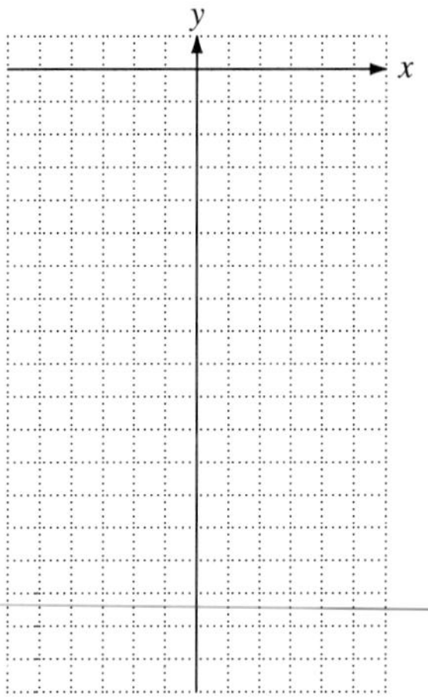

■

To Think About What happens to the graph when the coefficient of x^2 is negative? What happens when the coefficient of x^2 is an integer greater than 1 or less than −1? Graph $y = 3x^2$ and $y = -3x^2$. What might happen if $-1 < a < 1$. Graph $y = \frac{1}{3}x^2$ and $y = -\frac{1}{3}x^2$.

The curves that you have been graphing are called *parabolas.* What happens to the graph of a parabola as the equation changes?

EXAMPLE 2 Graph. **(a)** $y = x^2 - 3$ **(b)** $y = x^2 + 2$

(a) $y = x^2 - 3$

x	y
-3	6
-2	1
-1	-2
0	-3
1	-2
2	1
3	6

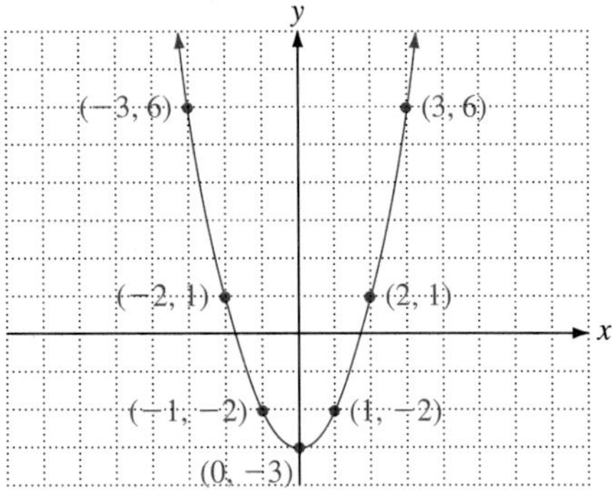

(b) $y = x^2 + 2$

x	y
-3	11
-2	6
-1	3
0	2
1	3
2	6
3	11

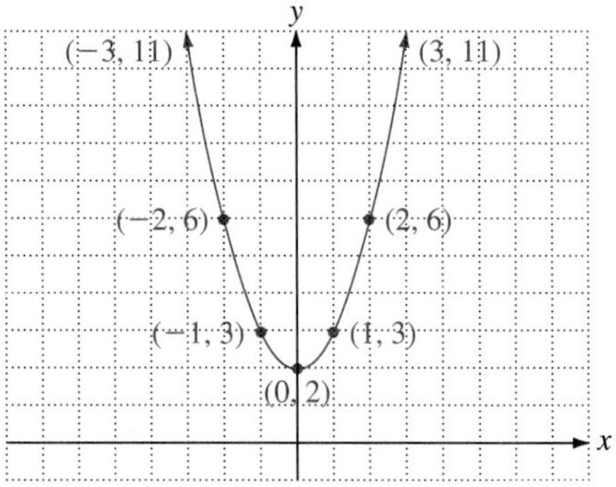

Practice Problem 2 Graph. **(a)** $y = (x - 1)^2$ **(b)** $y = (x + 2)^2$

$y = (x - 1)^2$

x	y
-2	
-1	
0	
1	
2	
3	
4	

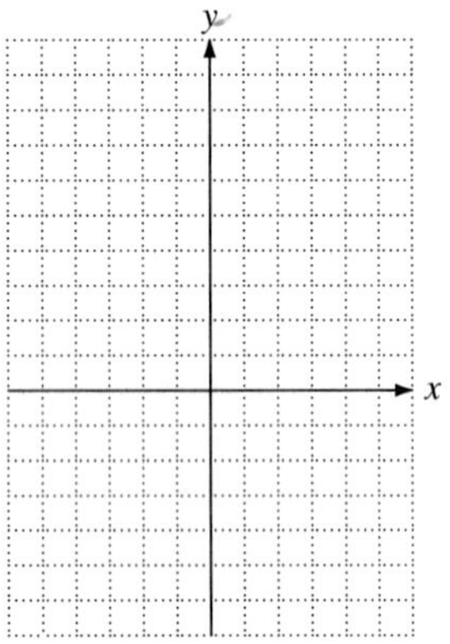

$y = (x + 2)^2$

x	y
-5	
-4	
-3	
-2	
-1	
0	
1	

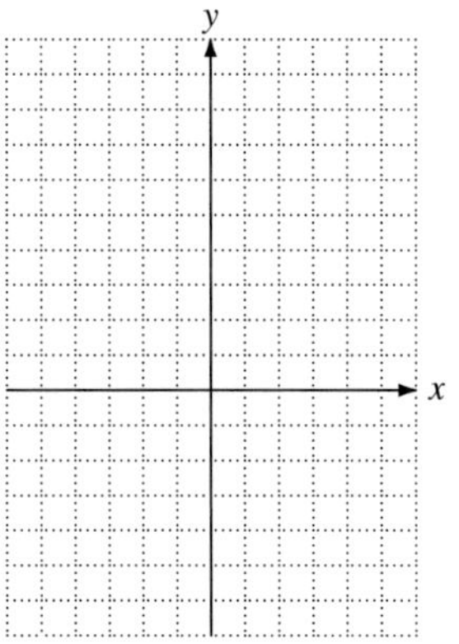

■

To Think About Look at the graphs for Example 2, describe how the constant c in the equation $y = ax^2 + c$ affects the graph of the parabola. Look at the graphs for Practice Problem 2. Describe how the constant c in the equation $y = (x + c)^2$ affects the graph of the parabola.

In the above examples the equations we worked with were relatively simple. They provided us with an understanding of the general shape of the graph of a quadratic equation. We have seen how changing the equation affects the graph of that equation. Let's look at a slightly more difficult equation to graph.

We introduce a new word, vertex. The **vertex** is the lowest point on a parabola that opens upward. The **vertex** is the highest point on a parabola that opens downward.

EXAMPLE 3 Graph $y = x^2 - 2x$. Identify the coordinates of the vertex.

x	y	
-2	8	
-1	3	
0	0	
1	-1	vertex
2	0	
3	3	
4	8	

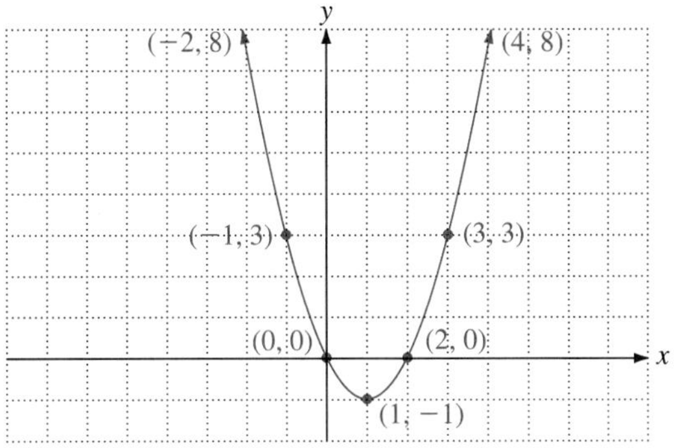

The vertex is at $(1, -1)$.

Notice the x-intercepts are at $(0, 0)$ and $(2, 0)$. The x-coordinates are the solutions to the equation.

$$x^2 - 2x = 0$$

$$x(x - 2) = 0$$

$$x = 0, \qquad x = 2$$

Practice Problem 3 Graph. **(a)** $y = x^2 + 2x$ **(b)** $y = -x^2 - 2x$ For each table of values, use x from -4 to 2. Identify the coordinates of the vertex.

$y = x^2 + 2x$

x	y
-4	
-3	
-2	
-1	
0	
1	
2	

$y = -x^2 - 2x$

x	y
-4	
-3	
-2	
-1	
0	
1	
2	

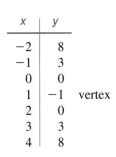

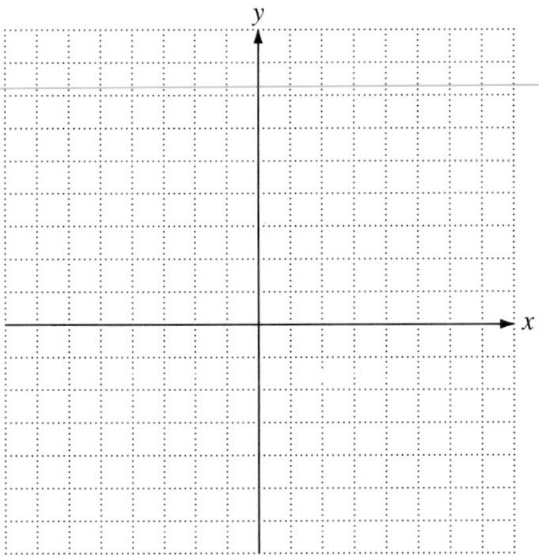

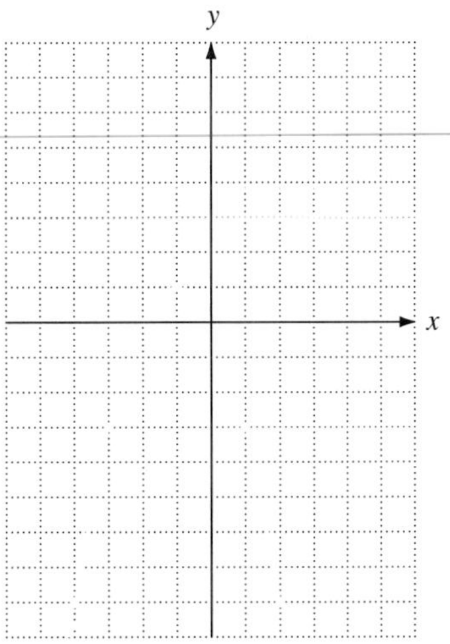

2 *Graphing Equations of the Form $y = ax^2 + bx + c$ by Using the Vertex*

Since the general shape of the parabola will be the same if the coefficient of the squared term is 1, the vertex is the key to graphing parabolas. Let's take a look at the previous graphs and identify the vertex of each parabola.

Equation:	$y = x^2$	$y = x^2 - 3$	$y = (x - 1)^2$ $y = x^2 - 2x + 2$	$y = (x + 2)^2$ $y = x^2 + 4x + 4$	$y = x^2 - 2x$
Vertex:	$(0, 0)$	$(0, -3)$	$(1, 0)$	$(-2, 0)$	$(1, -1)$

Notice that the x-coordinate of the vertex of the parabola for the first two equations is 0. You may also notice that these two equations do not have a middle term. That is, for $y = ax^2 + bx + c$, $b = 0$. This is not true of the last three equations. As you may have already guessed, there is a relationship between the equation and the x-coordinate of the vertex of the parabola.

Vertex of a Parabola

The x-coordinate of the vertex is

$$x = \frac{-b}{2a}$$

where $y = ax^2 + bx + c$.

Once we have the x-coordinate, we can find the y-coordinate using the equation. We can plot the point and draw a general sketch of the curve. The sign of the squared term will tell us if the graph opens upward or downward.

EXAMPLE 4 $y = -x^2 - 2x + 3$. Determine the vertex and the x-intercepts. Then sketch the graph.

We will first determine the coordinates of the vertex. We begin by finding the x-coordinate.

$$x = \frac{-b}{2a} = \frac{-(-2)}{2(-1)} = \frac{2}{-2} = -1$$

$y = -(-1)^2 - 2(-1) + 3$ *Determine y when x is -1.*

$= 4$ *The point $(-1, 4)$ is the vertex of the parabola.*

Now we will determine the x-intercepts, those points where the graph crosses the x-axis. This occurs when $y = 0$. These are the solutions to the equation. Note that this equation is factorable.

$y = -x^2 - 2x + 3$ *Find the solutions to the equation.*

$0 = (-x + 1)(x + 3)$ *Factor.*

$-x + 1 = 0 \quad x + 3 = 0$ *Solve for x.*

$x = 1 \quad x = -3$ *The x-intercepts are at $(1, 0)$ and at $(-3, 0)$.*

We now have enough information to sketch the graph. Since $a = -1$, the parabola opens downward and the vertex $(-1, 4)$ is the highest point. The graph crosses the x-axis at $(-3, 0)$ and at $(1, 0)$.

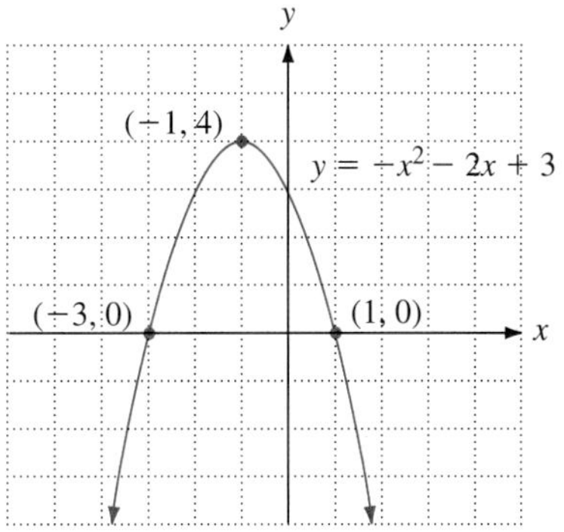

Practice Problem 4 $y = x^2 - 4x - 5$. Determine the vertex and the x-intercepts. Then sketch the graph.

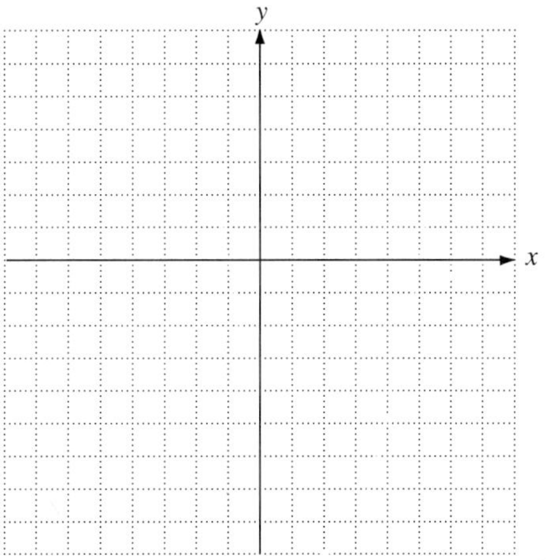

■

EXAMPLE 5 The height of a ball in feet is given by the equation $h = -16t^2 + 64t + 5$, where t is measured in seconds.

(a) Graph the equation $h = -16t^2 + 64t + 5$.

(b) What physical significance does the vertex of the graph have?

(c) After how many seconds will the ball hit the ground?

(a) Since the equation is a quadratic equation, the graph is a parabola. a is negative. Thus the parabola opens downward and the vertex is the highest point. We begin by finding the coordinates of the vertex. Our equation has ordered pairs (t, h) instead of (x, y). The t-coordinate of the vertex is

$$t = \frac{-b}{2a} = \frac{-64}{2(-16)} = \frac{-64}{-32} = 2$$

Substitute $t = 2$ into the equation to find the h-coordinate of the vertex.

$$h = -16(2)^2 + 64(2) + 5 = -64 + 128 + 5 = 69$$

The vertex (t, h) is $(2, 69)$.

To sketch this graph, it would be helpful to know the h-intercept, that is, the point where $t = 0$.

$$h = -16(0)^2 + 64(0) + 5$$
$$= 5$$

The h-intercept is at $(0, 5)$.

The t-intercept is where $h = 0$. We will use the quadratic formula to find t when h is 0, that is, to find the roots of the quadratic equation.

$$0 = -16t^2 + 64t + 5$$
$$t = \frac{-64 \pm \sqrt{(64)^2 - 4(-16)(5)}}{2(-16)}$$
$$t = -\frac{64 \pm \sqrt{4096 + 320}}{-32} = -\frac{64 \pm \sqrt{4416}}{-32}$$

Using a calculator or a square root table, we have

$$t = -\frac{64 - 66.453}{-32} = 4.077, \qquad t \approx 4.1 \quad \text{(to the nearest tenth)}$$
$$t = -\frac{64 + 66.453}{-32} = -0.077$$

We will not use negative t values. We are studying the height of the ball from the time of throwing ($t = 0$ seconds) to the time that it hits the ground (approximately $t = 4.1$ seconds). In this situation it is not useful to find points for negative values of t because they have no meaning in this situation.

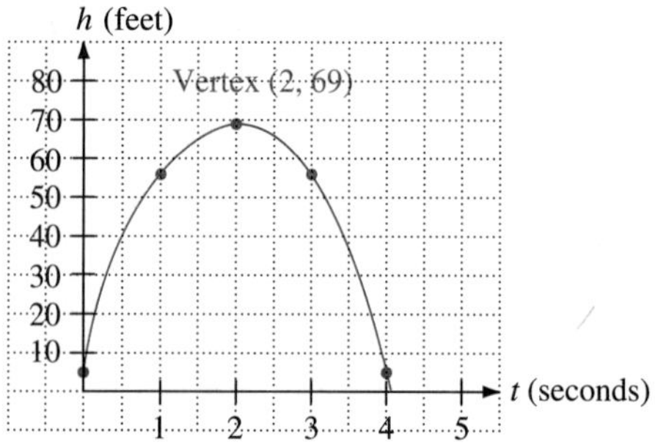

(b) The vertex represents the greatest height of the ball. The ball is 69 feet above the ground 2 seconds after it is thrown.

(c) The ball will hit the ground at approximately 4.1 seconds after it is thrown. This is the point at which the graph intersects the t-axis, the positive root of the equation.

Practice Problem 5 A ball is thrown upward with a speed of 32 feet per second from a distance 10 feet above the ground. The height of the ball in feet is given by the equation $h = -16t^2 + 32t + 10$, where t is measured in seconds.

(a) Graph the equation $h = -16t^2 + 32t + 10$.

(b) What is the greatest height of the ball?

(c) After how many seconds will the ball hit the ground? ■

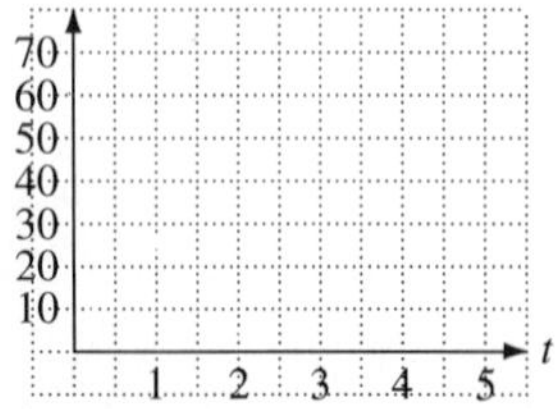

(b) 26

(c) 2.1 seconds

10.4 Exercises

Graph each equation.

1. $y = 2x^2 - 1$

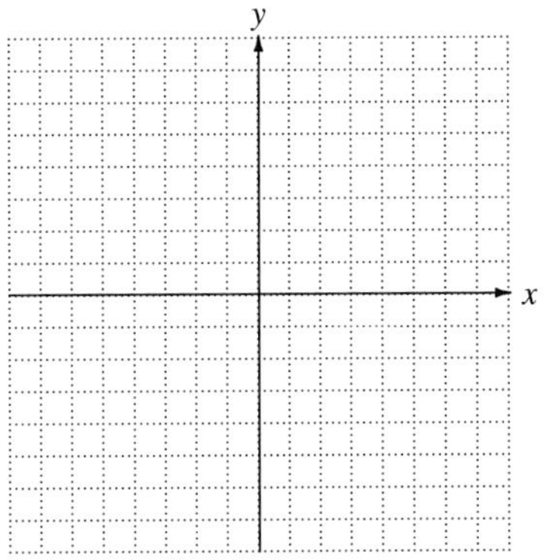

2. $y = x^2 + 2$

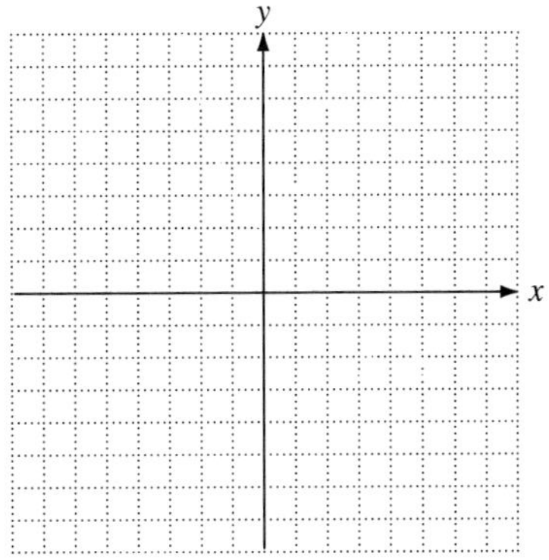

3. $y = -\frac{1}{3}x^2$

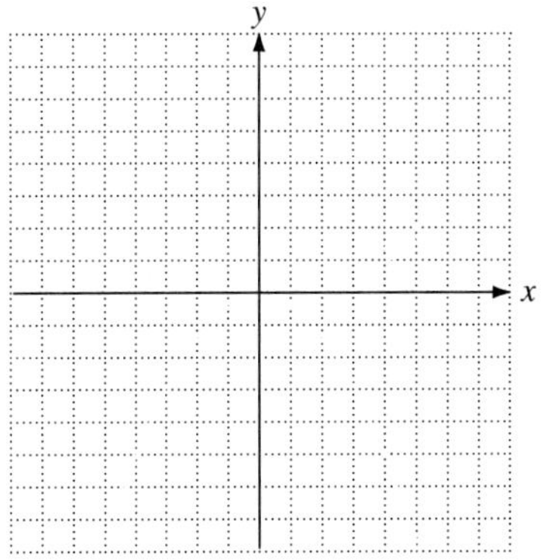

4. $y = \frac{1}{2}x^2$

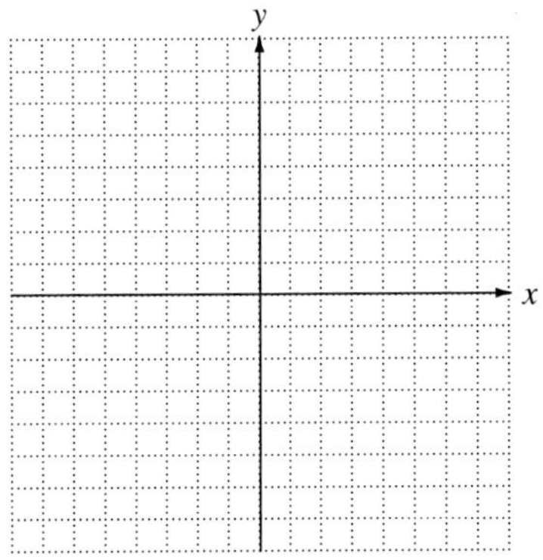

5. $y = x^2 - 2$

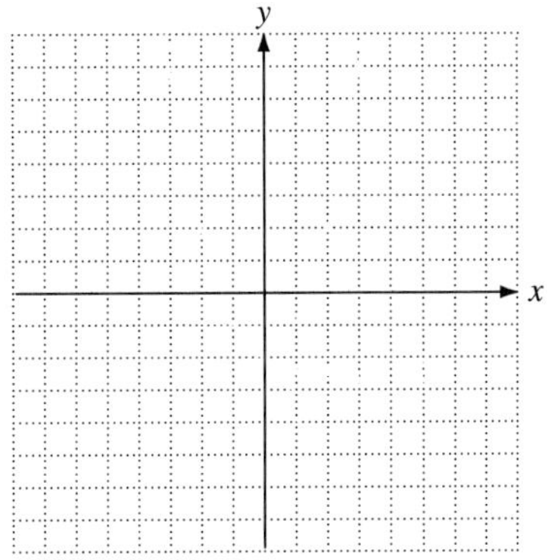

6. $y = x^2 + 3$

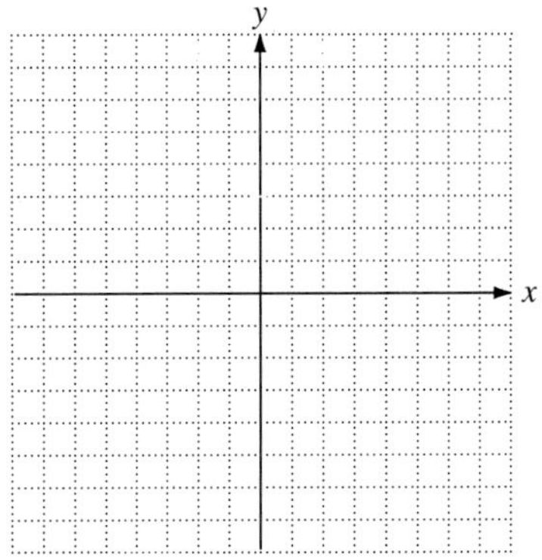

7. $y = (x - 2)^2$

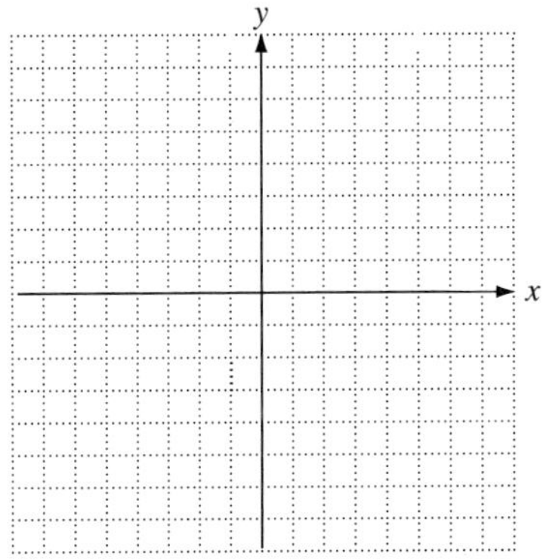

8. $y = (x + 1)^2$

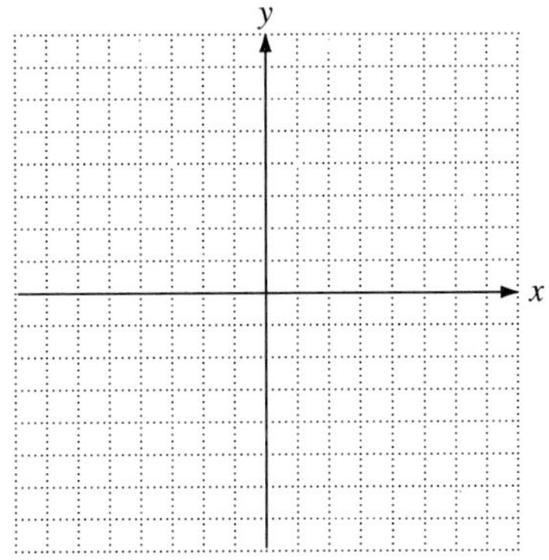

9. $y = -\frac{1}{2}x^2 + 4$

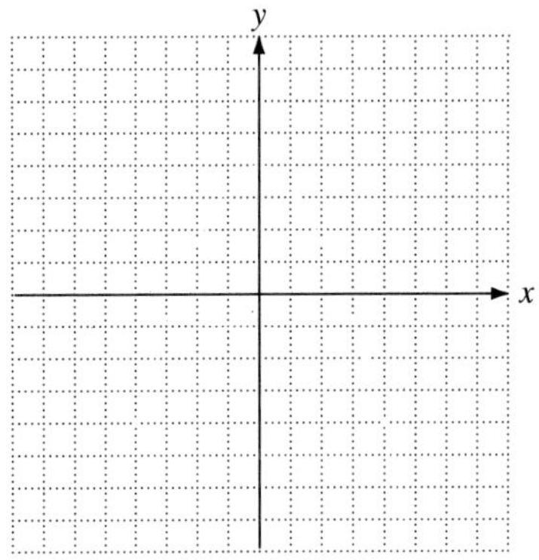

10. $y = \frac{1}{3}x^2 - 3$

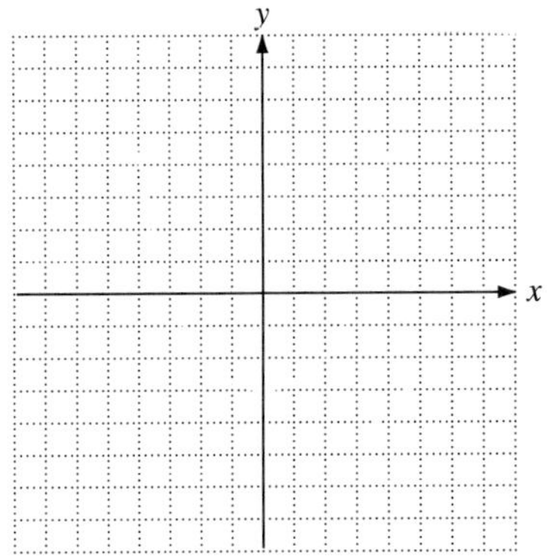

11. $y = \frac{1}{2}(x - 3)^2$

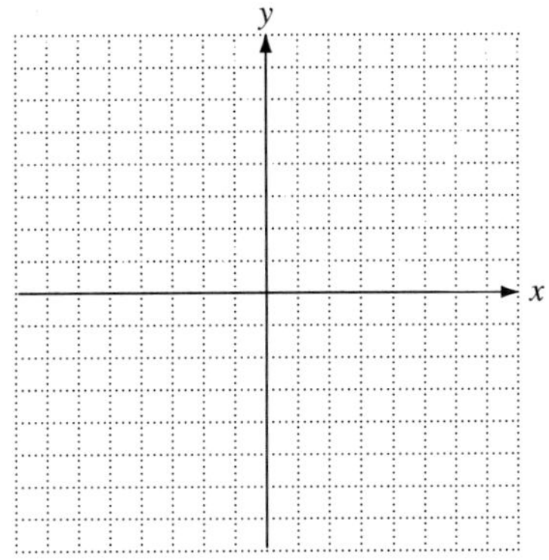

12. $y = \frac{1}{2}(x + 2)^2$

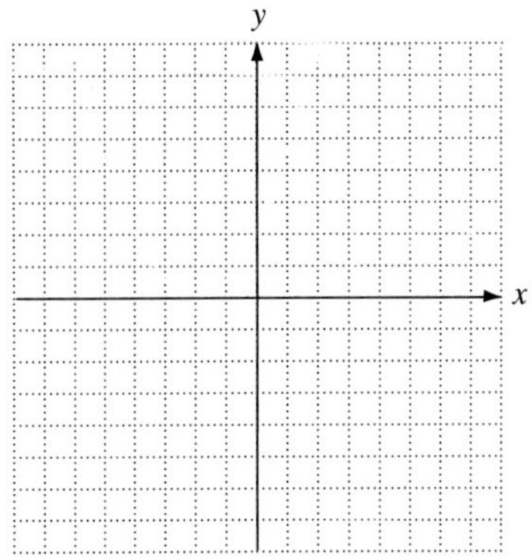

Graph each equation. Identify the coordinates of the vertex.

13. $y = x^2 + 4x$

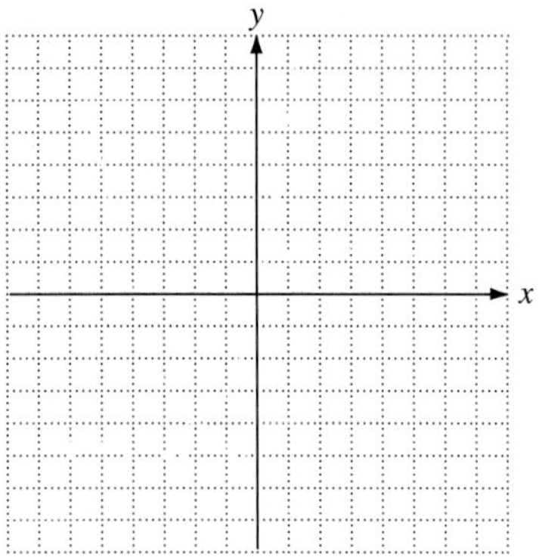

14. $y = -x^2 + 4x$

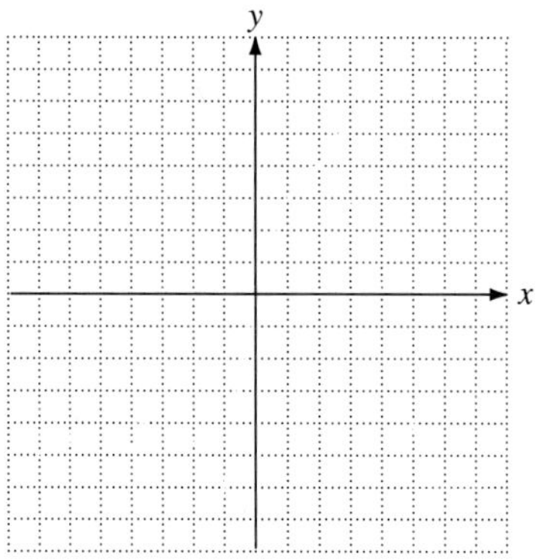

15. $y = x^2 - 4x$

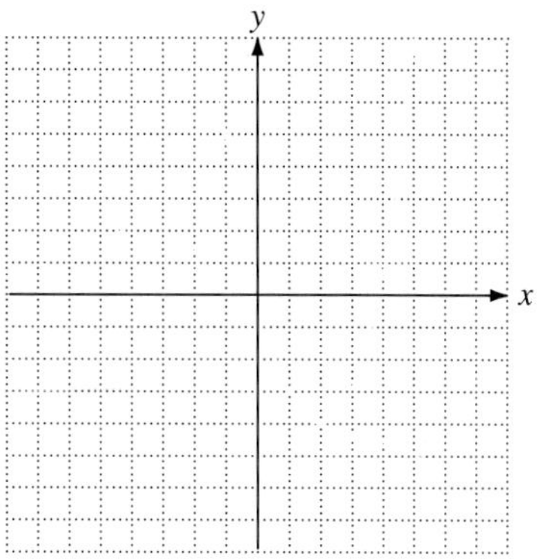

16. $y = -x^2 - 4x$

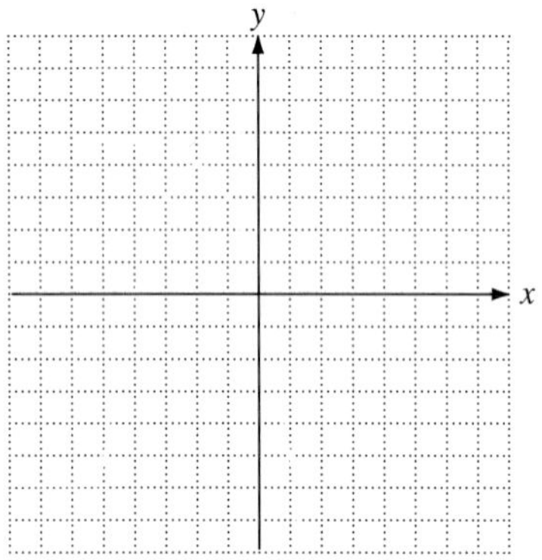

17. $y = -x^2 + 6x$

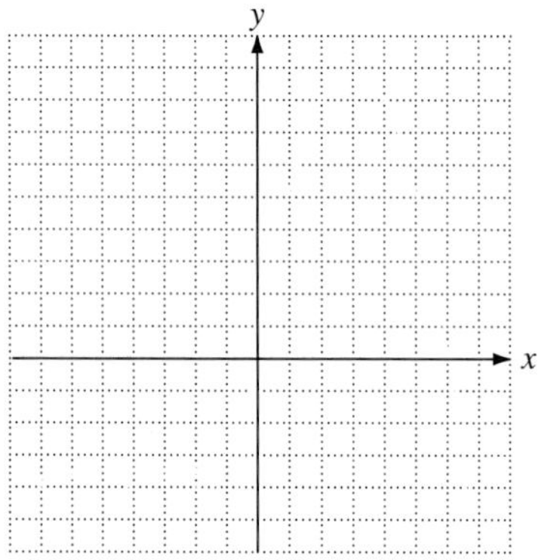

18. $y = x^2 + 6x$

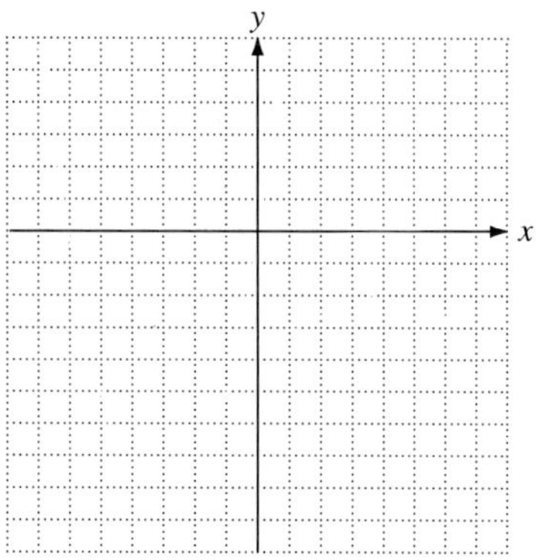

Determine the vertex and the x-intercepts. Then sketch the graph.

19. $y = x^2 - 4x + 3$

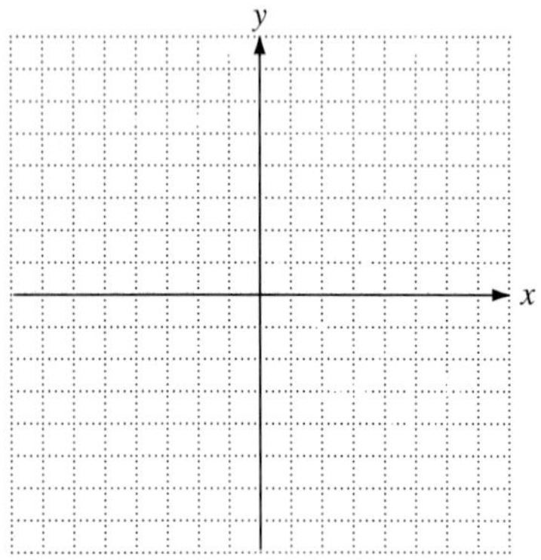

20. $y = x^2 + 2x - 3$

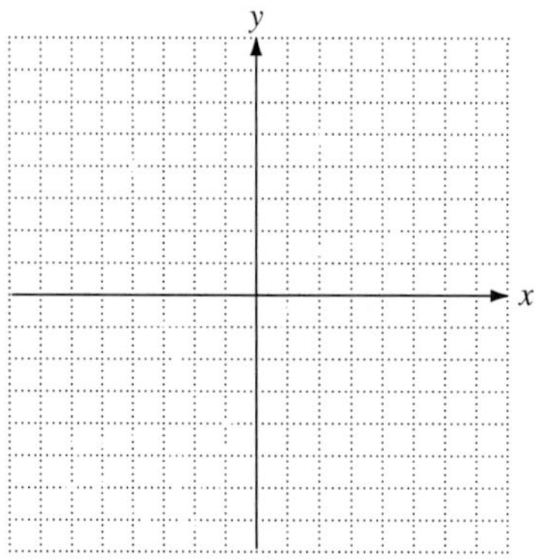

21. $y = -x^2 - 6x - 5$

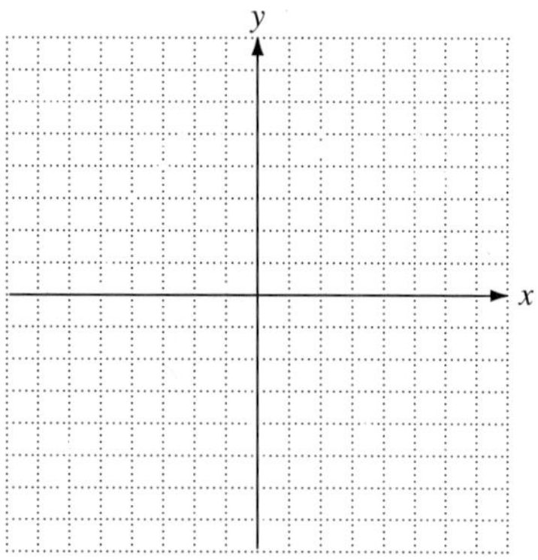

22. $y = -x^2 + 2x + 3$

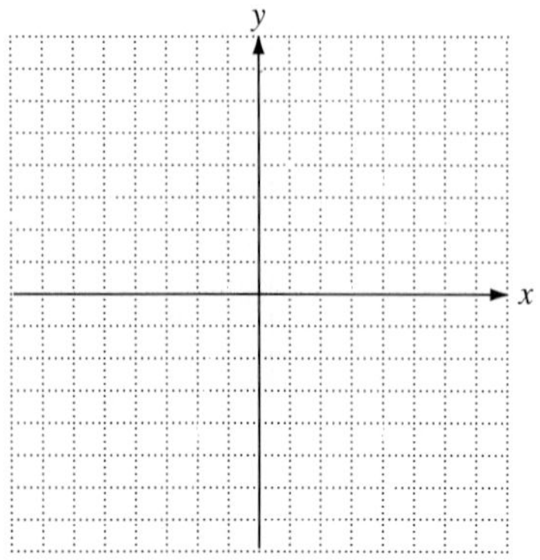

23. $y = -x^2 + 4x - 4$

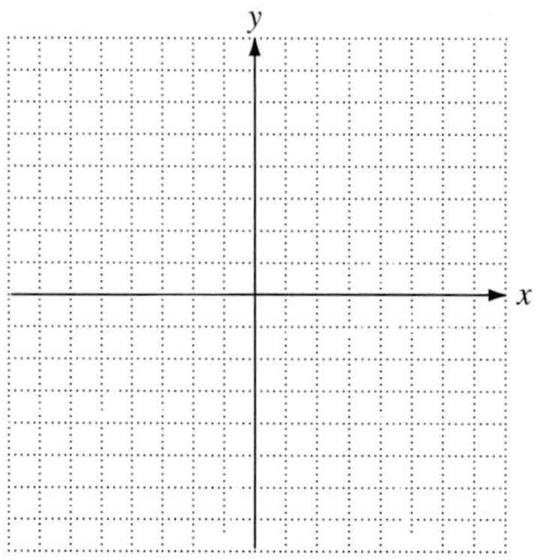

24. $y = x^2 - 6x + 8$

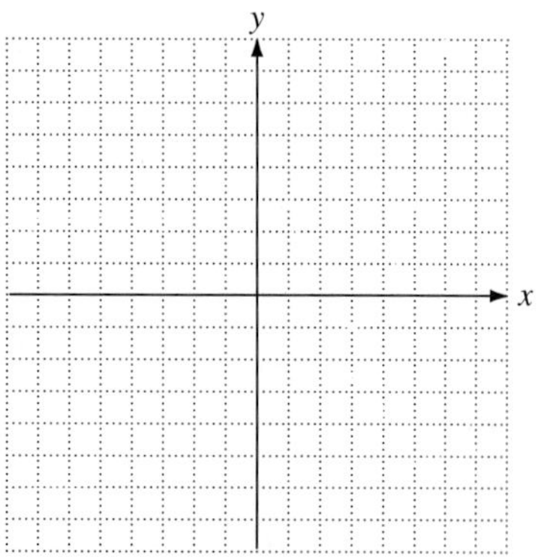

Applications

25. A miniature rocket is launched so that its height h in meters after t seconds is given by the equation $h = -4.9t^2 + 19.6t + 10$.
(a) Draw a graph of the equation.
(b) How high is the rocket after 3 seconds?
(c) What is the highest the rocket will travel?
(d) How long after the launch will it take for the rocket to strike the earth?

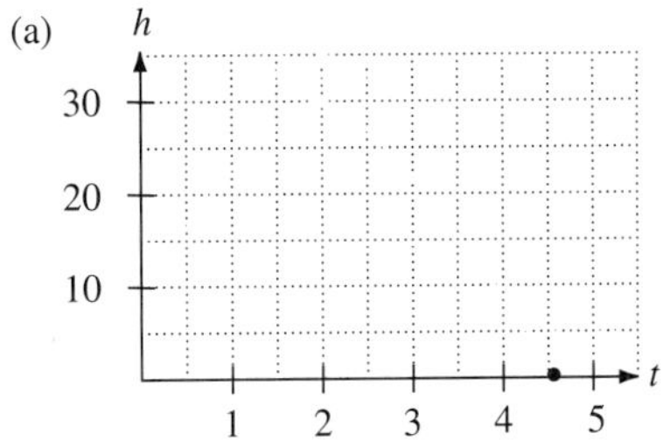

26. A miniature rocket is launched so that its height h in meters after t seconds is given by the equation $h = -4.9t^2 + 39.2t + 4$.
(a) Draw a graph of the equation.
(b) How high is the rocket after 2 seconds?
(c) What is the highest the rocket will travel?
(d) How long after the launch will it take for the rocket to strike the earth?

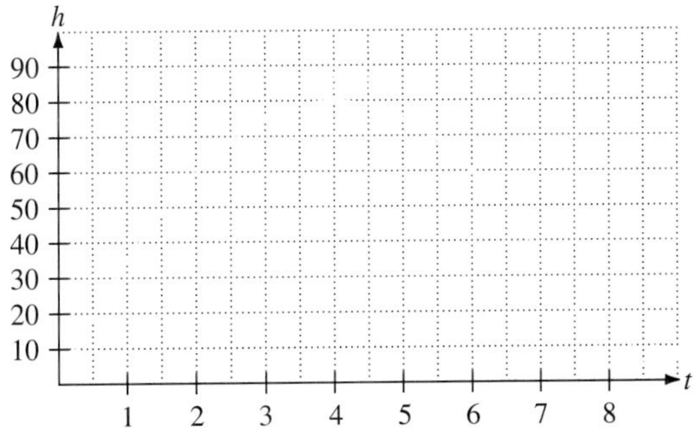

To Think About

Find the exact values of the vertex, the y-intercept, and the x-intercepts.

27. $y = -2x^2 + 4x + 5$

28. $y = -2x^2 - 4x + 1$

Cumulative Review Problems

29. Solve and check. $\sqrt{2x + 3} + 3 = 7$

30. Solve and check. $x - 3 = \sqrt{2x - 3}$

31. If y varies inversely as x and $y = \frac{1}{3}$ when $x = 21$, find y when $x = 7$.

32. If y varies directly as x^2, and $y = 12$ when $x = 2$, find y when $x = 5$.

Putting Your Skills to Work

HELPING THE FOOTBALL COACH PLAN THE DEFENSE

A football is thrown a distance of 40 yards by a quarterback. The ball is thrown and caught 5 feet above ground. The football is 15 feet above ground at the peak of the throw. The ball follows a path that is part of a parabola. The vertex of the parabola is the highest point the ball reaches. The height h of the football in feet at a distance of x yards from the quarterback is given by the equation

$$h = -\frac{1}{40}(x - 20)^2 + 15$$

A graph of the equation is shown below.

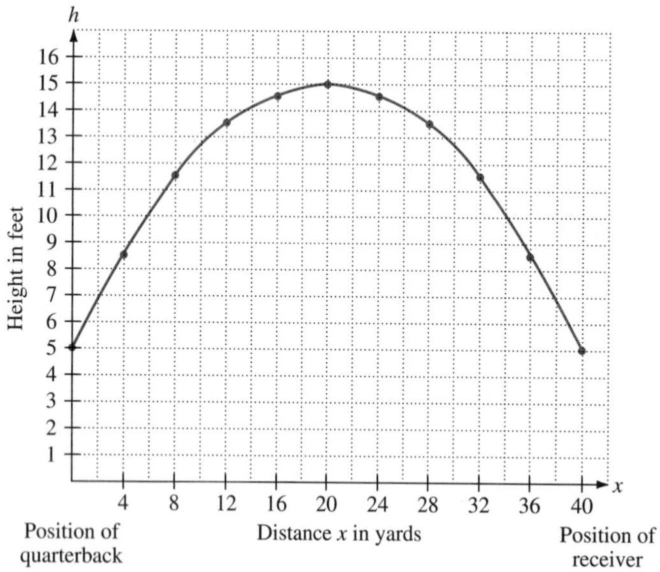

Approximate the answers to the following questions based on the graph.

1. How high is the football above ground when it is 8 yards from the quarterback?

2. How high is the football above ground when it is 32 yards from the quarterback?

3. If a member of the opposing team is rushing toward the quarterback and can leap up a distance of 8 feet to intercept or deflect the ball, how close does he have to be to impede the throw of the ball?

4. If a member of the opposing team is covering the receiver and he can leap up a distance of 9 feet to intercept or deflect the ball, how close does he need to be to the receiver to keep the receiver from catching the ball?

GROUP DISCUSSION AND ACTIVITY

Suppose that the equation $h = -\frac{1}{35}(x - 15)^2 + 12$ describes the most commonly used throw by an opposing quarterback. Graph the equation from $x = 0$ to $x = 30$.

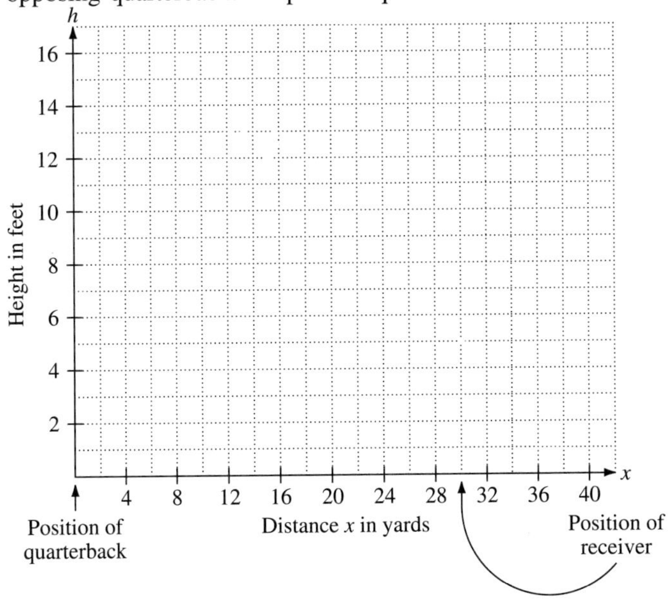

Answer the following questions based on your graph.

5. What is the maximum height of the football when it is thrown?

6. How far will the football travel to the receiver if the receiver catches it at a height of 5 feet above ground?

7. If a person can leap 7 feet up to intercept or deflect the ball, how close does he have to be to the quarterback to impede the throw of the ball?

8. If a person can leap 9 feet up to intercept or deflect the ball, how close does he have to be to the receiver to keep the receiver from catching the ball?

Note: Using videocameras and computer technology, it is possible to find an equation for the path of a football when thrown by a quarterback.

10.5 Formulas and Applied Problems

After studying this section, you will be able to:

1 *Solve applied problems requiring the use of a quadratic equation.*

1 *Solving Applied Problems Using a Quadratic Equation*

We can use what we have learned about quadratic equations to solve word problems. Some word problems will involve geometry. Recall that the Pythagorean theorem states that in a right triangle $a^2 + b^2 = c^2$, where a and b are legs and c is the hypotenuse. We will use this theorem to solve word problems involving right triangles.

EXAMPLE 1 The hypotenuse of a right triangle is 25 meters in length. One leg is 17 meters longer than the other. Find the length of each leg.

1. *Understand the problem.*
   Draw a picture.
2. *Write an equation.*
   The problem involves a right triangle.
   Use the Pythagorean theorem.

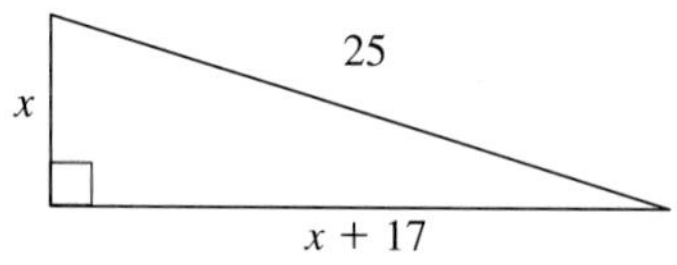

$$a^2 + b^2 = c^2$$

$$x^2 + (x + 17)^2 = 25^2$$

3. *Solve the equation and state the answer.*

$$x^2 + (x + 17)^2 = 25^2$$
$$x^2 + x^2 + 34x + 289 = 625$$
$$2x^2 + 34x - 336 = 0$$
$$x^2 + 17x - 168 = 0$$
$$(x + 24)(x - 7) = 0$$
$$x + 24 = 0 \qquad x - 7 = 0$$
$$x = -24 \qquad x = 7$$

(Note that $x = -24$ is not a valid solution.)
One leg is 7 meters in length. The other leg is $x + 17$ or $7 + 17 = 24$ meters in length.

4. *Check.*
   Check the conditions in the original problem. Do the two legs differ by 17 meters?

$$24 - 7 \stackrel{?}{=} 17 \qquad 17 = 17 \checkmark$$

Do the sides of the triangle satisfy the Pythagorean theorem?

$$7^2 + 24^2 \stackrel{?}{=} 25^2 \qquad 49 + 576 \stackrel{?}{=} 625 \qquad 625 = 625 \checkmark$$

Practice Problem 1 The hypotenuse of a right triangle is 30 meters in length. One leg is 6 meters shorter than the other leg. Find the length of each leg. ■

Sometimes it may help to use two variables in the initial part of the problem. Then one variable can be eliminated by substitution.

EXAMPLE 2 The Ski Club is renting a bus to travel to Mount Snow. The members agreed to share the cost of $180 equally. On the day of the ski trip three members were sick with the flu and could not go. This raised by $10 the share of each person going on the trip. How many people originally planned to attend?

1. *Understand the problem.*
Let s = the number of students in the ski club
c = the cost for each student in the original group

2. *Write an equation(s).*

number of students × cost per student = total cost

$$s \cdot c = 180 \qquad \textbf{(1)}$$

Now if 3 people are sick, the number of students drops by 3, but the cost for each increases by $10. The total is still $180. Therefore,

$$(s - 3)(c + 10) = 180 \qquad \textbf{(2)}$$

3. *Solve the equation(s) and state the answer.*

$$c = \frac{180}{s} \qquad \textit{Solve each } \textbf{(1)} \textit{ for } c.$$

$$(s - 3)\left(\frac{180}{s} + 10\right) = 180 \qquad \textit{Substitute } \tfrac{180}{s} \textit{ for } c \textit{ in equation } \textbf{(2)}.$$

$$(s - 3)\left(\frac{180 + 10s}{s}\right) = 180 \qquad \textit{Add } \tfrac{180}{s} + 10. \textit{ Use the LCD} = s.$$

$$(s - 3)(180 + 10s) = 180s \qquad \textit{Multiply both sides of the equation by } s.$$

$$10s^2 + 150s - 540 = 180s \qquad \textit{Use FOIL to multiply the binomials.}$$

$$10s^2 - 30s - 540 = 0$$

$$s^2 - 3s - 54 = 0$$

$$(s - 9)(s + 6) = 0$$

$$s = 9 \qquad s = -6$$

The number of students originally in the ski club was 9.

4. *Check.*
Was the cost increased by $10 if the number dropped from 9 students to 6 students? 9 people in the club would mean that each would pay $20 (180 ÷ 9 = 20). If the number is reduced by 3, there were 6 people who took the trip. Their cost was $30 (180 ÷ 6 = 30). The increase is $10.

$$20 + 10 \stackrel{?}{=} 30 \qquad 30 = 30 \checkmark$$

Practice Problem 2 Several friends from the University of New Mexico decided to charter a sport fishing boat for the afternoon while on a vacation on the coast. They agreed to share the rental cost of $240 equally. At the last minute, two of them could not go on the trip. This raised the cost by $4 for each person in the group. How many people originally planned to charter the boat? ■

EXAMPLE 3 Minette is fencing a garden that borders the back of a large barn. She has 120 feet of fencing. She would like a rectangular garden that measures 1350 square feet in area. She wants to use the back of the barn, so she only needs to use fencing on three sides. What dimensions should she use for her garden?

1. *Understand the problem.*
Draw a picture.

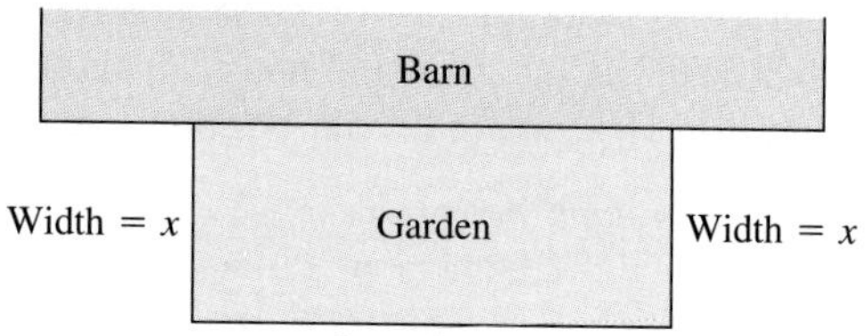

Let x = the width of the garden in feet

y = the length of the garden in feet

2. *Write an equation(s).*
Look at the drawing. The 120 feet of fencing will be needed for the width twice (two sides) and the length once (one side).

$$120 = 2x + y \qquad \textbf{(1)}$$

The area formula is A = (width)(length).

$$1350 = x(y)$$

3. *Solve the equation(s) and state the answer.*

$$120 = 2x + y \qquad \textbf{(1)}$$

$$y = 120 - 2x \qquad \textbf{(2)} \qquad \textit{Solve for } y \textit{ in equation } \textbf{(1)}.$$

$$1350 = x(120 - 2x) \qquad \textit{Substitute for } y \textit{ in equation } \textbf{(2)}.$$

$$1350 = 120x - 2x^2 \qquad \textit{Solve for } x.$$

$$2x^2 - 120x + 1350 = 0$$

$$x^2 - 60x + 675 = 0$$

$$(x - 15)(x - 45) = 0$$

$$x = 15 \qquad x = 45$$

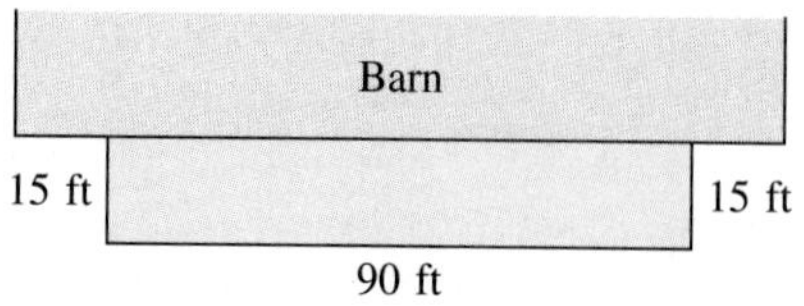

First solution

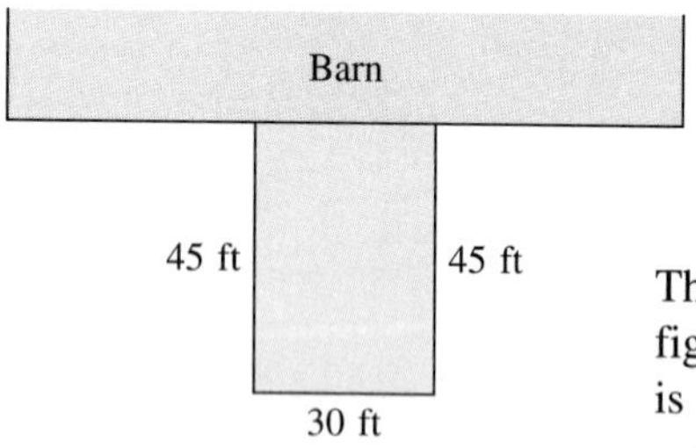

Second solution

First solution: If the width is 15 feet, the length is 90 feet.

Check $15 + 15 + 90 = 120$ ✓

$(15)(90) = 1350$ ✓

Second solution: If the width is 45 feet, the length is 30 feet.

Check $45 + 45 + 30 = 120$ ✓

$(45)(30) = 1350$ ✓

Thus we see that Minette has two choices that satisfy the given requirements (see the figure). She would probably choose the shape that is the most practical for her. (If her barn is not 90 feet long, she could not use this first solution!)

Remark: Isn't there another way to do this problem? Yes. Use only the variable x to represent the length and width. If you let x = the width and $120 - 2x$ = the length, you can immediately write equation (3), which is

$$1350 = x(120 - 2x) \qquad \textbf{(3)}$$

However, many students find that this is difficult to do, and they would prefer to use two variables. Try to work these problems in a way that seems fairly easy to you.

Practice Problem 3 An Arizona rancher has a small fenced-in area against a canyon wall. Therefore, he only needs three sides of fencing to make a rectangular holding area for cattle. He has available 160 feet of fencing. He wants to make the rectangular area with dimensions that provide him with an area of 3150 square feet. What dimensions should he use for the rectangular-shaped holding area? ■

10.5 Exercises

Applications

Solve each word problem.

1. The hypotenuse of a right triangle is 40 yards long. One leg of the triangle is 8 yards longer than the other leg. Find the length of each leg of the triangle.

2. The hypotenuse of a right triangle is 15 yards long. One leg of the triangle is 3 yards shorter than the other leg. Find the length of each leg of the triangle.

3. The diagonal of a rectangle is 13 meters. The width of the rectangle is 7 meters shorter than its length. Find the length and the width of the rectangle.

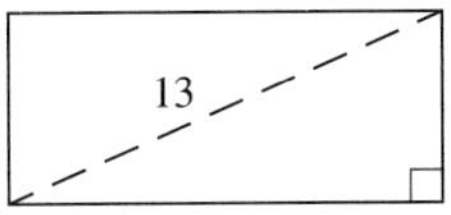

4. Manuel placed a ladder 20 feet long against the house. The distance from the top of the ladder to the bottom of the house is 4 feet greater than the distance from the bottom of the house to the foot of the ladder. How far is the foot of the ladder from the house? How far up on the house does the top of the ladder touch the building?

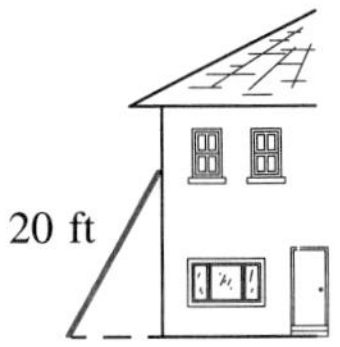

5. Several students at the college gave Professor Harmeling a Silver Service that cost \$420 at his retirement party. The students agreed to share the cost equally. At the last minute, seven former students joined the party and asked to share in the gift. This lowered the cost per student by \$5. How many students originally planned to give Professor Harmeling the gift?

6. All the families on Forest Street agreed to equally share in the cost of a new \$576 street light. However, before the light was installed, three new houses were built and the three new families volunteered to contribute their share. This lowered the contribution of each family by \$16. How many families originally agreed to share the cost of the new street light?

7. The dean of students needed to raise \$18,000 in student activity fees this semester. The fees were assigned equally to each full-time student. Due to an increase of 90 full-time students, the cost of the student activity fee went down \$10 for each student. How many full-time students were originally planned for at the college? After the increase, how many actually came to the college?

8. The members of the varsity football team decided to give the coach a gift that cost \$80. On the day that they collected money, 10 people were absent. They were in a hurry, so they collected an equal amount from each member and bought the gift. Later the team captain said that, if everyone had been present to chip in, each person would have contributed \$4 less. How many people actually contributed? How many people in total were on the football team?

9. The botany class at Golden West College planted an experimental garden at the west edge of the central classroom building. They planted a rectangular area, but only needed three sides of fencing since the building served as the fourth side of the rectangle. They had 140 feet of fencing available. The garden was 2000 square feet in area. What were the possible dimensions for the garden?

10. A security fence encloses a rectangular area on one side of Fenway Park in Boston. Three sides of fencing are used, since the fourth side of the area is formed by a building. The enclosed area measures 1250 square feet. Exactly 100 feet of fencing is used to fence in three sides of this rectangle. What are the possible dimensions that could have been used to construct this area?

★ 11. A pilot is testing a new experimental craft. The cruising speed is classified information. In a test the jet traveled 2400 miles. The pilot revealed that if he had increased his speed 200 mph the total trip would have taken 1 hour less. Can you determine the cruising speed of the jet?

★ 12. A gymnasium floor is being covered by square shock-absorbing tiles. The old gym floor required 864 square tiles. The new tiles are 2 inches larger in both length and width than the old tiles. The new flooring will require only 600 tiles. What is the length of a side of one of the new shock-absorbing tiles?
Hint: Since the area is the same in each case, write an expression for area with old tiles and one with new tiles. Set the two expressions equal to each other.

To Think About

The sum of all the counting numbers from 1 to 6 is $1 + 2 + 3 + 4 + 5 + 6 = 21$. A shortcut to the answer is $6(7)(\frac{1}{2}) = 21$. In general the sum of the numbers from 1 up to n, which we can write as $1 + 2 + \cdots + n$, can be computed by the formula

$$\frac{n(n+1)}{2} = s \qquad \text{where } s \text{ is the sum}$$

13. If the sum of the first n counting numbers is 91, find n.

14. If the sum of the first n counting numbers is 136, find n.

The Springfield Boot Company has determined that the daily profit of the company can be predicted by the equation $P = -2x^2 + 360x - 14{,}400$, where x is the number of pairs of boots produced each day.

15. What is the daily profit if 85 pairs of boots are produced each day?

16. (a) How many pairs of boots each day need to be produced to obtain the maximum daily profit?

(b) What is the maximum daily profit?

Cumulative Review Problems

17. Graph on a number line the solution to $3x + 11 \geq 9x - 4$.

18. Rationalize the denominator. $\dfrac{\sqrt{3}}{\sqrt{5} + 2}$

Chapter Organizer

Topic	Procedure	Examples
Solve quadratic equations by factoring, p. 560.	1. Clear the equation of fractions, if necessary. 2. Write as $ax^2 + bx + c = 0$. 3. Factor. 4. Set each factor equal to 0. 5. Solve the resulting equations.	Solve. $x + \frac{1}{2} = \frac{3}{2x}$ Multiply each term by LCD = $2x$. $2x(x) + 2x\left(\frac{1}{2}\right) = 2x\left(\frac{3}{2x}\right)$ $2x^2 + x = 3$ $2x^2 + x - 3 = 0$ $(2x + 3)(x - 1) = 0$ $2x + 3 = 0 \qquad x - 1 = 0$ $2x = -3 \qquad x = 1$ $x = -\frac{3}{2}$
Solving quadratic equations: taking the square root of each side of the equation, p. 569.	A quadratic equation of the form $ax^2 - c = 0$ can be solved by solving for x^2. Then use the property that if $x^2 = a$ then $x = \pm\sqrt{a}$. This amounts to taking the square root of each side of the equation.	$2x^2 + 1 = 99$ $2x^2 = 99 - 1$ $2x^2 = 98$ $\frac{2x^2}{2} = \frac{98}{2}$ $x^2 = 49$ $x = \pm\sqrt{49} = \pm 7$
Solving quadratic equations: completing the square, p. 570.	We can write a quadratic equation in the form $(ax + b)^2 = c$ by completing the square. 1. Write the equation as $ax^2 + bx = -c$. 2. Divide by a. 3. Take half the x coefficient, square it, and add this to both sides. 4. Factor the perfect square on the left. 5. Take the square root of both sides. 6. Solve for x.	Solve. $5x^2 - 3x - 7 = 0$ $5x^2 - 3x = 7$ $x^2 - \frac{3}{5}x = \frac{7}{5}$ $\left[\frac{1}{2} \cdot \left(-\frac{3}{5}\right)\right]^2 = \frac{9}{100}$ $x^2 - \frac{3}{5}x + \frac{9}{100} = \frac{7}{5} + \frac{9}{100}$ $\left(x - \frac{3}{10}\right)^2 = \frac{149}{100}$ $x - \frac{3}{10} = \pm\frac{\sqrt{149}}{10}$ $x = \frac{3}{10} \pm \frac{\sqrt{149}}{10}$ $x = \frac{3 \pm \sqrt{149}}{10}$
Solving quadratic equations: quadratic formula, p. 575.	1. Put the equation in standard form: $ax^2 + bx + c = 0$. 2. Carefully determine the values of a, b, and c. 3. Substitute these into the formula: $x = \frac{-b \pm \sqrt{b^2 - 4ac}}{2a}$ 4. Simplify.	Solve. $3x^2 - 4x - 8 = 0$ $a = 3, b = -4, c = -8$ $x = \frac{-(-4) \pm \sqrt{(-4)^2 - 4(3)(-8)}}{2(3)}$ $= \frac{4 \pm \sqrt{16 + 96}}{6}$ $= \frac{4 \pm \sqrt{112}}{6} = \frac{4 \pm \sqrt{16 \cdot 7}}{6}$ $= \frac{4 \pm 4\sqrt{7}}{6} = \frac{\cancel{2}(2 \pm 2\sqrt{7})}{\cancel{2} \cdot 3} = \frac{2 \pm 2\sqrt{7}}{3}$
The discriminant, p. 578.	The *discriminant* is $b^2 - 4ac$. If negative, there are no real solutions (since it is not possible to take the square root of a negative number and get a real number).	What kinds of solutions does this equation have? $3x^2 + 2x + 1 = 0$ $b^2 - 4ac = 2^2 - 4 \cdot 3 \cdot 1$ $= 4 - 12 = -8$ This equation has no real solution.

Chapter Organizer

Topic	Procedure	Examples
Properties of parabolas, p. 581.	**1.** The graph of $ax^2 + bx + c = y$ is a parabola. It opens up if $a > 0$ and down if $a < 0$. **2.** The vertex of a parabola is its lowest ($a > 0$) or highest ($a < 0$) point. Its x-coordinate is $x = \frac{-b}{2a}$.	$y = 3x^2 + 4x - 11$ has vertex at $x = \frac{-b}{2a} = \frac{-4}{6} = -\frac{2}{3}$ The parabola opens upward.
Graphing parabolas, p. 581.	Graph the vertex of a parabola and several other points to get a good idea of its graph. Usually, we find the y-intercept and the x-intercepts if any exist. The following procedure is helpful. **1.** Determine if $a < 0$ or $a > 0$. **2.** Find the y-intercept. **3.** Find the x-intercepts. **4.** Find the vertex. **5.** Plot one or two extra points if necessary. **6.** Draw the graph.	Graph. $y = x^2 - 2x - 8$ $a = 1, b = -2, c = -8$; $a > 0$, so the parabola opens upward. Let $x = 0$. $y = 0^2 - 2(0) - 8 = -8$ The y-intercept is $(0, -8)$. $0 = x^2 - 2x - 8$ $0 = (x - 4)(x + 2)$ $x = 4, \quad x = -2$ The x-intercepts are $(4, 0)$ and $(-2, 0)$. The vertex is at $x = \frac{-b}{2a} = -\frac{-2}{2} = 1$. If $x = 1$, $y = (1)^2 - 2(1) - 8 = -9$. The vertex is at $(1, -9)$. 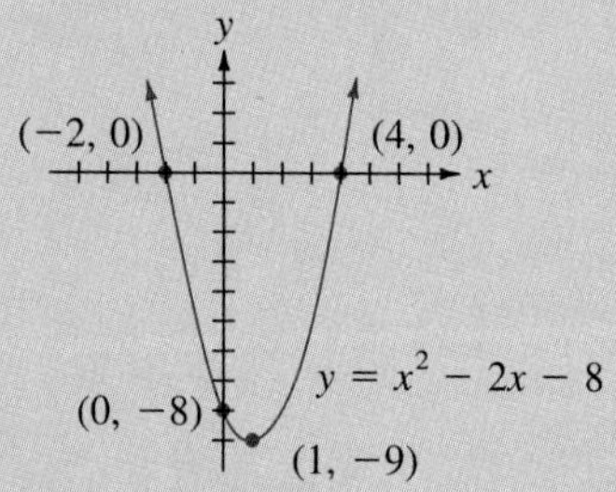
Formulas and applied problems, p. 592.	Some word problems require the use of a quadratic equation. You may want to follow this four-step procedure for problem solving. **1.** Understand the problem. **2.** Write an equation. **3.** Solve the equation and state the answer. **4.** Check.	The area of a rectangle is 48 square inches. The length is three times the width. Find the dimensions of the rectangle. **1.** Draw a picture. 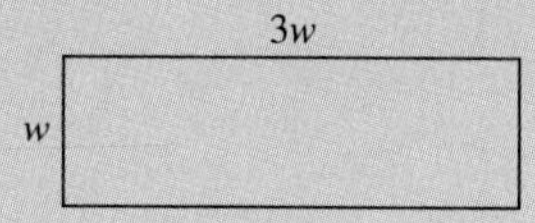 **2.** $(3w)w = 48$ **3.** $3w^2 = 48$ $w^2 = 16$ $w = 4, l = 12$ **4.** $(4)(12) = 48$

Putting Your Skills to Work

THE GOLDEN RATIO

Which of the following rectangles is most pleasing to you?

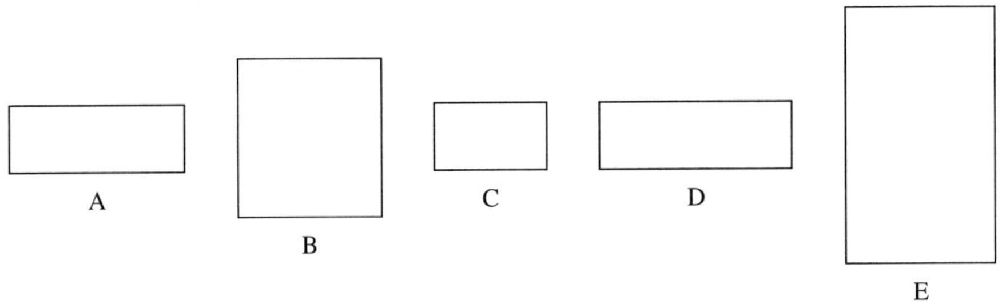

Although it would seem that everyone would have their own opinion and each person would tend to pick different choices, such is not the case. The overwhelming majority of people tend to pick choice C. This was true with the ancient Greeks and Renaissance painters, and recently scholars have shown that twentieth-century Americans tend to prefer choice C. The proportion of these sides is called the *golden ratio.*

If the shortest side of the rectangle is 1, what value should x have to be the most pleasing to the majority of people?

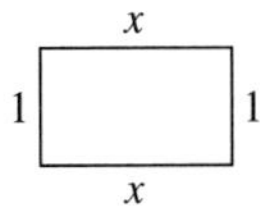

It turns out to be a number x such that the number is equal to its reciprocal added to 1. In other words x has the following value.

$$x = \frac{1}{x} + 1$$

What is the value of x?

$$x = \frac{1}{x} + 1$$

$$x(x) = x\left(\frac{1}{x}\right) + x \qquad \textit{Multiply each side by } x.$$

$$x^2 = 1 + x \qquad \textit{Simplify.}$$

$$x^2 - x - 1 = 0 \qquad \textit{Place in standard form.}$$

$$a = 1, \quad b = -1, \quad c = -1$$

$$x = \frac{-(-1) \pm \sqrt{(-1)^2 - 4(1)(-1)}}{2(1)}$$

$$x = \frac{1 \pm \sqrt{1 + 4}}{2} = \frac{1 \pm \sqrt{5}}{2}$$

Answer each question. Round your answer to the nearest thousandth.

1. A picture frame is 26 inches wide. Approximately how tall should it be if the frame has the golden ratio?

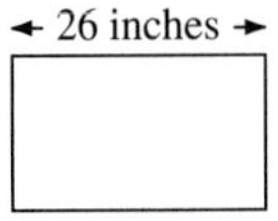

2. A window is 3.5 feet tall. Approximately how wide should it be if the window has the golden ratio?

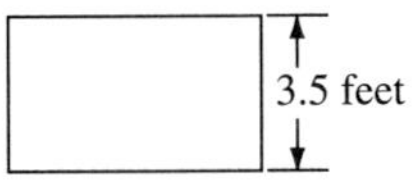

3. A color television screen is rectangular. It measures 21 inches diagonally. How tall and how wide should it be if the television screen has the golden ratio?

Chapter 10 Review Problems

10.1 *Write the quadratic equation in standard form. Determine the value of a, b, and c.* **Do not solve.**

1. $5x^2 = -2x + 3$

2. $7x^2 + 6x = -8x^2 - 4$

3. $(x - 4)(2x - 1) = x^2 + 7$

4. $x(3x + 1) = -5x + 10$

5. $\frac{1}{x^2} - \frac{3}{x} + 5 = 0$

6. $\frac{x}{x + 2} - \frac{3}{x - 2} = 5$

7. $x^2 + 7x + 8 = 4x^2 + x + 8$

8. $12x^2 - 9x + 1 = 15 - 9x$

Place each equation in standard form. Solve by factoring.

9. $x^2 + 21x + 20 = 0$

10. $x^2 + 10x + 25 = 0$

11. $x^2 + 6x - 27 = 0$

12. $4x^2 - 12x + 9 = 0$

13. $6x^2 + 23x = -15$

14. $4x^2 = 5x - 1$

15. $12x^2 = 4 - 13x$

16. $9x^2 = -27x - 8$

17. $x^2 + \frac{5}{6}x - 1 = 0$

18. $x^2 + \frac{7}{6}x - \frac{1}{2} = 0$

19. $x^2 + \frac{2}{15}x = \frac{1}{15}$

20. $1 + \frac{13}{12x} - \frac{1}{3x^2} = 0$

21. $5x^2 + 16x = 0$

22. $-2x^2 + 9x = 0$

23. $x^2 + 12x - 3 = 8x - 3$

24. $6x^2 - 13x + 2 = 5x + 2$

25. $1 + \frac{2}{3x + 4} = \frac{3}{3x + 2}$

26. $2 - \frac{5}{x + 1} = \frac{3}{x - 1}$

27. $5 + \frac{24}{2 - x} = \frac{24}{2 + x}$

28. $1 + \frac{14}{x^2} = \frac{9}{x}$

29. $3 = \frac{10}{x} + \frac{8}{x^2}$

30. $\frac{4}{9} = \frac{5x}{3} - x^2$

10.2 *Solve by taking the square root of each side.*

31. $x^2 = 100$

32. $x^2 = 121$

33. $x^2 + 3 = 28$

34. $x^2 - 4 = 32$

35. $x^2 - 5 = 17$

36. $x^2 + 11 = 50$

37. $2x^2 - 1 = 15$

38. $3x^2 + 1 = 61$

39. $3x^2 + 6 = 60$

40. $2x^2 - 5 = 43$

41. $(x - 4)^2 = 7$

42. $(x - 2)^2 = 3$

43. $(5x + 6)^2 = 20$

44. $(3x + 8)^2 = 12$

Solve by completing the square.

45. $x^2 + 6x + 5 = 0$

46. $x^2 + 8x + 15 = 0$

47. $x^2 - 4x + 2 = 0$

48. $x^2 + 2x - 6 = 0$

49. $2x^2 - 8x - 90 = 0$

50. $3x^2 - 36x - 60 = 0$

51. $3x^2 + 6x - 6 = 0$

52. $2x^2 + 10x - 3 = 0$

10.3 *Solve by the quadratic formula.*

53. $x^2 + 3x - 40 = 0$

54. $x^2 - 9x + 18 = 0$

55. $x^2 + 4x - 6 = 0$

56. $x^2 + 4x - 8 = 0$

57. $2x^2 - 7x + 4 = 0$

58. $2x^2 + 5x - 6 = 0$

59. $2x^2 + 4x - 5 = 0$

60. $2x^2 + 6x - 5 = 0$

61. $3x^2 - 5x = 4$

62. $4x^2 + 3x = 2$

63. $4x^2 = 1 - 4x$

64. $3x^2 - 8x = -2$

Solve by any method. If there is no real number solution, so state.

65. $2x^2 - 9x + 10 = 0$

66. $4x^2 - 4x - 3 = 0$

67. $9x^2 - 6x + 1 = 0$

68. $2x^2 - 11x + 12 = 0$

69. $3x^2 - 6x + 2 = 0$

70. $5x^2 - 7x = 8$

71. $4x^2 + 4x = x^2 + 5$

72. $5x^2 + 7x + 1 = 0$

73. $x^2 + 5x = -1 + 2x$

74. $3x^2 = 6 - 7x$

75. $2y^2 = -3y - 1$

76. $4y^2 = 5y - 1$

77. $3x^2 + 1 = 73 + x^2$

78. $2x^2 - 1 = 35$

79. $\dfrac{(y - 2)^2}{20} + 3 + y = 0$

80. $\dfrac{(y + 2)^2}{5} + 2y = -9$

81. $3x^2 + 1 = 6 - 8x$

82. $2x^2 + 10x = 2x - 7$

83. $2y - 10 = 10y(y - 2)$

84. $\dfrac{3y - 2}{4} = \dfrac{y^2 - 2}{y}$

85. $\dfrac{y^2 + 5}{2y} = \dfrac{2y - 1}{3}$

86. $\dfrac{5x^2}{2} = x - \dfrac{7x^2}{2}$

10.4 *Graph each quadratic equation. Label the vertex.*

87. $y = 2x^2$

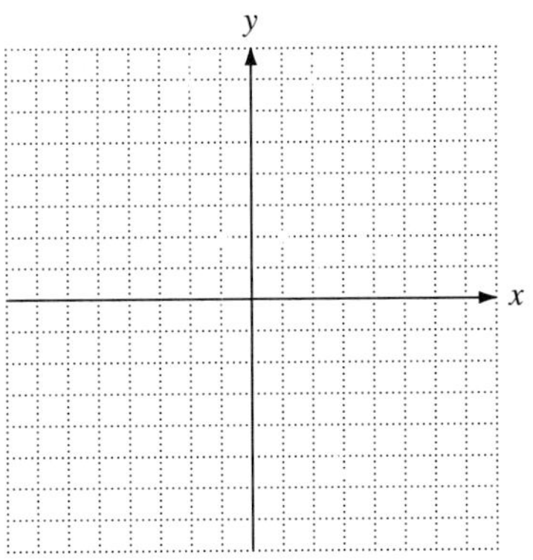

88. $y = x^2 + 4$

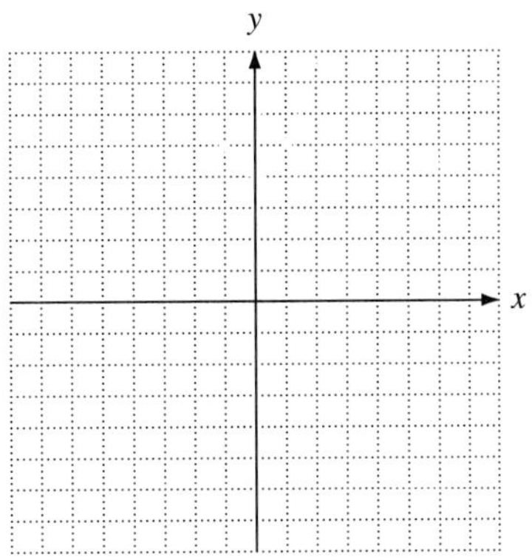

89. $y = x^2 - 3$

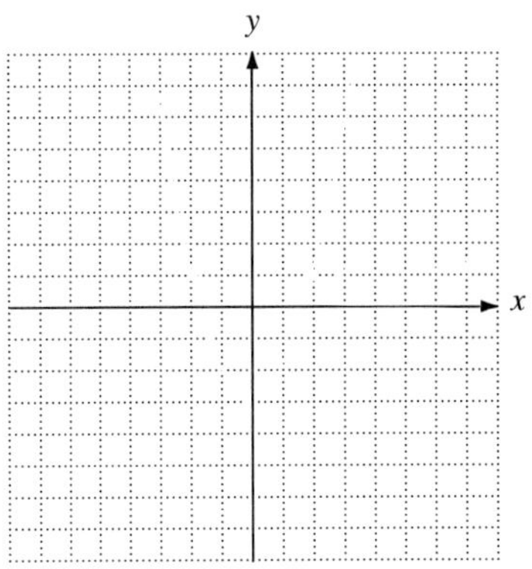

90. $y = -\frac{1}{2}x^2$

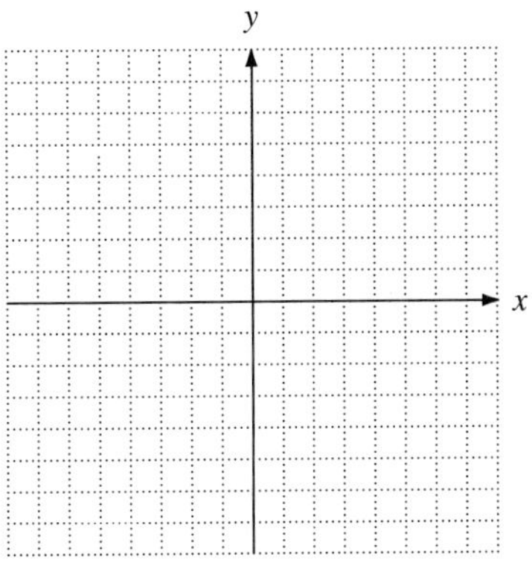

91. $y = x^2 - 3x - 4$

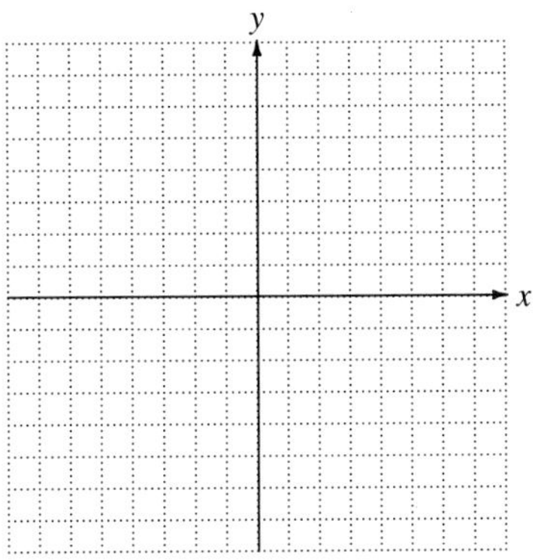

92. $y = \frac{1}{2}x^2 - 2$

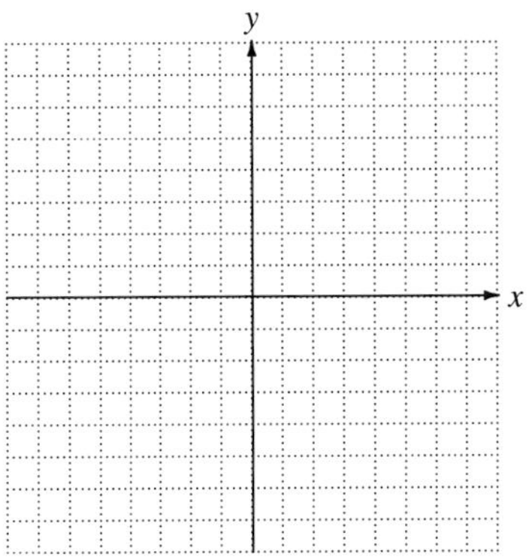

93. $y = -2x^2 + 12x - 17$

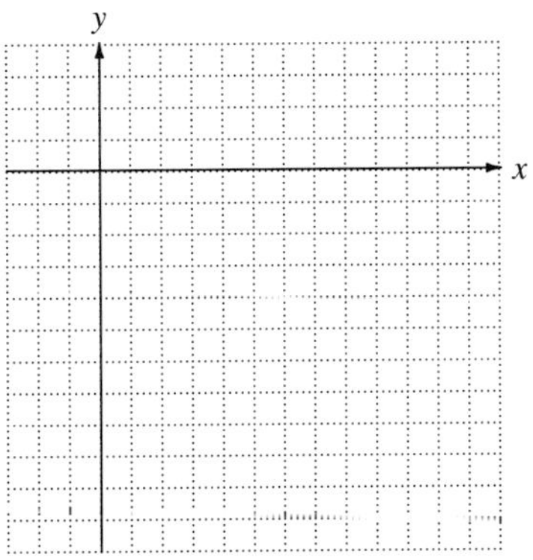

94. $y = -3x^2 - 2x + 4$

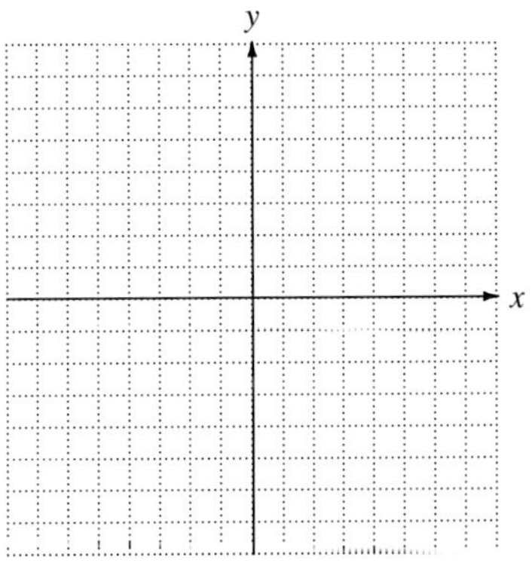

10.5 *Solve each word problem.*

95. Jon is building a dog pen against the house. He has 11 feet of fencing and will pen in an area of 15 square feet. What are the dimensions of the pen?

96. The hypotenuse of a right triangle is 20 centimeters. One leg of the right triangle is 4 centimeters less than the other. What is the length of each leg of the right triangle?

97. Last year the Golf Club on Campus raised \$720 through dues assigned equally to each member. This year there were four less members. As a result, the dues went up by \$6 for each member. How many members were in the Golf Club last year?

98. One positive number is 2 more than twice another positive number. The difference of the squares of these numbers is 119. Find the numbers that satisfy these conditions.

99. The length of a room exceeds its width by 4 feet. A rug covers the floor area except for a border 2 feet wide all around it. The area of the border is 68 square feet. What is the area of the rug?

★**100.** Alice drove 90 miles to visit her cousin. Her average speed on the trip home was 15 mph faster than her speed on the trip going. Her total travel time for both going and returning was 3.5 hours. What was her average rate of speed on each trip?

Chapter 10 Test

Solve by any desired method. If there is no real number solution, say so.

1. $5x^2 + 7x = 4$

2. $3x^2 + 13x = 10$

3. $2x^2 = 2x - 5$

4. $6x^2 - 7x - 5 = 0$

5. $12x^2 + 11x = 5$

6. $18x^2 + 32 = 48x$

7. $2x^2 - 11x + 3 = 5x + 3$

8. $2x^2 + 5 = 37$

9. $2x(x - 6) = 6 - x$

10. $x^2 - x = \frac{3}{4}$

1. ______

2. ______

3. ______

4. ______

5. ______

6. ______

7. ______

8. ______

9. ______

10. ______

11.

Graph each quadratic equation. Locate the vertex.

11. $y = 3x^2 - 6x$

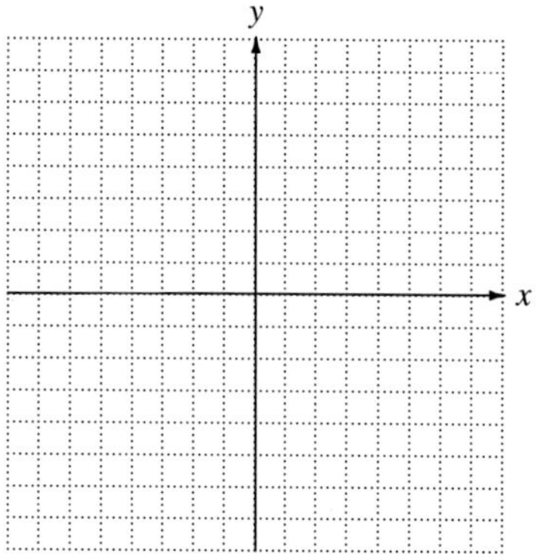

12. $y = -x^2 + 8x - 12$

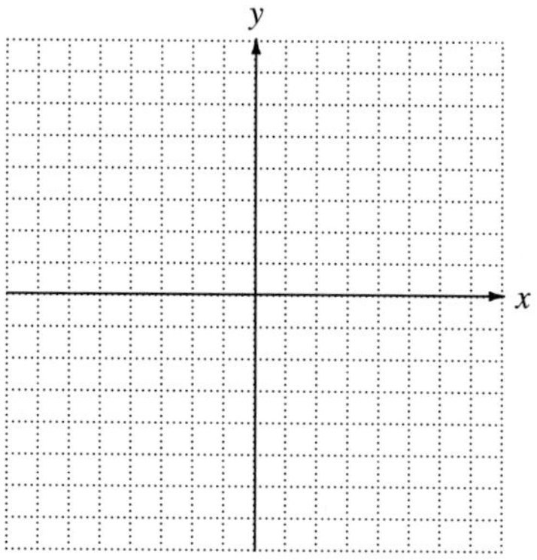

12.

Solve each word problem.

13. The hypotenuse of a right triangle is 15 meters in length. One leg is 3 meters longer than the other leg. Find the length of each leg.

13.

14. When an object is thrown upward, its height (S) in meters is given (approximately) by the quadratic equation

$$S = -5t^2 + vt + h$$

where v = the upward initial velocity in meters per second

t = the time of flight in seconds

h = the height above level ground from which the object is thrown

Suppose that a ball is thrown upward with a velocity of 33 meters/second at a height of 14 meters above the ground. When will it strike the ground?

14.

Cumulative Test for Chapters 1–10

Approximately one-half of this test covers Chapters 0–9. The remainder covers the content of Chapter 10.

1. Simplify. $3x\{2y - 3[x + 2(x + 2y)]\}$

2. Evaluate $2x^2 - 3xy + y^2$ when $x = 2$ and $y = -3$.

3. Solve for x. $\frac{1}{2}(x - 2) = \frac{1}{3}(x + 10) - 2x$

4. Factor completely. $16x^4 - 1$

5. Multiply. $(3x + 1)(2x - 3)(x + 4)$

6. Solve for x. $\frac{3x}{x^2 - 4} = \frac{2}{x + 2} + \frac{4}{2 - x}$

7. Graph the line $y = -\frac{3}{4}x + 2$. Identify the slope and the y-intercept.

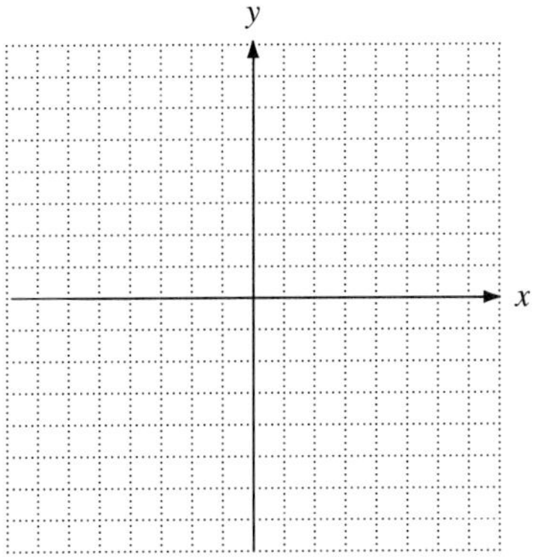

8. Find the equation of a line with slope $= -3$ that passes through the point $(2, -1)$.

9. Solve for (x, y).

$$\begin{aligned} 3x + 2y &= 5 \\ 7x + y &= 19 \end{aligned}$$

10. Simplify. $(-3x^3y^2)^4$

11. Simplify. $\sqrt{18x^5y^6z^3}$

12. Multiply. $(\sqrt{2} + \sqrt{3})(2\sqrt{2} - 4\sqrt{3})$

1. ______

2. ______

3. ______

4. ______

5. ______

6. ______

7. ______

8. ______

9. ______

10. ______

11. ______

12. ______

13. Rationalize the denominator. $\dfrac{\sqrt{5} - 3}{\sqrt{5} + 2}$

14. Solve for r. $A = 3r^2 - 2Ar$

15. Solve for b. $H = 25b^2 - 6$

Solve.

16. $2x^2 + 3x = 35$

17. $(2x + 1)^2 = 20$

18. $\dfrac{60}{R} = \dfrac{60}{R + 8} + 10$

19. $3x^2 + 11x + 2 = 0$

20. $7x^2 + 36x + 5 = 0$

21. $x^2 = 98$

22. $3x^2 + 4 = 79$

23. Graph $y = x^2 + 6x + 10$. Locate the vertex.

24. The length of a rectangle is 2 meters less than three times its width. The area is 96 square meters. Find its length and width.

13. ______

14. ______

15. ______

16. ______

17. ______

18. ______

19. ______

20. ______

21. ______

22. ______

23. ______

24. ______

Practice Final Examination

The following problems cover the content of Chapters 0–10. Follow the directions for each problem and simplify your answers.

1. Simplify.
$-2x + 3y\{7 - 2[x - (4x + y)]\}$.

2. Evaluate if $x = -2$ and $y = 3$.
$2x^2 - 3xy - 4y$

3. Simplify. $(-3x^2y)(-6x^3y^4)$

4. Combine like terms.
$5x^2y - 6xy + 8xy - 3x^2y - 10xy$

5. Solve for x.
$\frac{1}{2}(x + 4) - \frac{2}{3}(x - 7) = 4x$

6. Solve for b. $p = 2a + 2b$

7. Solve for x and graph the resulting inequality on a number line.
$5x + 3 - (4x - 2) \leq 6x - 8$

8. Multiply. $(2x + y)(x^2 - 3xy + 2y^2)$

9. Factor completely. $4x^2 - 18x - 10$

10. Factor completely. $25x^4 - 25$

11. Combine.
$\frac{2}{x - 3} - \frac{3}{x^2 - x - 6} + \frac{4}{x + 2}$

12. Simplify. $\dfrac{\frac{3}{x} + \frac{5}{2x}}{1 + \frac{2}{x + 2}}$

13. Solve for x.
$\frac{2}{x + 2} = \frac{4}{x - 2} + \frac{3x}{x^2 - 4}$

14. Find the slope of the line and then graph the line. $5x - 2y - 3 = 0$

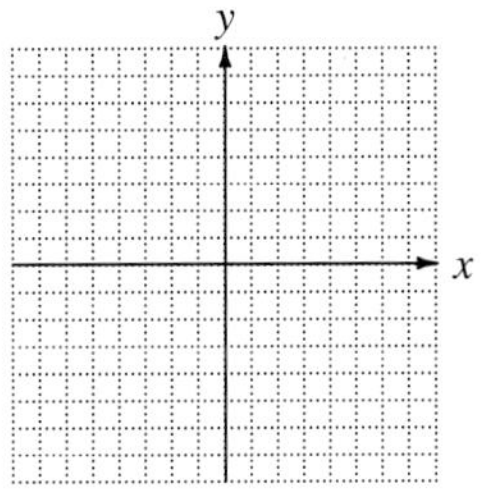

15. Find the equation of a line that has a slope of $-\frac{3}{4}$ and passes through the point $(-2, 5)$.

16. Two quarter-circles are attached to a rectangle. The dimensions are indicated in the figure. Find the area of the region. Use 3.14 as an approximation for π.

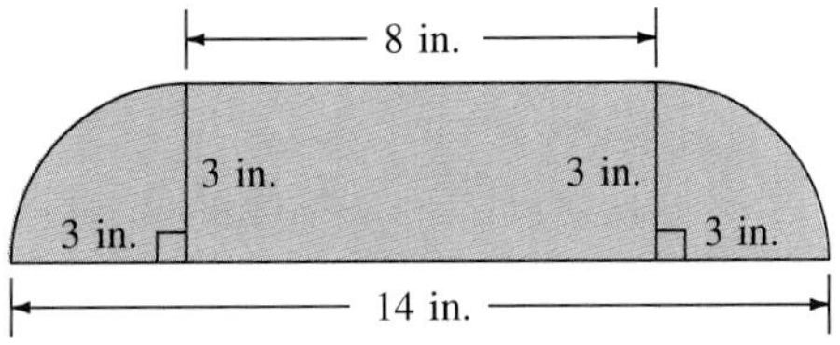

1. ______
2. ______
3. ______
4. ______
5. ______
6. ______
7. ______
8. ______
9. ______
10. ______
11. ______
12. ______
13. ______
14. ______
15. ______
16. ______

17. ______

18. ______

19. ______

20. ______

21. ______

22. ______

23. ______

24. ______

25. ______

26. ______

27. ______

28. ______

29. ______

30. ______

17. Solve for x and y.

$$2x + 3y = 8$$
$$3x - 5y = 31$$

18. Solve for a and b.

$$a - \frac{3}{4}b = \frac{1}{4}$$
$$\frac{3}{2}a + \frac{1}{2}b = -\frac{9}{2}$$

19. Simplify and combine.
$2x\sqrt{50} + \sqrt{98x^2} - 3x\sqrt{18}$

20. Multiply and simplify.
$\sqrt{6}(3\sqrt{2} - 2\sqrt{6} + 4\sqrt{3})$

21. Simplify by rationalizing the denominator.

$$\frac{\sqrt{3} + \sqrt{7}}{\sqrt{5} - \sqrt{7}}$$

22. Solve for x. $3x^2 - 2x - 5 = 0$

23. Solve for y. $2y^2 = 6y - 1$

24. Solve for x. $4x^2 + 3 = 19$

25. In the right triangle shown, sides b and c are given. Find the length of side a.

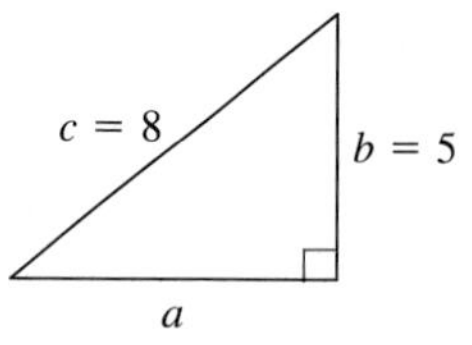

26. A number is tripled and then increased by 6. The result is 21. What is the original number?

27. A rectangular region has a perimeter of 38 meters. The length of the rectangle is 2 meters shorter than double the width. What are the dimensions of the rectangle?

28. Margaret invested \$7000 for one year. She placed part of the money in a tax-free bond earning 10% interest. She placed the rest in a money market earning 14% interest. At the end of one year she had earned \$860 in interest. How much did she invest at each interest rate?

29. A benefit concert was held on the campus to raise scholarship money. A total of 360 tickets were sold. Admission prices were \$5 for reserved seats and \$3 for general admission. Total receipts were \$1480. How many reserved tickets were sold? How many general admission tickets were sold?

30. The area of a triangle is 68 square meters. The altitude of the triangle is one meter longer than double the base of the triangle. Find the altitude and base of the triangle.

Glossary

Absolute Value of a Number (1.1) The absolute value of a number x is the distance between 0 and the number x on the number line. It is written as $|x|$. $|x| = x$ if $x \geq 0$, but $|x| = -x$ if $x < 0$.

Altitude of a Geometric Figure (3.5) The height of the geometric figure. In the three figures shown the altitude is labeled by the distance a.

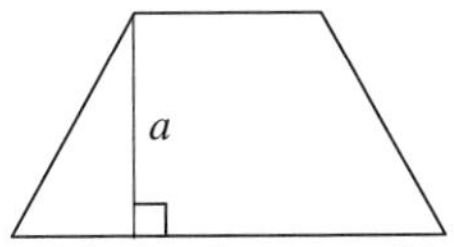

Altitude of a trapezoid

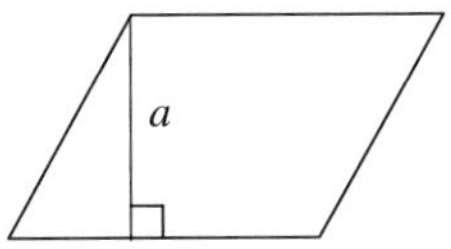

Altitude of a parallelogram

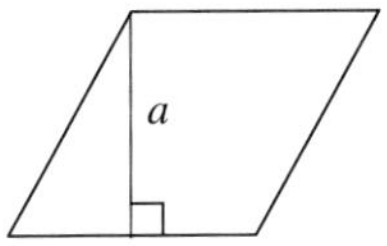

Altitude of a rhombus

Altitude of a Triangle (3.5) The height of any given triangle. In the three triangles shown the altitude is labeled by the distance a.

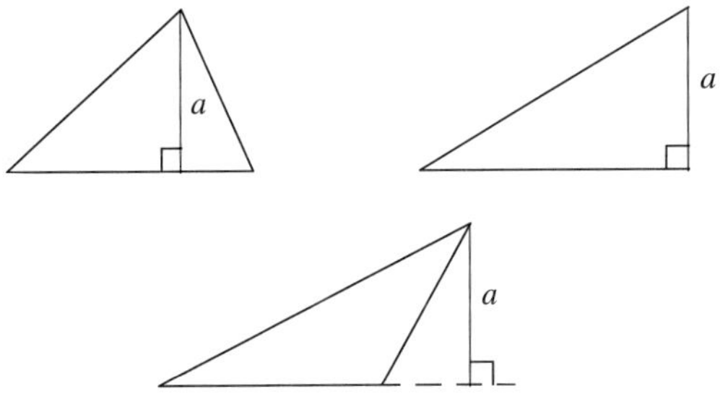

Associative Property of Addition (1.1) If a, b, and c are real numbers, then

$$a + (b + c) = (a + b) + c$$

This property states that if three numbers are added it does not matter *which two numbers* are added first, the result will be the same.

Associative Property of Multiplication (1.3) If a, b, and c are real numbers, then

$$a \times (b \times c) = (a \times b) \times c$$

This property states that if three numbers are multiplied it does not matter *which two numbers* are multiplied first, the result will be the same.

Base (1.4) The number or variable that is raised to a power. In the expression 2^6, the number 2 is the base.

Base of a triangle (7.5) The side of a triangle that is perpendicular to the altitude.

Binomial (1.5) A polynomial of two terms. The expressions $a + 2b$, $6x^3 + 1$, and $5a^3b^2 + 6ab$ are all binomials.

Circumference of a Circle (3.5) The distance around a circle. The circumference of a circle is given by the formula $C = \pi d$ or $C = 2\pi r$, where d is the diameter of the circle and r is the radius of the circle.

Coefficient (4.1) A coefficient is a factor or a group of factors in a product. In the term $4xy$ the coefficient of y is $4x$, but the coefficient of xy is 4. In the term $-5x^3y$ the coefficient of x^3y is -5.

Commutative Property for Addition (1.1) If a and b are any real numbers, then $a + b = b + a$.

Commutative Property for Multiplication (1.3) If a and b are any real numbers, then $ab = ba$.

Complex Fraction (6.4) A fraction that contains at least one fraction in the numerator or in the denominator or both. These three fractions are complex fractions:

$$\frac{7 + \frac{1}{x}}{x^2 + 2}, \quad \frac{1 + \frac{1}{5}}{2 - \frac{1}{7}}, \quad \text{and} \quad \frac{1}{\frac{3}{4}}$$

Constant (2.5) Symbol or letter that is used to represent exactly one single number during a particular problem or discussion.

Coordinates of a Point (7.1) An ordered pair of numbers (x, y) that specifies the location of a point on a rectangular coordinate system.

Degree of a Polynomial (3.1) The degree of the highest-degree term of a polynomial. The degree of the polynomial $5x^3 + 2x^2 - 6x + 8$ is 3. The degree of the polynomial $5x^2y^2 + 3xy + 8$ is 4.

Degree of a Term of a Polynomial (4.6) The sum of the exponents of the variables in the term. The degree of $3x^3$ is 3. The degree of $4x^5y^2$ is 7.

Denominator (0.1) and (6.1) The bottom number or algebraic expression in a fraction. The denominator of

$$\frac{3x - 2}{x + 4}$$

is $x + 4$. The denominator of $\frac{3}{7}$ is 7. The denominator of a fraction may not be zero.

Dependent Equations (8.1) Two equations are dependent if every value that satisfies one equation satisfies the other. A system of two dependent equations in two variables will not have a unique solution.

Diagonal of a Four-sided Figure (3.5) A line connecting two nonadjacent corners of the figure. In each of the figures shown, line AC is a diagonal.

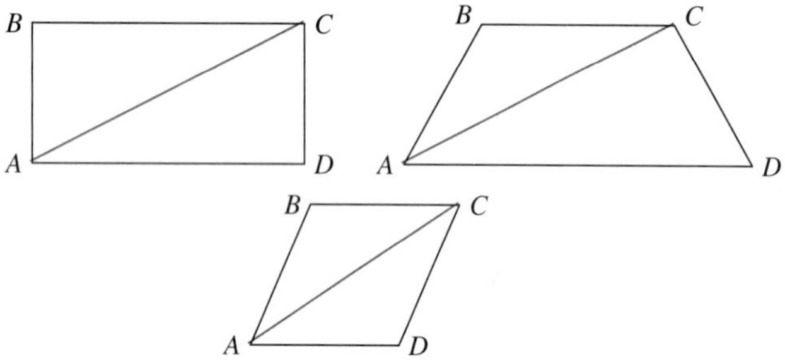

Difference (3.1) The result of subtracting one number or expression from another. The mathematical expression $x - 6$ can be written in words as the difference between x and 6.

Difference of Two Squares Polynomial (5.5) A polynomial of the form $a^2 - b^2$ that may be factored by using the formula

$$a^2 - b^2 = (a + b)(a - b).$$

Distributive Property (1.5) For all real numbers a, b, and c, it is true that $a(b + c) = ab + ac$.

Dividend (0.4) The number that is to be divided by another. In the problem $30 \div 5 = 6$, the three parts are as follows:

5 is the divisor → $5\overline{)30}$ with quotient 6

30 is the dividend

6 is the quotient

Divisor (0.4) The number you divide into another.

Domain of a Relation (7.6) In any relation, the set of values that can be used for the independent variable is called its domain. This is the set of all the first coordinates of the ordered pairs that define the relation.

Equilateral Triangle (3.5) A triangle with three sides equal in length and three angles that measure 60°. Triangle ABC is an equilateral triangle.

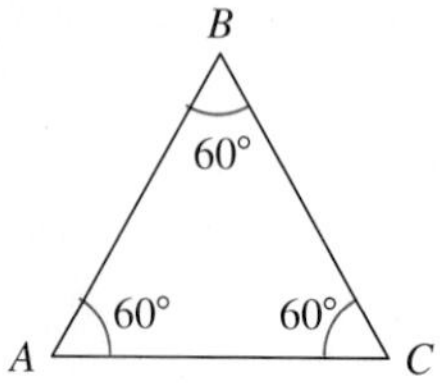

Even Integers (1.3) Integers that are exactly divisible by 2, such as $\ldots, -4, -2, 0, 2, 4, 6, \ldots$.

Exponent (4.1) The number that indicates the power of a base. If the number is a positive integer it indicates how many times the base is multiplied. In the expression 2^6, the exponent is 6.

Expression (3.1) A mathematic expression is any quantity using numbers and variables. Therefore, $2x$, $7x + 3$, and $5x^2 + 6x$ are all mathematical expressions.

Extraneous Solution (6.5) and (9.6) An obtained solution to an equation that when substituted back into the original equation, does *not* yield an identity. $x = 2$ is an extraneous solution to the equation

$$\frac{x}{x - 2} - 4 = \frac{2}{x - 2}$$

An extraneous solution is also called an extraneous root.

Factor (0.1) and (5.1) When two or more numbers, variables, or algebraic expressions are multiplied, each is called a factor. If we write $3 \cdot 5 \cdot 2$, the factors are 3, 5, and 2. If we write $2xy$, the factors are 2, x, and y. In the expression $(x - 6)(x + 2)$, the factors are $(x - 6)$ and $(x + 2)$.

Fractions

Algebraic Fractions (6.1) The indicated quotient of two algebraic expressions.

$$\frac{x^2 + 3x + 2}{x - 4} \quad \text{and} \quad \frac{y - 6}{y + 8}$$

are algebraic fractions. In these fractions the value of the denominator cannot be zero.

Numerical Fractions (0.1) A set of numbers used to describe parts of whole quantities. A numerical fraction can be represented by the quotient of two integers for which the denominator is not zero. The numbers $\frac{1}{5}$, $-\frac{2}{3}$, $\frac{8}{2}$, $-\frac{4}{31}$, $\frac{8}{1}$, and $-\frac{12}{1}$ are all numerical fractions. The set of rational numbers can be represented by numerical fractions.

Function (7.6) A relation in which no two different ordered pairs have the same first coordinates.

Hypotenuse of a Right Triangle (9.6) The side opposite the right angle in any right triangle. The hypotenuse is always the longest side of a right triangle. Side AB is the hypotenuse of triangle ABC.

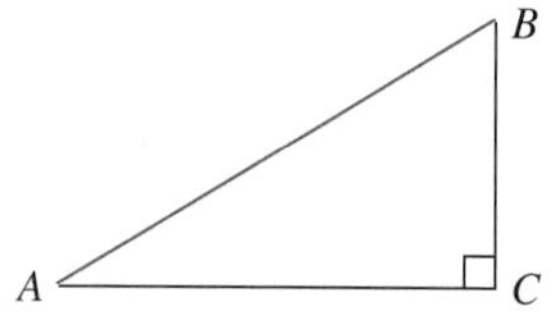

Identity (2.1) A statement that is always true. The equations $5 = 5$, $7 + 4 = 7 + 4$, and $x + 8 = x + 8$ are examples of identities.

Imaginary Number (9.1) A number that is not a real number. Imaginary numbers can be created by taking the square root of a negative real number. The numbers $\sqrt{-9}$, $\sqrt{-7}$, and $\sqrt{-12}$ are all imaginary numbers.

Improper Fraction (0.1) A numerical fraction whose numerator is larger than or equal to its denominator. $\frac{8}{3}$, $\frac{5}{2}$, and $\frac{7}{7}$ are improper fractions.

Inconsistent System of Equations (8.1) A system of equations that does not have a common solution.

Independent Equations (8.1) Two equations that are not dependent are said to be independent.

Index of a Radical (9.7) The number indicating the order of a root. The index number for $\sqrt[3]{x}$ is 3. The index number of $\sqrt[4]{5}$ is 4. The index number for a square root is not written, but is understood to be 2. The index number for $\sqrt{x}$ is 2.

Inequality (2.7) and (7.5) A mathematical relationship between quantities that are not equal. $x \leq -3$, $w > 5$, and $x < 2y + 1$ are mathematical inequalities.

Integers (1.1) The set of numbers . . . , $-5, -4, -3, -2, -1, 0, 1, 2, 3, 4, 5,$ . . . .

Intercepts of an Equation (7.2) The point or points where the graph of the equation crosses the x-axis or the y-axis or both. (*See* x-intercept.)

Irrational Numbers (9.1) A real number that cannot be expressed in the form $\frac{a}{b}$, where a and b are integers and $b \neq 0$. $\sqrt{2}$, $\sqrt[3]{3}$, π, $5 + 3\sqrt{2}$, and $-4\sqrt{7}$ are irrational numbers.

Isosceles Triangle (3.5) A triangle with two equal sides and two equal angles. Triangle ABC is an isosceles triangle. Angle BAC is equal to angle ACB. Side AB is equal in length to side BC.

Least Common Denominator of Numerical Fractions (0.2) The smallest whole number that is exactly divisible by all denominators of a group of fractions. The least common denominator (LCD) of $\frac{1}{6}$, $\frac{2}{3}$, and $\frac{3}{5}$ is 30. The least common denominator is also called the lowest common denominator.

Leg of a Right Triangle (9.6) One of the two shortest sides of a right triangle. Side AC and side BC are legs of triangle ABC.

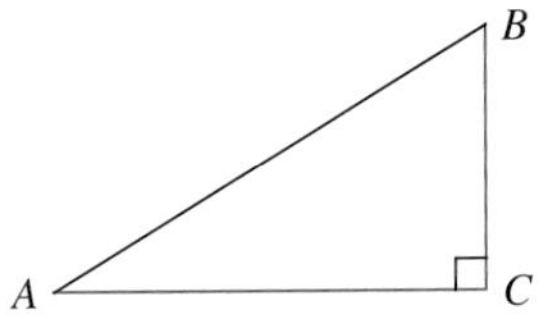

Like Terms (1.6) Terms that have identical variables and exponents. In the expression $5x^3 + 2xy^2 + 6x^2 - 3xy^2$, the term $2xy^2$ and the term $-3xy^2$ are like terms.

Linear Equation in Two Variables (7.2) An equation of the form $Ax + By = C$, where A, B, and C are real numbers. The graph of a linear equation in two variables is a straight line.

Line of Symmetry of a Parabola (10.4) A line that can be drawn through a parabola such that, if the graph were folded on this line, the two halves of the curve would correspond exactly. The line of symmetry through the parabola formed by $y = ax^2 + bx + c$ is given by $x = \frac{-b}{2a}$. The line of symmetry of a parabola always passes through the vertex of a parabola.

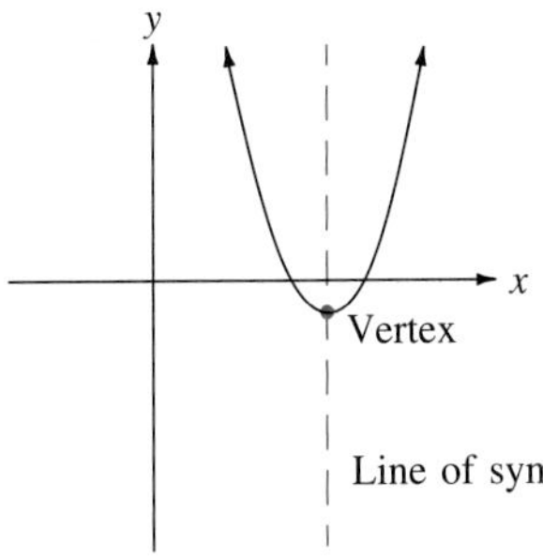

Mixed Fraction (0.1) A fraction that consists of an integer written next to a proper fraction. $2\frac{1}{3}$, $4\frac{6}{7}$, and $3\frac{3}{8}$ are all mixed fractions. Mixed fractions are sometimes called mixed numbers or mixed numerals.

Monomial (1.6) A polynomial of one term. The expressions $3xy$, $5a^2b^3cd$, and -6 are all monomials.

Natural Numbers (0.1) The set of numbers 1, 2, 3, 4, 5, ... This set is also called the set of counting numbers.

Numeral (0.1) The symbol used to describe a number.

Numerator (0.1) The top number or algebraic expression in a fraction. The numerator of

$$\frac{x + 3}{5x - 2}$$

is $x + 3$. The numerator of $\frac{12}{13}$ is 12.

Numerical Coefficient (4.1) The number that is multiplied by a variable or a group of variables. The numerical coefficient of $5x^3y^2$ is 5. The numerical coefficient of $-6abc$ is -6. The numerical coefficient of x^2y is 1. A numerical coefficient of 1 is not usually written.

Odd Integers (1.3) Integers that are not exactly divisible by 2, such as $-3, -1, 1, 3, 5, 7, 9,$ . . . .

Opposite of a Number (1.1) Two numbers that are the same distance from zero on the number line but lie on different sides of it are considered opposites. The opposite of -6 is 6. The opposite of $\frac{22}{7}$ is $-\frac{22}{7}$.

Ordered Pair (7.1) A pair of numbers represented in a specified order. An ordered pair is often used to specify a location on a graph. Every point on a rectangular coordinate system can be represented by an ordered pair (x, y).

Origin (7.1) The point (0, 0) in the rectangular coordinate system.

Parabola (9.5) A curve created by graphing a quadratic equation. The curves shown are all parabolas. The graph of the equation $y = ax^2 + bx + c$, where $a \neq 0$, will always yield a parabola.

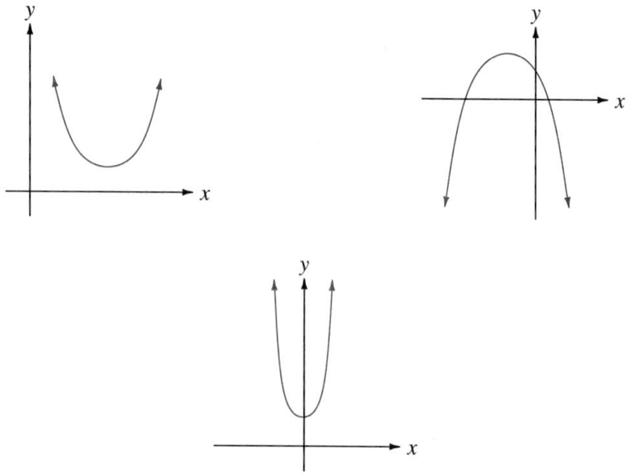

Parallel Lines (8.1) Two straight lines that never intersect. The graph of an inconsistent system of two equations in two variables will result in parallel lines.

Parallelogram (3.5) A four-sided figure with opposite sides parallel. Figure $ABCD$ is a parallelogram.

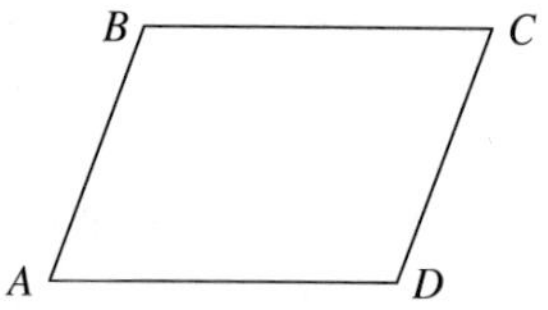

Percent (0.5) Hundredths or "per one hundred"; indicated by the % symbol. Thirty-seven hundredths $\left(\frac{37}{100}\right) = 37\%$ (thirty-seven percent).

Perfect Square Number (5.5) A number that is the square of an integer. The numbers 1, 4, 9, 16, 25, 36, 49, 64, 81, 100, 121, 144, . . . are considered perfect square numbers.

Perfect Square Trinomial (5.5) A polynomial of the form $a^2 + 2ab + b^2$ or $a^2 - 2ab + b^2$ that may be factored using one of the following formulas:

$$a^2 + 2ab + b^2 = (a + b)^2$$

or

$$a^2 - 2ab + b^2 = (a - b)^2$$

Perimeter (3.5) The distance around any closed geometric figure. The perimeter of this triangle is 13. The perimeter of this rectangle is 20.

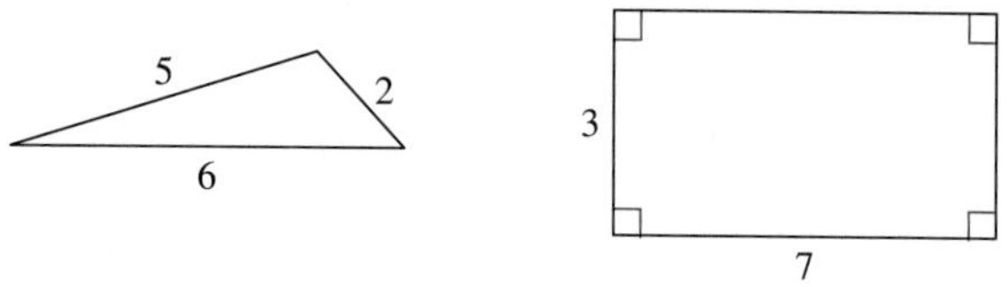

Pi (3.5) An irrational number, denoted by the symbol π, that is approximately equal to 3.141592654. In most cases 3.14 can be used as a sufficiently accurate approximation for π.

Polynomial (1.6) Expressions that contain terms with nonnegative integer exponents. The expressions $5ab + 6$, $x^3 + 6x^2 + 3$, -12, and $x + 3y - 2$ *are all polynomials.* The expressions $x^{-2} + 2x^{-1}$, $2\sqrt{x} + 6$, and $\frac{5}{x} + 2x^2$ *are not polynomials.*

Prime Numbers (0.1) Any natural number greater than 1 whose only natural number factors are 1 and itself. The first eight prime numbers are 2, 3, 5, 7, 11, 13, 17, and 19.

Prime Polynomial (5.6) A prime polynomial is a polynomial that cannot be factored by the methods of elementary algebra. $x^2 + x + 1$ is a prime polynomial.

Principal (3.4) In monetary problems, the principal is the original amount of money invested or borrowed.

Principal Square Root (9.1) For any given nonnegative number N, the principal square root of N (written $\sqrt{N}$) is the nonnegative number a if and only if $a^2 = N$. The principal square root of 25 $(\sqrt{25})$ is 5.

Proper Fraction (0.1) A numerical fraction whose numerator is less than its denominator; $\frac{3}{7}$, $\frac{2}{5}$, and $\frac{8}{9}$ are proper fractions.

Proportion (6.6) A proportion is an equation stating that two ratios are equal.

$$\frac{a}{b} = \frac{c}{d}$$

is a proportion.

Pythagorean Theorem (9.6) In any right triangle, if c is the length of the hypotenuse and a and b are the lengths of the two legs, then $c^2 = a^2 + b^2$.

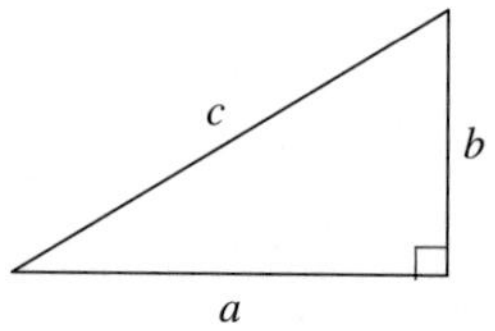

Quadratic Equation (5.7) A quadratic equation is a polynomial equation of degree 2. It is an equation with one variable that contains at least one term with the variable squared, but no term with the variable raised to a higher power. $5x^2 + 6x - 3 = 0$, $x^2 = 7$, and $5x^2 = 2x$ are all quadratic equations.

Quadratic Formula (10.3) If $ax^2 + bx + c = 0$ and $a \neq 0$, then the solutions to the quadratic equation are given by

$$x = \frac{-b \pm \sqrt{b^2 - 4ac}}{2a}$$

Quotient (0.4) The result of dividing one number or expression by another. In the problem $12 \div 4 = 3$, the quotient is 3.

Radical (9.1) An expression composed of a radical sign and a radicand. The expressions $\sqrt{5x}$, $\sqrt{\frac{3}{5}}$, $\sqrt{5x + b}$, and $\sqrt[3]{10}$ are called radicals.

Radical Equation (9.6) An equation that contains one or more radicals. $\sqrt{x + 4} = 12$, $\sqrt{x} = 3$, and $\sqrt{3x + 4} = x + 2$ are all radical equations.

Radical Sign (9.1) The symbol $\sqrt{\ }$ used to indicate the root of a number.

Radicand (9.1) The expression beneath the radical sign. The radicand of $\sqrt{5ab}$ is $5ab$.

Range of a Relation (7.6) In any relation, the set of values that represents the dependent variable is called its range. This is the set of all the second coordinates of the ordered pairs that define the relation.

Ratio (6.6) The ratio of one number a to another number b is the quotient $a \div b$ or $\frac{a}{b}$.

Rational Numbers (1.1) and (9.1) A number that can be expressed in the form $\frac{a}{b}$, where a and b are integers and $b \neq 0$. $\frac{7}{3}$, $-\frac{2}{5}$, $\frac{7}{-8}$, $\frac{5}{1}$, 1.62 and 2.7156 are rational numbers.

Rationalizing the Denominator (9.5) The process of transforming a fraction that contains a radical in the denominator to an equivalent fraction that does not contain a radical in the denominator.

Original Fraction	***Equivalent Fraction That Has Been Rationalized***
$\frac{3}{\sqrt{2}}$	$\frac{3\sqrt{2}}{2}$

Rationalizing the Expression (9.5) The process of transforming a radical that contains a fraction to an equivalent expression that does not contain a fraction inside a radical.

Original Radical	***Equivalent Expression after It Has Been Rationalized***
$\sqrt{\frac{3}{5}}$	$\frac{\sqrt{15}}{5}$

Real Number (9.1) Any number that is rational or irrational. 2, 7, $\sqrt{5}$, $\frac{3}{8}$, π, $-\frac{7}{5}$, and $-3\sqrt{5}$ are all real numbers.

Rectangle (3.5) A four-sided figure with opposite sides parallel and all interior angles measuring 90°. The opposite sides of a rectangle are equal.

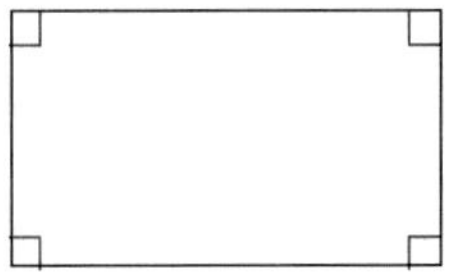

Relation (7.6) A relation is any set of ordered pairs.

Rhombus (3.5) A parallelogram with four equal sides. Figure *ABCD* is a rhombus.

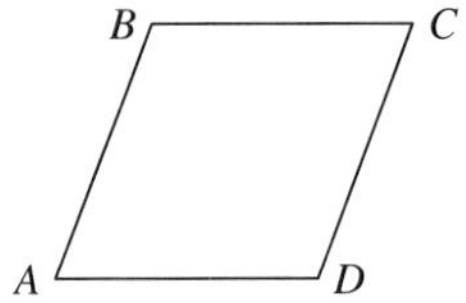

Right Angle (3.5) An angle that measures 90°. Right angles are usually labeled in a sketch by using a small square to indicate that it is a right angle. Here angle *ABC* is a right angle.

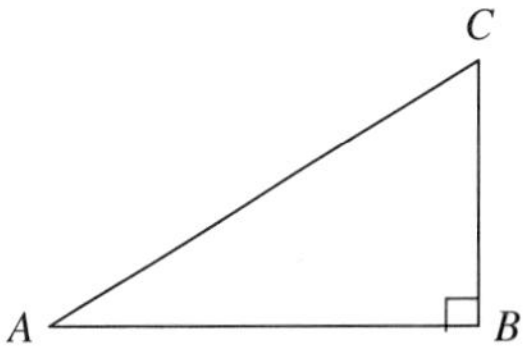

Right Triangle (3.5) A triangle that contains a right angle.

Root of an Equation (2.1) and (5.7) A value of the variable that makes an equation into a true statement. The root of an equation is also called the solution of an equation.

Scientific Notation (4.4) A positive number is written in scientific notation if it is in the form $a \times 10^n$, where $1 \leq a < 10$ and n is an integer.

Slope-Intercept Form (7.3) The equation of a line that has slope m and the y intercept at $(0, b)$ is given by $b = mx + b$.

Slope of a Line (7.3) The ratio of change in y over the change in x for any two different points on a nonvertical line. The slope m is determined by

$$m = \frac{y_2 - y_1}{x_2 - x_1}$$

where $x_2 \neq x_1$ for any two points (x_1, y_1) and (x_2, y_2) on a nonvertical line.

Solution of an Equation (2.1) A number that, when substituted into a given equation, yields an identity. The solution of an equation is also called the root of an equation.

Solution of a Linear Inequality (2.3) The possible values that make a linear inequality true.

Solution of an Inequality in Two Variables (7.5) The set of all possible ordered pairs that when substituted into the inequality will yield an identity.

Solution to a System of Two Equations in Two Variables (8.2) An ordered pair that can be substituted into each equation and an identity will be obtained in each case.

Square (3.7) A rectangle with four equal sides.

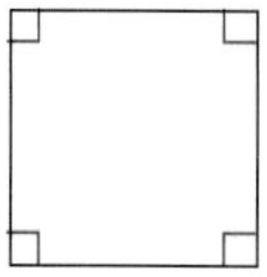

Square Root (9.1) For any given nonnegative number N, the square root of N is the number a if $a^2 = N$. One square root of 16 is 4 since $(4)^2 = 16$. Another square root of 16 is -4 since $(-4)^2 = 16$.

Standard Form of a Quadratic Equation (5.7) A quadratic equation that is placed in the form $ax^2 + bx + c = 0$.

Subscript of a Variable (3.3) A small number or letter written slightly below and to the right of a variable. In the expressions $5 = 2(x - x_0)$, the subscript of x is 0. In the expression $t_f = 5(t_a - b)$ the subscript of the first t is f. The subscript of the second t is a. A subscript is used to indicate a different value of the variable.

System of Equations (8.2) A set of two or more equations that must be considered together.

Term (1.5) A number, a variable, or a product of numbers and variables. In the expression $a^3 - 3a^2b + 4ab^2 + 6b^3 + 8$, there are five terms. They are a^3, $-3a^2b$, $4ab^2$, $6b^3$, and 8. The terms of a polynomial are separated by plus and minus signs.

Trapezoid (3.5) A four-sided figure with two sides parallel. The parallel sides are called the bases of the trapezoid. Figure $ABCD$ is a trapezoid.

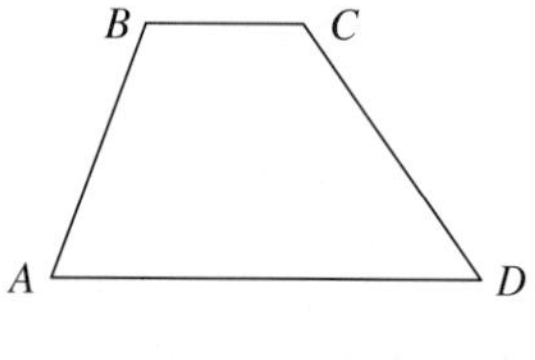

Trinomial (1.5) A polynomial of three terms. The expressions $x^2 + 6x - 8$ and $a + 2b - 3c$ are trinomials.

Variable (4.1) A letter that is used to represent a number or a set of numbers.

Variation (9.7) An equation relating values of one variable to those of other variables. An equation of the form $y = kx$, where k is a constant, is a *direct variation*. An equation of the form $y = \frac{k}{x}$, where k is a constant, is an *inverse variation*. k is called the *constant of variation*.

Vertex of a Parabola (10.4) The lowest point on a parabola opening upward or the highest point on a parabola opening downward. The x-coordinate of the vertex of the parabola formed by the equation $y = ax^2 + bx + c$ is given by $x = \frac{-b}{2a}$.

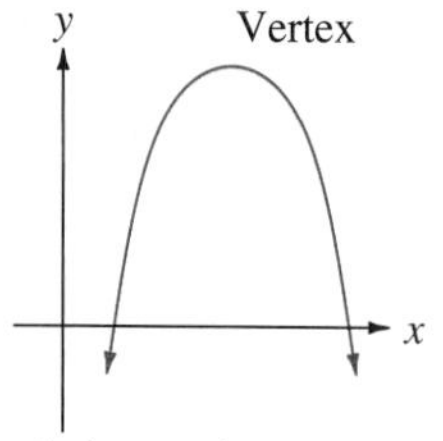

Parabola opening downward

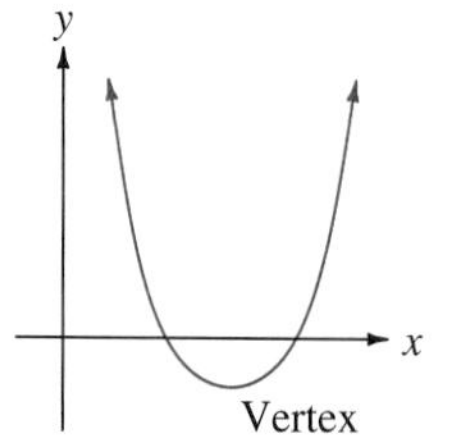

Parabola opening upward

Vertical Line Test (7.6) If a vertical line can intersect the graph of a relation more than once, the relation is not a function.

Whole Numbers (0.1) The set of numbers 0, 1, 2, 3, 4, 5, ...

x-Intercept (7.2) The number a is the x-intercept of a line if the line crosses the x-axis at $(a, 0)$. The x-intercept of line l on the graph shown is 4.

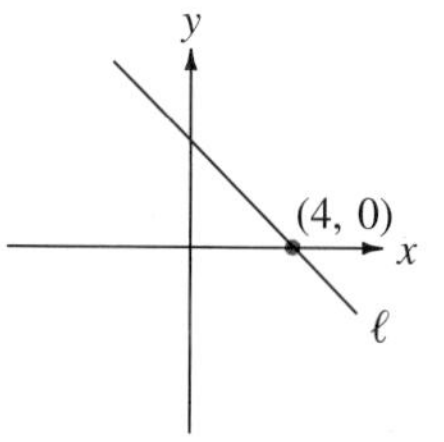

y-Intercept (7.2) The number b is the y-intercept of a line if the line crosses the y-axis at $(0, b)$. The y-intercept of line p on the accompanying graph is 3.

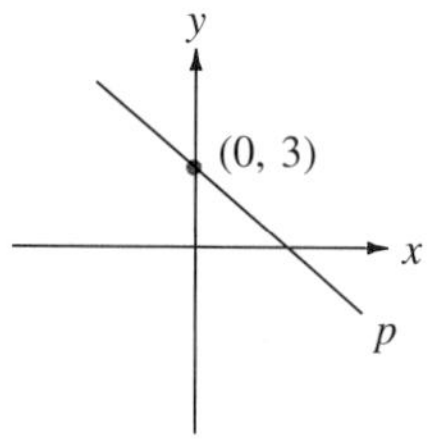

APPENDIX A
Table of Square Roots

x	$\sqrt{x}$	x	$\sqrt{x}$	x	$\sqrt{x}$	x	$\sqrt{x}$	x	$\sqrt{x}$
1	1.000	41	6.403	81	9.000	121	11.000	161	12.689
2	1.414	42	6.481	82	9.055	122	11.045	162	12.728
3	1.732	43	6.557	83	9.110	123	11.091	163	12.767
4	2.000	44	6.633	84	9.165	124	11.136	164	12.806
5	2.236	45	6.708	85	9.220	125	11.180	165	12.845
6	2.449	46	6.782	86	9.274	126	11.225	166	12.884
7	2.646	47	6.856	87	9.327	127	11.269	167	12.923
8	2.828	48	6.928	88	9.381	128	11.314	168	12.961
9	3.000	49	7.000	89	9.434	129	11.358	169	13.000
10	3.162	50	7.071	90	9.487	130	11.402	170	13.038
11	3.317	51	7.141	91	9.539	131	11.446	171	13.077
12	3.464	52	7.211	92	9.592	132	11.489	172	13.115
13	3.606	53	7.280	93	9.644	133	11.533	173	13.153
14	3.742	54	7.348	94	9.695	134	11.576	174	13.191
15	3.873	55	7.416	95	9.747	135	11.619	175	13.229
16	4.000	56	7.483	96	9.798	136	11.662	176	13.266
17	4.123	57	7.550	97	9.849	137	11.705	177	13.304
18	4.243	58	7.616	98	9.899	138	11.747	178	13.342
19	4.359	59	7.681	99	9.950	139	11.790	179	13.379
20	4.472	60	7.746	100	10.000	140	11.832	180	13.416
21	4.583	61	7.810	101	10.050	141	11.874	181	13.454
22	4.690	62	7.874	102	10.100	142	11.916	182	13.491
23	4.796	63	7.937	103	10.149	143	11.958	183	13.528
24	4.899	64	8.000	104	10.198	144	12.000	184	13.565
25	5.000	65	8.062	105	10.247	145	12.042	185	13.601
26	5.099	66	8.124	106	10.296	146	12.083	186	13.638
27	5.196	67	8.185	107	10.344	147	12.124	187	13.675
28	5.292	68	8.246	108	10.392	148	12.166	188	13.711
29	5.385	69	8.307	109	10.440	149	12.207	189	13.748
30	5.477	70	8.367	110	10.488	150	12.247	190	13.784
31	5.568	71	8.426	111	10.536	151	12.288	191	13.820
32	5.657	72	8.485	112	10.583	152	12.329	192	13.856
33	5.745	73	8.544	113	10.630	153	12.369	193	13.892
34	5.831	74	8.602	114	10.677	154	12.410	194	13.928
35	5.916	75	8.660	115	10.724	155	12.450	195	13.964
36	6.000	76	8.718	116	10.770	156	12.490	196	14.000
37	6.083	77	8.775	117	10.817	157	12.530	197	14.036
38	6.164	78	8.832	118	10.863	158	12.570	198	14.071
39	6.245	79	8.888	119	10.909	159	12.610	199	14.107
40	6.325	80	8.944	120	10.954	160	12.649	200	14.142

Unless the value of $\sqrt{x}$ ends in 000, all values are rounded to the nearest thousandth.

APPENDIX B
Metric Measurement and Conversion of Units

After studying this section, you will be able to:

1. *Convert from one metric unit of measurement to another.*
2. *Convert between metric and U.S. units of measure.*
3. *Convert between one type of unit of measure to another.*

1 *Converting from One Metric Unit of Measurement to Another*

Metric measurements are becoming more common in the United States. The metric system is used in many parts of the world and in the sciences. The basic unit of length in the metric system is the meter. For smaller measurements, centimeters are commonly used. There are 100 centimeters in 1 meter. For larger measurements, kilometers are commonly used. There are 1000 meters in 1 kilometer. The following tables give the metric units of measurement for length.

Metric Length

*1 kilometer (km)	=	1000 meters
1 hectometer (hm)	=	100 meters
1 dekameter (dam)	=	10 meters
*1 meter (m)	=	1 meter
1 decimeter (dm)	=	0.1 meter
*1 centimeter (cm)	=	0.01 meter
*1 millimeter (mm)	=	0.001 meter

The four most common units of metric measurement are indicated by an asterisk.

Metric Weight

1 kilogram (kg)	=	1000 grams
1 gram (g)	=	1 gram
1 milligram (mg)	=	0.001 gram

Metric Volume

1 kiloliter (kL)	=	1000 liters
1 liter (L)	=	1 liter
1 milliliter (mL)	=	0.001 liter

One way to convert from one metric unit to another is to multiply by 1. We know mathematically that to multiply by 1 will yield an equivalent expression since $(a)(1) = a$. In our calculations when solving word problems, we treat dimension symbols much like we treat variables in algebra.

EXAMPLE 1 Fred drove a distance of 5.4 kilometers. How many meters is that?

$$5.4 \text{ km} \cdot 1 = 5.4 \text{ km} \cdot \frac{1000 \text{ m}}{1 \text{ km}} = \frac{(5.4)(1000)}{1} \cdot \frac{\text{km}}{\text{km}} \cdot \text{m}$$

$$= 5400 \text{ m}$$

The distance is 5400 meters.

Practice Problem 1 June's room is 3.4 meters wide. How many centimeters is that? ■

EXAMPLE 2 The chemist measured 67 milliliters solution. How many liters is that?

$$67 \text{ mL} \cdot \frac{0.001 \text{ L}}{1 \text{ mL}} = 0.067 \text{ L}$$

The chemist measured 0.067 liters of the solution.

Practice Problem 2 The container has 125 liters of water. How many kiloliters is that? ■

2 *Converting between Metric and U.S. Units of Measure*

The most common relationships between the U.S. and metric systems are listed in the following table. These values are approximate.

Metric Conversion Ratios

LENGTH:	1 inch = 2.54 centimeters
	39.37 inches = 1 meter
	1 mile = 1.61 kilometers
	0.62 mile = 1 kilometer
WEIGHT:	1 pound = 454 grams
	2.20 pounds = 1 kilogram
	1 ounce = 28.35 grams
	0.0353 ounce = 1 gram
LIQUID CAPACITY:	1 quart = 946 milliliters
	1.06 quarts = 1 liter
	1 gallon = 3.785 liters

EXAMPLE 3 A box weighs 190 grams. How many ounces is that? Round your answer to the nearest hundredth.

$$190 \text{ g} \cdot \frac{0.0353 \text{ oz}}{1 \text{ g}} = 6.707 \text{ oz}$$

$$\approx 6.71 \text{ oz} \quad \textit{rounding to the nearest hundredth}$$

The box weighs approximately 6.71 ounces.

Practice Problem 3 A bag of groceries weighs 5.72 pounds. How many kilograms is that? ■

EXAMPLE 4 Juanita drives 23 kilometers to work each day. How many miles is the trip? Round your answer to the nearest mile.

$$23 \text{ km} \cdot \frac{0.62 \text{ mi}}{1 \text{ km}} = 14.26 \text{ mi} \approx 14 \text{ mi} \quad \text{(rounded to the nearest mile)}$$

Juanita drives approximately 14 miles to work each day.

Practice Problem 4 Carlos installed an electrical connection that is 8.00 centimeters long. How many inches long is the connection? Round your answer to the nearest hundredth. ■

EXAMPLE 5 Anita purchased 42.0 gallons of gasoline for her car last month. How many liters did she purchase? Round your answer to the nearest liter.

$$42.0 \text{ gal} \cdot \frac{3.785 \text{ L}}{1 \text{ gal}} = 158.97 \text{ L} \approx 159 \text{ L} \quad \text{(rounded to the nearest liter)}$$

Anita purchased approximately 159 liters of gasoline last month.

Practice Problem 5 Warren purchased a 3.00-liter bottle of Coca-Cola. How many quarts of Coca-Cola is that? Round your answer to the nearest hundredth of a quart. ■

3 *Converting between One Type of Unit of Measure to Another Type*

Sometimes you will need to convert form one type of unit of measure to another. For example you may need to convert days to minutes or from miles per hour to feet per second. Recall the U.S. units of measure.

Length	Time
12 inches = 1 foot	60 seconds = 1 minute
3 feet = 1 yard	60 minutes = 1 hour
5280 feet = 1 mile	24 hours = 1 day
1760 yards = 1 mile	7 days = 1 week
Weight	**Volume**
16 ounces = 1 pound	2 cups = 1 pint
2000 pounds = 1 ton	2 pints = 1 quart
	4 quarts = 1 gallon

EXAMPLE 6 A car was traveling at 50.0 miles per hour. How many feet per second was the car traveling? Approximate your answer to the nearest tenth of a foot per second.

$$\frac{50 \text{ mi}}{\text{hr}} \cdot \frac{5280 \text{ ft}}{1 \text{ mi}} \cdot \frac{1 \text{ hr}}{60 \text{ min}} \cdot \frac{1 \text{ min}}{60 \text{ sec}} = \frac{(50)(5280) \text{ ft}}{(60)(60) \text{ sec}} = \frac{73.333\ldots \text{ ft}}{\text{sec}}$$

Rounding to the nearest tenth of a foot per second, the car was traveling at 73.3 feet per second.

Practice Problem 6 A speeding car was traveling at 70.0 miles per hour. How many feet per second was the car traveling? Approximate your answer to the nearest tenth of a foot per second. ■

B Exercises

1. How many meters are in 34 km?
2. How many meters are in 128 km?
3. How many centimeters are in 57 m?
4. How many centimeters are in 46 m?
5. How many millimeters are in 25 cm?
6. How many millimeters are in 63 cm?
7. How many meters are in 563 mm?
8. How many meters are in 831 mm?
9. How many milligrams are in 29.4 g?
10. How many milligrams are in 75.2 g?
11. How many kilograms are in 98.4 g?
12. How many kilograms are in 62.7 g?
13. How many milliliters are in 7 L?
14. How many milliliters are in 12 L?
15. How many kiloliters are in 4 mL?
16. How many kiloliters are in 3 mL?

Use the table of metric conversion ratios to find each of the following. Round all answers to the nearest tenth.

17. How many inches are in 4.2 cm?
18. How many inches are in 3.8 cm?
19. How many kilometers are in 14 mi?
20. How many kilometers are in 13 mi?
21. How many meters are in 110 in.?
22. How many meters are in 150 in?
23. How many centimeters are in 7 in.?
24. How many centimeters are in 9 in.?
25. A box weighing 2.4 lb would weigh how many grams?
26. A box weighing 1.6 lb would weigh how many grams?
27. A man weighs 78 kg. How many pounds does he weigh?
28. A woman weighs 52 kg. How many pounds does she weigh?
29. Ferrante purchased 3 qt of milk. How many liters is that?
30. Wong Tin purchased 1 gal of milk. How many liters is that?

Answer the following questions. Round all answers to the nearest hundredth.

31. How many inches are in 3050 miles?
32. How many inches are in 4500 miles?
33. A truck traveled at a speed of 40 feet per second. How many miles per hour is that?
34. A car traveled at a speed of 55 feet per second. How many miles per hour is that?
35. How many years are in 3,500,000 seconds? (Use the approximate value that 365 days = 1 year.)
36. How many years are in 2,800,000 seconds? (Use the approximate value that 365 days = 1 year.)

APPENDIX C
Inductive and Deductive Reasoning

After studying this section you will be able:

1 *Use inductive reasoning to reach a conclusion.*
2 *Use deductive reasoning to reach a conclusion.*

1 *Inductive Reasoning*

When we reach a conclusion based on specific observations, we are using **inductive reasoning.** Much of our early learning is based on simple cases of inductive reasoning. If a child touches a hot stove or other appliance several times and each time he gets burned, he is likely to conclude, "If I touch something that is hot, I will get burned." This is inductive reasoning. The child has thought about several actions and their outcomes and has made a conclusion or generalization.

The next few examples show how inductive reasoning can be used in mathematics.

EXAMPLE 1 Find the next number in the sequence 10, 13, 16, 19, 22, 25, 28, ...

Solution: We observe a pattern that each number is 3 more than the preceding number. $10 + 3 = 13$; $13 + 3 = 16$, and so on. Therefore, if we add 3 to 28, we conclude that the next number in the sequence is 31.

Practice Problem 1 Find the next number in the sequence 24, 31, 38, 45, 52, 59, 66, ... ■

EXAMPLE 2 Find the next number in the sequence 1, 8, 27, 64, 125, ...

Solution: The sequence can be written as $1^3, 2^3, 3^3, 4^3, 5^3, \ldots$.
Each successive integer is cubed. The next number would be 6^3 or 216.

Practice Problem 2 Find the next number in the sequence.

3, 8, 15, 24, 35, 48, 63, 80, . . . ■

EXAMPLE 3 Guess the next seven digits in the following irrational number:

5.636336333633336 . . .

Solution: Between the 6's there are the digits 3, 33, 333, 3333, and so on. The pattern is that the number of 3's keeps increasing by 1 each time. Thus the next seven digits are 3333363.

Practice Problem 3 Guess the next seven digits in the following irrational number:

6.1213314441 . . . ■

EXAMPLE 4 Find the next two figures that would appear in the sequence.

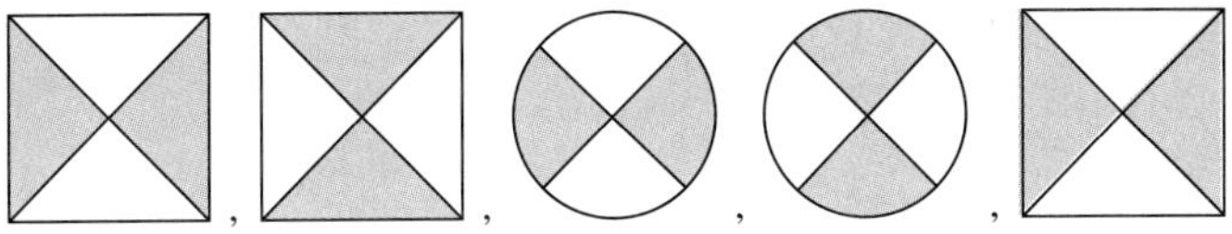

Solution: We notice an alternating pattern: square, square, circle, circle, square . . . . We would next expect a square followed by a circle.

We notice a shading pattern of horizontal, vertical, horizontal, vertical, horizontal . . . . We would next expect vertical, then horizontal. Thus, the next two figures are

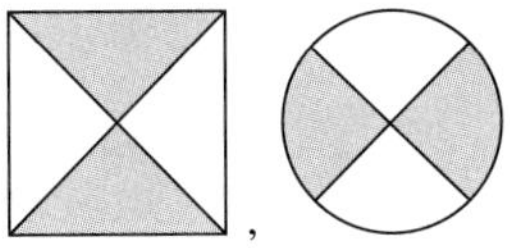

Practice Problem 4 Find the next two figures that would appear in the sequence.

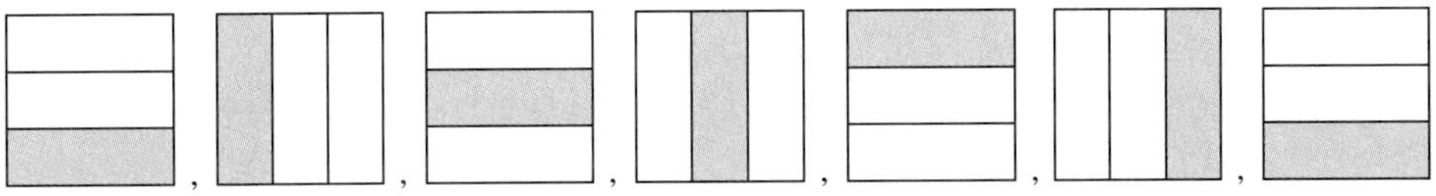

How accurate is inductive reasoning? Do we always come to the right conclusion? Conclusions arrived at by inductive reasoning are always tentative. They may require further investigation. When we use inductive reasoning we are using specific data to reach a general conclusion. However, we may have reached the wrong conclusion. To illustrate:

Take the sequence of numbers 20, 10, 5. What is the next number? You might say 2.5. In the sequence 20, 10, 5, each number appears to be one–half of the preceding number. Thus we would predict 20, 10, 5, 2.5, ...

But wait! There is another possibility. Maybe the sequence is 20, 10, 5, -5, -10, -20, -25, -35, -40, ...

To know for sure which answer is correct, we would need more information, such as more numbers in the sequence to verify the pattern. **You should always treat inductive reasoning conclusions as tentative, requiring further verification.**

2 *Deductive Reasoning*

Deductive reasoning requires us to take general facts, postulates, or accepted truths and use them to reach a specific conclusion. Suppose we know the following rules or "facts" of algebra:

For all numbers a, b, c and for all numbers $d \neq 0$:

1. *Addition principle of equations:* If $a = b$, then $a + c = b + c$.
2. *Division principle of equations:* If $a = b$, then $\frac{a}{d} = \frac{b}{d}$.
3. *Multiplication Principle of Equations:* If $a = b$, then $ac = bc$.
4. *Distributive Property:* $a(b + c) = ab + ac$.

EXAMPLE 5 Use deductive reasoning and the four properties listed above to justify each step in solving the equation.

$$2(7x - 2) = 38$$

Solution:

STATEMENT	REASON
1. $14x - 4 = 38$	**1.** Distributive property: $a(b + c) = ab + ac$ Here we distributed the 2.
2. $14x = 42$	**2.** Addition principle of equations: If $a = b$ then $a + c = b + c$ Here we added 4 to each side of the equation.
3. $\frac{14x}{14} = \frac{42}{14}$	**3.** Division principle of equations: If $a = b$ then $\frac{a}{d} = \frac{b}{d}$
$x = 3$	Here we divided each side by 14.

Practice Problem 5 Use deductive reasoning and the four properties listed above to justify each step in solving the equation.

$$\frac{1}{6}x = \frac{1}{3}x + 4 \quad ■$$

Sometimes we need to make conclusions about angles and lines in geometry. The following properties are useful. We will refer to angles 1, 2, 3, 4.

1. If two lines intersect, the opposite angles are equal. Here $\angle 1 = \angle 2$ and $\angle 4 = \angle 3$.

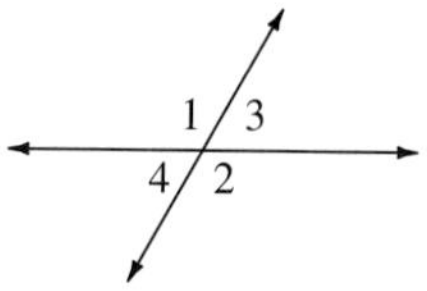

2. If two lines intersect, the adjacent angles are supplementary (they add up to 180°). Here, $\angle 4 + \angle 1 = 180°$, and $\angle 1 + \angle 3 = 180°$, also $\angle 3 + \angle 2 = 180°$, and $\angle 4 + \angle 2 = 180°$.

3. If a transversal (intersecting line) crosses two parallel lines, the alternate interior angles are equal. In the figure at right, line *P* is a transversal. If line *M* is parallel to line *N*, then $\angle 3 = \angle 4$ and $\angle 1 = \angle 2$.

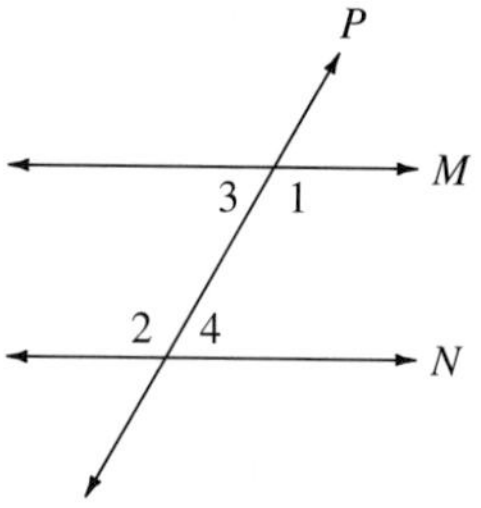

4. If two alternate interior angles on each side of a transversal cutting two straight lines are equal, the two straight lines are parallel. If $\angle 2 = \angle 4$, then line *M* is parallel to line *N*.

5. Supplements of equal angles are equal. In the figure at right, if $\angle 1$ and $\angle 2$ are supplementary, and $\angle 3$ and $\angle 4$ are supplementary, then if $\angle 1 = \angle 3$, then $\angle 2 = \angle 4$.

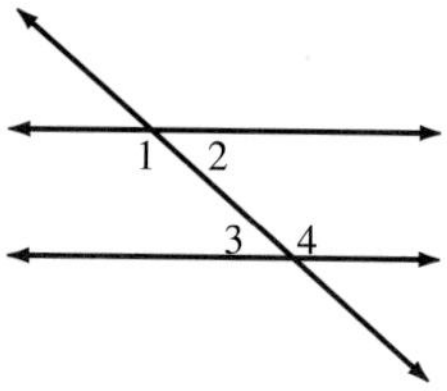

6. Transitive property of equality:

If $a = b$ and $b = c$, then $a = c$.

We will now use these facts to prove some geometric conclusions.

EXAMPLE 6 Prove that line M is parallel to line N if $\angle 1 = \angle 3$.

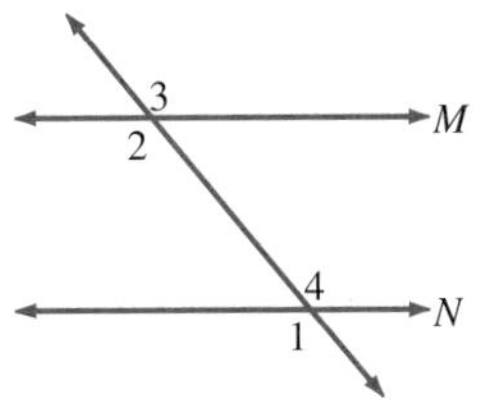

Solution:

STATEMENT	REASON
1. $\angle 3 = \angle 2$ and $\angle 4 = \angle 1$	**1.** If two lines intersect, the opposite angles are equal.
2. $\angle 2 = \angle 4$	**2.** Transitive property of equality.
3. Therefore, line M is parallel to line N. (We write this as $M \parallel N$.)	**3.** If two alternate interior angles on each side of a transversal cutting two straight lines are equal, the straight lines are parallel.

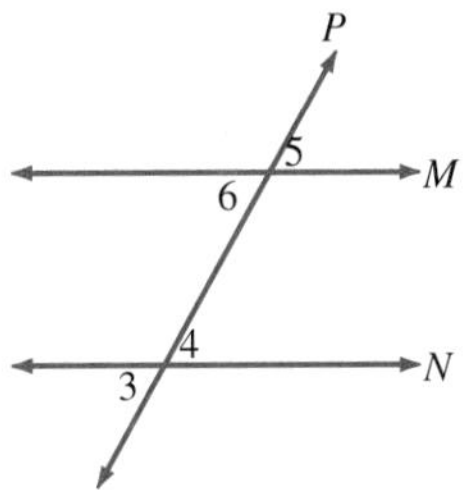

Practice Problem 6 Refer to the diagram in the margin. Line M is parallel to line N. $\angle 5 = 26°$. Prove $\angle 4 = 26°$. ■

Now let us see if we can use deductive reasoning to solve the following problem.

EXAMPLE 7 For next semester four professors from the psychology department have expressed the following desires for freshman courses. Each professor will teach only *one* freshman course next semester. The four freshman courses are General Psychology, Social Psychology, Psychology of Adjustment, and Educational Psychology.

1. Professors A and B don't want to teach General Psychology.
2. Professor C wants to teach Social Psychology.
3. Professor D will be happy to teach any course.
4. Professor B wants to teach Psychology of Adjustment.

Which professor will teach Educational Psychology, if all professors are given the courses they desire?

Solution: Let us organize the facts by listing the four freshman courses and each professor: A, B, C, and D.

General Psychology	Social Psychology	Psychology of Adjustment	Educational Psychology
A B C D	A B C D	A B C D	A B C D

STEP	REASON
1. We cross off Professors *A* and *B* from the General Psychology list.	**1.** Professors *A* and *B* don't want to teach General Psychology.

General Psychology			
~~*A*~~	*A*	*A*	*A*
~~*B*~~	*B*	*B*	*B*
C	*C*	*C*	*C*
D	*D*	*D*	*D*

STEP	REASON
2. We cross Professor *C* off every list (except Social Psychology) and mark that he will teach it.	**2.** Professor *C* wants to teach Social Psychology.

	Social Psychology		
~~*A*~~			
~~*B*~~			
~~*C*~~	[*C*]	~~*C*~~	~~*C*~~
D			

STEP	REASON
3. Professor *D* is thus the only person who can teach General Psychology. We cross him off every other list and mark that he will teach General Psychology.	**3.** Professor *D* is happy to teach any course.

~~*A*~~			
~~*B*~~			
~~*C*~~	[*C*]	~~*C*~~	~~*C*~~
[*D*]	~~*D*~~	~~*D*~~	~~*D*~~

STEP	REASON
4. We cross out all courses for Professor *B* except Psychology of Adjustment.	**4.** Professor *B* wants to teach Psychology of Adjustment.

		Psychology of Adjustment	
~~*A*~~	~~*A*~~	*A*	*A*
~~*B*~~	~~*B*~~	[*B*]	~~*B*~~
~~*C*~~	[*C*]	~~*C*~~	~~*C*~~
[*D*]	~~*D*~~	~~*D*~~	~~*D*~~

STEP	REASON
5. Professor *A* will teach Educational Psychology.	**5.** He is the only professor left. All others are assigned.

Educational Psychology

[*A*]

Practice Problem 7 A Honda, Toyota, Mustang, and a Camaro are parked side by side, but not in that order.

1. The Camaro is parked on the right end.
2. The Mustang is between the Honda and the Toyota.
3. The Honda is not next to the Camaro.

Which car is parked on the left end? ■

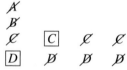

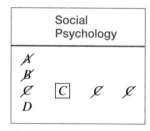

C Exercises

Find the next number in the sequence.

1. 2, 4, 6, 8, 10, 12 . . .

2. 0, 5, 10, 15, 20, 25 . . .

3. 7, 16, 25, 34, 43, . . .

4. 12, 25, 38, 51, 64, . . .

5. 1, 16, 81, 256, 625, . . .

6. 1, 6, 13, 22, 33, 46, . . .

7. $-7, 3, -6, 4, -5, 5, -4, 6, \ldots$

8. $2, -4, 8, -16, 32, -64, 128, \ldots$

9. $5x, 6x - 1, 7x - 2, 8x - 3, 9x - 4, \ldots$

10. $60x, 30x, 15x, 7.5x, 3.75x, 1.875x, \ldots$

Guess the next seven digits in the following irrational numbers.

11. 8.181181118 . . .

12. 3.043004300043 . . .

13. 12.98987987698765 . . .

14. 7.6574839201102938 . . .

15. 2.14916253649 . . .

16. 6.112223333 . . .

17. Find the next row in this triangular pattern.

1
1 1
1 2 1
1 3 3 1
1 4 6 4 1
1 5 10 10 5 1

18. Find the next two figures that would appear in the sequence.

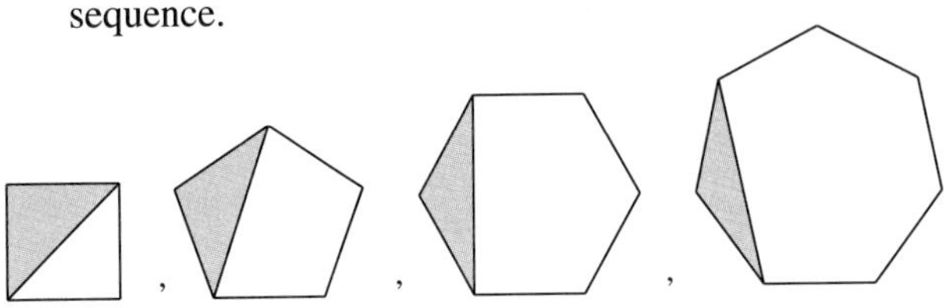

Use deductive reasoning and the four properties discussed in Example 5 to justify each step in solving each equation.

19. $12x - 30 = 6$

20. $3x + 7 = 4$

21. $4x - 3 = 3x - 5$

22. $2x - 9 = 4x + 5$

23. $8x - 3(x - 5) = 30$

24. $3x - 5(x - 1) = -5$

25. $\frac{1}{2}x + 6 = \frac{3}{2} + 6x$

26. $\frac{4}{5}x - 3 = \frac{1}{10}x + \frac{3}{5}$

Use deductive reasoning and the properties of geometry discussed in Example 6 to prove the following.

27. If $\angle 1 = \angle 5$,
prove that line P is parallel to line S.

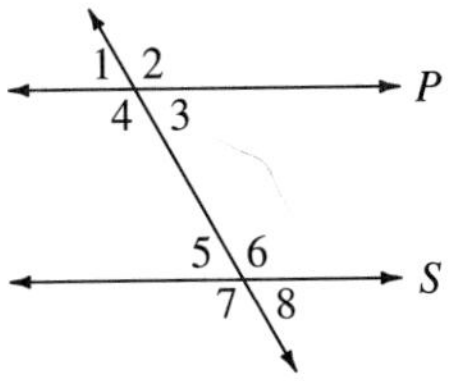

28. If line R is parallel to line S,
prove that $\angle 6 = \angle 3$.

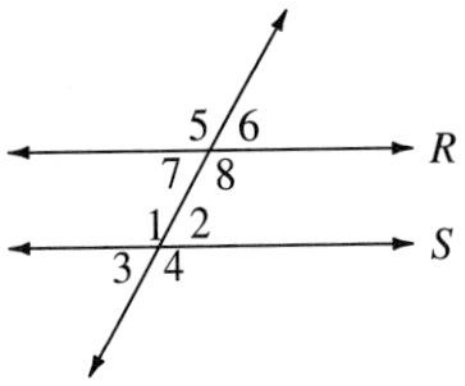

Solutions to Practice Problems

Chapter 0

0.1 Practice Problems

1. **(a)** $\frac{10}{16} = \frac{2 \times 5}{2 \times 8} = \frac{5}{8}$ **(b)** $\frac{24}{36} = \frac{12 \times 2}{12 \times 3} = \frac{2}{3}$

(c) $\frac{42}{36} = \frac{6 \times 7}{6 \times 6} = \frac{7}{6}$

2. **(a)** $\frac{4}{12} = \frac{2 \times 2 \times 1}{2 \times 2 \times 3} = \frac{1}{3}$ **(b)** $\frac{25}{125} = \frac{5 \times 5 \times 1}{5 \times 5 \times 5} = \frac{1}{5}$

(c) $\frac{73}{146} = \frac{73 \times 1}{73 \times 2} = \frac{1}{2}$

3. **(a)** $\frac{18}{6} = \frac{3 \times 6}{6} = 3$ **(b)** $\frac{146}{73} = \frac{73 \times 2}{73} = 2$

(c) $\frac{28}{7} = \frac{7 \times 4}{7} = 4$

4. 56 out of 154 $= \frac{56}{154} = \frac{2 \times 7 \times 4}{2 \times 7 \times 11} = \frac{4}{11}$

5. **(a)** $\frac{12}{7} = 12 \div 7 = 7\overline{)12} = 1\frac{5}{7}$

$$\begin{array}{r} 1 \\ 7\overline{)12} \\ \underline{-7} \\ 5 \end{array}$$ Remainder

(b) $\frac{20}{5} = 20 \div 5 = 5\overline{)20} = 4$

$$\begin{array}{r} 4 \\ 5\overline{)20} \\ \underline{20} \\ 0 \end{array}$$ Remainder

6. **(a)** $3\frac{2}{5} = \frac{(3 \times 5) + 2}{5} = \frac{15 + 2}{5} = \frac{17}{5}$

(b) $1\frac{3}{7} = \frac{(1 \times 7) + 3}{7} = \frac{7 + 3}{7} = \frac{10}{7}$

(c) $2\frac{6}{11} = \frac{(2 \times 11) + 6}{11} = \frac{22 + 6}{11} = \frac{28}{11}$

(d) $4\frac{2}{3} = \frac{(4 \times 3) + 2}{3} = \frac{12 + 2}{3} = \frac{14}{3}$

7. **(a)** $\frac{3}{8} = \frac{?}{24}$ Observe $8 \times 3 = 24$

$\frac{3 \times}{8 \times}\left|\frac{3}{3}\right| = \frac{9}{24}$

(b) $\frac{5}{6} = \frac{?}{30}$

$\frac{5 \times}{6 \times}\left|\frac{5}{5}\right| = \frac{25}{30}$

(c) $\frac{12}{13} = \frac{?}{26}$

$\frac{12 \times}{13 \times}\left|\frac{2}{2}\right| = \frac{24}{26}$

(d) $\frac{2}{7} = \frac{?}{56}$

$\frac{2 \times}{7 \times}\left|\frac{8}{8}\right| = \frac{16}{56}$

(e) $\frac{5}{9} = \frac{?}{27}$

$\frac{5 \times}{9 \times}\left|\frac{3}{3}\right| = \frac{15}{27}$

(f) $\frac{3}{10} = \frac{?}{60}$

$\frac{3 \times}{10 \times}\left|\frac{6}{6}\right| = \frac{18}{60}$

(g) $\frac{3}{4} = \frac{?}{28}$

$\frac{3 \times}{4 \times}\left|\frac{7}{7}\right| = \frac{21}{28}$

(h) $\frac{8}{11} = \frac{?}{55}$

$\frac{8 \times}{11 \times}\left|\frac{5}{5}\right| = \frac{40}{55}$

0.2 Practice Problems

1. **(a)** $\frac{3}{6} + \frac{2}{6} = \frac{3 + 2}{6} = \frac{5}{6}$

(b) $\frac{3}{11} + \frac{2}{11} + \frac{6}{11} = \frac{3 + 2 + 6}{11} = \frac{11}{11} = 1$

(c) $\frac{5}{8} + \frac{2}{8} + \frac{7}{8} = \frac{5 + 2 + 7}{8} = \frac{14}{8} = 1\frac{3}{4}$

2. **(a)** $\frac{11}{13} - \frac{6}{13} = \frac{11 - 6}{13} = \frac{5}{13}$

(b) $\frac{8}{9} - \frac{2}{9} = \frac{8 - 2}{9} = \frac{6}{9} = \frac{2}{3}$

3. Find LCD of $\frac{1}{8}$ and $\frac{5}{12}$.

$8 = 2 \cdot 2 \cdot 2$

$12 = \downarrow\ 2 \cdot 2 \cdot 3$

$2 \cdot 2 \cdot 2 \cdot 3 = 24$

LCD = 24

4. Find the LCD using Prime Factors.

$\frac{8}{35}$ and $\frac{6}{15}$

$35 = 7 \cdot 5$

$15 = \quad 5 \cdot 3$

$7 \cdot 5 \cdot 3$ LCD = 105

5. Find LCD of $\frac{5}{12}$ and $\frac{7}{30}$.

$12 = 3 \cdot 2 \cdot 2$

$30 = 3 \quad 2 \cdot 5$

$3 \cdot 2 \cdot 2 \cdot 5$ LCD = 60

6. Find LCD of $\frac{1}{18}, \frac{2}{27}$ and $\frac{5}{12}$.

$12 = 2 \cdot 2 \cdot 3$

$18 = 2 \quad 3 \cdot 3$

$27 = \quad 3 \cdot 3 \cdot 3$

$2 \cdot 2 \cdot 3 \cdot 3 \cdot 3$ LCD = 108

7. Add $\frac{1}{8} + \frac{5}{12}$

First find the LCD.

$8 = 2 \cdot 2 \cdot 2$

$12 = 2 \cdot 2 \quad \cdot 3$

$2 \cdot 2 \cdot 2 \cdot 3$ LCD = 24

Then change to equivalent fractions and add.

$\frac{1}{8} \times \left|\frac{3}{3}\right| + \frac{5}{12} \times \left|\frac{2}{2}\right| = \frac{3}{24} + \frac{10}{24} = \frac{3 + 10}{24} = \frac{13}{24}$

8. $\frac{3}{5} + \frac{4}{25} + \frac{1}{10}$

First find the LCD.

$5 \quad 5$

$10 \quad 2 \cdot 5$

$25 \quad 5 \cdot 5$

$2 \cdot 5 \cdot 5$ LCD = 50

Then change to equivalent fractions and add.

$\frac{3}{5} \times \left|\frac{10}{10}\right| + \frac{4}{25} \times \left|\frac{2}{2}\right| + \frac{1}{10} \times \left|\frac{5}{5}\right| = \frac{30}{50} + \frac{8}{50} + \frac{5}{50}$

$= \frac{30 + 8 + 5}{50} = \frac{43}{50}$

9. $\frac{1}{49} + \frac{3}{14}$

First find the LCD.

$14 = 2 \cdot 7$

$49 = \quad 7 \cdot 7$

$2 \cdot 7 \cdot 7$ LCD = 98

Then change to equivalent fractions and add.

$\frac{1}{49} \times \left|\frac{2}{2}\right| + \frac{3}{14} \times \left|\frac{7}{7}\right| = \frac{2}{98} + \frac{21}{98} = \frac{2 + 21}{98} = \frac{23}{98}$

10. $\frac{1}{12} - \frac{1}{30}$

First find the LCD.

$12 = 2 \cdot 2 \cdot 3$

$30 = \quad 2 \cdot 3 \cdot 5$

$2 \cdot 2 \cdot 3 \cdot 5 \qquad \text{LCD} = 60$

Then change to equivalent fractions and subtract.

$$\frac{1}{12} \times \left|\frac{5}{5}\right| - \frac{1}{30} \times \left|\frac{2}{2}\right| = \frac{5}{60} - \frac{2}{60} = \frac{5-2}{60} = \frac{3}{60} = \frac{1}{20}$$

11. $\frac{2}{3} + \frac{3}{4} - \frac{3}{8}$

First find the LCD.

3   3

4   $2 \cdot 2$

8   $2 \cdot 2 \cdot 2$

$2 \cdot 2 \cdot 2 \cdot 3 \qquad \text{LCD} = 24$

Then change to equivalent fractions and add and subtract.

$$\frac{2}{3} \times \left|\frac{8}{8}\right| + \frac{3}{4} \times \left|\frac{6}{6}\right| - \frac{3}{8} \times \left|\frac{3}{3}\right|$$

$$= \frac{16}{24} + \frac{18}{24} - \frac{9}{24} = \frac{16 + 18 - 9}{24} = \frac{25}{24} = 1\frac{1}{24}$$

12. (a) $1\frac{2}{3} + 2\frac{4}{5} = \frac{5}{3} + \frac{14}{5} = \frac{5}{3} \times \left|\frac{5}{5}\right| + \frac{14}{5} \times \left|\frac{3}{3}\right| = \frac{25}{15} + \frac{42}{15}$

$$= \frac{25 + 42}{15} = \frac{67}{15} = 4\frac{7}{15}$$

(b) $5\frac{1}{4} - 2\frac{2}{3} = \frac{21}{4} - \frac{8}{3} = \frac{21}{4} \times \left|\frac{3}{3}\right| - \frac{8}{3} \times \left|\frac{4}{4}\right| = \frac{63}{12} - \frac{32}{12}$

$$= \frac{63 - 32}{12} = \frac{31}{12} = 2\frac{7}{12}$$

13.

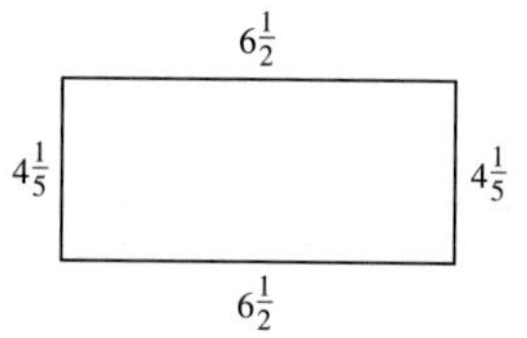

$$4\frac{1}{5} + 4\frac{1}{5} + 6\frac{1}{2} + 6\frac{1}{2}$$

$$= \frac{21}{5} + \frac{21}{5} + \frac{13}{2} + \frac{13}{2}$$

$= \text{LCD} = 10$

$$\frac{21}{5} \times \left|\frac{2}{2}\right| + \frac{21}{5} \times \left|\frac{2}{2}\right| + \frac{13}{2} \times \left|\frac{5}{5}\right| + \frac{13}{2} \times \left|\frac{5}{5}\right|$$

$$= \frac{42}{10} + \frac{42}{10} + \frac{65}{10} + \frac{65}{10} = \frac{42 + 42 + 65 + 65}{10} = \frac{214}{10} = 21\frac{2}{5}$$

0.3 Practice Problems

1. (a) $\frac{2}{7} \times \frac{5}{11} = \frac{2 \cdot 5}{7 \cdot 11} = \frac{10}{77}$

(b) $\frac{8}{9} \times \frac{3}{10} = \frac{8 \cdot 3}{9 \cdot 10} = \frac{24}{90} = \frac{4}{15}$

2. (a) $\frac{3}{5} \times \frac{4}{3} = \frac{3 \cdot 4}{5 \cdot 3} = \frac{4}{5}$

(b) $\frac{9}{10} \times \frac{5}{12} = \frac{3 \cdot 3}{2 \cdot 5} \times \frac{5}{2 \cdot 2 \cdot 3} = \frac{3}{8}$

3. (a) $4 \times \frac{2}{7} = \frac{4}{1} \times \frac{2}{7} = \frac{4 \cdot 2}{1 \cdot 7} = \frac{8}{7} = 1\frac{1}{7}$

(b) $12 \times \frac{3}{4} = \frac{12}{1} \times \frac{3}{4} = \frac{2 \cdot 2 \cdot 3}{1} \times \frac{3}{2 \cdot 2} = 9$

4. (a) $2\frac{1}{5} \times \frac{3}{7} = \frac{11}{5} \times \frac{3}{7} = \frac{11 \cdot 3}{5 \cdot 7} = \frac{33}{35}$

(b) $3\frac{1}{3} \times 1\frac{2}{5} = \frac{10}{3} \times \frac{7}{5} = \frac{2 \cdot 5}{3} \times \frac{7}{5} = \frac{14}{3} = 4\frac{2}{3}$

5. $3\frac{1}{2} \times \frac{1}{14} \times 4 = \frac{7}{2} \times \frac{1}{14} \times \frac{4}{1} = \frac{7}{2} \times \frac{1}{2 \cdot 7} \times \frac{2 \cdot 2}{1} = 1$

6. (a) $\frac{2}{5} \div \frac{1}{3} = \frac{2}{5} \times \frac{3}{1} = \frac{6}{5}$

(b) $\frac{12}{13} \div \frac{4}{3} = \frac{2 \cdot 2 \cdot 3}{13} \times \frac{3}{2 \cdot 2} = \frac{9}{13}$

7. (a) $\frac{3}{7} \div 6 = \frac{3}{7} \div \frac{6}{1} = \frac{3}{7} \times \frac{1}{6} = \frac{3}{42} = \frac{1}{14}$

(b) $8 \div \frac{2}{3} = \frac{8}{1} \times \frac{3}{2} = 12$

8. (a) $\dfrac{\frac{3}{11}}{\frac{5}{7}} = \frac{3}{11} \div \frac{5}{7} = \frac{3}{11} \times \frac{7}{5} = \frac{21}{55}$

(b) $\dfrac{\frac{12}{5}}{\frac{8}{15}} = \frac{12}{5} \div \frac{8}{15} = \frac{2 \cdot 2 \cdot 3}{5} \times \frac{3 \cdot 5}{2 \cdot 2 \cdot 2} = \frac{9}{2} = 4\frac{1}{2}$

9. (a) $1\frac{2}{5} \div 2\frac{1}{3} = \frac{7}{5} \div \frac{7}{3} = \frac{7}{5} \times \frac{3}{7} = \frac{3}{5}$

(b) $4\frac{2}{3} \div 7 = \frac{14}{3} \times \frac{1}{7} = \frac{2 \cdot 7}{3} \times \frac{1}{7} = \frac{2}{3}$

(c) $\dfrac{1\frac{1}{5}}{1\frac{2}{7}} = 1\frac{1}{5} \div 1\frac{2}{7} = \frac{6}{5} \div \frac{9}{7} = \frac{6}{5} \times \frac{7}{9} = \frac{2 \cdot 3}{5} \times \frac{7}{3 \cdot 3} = \frac{14}{15}$

10. $64 \div 5\frac{1}{3} = 64 \div \frac{16}{3} = 64 \times \frac{3}{16} = 12$

11. $25\frac{1}{2} \times 5\frac{1}{4} = \frac{51}{2} \times \frac{21}{4} = \frac{1071}{8} = 133\frac{7}{8}$

0.4 Practice Problems

1. (a) 1.371 $1\frac{371}{1000}$

One and three hundred seventy one thousandths.

(b) $\frac{9}{100}$ = Nine hundredths.

2. (a) $\frac{3}{8}$ $8\overline{)3.00}$ = .375

```
   .375
8)3.00
  2 4
    60
    56
     40
     40
      0
```

(b) $\frac{7}{200}$ $200\overline{)7.000}$ = 0.035

```
      0.035
200)7.000
    6 00
    1 000
    1 000
        0
```

(c) $\frac{33}{20}$ $20\overline{)33.00}$ = 1.65

```
     1.65
20)33.00
   20
   13 0
   12 0
    1 00
    1 00
       0
```

3. (a) $\frac{1}{6} = 6\overline{)1.00}$ $.166 = 0.1\overline{6}$

```
  .166
6)1.000
  6
  40
  36
   40
   36
    4
```

(b) $\frac{5}{11} = 11\overline{)5.000}$ $.4545 = .\overline{45}$

```
   .4545
11)5.0000
   4 4
     60
     55
      50
      44
       60
       55
        5
```

4. (a) $0.8 \quad \frac{8}{10} = \frac{2 \cdot 2 \cdot 2}{2 \cdot 5} = \frac{4}{5}$

(b) $0.88 \quad \frac{88}{100} = \frac{11 \cdot 2 \cdot 2 \cdot 2}{5 \cdot 5 \cdot 2 \cdot 2} = \frac{22}{25}$

(c) $0.45 \quad \frac{45}{100} = \frac{5 \cdot 3 \cdot 3}{5 \cdot 5 \cdot 2 \cdot 2} = \frac{9}{20}$

(d) $0.148 \quad \frac{148}{1000} = \frac{2 \cdot 2 \cdot 37}{5 \cdot 5 \cdot 5 \cdot 2 \cdot 2 \cdot 2} = \frac{37}{250}$

(e) $0.612 \quad \frac{612}{1000} = \frac{17 \cdot 3 \cdot 3 \cdot 2 \cdot 2}{5 \cdot 5 \cdot 5 \cdot 2 \cdot 2 \cdot 2} = \frac{153}{250}$

(f) $0.016 \quad \frac{16}{1000} = \frac{2 \cdot 2 \cdot 2 \cdot 2}{5 \cdot 5 \cdot 5 \cdot 2 \cdot 2 \cdot 2} = \frac{2}{125}$

5. (a) $\begin{array}{r} 3.12 \\ 5.08 \\ \underline{1.42} \\ 9.62 \end{array}$ **(b)** $\begin{array}{r} 152.003 \\ \underline{-\ 136.118} \\ 15.885 \end{array}$

(c) $\begin{array}{r} 1.1 \\ 3.16 \\ \underline{5.123} \\ 9.383 \end{array}$ **(d)** $\begin{array}{r} 1.0052 \\ \underline{-\ 0.1234} \\ 0.8818 \end{array}$

6. (a) $\begin{array}{r} 0.061 \\ 5.0008 \\ \underline{1.3} \\ 6.3618 \end{array}$ **(b)** $\begin{array}{r} 18.000 \\ \underline{-\ \ 0.126} \\ 17.874 \end{array}$

7. $\begin{array}{r} 0.5 \\ \underline{\times\ 0.3} \\ 0.15 \end{array}$ **8.** $\begin{array}{r} 0.12 \\ \underline{\times\ \ 0.4} \\ 0.048 \end{array}$

9. (a) $\begin{array}{r} 1.23 \\ \underline{\times\ 0.005} \\ 0.00615 \end{array}$ **(b)** $\begin{array}{r} 0.00002 \\ \underline{\times\ \ \ 0.003} \\ 0.00000006 \end{array}$

10. $0.18\overline{)0.11\,16}$ = 0.62

$\begin{array}{r} 10\,8 \\ \hline 3\,6 \\ \underline{3\,6} \\ 0 \end{array}$

11. $0.3\overline{)0.0.9}$ = 0.3

12. $.0.6.\overline{)1800.0.0.00}$ = 30,000.

13. $4.9.\overline{)0.0.1764}$

$49\overline{)0.1764}$ = .0036

$\begin{array}{r} 147 \\ \hline 294 \\ \underline{294} \\ 0 \end{array}$

14. (a) $0.0016 \times 100 = .16$
Move decimal point 2 places to the right.
(b) $2.34 \times 1000 = 2{,}340$
Move decimal point 3 places to the right.
(c) $56.75 \times 10{,}000 = 567{,}500$
Move decimal point 4 places to the right.

15. (a) $\frac{5.82}{10}$ = Move decimal point 1 place to the left. 0.582
(b) $\frac{123.4}{1000}$ = Move decimal point 3 places to the left. 0.1234
(c) $\frac{0.00614}{10{,}000}$ = Move decimal point 4 places to the left.
0.000000614

0.5 Practice Problems

1. (a) $0.92 = 92\%$ **(b)** $0.418 = 41.8\%$ **(c)** $0.7 = 70\%$
2. (a) $0.0019 = 0.19\%$ **(b)** $0.0736 = 7.36\%$
(c) $0.0003 = 0.03\%$
3. (a) 304% **(b)** 518.6% **(c)** 210%
4. (a) .82 **(b)** 3.47
5. (a) 0.07 **(b)** 0.093 **(c)** 0.002
6. (a) 1.31 **(b)** 3.016 **(c)** 0.0004
7. Change % to decimal and multiply.
(a) $.18 \times 50 = 9$ **(b)** $0.04 \times 64 = 2.56$
(c) $1.56 \times 35 = 54.6$ **(d)** $0.008 \times 60 = 0.48$
(e) $0.013 \times 82 = 1.066$ **(f)** $0.00002 \times 564 = 0.01128$
8. $4.2\% \times 18{,}000 = 0.042 \times 18{,}000$
(a) 756.00 **(b)** $18{,}000 + 756.00 = \$18{,}756.00$
9. $\frac{37}{148}$ reduce to $\frac{37 \cdot 1}{37 \cdot 2 \cdot 2} = \frac{1}{4} = .25 = 25\%$
10. (a) $\frac{24}{48} = \frac{2 \cdot 2 \cdot 2 \cdot 3}{2 \cdot 2 \cdot 2 \cdot 2 \cdot 3} = \frac{1}{2} = 0.5 = 50\%$
(b) $\frac{4}{25} = 0.16 = 16\%$
11. $\frac{430}{1256} = \frac{215}{628} = 0.342 = 34\%$

0.6 Practice Problems

1. $100{,}000 \times 400 = 40{,}000{,}000$
2. $12\frac{1}{2}$ 10; $9\frac{3}{4} \rightarrow 10$; First room $10 \times 10 = 100$ square feet.
$14\frac{1}{4}$ 15; $18\frac{1}{2} \rightarrow 20$; Second room $15 \times 20 = 300$ square feet.
Both rooms $100 + 300 = 400$ square feet.
3. (a) $422.8 \rightarrow 400$ miles
$\frac{400}{20} = 20$ miles/gallon
$19.3 \rightarrow 20$ gallons of gas
(b) $1.29\frac{9}{10} \rightarrow \$1.00$ per gallon
$\frac{4{,}000 \text{ miles}}{20 \text{ m/gallon}} = 200$ gallons
$3862 \rightarrow 4{,}000$ miles
200 gallons $\times$ \$1.00 per gallon = \$200.00
4. (a) $\frac{3{,}580{,}000{,}000}{43{,}300}$ estimate $\rightarrow$ $\frac{4{,}000{,}000{,}000 \text{ miles}}{40{,}000 \text{ m/hour}}$
100,000 hours
24 hours/day estimate $\rightarrow \frac{100{,}000}{20} = 5{,}000$ days
5. 56.93% estimate $\rightarrow 60\% \rightarrow 0.6$
293,567.12 estimate $\rightarrow$ 300,000
$0.6 \times 300{,}000 = 180{,}000$

0.7 Practice Problems

1.

Gather the Facts	What Am I Solving for	What Must I Calculate	Key Points to Remember
Living Room measures $16\frac{1}{2}$ ft × $10\frac{1}{2}$ ft.	Area of room in square feet	Multiply $16\frac{1}{2}$ by $10\frac{1}{2}$ ft. to get the area in square feet.	9 sq. feet = 1 square yard
The carpet cost \$20.00 per square yard.	Area of room in square yards. Cost of the carpet.	Divide the number of square feet by 9 to get the number of square yards. Multiply the number of square yards by \$20.00.	

$16\frac{1}{2} \times 10\frac{1}{2} = 173\frac{1}{4}$ square feet
$173\frac{1}{4} \div 9 = 19\frac{1}{4}$ square yards
$19\frac{1}{4} \times 20 = \$385.00$ total cost of carpet
CHECK:
Estimate area of room $16 \times 10 = 160$ square feet.
Estimate area in square yards $160 \div 10 = 16$
Estimate the cost $16 \times 20 = \$320.00$
This is close to our answer of \$385.00. Our answer seems reasonable.

2. (a) $\frac{10}{55} = 0.181 = 18\%$ **(b)** $\frac{3{,}660{,}000}{13{,}240{,}000} = 0.276 = 28\%$
(c) $\frac{15}{55} = 0.272 = 27\%$ **(d)** $\frac{3{,}720{,}000}{13{,}240{,}000} = 0.280 = 28\%$
(e) We notice that 18% of the company's sales force is located in the Northwest, and they were responsible for 28% of the sales volume. The percent of sales compared to the percent of sales force is about 150%. 27% of the company's sales force is located in the Southwest, and they were responsible for 28% of the sales volume. The percent of sales compared to the percent of sales force is approximately 100%. It would appear that the Northwest sales force is more effective.

Chapter 1

1.1 Practice Problems

1.

	Number	Integer	Rational Number	Irrational Number
(a)	$-\frac{2}{5}$		X	
(b)	1.515151...		X	
(c)	−8	X	X	
(d)	π			X

2. (a) Population growth of 1,259 is +1,259.
   (b) Depreciation of \$763 is −\$763.00.
   (c) Wind chill factor of minus 10 is −10.
3. (a) The opposite of $+\frac{2}{5}$ is $-\frac{2}{5}$.
   (b) The opposite of −1.92 is +1.92.
4. $(-23) + (-35)$; $23 + 35$; -58
5. $\left(\frac{-3}{5}\right) + \left(\frac{-4}{7}\right)$; $\frac{3}{5} + \frac{4}{7}$; $\frac{21}{35} + \frac{20}{35}$; $-\frac{41}{35}$
6. $(-12.7) + (-9.38)$; $12.7 + 9.38$; -22.08
7. $(-7) + (-11) + (-33)$; $7 + 11 + 33$; -51
8. $(-9) + (+3)$; $9 - 3$; -6
9. $(+37) + (-40)$; $40 - 37$; -3
10. $\left(-\frac{5}{12}\right) + \left(+\frac{7}{12}\right) + \left(-\frac{11}{12}\right)$
    $\frac{7}{12} - \frac{5}{12} = \frac{2}{12}$
    $\frac{2}{12} + \left(-\frac{11}{12}\right)$; $\frac{11}{12} - \frac{2}{12}$; $-\frac{9}{12} = -\frac{3}{4}$
11. $(-6) + (-8) + (+2)$
    $6 + 8 = 14$; $-14 + 2 = -12$
12. $(-6) + (+5) + (-7) + (-2) + (+5) + (+3)$
    $$\begin{array}{rr} -6 & +5 \\ -7 & +5 \\ -2 & +3 \\ \hline -15 & 13 \end{array} \qquad -15 + (+13) = -2$$
13. (a) $-2.9 + (-5.7) = -8.6$
    (b) $\frac{2}{3} + \left(-\frac{1}{4}\right)$
    $\frac{8}{12} + \left(-\frac{3}{12}\right) = \frac{5}{12}$
    (c) $-10 + (-3) + 15 + 4$
    $-13 + 19 = 6$

1.2 Practice Problems

1. $(-9) - (+3)$
   $-9 + (-3)$
   -12
2. $(+9) - (+12)$
   $+9 + (-12)$
   -3
3. $\left(-\frac{1}{5}\right) - \left(+\frac{1}{4}\right) = -\frac{1}{5} + \left(-\frac{1}{4}\right) = -\frac{9}{20}$
4. $\left(-\frac{3}{17}\right) - \left(-\frac{3}{17}\right) = -\frac{3}{17} + \left(+\frac{3}{17}\right) = 0$
5. Subtract −18 from 30
   $30 - (-18)$
   $30 + (+18)$
   48
6. (a) $-21 - 9$
   $-21 + (-9)$
   -30
   (b) $17 - 36$
   $17 + (-36)$
   -19
7. $350 - (-186)$
   $350 + (186)$
   536 The helicopter is 536 feet from the sunken vessel.

1.3 Practice Problems

1. (a) $(-6)(-2) = +12$ (b) $(7)(9) = +63$
   (c) $\left(-\frac{3}{5}\right)\left(\frac{2}{7}\right) = -\frac{6}{35}$ (d) $(40)(-20) = -800$
2. $(-3)(-4)(-3)$
   $(+12)(-3) = -36$
3. (a) $(-2)(-3)(-4)(-1)$
   $(2)(3)(4)(1)$
   $(6)(4)(1)$
   $(24)(1) = +24$
   (b) $(-1)(-3)(-2)$
   $-(1)(3)(2)$
   $-(3)(2)$
   -6
   (c) $(-4)(-\frac{1}{4})(-2)(-6)$
   $(4)(\frac{1}{4})(2)(6)$
   $(1)(2)(6)$
   $(2)(6) = +12$
4. $(-2)(3)(-4)(-1)(-2)$
   $(2)(3)(4)(1)(2)$
   $(6)(4)(1)(2)$
   $(24)(1)(2)$
   $(24)(2) = +48$
5. (a) $(-36) \div (-2) = +18$ (b) $18 \div 9 = +2$
   (c) $\frac{50}{-10} = -5$ (d) $(-49) \div 7 = -7$
6. (a) $(-1.242) \div (-1.8)$
   $$\begin{array}{r} .69 \\ 18.\overline{)12.42} \\ 10\,8 \\ \underline{8} \\ 1\,62 \\ \underline{1\,62} \\ 0 \end{array}$$
   Thus $(-1.242) \div (-1.8) = 0.69$
   (b) $(0.235) \div (-0.0025)$
   $$\begin{array}{r} 94. \\ 25\overline{)2350.00} \\ \underline{225} \\ 100 \\ \underline{100} \\ 0 \end{array}$$
   Thus $(0.235) \div (-0.0025) = -94$
7. $\left(-\frac{5}{16}\right) \div \left(-\frac{10}{13}\right) = \left(-\frac{5}{16}\right) \times \left(-\frac{13}{10}\right) = \frac{13}{32}$
8. (a) $\dfrac{-12}{-\frac{4}{5}} = -12 \div \left(-\frac{4}{5}\right) = -12 \times \left(-\frac{5}{4}\right) = 15$
   (b) $\dfrac{-\frac{2}{9}}{\frac{8}{13}} = \left(-\frac{2}{9}\right) \times \left(\frac{13}{8}\right) = -\frac{13}{36}$

1.4 Practice Problems

1. (a) 6^4 (b) $(-2)^5$ (c) 108^3 (d) $(-11)^6$
2. (a) $5^3 = (5)(5)(5) = 125$
   (b) $3^5 = (3)(3)(3)(3)(3) = 243$
   (c) $9^4 = (9)(9)(9)(9) = 6561$
3. (a) $(-3)^3 = -27$
   (b) $-2^4 = -(2^4) = -16$
   (c) $(-2)^6 = 64$
   (d) $(-1)^{10} = 1$
4. (a) $(\frac{1}{3})^3 = (\frac{1}{3})(\frac{1}{3})(\frac{1}{3}) = \frac{1}{27}$
   (b) $(0.3)^4 = (0.3)(0.3)(0.3)(0.3) = 0.0081$
   (c) $(\frac{3}{2})^4 = (\frac{3}{2})(\frac{3}{2})(\frac{3}{2})(\frac{3}{2}) = \frac{81}{16} = 5\frac{1}{16}$

1.5 Practice Problems

1. (a) $-3(x + 2y) = -3x - 6y$
   (b) $-a(a - 3b) = -a^2 + 3ab$
   (c) $5x(x + 4y) = 5x^2 + 20xy$
2. (a) $-(-3x + y) = 3x - y$
   (b) $-(5r - 3s - 1) = -5r + 3s + 1$

3. (a) $\frac{3}{5}(a^2 - 5a + 25) = \frac{3}{5}a^2 - 3a + 15$
(b) $2.5(x^2 - 3.5x + 1.2) = 2.5x^2 - 8.75x + 3$
4. (a) $(3x^2 - 2x)(-4) = -12x^2 + 8x$
(b) $(4y - 7x^2 + 1)(4y) = 16y^2 - 28x^2y + 4y$

1.6 Practice Problems

1. (a) $5a + 2b + 8a - 4b$
$5a$ and $8a$ are like terms.
$2b$ and $-4b$ are like terms.
(b) $x^2 + y^2 + 3x - 7y^2$
y^2 and $-7y^2$ are like terms.
Note that x^2 and $3x$ are not like terms.
2. (a) List the terms: $5x^2 - 6x + 8 - 4x - 9x^2 + 2$
They are $5x^2$, $-6x$, 8, $-4x$, $-9x^2$, and 2.
(b) $5x^2$ and $-9x^2$ are like terms.
$-6x$ and $-4x$ are like terms.
8 and 2 are like terms.
3. (a) $5a + 7a + 4a = (5 + 7 + 4)a = 16a$
(b) $16y^3 + 9y^3 = (16 + 9)y^3 = 25y^3$
4. $-8y^2 - 9y^2 + 4y^2 = (-8 - 9 + 4)y^2 = -13y^2$
5. (a) $-x + 3a - 9x + 2a = -x - 9x + 3a + 2a = -10x + 5a$
(b) $5ab - 2ab^2 - 3a^2b + 6ab = 5ab + 6ab - 2ab^2 - 3a^2b$
$= 11ab - 2ab^2 - 3a^2b$
(c) $7x^2y - 2xy^2 - 3x^2y - 4xy^2 + 5x^2y$
$= 7x^2y - 3x^2y + 5x^2y - 2xy^2 - 4xy^2 = 9x^2y - 6xy^2$
6. $5xy - 2x^2y + 6xy^2 - xy - 3xy^2 - 7x^2y$
$= 5xy - xy - 7x^2y - 2x^2y + 6xy^2 - 3xy^2$
$= 4xy - 9x^2y + 3xy^2$
7. $5a(2 - 3b) - 4(6a + 2ab) = 10a - 15ab - 24a - 8ab$
$= 10a - 24a - 15ab - 8ab = -14a - 23ab$
8. $\frac{1}{7}a^2 - \frac{5}{12}b + 2a^2 - \frac{1}{3}b$
$= \left(\frac{1}{7} + 2\right)a^2 + \left[\left(-\frac{5}{12}\right) + \left(-\frac{1}{3}\right)\right]b$
$= \frac{15}{7}a^2 + \left[\left(-\frac{5}{12}\right) + \frac{(-1)(4)}{(3)(4)}\right]b = \frac{15}{7}a^2 - \frac{9}{12}b = \frac{15}{7}a^2 - \frac{3}{4}b$

1.7 Practice Problems

1. $18 - (-3)^3$
$18 - (-27)$
$18 + (+27)$
45

2. $12 \div (-2)(-3) - 2^4$
$12 \div (-2)(-3) - 16$
$(-6)(-3) - 16$
$18 - 16 = 2$

3. $6 - (8 - 12)^2 + 8 \div 2$
$6 - (-4)^2 + 8 \div 2$
$6 - (16) + 4$
$-10 + 4$
-6

4. $\left(-\frac{1}{7}\right)\left(-\frac{14}{5}\right) + \left(-\frac{1}{2}\right) \div \left(\frac{3}{4}\right)$
$\left(-\frac{1}{7}\right)\left(-\frac{14}{5}\right) + \left(-\frac{1}{2} \times \frac{4}{3}\right)$
$\left(\frac{2}{5}\right) + \left(-\frac{2}{3}\right)$
$\left(\frac{6}{15}\right) + \left(-\frac{10}{15}\right) = -\frac{4}{15}$

1.8 Practice Problems

1. Evaluate $4 - \frac{1}{2}x$, for $x = -8$.
$4 - \frac{1}{2}(-8)$
$4 + 4$
8

2. Evaluate for $x = -3$.
(a) $-x^4 = -(-3)^4$
$= -(81) = -81$
(b) $(-x)^4 = [-(-3)]^4$
$= (3)^4 = 81$

3. Evaluate $(5x)^3 + 2x$, for $x = -2$.
$[5(-2)]^3 + 2(-2)$
$(-10)^3 + (-4)$
$(-1000) + (-4)$
-1004

4. Area of a triangle is
$A = \frac{1}{2}bh$
height = 3 meters (m)
base = 7 meters (m)
$A = \frac{1}{2}(7\text{m})(3\text{m})$
$A = \frac{21}{2}(\text{m})^2$
$A = 10.5$ square meters

5. Area of a circle is
$A = \pi r^2 \qquad r = 3$ meters
$A = \pi(3\text{ m})^2$
$A = \pi(9)(\text{m})^2$
$A = 28.26$ square meters

6. Formula $\quad C = \frac{5}{9}(F - 32)$
$F = 68°$
$C = \frac{5}{9}(68 - 32)$
$C = \frac{5}{9}(36)$
$C = 20°$ Celsius

7. Use the formula.
$K = 1.61$ (m) $\qquad$ Replace m by 35
$K = 1.61$ (35)
$K = 56.35$

1.9 Practice Problems

1. $5[4x - 3(x - 2)]$
$5[4x - 3x + 6]$
$5[x + 6]$
$5x + 30$

2. $3ab - [2ab - (2 - a)]$
$3ab - [2ab - 2 + a]$
$3ab - 2ab + 2 - a$
$ab + 2 - a$

3. $3[4x - 2(1 - x)] - [3x + (x - 2)]$
$3[4x - 2 + 2x] - [3x + x - 2]$
$12x - 6 + 6x - 3x - x + 2 = 14x - 4$
4. $-2\{5x - 3x[2x - (x^2 - 4x)]\}$
$-2\{5x - 3x[2x - x^2 + 4x]\}$
$-2\{5x - 6x^2 + 3x^3 - 12x^2\}$
$-2\{3x^3 - 12x^2 - 6x^2 + 5x\}$
$-2\{3x^3 - 18x^2 + 5x\} = -6x^3 + 36x^2 - 10x$

Chapter 2

2.1 Practice Problems

1. $x + 0.3 = 1.2$
$-0.3 - 0.3$
$x = 0.9$
Check. $0.9 + 0.3 = 1.2$
$1.2 = 1.2$
2. $17 = x - 5$
$+5 \quad +5$
$22 = x$
Check. $17 = 22 - 5$
$17 = 17$
3. $5 - 12 = x - 3$
$-7 = x - 3$
$+3 \quad +3$
$-4 = x$
Check. $5 - 12 = -4 - 3$
$-7 = -7$
4. $x + 8 = -22 + 6$
$x = -2$
$-2 + 8 = -22 + 6$
$6 \neq -16$ $\qquad$ This is not true.
Thus $x = -2$ is not a solution. Solve to find the solution.
$x + 8 = -22 + 6 = -16$
$x = -16 - 8$
$x = -24$
5. $\frac{1}{20} - \frac{1}{2} = x + \frac{3}{5}$
$\frac{1}{20} - \frac{1 \cdot 10}{2 \cdot 10} = x + \frac{3 \cdot 4}{5 \cdot 4}$
$\frac{1}{20} - \frac{10}{20} = x + \frac{12}{20}$
$x = \frac{1}{20} - \frac{10}{20} - \frac{12}{20} = -\frac{21}{20} = -1\frac{1}{20}$
Check.
$\frac{1}{20} - \frac{1}{2} \stackrel{?}{=} -1\frac{1}{20} + \frac{3}{5}$
$\frac{1}{20} - \frac{10}{20} \stackrel{?}{=} -\frac{21}{20} + \frac{12}{20}$
$-\frac{9}{20} = -\frac{9}{20}$

2.2 Practice Problems

1. $(8)\frac{1}{8}x = -2(8)$
$x = -16$

2. $\frac{9x}{9} = \frac{72}{9}$
$x = 8$

3. $\frac{6x}{6} = \frac{50}{6}$
$x = \frac{25}{3}$

4. $\frac{-27x}{-27} = \frac{54}{-27}$
$x = -2$

5. $\frac{-x}{-1} = \frac{36}{-1}$
$x = -36$

6. $\frac{-51}{-6} = \frac{-6x}{-6}$
$\frac{17}{2} = x$

7. $\dfrac{21}{4.2} = \dfrac{\cancel{4.2}x}{\cancel{4.2}}$
$5 = x$

2.3 Practice Problems

1. $$\begin{aligned} 9x + 2 &= 38 \\ -2 \quad &\; -2 \\ \hline \frac{9x}{9} &= \frac{36}{9} \\ x &= 4 \end{aligned}$$ $$\begin{aligned} 9(4) + 2 &= 38 \\ 36 + 2 &= 38 \\ 38 &= 38 \checkmark \end{aligned}$$

2. $$\begin{aligned} 13x &= 2x - 66 \\ -2x \quad & -2x \\ \hline \frac{11x}{11} &= \frac{-66}{11} \\ x &= -6 \end{aligned}$$ $$\begin{aligned} 13(-6) &= 2(-6) - 66 \\ -78 &= -12 - 66 \\ -78 &= -78 \checkmark \end{aligned}$$

3. $$\begin{aligned} 3x + 2 &= 5x + 2 \\ -2 \quad & \quad -2 \\ \hline 3x &= 5x \\ -5x \quad & -5x \\ \hline -2x &= 0 \\ x &= 0 \end{aligned}$$ $$\begin{aligned} 3(0) + 2 &= 5(0) + 2 \\ 2 &= 2 \checkmark \end{aligned}$$

4. $$\begin{aligned} -z + 8 - z &= 3z + 10 - 3 \\ -2z + 8 &= 3z + 7 \\ -3z \quad & -3z \\ \hline -5z + 8 &= 7 \\ -8 &= -8 \\ \hline \frac{-5z}{-5} &= \frac{-1}{-5} \\ z &= \frac{1}{5} \end{aligned}$$

5. $$\begin{aligned} 2x^2 - 6x + 3 &= -4x - 7 + 2x^2 \\ -2x^2 \quad & \quad - 2x^2 \\ \hline -6x + 3 &= -4x - 7 \\ +4x \quad & +4x \\ \hline -2x + 3 &= -7 \\ -3 &= -3 \\ \hline \frac{-2x}{-2} &= \frac{-10}{-2} \\ x &= 5 \end{aligned}$$

6. $$\begin{aligned} 4x - (x + 3) &= 12 - 3(x - 2) \\ 4x - x - 3 &= 12 - 3x + 6 \\ 3x - 3 &= -3x + 18 \\ +3x \quad & + 3x \\ \hline 6x - 3 &= 18 \\ +3 \quad & +3 \\ \hline \frac{6x}{6} &= \frac{21}{6} \\ x &= \frac{21}{6} = \frac{7}{2} \end{aligned}$$

Check. $$\begin{aligned} 4\left(\frac{7}{2}\right) - \left(\frac{7}{2} + 3\right) &= 12 - 3\left(\frac{7}{2} - 2\right) \\ 14 - \left(\frac{13}{2}\right) &= 12 - 3\left(\frac{3}{2}\right) \\ \frac{28}{2} - \frac{13}{2} = \frac{24}{2} - \frac{9}{2}&;\ \frac{15}{2} = \frac{15}{2} \end{aligned}$$

7. $$\begin{aligned} 4(-2x - 3) &= -5(x - 2) + 2 \\ -8x - 12 &= -5x + 10 + 2 \\ -8x - 12 &= -5x + 12 \\ +5x \quad & +5x \\ \hline -3x - 12 &= 12 \\ +12 &= +12 \\ \hline \frac{-3x}{-3} &= \frac{24}{-3} \\ x &= -8 \end{aligned}$$

8. $$\begin{aligned} 0.3x - 2(x + 0.1) &= 0.4(x - 3) - 1.1 \\ 0.3x - 2x - 0.2 &= 0.4x - 1.2 - 1.1 \\ -1.7x - 0.2 &= 0.4x - 2.3 \\ -0.4x \quad & -0.4x \\ \hline -2.1x - 0.2 &= -2.3 \\ +0.2 \quad & +0.2 \\ \hline \frac{-2.1x}{-2.1} &= \frac{-2.1}{-2.1} \\ x &= 1 \end{aligned}$$

9. $$\begin{aligned} 5(2z - 1) + 7 &= 7z - 4(z + 3) \\ 10z - 5 + 7 &= 7z - 4z - 12 \\ 10z + 2 &= 3z - 12 \\ -3z \quad & -3z \\ \hline 7z + 2 &= -12 \\ -2 \quad & \quad -2 \\ \hline \frac{7z}{7} &= -\frac{14}{7} \\ z &= -2 \end{aligned}$$

2.4 Practice Problems

1. $$\begin{aligned} \frac{3}{8}x - \frac{3}{2} &= \frac{1}{4}x \\ 3x - 12 &= 2x \\ -2x \quad & -2x \\ \hline x - 12 &= 0 \\ +12 \quad & +12 \\ \hline x &= 12 \end{aligned}$$

2. $$\begin{aligned} \frac{5x}{4} - 1 &= \frac{3x}{4} + \frac{1}{2} \\ 5x - 4 &= 3x + 2 \\ -3x \quad & -3x \\ \hline 2x - 4 &= +2 \\ +4 &= +4 \\ \hline \frac{2x}{2} &= \frac{6}{2} \\ x &= 3 \end{aligned}$$

Check. $$\begin{aligned} \frac{5(3)}{4} - 1 &= \frac{3(3)}{4} + \frac{1}{2} \\ \frac{15}{4} - 1 &= \frac{9}{4} + \frac{1}{2} \\ \frac{15}{4} - \frac{4}{4} &= \frac{9}{4} + \frac{2}{4} \\ \frac{11}{4} &= \frac{11}{4} \end{aligned}$$

3. $$\begin{aligned} \frac{5x}{6} - \frac{5}{8} &= \frac{3x}{4} - \frac{1}{3} \\ 20x - 15 &= 18x - 8 \\ -18x \quad & -18x \\ \hline 2x - 15 &= -8 \\ +15 &= +15 \\ \hline \frac{2x}{2} &= \frac{7}{2} \\ x &= \frac{7}{2} \end{aligned}$$

4. $$\begin{aligned} \frac{1}{3}(x - 2) &= \frac{1}{4}(x + 5) - \frac{5}{3} \\ \frac{1}{3}x - \frac{2}{3} &= \frac{1}{4}x + \frac{5}{4} - \frac{5}{3} \\ 4x - 8 &= 3x + 15 - 20 \\ 4x - 8 &= 3x - 5 \\ -3x \quad & -3x \\ \hline x - 8 &= -5 \\ +8 \quad & +8 \\ \hline x &= 3 \end{aligned}$$

5. $$\begin{aligned} 2.8 &= 0.3(x - 2) + 2(0.1x - 0.3) \\ 2.8 &= 0.3x - 0.6 + 0.2x - 0.6 \\ 10(2.8) &= 10(0.3x) - 10(0.6) + 10(0.2x) - 10(0.6) \\ 28 &= 3x - 6 + 2x - 6 \\ 28 &= 5x - 12 \\ +12 \quad & +12 \\ \hline 40 &= 5x \\ 8 &= x \end{aligned}$$

2.5 Practice Problems

1. $$\begin{aligned} d &= rt \\ 240 &= r(5) \\ \frac{240}{5} &= r \\ 48 &= r \end{aligned}$$

2. Solve for m. $\dfrac{E}{c^2} = \dfrac{mc^2}{c^2}$ $\quad \dfrac{E}{c^2} = m$

3. Solve for y.

$$\begin{aligned} 8 - 2y + 3x &= 0 \\ -3x &= -3x \\ \hline 8 - 2y &= -3x \\ -8 &\quad -8 \\ \hline \frac{-2y}{-2} &= \frac{-3x - 8}{-2} \\ y &= \frac{3x + 8}{2} \end{aligned}$$

4. Solve for d.

$$c = \pi d$$
$$\frac{c}{\pi} = \frac{\pi d}{\pi}$$
$$\frac{c}{\pi} = d$$

2.6 Practice Problems

1. (a) $7 > 2$ **(b)** $-3 > -4$ **(c)** $-1 < 2$

(d) $-8 < -5$ **(e)** $0 > -2$ **(f)** $\frac{2}{5} > \frac{3}{8}$

2. (a) $x > 5$; x is greater than 5

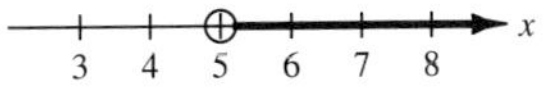

(b) $x \le -2$; x is less than or equal to -2

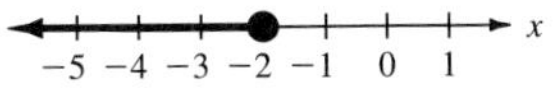

(c) $3 > x$; 3 is greater than x (or x is less than 3)

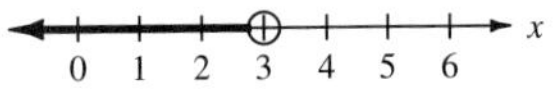

(d) $x \ge -\frac{3}{2}$; x is greater than or equal to $-\frac{3}{2}$

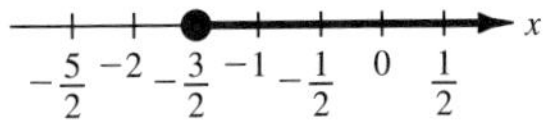

3. (a) $t \le 180$ **(b)** $d < 15{,}000$

2.7 Practice Problems

1. (a) $9 > 6$, $3 > 2$ **(b)** $2 < 5$, $4 < 10$ **(c)** $-8 < -3$, $-4 < 1$ **(d)** $-3 < 7$, $-10 < 0$

2. (a) $7 > 2$, $-14 < -4$ **(b)** $-3 < -1$, $3 > 1$

(c) $-10 > -20$, $1 < 2$ **(d)** $-15 < -5$, $3 > 1$

3.

$$\begin{aligned} 8x - 2 &< 3 \\ +2 &\quad +2 \\ \hline \frac{8x}{8} &< \frac{5}{8} \\ x &< \frac{5}{8} \end{aligned}$$

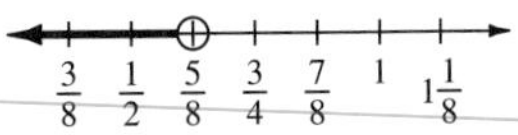

4.

$$\begin{aligned} 7 + x &< 28 \\ -7 &\quad -7 \\ \hline x &< 21 \end{aligned}$$

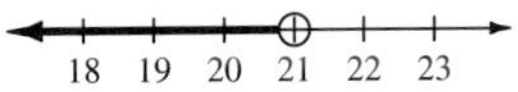

5.

$$\begin{aligned} \frac{1}{2}x + 3 &< \frac{2}{3}x \\ 3x + 18 &< 4x \\ -4x &\quad -4x \\ \hline -x + 18 &< 0 \\ -18 &\quad -18 \\ \hline \frac{-x}{-1} &> \frac{-18}{-1} \\ x &> 18 \end{aligned}$$

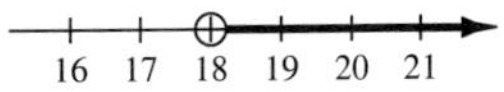

6.

$$\begin{aligned} \frac{1}{2}(3 - x) &\le 2x + 5 \\ \frac{3}{2} - \frac{1}{2}x &\le 2x + 5 \\ 3 - x &\le 4x + 10 \\ -5x &\le 7 \\ x &\ge -\frac{7}{5} \end{aligned}$$

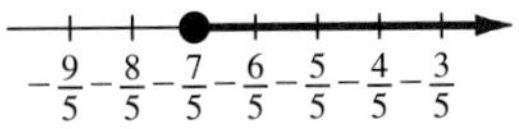

7.

$$\begin{aligned} 2000\text{N} - 700{,}000 &\ge 2{,}500{,}000 \\ 2{,}000\text{N} &\ge 3{,}200{,}000 \\ \text{N} &\ge 1{,}600 \end{aligned}$$

Chapter 3

3.1 Practice Problems

1. (a) $x + 4$ **(b)** $3x$

(c) $x - 8$ **(d)** $\frac{1}{4}x$

2. (a) $3x + 8$ **(b)** $3(x + 8)$

3. (a) $3x - 4$ **(b)** $\frac{2}{3}(x + 5)$

4. Let a = Ann's hours per week
Then $a - 17$ = Maria's hours per week

5. width = w
length = $2w + 5$

6. 1st angle = $s - 16$
2nd angle = s
3rd angle = $2s$

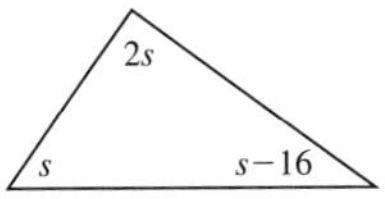

7. Let x = the number of students in the fall
$\frac{2}{3}x$ = the number of students in the spring
$\frac{1}{5}x$ = the number of students in the summer

3.2 Practice Problems

1.

$$\frac{3}{4}x = -81$$
$$x = -108$$

2.

$$\begin{aligned} 3x - 2 &= 49 \\ 3x &= 51 \\ x &= 17 \end{aligned}$$

3. $(x) + (3x - 12) = 24$
First number $x = 9$, Second number $= 3(9) - 12 = 15$

4. Let the precipitation for
S. Dakota = x
then Utah = $x - 2$

$$\begin{aligned} x + x - 2 &= 14 \\ 2x - 2 &= 14 \\ 2x &= 16 \\ x &= 8 \end{aligned}$$

therefore
S. Dakota had 8 inches of rain
Utah had 6 inches

5. (a) $\frac{220}{4} = 55$ mph **(b)** $\frac{225}{4.5} = 50$ mph

(c) The trip leaving the city by 5 mph.

6.

$$\frac{x + x + 78 + 80 + 100 + 96}{6} = 90$$
$$\begin{aligned} 354 + 2x &= 540 \\ 2x &= 186 \\ x &= 93 \end{aligned}$$

3.3 Practice Problems

1. short piece = x
long piece = $x + 17$

$$\begin{aligned} (x) + (x + 17) &= 89 \\ 2x + 17 &= 89 \\ 2x &= 72 \\ x &= 36 \end{aligned}$$

Therefore Short piece $x = 36$
Long piece $(36) + 17 = 53$

2. Family 1 = $x + 360$
Family 2 = x
Family 3 = $2x - 200$

$$\begin{aligned} (x) + (x + 360) + (2x - 200) &= 3960 \\ 4x + 160 &= 3960 \\ x &= 950 \end{aligned}$$

Therefore Family #2 = \$950.00
Family #3 = 2(950) − 200 = \$1700.00
Family #1 = 950 + 360 = \$1310.00

3. Let the longest side $= x$
other side $= x - 30$
shorter side $= \frac{1}{2}x$
$x + x - 30 + \frac{1}{2}x = 720$
$5x - 60 = 1440$
$5x = 1500$
$x = 300$

Therefore the
longest side = 300 meters
other side = 270 meters
shorter side = 150 meters

3.4 Practice Problems

1. $3(22) + (0.18)M = 363$
$66 + 0.18M = 363$
$0.18M = 297$
$M = 1650$ miles

2. $0.38x = 4560$
$x = \$12{,}000.00$

3. $x + 0.07x = 13{,}910$
$1.07x = 13{,}910$
$x = \$13{,}000.00$

4. $x = 7000(0.12)(1)$
$x = \$840.00$

5. $0.15x + 0.12(8000 - x) = 1155$
$0.15x + 960 - 0.12x = 1155$
$0.03x + 960 = 1155$
$0.03x = 195$
$x = \$6{,}500.00$
Therefore she invested \$6,500.00 @ 15%
(8,000 − 6,500) = \$1,500.00 @ 12%

6. x = Dimes
$x + 5$ = Quarters
$0.10x + 0.25(x + 5) = 5.10$
$0.10x + 0.25x + 1.25 = 5.10$
$0.35x + 1.25 = 5.10$
$0.35x = 3.85$
$x = 11$
Therefore she has 11 dimes, 16 quarters

7. Nickels $= 2x$
Dimes $= x$
Quarters $= x + 4$
$0.05(2x) + 0.10(x) + 0.25(x + 4) = 2.35$
$0.1x + 0.10x + 0.25x + 1 = 2.35$
$0.45x + 1 = 2.35$
$0.45x = 1.35$
$x = 3$
Therefore the boy has 3 dimes, 6 nickels, 7 quarters

3.5 Practice Problems

1. $A = \frac{1}{2}ab$
$= \frac{1}{2}(20)(14)$
$= (10)(14)$
$= 140$ square inches

2. $A = LW$
$120 = L(8)$
$\frac{120}{8} = L$
$L = 15$ yards

3. $A = \frac{1}{2}a(b_1 + b_2)$
$256 = \frac{1}{2}a(12 + 20)$
$256 = \frac{1}{2}(a)(32)$
$256 = 16a$
$\frac{256}{16} = a$
$a = 16$ feet

4. $C = 2\pi r$
$C = 2(3.14)(15)$
$C = 94.2$
$C = 94$ meters

5. $P = a + b + c$
$P = 15 + 15 + 15$
$P = 45$ cm

6. Surface Area $= 4\pi r^2$
$= 4(3.14)(5)^2$
$= 314 \text{ m}^2$

7. $V = \pi r^2 h$
$V = 3.14(3)^2(4) = 113 \text{ ft}^3$

8. Calculate the area of the pool.
$A = L \times W$
$A = 8 \times 12 = 96$ sq. ft.
Now add 6 feet to the length and 6 feet to the width of the pool and calculate the area.
$A = L \times W$
$A = 18 \times 14 = 252$ sq. ft.
Now subtract the areas.
$252 - 96 = 156$ sq. ft.
at \$12.00 a square foot.
$156 \times 12 = \$1{,}872.00$

9. $V = LWH$
$V = (6)(5)(8) = 240 \text{ ft}^3$
Weight $= 240 \text{ ft}^3 \times \frac{62.4 \text{ lb}}{1 \text{ ft}^3} = 14{,}976 \text{ lb} \approx 15{,}000 \text{ lb}$

3.6 Practice Problems

1. **(a)** person ≤ 6 **(b)** car > 65
(c) area ≥ 560 **(d)** profit ≥ 50

2. $\frac{1050 + 1250 + 950 + x}{4} \geq 1{,}100$
$\frac{x + 3250}{4} \geq 1{,}100$
$x + 3250 \geq 4{,}400$
$x \geq 1{,}150$

3. $1400 + 0.02x > 2{,}200$
$0.02x > 800$
$x > \$40{,}000.00$

Chapter 4

4.1 Practice Problems

1. **(a)** $a^7 \cdot a^5 = a^{12}$ **(b)** $w^{10} \cdot w = w^{11}$

2. **(a)** $x^3 \cdot x^9 = x^{12}$ **(b)** $3^7 \cdot 3^4 = 3^{11}$
(c) $a^3 \cdot b^2 = a^3 \cdot b^2$ (cannot be simplified)

3. **(a)** $(-a^8)(a^4) = -a^{12}$ **(b)** $(3y^2)(-2y^3) = -6y^5$
(c) $(-4x^3)(-5x^2) = 20x^5$

4. **(a)** $(-5a^3b)(-2ab^4) = 10a^4b^5$
(b) $(2xy)\left(-\frac{1}{4}x^2y\right)(6xy^3) = -3x^4y^5$

5. **(a)** $\frac{10^{13}}{10^7} = 10^6$ **(b)** $\frac{x^{11}}{x} = x^{10}$

6. **(a)** $\frac{c^3}{c^4} = \frac{1}{c}$ **(b)** $\frac{10^{31}}{10^{56}} = \frac{1}{10^{25}}$

7. **(a)** $\frac{-7x^7}{-21x^9} = \frac{1}{3x^2}$ **(b)** $\frac{15x^{11}}{-3x^4} = -5x^7$

8. **(a)** $\frac{x^7y^9}{y^{10}} = \frac{x^7}{y}$ **(b)** $\frac{12x^5y^6}{-24x^3y^8} = \frac{x^2}{-2y^2}$

9. **(a)** $\frac{10^7}{10^7} = 1$ **(b)** $\frac{12a^4}{15a^4} = \frac{4}{5}$

10. $\frac{-20a^3b^8c^4}{28a^3b^7c^5} = \frac{-5b}{7c}$

11. $\frac{(-6ab^5)(3a^2b^4)}{16a^5b^7} = \frac{-9b^2}{8a^2}$

12. **(a)** $(a^4)^3 = a^{12}$ **(b)** $(10^5)^2 = 10^{10}$

13. **(a)** $(3xy)^3 = 27x^3y^3$ **(b)** $(yz)^{37} = y^{37}z^{37}$

14. $\left(\frac{4a}{b}\right)^6 = \frac{4^6a^6}{b^6} = \frac{4096a^6}{b^6}$

15. **(a)** $\left(\frac{-2x^3y^0z}{4xz^2}\right)^5 = \left(\frac{-x^2}{2z}\right)^5 = \frac{(-1)^5(x^2)^5}{2^5z^5} = \frac{-x^{10}}{32z^5}$
(b) $\left(\frac{4x^8y^5}{-3x^4z}\right)^3 = \left(\frac{4x^4y^5}{-3z}\right)^3 = \frac{4^3x^{12}y^{15}}{(-3)^3z^3} = -\frac{64x^{12}y^{15}}{27x^3}$

4.2 Practice Problems

1. **(a)** $x^{-12} = \frac{1}{x^{12}}$ **(b)** $w^{-5} = \frac{1}{w^5}$ **(c)** $z^{-2} = \frac{1}{z^2}$
(d) $y^{-7} = \frac{1}{y^7}$

2. **(a)** $4^{-3} = \frac{1}{4^3} = \frac{1}{64}$ **(b)** $6^{-2} = \frac{1}{6^2} = \frac{1}{36}$

3. **(a)** $\frac{3}{w^{-4}} = 3w^4$ **(b)** $\frac{x^{-6}y^4}{z^{-2}} = \frac{y^4z^2}{x^6}$ **(c)** $x^{-6}y^{-5} = \frac{1}{x^6y^5}$

4. **(a)** $(2x^4y^{-5})^{-2} = 2^{-2}x^{-8}y^{10} = \frac{y^{10}}{4x^8}$
(b) $\frac{y^{-3}z^{-4}}{y^2z^{-6}} = \frac{z^6}{y^2y^3z^4} = \frac{z^6}{y^5z^4} = \frac{z^2}{y^5}$

5. **(a)** 7.82×10^4 **(b)** 4.786×10^6

6. **(a)** 9.8×10^{-1} **(b)** 7.63×10^{-4} **(c)** 9.2×10^{-5}

7. **(a)** 2960.0 **(b)** 1,930,000.0 **(c)** 0.0543 **(d)** 0.00008562

8. 3.09×10^{16} m

9. **(a)** $(5.6 \times 10^4)(1.4 \times 10^9)$
$(5.6)(1.4)(10^4)(10^9)$
7.84×10^{13}

(b) $\dfrac{1.11 \times 10^{-4}}{3.7 \times 10^{-7}} = \dfrac{1.11}{3.7} \times \dfrac{10^7}{10^4} = 0.3 \times 10^3 = 3.0 \times 10^2$

10. $(159 \text{ parsecs})\dfrac{(3.09 \times 10^{13} \text{ kilometers})}{1 \text{ parsec}} = 491.31 \times 10^{13} \text{ kilometers}$

$d = r \times t$

$491.31 \times 10^{13} \text{ km} = \dfrac{50{,}000 \text{ km}}{1 \text{ hr}} \times t$

$\dfrac{491.31 \times 10^{13} \text{ km}(1 \text{ hr})}{50{,}000 \text{ km}} = t$

$\dfrac{4.9131 \times 10^{15} \text{ km}(1 \text{ hr})}{5.0 \times 10^4 \text{ km}} = t$

$0.9826 \times 10^{11} \text{ hrs.} = t$

9.83×10^{10} hours

4.3 Practice Problems

1. $(-8x^3 + 3x^2 + 6) + (2x^3 - 7x^2 - 3)$
$[-8x^3 + 2x^3] + [3x^2 - 7x^2] + [6 - 3] = -6x^3 - 4x^2 + 3$

2. $\left(-\dfrac{1}{3}x^2 - 6x - \dfrac{1}{12}\right) + \left(\dfrac{1}{4}x^2 + 5x - \dfrac{1}{3}\right)$

$\left[-\dfrac{1}{3}x^2 + \dfrac{1}{4}x^2\right] + [-6x + 5x] + \left[-\dfrac{1}{12} - \dfrac{1}{3}\right]$

$-\dfrac{1}{12}x^2 - x - \dfrac{5}{12}$

3. $(3.5x^3 - 0.02x^2 + 1.56x - 3.5) + (-0.08x^2 - 1.98x + 4)$
$3.5x^3 + [-0.02x^2 - 0.08x^2] + [1.56x - 1.98x] + [-3.5 + 4]$
$3.5x^3 - 0.10x^2 - 0.42x + 0.5$

4. $(-3a^3 + 2ab + 7b) + (-4a^3 - 6ab + 3b^3)$
$[-3a^3 - 4a^3] + [2ab - 6ab] + 7b + 3b^3$
$-7a^3 - 4ab + 7b + 3b^3$

5. $(5x^3 - 15x^2 + 6x - 3) - (-4x^3 - 10x^2 + 5x + 13)$
$[5x^3 + 4x^3] + [-15x^2 + 10x^2] + [6x - 5x] + [-3 - 13]$
$9x^3 - 5x^2 + x - 16$

6. $(x^3 - 7x^2y + 3xy^2 - 2y^3) - (2x^3 + 4xy - 6y^3)$
$[x^3 - 2x^3] - 7x^2y + 3xy^2 - 4xy + [-2y^3 + 6y^3]$
$-x^3 - 7x^2y + 3xy^2 - 4xy + 4y^3$

4.4 Practice Problems

1. $-8x^5 + 12x^4$

2. **(a)** $21 - 14x$ **(b)** $-3x^3 - 6x^2 + 12x$
(c) $6x^4y + 12x^3y^2 - 6xy^3$

3. **(a)** $12x^3 - 28x^2 + 4x$ **(b)** $18x^4y - 12x^3y + 6x^2y$

4. $5x^2 - 10x - x + 2 = 5x^2 - 11x + 2$

5. $24ac - 8ad - 15bc + 5bd$

6. $3x^2 - 3x + 2x - 2 = 3x^2 - x - 2$

7. $6a^2 - 9ab + 4ab - 6b^2 = 6a^2 - 5ab - 6b^2$

8. $(3x - 2y)(3x - 2y) = 9x^2 - 6xy - 6xy + 4y^2 = 9x^2 - 12xy + 4y^2$

9. $10x^4 + 12x^2y^2 + 15x^2y^2 + 18y^4 = 10x^4 + 27x^2y^2 + 18y^4$

10. $(2x - 1)(7x + 3)$
$14x^2 + 6x - 7x - 3$
$14x^2 - x - 3$ square feet

4.5 Practice Problems

1. $(6x)^2 - (7)^2 = 36x^2 - 49$

2. $(3x)^2 - (5y)^2 = 9x^2 - 25y^2$

3. $(4x^2)^2 - (1)^2 = 16x^4 - 1$

4. **(a)** $(5x)^2 + 2(5x)(4) + (4)^2 = 25x^2 + 40x + 16$
(b) $(4a)^2 - 2(4a)(9b) + (9b)^2 = 16a^2 - 72ab + 81b^2$

5.
$$\begin{array}{r} 3x^2 - 2xy + 4y^2 \\ x - 2y \\ \hline -6x^2y + 4xy^2 - 8y^3 \\ +3x^3 - 2x^2y + 4xy^2 \\ \hline 3x^3 - 8x^2y + 8xy^2 - 8y^3 \end{array}$$

6. $(2x^2 - 5x + 3)(x^2 + 3x - 4)$
$2x^2(x^2 + 3x - 4) - 5x(x^2 + 3x - 4) + 3(x^2 + 3x - 4)$
$2x^4 + 6x^3 - 8x^2 - 5x^3 - 15x^2 + 20x + 3x^2 + 9x - 12$
$2x^4 + x^3 - 20x^2 + 29x - 12$

7. $(3x - 2)(3x + 2)(2x + 3)$
$(9x^2 - 4)(2x + 3)$
$$\begin{array}{r} 9x^2 - 4 \\ 2x + 3 \\ \hline 27x^2 + 0x - 12 \\ 18x^3 + 0x^2 - 8x \\ \hline 18x^3 + 27x^2 - 8x - 12 \end{array}$$

4.6 Practice Problems

1. $\dfrac{14y^2 - 49y}{7y} = \dfrac{14y^2}{7y} - \dfrac{49y}{7y} = 2y - 7$

2. $\dfrac{15y^4 - 27y^3 - 21y^2}{3y^2} = 5y^2 - 9y - 7$

3.
$$\begin{array}{r} x^2 + 6x + 7 \\ x + 4 \overline{)\, x^3 + 10x^2 + 31x + 35} \\ \underline{x^3 + 4x^2} \\ 6x^2 + 31x \\ \underline{6x^2 + 24x} \\ 7x + 35 \\ \underline{7x + 28} \\ 7 \end{array}$$
$= x^2 + 6x + 7 + \dfrac{7}{x + 4}$

4.
$$\begin{array}{r} 2x^2 + x + 1 \\ x - 1 \overline{)\, 2x^3 - x^2 + 0x + 1} \\ \underline{2x^3 - 2x^2} \\ x^2 + 0x \\ \underline{x^2 - x} \\ x + 1 \\ \underline{x - 1} \\ 2 \end{array}$$
$= 2x^2 + x + 1 + \dfrac{2}{x - 1}$

5.
$$\begin{array}{r} 5x^2 + x - 2 \\ 4x - 3 \overline{)\, 20x^3 - 11x^2 - 11x + 6} \\ \underline{20x^3 - 15x^2} \\ 4x^2 - 11x \\ \underline{4x^2 - 3x} \\ -8x + 6 \\ \underline{-8x + 6} \\ 0 \end{array}$$
$= 5x^2 + x - 2$

Check. $(4x - 3)(5x^2 + x - 2)$
$20x^3 + 4x^2 - 8x - 15x^2 - 3x + 6$
$20x^3 - 11x^2 - 11x + 6$

Chapter 5

5.1 Practice Problems

1. **(a)** $7(3a - b)$ **(b)** $p(1 + rt)$

2. $4a(3a + 4b^2 - 3ab)$

3. **(a)** $8(2a^3 - 3b^3)$ **(b)** $r^3(s^2 - 4rs + 7r^2)$

4. $9ab^2c(2a^2 - 3bc - 5ac)$

5. $6xy^2(5x^2 - 4x + 1) = 30x^3y^2 - 24x^2y^2 + 6xy^2$

6. $(a + b)(3 + x)$ **7.** $(8y - 1)(9y^2 - 2)$

8. $\pi b^2 - \pi a^2 = \pi(b^2 - a^2)$

5.2 Practice Problems

1. $(2x - 7)(3y - 8)$

2. $3x(2x - 5) + 2(2x - 5)$
$(2x - 5)(3x + 2)$

3. $a(x + 2) + 4b(x + 2)$
$(x + 2)(a + 4b)$

4. $4(3x + y) + 7a(3x + y)$
$(3x + y)(4 + 7a)$

5. $6a^2 + 3ac + 10ab + 5bc$
$3a(2a + c) + 5b(2a + c)$
$(2a + c)(3a + 5b)$

6. $6xy + 14x - 15y - 35$
$2x(3y + 7) - 5(3y + 7)$
$(3y + 7)(2x - 5)$

7. $3x + 6y - 5ax - 10ay$
$3(x + 2y) - 5a(x + 2y)$
$(3 - 5a)(x + 2y)$

8. $10ad + 27bc - 6bd - 45ac$
$10ad - 6bd - 45ac + 27bc$
$2d(5a - 3b) - 9c(5a - 3b)$
$(2d - 9c)(5a - 3b)$
Check: $10ad - 6bd - 45ac + 27bc$

5.3 Practice Problems

1. $(x + 6)(x + 2)$
2. $(x + 2)(x + 15)$
3. $(x + 1)(x + 6)$
4. $(x - 9)(x - 2)$
5. $(y - 3)(y - 8)$
6. $(x + 6)(x - 1)$
7. $(a + 3)(a - 8)$
8. $(x + 20)(x - 3)$
Check: $x^2 - 3x + 20x - 60$
$= x^2 + 17x - 60$
9. $(x + 5)(x - 12)$
10. $(a^2 + 7)(a^2 - 6)$
11. $3x^2 + 45x + 150$
$3(x^2 + 15x + 50)$
$3(x + 5)(x + 10)$
12. $4x^2 - 8x - 140$
$4(x^2 - 2x - 35)$
$4(x + 5)(x - 7)$

5.4 Practice Problems

1. $(2x - 5)(x - 1)$
2. $(9x - 1)(x - 7)$
3. $(3x - 2)(x + 2)$
4. $(3x - 7)(x + 2)$
5. $2x^2 - 2x - 5x + 5$
$2x(x - 1) - 5(x - 1)$
$(2x - 5)(x - 1)$
6. $9x^2 - 63x - x + 7$
$9x(x - 7) - 1(x - 7)$
$(x - 7)(9x - 1)$
7. $3x^2 + 6x - 2x - 4$
$3x(x + 2) - 2(x + 2)$
$(x + 2)(3x - 2)$
8. $8x^2 - 8x - 6$
$2(4x^2 - 4x - 3)$
$2(2x + 1)(2x - 3)$
9. $24x^2 - 38x + 10$
$2(12x^2 - 19x + 5)$
$2(12x^2 - 4x - 15x + 5)$
$2[4x(3x - 1) - 5(3x - 1)]$
$2(4x - 5)(3x - 1)$

5.5 Practice Problems

1. $(1 + 8x)(1 - 8x)$
2. $(6x + 7)(6x - 7)$
3. $(10x + 9y)(10x - 9y)$
4. $(x^4 + 1)(x^4 - 1)$
$(x^4 + 1)(x^2 + 1)(x^2 - 1)$
$(x^4 + 1)(x^2 + 1)(x + 1)(x - 1)$
5. $(4x + 1)(4x + 1) = (4x + 1)^2$
6. $(5x - 3)(5x - 3) = (5x - 3)^2$
7. **(a)** $(5x - 6y)(5x - 6y) = (5x - 6y)^2$
(b) $(8x^3 - 3)(8x^3 - 3) = (8x^3 - 3)^2$
8. $9x^2 - 12x - 3x + 4$
$3x(3x - 4) - 1(3x - 4)$
$(3x - 1)(3x - 4)$
9. $20x^2 - 45$
$5(4x^2 - 9)$
$5(2x - 3)(2x + 3)$
10. $3(25x^2 - 20x + 4)$
$3(5x - 2)(5x - 2)$
$3(5x - 2)^2$

5.6 Practice Problems

1. **(a)** $9y^2(x^4 - 1)$
$9y^2(x^2 + 1)(x^2 - 1)$
$9y^2(x^2 + 1)(x + 1)(x - 1)$
(b) $-(4x^2 - 12x + 9)$
$-(2x - 3)(2x - 3)$
$-(2x - 3)^2$
(c) $3(x^2 - 12x + 36)$
$3(x - 6)(x - 6)$
$3(x - 6)^2$
(d) $5x(x^2 - 3xy + 2x - 6y)$
$5x[x(x - 3y) + 2(x - 3y)]$
$5x(x + 2)(x - 3y)$
2. $x^2 - 9x - 8$
The factors of -8 are $(-2)(4) = -8$
$(2)(-4) = -8$
$(-8)(1) = -8$
$(-1)(8) = -8$
None of these pairs will add up to be the coefficient of the middle term. Thus the problem cannot be factored.
It is prime.
3. $25x^2 + 82x + 4$
Check to see if this is a perfect square trinomial.
$2[(5)(2)] = 2(10) = 20$ This is not the coefficient of the middle term. The grouping number equals 100. No factors.
It is prime.

5.7 Practice Problems

1. $10x^2 - x - 2 = 0$
$(5x + 2)(2x - 1) = 0$

$5x + 2 = 0$
$5x = -2$
$x = -\frac{2}{5}$

$2x - 1 = 0$
$2x = 1$
$x = \frac{1}{2}$

Check.
$10\left(-\frac{2}{5}\right)^2 - \left(-\frac{2}{5}\right) - 2 = 0$
$10\left(\frac{4}{25}\right) + \frac{2}{5} - 2 = 0$
$\frac{8}{5} + \frac{2}{5} - 2 = 0$
$\frac{10}{5} - \frac{10}{5} = 0$
$0 = 0$

$10\left(\frac{1}{2}\right)^2 - \frac{1}{2} - 2 = 0$
$10\left(\frac{1}{4}\right) - \frac{1}{2} - 2 = 0$
$\frac{5}{2} - \frac{1}{2} - 2 = 0$
$\frac{4}{2} - 2 = 0$
$2 - 2 = 0$
$0 = 0$

Thus $x = -\frac{2}{5}$ and $x = \frac{1}{2}$ are both roots for the equation.

2. $3x^2 - 5x + 2 = 0$
$(3x - 2)(x - 1) = 0$
$3x - 2 = 0$
$3x = 2$
$x = \frac{2}{3}$
$x - 1 = 0$
$x = 1$

3. $7x^2 + 11x = 0$
$x(7x + 11) = 0$
$x = 0$
$7x + 11 = 0$
$7x = -11$
$x = \frac{-11}{7}$

4. $x^2 - 6x + 4 = -8 + x$
$x^2 - 7x + 12 = 0$
$(x - 3)(x - 4) = 0$
$x - 3 = 0$
$x = 3$
$x - 4 = 0$
$x = 4$

5. $x(3x - 7) = 20$
$3x^2 - 7x = 20$
$3x^2 - 7x - 20 = 0$
$(3x + 5)(x - 4) = 0$
$3x + 5 = 0$
$x = -\frac{5}{3}$
$x - 4 = 0$
$x = 4$

6. $\frac{2x^2 - 7x}{3} = 5$
$2x^2 - 7x = 15$
$2x^2 - 7x - 15 = 0$
$(2x + 3)(x - 5) = 0$
$2x + 3 = 0$
$x = -\frac{3}{2}$
$x - 5 = 0$
$x = 5$

7. Let w = width, then
$3w + 2$ = length
$(3w + 2)w = 85$
$3w^2 + 2w = 85$
$3w^2 + 2w - 85 = 0$
$(3w + 17)(w - 5) = 0$
$3w + 17 = 0$
$w = -\frac{17}{3}$
$w - 5 = 0$
$w = 5$
The only valid answer is width = 5 meters
length = 3(5) + 2 = 17 meters

8. Let b = base
$b - 3$ = altitude
$\frac{b(b - 3)}{2} = 35$
$b^2 - 3b = 70$
$b^2 - 3b - 70 = 0$
$(b + 7)(b - 10) = 0$
$b + 7 = 0$
$b = -7$
This is not a valid answer.
$b - 10 = 0$
$b = 10$
Thus the base = 10 centimeters
altitude = 10 − 3 = 7 centimeters

9. $-5t^2 + 45$
$-5(t^2 - 9)$
$t^2 - 9 = 0$
$(t + 3)(t - 3) = 0$
$t = 3 \qquad t = -3$ This is not a valid answer.
Thus it will be 3 seconds before she breaks the water's surface.

Chapter 6

6.1 Practice Problems

1. **(a)** $\dfrac{7}{3} = \dfrac{?}{18}$ **(b)** $\dfrac{28}{63} = \dfrac{7 \times 2 \times 2}{7 \times 3 \times 3} = \dfrac{4}{9}$

$\dfrac{7 \cdot 6}{3 \cdot 6} = \dfrac{42}{18}$

2. $\dfrac{12x - 6}{14x - 7} = \dfrac{6(2x - 1)}{7(2x - 1)} = \dfrac{6}{7}$

3. $\dfrac{4x - 6}{2x^2 - x - 3} = \dfrac{2(2x - 3)}{(2x - 3)(x + 1)} = \dfrac{2}{x + 1}$

4. $\dfrac{7x^2 - 14xy}{14x^2 - 28xy} = \dfrac{7x(x - 2y)}{14x(x - 2y)} = \dfrac{7x}{14x} = \dfrac{1}{2}$

5. $\dfrac{x^3 + 11x^2 + 30x}{3x^3 + 17x^2 - 6x} = \dfrac{x(x^2 + 11x + 30)}{x(3x^2 + 17x - 6)} = \dfrac{(x + 5)(x + 6)}{(3x - 1)(x + 6)}$

$= \dfrac{x + 5}{3x - 1}$

6. $\dfrac{2x - 5}{5 - 2x} = \dfrac{-1(-2x + 5)}{(5 - 2x)} = \dfrac{-1(5 - 2x)}{(5 - 2x)} = -1$

7. $\dfrac{4x^2 + 3x - 10}{25 - 16x^2} = \dfrac{(4x - 5)(x + 2)}{(5 + 4x)(5 - 4x)} = \dfrac{(4x - 5)(x + 2)}{-1(4x - 5)(5 + 4x)}$

$= \dfrac{x + 2}{-1(5 + 4x)} = -\dfrac{x + 2}{5 + 4x}$

8. $\dfrac{4x^2 - 9y^2}{4x^2 + 12xy + 9y^2} = \dfrac{(2x + 3y)(2x - 3y)}{(2x + 3y)(2x + 3y)} = \dfrac{2x - 3y}{2x + 3y}$

9. $\dfrac{12x^3 - 48x}{6x - 3x^2} = \dfrac{12x(x^2 - 4)}{3x(2 - x)} = \dfrac{12x(x + 2)(x - 2)}{3x(2 - x)} = -4(x + 2)$

6.2 Practice Problems

1. $\dfrac{10x}{x^2 - 7x + 10} \cdot \dfrac{x^2 + 3x - 10}{25x} = \dfrac{10x}{(x - 5)(x - 2)} \cdot \dfrac{(x - 2)(x + 5)}{25x}$

$= \dfrac{2(x + 5)}{5(x - 5)}$

2. $\dfrac{2y^2 - 6y - 8}{y^2 - y - 2} \cdot \dfrac{y^2 - 5y + 6}{2y^2 - 4y - 6}$

$= \dfrac{2(y + 1)(y - 4)}{(y + 1)(y - 2)} \cdot \dfrac{(y - 2)(y - 3)}{2(y + 1)(y - 3)} = \dfrac{y - 4}{y + 1}$

3. $\dfrac{x^2 + 5x + 6}{x^2 + 8x} \div \dfrac{2x^2 + 5x + 2}{2x^2 + x}$

$= \dfrac{(x + 2)(x + 3)}{x(x + 8)} \cdot \dfrac{x(2x + 1)}{(2x + 1)(x + 2)} = \dfrac{x + 3}{x + 8}$

4. $\dfrac{x + 3}{x - 3} \div (9 - x^2) = \dfrac{x + 3}{x - 3} \cdot \dfrac{1}{(3 + x)(3 - x)} = \dfrac{1}{(x - 3)(3 - x)}$

6.3 Practice Problems

1. $\dfrac{2s + t}{2s - t} + \dfrac{s - t}{2s - t} = \dfrac{2s + t + s - t}{2s - t} = \dfrac{3s}{2s - t}$

2. $\dfrac{b}{(a - 2b)(a + b)} - \dfrac{2b}{(a - 2b)(a + b)} = \dfrac{b - 2b}{(a - 2b)(a + b)}$

$= \dfrac{-b}{(a - 2b)(a + b)}$

3. $\dfrac{7}{6x + 21}, \dfrac{13}{10x + 35}$ $\quad 6x + 21 = 3(2x + 7)$
$10x + 35 = 5(2x + 7)$
$\text{LCD} = 3 \cdot 5 \cdot (2x + 7) = 15(2x + 7)$

4. **(a)** $\dfrac{3}{50xy^2z}, \dfrac{19}{40x^3yz}$ $\quad 50xy^2 = 2 \cdot 5^2 \cdot x \cdot y^2 \cdot z$
$40x^3yz = 2^3 \cdot 5 \cdot x^3 \cdot y \cdot z$
$\text{LCD} = 2^3 \cdot 5^2 \cdot x^3 \cdot y^2 \cdot z$
$\text{LCD} = 200x^3y^2z$

(b) $\dfrac{12}{x^2 + 5x}, \dfrac{16}{x^2 + 10x + 25}$ $\quad x^2 + 5x = x(x + 5)$
$x^2 + 10x + 25 = (x + 5)(x + 5)$
$\text{LCD} = x(x + 5)^2$

(c) $\dfrac{2}{x^2 + 5x + 6}, \dfrac{6}{3x^2 + 5x - 2}$
$x^2 + 5x + 6 = (x + 3)(x + 2)$
$3x^2 + 5x - 2 = (3x - 1)(x + 2)$
$\text{LCD} = (x + 2)(x + 3)(3x - 1)$

5. $\dfrac{7}{a} + \dfrac{3}{abc}$ LCD $= abc$ $\quad \dfrac{7bc + 3}{abc}$

6. $\dfrac{2a - b}{(a + 2b)(a - 2b)} + \dfrac{2}{(a + 2b)}$ $\quad \text{LCD} = (a + 2b)(a - 2b)$

$\dfrac{2a - b}{(a + 2b)(a - 2b)} + \dfrac{2(a - 2b)}{(a + 2b)(a - 2b)} = \dfrac{2a - b + (2a - 4b)}{(a + 2b)(a - 2b)}$

$= \dfrac{4a - 5b}{(a + 2b)(a - 2b)}$

7. $\dfrac{3}{x^2} + \dfrac{4x}{xy} + \dfrac{7}{y}$ $\quad \text{LCD} = x^2y$

$\dfrac{3y}{x^2y} + \dfrac{4x^2}{x^2y} + \dfrac{7x^2}{x^2y} = \dfrac{3y + 4x^2 + 7x^2}{x^2y} = \dfrac{11x^2 + 3y}{x^2y}$

8. $\dfrac{7a}{a^2 + 2ab + b^2} + \dfrac{4}{a^2 + ab}$

$\dfrac{7a}{(a + b)(a + b)} + \dfrac{4}{a(a + b)}$ $\quad \text{LCD} = a(a + b)^2$

$\dfrac{7a^2}{a(a + b)(a + b)} + \dfrac{4(a + b)}{a(a + b)(a + b)} = \dfrac{7a^2 + 4a + 4b}{a(a + b)^2}$

9. $\dfrac{x + 7}{3x - 9} - \dfrac{x - 6}{x - 3} = \dfrac{x + 7}{3(x - 3)} - \dfrac{x - 6}{x - 3}$ $\quad \text{LCD} = 3(x - 3)$

$\dfrac{x + 7 - 3(x - 6)}{3(x - 3)} = \dfrac{x + 7 - 3x + 18}{3(x - 3)} = \dfrac{-2x + 25}{3(x - 3)}$

10. $\dfrac{x - 2}{x^2 - 4} - \dfrac{x + 1}{2x^2 + 4x} = \dfrac{x - 2}{(x + 2)(x - 2)} - \dfrac{x + 1}{2x(x + 2)}$

$\text{LCD} = 2x(x + 2)(x - 2)$

$\dfrac{2x(x - 2)}{2x(x + 2)(x - 2)} - \dfrac{(x - 2)(x + 1)}{2x(x + 2)(x - 2)}$

$= \dfrac{2x^2 - 4x - (x^2 - x - 2)}{2x(x + 2)(x - 2)} = \dfrac{2x^2 - 4x - x^2 + x + 2}{2x(x + 2)(x - 2)}$

$= \dfrac{x^2 - 3x + 2}{2x(x + 2)(x - 2)} = \dfrac{\cancel{(x - 2)}(x - 1)}{2x(x + 2)\cancel{(x - 2)}} = \dfrac{x - 1}{2x(x + 2)}$

6.4 Practice Problems

1. $\dfrac{\dfrac{1}{a} + \dfrac{1}{b}}{\dfrac{2}{ab^2}} = \dfrac{\dfrac{b + a}{ab}}{\dfrac{2}{ab^2}}$

$\dfrac{b + a}{ab} \div \dfrac{2}{ab^2} = \dfrac{b + a}{ab} \cdot \dfrac{ab^2}{2} = \dfrac{\cancel{ab^2}(a + b)}{2\cancel{ab}} = \dfrac{b(a + b)}{2}$

2. $\dfrac{\dfrac{1}{a} + \dfrac{1}{b}}{\dfrac{1}{a} - \dfrac{1}{b}} = \dfrac{\dfrac{b + a}{ab}}{\dfrac{b - a}{ab}} = \dfrac{b + a}{\cancel{ab}} \cdot \dfrac{\cancel{ab}}{b - a} = \dfrac{b + a}{b - a}$

3. $\dfrac{\dfrac{x}{x^2 + 4x + 3} + \dfrac{2}{x + 1}}{x + 1} = \dfrac{\dfrac{x}{(x + 1)(x + 3)} + \dfrac{2}{x + 1}}{x + 1}$

$= \dfrac{\dfrac{x + 2(x + 3)}{(x + 1)(x + 3)}}{x + 1} = \dfrac{\dfrac{x + 2x + 6}{(x + 1)(x + 3)}}{(x + 1)} = \dfrac{3x + 6}{(x + 1)(x + 3)} \cdot \dfrac{1}{(x + 1)}$

$= \dfrac{3(x + 2)}{(x + 1)^2(x + 3)}$

4. $\dfrac{\dfrac{6}{x^2-y^2}}{\dfrac{1}{x-y}+\dfrac{3}{x+y}} = \dfrac{\dfrac{6}{(x+y)(x-y)}}{\dfrac{(x+y)+3(x-y)}{(x+y)(x-y)}} = \dfrac{6}{(x+y)(x-y)}$

$\div \dfrac{x+y+3x-3y}{(x+y)(x-y)} = \dfrac{6}{(x+y)(x-y)} \cdot \dfrac{(x+y)(x-y)}{4x-2y}$

$= \dfrac{6}{\cancel{(x+y)(x-y)}} \cdot \dfrac{\cancel{(x+y)(x-y)}}{2(2x-y)} = \dfrac{3}{2x-y}$

5. $\dfrac{\dfrac{2}{3x^2}-\dfrac{3}{y}}{\dfrac{5}{xy}-4}$ LCD $= 3x^2y$

$\dfrac{3x^2y\left(\dfrac{2}{3x^2}\right)-3x^2y\left(\dfrac{3}{y}\right)}{3x^2y\left(\dfrac{5}{xy}\right)-3x^2y(4)} = \dfrac{2y-3x^2(3)}{3x(5)-12x^2y} = \dfrac{2y-9x^2}{15x-12x^2y}$

6. $\dfrac{\dfrac{6}{x^2-y^2}}{\dfrac{7}{x-y}+\dfrac{3}{x+y}} = \dfrac{\dfrac{6}{(x+y)(x-y)}}{\dfrac{7}{x-y}+\dfrac{3}{x+y}}$ LCD $= (x+y)(x-y)$

$\dfrac{(x+y)(x-y)\left(\dfrac{6}{(x+y)(x-y)}\right)}{(x+y)(x-y)\left(\dfrac{7}{(x-y)}\right)+(x+y)(x-y)\left(\dfrac{3}{x+y}\right)}$

$= \dfrac{6}{7(x+y)+3(x-y)} = \dfrac{6}{7x+7y+3x-3y} = \dfrac{6}{10x+4y}$

$= \dfrac{3}{5x+2y}$

6.5 Practice Problems

1. $\dfrac{2}{x}+\dfrac{4}{x}=3-\dfrac{1}{x}-\dfrac{17}{8}$

$\text{LCD} = 8x$
$16+32=24x-8-17x$
$48=7x-8$
$56=7x$
$x=8$

Check. $\dfrac{2}{8}+\dfrac{4}{8}=3-\dfrac{1}{8}-\dfrac{17}{8}$

$\dfrac{6}{8}=3-\dfrac{18}{8}$

$\dfrac{6}{8}=\dfrac{24}{8}-\dfrac{18}{8}$

$\dfrac{6}{8}=\dfrac{6}{8}$ ✓

2. $\dfrac{4}{2x+1}=\dfrac{6}{2x-1}$ LCD $= (2x+1)(2x-1)$

$(2x+1)(2x-1)\left[\dfrac{4}{2x+1}\right]=(2x+1)(2x-1)\left[\dfrac{6}{2x-1}\right]$

$4(2x-1)=6(2x+1)$
$8x-4=12x+6$
$-4x=10$
$x=-\dfrac{5}{2}$

Check. $\dfrac{4}{2\left(-\dfrac{5}{2}\right)+1} \stackrel{?}{=} \dfrac{6}{2\left(-\dfrac{5}{2}\right)-1}$

$\dfrac{4}{-5+1} \stackrel{?}{=} \dfrac{6}{-5-1}$

$\dfrac{4}{-4} \stackrel{?}{=} \dfrac{6}{-6}$

$-1=-1$ ✓

3. $\dfrac{x-1}{x^2-4}=\dfrac{2}{x+2}+\dfrac{4}{x-2}$

$\dfrac{x-1}{(x+2)(x-2)}=\dfrac{2}{x+2}+\dfrac{4}{x-2}$ LCD $= (x+2)(x-2)$

$\cancel{(x+2)(x-2)}\left[\dfrac{x-1}{\cancel{(x+2)(x-2)}}\right]=\cancel{(x+2)}(x-2)\left[\dfrac{2}{\cancel{x+2}}\right]$
$+(x+2)\cancel{(x-2)}\left[\dfrac{4}{\cancel{x-2}}\right]$

$x-1=2(x-2)+4(x+2)$
$x-1=2x-4+4x+8$
$x-1=6x+4$
$-5x=5$
$x=-1$

Check. $\dfrac{-1-1}{(-1)^2-4}=\dfrac{2}{-1+2}+\dfrac{4}{-1-2}$

$\dfrac{-2}{-3}=\dfrac{2}{1}+\dfrac{4}{-3}$

$\dfrac{2}{3}=\dfrac{6}{3}-\dfrac{4}{3}$

$\dfrac{2}{3}=\dfrac{2}{3}$

4. $\dfrac{2x}{x+1}=\dfrac{-2}{x+1}+1$ LCD $= (x+1)$

$(x+1)\left[\dfrac{2x}{x+1}\right]=(x+1)\left[\dfrac{-2}{x+1}\right]+(x+1)[1]$

$2x=-2+x+1$
$2x=x-1$
$x=-1$

Check. $\dfrac{2(-1)}{-1+1}=\dfrac{-2}{-1+1}+1$

$\dfrac{-2}{0}=\dfrac{-2}{0}+1$

These expressions are not defined, therefore, there is no solution to this problem.

6.6 Practice Problems

1. $\dfrac{8}{420}=\dfrac{x}{315}$
$8(315)=420x$
$2{,}520=420x$
$x=6$

It would take Brenda 6 hours to drive 315 miles.

2. $\dfrac{\dfrac{5}{8}}{30}=\dfrac{2\frac{1}{2}}{x}$

$\dfrac{5}{8}x=30\left(2\dfrac{1}{2}\right)$

$\dfrac{5}{8}x=75$

$x=120$

Therefore $2\frac{1}{2}$ inches would represent 120 miles.

3. $\dfrac{13}{x}=\dfrac{16}{18}$
$13(18)=16x$
$234=16x$
$x=14\dfrac{5}{8}$

4. $\dfrac{6}{7}=\dfrac{x}{38.5}$
$6(38.5)=7x$
$231=7x$
$x=33$ feet

5. Train A $\dfrac{180}{x+10}$ Train B $\dfrac{150}{x}$

$\dfrac{180}{x+10}=\dfrac{150}{x}$
$180x=150(x+10)$
$180x=150x+1500$
$30x=1500$
$x=50$

Train B travels 50 kilometers per hour. Train A travels $50+10=60$ kilometers per hour.

6.

	Number of Hours	Part of the Job Done in One Hour
John	6 hours	$\dfrac{1}{6}$
Dave	7 hours	$\dfrac{1}{7}$
John & Dave Together	x	$\dfrac{1}{x}$

$$\frac{1}{6} + \frac{1}{7} = \frac{1}{x} \qquad \text{LCD} = 42x$$
$$7x + 6x = 42$$
$$13x = 42$$
$$x = 3\frac{3}{13} \qquad \frac{3}{13}\text{ hour} \times \frac{60\text{ min}}{1\text{ hour}} = \frac{180}{13}\text{ min} = 13.846\text{ min}$$
Thus, doing the job together will take 3 hours and 14 minutes.

Chapter 7

7.1 Practice Problems

1. Point B is three units to the right of the origin and 4 units up.

2. Point D is one unit to the right and 4 units up.
Point E is four units to the right and one unit up.

3.

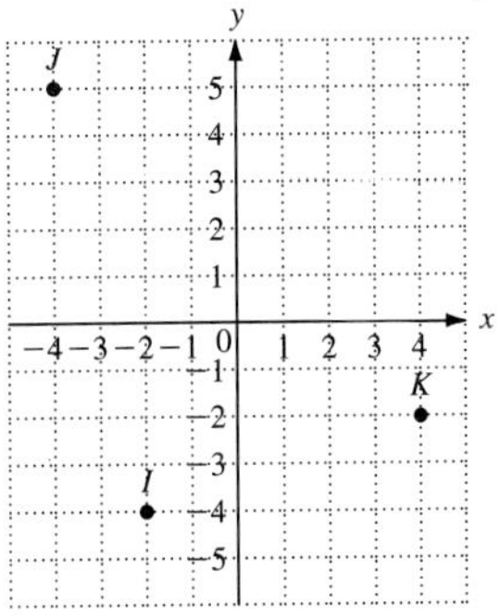

4.

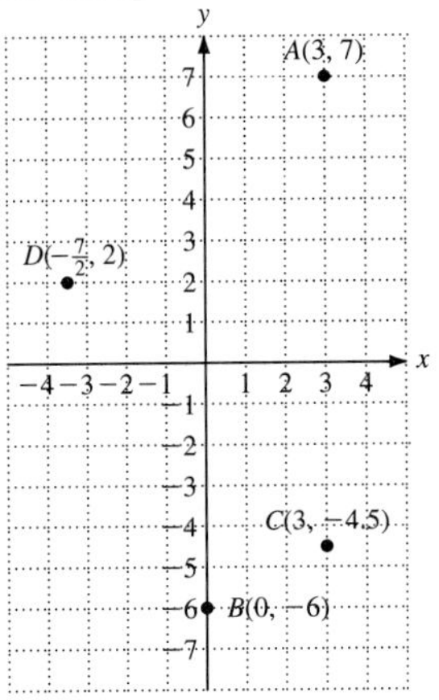

5. $B = (2, 7)$

6. $A = (-2, -1)$; $B = (-1, 3)$; $C = (0, 0)$; $D = (2, -1)$; $E = (3, 1)$

8. $y = -2x - 5$

(a) $y = -2(3) - 5$
$y = -6 - 5$
$y = -11 \qquad (3, -11)$

(b) $y = -2(-4) - 5$
$y = 8 - 5$
$y = 3 \qquad (-4, 3)$

7. $3x - 4y = 12$

(a) $3(0) - 4y = 12$
$-4y = 12$
$y = -3 \qquad (0, -3)$

(b) $3x - 4(3) = 12$
$3x - 12 = 12$
$3x = 24$
$x = 8 \qquad (8, 3)$

(c) $3x - 4(-6) = 12$
$3x + 24 = 12$
$3x = -12$
$x = -4 \qquad (-4, -6)$

7.2 Practice Problems

1. Graph $x + y = 10$.
Let $x = 0$ then
$0 + y = 10$
$y = 10$

Let $x = 5$ then
$5 + y = 10$
$y = 5$

Let $x = 2$ then
$2 + y = 10$
$y = 8$

Plot the ordered pairs.
(0, 10), (5, 5) and (2, 8)

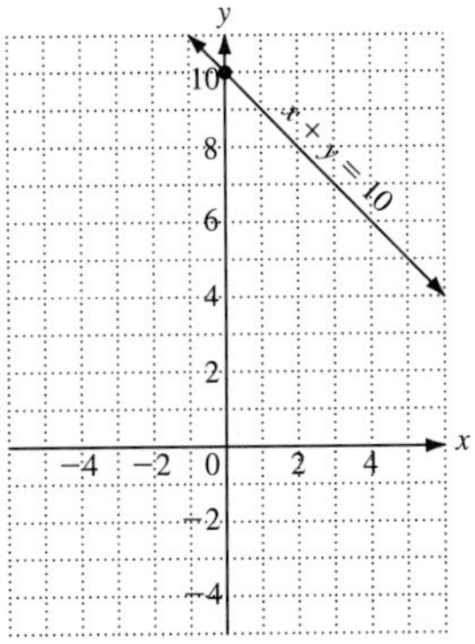

2. $7x + 3 = -2y + 3$
$7x + 2y = 0$
Let $x = 0$ then
$7(0) + 2y = 0$
$2y = 0$
$y = 0$
Let $x = -2$ then

$7(-2) + 2y = 0$
$-14 + 2y = 0$
$2y = 14$
$y = 7$
Let $x = 2$ then
$7(2) + 2y = 0$
$14 + 2y = 0$
$2y = -14$
$y = -7$
Graph the ordered pairs. (0, 0), (2, −7) and (−2, 7)

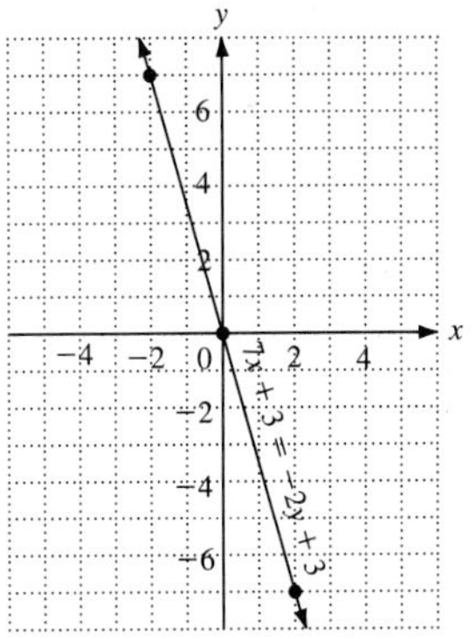

3. $3x - 2y = 6$
Let $y = 0$ then
$3x - 2(0) = 6$
$3x = 6$
$x = 2$
Let $y = 3$ then
$3x - 2(3) = 6$
$3x - 6 = 6$
$3x = 12$
$x = 4$
Let $y = -3$ then
$3x - 2(-3) = 6$
$3x + 6 = 6$
$3x = 0$
$x = 0$
Graph the ordered pairs. (2, 0), (4, 3) and (0, −3)

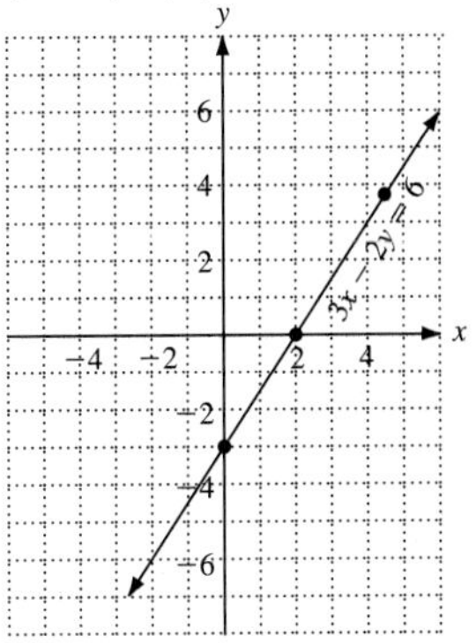

4. $2y - x = 6$
Find the two intercepts.

Let $x = 0$
$2y - 0 = 6$
$2y = 6$
$y = 3$

Let $y = 0$
$2(0) - x = 6$
$-x = 6$
$x = -6$

Third point
Let $y = 1$
$2(1) - x = 6$
$2 - x = 6$
$-x = 4$
$x = -4$
Graph the ordered pairs.
(0, 3), (−6, 0), and (−4, 1)

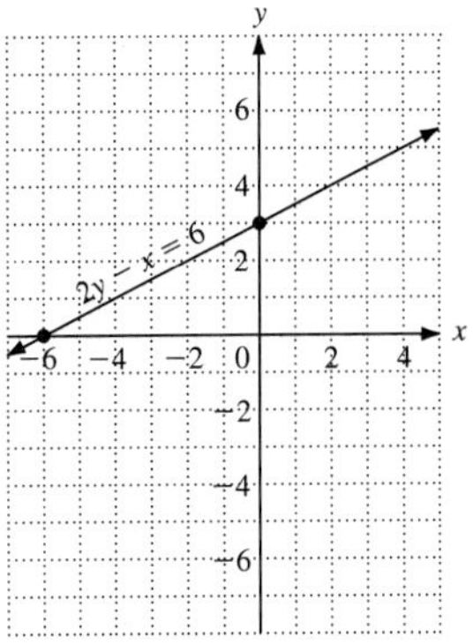

5. $2y - 3 = 0$
Solve for y.
$2y = 3$
$$y = \frac{3}{2}$$
This line is parallel to the x-axis. It is a horizontal line $1\frac{1}{2}$ units above the x-axis.

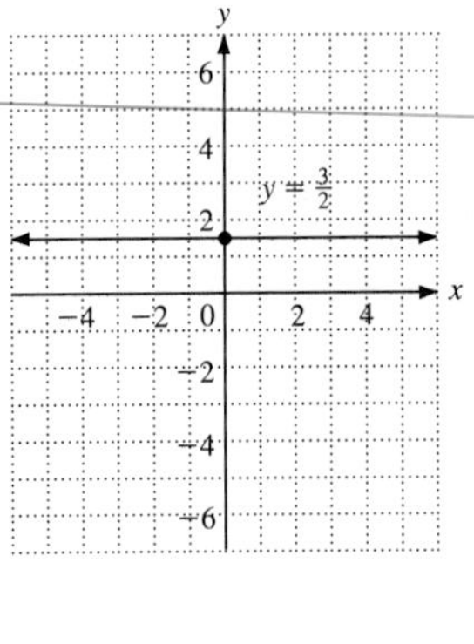

6. $x + 3 = 0$
Solve for x.
$x = -3$
This line is parallel to the y-axis. It is a vertical line 3 units to the left of the y-axis.

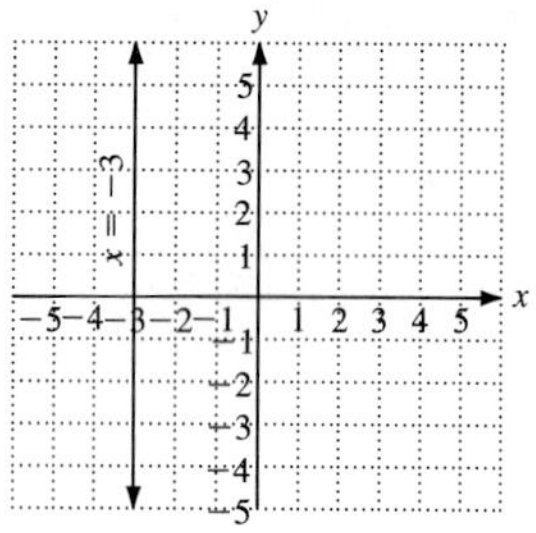

7.3 Practice Problems

1. $m = \dfrac{-1 - (1)}{-4 - (6)} = \dfrac{-2}{-10} = \dfrac{1}{5}$ $\quad m = \dfrac{1}{5}$

2. $m = \dfrac{6 - (-4)}{1 - (-1)} = \dfrac{10}{2} = 5$ $\quad m = 5$

3. $m = \dfrac{1 - 0}{-1 - 2} = \dfrac{1}{-3}$ $\quad m = -\dfrac{1}{3}$

4. **(a)** $m = \dfrac{4 - 0}{-4 - (-4)} = \dfrac{4}{0}$

$\dfrac{4}{0}$ is undefined. Therefore there is no slope and the line is a vertical line through $x = -4$.

(b) $m = \dfrac{-11 - (-11)}{3 - (-7)} = \dfrac{0}{10} = 0$

$m = 0$. The line is a horizontal line through $y = -11$.

5. $y = -2x + 6$
The slope $= -2$ and the y-intercept $= 6$.

6. $4x - 2y = -5$ $\quad$ Solve for y.
$-2y = -4x - 5$
$y = 2x + \dfrac{5}{2}$ $\quad$ Slope $= 2$ $\quad$ y-intercept $= \dfrac{5}{2}$

7. **(a)** $y = mx + b$
$m = -\dfrac{3}{7}$ $\quad$ y-intercept $= \dfrac{2}{7}$
$y = -\dfrac{3}{7}x + \dfrac{2}{7}$

(b) $y = -\dfrac{3}{7}x + \dfrac{2}{7}$
$7y = -3x + 2$
$7y + 3x = 2$
$3x + 7y = 2$

8. y = intercept $= -1$. Thus the coordinates of the y-intercept for this line is $(0, -1)$. Plot the point.
Slope is $\dfrac{\text{rise}}{\text{run}}$ since the slope for this line is $\dfrac{3}{4}$, we will go up(rise) 3 units and go over(run) 4 units to the right. This is the point $(4, 2)$.

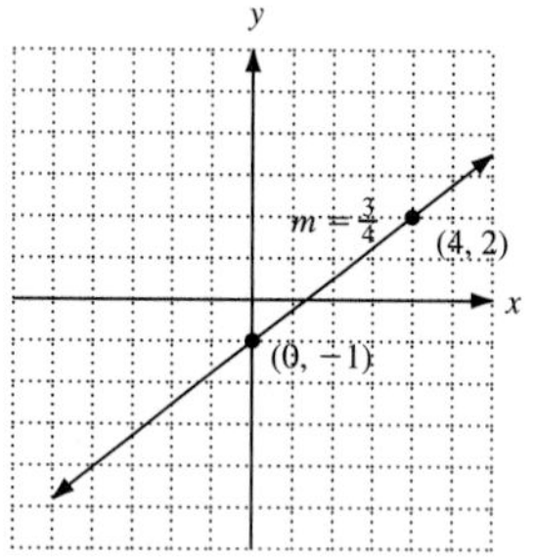

9. $y = -\dfrac{2}{3}x + 5$
y-intercept is 5, plot the point $(0, 5)$. The slope is $-\dfrac{2}{3}$. Begin at $(0, 5)$, go down 2 units and to the right 3 units. Draw a line that connects the points $(0, 5)$ and $(3, 3)$.

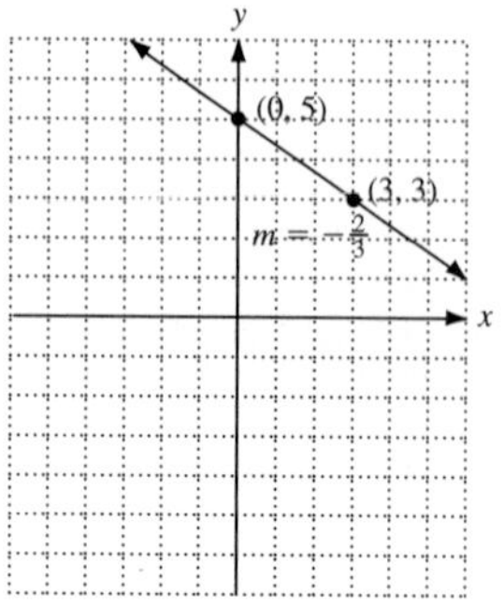

10. **(a)** $m = \dfrac{1}{4}$ **(b)** $m = -4$ $\quad$ 11. **(a)** $m = \dfrac{1}{4}$ **(b)** $m = -4$

7.4 Practice Problems

1. $y = mx + b$
$4 = \dfrac{1}{2}(-2) + b$ $\quad$ Solve for b.
$4 = -1 + b$
$5 = b$ $\quad$ Thus y-intercept is 5.
We can now write the equation of the line. $\quad y = \dfrac{1}{2}x + 5$

2. $y = mx + b$
$12 = -\dfrac{3}{4}(-8) + b$ $\quad$ Solve for b.
$12 = 6 + b$
$6 = b$ $\quad$ Thus y-intercept is 6.
We can now write the equation of the line. $\quad y = -\dfrac{3}{4}x + 6$

3. Find the slope.
$m = \dfrac{1 - 5}{-1 - 3} = \dfrac{-4}{-4} = 1$
Using either of the two points given substitute x and y values into the equation.
$m = 1 \quad x = 3$ and $y = 5$.
$y = mx + b$
$5 = 1(3) + b$ $\quad$ Solve for b.
$5 = 3 + b$
$2 = b$ $\quad$ Thus y-intercept is 2.
$y = x + 2$

4. y-intercept is $(0, 1)$. Look for another point in the line. We choose $(6, 2)$. Count the number of units from 1 to 2 (rise). Count the number of units from 0 to 6 (run).
$m = \dfrac{1}{6}$ $\quad$ Now we can write the equation of the line.
$y = mx + b$
$y = \dfrac{1}{6}x + 1$

7.5 Practice Problems

1. Graph $x - y \geq -10$.
Begin by graphing the line $x - y = -10$. You may use any method discussed previously. The easiest test point is $(0, 0)$. Substitute $x = 0$, $y = 0$ in the inequality.
$x - y \geq -10$
$0 - 0 \geq -10$
$0 \geq -10$ $\quad$ True
Therefore shade the side of the line that includes the point $(0, 0)$.

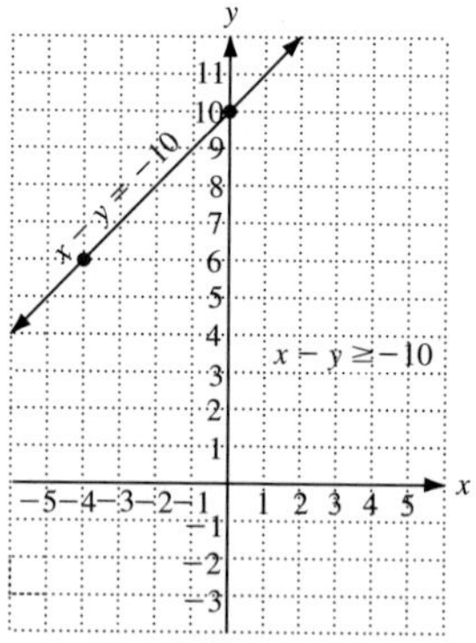

2. $y > \dfrac{1}{2}x$
Since the line goes through the origin $(0, 0)$ choose another test point. We will choose $(-1, 1)$.
$y > \dfrac{1}{2}x \qquad 1 > \dfrac{1}{2}(-1)$
$1 > -\dfrac{1}{2}$ $\quad$ True
Therefore shade the side of the line that includes the point $(-1, 1)$.

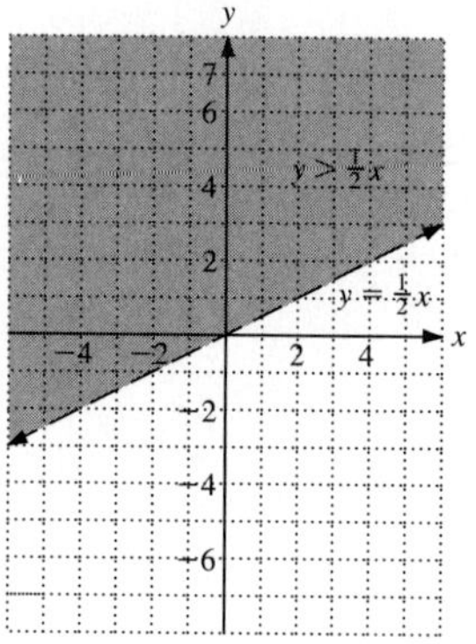

3. $y \geq -3$
A test point is not needed in the case of horizontal and vertical lines. Usually the region can be determined by inspection.

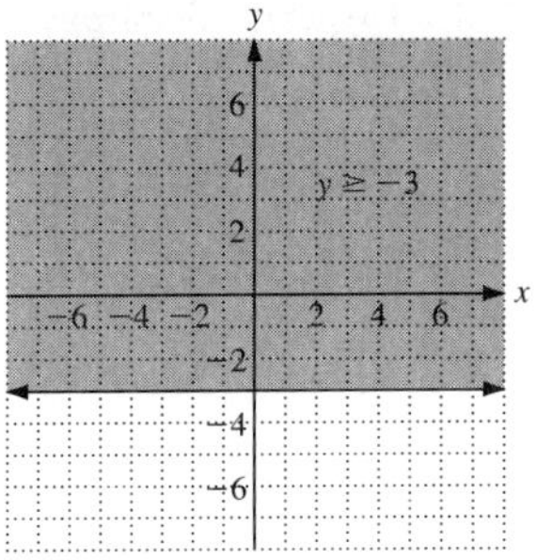

7.6 Practice Problems

1. The domain is $\{-3, 3, 0, 20\}$. The range is $\{-5, 5\}$.
2. **(a)** Function **(b)** Not a function
3. **(a)** Function
The domain is {all nonnegative real numbers}.
The range is {all nonnegative real numbers}.
(b) Not a function
4. $y = x^2 - 2$

x	$y = x^2 - 2$	y
-2	$y = (-2)^2 - 2 = 2$	2
-1	$y = (-1)^2 - 2 = -1$	-1
0	$y = 0 - 2 = -2$	-2
1	$y = (1)^2 - 2 = -1$	-1
2	$y = (2)^2 - 2 = 2$	2

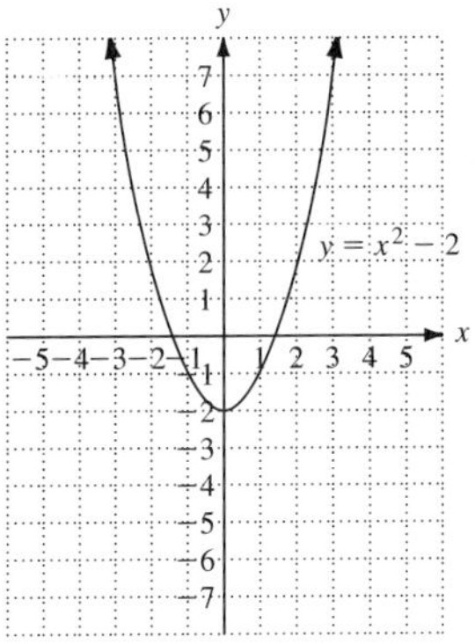

5. $x = y^2 - 1$

y	$x = y^2 - 1$	x	y
-2	$x = (-2)^2 - 1 = 3$	3	-2
-1	$x = (-1)^2 - 1 = 0$	0	-1
0	$x = 0 - 1 = -1$	-1	0
1	$x = 1^2 - 1 = 0$	0	1
2	$x = 2^2 - 1 = 3$	3	2

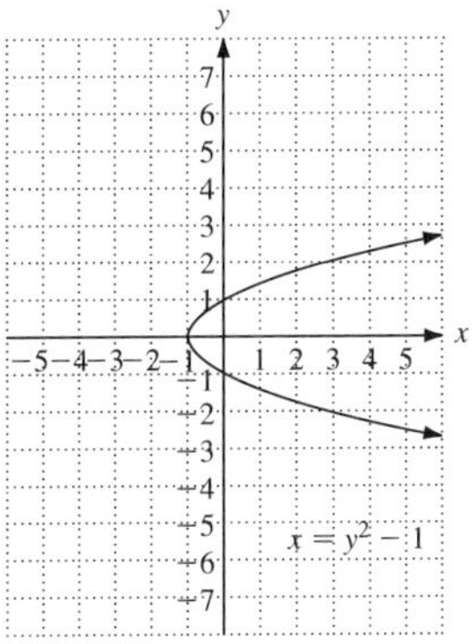

6. $y = \dfrac{6}{x}$

x	$y = \dfrac{6}{x}$	y
-3	$y = \dfrac{6}{-3} = -2$	-2
-2	$y = \dfrac{6}{-2} = -3$	-3
-1	$y = \dfrac{6}{-1} = -6$	-6
0	We cannot divide by 0.	
1	$y = \dfrac{6}{1} = 6$	6
2	$y = \dfrac{6}{2} = 3$	3
3	$y = \dfrac{6}{3} = 2$	2

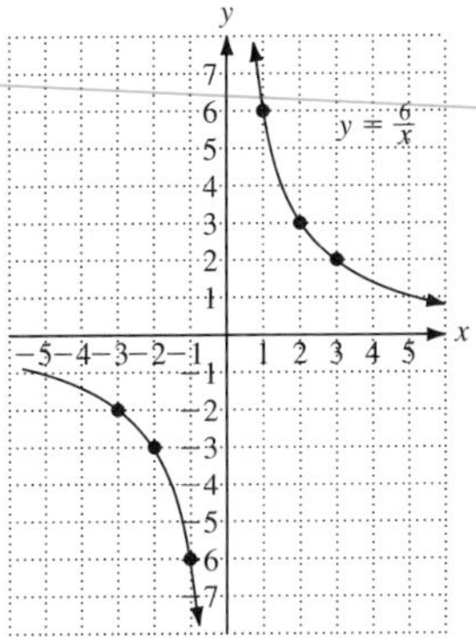

7. **(a)** Not a function **(b)** Function **(c)** Not a function
8. $f(x) = -3x + 4$
$f(5) = -3(5) + 4$
$f(5) = -15 + 4$
$f(5) = -11$
9. $f(x) = -2x^2 + 3x - 8$
(a) $f(-1) = -2(-1)^2 + 3(-1) - 8$
$f(-1) = -2 - 3 - 8$
$f(-1) = -13$
(b) $f(2) = -2(2)^2 + 3(2) - 8$
$f(2) = -8 + 6 - 8$
$f(2) = -10$
(c) $f(-3) = -2(-3)^2 + 3(-3) - 8$
$f(-3) = -18 - 9 - 8$
$f(-3) = -35$
(d) $f(0) = -2(0)^2 + 3(0) - 8$
$f(0) = 0 + 0 - 8$
$f(0) = -8$

Chapter 8

8.1 Practice Problems

1. $x + y = 12$
$-x + y = 4$
The lines intersect at the point (4, 8). Thus the solution to the system of equation is $x = 4$, $y = 8$.

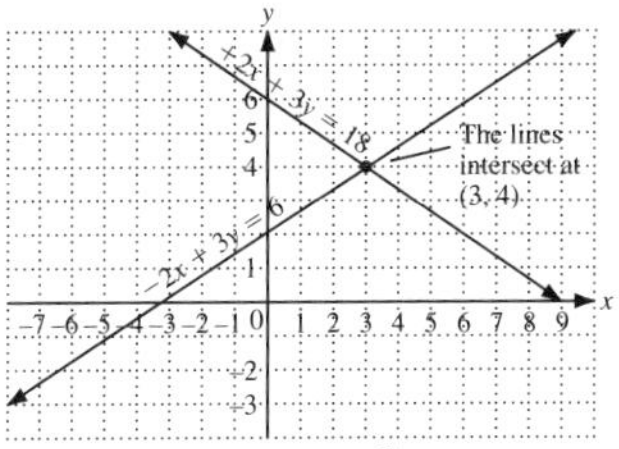

2. $4x + 2y = 8$
$-6x - 3y = 6$
These lines are parallel. They do not intersect. Hence there is no solution to this system of equations.

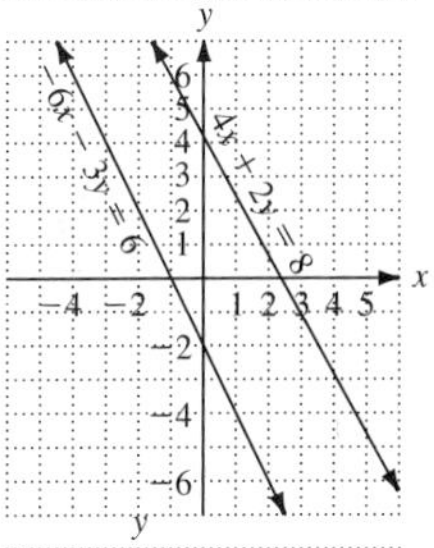

3. $3x - 9y = 18$
$-4x + 12y = -24$
Notice both equations represent the same line. Thus there is an infinite number of solutions to this system. The equations are said to be dependent.

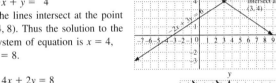

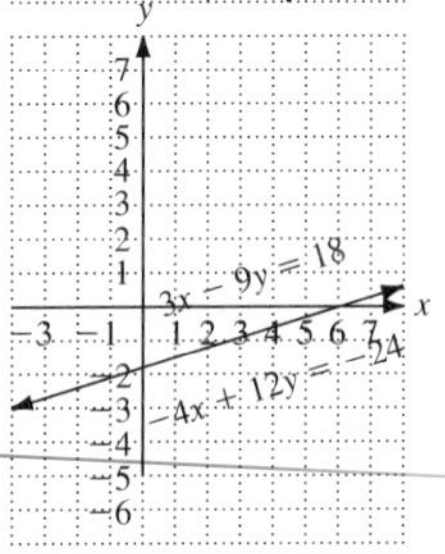

8.2 Practice Problems

1. $y = 3x + 10$
$2x - 3y = -16$
Substitute $y = 3x + 10$ into the equation.
$2x - 3y = -16$
$2x - 3(3x + 10) = -16$
$2x - 9x - 30 = -16$
$-7x = 14$
$x = -2$
Now we obtain the value of the second variable.

$y = 3(-2) + 10$
$y = -6 + 10$
$y = 4$

Check the solution $x = -2$ and $y = 4$ in both original equations.

$y = 3x + 10$	$2x - 3y = -16$
$4 = 3(-2) + 10$	$2(-2) - 3(4) = -16$
$4 = -6 + 10$	$-4 - 12 = -16$
$4 = 4$	$-16 = -16$

2. $5x + 3y = 19$
$2x - \ y = 12$
Solve the second equation for y.
$2x - y = 12$
$-y = -2x + 12$
$y = 2x - 12$
Now substitute $y = 2x - 12$ into the equation.
$5x + 3y = 19$
$5x + 3(2x - 12) = 19$
$5x + 6x - 36 = 19$
$11x - 36 = 19$
$11x = 55$
$x = 5$
Now obtain the value for the second variable.
$2x - y = 12$
$2(5) - y = 12$
$10 - y = 12$
$-y = 2$
$y = -2$
The solution is $x = 5$, $y = -2$. Check this result.

3. $\frac{x}{3} - \frac{y}{2} = 1$ Multiply by the LCD. $\rightarrow 2x - 3y = 6$
$x + 4y = -8$
Solve the second equation for x.
$x + 4y = -8$
$x = -4y - 8$
Substitute $x = -4y - 8$ into the equation $2x - 3y = 6$.
$2(-4y - 8) - 3y = 6$
$-8y - 16 - 3y = 6$
$-11y - 16 = 6$
$-11y = 22$
$y = -2$
Obtain the value of the second variable.
$x + 4y = -8$
$x + 4(-2) = -8$
$x - 8 = -8$
$x = 0$
The solution is $x = 0$, $y = -2$. Remember to check the solution in both original equations.

8.3 Practice Problems

1.
$$\begin{array}{r} 3x + 5y = 1 \\ -3x + y = 11 \\ \hline 6y = 12 \end{array}$$
$y = 2$ Substitute for y in equation 2.
$-3x + 2 = 11$
$-3x = 9$
$x = -3$
The solution is $x = -3$, $y = 2$. Check this result.

2. $3x + y = 7$ Multiply by 2. $\rightarrow$ $6x + 2y = 14$
$5x - 2y = 8$ $\qquad + \ 5x - 2y = 8$
$$\begin{array}{r} 6x + 2y = 14 \\ +\ 5x - 2y = 8 \\ \hline 11x \quad = 22 \\ x \quad = 2 \end{array}$$
Substitute $x = 2$ into one of the original equations.
$5(2) - 2y = 8$
$10 - 2y = 8$
$-2y = -2$
$y = 1$
The solution is $x = 2$, $y = 1$. Check this result.

3. $4x + 5y = 7$ Multiply by -3. $\rightarrow -12x - 15y = -21$
$3x + 7y = 2$ Multiply by 4. $\rightarrow$ $12x + 28y = 8$
$$\begin{array}{r} 13y = -13 \\ y = -1 \end{array}$$
Substitute $y = -1$ into one of the original equations.
$4x + 5(-1) = 7$
$4x - 5 = 7$
$4x = 12$
$x = 3$
The solution is $x = 3$, $y = -1$. Check this result.

4. $\frac{2}{3}x - \frac{3}{4}y = 3$ Multiply by the LCD. $\rightarrow 8x - 9y = 36$
Use the addition method to solve the system.
$8x - 9y = 36$ $\qquad 8x - 9y = 36$
$-2x + \ y = 6$ Multiply by 4. $\rightarrow -8x + 4y = 24$
$$\begin{array}{r} -5y = 60 \\ y = -12 \end{array}$$
Substitute $y = -12$ into one of the original equations.
$-2x - 12 = 6$
$-2x = 18$
$x = -9$
The solution is $x = -9$, $y = -12$. Check this result.

5. Multiply each term of each equation by 10.
$0.2x + 0.3y = -0.1 \Rightarrow 2x + 3y = -1$
$0.5x - 0.1y = -1.1 \Rightarrow 5x - \ y = -11$
$2x + 3y = -1$ $\qquad 2x + 3y = -1$
$5x - \ y = -11$ Multiply by 3. $\rightarrow 15x - 3y = -33$
$$\begin{array}{r} 17x \quad = -34 \\ x = \ -2 \end{array}$$
Substitute $x = -2$ into one of the equations.
$2(-2) + 3y = -1$
$-4 + 3y = -1$
$3y = 3$
$y = 1$
The solution is $x = -2$, $y = 1$. Check this result. Be sure to substitute $x = -2$ and $y = 1$ into both of the original equations.

8.4 Practice Problems

1. (a) $3x + 5y = 1485$
$x + 2y = 564$
Solve for x in equation 2 and solve using the substitution method.
$x = -2y + 564$
$3(-2y + 564) + 5y = 1485$
$-6y + 1692 + 5y = 1485$
$-y = -207$
$y = 207$
Substitute $y = 207$ into equation 2 and solve for x.
$x + 2(207) = 564$
$x + 414 = 564$
$x = 150$
The solution is $x = 150$, $y = 207$. Check this result.

(b) $7x + 6y = 45$
$6x - 5y = -2$
Using the addition method, multiply the first equation by -6 and the second equation by 7.
$$\begin{array}{r} -42x - 36y = -270 \\ 42x - 35y = -14 \\ \hline -71y = -284 \\ y = 4 \end{array}$$
Substitute $y = 4$ into one of the original equations and solve for x.
$7x + 6(4) = 45$
$7x + 24 = 45$
$7x = 21$
$x = 3$
The solution is $x = 3$, $y = 4$. Check this result.

2. $4x + 2y = 2$
$-6x - 3y = 6$
Use the addition method. Multiply the first equation by 3 and the second equation by 2.
$12x + 6y = 6$
$-12x - 6y = 12$
$0 = 18$
The statement $0 = 18$ is inconsistent. There is no solution to this system of equations.

3. $3x - 9y = 18$
$-4x + 12y = -24$
Use the addition method. Multiply the first equation by 4 and the second equation by 3.
$12x - 36y = 72$
$-12x + 36y = -72$
$0 = 0$
The statement $0 = 0$ is true and can identify the equations as dependent. There are an unlimited number of solutions to the system.

8.5 Practice Problems

1. Let x = Math Score
y = Verbal Score
$x + y = 1153$
$x - y = 409$
$2x = 1562$
$x = 781$
Substitute $x = 781$ into one of the original equations.
$781 + y = 1153$
$y = 372$
Therefore she scored 781 on Math and 372 on Verbal.

2. Let x = Smaller pump
y = Larger pump
1st day $8x + 5y = 49{,}000$
2nd day $5x + 3y = 30{,}000$
Multiply the first equation by 5 and the second equation by -8.
$40x + 25y = 245{,}000$
$-40x - 24y = -240{,}000$
$y = 5{,}000$
Substitute $y = 5{,}000$ into one of the original equations and solve for x.
$8x + 5(5000) = 49{,}000$
$8x + 25{,}000 = 49{,}000$
$8x = 24{,}000$
$x = 3{,}000$
The smaller pump removes 3000 gallons per hour.
The larger pump removes 5000 gallons per hour.

3. Let x = cost per gallon of unleaded premium gasoline.
y = cost per gallon of leaded regular gasoline.
Last week's purchase $7x + 8y = 27.22$
This week's purchase $8x + 4y = 22.16$
Use the addition method. Multiply the first equation by 8 and the second equation by -7.
$56x + 64y = 217.76$
$-56x - 28y = -155.12$
$36y = 62.64$
$y = 1.74$
Substitute $y = 1.74$ into one of the original equations and solve for x.
$7x + 8(1.74) = 27.22$
$7x + 13.92 = 27.22$
$7x = 13.30$
$x = 1.90$
Unleaded premium gasoline costs \$1.90 per gallon.
Leaded regular gasoline costs \$1.74 per gallon.

4. Let x = the amount of 20% solution
y = the amount of 80% solution
$x + y = 4000$
$.20x + .80y = 2600$
Solve for x in the first equation.
$x + y = 4000$
$x = 4000 - y$
Substitute $x = 4000 - y$ into the second equation.
$.20(4000 - y) + .80y = 2600$
$800 - .2y + .8y = 2600$
$800 + 0.6y = 2600$
$0.6y = 1800$
$y = 3000$
Now substitute $y = 3000$ into the first equation.
$x + y = 4000$
$x + 3000 = 4000$
$x = 1000$
Therefore he will need 1,000 liters of 20% solution and 3,000 liters of 80% solution.

5. To go downstream with the current we have as the rate of travel
$r = \frac{d}{t} = \frac{72}{3} = 24$ mph
To go upstream against the current we have as the rate of travel
$r = \frac{d}{t} = \frac{72}{4} = 18$ mph
Let x = the speed of boat in mph.
y = the speed of the current in mph.
$x + y = 24$
$x - y = 18$
$2x = 42$
$x = 21$
Substitute $x = 21$ into the original equation and solve for y.
$21 + y = 24$
$y = 3$
Thus the boat speed is 21 mph and the speed of the current is 3 mph.

Chapter 9

9.1 Practice Problems

1. **(a)** $\sqrt{64} = 8$ because $8^2 = 64$
(b) $-\sqrt{121} = -11$ because $(-11)^2 = 121$
2. **(a)** $\sqrt{\frac{9}{25}} = \frac{\sqrt{9}}{\sqrt{25}} = \frac{3}{5}$ **(b)** $-\sqrt{\frac{121}{169}} = \frac{-\sqrt{121}}{\sqrt{169}} = \frac{-11}{13}$
3. **(a)** $-\sqrt{0.0036} = -0.06$ **(b)** $\sqrt{2500} = 50$
4. **(a)** 1.732 **(b)** 5.916 **(c)** 11.269

9.2 Practice Problems

1. **(a)** 6
(b) 13^2
(c) 18^6
2. **(a)** x^2
(b) y^9
(c) x^{15}
3. **(a)** $25y^2$
(b) x^8y^{11}
(c) $11x^6y^3$
4. **(a)** $\sqrt{49} \cdot \sqrt{2} = 7\sqrt{2}$
(b) $\sqrt{4} \cdot \sqrt{3} = 2\sqrt{3}$
(c) $\sqrt{25} \cdot \sqrt{3} = 5\sqrt{3}$
5. **(a)** $\sqrt{x^{11}} = \sqrt{x^{10}} \cdot \sqrt{x} = x^5\sqrt{x}$
(b) $\sqrt{x^5y^3} = \sqrt{x^4y^2} \cdot \sqrt{xy} = x^2y\sqrt{xy}$
6. **(a)** $\sqrt{48x^{11}} = \sqrt{16x^{10}} \cdot \sqrt{3x} = 4x^5\sqrt{3x}$
(b) $\sqrt{72x^8y^9} = \sqrt{36x^8y^8} \cdot \sqrt{2y} = 6x^4y^4\sqrt{2y}$
(c) $\sqrt{121x^6y^7z^8} = \sqrt{121x^6y^6z^8} \cdot \sqrt{y} = 11x^3y^3z^4\sqrt{y}$

9.3 Practice Problems

1. **(a)** $11\sqrt{11}$ **(b)** $\sqrt{t}$
2. $3\sqrt{x} - 2\sqrt{xy} + 7\sqrt{xy} - 5\sqrt{y}$
$3\sqrt{x} + 5\sqrt{xy} - 5\sqrt{y}$

3. (a) $\sqrt{50} - \sqrt{18} + \sqrt{98}$
$= \sqrt{25}\cdot\sqrt{2} - \sqrt{9}\cdot\sqrt{2} + \sqrt{49}\cdot\sqrt{2}$
$= 5\sqrt{2} - 3\sqrt{2} + 7\sqrt{2} = 2\sqrt{2} + 7\sqrt{2} = 9\sqrt{2}$
(b) $\sqrt{12} + \sqrt{18} - \sqrt{50} + \sqrt{27}$
$= \sqrt{4}\cdot\sqrt{3} + \sqrt{9}\cdot\sqrt{2} - \sqrt{25}\cdot\sqrt{2} + \sqrt{9}\cdot\sqrt{3}$
$= 2\sqrt{3} + 3\sqrt{3} + 3\sqrt{2} - 5\sqrt{2} = 5\sqrt{3} - 2\sqrt{2}$

4. $\sqrt{9x} + \sqrt{8x} - \sqrt{4x} + \sqrt{50x}$
$= 3\sqrt{x} + 2\sqrt{2x} - 2\sqrt{x} + 5\sqrt{2x}$
$= \sqrt{x} + 7\sqrt{2x}$

5. (a) $2\sqrt{18} + 3\sqrt{8}$
$= 2\sqrt{9}\sqrt{2} + 3\sqrt{4}\sqrt{2}$
$= 2(3)\sqrt{2} + 3(2)\sqrt{2}$
$= 6\sqrt{2} + 6\sqrt{2} = 12\sqrt{2}$
(b) $\sqrt{27} - 4\sqrt{3} + 2\sqrt{75}$
$= \sqrt{9}\sqrt{3} - 4\sqrt{3} + 2\sqrt{25}\sqrt{3}$
$= 3\sqrt{3} - 4\sqrt{3} + 10\sqrt{3} = 9\sqrt{3}$

6. $2\sqrt{12x} - 3\sqrt{45x} - 3\sqrt{27x} + \sqrt{20}$
$= 2\sqrt{4}\cdot\sqrt{3x} - 3\sqrt{9}\cdot\sqrt{5x} - 3\sqrt{9}\cdot\sqrt{3x} + \sqrt{4}\sqrt{5}$
$= 2(2)\sqrt{3x} - 3(3)\sqrt{5x} - 3(3)\sqrt{3x} + 2\sqrt{5}$
$= 4\sqrt{3x} - 9\sqrt{3x} - 9\sqrt{5x} + 2\sqrt{5}$
$= -5\sqrt{3x} - 9\sqrt{5x} + 2\sqrt{5}$

9.4 Practice Problems

1. $\sqrt{3a}\sqrt{6a} = \sqrt{18a^2} = 3a\sqrt{2}$
2. (a) $(2\sqrt{3})(5\sqrt{5}) = 10\sqrt{15}$
(b) $(2x\sqrt{6})(3\sqrt{6}) = 6x\sqrt{36} = 36x$
(c) $(4\sqrt{3x})(2x\sqrt{6x}) = 8x\sqrt{18x^2} = 24x^2\sqrt{2}$
3. $2\sqrt{3}(\sqrt{3} + \sqrt{5} - \sqrt{12})$
$= 2\sqrt{9} + 2\sqrt{15} - 2\sqrt{36}$
$= 2(3) + 2\sqrt{15} - 2(6) = 6 + 2\sqrt{15} - 12 = -6 + 2\sqrt{15}$
4. $2\sqrt{x}(4\sqrt{x} - x\sqrt{2})$
$= 8\sqrt{x^2} - 2x\sqrt{2x}$
$= 8x - 2x\sqrt{2x}$
5. $\sqrt{6x}(\sqrt{15x} + \sqrt{10x^2})$
$= \sqrt{90x^2} + \sqrt{60x^3}$
$= 3x\sqrt{10} + 2x\sqrt{15x}$
6. $(\sqrt{2} + \sqrt{6})(2\sqrt{2} - \sqrt{6})$
$= 2\sqrt{4} - \sqrt{12} + 2\sqrt{12} - \sqrt{36}$
$= 2(2) + \sqrt{12} - 6 = 4 + 2\sqrt{3} - 6 = -2 + 2\sqrt{3}$
7. $(\sqrt{6} + \sqrt{5})(\sqrt{2} + 2\sqrt{5})$
$= \sqrt{12} + 2\sqrt{30} + \sqrt{10} + 2\sqrt{25}$
$= 2\sqrt{3} + 2\sqrt{30} + \sqrt{10} + 2(5)$
$= 2\sqrt{3} + 2\sqrt{30} + \sqrt{10} + 10$
8. $(3\sqrt{5} - \sqrt{10})^2$
$= (3\sqrt{5} - \sqrt{10})(3\sqrt{5} - \sqrt{10})$
$= 9\sqrt{25} - 3\sqrt{50} - 3\sqrt{50} + \sqrt{100}$
$= 45 - 6\sqrt{25}\sqrt{2} + 10 = 55 - 30\sqrt{2}$

9.5 Practice Problems

1. (a) $\sqrt{\dfrac{x^3}{x}} = \sqrt{x^2} = x$ (b) $\sqrt{\dfrac{49}{x^2}} = \dfrac{7}{x}$
2. $\sqrt{\dfrac{50}{a^4}} = \dfrac{5\sqrt{2}}{a^2}$
3. (a) $\dfrac{9}{\sqrt{7}}\cdot\dfrac{\sqrt{7}}{\sqrt{7}} = \dfrac{9\sqrt{7}}{7}$ (b) $\dfrac{\sqrt{5}}{\sqrt{6}}\cdot\dfrac{\sqrt{6}}{\sqrt{6}} = \dfrac{\sqrt{30}}{6}$
4. (a) $\dfrac{\sqrt{2}}{\sqrt{12}}\cdot\dfrac{\sqrt{3}}{\sqrt{3}} = \dfrac{\sqrt{6}}{\sqrt{36}} = \dfrac{\sqrt{6}}{6}$
(b) $\dfrac{6a}{\sqrt{a^7}}\cdot\dfrac{\sqrt{a}}{\sqrt{a}} = \dfrac{6a\sqrt{a}}{\sqrt{a^8}} = \dfrac{6a\sqrt{a}}{a^4} = \dfrac{6\sqrt{a}}{a^3}$
5. $\dfrac{\sqrt{2x}}{\sqrt{8x}} = \dfrac{\sqrt{2x}}{2\sqrt{2x}}\cdot\dfrac{\sqrt{2x}}{\sqrt{2x}} = \dfrac{\sqrt{4x^2}}{2\sqrt{4x^2}} = \dfrac{2x}{4x} = \dfrac{1}{2}$
6. (a) $\dfrac{4}{\sqrt{3} + \sqrt{5}}\cdot\dfrac{\sqrt{3} - \sqrt{5}}{\sqrt{3} - \sqrt{5}}$
$= \dfrac{4(\sqrt{3} - \sqrt{5})}{3 - 5} = \dfrac{4(\sqrt{3} - \sqrt{5})}{-2}$
$= \dfrac{2(\sqrt{3} - \sqrt{5})}{-1}$ or $-2(\sqrt{3} - \sqrt{5})$
(b) $\dfrac{\sqrt{a}}{(\sqrt{10} - 3)}\cdot\dfrac{(\sqrt{10} + 3)}{(\sqrt{10} + 3)}$
$= \dfrac{\sqrt{a}(\sqrt{10} + 3)}{\sqrt{100} + 3\sqrt{10} - 3\sqrt{10} - 9}$
$= \dfrac{\sqrt{a}(\sqrt{10} + 3)}{10 - 9} = \sqrt{a}(\sqrt{10} + 3)$
7. $\dfrac{\sqrt{7} - \sqrt{x}}{\sqrt{7} + \sqrt{x}}\cdot\dfrac{\sqrt{7} - \sqrt{x}}{\sqrt{7} - \sqrt{x}}$
$= \dfrac{7 - 2\sqrt{7x} + x}{7 - x} = \dfrac{7 + x - 2\sqrt{7x}}{7 - x}$

9.6 Practice Problems

1. $c^2 = a^2 + b^2$
$c^2 = 9^2 + 12^2$
$c^2 = 81 + 144$
$c^2 = 225$
$c = \pm\sqrt{225}$
$c = \pm 15$
$c = +15$ because length is not negative.
2. $(\sqrt{17})^2 = (1)^2 + b^2$
$17 = 1 + b^2$
$16 = b^2$
$\sqrt{16} = b$
$4 = b$
The leg is 4 meters long.
3. $c^2 = 3^2 + 8^2$
$c^2 = 9 + 64$
$c^2 = 73$
$c = \sqrt{73}$
$c = 8.5$ meters
4. $\sqrt{2x - 5} = 6$
$(\sqrt{2x - 5})^2 = (6)^2$
$2x - 5 = 36$
$2x = 41$
$x = 20.5$
Check.
$\sqrt{2(20.5) - 5} \stackrel{?}{=} 6$
$\sqrt{41 - 5} \stackrel{?}{=} 6$
$\sqrt{36} \stackrel{?}{=} 6$
$6 = 6$ ✓
5. $\sqrt{3x - 2} - 7 = 0$
$\sqrt{3x - 2} = 7$
$(\sqrt{3x - 2})^2 = (7)^2$
$3x - 2 = 49$
$3x = 51$
$x = 17$
Check.
$\sqrt{3(17) - 2} - 7 \stackrel{?}{=} 0$
$\sqrt{51 - 2} - 7 \stackrel{?}{=} 0$
$\sqrt{49} - 7 \stackrel{?}{=} 0$
$7 - 7 \stackrel{?}{=} 0$
$0 = 0$ ✓
6. $\sqrt{5x + 4} + 2 = 0$
$\sqrt{5x + 4} = -2$
$(\sqrt{5x + 4})^2 = (-2)^2$
$5x + 4 = 4$
$5x = 0$
$x = 0$
Check.
$\sqrt{5(0) + 4} + 2 \stackrel{?}{=} 0$
$\sqrt{4} + 2 \stackrel{?}{=} 0$
$2 + 2 \stackrel{?}{=} 0$
$4 \neq 0$
No, this does not check. There is no solution.
7. $-2 + \sqrt{6x - 1} = 3x - 2$
$\sqrt{6x - 1} = 3x$
$(\sqrt{6x - 1})^2 = (3x)^2$
$6x - 1 = 9x^2$
$0 = 9x^2 - 6x + 1$ *Factor.*
$0 = (3x - 1)^2$
$0 = 3x - 1$
$x = \dfrac{1}{3}$ is a double root.

Check.
$-2 + \sqrt{6\left(\dfrac{1}{3}\right) - 1} \stackrel{?}{=} 3\left(\dfrac{1}{3}\right) - 2$
$-2 + \sqrt{1} \stackrel{?}{=} 1 - 2$
$-2 + 1 \stackrel{?}{=} 1 - 2$
$-1 = -1$ ✓

8. $2 - x + \sqrt{x+4} = 0$

$$\sqrt{x+4} = x - 2$$
$$(\sqrt{x+4})^2 = (x-2)^2$$
$$x + 4 = x^2 - 4x + 4$$
$$0 = x^2 - 5x$$
$$0 = x(x-5)$$
$$x = 0 \qquad x - 5 = 0$$
$$x = 5$$

Check.

$$2 - 0 + \sqrt{0+4} \stackrel{?}{=} 0$$
$$2 + \sqrt{4} \stackrel{?}{=} 0$$
$$2 + 2 \stackrel{?}{=} 0$$
$$4 \neq 0$$

$x = 0$ does not check

$$2 - 5 + \sqrt{5+4} \stackrel{?}{=} 0$$
$$-3 + \sqrt{9} \stackrel{?}{=} 0$$
$$-3 + 3 \stackrel{?}{=} 0$$
$$0 = 0 \checkmark$$

The only solution is $x = 5$.

9. $\sqrt{2x+1} = \sqrt{x-10}$

$$(\sqrt{2x+1})^2 = (\sqrt{x-10})^2$$
$$2x + 1 = x - 10$$
$$x + 11 = 0$$
$$x = -11$$

Check.

$$\sqrt{2(-11)+1} \stackrel{?}{=} \sqrt{-11-10}$$
$$\sqrt{-22+1} \stackrel{?}{=} \sqrt{-21}$$
$$\sqrt{-21} = \sqrt{-21} \checkmark$$

$x = -11$ produces a negative radicand and thus is extraneous. No solution.

9.7 Practice Problems

1. Change of 100°C = change of 180°F.
Let C = the change in C
f = the change in F
k = the variation constant

$$C = kf$$
$$100 = k \cdot 180$$
$$k = \frac{5}{9}$$

if $C = kf$ then

$$C = \frac{5}{9}f$$

if $f = -20$ then

$$C = \frac{5}{9}(-20)$$
$$C = -11\frac{1}{9}$$

C changes $-11\frac{1}{9}$ degrees.

2. $y = kx$
To find K we substitute $y = 18$ and $x = 5$.

$$18 = k(5)$$
$$\frac{18}{5} = k$$

We now write the variation equation with k replaced by $\frac{18}{5}$.

$$y = \frac{18}{5}x$$

Replace x by $\frac{20}{23}$ and solve for y.

$$y = \frac{18}{5} \cdot \frac{20}{23}$$
$$y = \frac{72}{23}$$

Thus $y = \frac{72}{23}$ when $x = \frac{20}{23}$.

3. Let d = the distance to stop the car
s = speed of the car
k = the constant of variation
Since the distance varies directly as the square of the speed, we have

$$d = ks^2$$

To evaluate k we substitute the known distance and speed.

$$60 = k(20)^2$$
$$60 = k(400)$$
$$k = \frac{3}{20}$$

Now write the variation equation with the constant evaluated.

$$d = \frac{3}{20}s^2$$
$$d = \frac{3}{20}(40)^2 \quad \textit{Substitute } s = 40.$$
$$d = \frac{3}{20}(1600)$$
$$d = 240$$

The distance to stop the car going 40 mph on an ice covered road is 240 feet.

4. $y = \frac{k}{x}$

$$8 = \frac{k}{15} \quad \textit{Substitute known values of } x \textit{ and } y \textit{ to find } k.$$
$$120 = k$$
$$y = \frac{120}{x} \quad \textit{Write the variation equation with } k \textit{ replaced by } 120.$$
$$y = \frac{120}{\frac{3}{5}} \quad \textit{Find } y \textit{ when } x = \frac{3}{5}.$$
$$y = 200 \quad \textit{Thus } y = 200 \textit{ when } x = \frac{3}{5}.$$

5. Let V = the volume of sales
p = price of calculators
k = the constant variation
Since volume varies inversely as the price then

$$V = \frac{k}{p} \quad \textit{Evaluate } k \textit{ by substituting known values for } V \textit{ and } p.$$
$$120{,}000 = \frac{k}{30}$$
$$3{,}600{,}000 = k$$
$$V = \frac{3{,}600{,}000}{24} \quad \textit{Write the variation equation with the evaluated constant and substitute } p = \$24.00.$$
$$V = 150{,}000$$

Thus the volume increased to 150,000 sales per year when the price was reduced to $24 per calculator.

6. Let R = the resistance in the circuit
i = the amount of current
k = the constant of variation
Since the resistance in the circuit varies inversely as the square of the amount of current

$$R = \frac{k}{i^2}$$

To evaluate k, substitute the known values for R and i.

$$800 = \frac{k}{(0.01)^2}$$
$$k = 0.08$$
$$R = \frac{0.08}{(0.02)^2} \quad \textit{Write the variation equation with the evaluated constant of } k = 0.08 \textit{ and substitute } i = 0.02.$$
$$R = \frac{0.08}{0.0004}$$
$$R = 200$$

Thus the resistance is 200 ohms if the current is increased to 0.02 ampere.

Chapter 10

10.1 Practice Problems

1. (a) $2x^2 + 12x - 9 = 0$
$a = 2,\ b = 12,\ c = -9$

(b) $7x^2 = 6x - 8$

$7x^2 - 6x + 8 = 0$

$a = 7,\ b = -6,\ c = 8$

(c) $-x^2 - 6x + 3 = 0$

$x + 6x - 3 = 0$

$a = 1,\ b = 6,\ c = -3$

(d) $10x^2 - 12x = 0$

$a = 10,\ b = -12,\ c = 0$

(e) $5x^2 - 4x + 20 = 4x^2 - 3x - 10$

$x^2 - x + 30 = 0$

$a = 1,\ b = -1,\ c = 30$

2. (a) $3x^2 + 18x = 0$

$3x(x + 6) = 0$

$3x = 0 \qquad x + 6 = 0$

$x = 0 \qquad x = -6$

The two roots are 0 and -6.

(b) $2x^2 - 7x - 6 = 4x - 6$

$2x^2 - 11x = 0$

$x(2x - 11) = 0$

$x = 0 \qquad 2x - 11 = 0$

$x = \frac{11}{2}$

The two roots are 0 and $\frac{11}{2}$.

3. $3x + 10 - \frac{8}{x} = 0$

$3x^2 + 10x - 8 = 0$

$(3x - 2)(x + 4) = 0$

$3x - 2 = 0 \qquad x + 4 = 0$

$x = \frac{2}{3} \qquad x = -4$

Check. $3\left(\frac{3}{2}\right) + 10 - \frac{8}{\frac{2}{3}} \stackrel{?}{=} 0 \qquad 3(-4) + 10 - \frac{8}{-4} \stackrel{?}{=} 0$

$2 + 10 - 12 \stackrel{?}{=} 0 \qquad -12 + 10 + 2 \stackrel{?}{=} 0$

$2 - 2 \stackrel{?}{=} 0 \qquad -2 + 2 \stackrel{?}{=} 0$

$0 = 0$ ✓ $\qquad 0 = 0$ ✓

Both roots check so $x = \frac{2}{3}$ and $x = -4$ are the two roots that satisfy the equation $3x + 10 - \frac{8}{x} = 0$.

4. $2x^2 + 9x - 18 = 0$

$(2x - 3)(x + 6) = 0$

$2x - 3 = 0 \qquad x + 6 = 0$

$x = \frac{3}{2} \qquad x = -6$

Check. $2\left(\frac{3}{2}\right)^2 + 9\left(\frac{3}{2}\right) - 18 \stackrel{?}{=} 0 \qquad 2(-6)^2 + 9(-6) - 18 \stackrel{?}{=} 0$

$2\left(\frac{9}{4}\right) + \frac{27}{2} - 18 \stackrel{?}{=} 0 \qquad 2(36) - 54 - 18 \stackrel{?}{=} 0$

$\frac{9}{2} + \frac{27}{2} - 18 \stackrel{?}{=} 0 \qquad 72 - 54 - 18 \stackrel{?}{=} 0$

$\frac{36}{2} - 18 \stackrel{?}{=} 0 \qquad 18 - 18 \stackrel{?}{=} 0$

$18 - 18 \stackrel{?}{=} 0 \qquad 0 = 0$ ✓

$0 = 0$ ✓

Both roots check so $x = \frac{3}{2}$ and $x = -6$ are the two roots that satisfy the equation $2x^2 + 9x - 18 = 0$.

5. $-4x + (x - 5)(x + 1) = 5(1 - x)$

$-4x + x^2 - 4x - 5 = 5 - 5x$

$x^2 - 8x - 5 = 5 - 5x$

$x^2 - 3x - 10 = 0$

$(x + 2)(x - 5) = 0$

$x + 2 = 0 \qquad x - 5 = 0$

$x = -2 \qquad x = 5$

Check. $-4(-2) + (-2 - 5)(-2 + 1) \stackrel{?}{=} 5(1 + 2)$

$8 + (-7)(-1) \stackrel{?}{=} 5(3)$

$8 + 7 \stackrel{?}{=} 15$

$15 = 15$ ✓

$-4(5) + (5 - 5)(5 + 1) \stackrel{?}{=} 5(-4)$

$-20 + 0 \stackrel{?}{=} -20$

$-20 = -20$ ✓

Both roots check so $x = -2$ and $x = 5$ are the two roots that satisfy the equation $-4x + (x - 5)(x + 1) = 5(1 - x)$.

6. $\frac{10x + 18}{x^2 + x - 2} = \frac{3x}{x + 2} + \frac{2}{x - 1}$

$\frac{10x + 18}{(x - 1)(x + 2)} = \frac{3x}{x + 2} + \frac{2}{x - 1}$ *Multiply by the LCD.*

$10x + 18 = 3x(x - 1) + 2(x + 2)$

$10x + 18 = 3x^2 - 3x + 2x + 4$

$0 = 3x^2 - 11x - 14$

$0 = (3x - 14)(x + 1)$

$3x - 14 = 0 \qquad x + 1 = 0$

$x = \frac{14}{3} \qquad x = -1$

7. t = the number of truck routes

n = the number of cities they can service

$t = \frac{n^2 - n}{2}$

$28 = \frac{n^2 - n}{2}$

$56 = n^2 - n$

$0 = n^2 - n - 56$

$0 = (n + 7)(n - 8)$

$n + 7 = 0 \qquad n - 8 = 0$

$n = -7 \qquad n = 8$

We reject $n = -7$. We cannot have a negative number of cities. Thus our answer is 8 cities.

10.2 Practice Problems

1. (a) $x^2 = 1$

$x = \pm\sqrt{1}$

$x = \pm 1$

(b) $x^2 - 50 = 0$

$x^2 = 50$

$x = \sqrt{50}$

$x = \pm 5\sqrt{2}$

(c) $3x^2 = 81$

$x^2 = 27$

$x = \pm\sqrt{27} = \pm 3\sqrt{3}$

2. $4x^2 - 5 = 319$

$4x^2 = 324$

$x^2 = 81$

$x = \pm\sqrt{81} = \pm 9$

3. $(2x + 3)^2 = 25$

$\sqrt{(2x + 3)^2} = \pm\sqrt{25}$

$2x + 3 = \pm 5$

$2x + 3 = 5 \qquad 2x + 3 = -5$

$2x = 2 \qquad 2x = -8$

$x = 1 \qquad x = -4$

The two roots of the equation are $x = 1$ and $x = -4$.

4. $(2x - 3)^2 = 12$

$\sqrt{(2x - 3)^2} = \pm\sqrt{12}$

$2x - 3 = \pm 2\sqrt{3}$

$2x = 3 \pm 2\sqrt{3}$

$x = \frac{3 \pm 2\sqrt{3}}{2}$

5. (a) $x^2 - 8x = 20$

$x^2 - 8x + (4)^2 = 20 + 16$

$x^2 - 8x + 16 = 36$

$(x - 4)^2 = 36$

$\sqrt{(x - 4)^2} = \sqrt{36}$

$x - 4 = \pm 6$

$x = 4 \pm 6$

$x = 10 \qquad x = -2$

(b) $x^2 + 10x = 3$

$x^2 + 10x + (5)^2 = 3 + 25$

$(x+5)^2 = 28$

$\sqrt{(x+5)^2} = \pm\sqrt{28}$

$x + 5 = \pm\sqrt{28}$

$x = -5 \pm 2\sqrt{7}$

6. $x^2 - 5x = 7$

$x^2 - 5x + \left(\frac{5}{2}\right)^2 = 7 + \left(\frac{5}{2}\right)^2$

$\left(x - \frac{5}{2}\right)^2 = 7 + \frac{25}{4}$

$\left(x - \frac{5}{2}\right)^2 = \frac{28}{4} + \frac{25}{4}$

$\left(x - \frac{5}{2}\right)^2 = \frac{53}{4}$

$x - \frac{5}{2} = \pm\sqrt{\frac{53}{4}}$

$x = \frac{5}{2} \pm \frac{\sqrt{53}}{2}$

$x = \frac{5 \pm \sqrt{53}}{2}$

7. $4y^2 + 4y - 3 = 0$

$(2y + 3)(2y - 1) = 0$

$2y + 3 = 0 \qquad 2y - 1 = 0$

$y = -\frac{3}{2} \qquad y = \frac{1}{2}$

10.3 Practice Problems

1. $x^2 - 7x + 6 = 0$

$x = \frac{(7) \pm \sqrt{(-7)^2 - 4(1)(6)}}{2(1)}$

$x = \frac{7 \pm \sqrt{49 - 24}}{2}$

$x = \frac{7 \pm \sqrt{25}}{2}$

$x = \frac{7 \pm 5}{2}$

$x = \frac{7+5}{2} \qquad x = \frac{7-5}{2}$

$x = 6 \qquad x = 1$

2. $3x^2 - 8x + 3 = 0$

$x = \frac{-(-8) \pm \sqrt{(-8)^2 - 4(3)(3)}}{2(3)}$

$x = \frac{8 \pm \sqrt{64 - 36}}{6}$

$x = \frac{8 \pm \sqrt{28}}{6}$

$x = \frac{8 \pm 2\sqrt{7}}{6}$

$x = \frac{2(4 \pm \sqrt{7})}{6} = \frac{4 \pm \sqrt{7}}{3}$

3. $2x^2 = 3x + 1$

$2x^2 - 3x - 1 = 0$

$x = \frac{-(-3) \pm \sqrt{(-3)^2 - 4(2)(-1)}}{2(2)}$

$x = \frac{3 \pm \sqrt{9 + 8}}{4}$

$x = \frac{3 \pm \sqrt{17}}{4}$

4. $2x^2 = 13x + 5$

$2x^2 - 13x - 5 = 0$

$x = \frac{-(-13) \pm \sqrt{(-13)^2 - 4(2)(-5)}}{2(2)}$

$x = \frac{13 \pm \sqrt{169 + 40}}{4}$

$x = \frac{13 \pm \sqrt{209}}{4}$

$x = \frac{13 \pm \sqrt{209}}{4} \qquad x = \frac{13 - \sqrt{209}}{4}$

$x = 6.864$ and $x = -0.364$

5. $5x^2 + 2x = -3$

$5x^2 + 2x + 3 = 0$

$x = \frac{-2 \pm \sqrt{(2)^2 - 4(5)(3)}}{2(5)}$

$x = \frac{-2 \pm \sqrt{4 - 60}}{10}$

$x = \frac{-2 \pm \sqrt{-56}}{10}$

There is no real number that is $\sqrt{-56}$.

There is no solution to this problem.

6. (a) $5x^2 - 3x - 2 = 0$

$b^2 - 4ac = (-3)^2 - 4(5)(-2)$

$= 9 + 40 = 49$

Because $49 > 0$ there will be two real roots.

(b) $2x^2 - 4x - 5 = 0$

$b^2 - 4ac = (-4)^2 - 4(2)(-5)$

$= 16 + 40 = 56$

Because $56 > 0$ there will be two real roots.

10.4 Practice Problems

1. (a) $y = 2x^2$

x	y
−2	8
−1	2
0	0
1	2
2	8

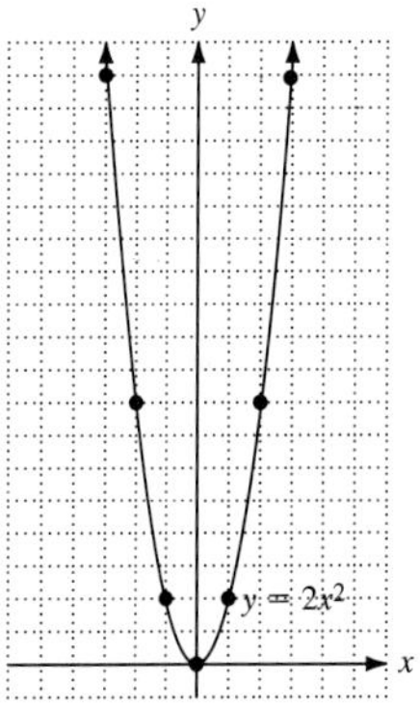

(b) $y = -2x^2$

x	y
−2	−8
−1	−2
0	0
1	−2
2	−8

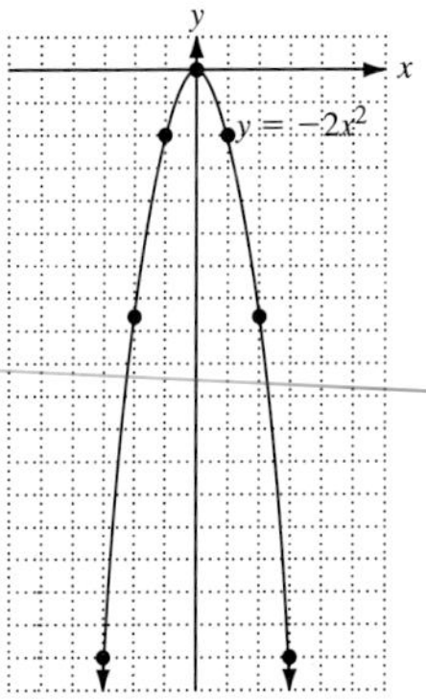

2. (a) $y = (x - 1)^2$

x	y
−2	9
−1	4
0	1
1	0
2	1
3	4
4	9

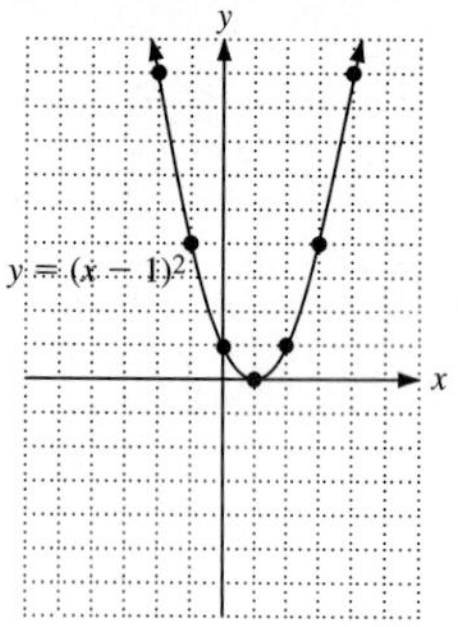

(b) $y = (x + 2)^2$

x	y
−5	9
−4	4
−3	1
−2	0
−1	1
0	4
1	9

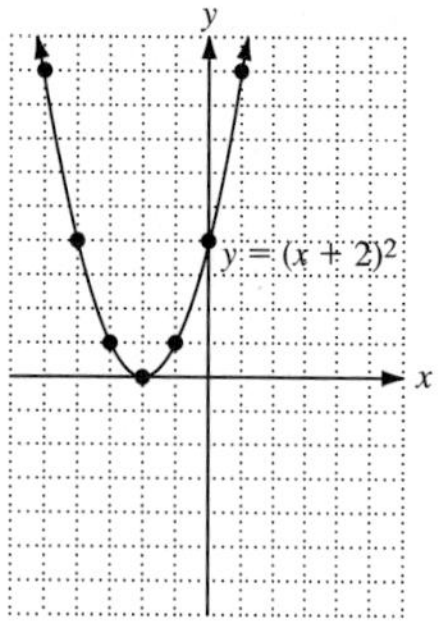

3. (a) $y = x^2 + 2x$

x	y	
−4	8	
−3	3	
−2	0	
−1	−1	vertex
0	0	
1	3	
2	8	

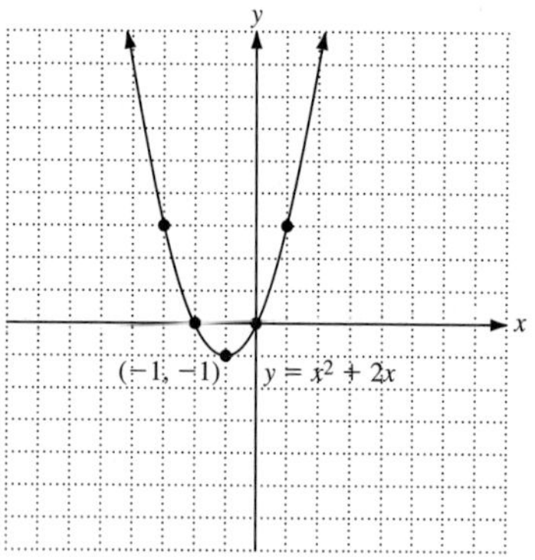

(b) $y = -x^2 - 2x$

x	y	
−4	−8	
−3	−3	
−2	0	
−1	1	vertex
0	0	
1	−3	
2	−8	

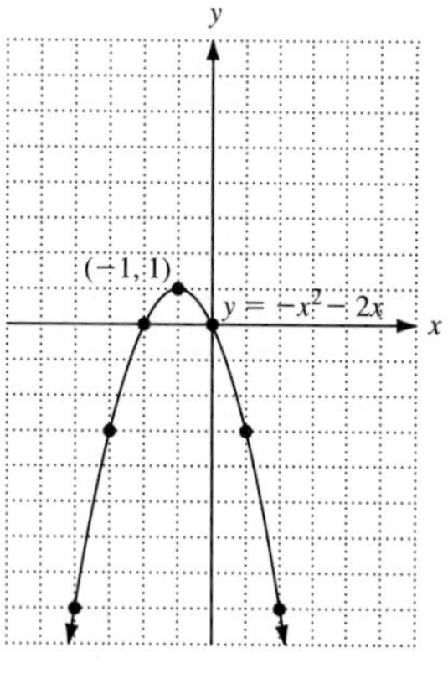

4. $y = x^2 - 4x - 5$

$x = \dfrac{-b}{2a} = \dfrac{-(-4)}{2(1)} = 2$

if $x = 2$ then

$y = (2)^2 - 4(2) - 5$

$y = 4 - 8 - 5$

$y = -9$

The point $(2, -9)$ is the vertex.

x-intercepts:

$-x^2 + 4x + 5 = 0$

$x^2 - 4x - 5 = 0$

$(x + 1)(x - 5) = 0$

$x + 1 = 0 \qquad x - 5 = 0$

$x = -1 \qquad x = 5$

The x-intercepts are the points $(-1, 0)$ and $(5, 0)$.

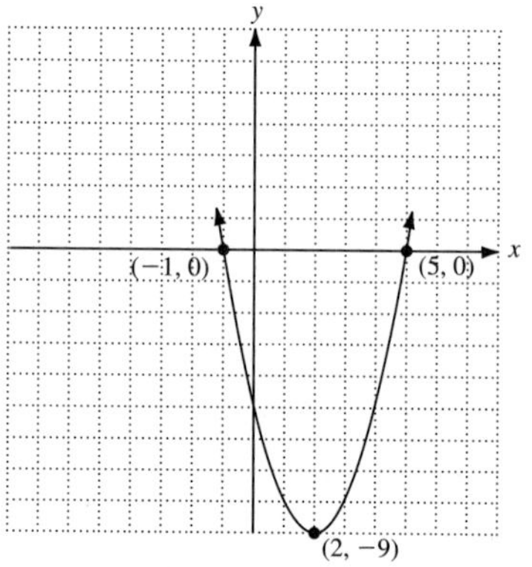

5. $h = -16t^2 + 32t + 10$

(a) Since the equation is a quadratic equation, the graph is a parabola. A is negative, therefore the parabola opens downward and the vertex is the highest point.

$$\text{Vertex} = t = \frac{-b}{2a} = \frac{-32}{2(-16)} = \frac{-32}{-32} = 1$$

if $t = 1$ then $h = -16(1)^2 + 32(1) + 10$

$h = -16 + 32 + 10$

$h = 26$

The vertex (t, h) is $(1, 26)$.

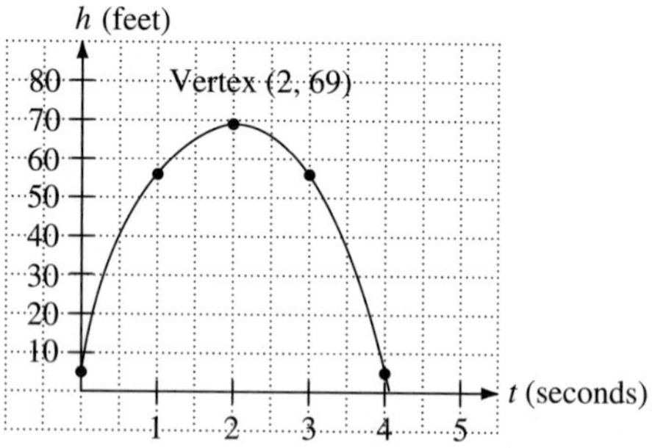

Now find the h-intercept. The point where $t = 0$.

$h = -16t^2 + 32t + 10$

$h = 10$

The h-intercept is at $(0, 10)$.

(b) The ball is 26 feet above the ground one second after it is thrown. This is the greatest height of the ball.

(c) The t-intercept is where $h = 0$.

$0 = -16t^2 + 32t + 10$

$t = \dfrac{-32 \pm \sqrt{(32)^2 - 4(-16)(10)}}{2(-16)}$

$t = \dfrac{-32 \pm \sqrt{1024 + 640}}{-32}$

$t = \dfrac{-32 \pm \sqrt{1664}}{-32}$

$t = \dfrac{-32 \pm 40.792}{-32}$

$t = \dfrac{-32 + 40.792}{-32}$ and $t = \dfrac{-32 - 40.792}{-32}$

$t = \dfrac{8.792}{-32}$ $\quad$ $t = \dfrac{-72.792}{-32}$

$t = -0.275$ $\quad$ $t = 2.275$

In this situation it is not useful to find points for negative values of t. Therefore $t = 2.3$ seconds. The ball will hit the ground approximately 2.3 seconds after it is thrown.

10.5 Practice Problems

1. $a^2 + b^2 = c^2$ $\quad$ Let $a = x$

$b = x - 6$

$x^2 + (x - 6)^2 = 30^2$

$x^2 + x^2 - 12x + 36 = 900$

$2x^2 - 12x + 36 = 900$

$2x^2 - 12x - 864 = 0$

$x^2 - 6x - 432 = 0$

$(x + 18)(x - 24) = 0$

$x + 18 = 0$ $\quad$ $x - 24 = 0$

$x = -18$ $\quad$ $x = 24$

One leg is 24 meters, the other leg is $x - 6$ or $24 - 6 = 18$ meters.

2. Let c = cost for each student in the original group

s = number of students

Number of students × cost per student = total cost

$s \times c = 240.00$

If 2 people cannot go, then the number of students drops by 2 but the cost of each increases by $4.00.

$(s - 2)(c + 4) = 240$

if $s \times c = 240$ then

$c = \dfrac{240}{s}$

$(s - 2)\left(\dfrac{240}{s} + 4\right) = 240$ *Multiply by the LCD = s.*

$(s - 2)(240 + 4s) = 240s$

$4s^2 + 232s - 480 = 240s$

$4s^2 - 8s - 480 = 0$

$s^2 - 2s - 120 = 0$

$(s - 12)(s + 10) = 0$

$s - 12 = 0$ $\quad$ $s + 10 = 0$

$s = 12$ $\quad$ $s = -10$

The number of students originally going on the trip was 12.

Check. If 12 students originally planned the trip then

$12 \times c = 240$

$c = \$20.00$ per student

If 2 students dropped out, does this mean they must increase the cost of the trip by $4.00?

$10 \times c = 240$

$c = \$24.00$ YES!

3. Let x = width

y = length

$160 = 2x + y$

The area of a rectangle is (width)(length).

$A = x(y)$ *Substitute* $y = 160 - 2x$.

$3150 = x(160 - 2x)$

$3150 = 160x - 2x^2$

$0 = -2x^2 + 160x - 3150$

$2x^2 - 160x + 3150 = 0$

$x^2 - 80x + 1575 = 0$

$(x - 45)(x - 35) = 0$

$x = 45$ $\quad$ $x = 35$

First Solution

If the width is 45 feet then the length is

$160 = 2(45) + y$

$160 = 90 + y$

$y = 70$ feet

Check. $45 + 45 + 70 \stackrel{?}{=} 160$

$90 + 70 \stackrel{?}{=} 160$

$160 = 160$ ✓

Second Solution

If the width is 35 feet then the length is

$160 = 2(35) + y$

$160 = 70 + y$

$y = 90$ feet

Check. $35 + 35 + 90 \stackrel{?}{=} 160$

$70 + 90 \stackrel{?}{=} 160$

$160 = 160$ ✓

Selected Answers

Chapter 0

Pretest Chapter 0

1. $\frac{7}{9}$ **2.** $\frac{2}{3}$ **3.** $3\frac{3}{4}$ **4.** $\frac{33}{7}$ **5.** 6 **6.** 35 **7.** 120 **8.** $\frac{13}{9}$ **9.** $\frac{5}{12}$ **10.** $8\frac{5}{12}$ **11.** $1\frac{33}{40}$ **12.** $\frac{10}{9}$ or $1\frac{1}{9}$ **13.** $\frac{21}{2}$ or $10\frac{1}{2}$ **14.** $\frac{7}{2}$ or $3\frac{1}{2}$ **15.** $\frac{12}{13}$ **16.** 0.625 **17.** $0.\overline{5}$ **18.** 19.651 **19.** 2.0664 **20.** 0.45 **21.** 2.246 **22.** 70% **23.** 0.6% **24.** 3.27 **25.** 0.02 **26.** 407.5 **27.** 7.2% **28.** 10,000 **29.** 21,000 **30.** 20 miles per gallon **31.** 28.4 miles per gallon

0.1 Exercises

7. $\frac{5}{7}$ **9.** $\frac{1}{5}$ **11.** 5 **13.** $5\frac{4}{7}$ **15.** $31\frac{1}{4}$ **17.** $14\frac{5}{6}$ **19.** $\frac{20}{7}$ **21.** $\frac{19}{4}$ **23.** $\frac{79}{6}$ **25.** $\frac{20}{28}$ **27.** $\frac{6}{21}$ **29.** $\frac{39}{51}$ **31.** $\frac{5}{7}$ **33.** 119 **35.** $\frac{3}{5}$ **37.** Aaron Dunsay $\frac{3}{5}$, Paul Banks $\frac{2}{5}$, Tom Re $\frac{4}{5}$

0.2 Exercises

3. 36 **5.** 20 **7.** 63 **9.** 90 **11.** $\frac{9}{11}$ **13.** $\frac{3}{17}$ **15.** $\frac{41}{30}$ or $1\frac{11}{30}$ **17.** $\frac{3}{28}$ **19.** $\frac{19}{18}$ or $1\frac{1}{18}$ **21.** $\frac{17}{40}$ **23.** $\frac{41}{24}$ or $1\frac{17}{24}$ **25.** $\frac{17}{36}$ **27.** $\frac{5}{4}$ or $1\frac{1}{4}$ **29.** $\frac{13}{20}$ **31.** $\frac{4}{75}$ **33.** $8\frac{5}{12}$ **35.** $7\frac{5}{12}$ **37.** $3\frac{5}{8}$ **39.** $2\frac{3}{4}$ **41.** $7\frac{10}{21}$ **43.** $1\frac{5}{9}$ **45.** $10\frac{7}{24}$ miles **47.** $4\frac{1}{12}$ hours **49.** $A = 12$ inches, $B = 15\frac{7}{8}$ inches **51.** $\frac{9}{11}$ **52.** $\frac{133}{5}$ **53.** $19\frac{3}{7}$ **54.** 12

0.3 Exercises

1. $\frac{6}{55}$ **3.** 10 **5.** $\frac{24}{7}$ or $3\frac{3}{7}$ **7.** $\frac{10}{3}$ or $3\frac{1}{3}$ **9.** 5 **11.** $\frac{2}{3}$ **13.** $\frac{1}{7}$ **15.** 14 **17.** $\frac{15}{4}$ or $3\frac{3}{4}$ **19.** $\frac{8}{35}$ **21.** $\frac{5}{3}$ or $1\frac{2}{3}$ **23.** 1 **25.** $\frac{5}{2}$ or $2\frac{1}{2}$ **27.** $\frac{219}{4}$ or $54\frac{3}{4}$ **29.** $\frac{17}{2}$ or $8\frac{1}{2}$ **31.** 15 **33.** $\frac{3}{16}$ **35.** $71\frac{1}{2}$ yds

Putting Your Skills to Work

1. $9\frac{3}{4} \times 100 = \$975$ **2.** $50 \times 17\frac{7}{8} = \$893\frac{3}{4}$ or \$893.75 **3.** $2\frac{3}{4}$ **4.** $38\frac{1}{2}$ **5.** 150 **6.** \$62.50

0.4 Exercises

5. 0.15 **7.** 0.9375 **9.** 0.3 **11.** $\frac{3}{20}$ **13.** $\frac{2}{25}$ **15.** $\frac{5}{8}$ **17.** 7.72 **19.** 0.9094 **21.** 24.356 **23.** 87.2 **25.** 18.512 **27.** 0.51191 **29.** 0.001428 **31.** 219.7 **33.** 3.69 **35.** 8.8 **37.** 6.2 **39.** 6.3 **41.** 0.0076 **43.** 2 **45.** 243 **47.** 0.73892 **49.** 0.005808 **51.** 24.36 **53.** 317325 **55.** 0.003472 **57.** $478.6 \div 26.8 = 17.9$ mpg **58.** $\frac{2}{3}$ **59.** $\frac{1}{6}$ **60.** $\frac{48 + 45}{100} = \frac{93}{100}$ **61.** $\frac{16 - 5}{10} = \frac{11}{10}$ or $1\frac{1}{10}$

Putting Your Skills to Work

1. \$11,000; \$8,000; \$10,000; \$600; \$4,800; \$34,400 **2.** \$3,600 lower **3.** \$286.67/day

0.5 Exercises

3. 62.4% **5.** 0.3% **7.** 156% **9.** 0.04 **11.** 0.002 **13.** 2.50 **15.** 358 **17.** 0.27 **19.** 296.8 **21.** 176 **23.** 5% **25.** 5% **27.** 150% **29.** 0.4% **31.** 137.5% **33.** 14,100 **35.** 53.89% **37.** 648 products were defective. **39.** 77% **41.** **(a)** 12,090 miles traveled on business. **(b)** \$3324.75 is the amount the company should pay him.

Putting Your Skills to Work

1. 5%   **2.** 13%   **3.** \$250   **4.** \$16

0.6 Exercises

1. 180,000   **3.** 120,000,000   **5.** 290   **7.** 30,000   **9.** 0.1   **11.** \$2400   **13.** 200 feet   **15.** 20 mpg   **17.** \$6400   **19.** \$20   **21.** \$600   **23.** \$200,000,000   **25.** \$240   **26.** \$300

0.7 Exercises

1. \$462.50   **3.** 20 cubic yards   **5.** Jog $2\frac{2}{15}$ miles; walk $3\frac{1}{9}$ miles; rest $4\frac{4}{9}$ minutes; walk $1\frac{7}{9}$ miles   **7.** Betty; Melinda increases each activity by $\frac{2}{3}$ by day 3 and Betty increases each activity by $\frac{7}{9}$ by day 3.   **8.** $3\frac{107}{135}$ **9.** $3\frac{3}{5}$ **10.** Answers will vary.   **11.** 27 miles per gallon   **13.** 142,560; Answers will vary. Samples include: improved schools, new job opportunities, availability of affordable housing.   **15.** 49%; 61%; business travelers; Answers may vary; A sample is: Men 21 or older spend more of their travel time on business than for pleasure.   **17.** 18%   **19.** 69%

Chapter 0 Review Problems

1. $\frac{3}{4}$ **2.** $\frac{3}{10}$ **3.** $\frac{1}{3}$ **4.** $\frac{3}{5}$ **5.** $\frac{19}{7}$ **6.** $6\frac{4}{5}$ **7.** $6\frac{3}{4}$ **8.** 15   **9.** 15   **10.** 12   **11.** 22   **12.** $\frac{11}{12}$ **13.** $\frac{11}{24}$ **14.** $\frac{4}{15}$ **15.** $\frac{17}{48}$ **16.** $\frac{173}{30}$ or $5\frac{23}{30}$ **17.** $\frac{79}{20}$ or $3\frac{19}{20}$ **18.** $\frac{8}{15}$ **19.** $\frac{23}{12}$ or $1\frac{11}{12}$ **20.** $\frac{3}{5}$ **21.** $\frac{2}{7}$ **22.** $\frac{30}{11}$ or $2\frac{8}{11}$ **23.** $\frac{21}{2}$ or $10\frac{1}{2}$ **24.** $\frac{19}{8}$ or $2\frac{3}{8}$ **25.** $\frac{20}{7}$ or $2\frac{6}{7}$ **26.** 2   **27.** 4   **28.** $\frac{3}{10}$ **29.** $\frac{2}{3}$ **30.** $\frac{3}{20}$ **31.** 6   **32.** 4.462   **33.** 7.737   **34.** 11.427   **35.** 4.814   **36.** 13.7   **37.** 111.1121   **38.** 0.00862   **39.** 362,341   **40.** 0.07956   **41.** 10.368   **42.** 0.00186   **43.** 0.07132   **44.** 0.002   **45.** 90   **46.** 0.07   **47.** 0.5   **48.** 37.5%   **49.** $\frac{6}{25}$ **50.** 0.014   **51.** 0.361   **52.** 0.0002   **53.** 1.253   **54.** 260   **55.** 90   **56.** 18   **57.** 12.5%   **58.** 60%   **59.** 97,200,000   **60.** 75%   **61.** 400,000,000,000   **62.** 2500   **63.** 600,000   **64.** 19   **65.** \$12,000   **66.** 30   **67.** \$480 or \$640   **68.** \$300   **69.** \$349.07   **70.** 6%

Chapter 0 Test

1. $\frac{8}{9}$ **2.** $\frac{7}{8}$ **3.** $\frac{45}{7}$ **4.** $3\frac{3}{11}$ **5.** 33   **6.** $\frac{13}{18}$ **7.** $\frac{17}{12}$ or $1\frac{5}{12}$ **8.** $\frac{1}{2}$ **9.** $\frac{39}{8}$ or $4\frac{7}{8}$ **10.** $\frac{5}{6}$ **11.** $\frac{5}{4}$ or $1\frac{1}{4}$ **12.** $\frac{4}{3}$ or $1\frac{1}{3}$ **13.** $\frac{21}{55}$ **14.** $\frac{7}{2}$ or $3\frac{1}{2}$ **15.** $\frac{65}{8}$ or $8\frac{1}{8}$ **16.** $\frac{159}{74}$ or $2\frac{11}{74}$ **17.** $\frac{18}{25}$ **18.** 0.4375   **19.** 14.64   **20.** 1806.62   **21.** 1.312   **22.** 16.32   **23.** 230   **24.** 19.3658   **25.** 7.3%   **26.** 1.965   **27.** 63   **28.** 0.336   **29.** 12.5%   **30.** 30%   **31.** 52%   **32.** 18   **33.** 100   **34.** 99   **35.** \$2700   **36.** 65%

Chapter 1

Pretest Chapter 1

1. 3   **2.** −4   **3.** −18   **4.** 20   **5.** −42   **6.** $\frac{8}{3}$ **7.** 28   **8.** −4   **9.** 16   **10.** 64   **11.** $-\frac{8}{27}$ **12.** −256   **13.** $-6x^2 + 4x^2y - 2xz$ **14.** $-3x + 4y + 12$ **15.** $5x^2 - 6x^2y - 11xy$ **16.** $7x - 2y + 7$ **17.** $5x - 7y$ **18.** $-x^2y + 17y^2$ **19.** 28   **20.** 197   **21.** $\frac{107}{12}$ **22.** 6.37   **23.** 18   **24.** 29   **25.** 25°C   **26.** $-12x - 3y$ **27.** $-x^2 + 6xy$

1.1 Exercises

11. −53   **13.** −85   **15.** +7   **17.** $-\frac{3}{4}$ **19.** $+\frac{5}{13}$ **21.** −4   **23.** −13   **25.** $\frac{1}{3}$ **27.** 55   **29.** −31   **31.** $-\frac{7}{13}$ **33.** −3.8   **35.** 0.4   **37.** −25   **39.** $\frac{1}{12}$ **41.** −11   **43.** 3   **45.** −2   **47.** 13   **49.** −3.2   **51.** 17   **53.** 5   **55.** $-\frac{3}{8}$ **57.** $-\frac{11}{12}$ **59.** −3.4   **61.** −32   **63.** 1   **65.** 13.73   **67.** −6°F   **69.** −265   **71.** 57   **73.** \$143   **74.** 18   **75.** 28   **76.** 0   **77.** Under one of the following cases: (a) $|x| > |y|$ with x negative and y positive, (b) $|y| > |x|$ with x positive and y negative, or (c) x and y both negative.   **78.** $x = y$ **79.** \$39.00

Putting Your Skills to Work

1. \$265.96   **2.** \$261.60

1.2 Exercises

1. 3   **3.** 10   **5.** −8   **7.** −26   **9.** −26   **11.** 8   **13.** 5   **15.** −5   **17.** 0   **19.** −4   **21.** −0.9   **23.** 4.47   **25.** 2   **27.** $1\frac{7}{20}$ **29.** $-\frac{23}{40}$ **31.** −53   **33.** −99   **35.** 7.1   **37.** $\frac{14}{5}$ or $2\frac{4}{5}$ **39.** $\frac{130}{9}$ or $14\frac{4}{9}$ **41.** $-\frac{21}{20}$ or $-1\frac{1}{20}$ **43.** −8.5

45. 0.0499 **47.** $-5\frac{4}{5}$ **49.** $-\frac{3}{10}$ **51.** 7 **53.** −48 **55.** −2 **57.** 11 **59.** −62 **61.** −34 **63.** −127 **65.** 38
67. $149 **69.** A drop of 16°F or −16°F **70.** −21 **71.** −51 **72.** −19 **73.** −8°C

1.3 Exercises

1. −36 **3.** 30 **5.** 60 **7.** −24 **9.** 0 **11.** 24 **13.** 0.264 **15.** 0.054 **17.** −24 **19.** $\frac{15}{16}$ **21.** −3 **23.** 0
25. −6 **27.** 36 **29.** −15 **31.** −2 **33.** 4 **35.** −0.9 **37.** $-\frac{3}{10}$ **39.** $1\frac{2}{7}$ **41.** −30 **43.** $\frac{9}{16}$ **45.** $-\frac{3}{4}$
47. −24 **49.** 36 **51.** −16 **53.** −18 **55.** $-\frac{1}{9}$ **57.** $\frac{2}{27}$ **59.** −4 **61.** −2 **63.** 17 **65.** −72 **67.** −1
69. $30 **71.** 150 or 150 m right of 0 **72.** −6.69 **73.** −72 **74.** 4.63 **75.** −88

1.4 Exercises

7. 8 **9.** 81 **11.** 343 **13.** −27 **15.** 64 **17.** −216 **19.** $\frac{1}{16}$ **21.** $\frac{8}{125}$ **23.** 0.09 **25.** 0.125 **27.** 4096
29. −4096 **31.** −0.07776 **33.** 0.000001 **35.** 75 **37.** −72,000 **39.** −1008 **41.** 3 **43.** 11 **45.** 20 **47.** 2
48. −19 **49.** $-\frac{5}{3}$ or $-1\frac{2}{3}$ **50.** −8 **51.** 2.52

1.5 Exercises

7. $2x^2 - xy$ **9.** $-6x^2 + 4$ **11.** $-15a^2 - 10ab^2$ **13.** $-3x^2 - 6xy + 3x$ **15.** $-5x^2y + 2xy^2 + xy$ **17.** $-15x - 45 + 35y$
19. $\frac{x^2}{4} + \frac{x}{2} - 2$ **21.** $4x^2 - 12x - \frac{12}{7}$ **23.** $y^2 - \frac{4xy}{3} - 2y$ **25.** $6a^2 + 3ab - 3ac - 12a$ **27.** $-12 + 6x$ **29.** $25x^3 - 5x$
31. $12x^2 - 15xy - 18x$ **33.** $-4x^2 - 4xy + 20x$ **35.** $-4a^2b + 2ab^2 + ab$ **37.** $-8x^2y + 4xy^2 - 12xy$ **39.** $3.75a^2 - 8.75a + 5$
41. $-0.12a^2 + 0.08ab - 0.008ab^2$ **43.** −16 **44.** 64 **45.** 1 **46.** 14

1.6 Exercises

7. $10x$ **9.** $-28x^3$ **11.** $18x^4 + 7x^2$ **13.** $5a + 2b - 7a^2$ **15.** $-4ab - 7$ **17.** $7.1x - 3.5y$ **19.** $-2x - 8.7y$ **21.** $\frac{3}{4}x^2 - \frac{10}{3}y$
23. $-\frac{1}{4}x - \frac{3}{8}y$ **25.** $-5x - 7y - 20$ **27.** $5x^2y - 10xy^6 - xy^2$ **29.** $5bc - 6ac$ **31.** $3x^2 - 2x + 1$ **33.** $-10y^2 - 16y + 12$
35. $5a - 3ab - 8b$ **37.** $-12x + 13y$ **39.** $3a^2 + 27ab$ **41.** $-39a^2 - 13b$ **43.** $71x - 27$ **44.** $-\frac{2}{15}$ **45.** $-\frac{5}{6}$ **46.** $\frac{23}{50}$
47. $-\frac{15}{98}$

1.7 Exercises

5. 24 **7.** 7 **9.** 6 **11.** −29 **13.** 14 **15.** 145 **17.** −6 **19.** 42 **21.** $\frac{9}{4}$ or $2\frac{1}{4}$ **23.** 0.848 **25.** $\frac{3}{10}$
27. 41.25 **29.** $\frac{1}{4}$ **31.** x **32.** $-\frac{19}{12}$ or $-1\frac{7}{12}$ **33.** −1 **34.** $\frac{72}{125}$

1.8 Exercises

1. −5 **3.** −2 **5.** $12\frac{1}{2}$ **7.** −26 **9.** 15 **11.** 3 **13.** −50 **15.** −5 **17.** 9 **19.** 39 **21.** −2 **23.** 34
25. 20 **27.** 29 **29.** **(a)** 8 **(b)** 4 **(c)** 14 **31.** **(a)** −10 **(b)** 6 **(c)** 5 **33.** 42 **35.** 29 **37.** 91 square feet
39. 78.5 square meters **41.** 14°F **43.** $53,851 **45.** −76.4°F to −22°F **47.** 12.4 miles **48.** 6 **49.** $-x^2 + 2x - 4y$
50. −34 **51.** −0.72

Putting Your Skills to Work

1. $1549.80 **2.** $1,196.58 **3.** $234 **4.** 15 years

1.9 Exercises

5. $-3x - 4y$ **7.** $5a + 3b$ **9.** $-x^4 - x^2y^2$ **11.** $-245 + 25x$ **13.** $4x^2 + 6$ **15.** $-12y^3 - 8y^2 + 32y$ **17.** $4x - 6y - 3$
19. $2a - 6ab - 8b^2$ **21.** $-x^3 + 2x^2 - 3x - 12$ **23.** $3a^2 + 16b + 12b^2$ **25.** $15x - 10y$ **27.** $4a^2 - 2b$ **29.** $12a^2 - 8b$
31. 97.52° **32.** 465,426.5 sq ft **33.** 35 **34.** 50

Putting Your Skills to Work

1. 28646 **2.** $6,330,700

Chapter 1 Review Problems

1. −8 **2.** −4.2 **3.** −9 **4.** 1.9 **5.** $-\frac{1}{3}$ **6.** $-\frac{7}{22}$ **7.** $\frac{1}{6}$ **8.** $\frac{22}{15}$ **9.** −13 **10.** −2 **11.** 8 **12.** 13
13. −33 **14.** 9.2 **15.** $-\frac{13}{8}$ **16.** $\frac{1}{2}$ **17.** −22.7 **18.** −27 **19.** −4 **20.** 16 **21.** −29 **22.** 1 **23.** −42
24. −3 **25.** −2 **26.** 27 **27.** −5 **28.** 2 **29.** $-\frac{2}{3}$ **30.** $-\frac{25}{7}$ **31.** 10 **32.** −72 **33.** 30 **34.** −30 **35.** $-\frac{1}{2}$
36. $-\frac{4}{7}$ **37.** −30 **38.** −5 **39.** −9.1 **40.** 0.9 **41.** 10.1 **42.** −1.2 **43.** −0.4 **44.** −2.3 **45.** 24 yards
46. −22° **47.** 7363 feet **48.** $2\frac{1}{4}$ point loss **49.** 81 **50.** 64 **51.** −125 **52.** $\frac{1}{8}$ **53.** −64 **54.** 0.49 **55.** $\frac{25}{36}$
56. $\frac{27}{64}$ **57.** $15x - 35y$ **58.** $6x^2 - 14xy + 8x$ **59.** $-7x^2 + 3x - 11$ **60.** $-6xy^2 - 3xy + 3y^2$ **61.** $-5a^2b + 3bc$
62. $14x - 2y$ **63.** $-5x^2 - 35x - 9$ **64.** $10x^2 - 8x - \frac{1}{2}$ **65.** −44 **66.** 30 **67.** 1 **68.** −19 **69.** 1 **70.** −4
71. −15 **72.** 10 **73.** −16 **74.** $\frac{32}{5}$ **75.** \$810 **76.** 86°F **77.** \$2119.50 **78.** \$8580.00 **79.** $x - 8$ **80.** $-19x - 30$
81. $-2 + 10x$ **82.** $-12x^2 + 63x$ **83.** $5xy^3 - 6x^3y - 13x^2y^2 - 6x^2y$ **84.** $-14y^2 - 4xy + x^2$ **85.** $x - 10y + 35 - 15xy$
86. $10x - 22y - 36$ **87.** $-4a - 24b + 32ab + 18a^2 - 6a^2b$ **88.** $-10a + 25ab - 15b^2 - 10ab^2$ **89.** $-3x - 9xy + 18y^2$

Chapter 1 Test

1. 4 **2.** $\frac{1}{3}$ **3.** 0.2 **4.** 96 **5.** −70 **6.** 4 **7.** −3 **8.** $-\frac{10}{3}$ or $-3\frac{1}{3}$ **9.** −64 **10.** 1.44 **11.** −125 **12.** $\frac{16}{81}$
13. $-5x^2 - 10xy + 35x$ **14.** $6a^2b^2 + 4ab^3 - 14a^2b^3$ **15.** $-6xy - 5x^2y - 4xy^2$ **16.** $2a^2b + \frac{15}{2}ab$ **17.** $8a^2 + 20ab$ **18.** $5a + 30$
19. $14x - 16y$ **20.** 8 **21.** −25 **22.** 122 **23.** 37 **24.** 1 **25.** 96.6 **26.** $-3a - 9ab + 3b^2 - 3ab^2$
27. $-69x + 78y - 51$

Chapter 2

Pretest Chapter 2

1. $x = 58$ **2.** $x = 48$ **3.** $x = -4$ **4.** $x = 4$ **5.** $x = -\frac{7}{2}$ **6.** $x = -\frac{1}{7}$ **7.** $x = \frac{5}{2}$ **8.** $x = \frac{17}{10}$ **9.** **(a)** $F = \frac{9C + 160}{5}$
(b) $F = 5°$ **10.** **(a)** $r = \frac{I}{Pt}$ **(b)** 6% **11.** < **12.** > **13.** > **14.** < **15.** 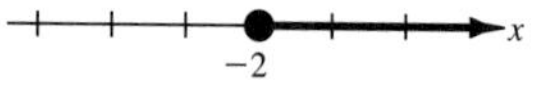

16. 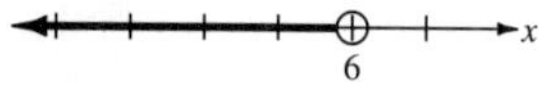

2.1 Exercises

7. 3 **9.** −1 **11.** 20 **13.** 7 **15.** 1 **17.** 0 **19.** −21 **21.** 66 **23.** 27 **25.** −14 **27.** no, $x = 26$
29. no, $x = 7$ **31.** no, $x = 8$ **33.** yes **35.** 5 **37.** 22 **39.** 3.5 **41.** −1.2 **43.** $\frac{1}{3}$ **45.** $\frac{4}{5}$ **47.** 1 **49.** $27\frac{1}{8}$
51. 7.2 **53.** −27.4976 **54.** $7x - 20$ **55.** $-2x - 4y$ **56.** $-7x^2 - 4x - 7$ **57.** $-2y^2 - 4y + 4$

2.2 Exercises

5. 63 **7.** −8 **9.** 80 **11.** −48 **13.** 4 **15.** $-\frac{8}{3}$ **17.** 50 **19.** 15 **21.** −7 **23.** $\frac{2}{5}$ **25.** no, $x = -7$ **27.** yes
29. −0.8 **31.** $\frac{9}{4}$ **33.** −0.6 **35.** 3 **37.** −4 **39.** $\frac{15}{11}$ **41.** $\frac{7}{6}$ **43.** −5.26 **45.** −1 **46.** 42 **47.** −37 **48.** 21

2.3 Exercises

1. 6 **3.** 11 **5.** −17 **7.** 11 **9.** 15 **11.** 28 **13.** −12 **15.** 8 **17.** 3 **19.** 5 **21.** −9 **22.** yes **23.** yes
24. no, $x = -11$ **25.** no, $x = 12$ **26.** 3; 3 **27.** 7; 7 **29.** $-\frac{5}{6}$ **31.** 5 **33.** −1 **35.** 6 **37.** 4 **39.** $\frac{1}{2}$ **41.** 1
43. 4 **45.** −33 **47.** 2 **49.** 4 **51.** 6 **53.** 5 **55.** −7 **57.** 2 **59.** $-\frac{1}{4}$ **61.** −0.25 or $-\frac{1}{4}$ **63.** 8 **65.** −6
67. $\frac{1}{2}$ **69.** −4.23 **71.** $14x^2 - 14xy$ **72.** $5x + 25y$ **73.** $28x + 54$ **74.** $-10x - 60$

2.4 Exercises

1. 2 **3.** $\frac{1}{2}$ **5.** 1 **7.** 24 **9.** 20 **11.** 7 **13.** −11 **15.** 3 **17.** yes **19.** no **21.** 1 **23.** 0 **25.** −7
27. −3 **29.** 4 **31.** −22 **33.** 3 **35.** −12 **37.** $-\frac{6}{13}$ **39.** $\frac{4}{113}$ **41.** $\frac{2}{7}$ **43.** $\frac{2}{3}$ **45.** $\frac{27}{14}$ or $1\frac{13}{14}$ **46.** $\frac{19}{20}$

47. $\frac{539}{40}$ [illegible] $\frac{19}{40}$ 48. $\frac{22}{5}$ or $4\frac{2}{5}$

2.5 Exercises

1. 50 miles per hour **3. (a)** 10 meters **(b)** 16 meters **5. (a)** $y = \frac{3}{5}x - 3$; -6 **7.** $h = \frac{2A}{b}$ **9.** $r = \frac{I}{Pt}$ **11.** $b = y - mx$ **13.** $r = \frac{A - P}{Pt}$ **15.** $y = -\frac{5}{9}x - 2$ **17.** $x = \frac{7}{6}y + 14$ **19.** $x = \frac{c - by}{a}$ **21.** $r^2 = \frac{s}{4\pi}$ **23.** $m = \frac{E}{c^2}$ **25.** $t^2 = \frac{2S}{g}$ **27.** $I = \frac{E}{R}$ **29.** $H = \frac{V}{LW}$ **31.** $B = \frac{3V}{h}$ **33.** $L = \frac{P - 2W}{2}$ **35.** $b^2 = c^2 - a^2$ **37.** $F = \frac{9}{5}C + 32$ **39.** $v = \frac{d - 16t^2}{t}$ **41.** I doubles **42.** I increases four times **43.** A increases four times **44.** A becomes $\frac{1}{4}A$ **45.** 31.2 **46.** 0.096 **47.** 40% **48.** 2.5%

Putting Your Skills to Work

1. $d = (1100)(1.4)$; $d = 1540$ feet **2.** $d = (1100)(2.5)$; $d = 2750$ feet **3.** You are closer to the storm; 7150 feet

2.6 Exercises

3. $>$ **5.** $<$ **7.** $<$ **9.** $<$ **11.** $<$ **13.** $>$ **15.** 0.0247 **17.**
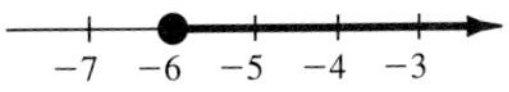

19.
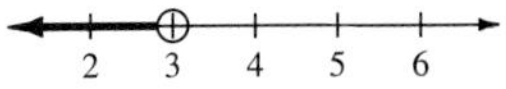

21.
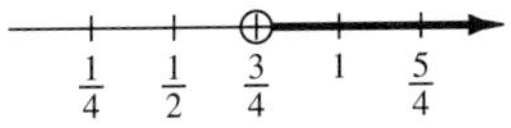

23.
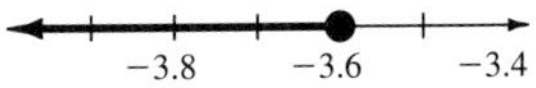

25.
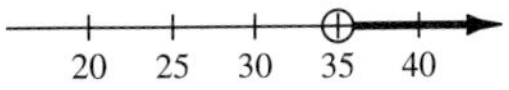

27. $x \geq -1$ **29.** $x < -20$ **31.** $x > 6.5$ **33.** $c < 56$ **35.** $h \geq 37$ **37.**
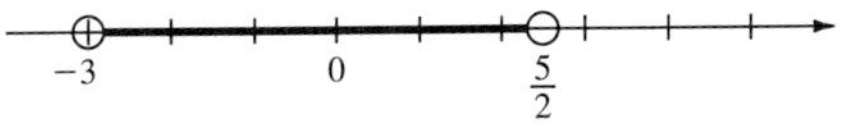

38.

39. 6.08 **40.** 16.25 **41.** 2% **42.** 37.5%

2.7 Exercises

1. $x > 8$
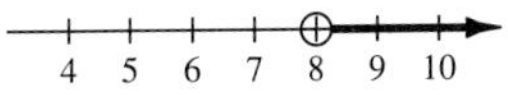

3. $x < 3$
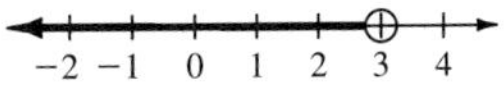

5. $x \geq 8$

7. $x < \frac{7}{2}$
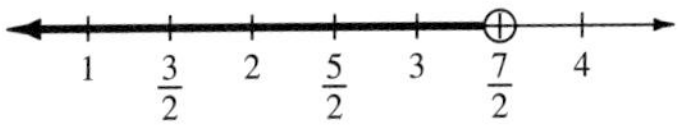

9. $x \leq -1$
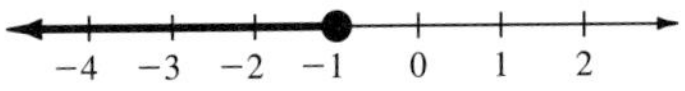

11. $x > -6$
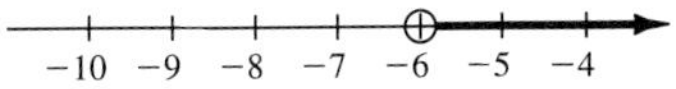

13. $x \leq -2$
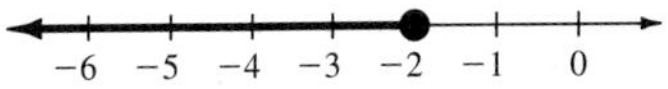

17. $x > 2$ **19.** $x \geq 4$ **21.** $x < -1$ **23.** $x \geq -6$ **25.** $x \geq -6$ **27.** $x > 0$ **29.** $x > -\frac{4}{11}$ **31.** $x > -1.46$ **33.** $x \geq 76$ **35. (a)** 24,000 chips **(b)** 76,000 chips; 61,600 chips

Putting Your Skills to Work

1. \$800 **2.** \$1580

Chapter 2 Review Problems

1. -10 **2.** -1 **3.** 5 **4.** 4 **5.** 10 **6.** -13 **7.** 40.4 **8.** -7 **9.** -2 **10.** -64 **11.** -11 **12.** $\frac{22}{3}$ or $7\frac{1}{3}$ **13.** -3 **14.** -1 **15.** 4 **16.** 3 **17.** 3 **18.** -6 **19.** 3 **20.** $-\frac{7}{2}$ or $-3\frac{1}{2}$ **21.** -3 **22.** 21 **23.** $-\frac{7}{3}$ or $-2\frac{1}{3}$ **24.** $-\frac{2}{7}$ **25.** 5 **26.** 0 **27.** 1 **28.** 20 **29.** $\frac{2}{3}$ **30.** 5 **31.** $\frac{35}{11}$ **32.** 4 **33.** -17 **34.** $\frac{2}{5}$ **35.** 32 **36.** $\frac{26}{7}$ **37.** -1 **38.** $-\frac{5}{2}$ or $-2\frac{1}{2}$ **39.** 4 **40.** -17 **41.** 4 **42.** 0 **43.** 4 **44.** -5 **45.** 3 **46.** -28 **47.** $-\frac{17}{5}$ **48.** 9 **49.** 0 **50.** 3 **51.** $y = 3x - 10$ **52.** $y = \frac{-5x - 7}{2}$ **53.** $r = \frac{A - P}{Pt}$ **54.** $h = \frac{A - 4\pi r^2}{2\pi r}$ **55.** $p = \frac{3H - a - 3}{2}$ **56. (a)** $T = \frac{1000\,C}{WR}$ **(b)** $T = 6000$ **57. (a)** $y = \frac{5}{3}x - 4$ **(b)** $y = 11$ **58. (a)** $R = \frac{E}{I}$ **(b)** $R = 5$

59. $x \le 2$

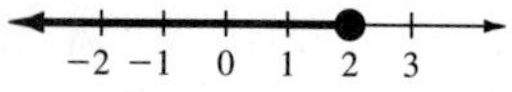

60. $x \ge 1$

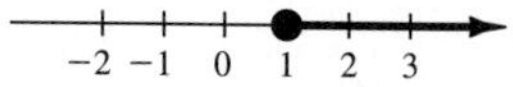

61. $x < -4$

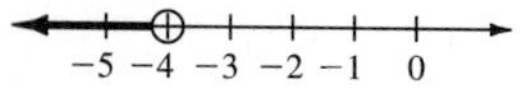

62. $x > -3$

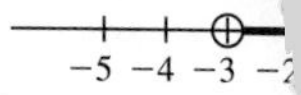

63. $x \ge 3$

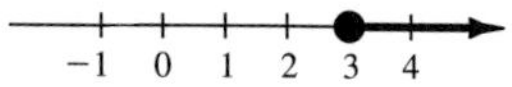

64. $x < 2$

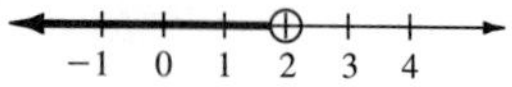

65. $x \le -8$

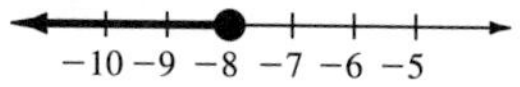

66. $x \ge 3$

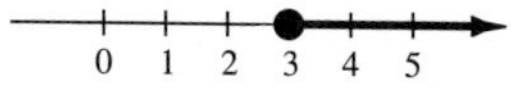

67. $x < 10$

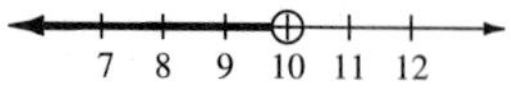

68. $x > -3$

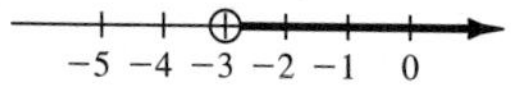

69. $x > \frac{17}{2}$

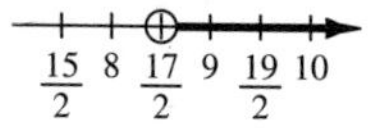

70. $x \ge \frac{19}{7}$

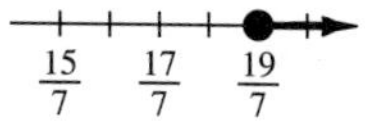

71. $x \ge -15$

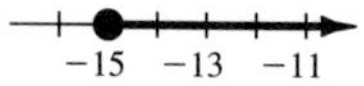

72. $x > \frac{7}{5}$

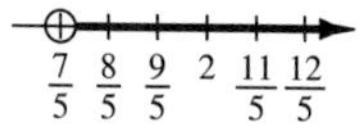

73. $n \le 46$ **74.** $n \le 17$

Chapter 2 Test

1. 2.1 **2.** 2 **3.** $-\frac{7}{2}$ **4.** −30 **5.** 2 **6.** −1.2 **7.** 7 **8.** $\frac{7}{3}$ **9.** 13 **10.** 125 **11.** 10 **12.** −4 **13.** 12 **14.** $-\frac{1}{5}$ **15.** 3 **16.** 2 **17.** 18 **18.** $w = \frac{A - 2P}{3}$ **19.** $w = \frac{6 - 3x}{4}$ **20.** $a = \frac{2A - hb}{h}$ **21.** $y = \frac{10ax - 5}{8ax}$ **22.** $x \le -\frac{1}{2}$

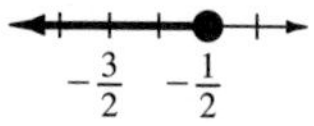

23. $x > -\frac{5}{4}$

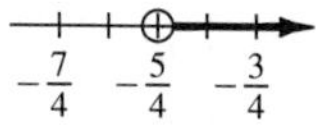

24. $x < 2$

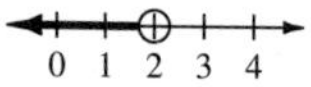

25. $x \ge \frac{1}{2}$

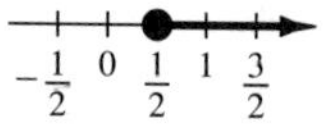

Cumulative Test for Chapters 0–2

1. $\frac{4}{21}$ **2.** $\frac{79}{20}$ or $3\frac{19}{20}$ **3.** $\frac{32}{15}$ or $2\frac{2}{15}$ **4.** 0.6888 **5.** 0.12 **6.** 61.2 **7.** 30 **8.** $12ab - 28ab^2$ **9.** $25x^4y^6$ **10.** $2x + 6y - 4xy + 16xy^2$ **11.** 5 **12.** 4 **13.** 1 **14.** $y = \frac{3x + 2}{7}$ **15.** $b = \frac{3H - 8a}{2}$ **16.** $t = \frac{I}{Pr}$ **17.** $a = \frac{2A}{h} - b$ **18.** $x < 5$

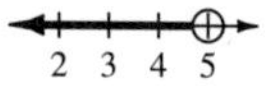

19. $x \le 3$

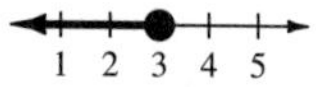

20. $x > -15$

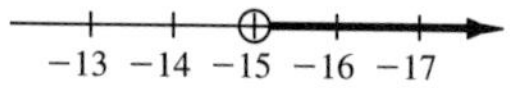

21. $x \le -1$

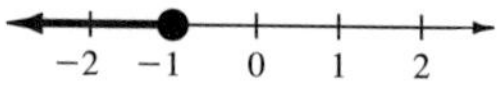

22. $x \ge -12$

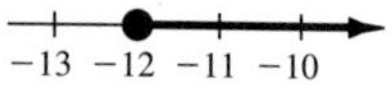

Chapter 3

Pretest Chapter 3

1. $2x - 30$ **2.** $0.40x + 150$ **3.** 5 **4.** 1st number = 12; 2nd number = 4; 3rd number = 7 **5.** $x = 8.6$ m; $2x - 3 = 14.2$ m; $2x + 1 = 18.2$ m **6.** $x = 7.5$ pounds; $x - 3.5 = 4$ pounds; $x - 2 = 5.5$ pounds **7.** width = 8 m; length = 29 m **8.** \$400 at 12%; \$600 at 9% **9.** 22,000 people **10.** 3 nickels; 7 dimes; 10 quarters **11.** 72 square inches **12.** 339.12 cubic meters **13.** 25.12 centimeters **14.** \$6859.80 **15.** $h \le 72$ **16.** $r \ge 450$ **17.** $a < 15$ **18.** $c > 8{,}900$ **19.** \$30,000 **20.** Length ≤ 16

3.1 Exercises

1. $x + 7$ **3.** $x - 5$ **5.** $\frac{x}{3}$ **7.** $3x$ **9.** $\frac{x}{8}$ **11.** $2x + 5$ **13.** $\frac{2}{7}x - 3$ **15.** $5(x + 12)$ **17.** $\frac{2}{3}(x - 6)$ **19.** $3x + \frac{1}{2}x$ **21.** $4x - \frac{1}{2}x$ **23.** $b + 15$ = no. of hours Alicia works; b = no. of hours Barbara works **25.** $c - 390$ = cost of the printer; c = cost of the computer **27.** $3w + 3$ = length of rectangle; w = width of rectangle **29.** s = attendance on Saturday; $s + 1600$ = attendance on Friday; $s - 783$ = attendance on Thursday **31.** 1st angle = $t + 19$; 2nd angle = $3t$; 3rd angle = t **33.** $3t - 200$ = attendance on Friday; t = attendance on Thursday **35.** 1st angle = x; 2nd angle = $x - 6$; 3rd angle = $2x$ **37.** February = $\frac{5}{6}x$; March = x; April = $\frac{3}{4}x$ **39.** p = last year's profit; $3p - 400 - 0.5p$ = this year's new profit **40.** 17 **41.** 10 **42.** 12 **43.** 7

3.2 Exercises

1. 45 **3.** 720 **5.** 290 **7.** 39 **9.** 51 **11.** −10 **13.** 72; 84 **15.** 17; 31 **17.** 80 **19.** 77.0973; 118.9605 **21.** 15 tables **23.** 25 months **25.** 9 hours **27.** 4 French fries **29.** 6 hours **31.** 6 miles per hour **33.** 5 miles **35.** 6 mph **37.** 96 **39.** 9 mpg **41. (a)** $F - 40 = \frac{x}{4}$; **(b)** 200 chirps; **(c)** 77° **42.** $10x^3 - 30x^2 - 15x$ **43.** $-2a^2b + 6ab - 10a^2$ **44.** $-5x - 6y$ **45.** $-4x^2y - 7xy^2 - 8xy$

3.3 Exercises

1. Melissa earned \$16,350; Heather earned \$12,750 **3.** short piece = 41.5 meters; long piece = 51.5 meters **5.** Monday attendance = 20,600; Tuesday attendance = 20,250; Wednesday attendance = 23,700 **7.** Margaret lost 66 lb; Rita lost 44 lb; Tony lost 46 lb **9.** first transistor = 0.0045 amp; second transistor = 0.009 amp; third transistor = 0.0135 amp **11.** width = 6 cm; length = 15 cm **13.** width = 17 cm; length = 31 cm **15.** first side = 16 m; second side = 8 m; third side = 13 m **17.** Longest side = 18 in.; Shortest side = 13 in.; Third side = 15 in. **19.** Length = 16.858 cm Width = 7.762 cm **21.** Original square was 11 m × 11 m **23.** $8x^3 + 12x^2 - 32x$ **24.** $5a^2b + 30ab - 10a^2$ **25.** $-19x + 2y - 2$ **26.** $9x^2y - 6xy^2 + 7xy$

3.4 Exercises

1. 360 miles **3.** 17 dictionaries **5.** 7 hours **7.** \$240 **9.** 800 **11.** \$14,500 **13.** \$5000 **15.** \$2600 at 12%; \$1400 at 14% **17.** \$1800 at 9%; \$4200 at 11% **19.** \$6500 at 14%; \$2500 at 9% **21.** 13 quarters; 9 nickels **23.** 18 nickels; 6 dimes; 9 quarters **25.** 7 nickels; 5 quarters; 15 half-dollars **27.** First angle measures 70°; Second angle measures 35°; Third angle measures 75° **29.** \$1328 at 7%, \$1641 at 11% **31.** 225 miles **33.** $15 - 3 = 12$ **34.** $2 + 8 + 5 = 15$ **35.** $8 - 18 - 18 = -28$ **36.** $-8 - 20 + 9 - 6 = -25$

Putting Your Skills to Work

1. \$982.40 **2.** \$128.17 **3.** \$755.40 **4.** \$66.24 **5.** \$150.60 **6.** 9% for 36 months

3.5 Exercises

7. 70 in.^{2} **9.** 10 feet **11.** 60 cm **13.** 28.26 m^2 **15.** 24 inches **17.** 70 meters **19.** 62.8 meters **21.** 9 inches **23.** equal sides = 12 feet; base = 8 feet **25.** 13° **27.** First angle = 100°; Second angle = 50°; Third angle = 30° **29.** 2512.6 in.^{3} **31.** 3 feet **33.** 552 square inches **35. (a)** 113.04 cm^3 **(b)** 113.04 cm^2 **37.** \$23.85 **39. (a)** 285.74 sq. in. **(b)** \$2143.05 **41. (a)** 36 cubic yd. **(b)** \$126 **42.** No. The perimeter may be 19 feet or it may be 15.5 feet. **43. (a)** 36π in.^{2} **(b)** 10π in.^{2}; the sphere **44.** 6.28 ft **45.** $3x^2 + 8x - 24$ **46.** $-11x - 17$ **47.** $-4x - 9$ **48.** $22x - 84$

Putting Your Skills to Work

1. 186 sq ft **2.** \$82 **3.** \$180 **4.** \$235 **5.** Depending on the circumstances you may give a low estimate of \$225 or a high estimate of \$250.

3.6 Exercises

1. $x > 67{,}000$ **3.** $x \leq 120$ **5.** $x < 34$ **7.** $x \geq 3500$ **9.** $x \leq 350$ **11.** $x \leq 224$ **13.** $x \geq 90$ **15.** $x \leq 17$ **17.** $x \geq 17$ feet **19.** $x \geq 15$ cars **21.** $x > 60{,}000$ **23.** $F < 230$ **25.** $x \leq 12$ **27.** $n > 454$ **29.** $x < -4$ **30.** $x \leq 3$ **31.** $x \geq 4\frac{2}{3}$ **32.** $x > -4$

Chapter 3 Review Problems

1. $x + 19$ **2.** $\frac{2}{3}x$ **3.** $x - 56$ **4.** $3x$ **5.** $2x + 7$ **6.** $2x - 3$ **7.** $5x - \frac{x}{3}$ **8.** $\frac{x + 6}{4}$ **9.** $h + 370$ = the speed of the jet; h = the speed of the helicopter **10.** $g - 2000$ = the cost of the truck; g = the cost of the garage **11.** $3w + 5$ = the length of the rectangle; w = the width of the rectangle **12.** r = the number of retired people; $4r$ = the number of working people; $0.5r$ = the number of unemployed people **13.** b = the number of degrees in angle B; $2b$ = the number of degrees in angle A; $b - 17$ = the number of degrees in angle C **14.** a = the number of students in algebra; $a + 29$ = the number of students in biology; $0.5a$ = the number of students in geology **15.** 12 **16.** 18 **17.** 44 **18.** 1 **19.** 16 years old **20.** \$40 **21.** 6.6 hours; 6 hours **22.** 90 **23.** $W = 10$ m; $L = 23$ m **24.** $W = 12$ ft; $L = 30$ ft **25.** 1st side = 8 yd; 2nd side = 15 yd; 3rd side = 17 yd **26.** 1st angle = 32°; 2nd angle = 96°; 3rd angle = 52° **27.** 31.25 yd; 18.75 yd **28.** Jon = \$30,000; Lauren = \$18,000 **29.** 280 miles **30.** 310 kilowatt-hours **31.** \$200.00 **32.** \$22,500 **33.** \$7000 at 12%; \$2000 at 8% **34.** \$2000 at $4\frac{1}{2}$%; \$3000 at 6% **35.** 18 nickels; 6 dimes; 9 quarters **36.** 7 nickels; 8 dimes; 10 quarters **37.** 8 m **38.** $16\frac{1}{2}$ in. **39.** 71° **40.** 42 sq. mi. **41.** 56 in. **42.** 168 square centimeters **43.** 1920 ft^3 **44.** 113.04 cm^3 **45.** 12 in. **46.** 254.34 cm^2 **47.** \$23,864 **48.** \$3440 **49.** 84 **50.** 80 dimes; 8 quarters **51.** 200 miles **52.** more than \$300,000 **53.** \$21.85 **54.** 3 hours **55.** $24\frac{4}{5}$ min **56.** 4 ounces **57.** 16 **58.** 14 games

Putting Your Skills to Work

1. $\frac{1}{2}$ hour or 30 minutes **2. (a)** 18,850 square miles; **(b)** 157 hours or 6.5 days

Chapter 3 Test

1. 35 **2.** 36 **3.** −12 **4.** first side = 20 m; second side = 30 m; third side = 16 m **5.** width = 20 m; length = 47 m **6.** 1st pollutant = 8 ppm; 2nd pollutant = 4 ppm; 3rd pollutant = 3 ppm **7.** 15 months **8.** \$8500 **9.** \$1400 at 14%; \$2600 at 11% **10.** 16 nickels; 7 dimes; 8 quarters **11.** 138.16 in. **12.** 192 square inches **13.** 4187 in.^{3} **14.** 70 cm^2 **15.** \$450 **16.** at least an 82 **17.** more than \$110,000

Cumulative Test for Chapters 0–3

1. 3.69 **2.** $\frac{31}{24}$ **3.** $-23y + 18$ **4.** 28 **5.** $\dfrac{2H - 3a}{5}$ **6.** $-3 \leq x$;  **7.** 40 **8.** Psych students = 50; Wrld Hist. students = 84 **9.** width = 7 cm; length = 32 cm **10.** \$135,000 **11.** \$3000 at 15%; \$4000 at 7% **12.** 7 nickels; 12 dimes; 4 quarters **13.** \$731.25 **14.** 169.56 pounds

Chapter 4

Pretest Chapter 4

1. 3^{16} **2.** $-10x^7$ **3.** $-24a^5b^4$ **4.** x^{23} **5.** $\dfrac{-2}{x^2y^2}$ **6.** $\dfrac{5bc}{3a}$ **7.** x^{50} **8.** $-8x^6y^3$ **9.** $\dfrac{9a^2b^4}{c^6}$ **10.** $\dfrac{3x^2}{y^3z^4}$ **11.** $\dfrac{-2b^2}{a^3c^4}$ **12.** $\frac{1}{9}$ **13.** 6.38×10^{-4} **14.** 1,894,000,000,000 **15.** $-2x^2 + 7x - 29$ **16.** $x^3 - 5x^2 + 11x - 12$ **17.** $-6x + 21x^2$ **18.** $2x^4 + 8x^3 - 2x^2$ **19.** $5x^4y^2 + 15x^2y^3 + 5xy^4$ **20.** $x^2 + 14x + 45$ **21.** $12x^2 + xy - 6y^2$ **22.** $2x^4 - 7x^2y^2 + 3y^4$ **23.** $64x^2 - 121y^2$ **24.** $36x^2 + 36xy + 9y^2$ **25.** $25a^2b^2 - 60ab + 36$ **26.** $4x^3 - 13x^2 + 11x - 2$ **27.** $16x^4 - 81$ **28.** $7x^3 - 6x^2 + 11x$ **29.** $5x^2 - 6x + 2 - \dfrac{3}{3x - 2}$

4.1 Exercises

5. 6; x, y; 11, and 1 **7.** $3x^3y^2$ **9.** $-3a^2b^2c^3$ **11.** 3^{15} **13.** 5^{26} **15.** $3^5 \cdot 8^2$ **17.** $-60x^3$ **19.** $26x^3$ **21.** $-12x^{11}$ **23.** $-10x^4y^3$ **25.** $\dfrac{10}{21}x^4y^5$ **27.** $-8.05x^5wy^4$ **29.** 0 **31.** $-24x^4y$ **33.** $72x^6y^{11}$ **35.** $-28a^5b^6$ **37.** $-3x^3y$ **39.** $12a^8b^6$ **41.** 0 **43.** $-24w^5xyz^6$ **45.** $\dfrac{1}{x^2}$ **47.** y^7 **49.** 3^{28} **51.** $\dfrac{1}{13^{10}}$ **53.** 2^{13} **55.** $\dfrac{a^8}{4}$ **57.** $\dfrac{x^7}{y^9}$ **59.** $\dfrac{1}{2b}$ **61.** $-4x^4y^5$ **63.** $\dfrac{-1}{2a^2b}$ **65.** -1 **67.** $\dfrac{-3x^2}{y^3z}$ **69.** $\dfrac{y^5}{10x^4}$ **71.** $6x^2$ **73.** $\dfrac{y^2}{16x^3}$ **75.** $\dfrac{3a}{4}$ **77.** $\dfrac{5x^6}{7y^8}$ **79.** $\dfrac{-xyz}{2}$ **81.** $2x$ **83.** $\dfrac{-2}{3a^3}$ **84.** $-27x^5yz^3$ **85.** $-7ab^5$ **86.–88.** Answers will vary. **89.** a^{5b} **90.** b^{3x} **91.** c^{y+2} **93.** w^{40} **95.** $a^{15}b^{10}$ **97.** $a^{24}b^{16}c^8$ **99.** $3^8x^4y^8$ **101.** $16a^{20}$ **103.** $\dfrac{12^5x^5}{y^{10}}$ **105.** $\dfrac{36x^2}{25y^6}$ **107.** $-32a^{25}b^{10}c^5$ **109.** $-64x^3z^{12}$ **111.** $\dfrac{a}{4b^4}$ **113.** $-8a^7b^{11}$ **115.** $\dfrac{16y^8}{9z^6}$ **117.** $\dfrac{y^{18}}{8}$ **119.** $\dfrac{a^{15}b^5}{c^{25}d^5}$ **121.** $16x^9y^3$ **123.** $\dfrac{81a^{12}b^4}{c^8}$ **124.** $-3x^3y^4z^7$ **125.** $\pm 2x^5y^4z^7$ **126.–128.** Answers will vary. **129.** 2^{17} **131.** -11 **132.** -46 **133.** $-\dfrac{7}{4}$ **134.** 5

4.2 Exercises

1. $\dfrac{3}{x^2}$ **3.** $\dfrac{1}{2xy}$ **5.** $\dfrac{3xz^3}{y^2}$ **7.** $3x$ **9.** $\dfrac{x^6}{y^2z^3w^{10}}$ **11.** $\dfrac{1}{64}$ **13.** $\dfrac{z^8}{9x^2y^4}$ **15.** $\dfrac{1}{x^6y}$ **17.** $-\dfrac{y^6}{8x^9}$ **19.** 1.2378×10^5 **21.** 7.42×10^{-4} **23.** 7.652×10^9 **25.** 56,300 **27.** 0.000033 **29.** 983,000 **31.** 300,000,000 meters per second **33.** 9.11×10^{-31} kilogram **35.** 1.0×10^1 **37.** 3.2×10^{-19} **39.** 4.5×10^5 **41.** \$$1.41 \times 10^4$ **43.** 1.9×10^{11} hours **45.** 1.13875×10^{-19} grams **47.** -0.8 **48.** -1 **49.** $-\dfrac{1}{28}$

Putting Your Skills to Work

1. 1.98×10^{20} newtons **2.** 8.39×10^{14} newtons **3.** 1.37×10^{19} newtons **4.** 4.46×10^{19} pounds

4.3 Exercises

1. $2x - 17$ **3.** $-x^2 - x + 12$ **5.** $x^3 - 2x^2 + 5x - 1$ **7.** $\frac{5}{6}x^2 + \frac{1}{2}x - 9$ **9.** $1.7x^3 - 3.4x^2 - 13.2x - 5.4$ **11.** $-11x - 2$ **13.** $17x^3 + x^2$ **15.** $-\frac{1}{6}x^2 + 7x - \frac{3}{10}$ **17.** $-2x^3 + 5x - 10$ **19.** $-4.7x^4 - 0.7x^2 - 1.6x + 0.4$ **21.** $10x + 10$ **23.** $-3x^2y + 6xy^2 - 4$ **25.** $-11a^4 - 5a^3 - 2$ **27.** $-x^3 - 2x^2y - 5xy^2 - 5y^3$ **29.** $-58x^2 + 27xy - 8y^2$ **30.** $-x^2y^2 + 7xy$ **31.** $7x^3 + 41x^2 + 22$ **32.** $-25x^3 - 20x + 21$ **33.** $\dfrac{3y - 2}{8}$ **34.** $\dfrac{3A}{2CD}$ **35.** b **36.** $\dfrac{5xy}{B}$

4.4 Exercises

1. $15x^3 - 3x^2$ **3.** $-4x^3 + 12x^2 + 24x$ **5.** $2x^4 - 6x^3 + 2x^2 - 12x$ **7.** $-5x^3y^3 + 15x^2y^2 - 25xy$ **9.** $3x^7 + 6x^5 - 9x^4 + 21x^3$ **11.** $6ab^2 - 4a^2b^3 - 10ab^5$ **13.** $-15x^4y^2 + 6x^3y^2 - 18x^2y^2$ **15.** $4b^4 + 6b^3 - 8b^2$ **17.** $3x^4 - 9x^3 + 15x^2 - 6x$

19. $-2x^3y^3 + 12x^2y^2 - 16xy$ **21.** $-28x^5y + 12x^4y + 8x^3y - 4x^2y$ **23.** $-15a^4b^2 - 3a^3b^2 + 21ab^2$ **25.** $12x^7 - 6x^5 + 18x^4 + 54x^3$ **27.** $-18x^7 - 4x^5 + 8x^4$ **29.** $x^2 + 13x + 30$ **31.** $x^2 + 8x + 12$ **33.** $x^2 - 3x - 18$ **35.** $x^2 - 11x + 30$ **37.** $-5x^2 - 16x - 3$ **39.** $3x^2 + 9xy - 5x - 15y$ **41.** $12y^2 - y - 6$ **43.** $12y^2 - 16y + 5$ **44.** The signs are incorrect. The result should be $-3x + 6$. **45.** The last term is incorrect. The result should be $-3x + 7$. **46.** $12x$ **47.** $49x^2 - 9$, difference of 2 perfect squares **49.** $-28x^2 + 2xy + 6y^2$ **51.** $15a^2 - 28ab^2 + 12b^4$ **53.** $4x^4 - 25y^4$ **55.** $64x^2 - 32x + 4$ **57.** $25x^4 + 20x^2y^2 + 4y^4$ **59.** $15t^2 + 8pt - 63p^2$ **61.** $6a^2b^2 - 31abd + 35d^2$ **63.** $30x^2 + 43xy - 8y^2$ **65.** $35x^2 + 21xz - 10xy - 6yz$ **67.** $4ac - 3ab - 32bc + 24b^2$ **69.** $\frac{1}{6}x^6y^8 + \frac{5}{126}x^4y^{10} - \frac{1}{21}x^2y^{12}$ **71.** -10 **72.** $-\frac{25}{7}$ **73.** 8 dimes; 11 quarters **74.** -5 **75.** $3.2076x^2 - 0.819xy - 17.0146y^2$ **76.** $57.1536x^2 - 130.788xy + 74.8225y^2$

4.5 Exercises

5. $x^2 - 16$ **7.** $x^2 - 81$ **9.** $64x^2 - 9$ **11.** $4x^2 - 49$ **13.** $4x^2 - 25y^2$ **15.** $100p^2 - 49q^2$ **17.** $144x^2 - 49$ **19.** $9x^4 - 16y^6$ **21.** $16x^2 - 8x + 1$ **23.** $64x^2 + 48x + 9$ **25.** $25x^2 - 70x + 49$ **27.** $4x^2 + 12xy + 9y^2$ **29.** $64x^2 - 80xy + 25y^2$ **31.** $64 + 48x + 9x^2$ **33.** $36w^2 + 60wz + 25z^2$ **35.** $49x^2 - 9y^2$ **37.** $16x^4 - 56x^2y + 49y^2$ **39.** $x^3 - 4x^2 + 5x - 2$ **41.** $9y^3 + 3y^2z - 8yz^2 - 4z^3$ **43.** $4x^4 - 7x^3 + 2x^2 - 3x - 1$ **45.** $2x^3 + 3x^2 - 17x - 30$ **47.** $3x^3 - 13x^2 - 6x + 40$ **49.** $2x^3 - 7x^2 - 32x + 112$ **51.** $x^4 + x^3 - 4x^2 - 7x - 3$ **53.** $2x^6 + 5x^5y - 6x^4y^2 - 18x^2y^2 - 45xy^3 + 54y^4$ **55.** \$11,000 at 7%; \$7,000 at 11% **56.** width = 7 m; length = 10 m

Putting Your Skills to Work

1. $x^2 + 2xy$ **2.** 112 square inches **3.** 481 square inches **4.** $x^2 - y^2$ **5.** 112 square inches; 481 square inches **6.** $(x - y)(x + y)$ **7.** 189 square inches

4.6 Exercises

1. $4x^2 - 10x + 2$ **3.** $2y^2 - 3y - 1$ **5.** $7x^4 - 3x^2 + 8$ **7.** $2x^3 - 9x + 3$ **9.** $10x^3 - 7x^2 - x + 5$ **11.** $3x + 5$ **13.** $x - 2$ **15.** $3x^2 - 4x + 8$ **17.** $x^2 - x + 2$ **19.** $2x^2 + x - 2$ **21.** $2y^2 - 9y + 10$ **23.** $y^2 - 4y - 1$ **25.** $2y^2 + y + 2$ **27.** $y^3 - 3y^2 + 6y - 16$ **29.** $y^3 + 2y^2 - 5y - 10$ **31.** $x \geq 3$ **32.** $x \leq -2$ **33.** $x < 6$ **34.** $x \leq -1$

Chapter 4 Review Problems

1. $-18a^7$ **2.** 25^{23} **3.** $6x^4y^6$ **4.** $40x^5y^5$ **5.** 8^{17} **6.** $\frac{1}{7^{12}}$ **7.** $\frac{1}{x^4}$ **8.** w^5 **9.** $\frac{x^4}{3}$ **10.** $-\frac{3}{5x^5y^4}$ **11.** $-\frac{2a}{3b}$ **12.** $\frac{5a^3b}{3c}$ **13.** x^{24} **14.** $125x^3y^6$ **15.** $9a^6b^4$ **16.** $27a^6b^6$ **17.** $\frac{2x^4}{3y^2}$ **18.** $\frac{25a^2b^4}{c^6}$ **19.** $\frac{1}{2xy^2}$ **20.** $\frac{y^9}{64w^{15}z^6}$ **21.** $\frac{1}{x^3}$ **22.** $\frac{1}{x^6y^8}$ **23.** $\frac{2y^3}{x^6}$ **24.** $\frac{x^5}{2y^6}$ **25.** $\frac{1}{4x^6}$ **26.** $\frac{3y^2}{x^3}$ **27.** $\frac{4w^2}{x^5y^6z^8}$ **28.** $\frac{b^5c^3d^4}{27a^2}$ **29.** 1.563402×10^{11} **30.** 8.8367×10^4 **31.** 1.79632×10^5 **32.** 7.8×10^{-3} **33.** 6.173×10^{-5} **34.** 1.3×10^{-1} **35.** 120,000 **36.** 83,670,000,000 **37.** 3,000,000 **38.** 0.25 **39.** 0.00000005708 **40.** 0.000000006 **41.** 2.0×10^{13} **42.** 9.36×10^{19} **43.** 9.6×10^{-10} **44.** 7.8×10^{-11} **45.** 3.504×10^8 kilometers **46.** 7.94×10^{14} cycles **47.** 6×10^9 **48.** $-5x^2 - 11x - 18$ **49.** $-5x^2 - 16x - 12$ **50.** $-x^3 + 2x^2 - x + 8$ **51.** $7x^3 - 3x^2 - 6x + 4$ **52.** $-11x^2 + 7xy - 20y^2$ **53.** $3x^2 - 6y + 20xy$ **54.** $5x^3y^3 - 2x^2y^2 + 10xy - 4$ **55.** $11x^2y^2 + xy - 2y - 3y^2$ **56.** $-x^2 - 2x + 10$ **57.** $-6x^2 - 6x - 3$ **58.** $15x^2 + 2x - 1$ **59.** $32x^2 - 32x + 6$ **60.** $20x^2 + 48x + 27$ **61.** $10x^3 - 30x^2 + 15x$ **62.** $-6x^3y^3 + 10x^2y^2 - 12xy$ **63.** $4x^4 - 12x^3 + 20x^2 - 8x$ **64.** $5a^2 - 8ab - 21b^2$ **65.** $8x^4 - 12x^2 - 10x^2y + 15y$ **66.** $-15x^6y^2 - 9x^4y + 6x^2y$ **67.** $9x^2 - 12x + 4$ **68.** $25x^2 - 9$ **69.** $49x^2 - 36y^2$ **70.** $25a^2 - 20ab + 4b^2$ **71.** $64x^2 + 144xy + 81y^2$ **72.** $4x^3 + 27x^2 + 5x - 3$ **73.** $2x^4 + x^3 - 8x^2 - 5x + 12$ **74.** $2x^3 - 7x^2 - 42x + 72$ **75.** $8x^3 + 36x^2 + 54x + 27$ **76.** $2y^2 + 3y + 4$ **77.** $6x^3 + 7x^2 - 18x$ **78.** $4x^2y - 6x + 8y$ **79.** $53x^3 - 12x^2 + 19x + 13$ **80.** $3x + 7$ **81.** $4x + 1$ **82.** $2x^2 - 5x + 13 + \frac{-27}{x + 2}$ **83.** $3x^2 + 2x + 4 + \frac{9}{2x - 1}$ **84.** $4x - 5 + \frac{1}{2x + 1}$ **85.** $4x + 1$ **86.** $x^2 + 3x + 8$ **87.** $2x^2 + 4x + 5 + \frac{11}{x - 2}$ **88.** $3x^3 - x^2 + x - 5$ **89.** $4x^2 + 5x + 1$

Chapter 4 Test

1. 3^{34} **2.** 3^{28} **3.** 8^{24} **4.** $6x^7y^7$ **5.** $\frac{-7x^3}{5}$ **6.** $-125x^3y^{18}$ **7.** $\frac{49a^{14}b^4}{9}$ **8.** $\frac{1}{4a^5b^6}$ **9.** $\frac{1}{125}$ **10.** $\frac{6c^5}{a^4b^3}$ **11.** $\frac{2w^6}{x^3y^4z^8}$ **12.** 1.765×10^{-7} **13.** 582,000,000 **14.** 2.4×10^{-6} **15.** $-2x^2 + 5x$ **16.** $3x^2 + 3xy - 6y - 7y^2$ **17.** $-21x^5 + 28x^4 - 42x^3 + 14x^2$ **18.** $15x^4y^3 - 18x^3y^2 + 6x^2y$ **19.** $10a^2 + 7ab - 12b^2$ **20.** $6x^3 - 11x^2 - 19x - 6$ **21.** $49x^4 + 28x^2y^2 + 4y^4$ **22.** $81x^2 - 4y^2$ **23.** $12x^4 - 14x^3 + 25x^2 - 29x + 10$ **24.** $3x^4 + 4x^3y - 15x^2y^2$ **25.** $3x^3 - x + 5$ **26.** $2y^2 + 3y - 1$

Cumulative Test for Chapters 0–4

1. $-\frac{11}{24}$ **2.** -0.74 **3.** $-\frac{6}{7}$ **4.** \$8.96 **5.** $2x^2 - 13x$ **6.** 35 **7.** $x = -\frac{9}{2}$ **8.** $x = 44$ **9.** $x > -1$ **10.** $f = \frac{2B}{3a} - \frac{c}{3}$ **11.** 12,400 employees **12.** \$199.20 **13.** $15x^2 - 47x + 28$ **14.** $9x^2 - 30x + 25$ **15.** $8x^3 + 12x^2 + 6x + 1$ **16.** $-20x^5y^8$ **17.** $-2\frac{x^3}{3y^9}$ **18.** $-27x^3y^{12}z^6$ **19.** $\frac{9z^8}{w^2x^3y^4}$ **20.** 1.36×10^{15} **21.** 4.0×10^{-35} **22.** $5x^3 + 7x^2 - 6x + 50$ **23.** $-36x^3y^2 + 18x^2y^3 - 48xy^4$ **24.** $2x^4 - 15x^3 + 28x^2 - 33x + 12$

Chapter 5

Pretest Chapter 5

1. $2x(x - 3y + 6y^2)$ **2.** $(3x + 4y)(a - 2b)$ **3.** $18ab(2b - 1)$ **4.** $(a - 2b)(5 - 3x)$ **5.** $(7x - 3)(3x - 2)$ **6.** $7x(x - 2) - 3(9x - 2)$ **7.** $(x - 24)(x + 2)$ **8.** $(x - 5)(x - 3)$ **9.** $(x + 5)(x + 1)$ **10.** $2(x + 6)(x - 2)$ **11.** $3(x - 9)(x + 7)$ **12.** $(5x - 2)(3x - 2)$ **13.** $(3y - 2z)(2y + 3z)$ **14.** $4(3x + 5)(x + 2)$ **15.** $(9x^2 + 4)(3x + 2)(3x - 2)$ **16.** $(7x - 2y)^2$ **17.** $(5x + 8)^2$ **18.** $3x(2x - 1)(x + 3)$ **19.** $2y^2(4x - 3)^2$ **20.** Cannot be factored **21.** $x = 1, x = -\frac{3}{2}$ **22.** $x = 5, x = 6$ **23.** $x = -\frac{2}{3}, x = 3$

5.1 Exercises

5. $5b(a - 1)$ **7.** $9wz(2 - 3w)$ **9.** $2x(4x^2 - 5x - 7)$ **11.** $r(9pq - 12p^2q - 14x)$ **13.** $a(a^3y - a^2y^2 + ay^3 - 2)$ **15.** $2x^5(3x^4 - 4x^2 + 2)$ **17.** $9ab(ab - 4)$ **19.** $8a(5a - 2b - 3)$ **21.** $17x(2x - 3y - 1)$ **23.** $6xy(3y + x + 4 - 2xy)$ **25.** $(x - 2y)(5 - z)$ **27.** $(x - 7)(5x + 3)$ **29.** $(c + 2d)(5b - 2a)$ **31.** $(bc - 1)(5a + b + c)$ **33.** $(x^2 + 8)(4 - 2y - w)$ **35.** $(x - 2y)(3x^2 - 1)$ **37.** $(9x + 5y)(2ab + 1)$ **39.** $y(5a - 3b + 1)(3x^2 - 19x - 1)$ **41.** $57{,}982(3x^3 - 6y^2 - 11z^2)$ **43.** $c = 25.95(a + b + c + d)$ **44.** $4x^2 - 9$ **45.** $16a^2 - 8ab + b^2$ **46.** 3rd number is 21 **47.** \$12,300

5.2 Exercises

1. $(x - 2)(3 + y)$ **3.** $(x + y)(a - 2b)$ **5.** $(x - 4)(x^2 + 3)$ **7.** $(3a + b)(x - 2)$ **9.** $(a + 2b)(5 + 6c)$ **11.** $(a - b)(5 - 2x)$ **13.** $(y - 2)(y - 3)$ **15.** $(x + y)(3a - 1)$ **17.** $(3x + y)(2a - 1)$ **19.** $(x + 6)(x - 8)$ **21.** $(4x + 3w)(7x + 2y^2)$ **23.** $(4a - 3b)(2a + 5e)$ **25.** $(4x + 7w)(11x^2 - 9y^2)$ **28.** $121x^2 - 49y^2$ **29.** trinomial **30.** 84 at \$1.75; 36 at \$3.25 **31.** 91.5

5.3 Exercises

3. $(x + 1)(x + 8)$ **5.** $(x + 6)(x + 2)$ **7.** $(x - 3)(x - 1)$ **9.** $(x - 11)(x - 2)$ **11.** $(x + 4)(x - 3)$ **13.** $(x - 14)(x + 1)$ **15.** $(x + 5)(x - 4)$ **17.** $(x - 8)(x + 3)$ **19.** $(x - 2)(x - 5)$ **21.** $(x - 2)(x + 1)$ **23.** $(x + 7)(x - 2)$ **25.** $(x - 4)(x - 5)$ **27.** $(x + 15)(x + 2)$ **29.** $(y - 5)(y + 1)$ **31.** $(a + 8)(a - 2)$ **33.** $(x - 4)(x - 8)$ **35.** $(x + 7)(x - 3)$ **37.** $(x + 5)(x + 8)$ **39.** $(x - 22)(x + 1)$ **41.** $(x + 12)(x - 3)$ **43.** $(x - 7y)(x + 6y)$ **45.** $(x - 7y)(x - 9y)$ **47.** $(x^2 - 8)(x^2 + 5)$ **49.** $(x^2 + 30)(x^2 + 2)$ **51.** $2(x - 2)(x - 4)$ **53.** $3(x + 4)(x - 6)$ **55.** $4(x + 5)(x + 1)$ **57.** $5(x - 1)(x + 3)$ **59.** $6(x + 1)(x + 2)$ **61.** $2(x - 7)(x - 9)$ **63.** $\frac{A - P}{Pr}$ **64.** $x \geq -\frac{5}{3}$

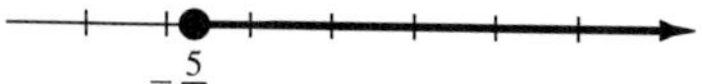

$-\frac{5}{3}$

65. 120 miles **66.** Sales must be at least \$130,000

5.4 Exercises

1. $(3x + 1)(x + 2)$ **3.** $(2x - 1)(x - 2)$ **5.** $(2x + 5)(x - 1)$ **7.** $(2x + 1)(x - 3)$ **9.** $(5x - 2)(x + 1)$ **11.** $(3x - 2)(2x - 3)$ **13.** $(2x - 5)(x + 4)$ **15.** $(3x + 1)(3x + 2)$ **17.** $(3x + 2)(2x - 3)$ **19.** $(4x - 1)(2x - 1)$ **21.** $(2x - 1)(2x + 9)$ **23.** $(9y - 4)(y - 1)$ **25.** $(5a + 2)(a - 3)$ **27.** $(6x - 1)(2x - 3)$ **29.** $(5x - 2)(3x + 2)$ **31.** $(5x + 3)(2x + 3)$ **33.** $(8x + 1)(2x - 1)$ **35.** $(2x^2 - 1)(x^2 + 8)$ **37.** $(2x - y)(2x + 5y)$ **39.** $(5x - 4y)(x + 4y)$ **41.** $2(5x + 6)(x + 1)$ **43.** $3(2x - 1)(2x - 3)$ **45.** $3(3x - 1)(2x + 5)$ **47.** $3(7x - 1)(x + 3)$ **49.** $(2x + 5)(6x - 7)$ **51.** $(4x - 3)(5x - 3)$ **53.** $\frac{1}{10}$ **54.** $(x + 9)(x - 4)$ **55.** 1st = 20°; 2nd = 35°; 3rd = 125° **56.** width = 4 m; length = 12 m

5.5 Exercises

1. $(3x + 4)(3x - 4)$ **3.** $(1 + 10s)(1 - 10s)$ **5.** $(5x - 4)(5x + 4)$ **7.** $(2x - 5)(2x + 5)$ **9.** $(6x - 5)(6x + 5)$ **11.** $(1 - 7x)(1 + 7x)$ **13.** $(9x - 7y)(9x + 7y)$ **15.** $(5 + 11x)(5 - 11x)$ **17.** $(9x + 10y)(9x - 10y)$ **19.** $(5a + 1)(5a - 1)$ **21.** $(7x + 5y)(7x - 5y)$ **23.** $(7x + 1)^2$ **25.** $(y - 3)^2$ **27.** $(3x - 4)^2$ **29.** $(7x + 2)^2$ **31.** $(x + 5)^2$ **33.** $(3x - 7)^2$ **35.** $(5x - 4)^2$ **37.** $(9x + 2y)^2$ **39.** $(3x - 7y)^2$ **41.** $(4a + 9b)^2$ **43.** $(3x^2 - y)^2$ **45.** $(7x + 1)(7x + 9)$ **47.** $(x^3 + 5)(x^3 - 5)$ **49.** $(x^5 + 6y^5)(x^5 - 6y^5)$ **51.** $(2x^4 + 3)^2$ **53.** Because no matter what combination you try, you cannot multiply two binomials to equal $9x^2 + 1$. **55.** There is one answer. It is 49. **57.** $3(3x - 1)(3x + 1)$ **59.** $5(5x - 1)(5x + 1)$ **61.** $3(2x - 3)(2x - 3)$ **63.** $2(7x + 3)(7x + 3)$ **65.** $(x - 2)(x - 7)$ **67.** $(2x - 1)(x + 3)$ **69.** $(4x - 11)(4x + 11)$ **71.** $(5x - 2)^2$ **73.** $3(x + 5)(x - 3)$ **75.** $7(x - 3)(x + 3)$ **77.** $5(x + 2)^2$ **79.** $2(x - 9)(x - 7)$ **81.** $2x^3 + 7x^2 - 10x - 24$ **82.** $x > \frac{5}{2}$

$\frac{5}{2}$

83. $x^2 + 3x + 4 + \frac{-3}{x - 2}$ **84.** $2x^2 + x - 5$

5.6 Exercises

1. $3(3x - 2y + 5)$ **3.** $(6x - 3y)(6x + 3y)$ **5.** $(3x - 2y)^2$ **7.** $(x - 4)(x + 7)$ **9.** $(5x - 3)(x - 2)$ **11.** $(x - 3y)(a + 2b)$ **13.** $3(x^2 + 2)(x^2 - 2)$ **15.** $2(x - 11)(x + 6)$ **17.** $(2x - 3)(x - 4)$ **19.** $(x - 10y)(x + 7y)$ **21.** $(a + 4)(x - 5)$ **23.** $5x(3 - x)(3 + x)$ **25.** $5xy^3(x - 1)^2$ **27.** $3xy(3z + 2)(3z - 2)$ **29.** $3(x + 7)(x - 5)$ **31.** $5(x - 2)(x - 4)$ **33.** $-1(2x^2 + 1)(x + 2)(x - 2)$ **35.** prime **37.** $3(x - 2y)(x - 3y)$ **39.** $6(3x + y)(2x - y)$ **41.** $4(2x - 1)(x + 4)$ **43.** $11x^2(3x - 1)(3x + 1)$ **45.** $a(x + 2y)(a - 3)$ **47.** prime **48.** The number d must be a perfect square such as 1, 4, 9, 16, 25, 36, 49, 64, 81, 100, . . . **49.** −3 **50.** \$30,000

5.7 Exercises

1. 5, −2 **3.** −5, −8 **5.** $-\frac{1}{2}, 3$ **7.** $-\frac{1}{4}, 3$ **9.** 8, −2 **11.** 2, 5 **13.** $\frac{2}{3}, \frac{3}{2}$ **15.** 0, 8 **17.** $0, \frac{2}{3}$ **19.** 0, 1 **21.** −5, 4 **23.** 6 **25.** $-\frac{1}{2}$ **27.** 0, −2 **29.** 9, −6 **31.** $-\frac{9}{2}, 4$ **33.** You can always factor out x. **34.** You must have the product of two or more factors equal to zero, not 14. Remember the zero factor property. **35.** $L = 14$ m; $W = 10$ m **37.** 4; 14

39. 12 meters above ground after 2 sec **41.** 5 stereo sets were produced. **43.** $-12x^3y^7$ **44.** $12a^{10}b^{13}$ **45.** $-\frac{3a^4}{2b^2}$ **46.** $\frac{1}{3x^5y^4}$

Putting Your Skills to Work

1. 30 ft × 75 ft **2.** 90 ft × 30 ft

Chapter 5 Review Problems

1. $3x^2y(5x - 3y)$ **2.** $7x(3x - 2)$ **3.** $7xy(x - 2y - 3x^2y^2)$ **4.** $25a^4b^4(2b - 1 + 3ab)$ **5.** $9x^2(3x - 1)$ **6.** $2(x - 2y + 3z + 6)$ **7.** $(a + 3b)(2a - 5)$ **8.** $3xy(5x^2 + 2y + 1)$ **9.** $(3x - 7)(a - 2)$ **10.** $(a + 5b)(a - 4)$ **11.** $(x^2 + 3)(y - 2)$ **12.** $(4a - 5)(d + 5c)$ **13.** $(5x - 1)(3x + 2)$ **14.** $(5w - 3)(6w + z)$ **15.** $(x - 6)(x + 3)$ **16.** $(x - 4)(x - 6)$ **17.** $(x + 6)(x + 8)$ **18.** $(x + 3y)(x + 5y)$ **19.** $(x^2 + 7)(x^2 + 6)$ **20.** $(x + 3)(x - 13)$ **21.** $5(x + 1)(x + 3)$ **22.** $3(x + 1)(x + 12)$ **23.** $2(x - 6)(x - 8)$ **24.** $4(x - 5)(x - 6)$ **25.** $(4x - 5)(x + 3)$ **26.** $(3x - 1)(4x + 5)$ **27.** $(5x + 4)(3x - 1)$ **28.** $(3x - 2)(2x - 3)$ **29.** $(2x - 3)(x + 1)$ **30.** $(3x - 1)(x + 5)$ **31.** $(10x - 1)(2x + 5)$ **32.** $(5x - 1)(4x + 5)$ **33.** $(4a + 1)(a - 3)$ **34.** $(4a + 3)(a - 1)$ **35.** $2(x - 1)(3x + 5)$ **36.** $2(x + 1)(3x - 5)$ **37.** $2(2x - 3)(x - 5)$ **38.** $4(x - 9)(x + 4)$ **39.** $2(x - 1)(5x - 6)$ **40.** $(2x + 11)(x - 4)$ **41.** $(3x - 1)(4x + 3)$ **42.** $2(x - 1)(8x + 15)$ **43.** $(3x - 2y)(2x - 5y)$ **44.** $2(3x - y)(x - 5y)$ **45.** $(3x^2 - 8)(x^2 + 1)$ **46.** $(3x^2 - 2)(x^2 + 4)$ **47.** $(7x + y)(7x - y)$ **48.** $4(2x - 3y)(2x + 3y)$ **49.** $(3x - 2)^2$ **50.** $(4x - 1)(4x + 1)$ **51.** $(5x - 6)(5x + 6)$ **52.** $(10x - 3)(10x + 3)$ **53.** $(1 - 7x)(1 + 7x)$ **54.** $(2 - 7x)(2 + 7x)$ **55.** $(6x + 1)^2$ **56.** $(5x - 2)^2$ **57.** $(4x - 3y)^2$ **58.** $(7x - 2y)^2$ **59.** $2(x - 3)(x + 3)$ **60.** $3(x - 5)(x + 5)$ **61.** $2(2x + 5)^2$ **62.** $2(5x + 6)^2$ **63.** $(2x + 3y)(2x - 3y)$ **64.** $(x + 3)^2$ **65.** $(x - 3)(x - 6)$ **66.** $(x + 15)(x - 2)$ **67.** $(x - 1)(x + 7)$ 68. $(5x - 2)(2x + 1)$ **69.** $4(3x + 4)$ **70.** $4xy(2xy - 1)$ **71.** $10x^2y^2(5x + 2)$ **72.** $13ab(2a^2 - b^2 + 4ab^3)$ **73.** $x(x - 8)^2$ **74.** $2(x + 10)^2$ **75.** $3(x - 3)^2$ **76.** $x(5x - 6)^2$ **77.** $(7x + 10)(x - 1)$ **78.** $(5x + 2)(x - 2)$ **79.** $xy(3x + 2y)(3x - 2y)$ **80.** $x^3a(3a + 4x)(a - 5x)$ **81.** $2(3a + 5b)(2a - b)$ **82.** $(4a - 5b)^2$ **83.** $(a - 1)(7 - b)$ **84.** $(3d - 4)(1 - c)$ **85.** $(2x - 1)(1 + b)$ **83.** $(b - 7)(5x + 4y)$ **87.** $x(2a - 1)(a - 7)$ **88.** $x(x + 4)(x - 4)(x + 1)(x - 1)$ **89.** $(x^2 + 9y^6)(x + 3y^3)(x - 3y^3)$ **90.** $(3x^2 - 5)(2x^2 + 3)$ **91.** $yz(14 - x)(2 - x)$ **92.** $x(3x + 2)(4x + 3)$ **93.** $(2w + 1)(8w - 5)$ **94.** $3(2w - 1)^2$ **95.** $2y(2y - 1)(y + 3)$ **96.** $(5y - 1)(2y + 7)$ **97.** $8y^8(y^2 - 2)$ **98.** $49(x^2 + 1)(x + 1)(x - 1)$ **99.** prime **100.** prime **101.** $4y(2y^2 - 5)(y^2 + 3)$ **102.** $3x(3y + 7)(y - 2)$ **103.** $(4x^2y - 7)^2$ **104.** $2xy(8x + 1)(8x - 1)$ **105.** $(2x + 5)(a - 2b)$ **106.** $(2x + 1)(x + 3)(x - 3)$ **107.** $-3, 6$ **108.** $2, -5$ **109.** $0, \frac{1}{6}$ **110.** $0, -\frac{11}{6}$ **111.** $-5, \frac{1}{2}$ **112.** $-8, -3$ **113.** $-5, -9$ **114.** $-\frac{3}{5}, 2$ **115.** -3 **116.** $-3, \frac{3}{4}$ **117.** $\frac{1}{5}, 2$ **118.** base = 10 cm altitude = 7 cm **119.** width = 7 feet, length = 15 feet **120.** 6 seconds **121.** 8 amperes, 12 amperes

Putting Your Skills to Work

1. \$125 **2.** You will get two answers. Which answer should you use? Why? $(7t - 840)(t - 280) = 0$; $t = 120$ items, $t = 280$ items; 120 items is a profit of \$180, 280 items is a loss of \$188 **3.** 200 items **4.** \$300 **5.** Answers may vary. Check students' work.

Chapter 5 Test

1. $(x + 14)(x - 2)$ **2.** $(3x + 10y)(3x - 10y)$ **3.** $(5x + 1)(2x + 5)$ **4.** $(3a - 5b)^2$ **5.** $x(7 - 9x + 14y)$ **6.** $(x + 2y)(3x - 2w)$ **7.** $2x(3x - 4)(x - 2)$ **8.** $c(5a - b)(a - 2b)$ **9.** $4(5x^2 + 2y^2)(5x^2 - 2y^2)$ **10.** $(3x - y)(3x - 4y)$ **11.** $7x(x - 6)$ **12.** prime **13.** prime **14.** $-5y(2x - 3y)^2$ **15.** $(4x + 1)(4x - 1)$ **16.** $(x^8 + 1)(x^4 + 1)(x^2 + 1)(x + 1)(x - 1)$ **17.** $(x + 3)(2a - 5)$ **18.** $(a + 2b)(w + 2)(w - 2)$ **19.** $3(x - 6)(x + 5)$ **20.** $x(2x + 5)(x - 3)$ **21.** $-5, -9$ **22.** $-\frac{7}{3}, -2$ **23.** $-\frac{5}{2}, 2$ **24.** width = 7 miles, length = 13 miles

Cumulative Test for Chapters 0–5

1. 15% **2.** 0.494 **3.** -10.36 **4.** $8x^4y^{10}$ **5.** -64 **6.** $14x^2 + x - 3$ **7.** $2x^3 - 12x^2 + 19x - 3$ **8.** $x \leq -3$ **9.** 2 **10.** -15 **11.** $\frac{2s - 2a}{3}$ **12.** $(3x - 1)(2x - 1)$ **13.** $(6x + 1)(x - 1)$ **14.** $(3x + 2)(3x - 1)$ **15.** $(11x + 8y)(11x - 8y)$ **16.** $-4(5x + 6)(4x - 5)$ **17.** prime **18.** $x(4x + 5)^2$ **19.** $(9x^2 + 4b^2)(3x + 2b)(3x - 2b)$ **20.** $(2x + 3)(a - 2b)$ **21.** $(x^2 + 5)(x^2 + 3)$ **22.** $-8, 3$ **23.** $\frac{5}{3}, 2$ **24.** base = 6 miles, altitude = 19 miles

Chapter 6

Pretest Chapter 6

1. -2 **2.** $\frac{3x - 2}{4x + 3}$ **3.** $\frac{b}{2a - b}$ **4.** $\frac{2a - b}{6}$ **5.** $\frac{x}{x + 5}$ **6.** $\frac{1}{x + 3}$ **7.** $\frac{y}{x - 1}$ **8.** 3 **9.** $\frac{-2y - 7}{(y - 1)(2y + 3)}$ **10.** $\frac{1}{x + 5}$ **11.** $\frac{5x^2 + 6x - 9}{3x(x - 3)^2}$ **12.** $\frac{2a - 3}{5a^2 + a}$ **13.** $\frac{a^2 - 2a - 2}{3a^2(a + 1)}$ **14.** $\frac{-x^2 - y^2}{x^2 + 2xy - y^2}$ **15.** $x = -5$ **16.** no solution **17.** 9.3 **18.** \$263.50

6.1 Exercises

1. 3 **3.** $\frac{1}{3 - x}$ **5.** $\frac{2}{x - 4}$ **7.** $x - 3y$ **9.** $\frac{3x}{x - 3y}$ **11.** $\frac{x + 2}{x}$ **13.** $\frac{x + 3}{4x - 1}$ **15.** $\frac{x(x - 4)}{x + 6}$ **17.** $\frac{x + 5}{2x - 1}$ **19.** $\frac{3x - 5}{4x - 1}$ **21.** $\frac{x - 4}{x - 2}$ **23.** $\frac{x + 5}{x - 3}$ **25.** $\frac{-3}{2}$ **27.** $\frac{-2x - 3}{x + 5}$ **29.** $\frac{4x + 5}{2x - 1}$ **31.** $\frac{-2y - 3}{y + 2}$ **33.** $\frac{a - b}{2a - b}$ **35.** $\frac{4x + 5y}{x + 2y}$ **37.** $\frac{x(x - 2)}{2(x + 3)}$ **39.** $9x^2 - 42x + 49$ **40.** $100x^2 - 81y^2$ **41.** $2x^3 - 9x^2 - 2x + 24$ **42.** $2x^4 - 5x^3 - 10x^2 - 14x - 15$

6.2 Exercises

3. $\frac{x+5}{x+7}$ **5.** $\frac{x^2}{2(x-2)}$ **7.** $\frac{x+4}{x-4}$ **9.** $\frac{(x+8)(x+7)}{7x(x-3)}$ **11.** $\frac{x+1}{(x+2)(x+3)}$ **13.** $\frac{4(x+y)}{3(x-y)}$ **15.** $\frac{y(x+y)}{(x+1)^2}$ **17.** $\frac{-3(x-2)}{(x+5)}$ **19.** $\frac{(x+4)(x-5)}{2x+3}$ **21.** 1 **23.** $\frac{(2x+1)(4x-1)}{(x+1)(x-6)}$ **25.** $\frac{3}{2}$ **26.** $7x^3-22x^2+2x+3$

6.3 Exercises

1. $\frac{2}{3x^3}$ **3.** $\frac{x+8}{x+2}$ **5.** $\frac{-1}{x-4}$ **7.** $\frac{2x-7}{5x+7}$ **9.** $\frac{2x-7}{5x-6}$ **11.** $5a^3$ **13.** $120x^4y^5$ **15.** x^2-16 **17.** $(x+2)(4x-1)^2$ **19.** $\frac{8y+2}{xy^2}$ **21.** $\frac{2x+9}{(x+3)(x+2)}$ **23.** $\frac{4y}{(y+1)(y-1)}$ **25.** $\frac{10a+5b+2ab}{2ab(2a+b)}$ **27.** $\frac{x^2-3x+24}{4x^2}$ **29.** $\frac{6x+20}{(x+5)^2(x-5)}$ **31.** $\frac{2x}{(x+1)(x-1)}$ **33.** $\frac{a+5}{6}$ **35.** $\frac{7x-12}{12x}$ **37.** $\frac{9x+2}{(3x-4)(4x-3)}$ **39.** $\frac{-4x-2}{x(x-2)^2}$ **41.** $\frac{x^2-5x-2}{(x+2)(x+3)(x-1)}$ **43.** $\frac{3a-3b+2ab+a^3-a^2b}{ab(a-b)}$ **45.** $\frac{3x}{x-5}$ **47.** $\frac{3}{y+4}$ **49.** $\frac{3x^2+3x+2}{(x+3)(x-3)}$ **51.** $\frac{(4y+1)(y-1)}{2y(3y+1)(y-3)}$ **53.** $\frac{9}{2(5x-1)}$ **55.** $\frac{-2x^2-x+8}{(x+2)^2(x+3)}$ **57.** $\frac{6x}{y-7}$ **59.** $\frac{9x^2-3x-5}{(x+3)(x-2)(x-1)}$ **60.** -7 **61.** $\frac{5ax+6bc}{2a}$ **62.** $x<\frac{3}{2}$ **63.** $81x^{12}y^{16}$

Putting Your Skills to Work

1. 37.5 miles per hour **2.** $\frac{2xy}{x+y}$

6.4 Exercises

1. $\frac{3y}{2x}$ **3.** $y+x$ **5.** $\frac{1}{xy(x-y)}$ **7.** $\frac{x-3}{x}$ **9.** $\frac{-2x}{3(x+1)}$ **11.** $\frac{3a^2+9}{a^2+2}$ **13.** $\frac{x+2y}{x-y}$ **15.** $\frac{x(x-2)}{x^2+4}$ **17.** $\frac{2x-5}{3(x+3)}$ **19.** $\frac{y+1}{-y+1}$ **21.** $\frac{y-1}{2y^3+2y^2-y-1}$ **23.** $\frac{P-2l}{2}$ **24.** $x>-1$

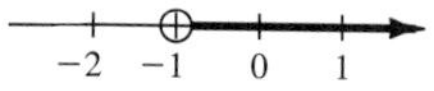

6.5 Exercises

1. 3 **3.** 2 **5.** $\frac{17}{2}$ **7.** -2 **9.** 4 **11.** $\frac{-14}{3}$ **13.** There is no solution. **15.** -3 **17.** -5 **19.** 12 **21.** 2 **23.** There is no solution. **25.** There is no solution. **27.** 3 **29.** $\frac{12}{5}$ **31.** 7 **33.** 2, 4 **34.** $-2, -3$ **35.** $(3x+4)(2x-3)$ **36.** $\frac{-10}{3}$ **37.** width = 7 m, length = 20 m **38.** $-\frac{1}{3}$

6.6 Exercises

1. 15 **3.** $40\frac{4}{5}$ **5.** $\frac{40}{3}$ **7.** $2\frac{1}{2}$ **9.** 114 cups **11.** 19.25 gallons **13.** 504 lb **15.** 56 miles per hour **17.** 46 gallons **19.** 29 miles **21.** $18\frac{17}{20}$ inches **23.** $114\frac{2}{7}$ meters **25.** 56 feet tall **27.** 86 meters **29.** Boat A traveled at 80 kilometers per hour. Boat B traveled at 70 kilometers per hour. **31.** Helicopter B travels at 250 kilometers per hour. Helicopter A travels at 210 kilometers per hour. **33.** $2\frac{2}{9}$ hours or 2 hours 13 minutes **35.** $8\frac{4}{7}$ hours or 8 hours 34 minutes **37.** 6.316×10^{-7} **38.** 582,000,000 **39.** $\frac{w^8}{x^3y^2z^4}$ **40.** $\frac{27}{8}$ or $3\frac{3}{8}$

Putting Your Skills to Work

1. 3.75 ohms **2.** 3 ohms

Chapter 6 Review Problems

1. $-\frac{4}{5}$ **2.** $\frac{x}{x-y}$ **3.** $\frac{x+3}{x-4}$ **4.** $\frac{2x+1}{x+1}$ **5.** $\frac{x+3}{x-7}$ **6.** $\frac{2(x+4)}{3}$ **7.** $\frac{2x+3}{2x}$ **8.** $\frac{x}{x-5}$ **9.** $\frac{2(x-4y)}{2x-y)}$ **10.** $\frac{2-y}{3y-1}$ **11.** $\frac{x-2}{5x^2+x-6}$ **12.** $4x+2y$ **13.** $\frac{x-25}{x+5}$ **14.** $\frac{x-4}{4}$ **15.** $\frac{2(x+3)(x-3)}{3y(y+3)}$ **16.** $\frac{(y+2)(2y+1)(y-2)}{(y+1)(2y-1)(y+1)}$ **17.** $\frac{3(x+2)}{x(x+5)}$ **18.** $\frac{4}{5(x-5)}$ **19.** $\frac{4y(3y-1)}{(3y+1)(2y+5)}$ **20.** $\frac{3y(4x+3)}{2(x-5)}$ **21.** $\frac{(x+2)}{2}$ **22.** $\frac{3}{16}$ **23.** $\frac{1}{x(4x-3)}$ **24.** $\frac{2(x+3y)}{x-4y}$ **25.** $\frac{9x+2}{x(x+1)}$ **26.** $\frac{5x^2+7x+1}{x(x+1)}$ **27.** $\frac{(x-1)(x-2)}{(x+3)(x-3)}$ **28.** $\frac{10x-22}{(x+2)(x-4)}$ **29.** $\frac{2xy+4x+5y+6}{2y(y+2)}$ **30.** $\frac{(2a+b)(a+4b)}{ab(a+b)}$ **31.** $\frac{3x-2}{3x}$ **32.** $\frac{2x^2+7x-2}{2x(x+2)}$ **33.** $\frac{1-2x-x^2}{(x+5)(x+2)}$ **34.** $\frac{7+15x-3x^2}{(x-4)(x-5)}$ **35.** $\frac{3(2x-1)(x+1)}{(x+2)(x+3)(x-3)}$ **36.** $\frac{x^2+13x+1}{(x+2)(x-1)(x+4)}$

37. $\frac{1}{11}$ **38.** $\frac{5}{3x^2}$ **39.** $w - 2$ **40.** 1 **41.** $-\frac{y^2}{2}$ **42.** $\frac{x + 2y}{y(x + y + 2)}$ **43.** $\frac{-1}{a(a + b)}$ **44.** $\frac{-3a - b}{b}$ **45.** $\frac{-3y}{2(x + 2y)}$ **46.** $\frac{5y(x + 5y)^2}{x(x - 6y)}$ **47.** 15 **48.** 2 **49.** $\frac{2}{7}$ **50.** -8 **51.** -4 **52.** -2 **53.** $\frac{1}{5}$ **54.** $\frac{1}{2}$ **55.** $-4, 6$ **56.** no solution **57.** 2 **58.** 6 **59.** no solution **60.** -2 **61.** 9 **62.** 0 **63.** 1.3 **64.** 2.8 **65.** 26.4 **66.** 2 **67.** 11 hours **68.** 40 miles per hour **69.** 8.3 gallons **70.** 167 cookies **71.** 46 gallons **72.** 91.5 miles **73.** train: 60 miles per hour; car: 40 miles per hour **74.** 3 hours 5 minutes **75.** 1200 feet tall **76.** 182 feet tall

Chapter 6 Test

1. $\frac{2}{3a}$ **2.** $\frac{2x^2(2 - y)}{(y + 2)}$ **3.** $\frac{5}{12}$ **4.** $\frac{1}{3b^2}$ **5.** $\frac{(a - 2)(4a + 7)}{(a + 2)^2}$ **6.** $\frac{23x - 64}{2(x - 2)(x - 3)}$ **7.** $\frac{x - a}{ax}$ **8.** $-\frac{x + 2}{x + 3}$ **9.** $\frac{x}{4}$ **10.** $\frac{x + 1}{x(x + 3)^2}$ **11.** $\frac{2x - 3y}{4x + y}$ **12.** $\frac{x}{(x + 2)(x + 4)}$ **13.** 3 **14.** 4 **15.** no solution **16.** $x = \frac{47}{6}$ **17.** $x = \frac{45}{13}$ **18.** 6 hours, 25 minutes **19.** \$368 **20.** 102 feet

Cumulative Test for Chapters 0–6

1. 0.006 cm **2.** \$247 **3.** \$49.98 **4.** 6 **5.** $\frac{A}{\pi r^2}$ **6.** $x > \frac{5}{4}$

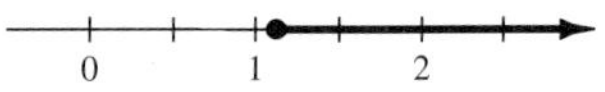

7. $x > 5$

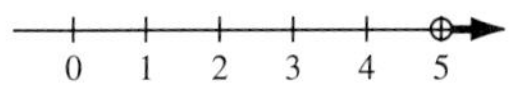

8. $(a + b)(3x - 2y)$ **9.** $2a(4a + b)(a - 5b)$ **10.** $-12x^5y^7$ **11.** $\frac{2x + 5}{x + 7}$ **12.** $3x + 1$ **13.** $\frac{11x - 3}{2(x + 2)(x - 3)}$ **14.** $\frac{x^2 + 2x + 6}{(x + 4)(x - 3)(x + 2)}$ **15.** -2 **16.** $\frac{-9}{2}$ **17.** $\frac{x + 8}{6x(x - 3)}$ **18.** $\frac{3ab^2 + 2a^2b}{5b^2 - 2a^2}$ **19.** $\frac{98}{3}$ **20.** 208 miles **21.** 484 phone calls

Chapter 7

Pretest Chapter 7

1.

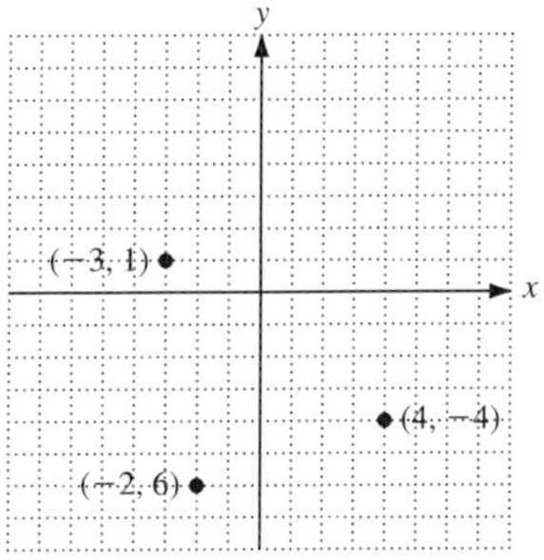

2. $A\ (-1, -2)$; $B\ (3, -2)$; $C\ (-1, 2)$
3. **(a)** $(-1, 10)$ **(b)** $(2, 1)$

4.

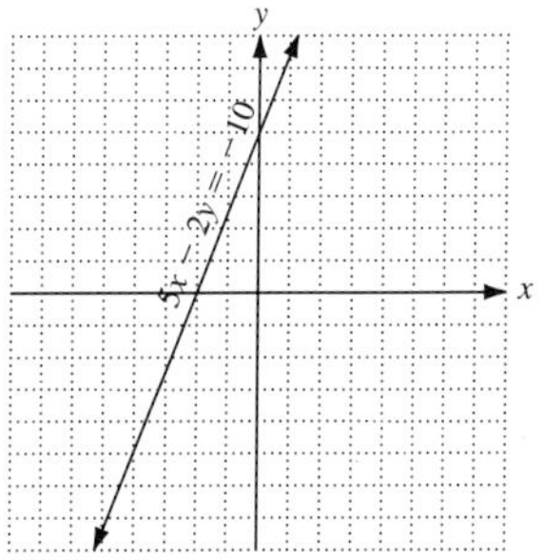

5.

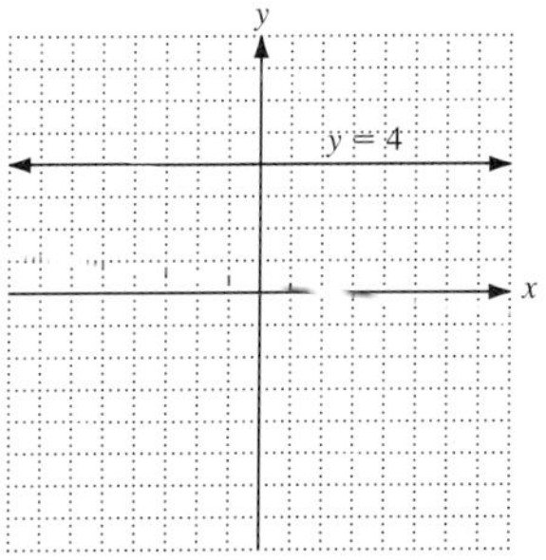

6. $\frac{2}{7}$ **7.** $\frac{4}{3}$
8. **(a)** $y = 4x - 5$ **(b)** $4x - y = 5$

9.

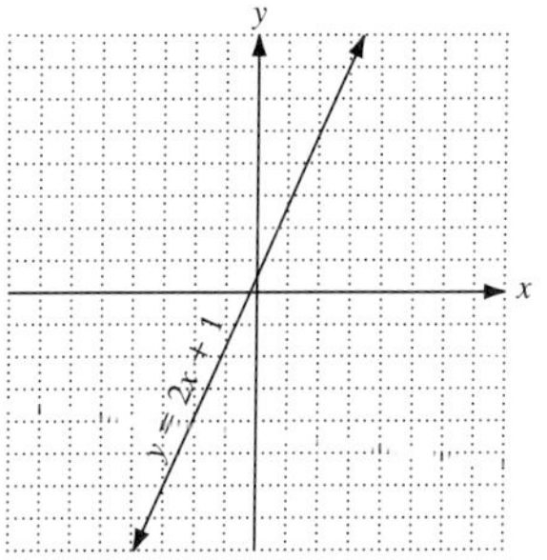

10. $y = -\frac{3}{2}x + 2$
11. **(a)** $-\frac{1}{4}$ **(b)** 4
12. $y = \frac{2}{3}x - 1$
13. $y = -x + 5$

14.

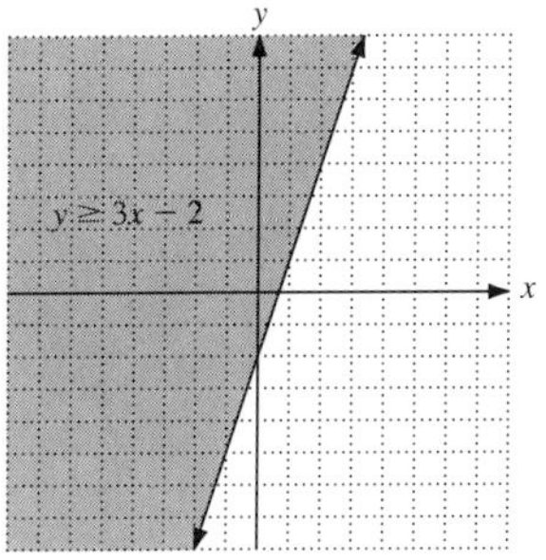

15.

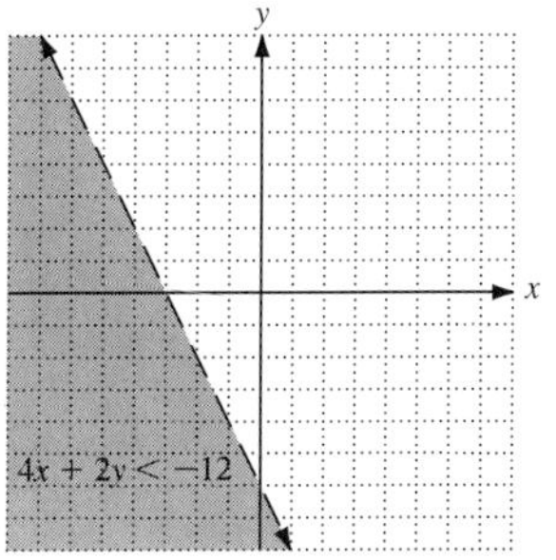

16. yes **17.** no **18.** **(a)** -13 **(b)** 5 **19.** **(a)** 72 **(b)** $\frac{1}{2}$

7.1 Exercises

1.

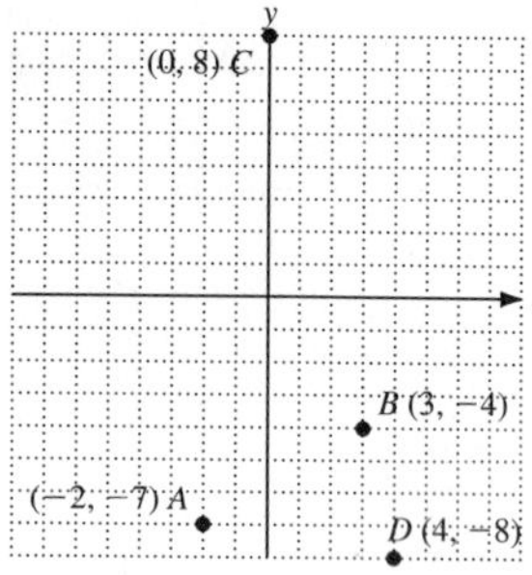

3.

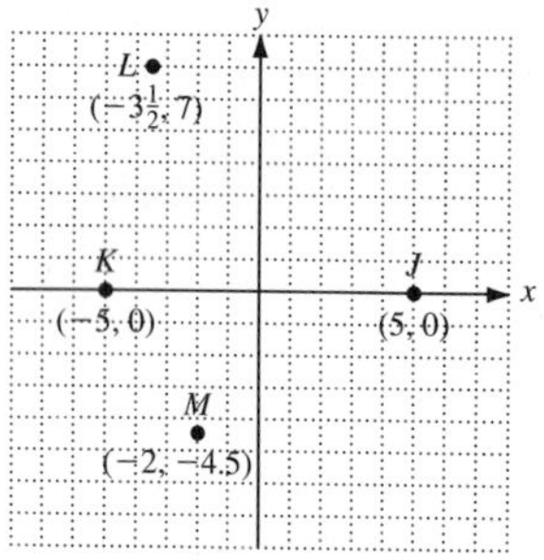

5. R: $(-3, -5)$; S: $\left(-4\frac{1}{2}, 0\right)$; X: $(3, -5)$, Y: $\left(2\frac{1}{2}, 6\right)$

13. **(a)** $(0, 8)$ **(b)** $(4, 20)$

15. **(a)** $(-6, 15)$ **(b)** $(3, -3)$

17. **(a)** $(8, 0)$ **(b)** $(-12, -5)$

19. **(a)** $(2, -15)$ **(b)** $(-2, -5)$

21. $\frac{8x + 8y}{x + y} = 8$

22. $\frac{(x + y)(x - y)}{(x + y)(x + 3y)} \cdot \frac{(x + 3y)(x - 3y)}{(x - 3y)(x - 3y)} = \frac{x - y}{x - 3y}$

23. Use $\pi = 3.14$.) 1256 square yards

24. \$24,000

7.2 Exercises

5. $(-2, 1)$, $(-1, -1)$, $(0, -3)$

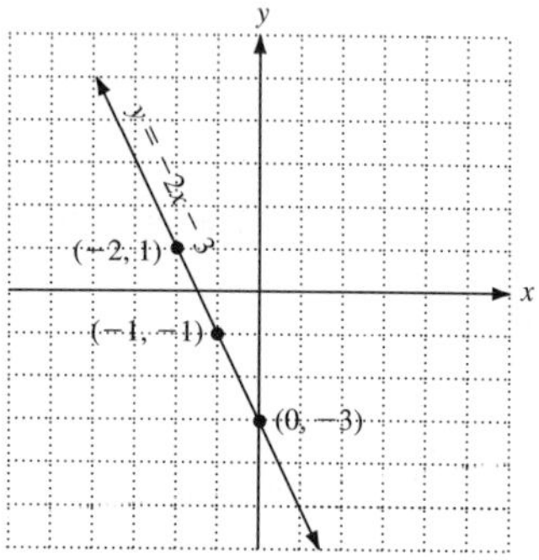

7. $(0, -4)$, $(2, -2)$, $(4, 0)$

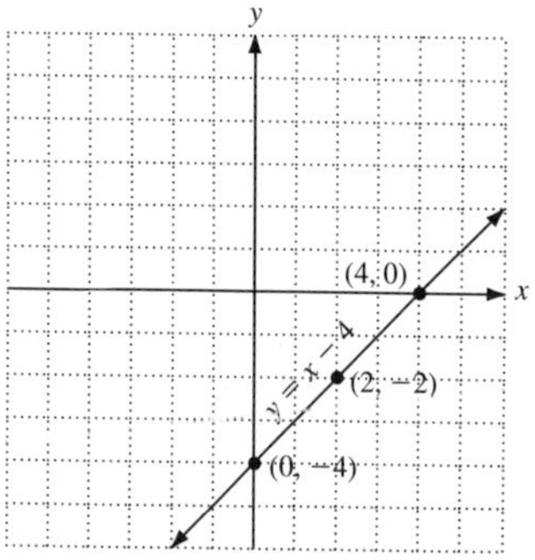

9.

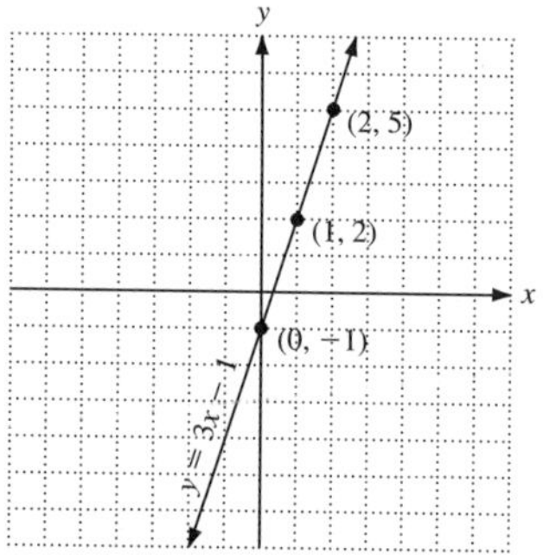

11.

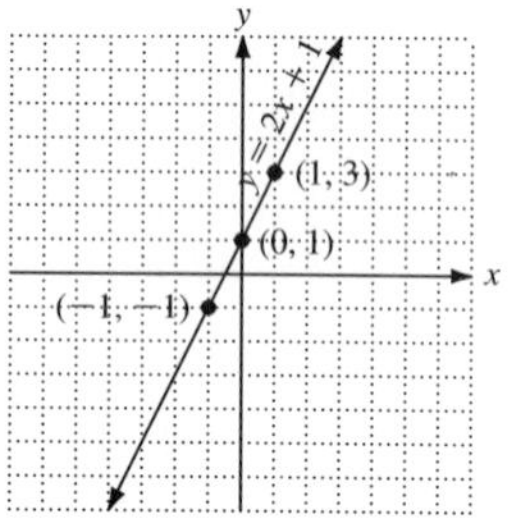

13.

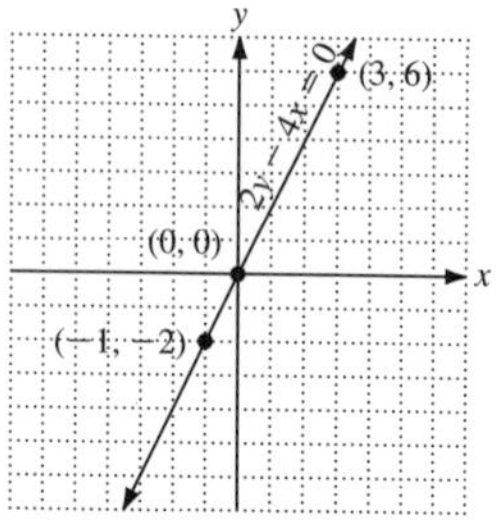

15.

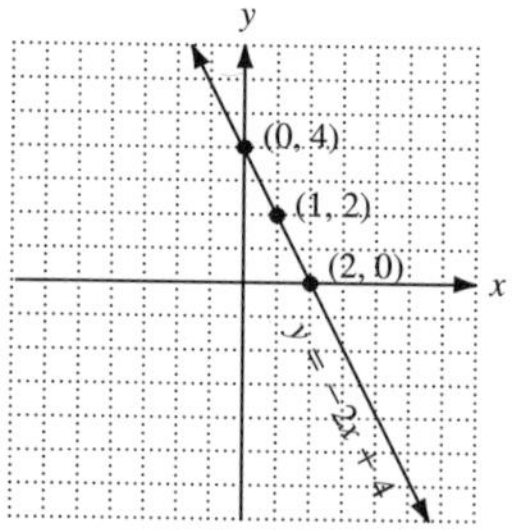

17.

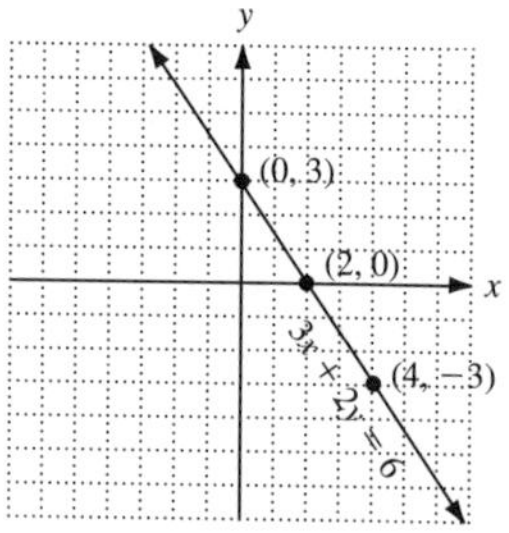

19.

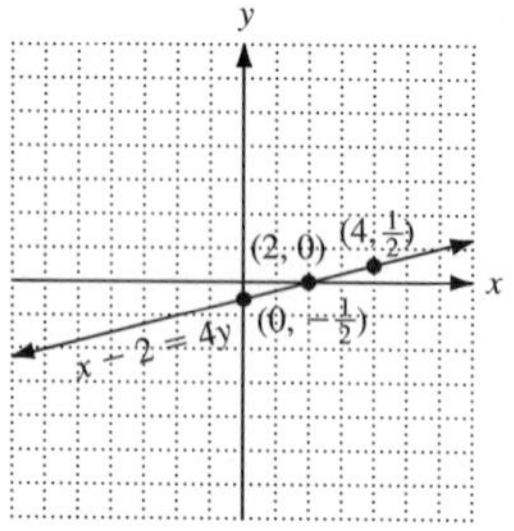

21.

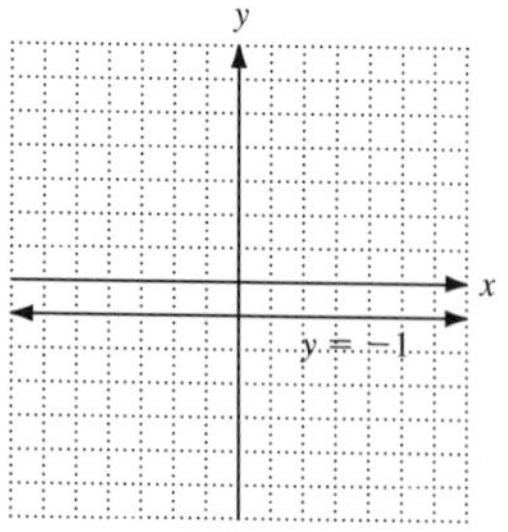

23. $x = 28$

24. $x \geq -\frac{14}{3}$

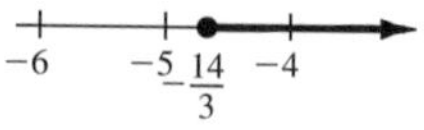

25. $L = 17$ meters; $W = 9\frac{1}{2}$ meters

26. 212 faculty members

7.3 Exercises

1. -6 **3.** $\frac{3}{5}$ **5.** 2 **7.** 0 **9.** $-\frac{4}{3}$ **11.** -9 **15.** $m = 8$; $b = 9$ **17.** $m = 3$; $b = -7$ **19.** $m = \frac{5}{6}$; $b = -\frac{2}{9}$ **21.** $m = -6$; $b = 0$ **23.** $m = -4$; $b = -\frac{1}{3}$ **25.** $m = 3$; $b = +2$ **27.** $m = -\frac{5}{2}$; $b = \frac{3}{2}$ **29.** $m = \frac{4}{3}$; $b = -\frac{7}{3}$ **31.** **(a)** $y = \frac{3}{4}x + 2$ **(b)** $3x - 4y = -8$ **33.** **(a)** $y = 6x - 3$ **(b)** $6x - y = 3$ **35.** **(a)** $y = \frac{-5}{2}x + 6$ **(b)** $5x + 2y = 12$ **37.** **(a)** $y = -2x + \frac{1}{2}$ **(b)** $4x + 2y = 1$

39.

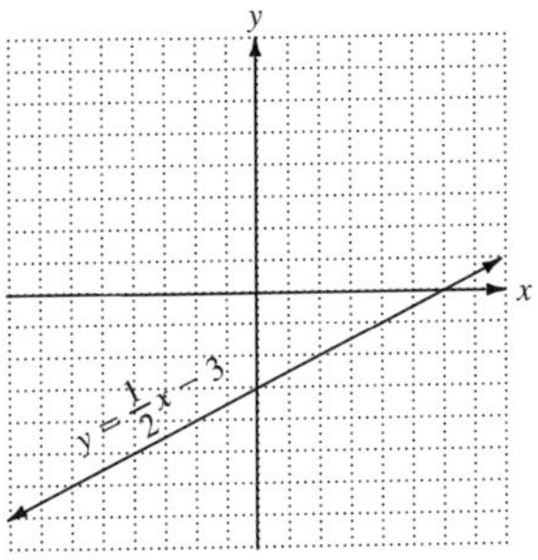

41.

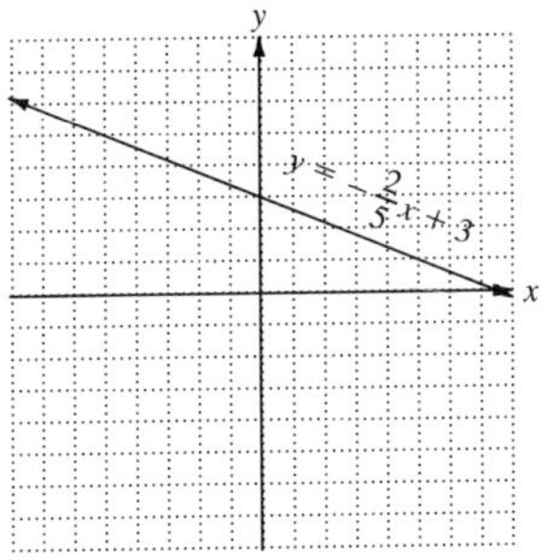

43.

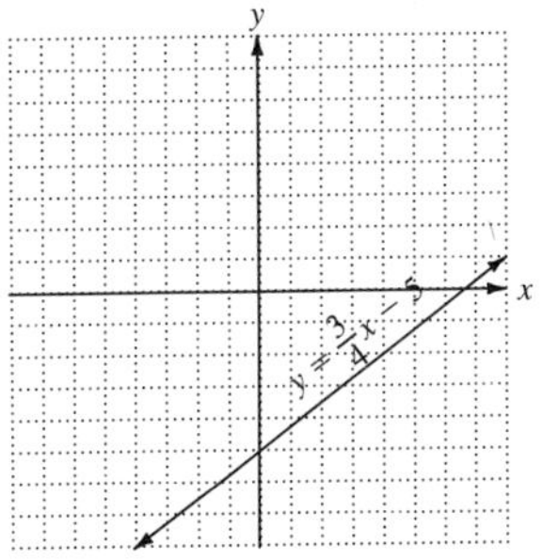

45.

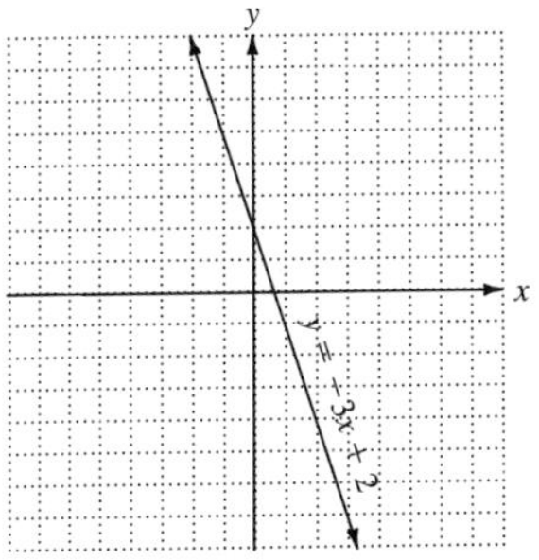

47.

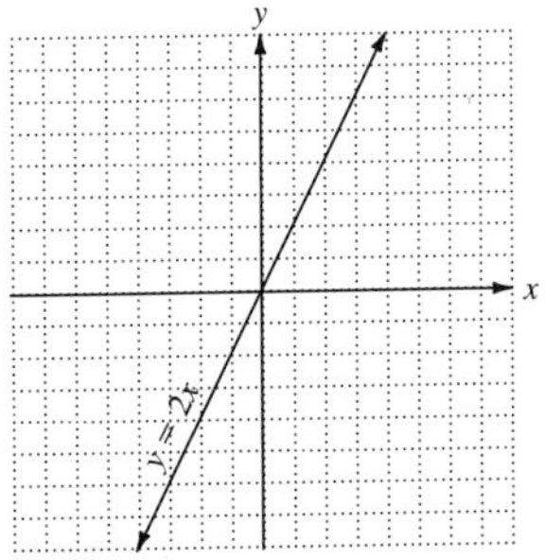

49. **(a)** $\frac{13}{5}$ **(b)** $-\frac{5}{13}$

51. **(a)** $\frac{1}{8}$ **(b)** -8

53. **(a)** -5 **(b)** $\frac{1}{5}$

55. **(a)** $-\frac{1}{2}$ **(b)** 2

56. $x > 4$ (number line: 3, 4, 5, 6, 7)

57. $x < \frac{12}{5}$ (number line: $\frac{11}{5}$, $\frac{12}{5}$, $\frac{13}{5}$, $\frac{14}{5}$, 3)

58. $x \le 24$ (number line: 22, 23, 24, 25, 26)

59. $x \le -22$ (number line: −25, −24, −23, −22, −21)

Putting Your Skills to Work

1. 25 pounds per square inch
2. 44 feet
3. $\frac{5}{11}$

4.

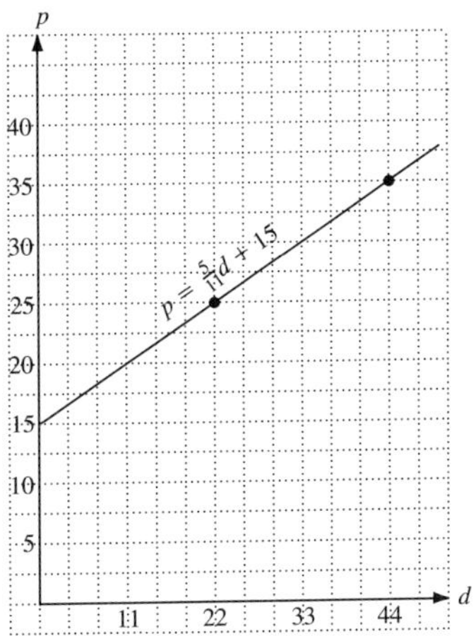

5. $\frac{5}{11}$; yes. You can determine the slope of a line using the coordinates of two points, from the equation in slope-intercept form, from the graph by counting.

6.

d	0	11	22	33	44
p	15	20	25	30	35

(a) 5 **(b)** 11 **(c)** Slope is the difference between the p values over the difference between the d values.

7.4 Exercises

1. $6x + 20$ **3.** $-2x + 11$ **5.** $\frac{2}{3}x$ **7.** $-3x + \frac{7}{2}$ **9.** $\frac{3}{4}x - 5$ **11.** -4 **13.** $y = 2x + 3$ **15.** $y = -3x + 1$ **17.** $y = -3x - 4$ **19.** $y = 5x - 10$ **21.** $y = 9x - 12$ **23.** $y = \frac{1}{3}x + \frac{1}{2}$ **25.** $y = -\frac{2}{3}x + 1$ **27.** $y = \frac{2}{3}x - 4$ **29.** $y = 3x$ **31.** $y = -2$ **33.** $x = 6$ **34.** $y = 5$ **35.** $x = -5$ **36.** $y = -3$ **37.** $y = 2$ **38.** $x = -1$ **39.** $y = 2x + 3$ **40.** $y = -\frac{1}{2}x + 3$ **41.** $y = 3x + 2$ **42.** $y = -\frac{1}{3}x + 1$ **43.** $\frac{24}{5}$ **44.** $\frac{1}{(2x - 5)^2}$ **45.** 1.03% **46.** \$3000 is invested in the risky venture.

7.5 Exercises

3.

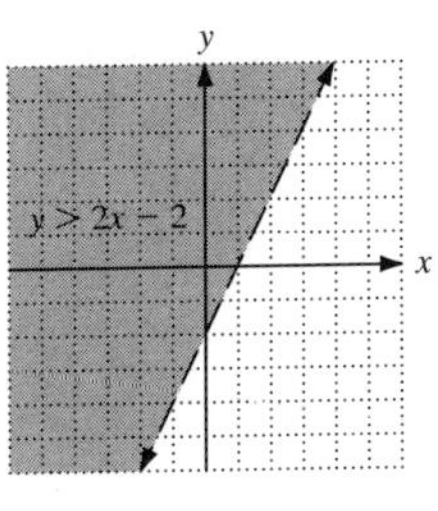

5.

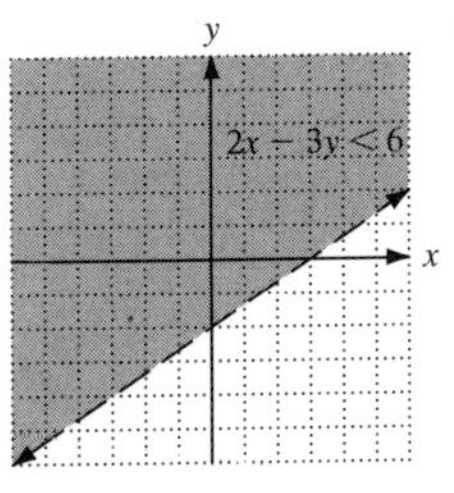

7.

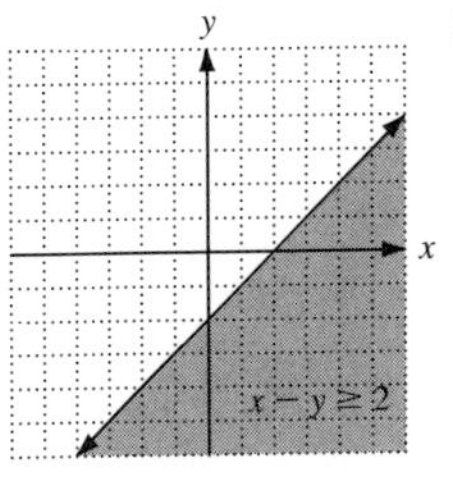

9.

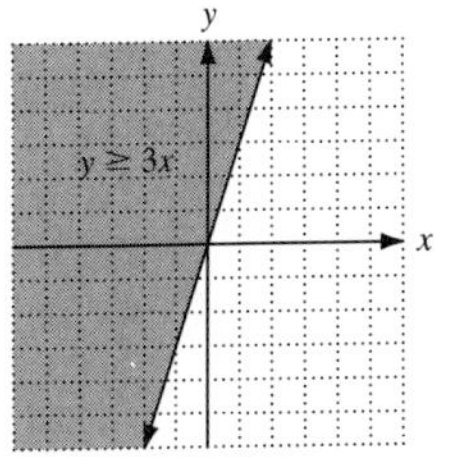

11.

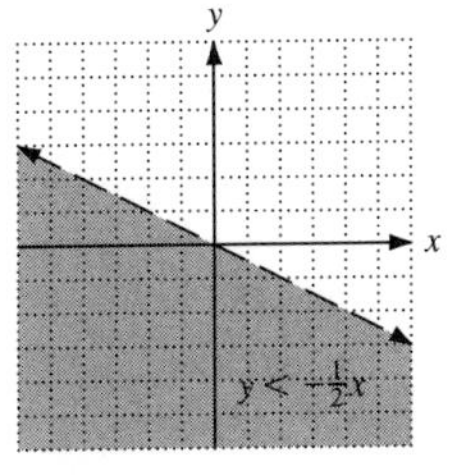

13.
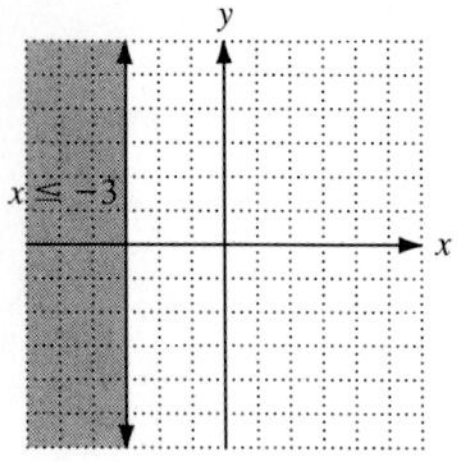

15.
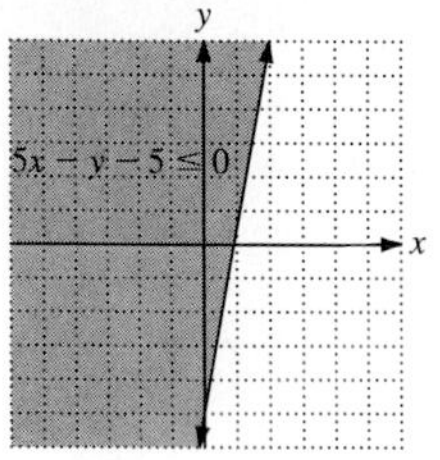

17.
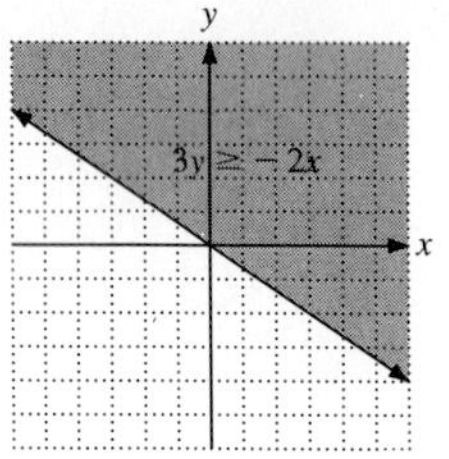

19.
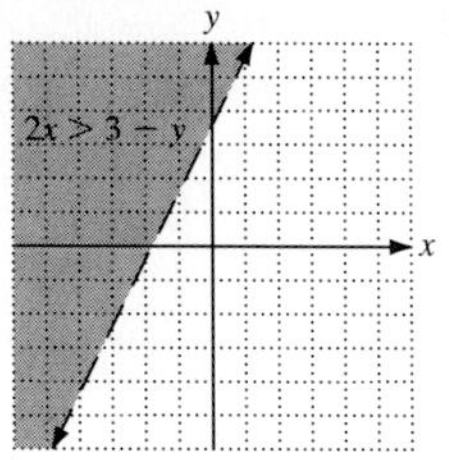

21.
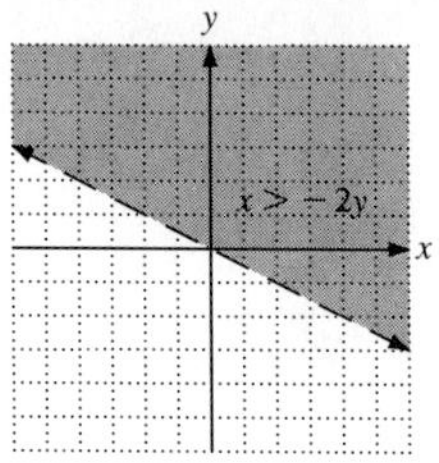

23. **(a)** $x < 3$ **(b)** $x < 3$ **(c)** $3x + 3$ will be greater than $5x - 3$ for those values of x where the graph of $y = 3x + 3$ lies *above* the graph of $y = 5x - 3$. **24.** $y = -\frac{7}{3}x + 7$ **25.** $x = -\frac{3}{2}, x = -2$ **26.** 4 **27.** 83 is largest.

7.6 Exercises

7. Domain = {2, 6, 9}, Range = {2, 6, 8, 9}, Not a function **9.** Domain = $\{\frac{1}{4}, \frac{1}{2}, \frac{3}{4}\}$, Range = {5, 6, 10}, Not a function
11. Domain = {1, 3, 8, 10}, Range = {1, 3, 8, 10}, Function **13.** Domain = {5, 5.6, 5.8, 6}, Range = {5.8, 6, 8}, Function
15. Domain = {30, 40, 50}, Range = {2, 30, 60, 80}, Not a function

17.
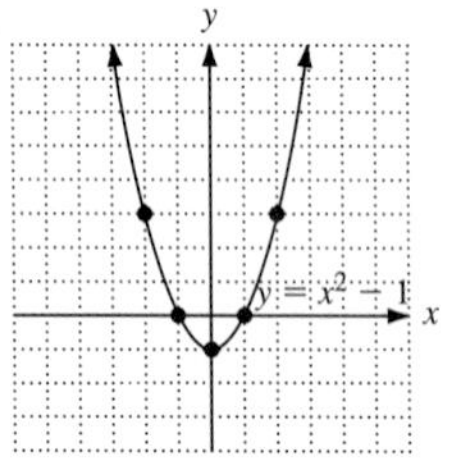

19.
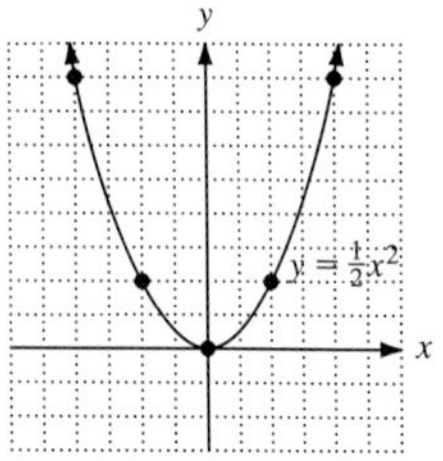

21.
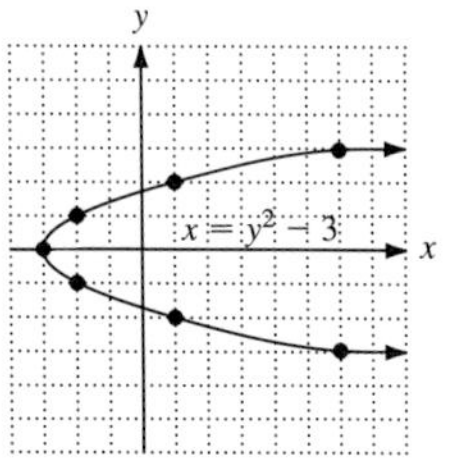

23.
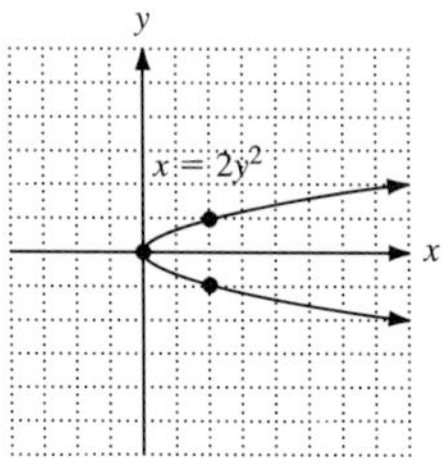

25.
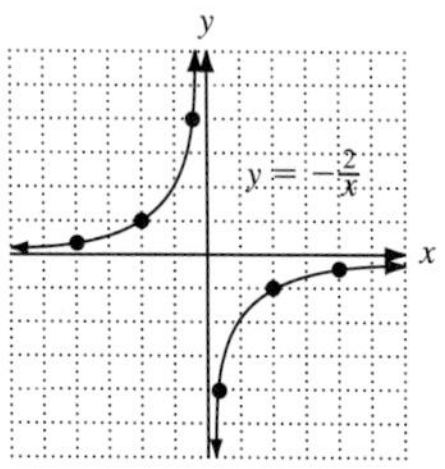

27.
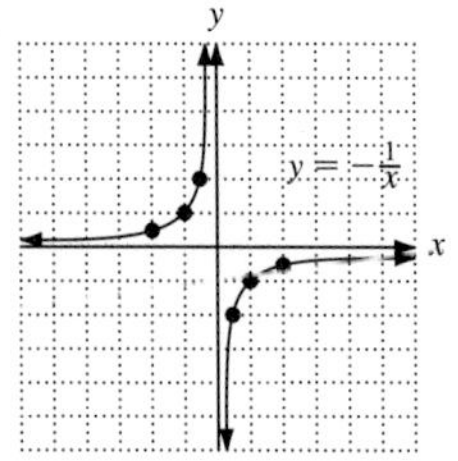

29.
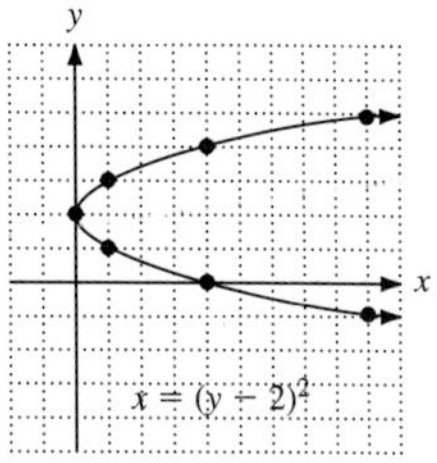

31.
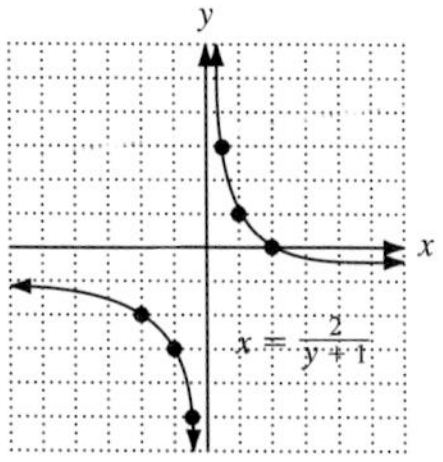

33. Function **35.** Not a function **37.** Function **39.** Not a function **41.** **(a)** −4 **(b)** −13 **(c)** 2 **43.** **(a)** −4 **(b)** 2 **(c)** 29
45. **(a)** −4 **(b)** −3 **(c)** 12 **47.** **(a)** 3 **(b)** −13 **(c)** −33 **49.** **(a)** −1 **(b)** −4 **(c)** $-2\frac{1}{2}$ **50.** $-2x^2 - 14x + 6$ **51.** $-3x^2 - 7x + 5$
52. $2x^2 - x + 3$ **53.** $2x + 4 + \frac{-1}{3x - 1}$

Putting Your Skills to Work

1. \$14.25; \$151,050; \$14.00; \$151,200; \$13.75; \$151,250; \$13.50; \$151,200; Yes. Monthly income is the dependent variable. If either number of subscribers or cost per month is the independent variable, no two ordered pairs in the table have the same first coordinate. **2.** \$13.75
3. The curve goes up and then down. The maximum monthly income is the highest point on the curve.

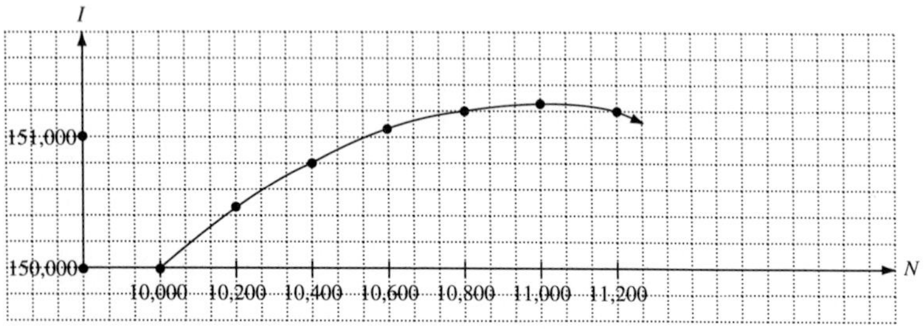

Chapter 7 Review Problems

1.

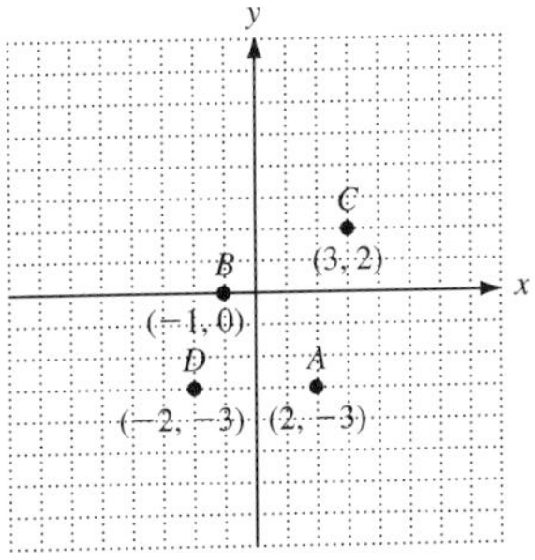

2.

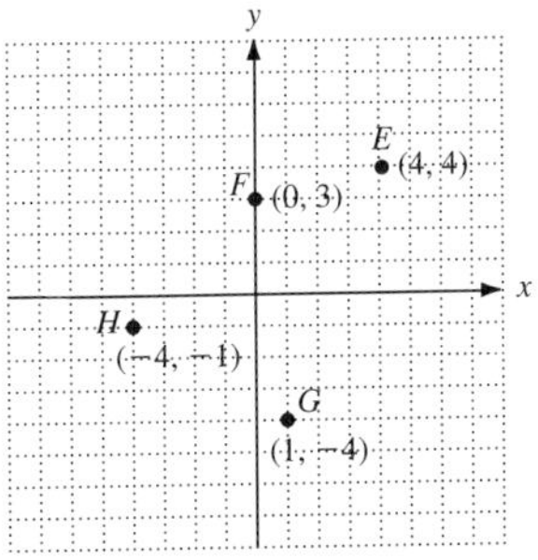

3. **(a)** (0, −5) **(b)** (3, 4)
4. **(a)** (1, 2) **(b)** (−4, 4)
5. **(a)** (6, −1) **(b)** (6, 3)

6.

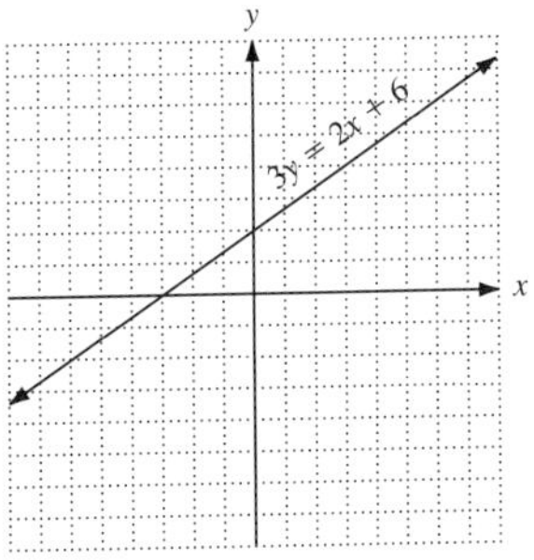

7.

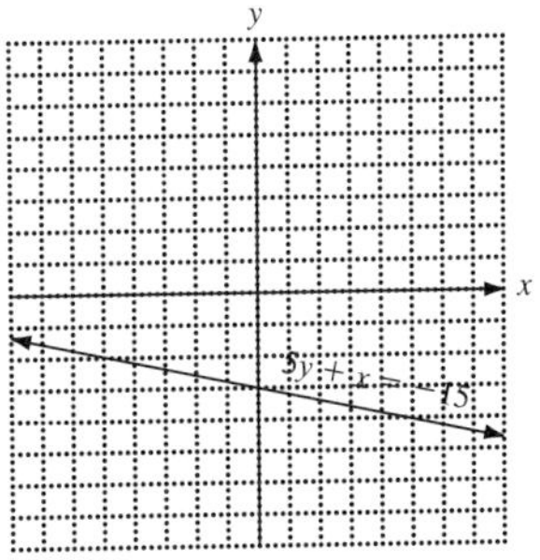

8.

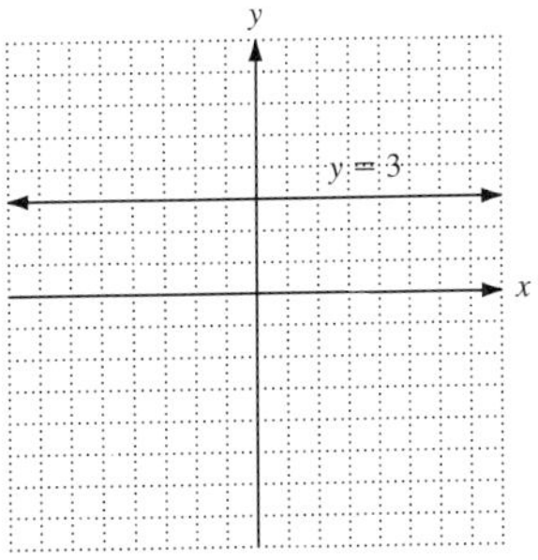

9. $-\frac{5}{6}$ **10.** $\frac{5}{8}, \frac{3}{2}$ **11.** $y = -\frac{1}{2}x + 3$ **12.** $y = 7$ **13.** $\frac{3}{2}$ **14.** $\frac{1}{7}$

15.

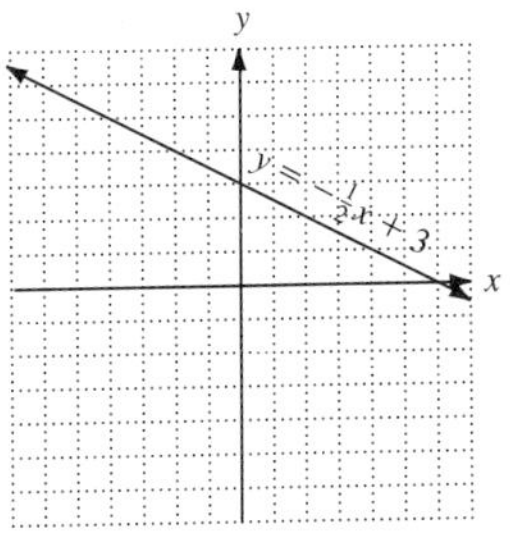

16.

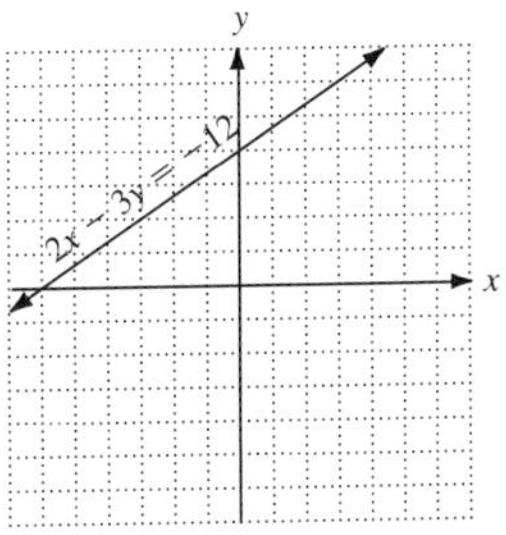

17.

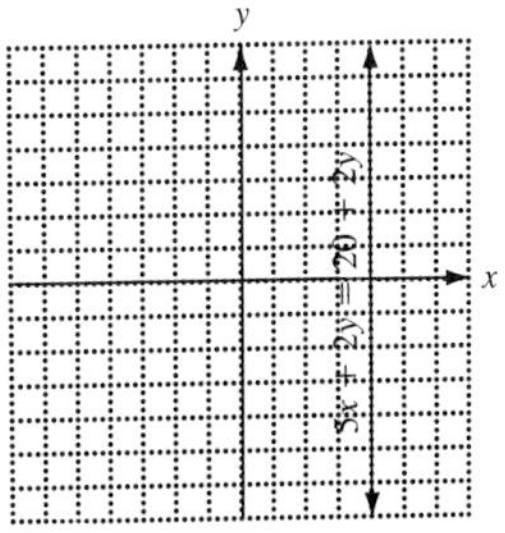

18. $y = 2x - 4$ **19.** $y = -6x + 14$ **20.** $y = \frac{1}{3}x - 5$ **21.** $y = -\frac{1}{2}x + \frac{9}{2}$ **22.** $y = x - 5$ **23.** $y = 5$ **24.** $y = \frac{2}{3}x - 3$
25. $y = -3x + 1$ **26.** $x = -5$

27.

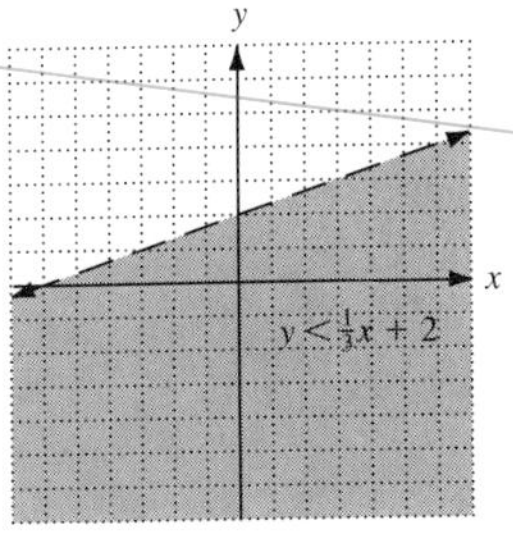

28.

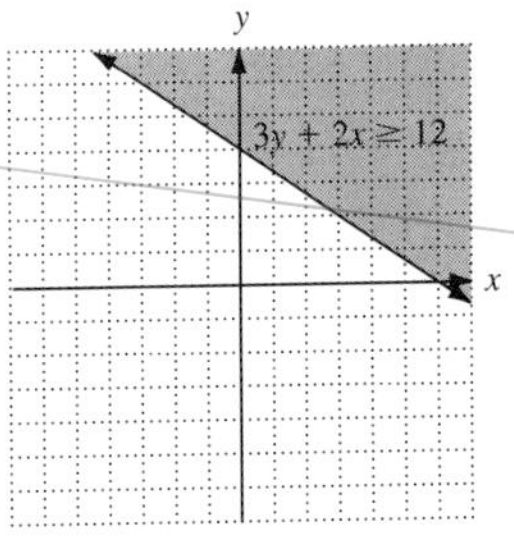

29.

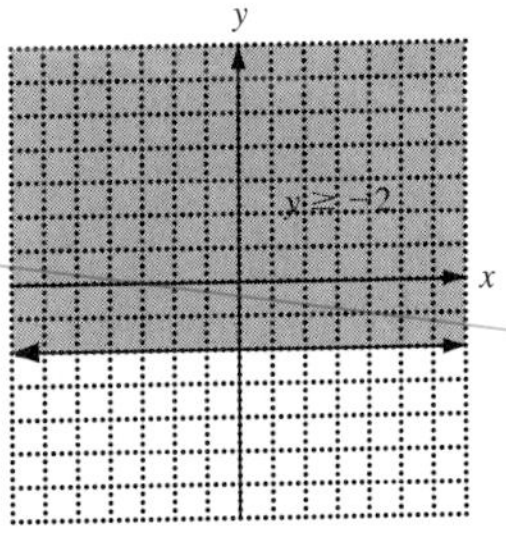

30. Domain: {−3, 0, 1, 6}, Range: {−2, 4, 7}, function **31.** Domain: {0, 1, 2}, Range: {6, 4, 8), function **32.** Domain: {5, −6, −5}, Range: {−6, 5}, not a function **33.** Domain: {3, −7, −3, 7}, Range: {−7, −3, 3, 7}, function **34.** function **35.** not a function **36.** function

37. 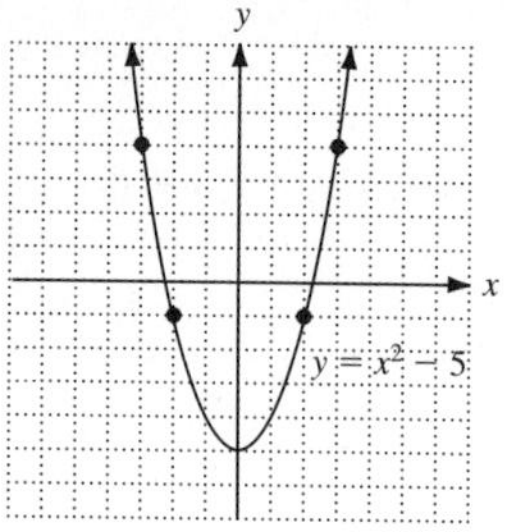

38. 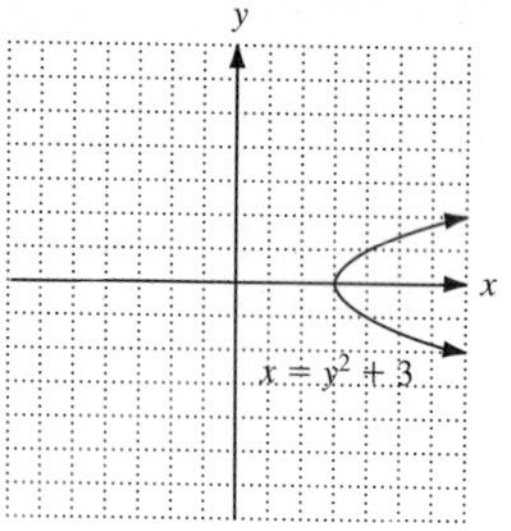

39. 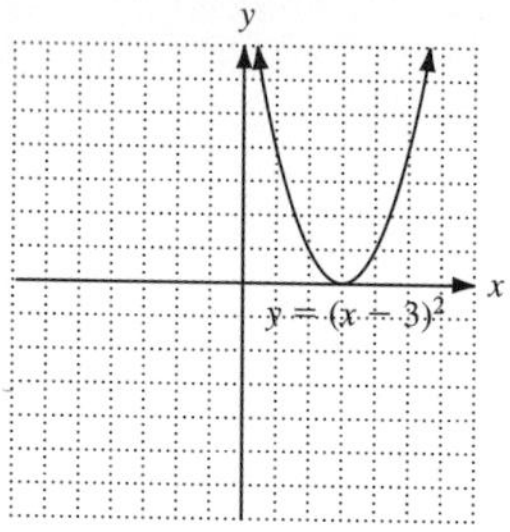

40. **(a)** 7 **(b)** 19 **41.** **(a)** 9 **(b)** 24 **42.** **(a)** -1 **(b)** 2 **43.** **(a)** 31 **(b)** 59 **44.** **(a)** 1 **(b)** -6 **45.** **(a)** 1 **(b)** $\frac{1}{5}$

Chapter 7 Test

1. 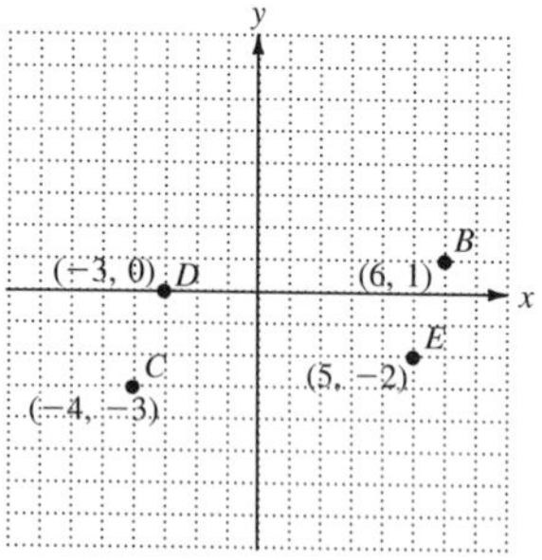

2. 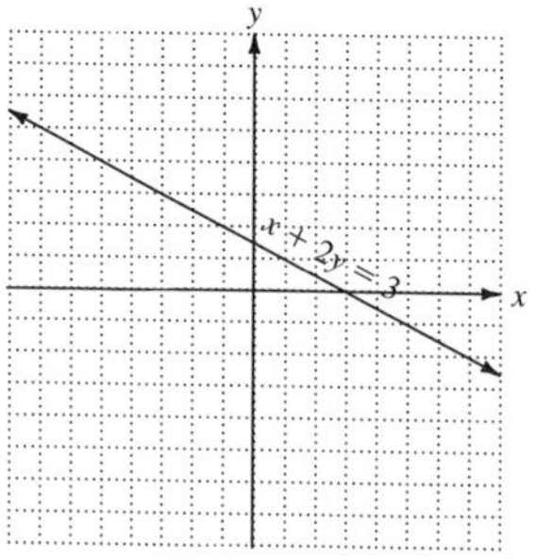

3. 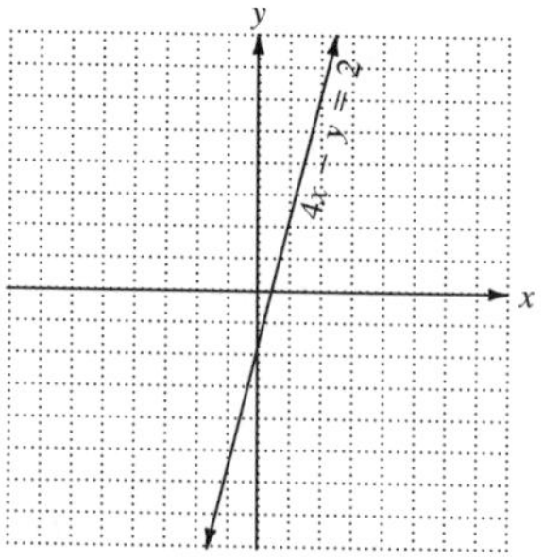

4. 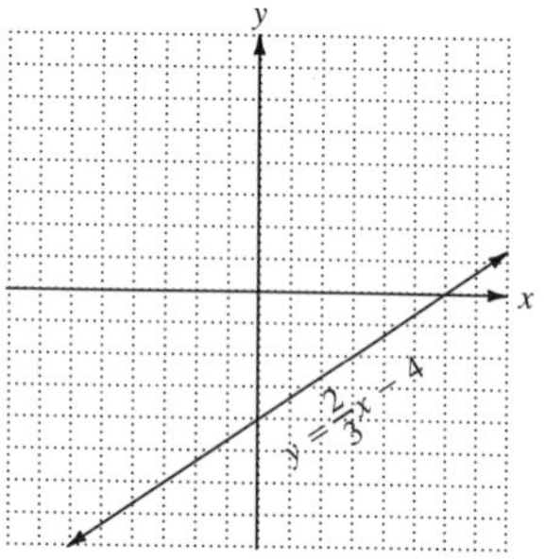

5. $-\frac{3}{2}, \frac{5}{2}$ **6.** $\frac{1}{2}$ **7.** $x - 2y = 8$ **8.** 0 **9.** $x + 3y = 17$ **10.** $x = 2$, no slope

11. 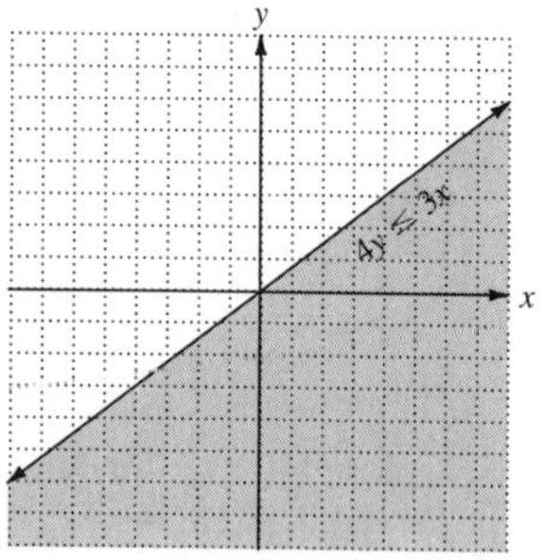

12. 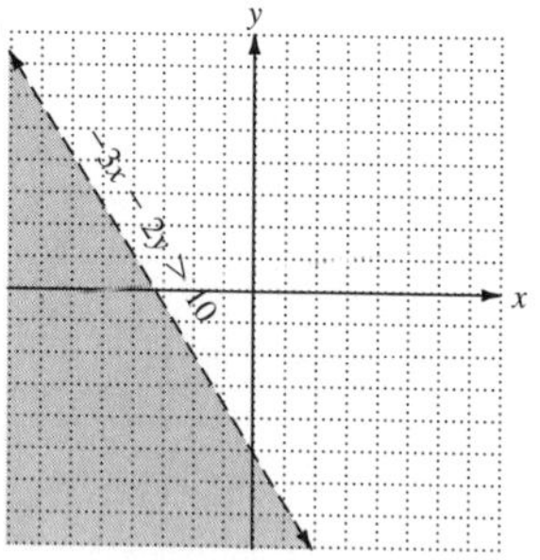

13. Yes. No two different ordered pairs have the same first coordinate. **14.** Yes. A vertical line passes through every point on the graph once.

15. 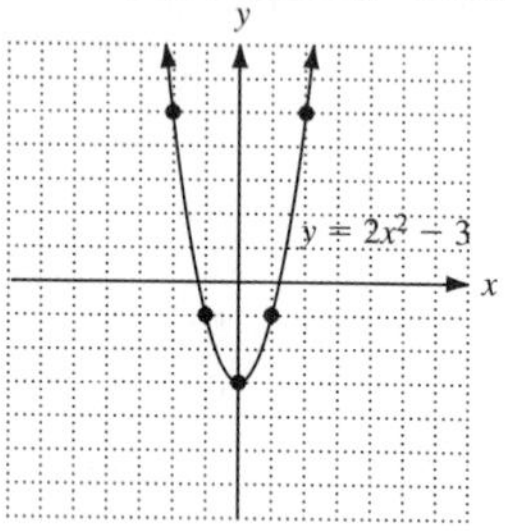

16. **(a)** -4 **(b)** -2 **17.** **(a)** -3 **(b)** $-\frac{3}{4}$

Cumulative Test for Chapters 0–7

1. $-27x^9y^{12}z^3$ **2.** $-\dfrac{3a^2}{4b^2}$ **3.** 7 **4.** 40 **5.** 5 **6.** $w = \dfrac{A}{l} - h$ **7.** $2(5a + 7b)(5a - 7b)$ **8.** $(2x + 5)(x + 4)$ **9.** $2(x - 1)$

10. $3x - y = 10$ **11.** $x = 7$ **12.** $x - 3y = -11$ **13.** 0 **14.** $\frac{3}{7}$

15. 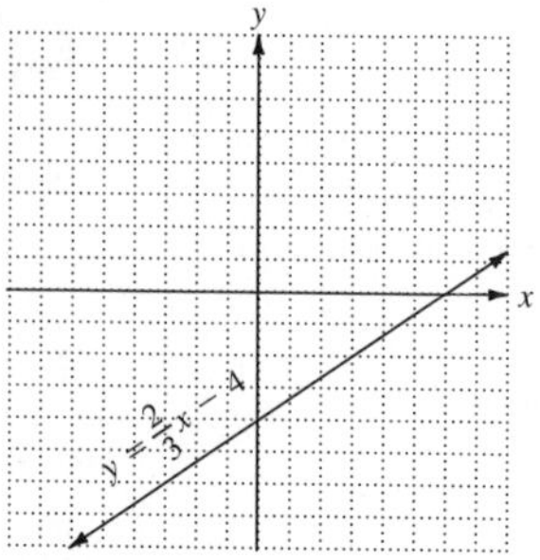

16. 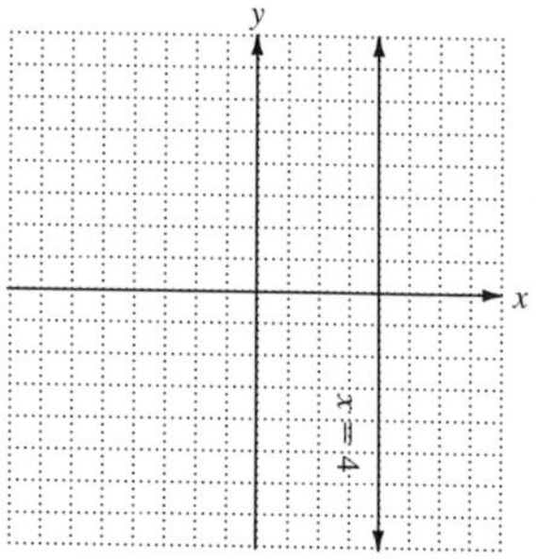

17. 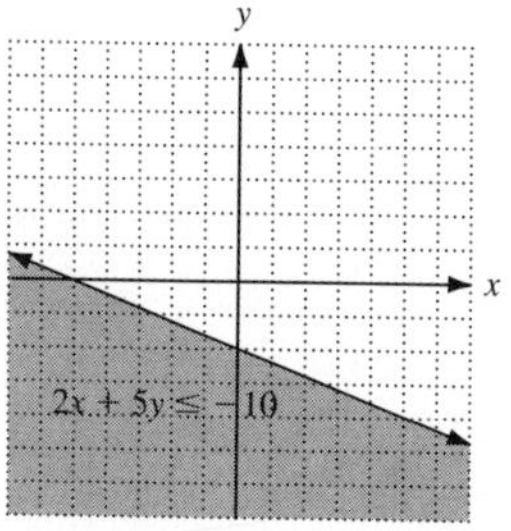

18. No

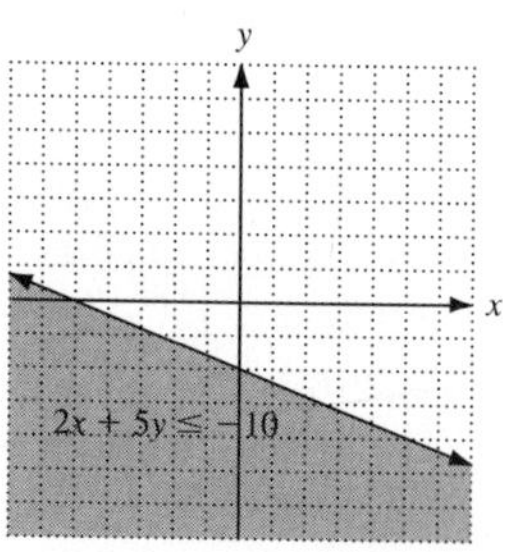

19. Yes **20.** **(a)** 5 **(b)** 3

Chapter 8

Pretest Chapter 8

1. $x = -3, y = 2$

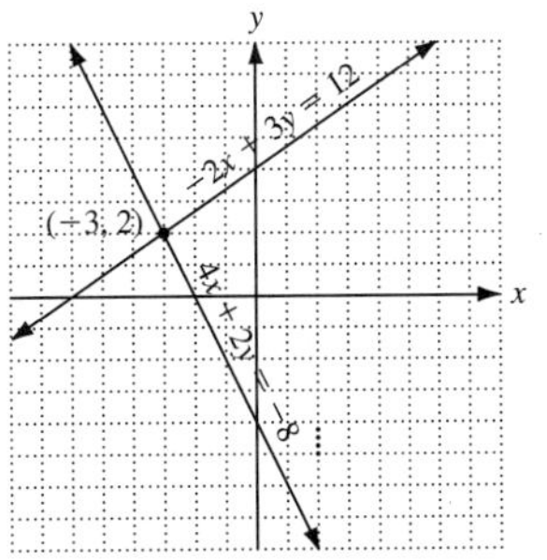

2. no solution

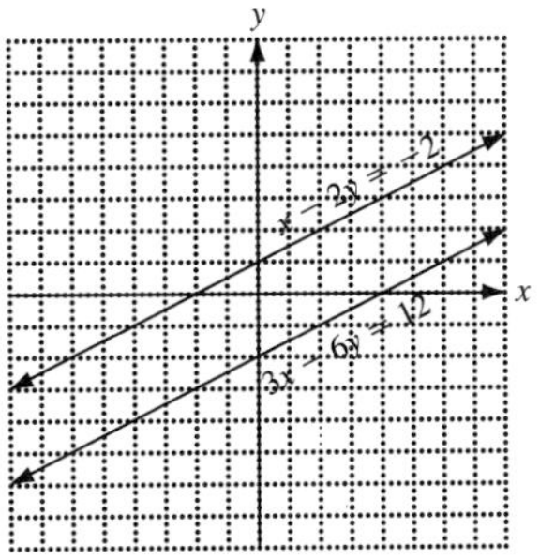

3. $x = -3, y = -2$ **4.** $x = \frac{3}{5}, y = \frac{3}{5}$ **5.** $x = 2, y = 0$ **6.** $x = 6, y = 2$ **7.** $x = 2, y = 3$ **8.** intersect **9.** are parallel **10.** coincide **11.** $x = 3, y = \frac{1}{2}$ **12.** $x = 0, y = -5$ **13.** $x = 2, y = -1$ **14.** James = 290, Michael = 690

8.1 Exercises

1.

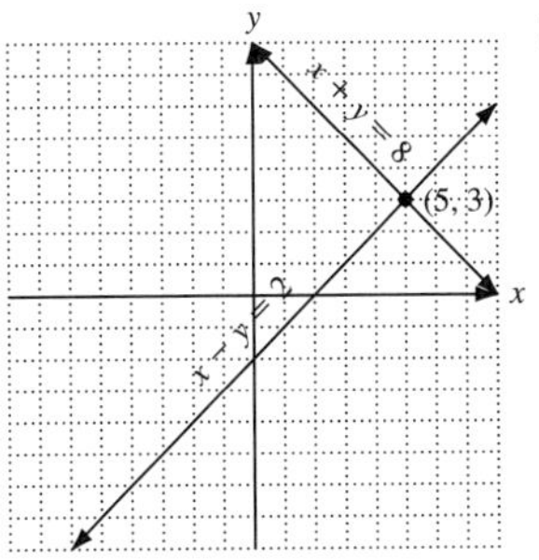

3.

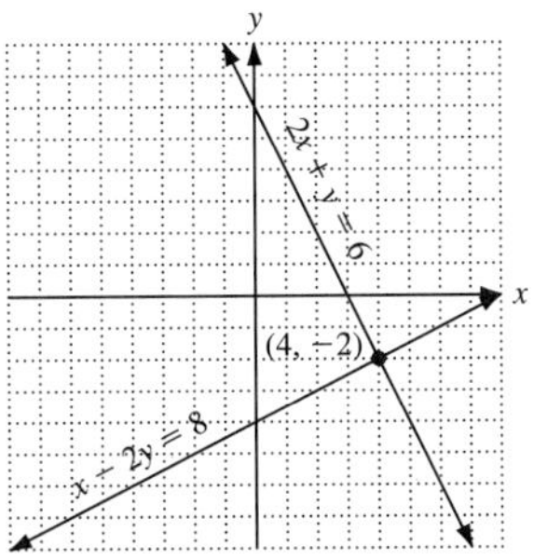

5.

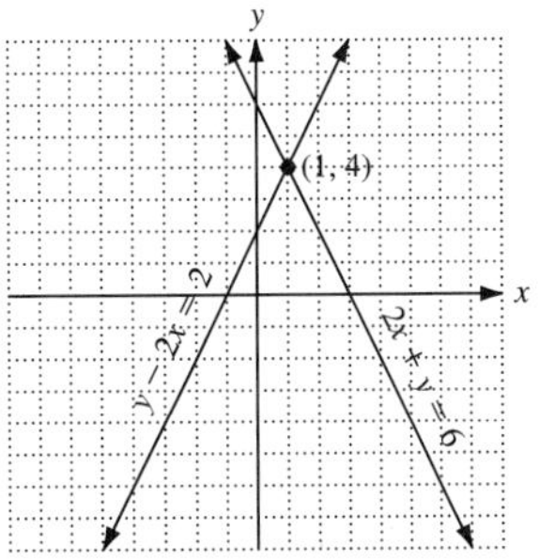

7.

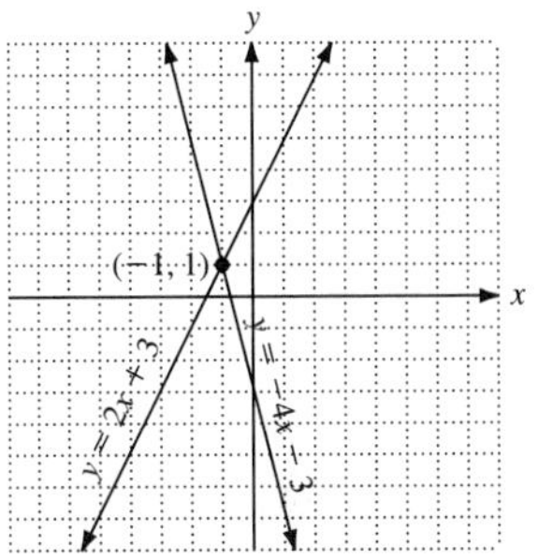

9.

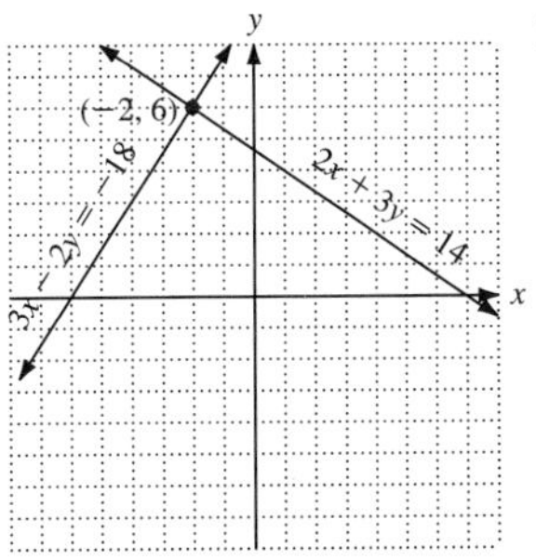

11.

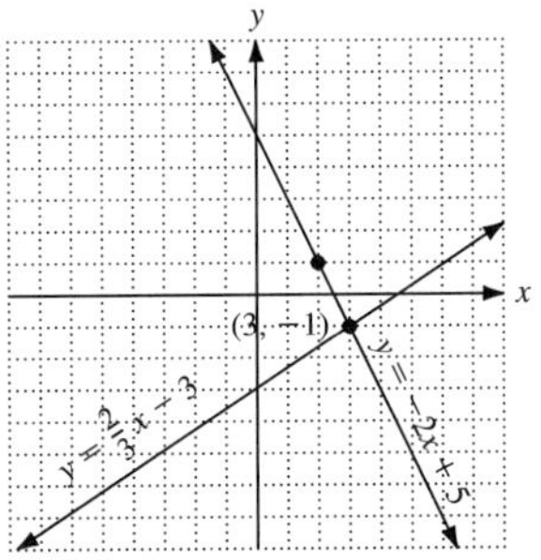

13.

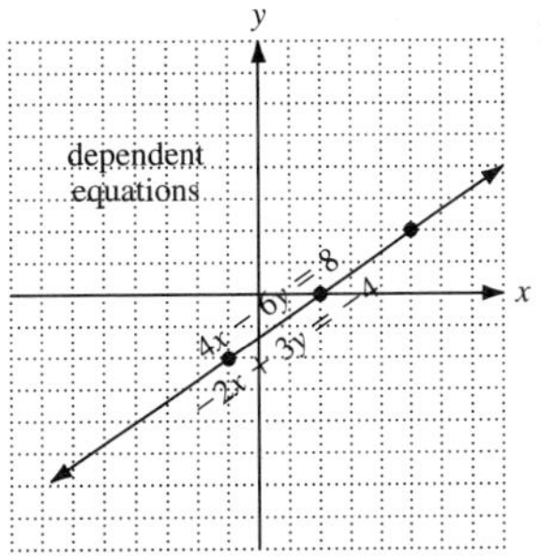

15.

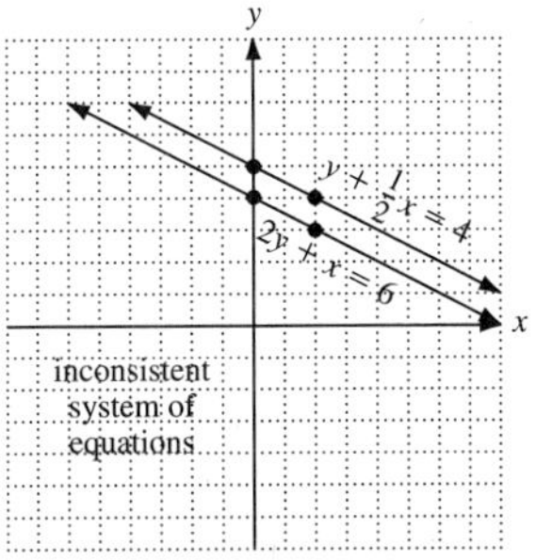

17.

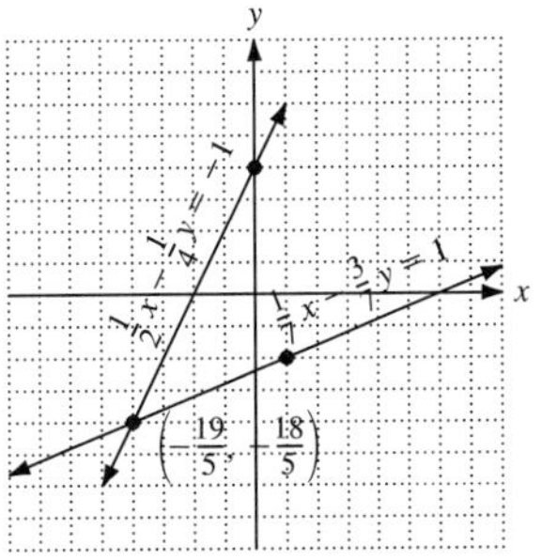

24. $\frac{1}{8} \leq x$ **25.** $d = \dfrac{5y + 16}{x}$ **26.** $9x^2 - 42x + 49$ **27.** $(5x - 6)^2$

8.2 Exercises

1. $x = 1; y = 1$ **3.** $y = -1; x = 3$ **5.** $x = 1; y = 2$ **7.** $x = \dfrac{8}{5}; y = \dfrac{16}{5}$ **9.** $y = -1; x = 5$ **11.** $y = 0; x = 3$ **13.** $b = 5; a = 9$ **15.** $x = 2; y = 0$ **17.** $c = 3; d = -2$ **19.** $x = 2; y = -3$ **21.** $x = -3; y = 6$ **23.** $x = -3; y = 1$ **25.** $x = 7; y = -2$ **27.** $x = -3; y = 3$ **29.** $x = 2; y = 1$ **31.** $x = 3; y = \dfrac{1}{4}$ **33.** $x = 2.9346; y = 1.0994$ **40.** $m = -\dfrac{7}{11}; b = \dfrac{19}{11}$

41.

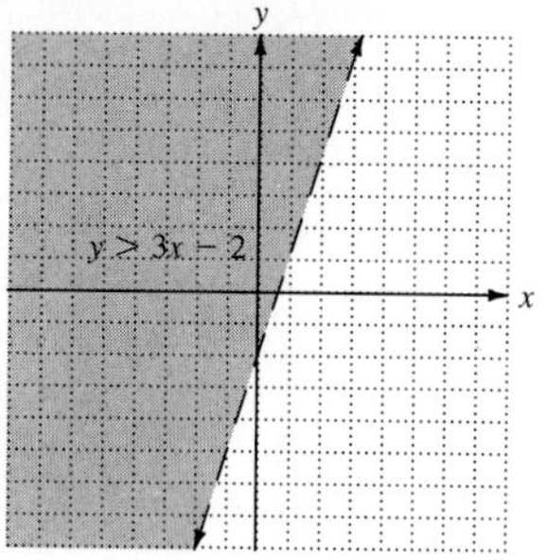

42. Larger number is 40, smaller number is 10 **43.** 124 quarters, 30 dimes

8.3 Exercises

3. $x = -1; y = 4$ **5.** $x = 1; y = 2$ **7.** $x = 0; y = 5$ **9.** $x = -3; y = 4$ **11.** $x = 8; y = -1$ **13.** $b = 2; a = -6$ **15.** $y = 1; x = -\frac{9}{4}$ **17.** $x = 2; y = 2$ **19.** $x = 1; y = -2$ **21.** $x = 6; y = 3$ **23.** $x = 2; y = \frac{1}{5}$ **25.** $x = -3; y = 5$ **27.** $x = -2; y = \frac{4}{3}$ **29.** $x = 9; y = 4$ **33.** $x = 1; y = 0$ **35.** $x = -4; y = -7$ **37.** $x = 8; y = -10$ **39.** $x = 7; y = -2$ **40.** $x = 5; y = -3$ **41.** $x = -3; y = -4$ **42.** $x = 3; y = 5$ **43.** $x = 5; y = 4$ **44.** $x = a - b; y = a + b$ **45.** $x = 4a; y = 2b$ **46.** $5 = x$ **47.** $y = -3; x = 8$ **48.** $3(5x - 1)(5x + 1)$ **49.** $(2x - 5)(3x - 5)$

Putting Your Skills to Work

1. $x + 5y = 6.50, 2x + 9y = 11.90$ **2.** The cost of 1 gallon of gasoline is \$1.10. **3.** The cost of 1 quart of oil is \$2.10. The cost of 1 gallon of gasoline is \$1.40.

8.4 Exercises

5. $x = -14, y = -5$ **7.** infinite number of solutions **9.** $x = 3, y = -7$ **11.** no solution **13.** $x = 0, y = -3$ **15.** $x = 3, y = 3$ **17.** $y = 1, x = 3$ **19.** $x = 3, y = 1$ **21.** $a = -5, b = -7$ **23.** $x = 5, y = -2$ **25.** $c = \frac{3}{5}$ **26.** $\frac{1}{2}$ **27.** -1 **28.** 25 **29.** 2

8.5 Exercises

1. $x = 13, y = -17$ **3.** $x = 42$ mph (highway), $y = 30$ mph (city) **5.** 32 attended college, 14 did not attend college **7.** 250 orchestra tickets, 170 mezzanine tickets **9.** width = 11 feet, length = 13 feet **11.** 8 qts before, 8 qts added **13.** 30 lb of nuts, 20 lb of raisins **15.** $r = 195$ mph (still air), $w = 15$ mph (wind speed) **17.** $r = 13.125$ mph in still water, $c = 1.875$ mph speed of current **19.** value of dictionaries = \$500, value of encyclopedias = \$4000 **20.** $\frac{2x^2 - 12}{x^2 - 9}$ **21.** 35 **22.** $-64x^9y^6$

Putting Your Skills to Work

1. $D = 80, S = 90$ **2.** $D = 90, S = 60$

Chapter 8 Review Problems

1.

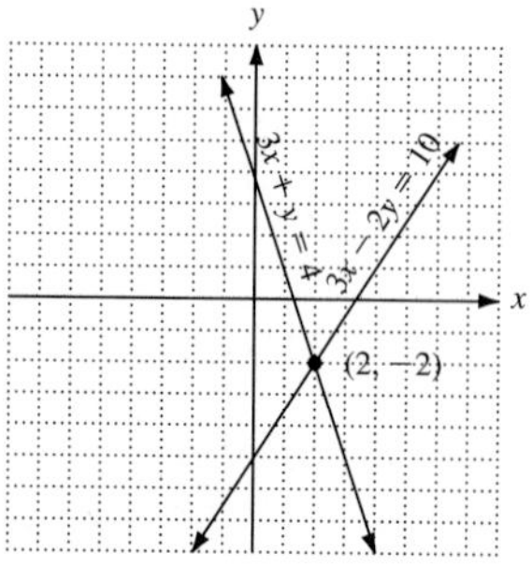

2.

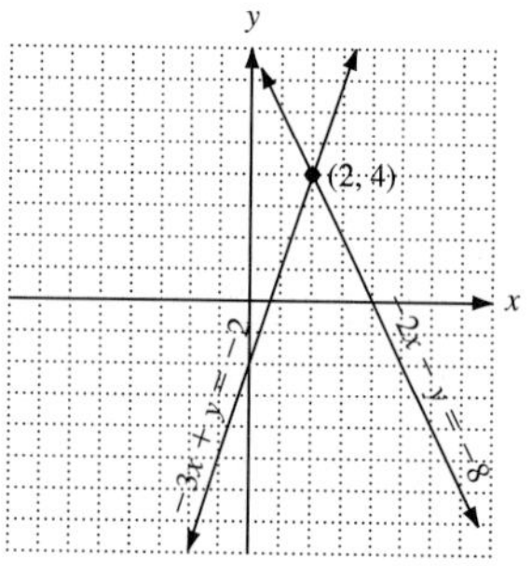

3.

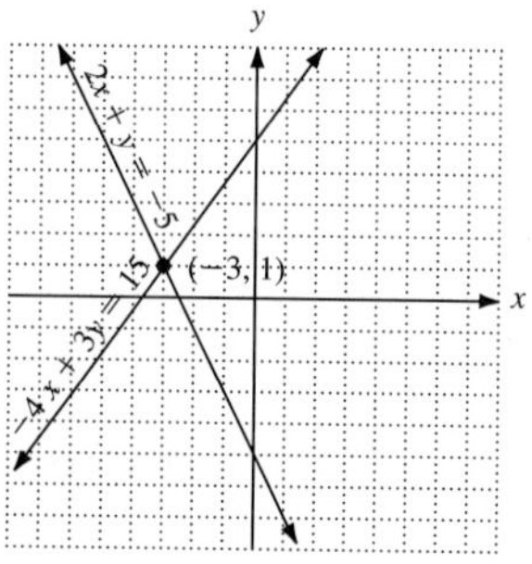

4.

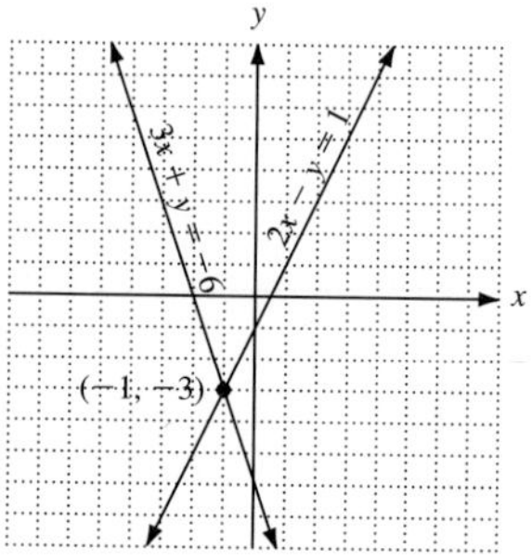

5. $x = 3, y = 5$ **6.** $x = 3, y = 3$ **7.** $x = 2, y = -3$ **8.** $x = 14, y = 66$ **9.** $x = 2, y = 1$ **10.** $x = -1, y = 0$ **11.** $x = -4, y = -2$ **12.** $x = 3, y = 1$ **13.** (7, 2) **14.** (4, 1) **15.** (3, −7) **16.** (−1, −3) **17.** (4, −4) **18.** (10, 2) **19.** $\left(\frac{4}{5}, -\frac{6}{5}\right)$ **20.** (−3, 2) **21.** (−3, −1) **22.** (2, −3) **23.** No solution, Inconsistent system **24.** (9, 4) **25.** $\left(-3, -\frac{2}{3}\right)$ **26.** Infinite number of solutions, dependent equations **27.** (−2, −2) **28.** (−1, −2) **29.** (4, −3) **30.** $\left(-\frac{24}{41}, \frac{2}{41}\right)$

31. Infinite number of solutions. Dependent equations. **32.** (1960, 1040) **33.** $(-4, -7)$ **34.** $(8, -5)$ **35.** $\left(\frac{29}{28}, \frac{1}{7}\right)$ **36.** (4, 10) **37.** (10, 8) **38.** $(-12, -15)$ **39.** $\left(\frac{1}{2}, -\frac{2}{3}\right)$ **40.** $\left(\frac{15}{13}, \frac{36}{13}\right)$ **41.** Infinite number of solutions, dependent equations **42.** Infinite number of solutions, dependent equations **43.** $(-5, 4)$ **44.** $(9, -8)$ **45.** (2, 4) **46.** (3, 1) **47.** (3, 2) **48.** $(12, -12)$ **49.** $\left(a + b, \frac{a^2 - b^2}{b}\right)$ **50.** $\left(\frac{a + b}{a}, \frac{a - b}{b}\right)$ **51.** 120 reserved seats, 640 general admission seats **52.** 7 pounds of apples, 5 pounds of oranges **53.** 275 mph is the speed of the plane in still air; 25 mph is the speed of the wind **54.** 32 liters **55.** 44 regular sized cars, 42 compact cars **56.** 8 tons of 15% salt, 16 tons of 30% salt **57.** Speed of the boat is 19 kilometers per hour, speed of the current is 4 kilometers per hour. **58.** It costs \$3 for a gram of copper. The hourly labor rate is \$4.

Putting Your Skills to Work

1. **(a)** \$700 **(b)** \$600 **(c)** Loss because it costs \$100 more than you make. **2.** \$325 profit **3.** \$25 loss

Chapter 8 Test

1. $(-3, -4)$ **2.** $x = -3$; $y = 4$ **3.**

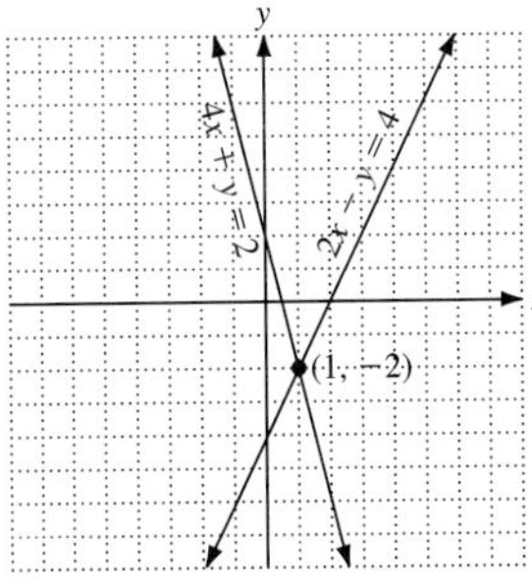

4. $(-6, -2)$ **5.** (4, 3) **6.** (1, 2) **7.** (3, 0) **8.** no solution, inconsistent system **9.** dependent equations, infinite number of solutions **10.** $(-2, 3)$ **11.** $(-5, 3)$ **12.** no solution, inconsistent system **13.** $(5, -3)$ **14.** 12,896 of the \$12 tickets, 17,894 of the \$8 tickets **15.** A shirt costs \$20 and a pair of slacks costs \$24. **16.** 500 per hr by 1st machine; 400 per hr by 2nd machine **17.** 450 k/hr speed of jet; 50 k/hr speed of wind

Cumulative Test for Chapters 0–8

1. \$4.09 **2.** -8 **3.** $15x^3 - 19x^2 + 11x - 3$ **4.** $\frac{1}{(x + 3)(x - 4)}$ **5.** $x \leq \frac{9}{5}$ **6.** $x = -5, x = 9$ **7.** 3000 **8.** $x = \frac{7}{3}$ **9.** $5a(a - 4)(a + 3)$ **10.** $x = 3$ **11.** $x = 8, y = -8$ **11.** $x = 24, y = 8$ **13.** infinite number of solutions, dependent equations **14.** (1, 1) **15.** inconsistent system of equations, no solution **16.** 4000 liters of 8%; 8000 liters of 20% **17.** 1 mph **18.** 1200 and 1800

Chapter 9

Pretest Chapter 9

1. 11 **2.** $\frac{5}{8}$ **3.** 13 **4.** -10 **5.** 2.236 **6.** not real **7.** x^2 **8.** $5x^2y^3$ **9.** $7\sqrt{2}$ **10.** $x^2\sqrt{x}$ **11.** $6x\sqrt{x}$ **12.** $2a^2b\sqrt{2b}$ **13.** $4\sqrt{2}$ **14.** $\sqrt{2} + \sqrt{3}$ **15.** $4\sqrt{2}$ **16.** $13\sqrt{7} - 7$ **17.** $11\sqrt{2} - 18\sqrt{3}$ **18.** $18\sqrt{2}$ **19.** $4 + 2\sqrt{6}$ **20.** $23 + 8\sqrt{7}$ **21.** 1 **22.** -1 **23.** $17 - 4\sqrt{15}$ **24.** $\frac{6}{x}$ **25.** $\frac{5\sqrt{3}}{3}$ **26.** $3\sqrt{5} - 6$ **27.** $13 - 2\sqrt{42}$ **28.** $2\sqrt{13}$ **29.** $\sqrt{11}$ **30.** 7 **31.** $\frac{35}{3}$ **32.** $\frac{1}{2}$ **33.** 7

9.1 Exercises

5. ± 3 **7.** ± 8 **9.** ± 9 **11.** ± 11 **13.** 5 **15.** 0 **17.** 7 **19.** 10 **21.** -6 **23.** 0.9 **25.** 0.1 **27.** $\frac{4}{5}$ **29.** $\frac{7}{8}$ **31.** 20 **33.** -100 **35.** 13 **37.** $-\frac{1}{8}$ **39.** $\frac{5}{7}$ **41.** 0.6 **43.** 120 **45.** 21 **47.** 0.08 **49.** 6.481 **51.** 7.681 **53.** -11.533 **55.** -13.964 **57.** -18.97 **59.** 6.75 **61.** 7.5 seconds **63.** 9.5 seconds **65.** 3 **66.** 2 **67.** $x = 2$; $y = 1$ **68.** no solution, inconsistent system of equations **69.** Each tire costs \$80. Each battery costs \$60. **70.** 600 mph

9.2 Exercises

1. 7 **3.** 10^2 **5.** 5^5 **7.** 56^3 **9.** 3^{75} **11.** w^6 **13.** x^{12} **15.** y^{13} **17.** $7x^3$ **19.** $12x$ **21.** $20x^4$ **23.** xy^3 **25.** $5x^4y^2$ **27.** $8xy^2$ **29.** $11x^3y^7$ **31.** $2\sqrt{5}$ **33.** $2\sqrt{6}$ **35.** $3\sqrt{2}$ **37.** $5\sqrt{2}$ **39.** $6\sqrt{2}$ **41.** $3\sqrt{10}$ **43.** $7\sqrt{2}$ **45.** $5\sqrt{6}$ **47.** $2y^2\sqrt{3y}$ **49.** $3y^6\sqrt{2y}$ **51.** $7x^2\sqrt{x}$ **53.** $5x^3\sqrt{2xy}$ **55.** $2xy\sqrt{3y}$ **57.** $3a^4y^3\sqrt{3y}$ **59.** $2x^3y^5\sqrt{35xy}$ **61.** $3bc^2\sqrt{5ab}$ **63.** $3a^2b^3c\sqrt{7c}$ **65.** $9x^6y^5w^2\sqrt{yw}$ **66.** $x + 3$ **67.** $x + 6$ **68.** $2x + 3$ **69.** $3x + 2$ **70.** $y(x + 7)$ **71.** $2(2x + 1)$ **72.** $x = 4$; $y = -1$ **73.** -4 **74.** -3 **75.** $x = \frac{51}{4}$; $y = \frac{29}{4}$

9.3 Exercises

3. $-2\sqrt{3}$ **5.** $5\sqrt{2} + 3\sqrt{3}$ **7.** $-4\sqrt{2x} + 5\sqrt{x}$ **9.** $\sqrt{3} - 2\sqrt{3} = -\sqrt{3}$ **11.** $17\sqrt{2}$ **13.** $\sqrt{2}$ **15.** 0 **17.** $4\sqrt{3} + 2\sqrt{5} + 6$ **19.** $-2\sqrt{5} + 12\sqrt{3}$ **21.** $5\sqrt{3x}$ **23.** $\sqrt{2xy}$ **25.** $0.2\sqrt{3x}$ **27.** $7\sqrt{5y}$ **29.** $5y\sqrt{y}$ **31.** $4x\sqrt{2x} + 15\sqrt{2x}$ **33.** $-5x\sqrt{2x}$ **35.** $x\sqrt{3}$ **37.** $2y\sqrt{6y} - 6y\sqrt{6}$ **39.** $-40x^3\sqrt{2}$ **41.** $-110x\sqrt{2x}$ **43.** $88 + 16\sqrt{5} + 8\sqrt{13}$ **44.** $5\sqrt{x^2 + 1}$ **45.** $3x\sqrt{x^3 - 2}$

46. $2x$ **47.** 2 **48.** $\frac{4}{3}\sqrt{3}$ **49.** $\frac{3}{2}\sqrt{2}$ **50.** $2\frac{1}{2}$ **51.**

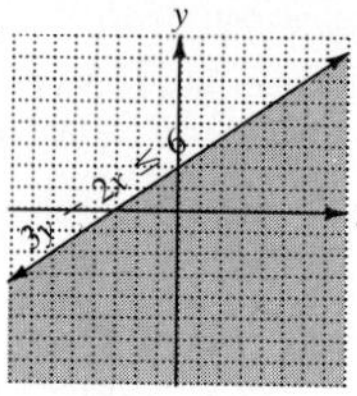

52.

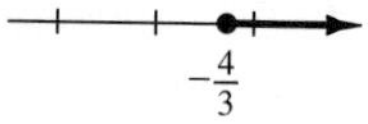

53. 55 of the 15¢ stamps; 77 of the 25¢ stamps
54. \$5000 at 5%; \$3000 at 8%

9.4 Exercises

1. $\sqrt{35}$ **3.** $2\sqrt{11}$ **5.** $a\sqrt{15}$ **7.** $5\sqrt{2x}$ **9.** $6\sqrt{30}$ **11.** $6x^3\sqrt{5}$ **13.** $-6b\sqrt{a}$ **15.** $\sqrt{6} + 2\sqrt{15}$ **17.** $2\sqrt{5} - 6\sqrt{3}$ **19.** $2x - 16\sqrt{5x}$ **21.** $2\sqrt{3} - 18 + 4\sqrt{15}$ **23.** $6\sqrt{ab} + 2a\sqrt{b} - 4a$ **25.** $16 + 3\sqrt{14}$ **27.** $22 + 18\sqrt{3}$ **29.** $5 - \sqrt{21}$ **31.** $-6 + 9\sqrt{2}$ **33.** $2\sqrt{14} + 34$ **35.** $29 - 12\sqrt{5}$ **37.** $53 + 10\sqrt{6}$ **39.** $18 - 12\sqrt{2}$ **41.** $a - 2a\sqrt{b} + ab$ **43.** $5a + 2a\sqrt{3a} - \sqrt{3a}$ **45.** $5x\sqrt{5x} - 5x^2 - 10\sqrt{x} + 2x\sqrt{5}$ **47.** **(a)** It is not true when a is negative. **(b)** In order to deal with real numbers. **48.** Hank's final answer does not have a middle term. $(\sqrt{a} + \sqrt{b})^2 = a + 2\sqrt{ab} + b$. **49.** $(6x + 7y)(6x - 7y)$ **50.** $(x^4 + 1)(x^2 + 1)(x + 1)(x - 1)$ **51.** \$12,500 at 6%; \$27,500 at 11% **52.** \$55

9.5 Exercises

1. 4 **3.** $\frac{1}{2}$ **5.** 3 **7.** $\frac{\sqrt{6}}{x^2}$ **9.** 5 **11.** $\frac{2\sqrt{2}}{x^3}$ **13.** $\frac{3\sqrt{7}}{7}$ **15.** $\frac{5\sqrt{6}}{6}$ **17.** $\frac{x\sqrt{3}}{3}$ **19.** $\frac{2\sqrt{2x}}{x}$ **21.** $\frac{\sqrt{3}}{2}$ **23.** $\frac{7\sqrt{a}}{a^2}$ **25.** $\frac{\sqrt{2x}}{2x^2}$ **27.** $\frac{\sqrt{x}}{2x^4}$ **29.** $\frac{\sqrt{14}}{7}$ **31.** $\frac{\sqrt{3a}}{ab}$ **33.** $\frac{9\sqrt{2x}}{8x}$ **35.** $3\sqrt{2} + 3$ **37.** $\sqrt{5} - \sqrt{2}$ **39.** $2 + \sqrt{2}$ **41.** $x(2\sqrt{2} - \sqrt{7})$ **43.** $\frac{x(\sqrt{3} + 2x)}{3 - 4x^2}$ **45.** $2\sqrt{14} - 7$ **47.** $\frac{7 - 2\sqrt{10}}{3}$ **49.** $3\sqrt{2} + 4$ **51.** $-3\sqrt{35} - 6\sqrt{10} - 6\sqrt{7} - 12\sqrt{2}$ **53.** $\sqrt{x} - 2$ **55.** $\frac{x + \sqrt{2x} + 2\sqrt{2}}{x - 2} + 2\sqrt{x}$ **57.** **(a)** $x^2 - 2$ **(b)** $(x + \sqrt{2})(x - \sqrt{2})$ **(c)** $(x + 2\sqrt{3})(x - 2\sqrt{3})$ **58.** 0.904534033 The values are the same. They are equivalent expressions. **59.** 1.183215957 The values are the same. They are equivalent expressions. **60.** 38.49 **61.** 448 **62.** 6 cm **63.** He must score 190 on his final.

9.6 Exercises

1. 5 **3.** 5 **5.** $\sqrt{74}$ **7.** $\sqrt{34}$ **9.** $2\sqrt{19}$ **11.** $\sqrt{7}$ **13.** 9.91 **15.** 9.9 ft **17.** 127.3 ft **19.** 6.7 mi **21.** 121 **23.** 17 **25.** $-\frac{2}{3}$ **27.** 7 **29.** $\frac{81}{2}$ **31.** 5 only **33.** $y = 0, y = 3$ **35.** 6 only **37.** $y = 0, y = 1$ **39.** x only **41.** 5 only **43.** $\frac{2}{9}$ only **44.** $x = 2, x = 6$ **45.** no solution **46.** $9x^2 - 6x + 1$ **47.** 20 **48.** relation

9.7 Exercises

1. $y = \frac{63}{4}$ **3.** $y = 70$ **5.** $y = \frac{1029}{2}$ **7.** $y = \frac{625}{2}$ **9.** $y = 373.03$ **11.** 169 lb/sq in. **13.** 56 min **15.** $y = \frac{48}{7}$ **17.** $y = 2$ **19.** $y = \frac{2500}{9}$ **21.** $y = \frac{4}{9}$ **23.** $y = 10$ **25.** 52.5 amperes **27.** $444\frac{4}{9}$ pounds **31.** 5000 Btu/hr **32.** 7467 lb **33.** $a = 160$

34.

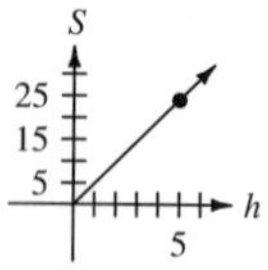

35. $4(3x + 4)(x - 3)$ **36.** $x = \frac{13}{4}$

Chapter 9 Review Problems

1. 3 **2.** 5 **3.** 6 **4.** 7 **5.** 10 **6.** 2 **7.** not a real number **8.** not a real number **9.** 4 **10.** 11 **11.** 15 **12.** 13 **13.** 0.9 **14.** 0.6 **15.** $\frac{1}{5}$ **16.** $\frac{6}{7}$ **17.** -0.02 **18.** 0.03 **19.** not a real number **20.** -0.09 **21.** 10.247 **22.** 14.071 **23.** 8.775 **24.** 9.381 **25.** $5\sqrt{2}$ **26.** $4\sqrt{3}$ **27.** $7\sqrt{2}$ **28.** $6\sqrt{2}$ **29.** $2\sqrt{10}$ **30.** $4\sqrt{5}$ **31.** x^4 **32.** y^5 **33.** $x^2y^3\sqrt{x}$ **34.** $ab^2\sqrt{a}$ **35.** $4xy^2\sqrt{xy}$ **36.** $7x^2y^3\sqrt{2}$ **37.** $2x^2\sqrt{3x}$ **38.** $3x^3\sqrt{3x}$ **39.** $5x^5\sqrt{3}$ **40.** $5x^6\sqrt{5}$ **41.** $2ab^2c^2\sqrt{30ac}$ **42.** $11a^3b^2\sqrt{c}$ **43.** $2x^3y^4\sqrt{14xy}$ **44.** $3x^6y^3\sqrt{11xy}$ **45.** $3\sqrt{2}$ **46.** $-2\sqrt{5}$ **47.** $7x\sqrt{3}$ **48.** $8a\sqrt{2} + 2a\sqrt{3}$ **49.** $-7\sqrt{5} + 2\sqrt{10}$ **50.** $9\sqrt{6} - 15\sqrt{2}$ **51.** $-5\sqrt{7} + 2x\sqrt{7}$ **52.** $5a\sqrt{3} - 26\sqrt{3}$ **53.** $6x^2$ **54.** $-10a\sqrt{b}$ **55.** $4ab\sqrt{a}$ **56.** $-15x^4$ **57.** 5 **58.** -18 **59.** $\sqrt{10} - \sqrt{6} - 4$ **60.** $\sqrt{30} - 10 + 5\sqrt{2}$ **61.** $20 + 3\sqrt{11}$ **62.** $27 + 8\sqrt{10}$ **63.** $8 - 4\sqrt{3} + 12\sqrt{6} - 18\sqrt{2}$ **64.** $15 - 3\sqrt{2} - 10\sqrt{3} + 2\sqrt{6}$ **65.** $66 + 36\sqrt{2}$ **66.** $74 - 40\sqrt{3}$ **67.** $ab - 3a\sqrt{ab} - 2\sqrt{ab} + 6a$ **68.** $3xy + x\sqrt{xy} - 6\sqrt{xy} - 2x$ **69.** $\frac{\sqrt{3x}}{3x}$ **70.** $\frac{2y\sqrt{5}}{5}$ **71.** $\frac{x^2y\sqrt{2}}{4}$ **72.** $\frac{3a\sqrt{2b}}{2}$ **73.** $\frac{\sqrt{21}}{7}$ **74.** $\frac{\sqrt{2}}{3}$ **75.** 6 **76.** $\sqrt{6}$ **77.** $\frac{a^2\sqrt{2}}{2}$ **78.** $\frac{x\sqrt{3}}{3}$ **79.** $\frac{\sqrt{ab}}{b^2}$ **80.** $\frac{x^2\sqrt{xy}}{y^2}$ **81.** $\sqrt{5} - \sqrt{2}$ **82.** $\frac{2(\sqrt{6} + \sqrt{3})}{3}$ **83.** $-7 + 3\sqrt{5}$ **84.** $\frac{3 - 2\sqrt{3}}{3}$ **85.** $c = \sqrt{89}$ **86.** $a = \sqrt{2}$ **87.** $b = \frac{\sqrt{51}}{2}$ **88.** $c = 3.124$ **89.** 30 meters **90.** 45.8 miles **91.** 7.1 inches **92.** $x = 13$ **93.** $x = 42$ **94.** $x = -1$ **95.** $x = 6$ **96.** $x = 4$ **97.** $x = 7$ **98.** $x = 2$ **99.** $x = 7$ **100.** $y = 77$ **101.** $y = \frac{2}{15}$ **102.** $y = \frac{1}{2}$ **103.** 400 lb **104.** 134 ft **105.** 8 times as much

Putting Your Skills to Work

1. $r = 10$ miles per hour **2.** $r = 30$ miles per hour **3.** $r = 50$ miles per hour **4.** $r = 70$ miles per hour **5.** $L = 20$ feet **6.** $L = 80$ feet **7.** $L = 320$ feet **8.** $L = 500$ feet

Chapter 9 Test

1. 9 **2.** $\frac{3}{10}$ **3.** $4xy^3\sqrt{3y}$ **4.** $10xz^2\sqrt{xy}$ **5.** $\sqrt{5}$ **6.** $8\sqrt{a} + 5\sqrt{2a}$ **7.** $12ab$ **8.** $3\sqrt{2} - \sqrt{6} + 45$ **9.** $10 - 4\sqrt{6}$ **10.** $19 + \sqrt{10}$ **11.** $\frac{\sqrt{5x}}{5}$ **12.** $\frac{\sqrt{3}}{2}$ **13.** $\frac{16-3\sqrt{2}}{-4}$ or $\frac{3\sqrt{2}-16}{4}$ **14.** $a(\sqrt{5} - \sqrt{2})$ **15.** 12.49 **16.** $x = \sqrt{95}$ **17.** $x = 3\sqrt{3}$ **18.** $x = \frac{35}{2}$ **19.** $x = 9$ **20.** 16 inches **21.** \$912.05 **22.** $A = 21.22\text{ cm}^2$

Cumulative Test for Chapters 0–9

1. $\frac{4}{5}$ **2.** $8\frac{11}{12}$ **3.** $\frac{15}{16}$ **4.** 0.0007 **5.** 3 **6.** 31 **7.** $-2x^3yz^2 + 4xyz$ **8.** $27x^3 - 54x^2 + 36x - 8$ **9.** $4x^2 - 12x + 9$ **10.** $9x^2 - 121$ **11.** $\frac{4x+4}{x+2}$ **12.** $\frac{(x-3)(x+1)}{(x-6)(x+2)}$ **13.** $7x^2y^3\sqrt{2x}$ **14.** $14\sqrt{3}$ **15.** $9 - 6\sqrt{2}$ **16.** $5 + 2\sqrt{6}$ **17.** $10 + 22\sqrt{3}$ **18.** $\frac{3+\sqrt{2}}{7}$ **19.** -4 **20.** not a real number **21.** $c = \sqrt{23}$ **22.** $b = 4\sqrt{5}$ **23.** $x = 5$ **24.** $y = 6$ **25.** $y = 10$ **26.** 78.5 m^2

Chapter 10

Pretest Chapter 10

1. $x = 14, -2$ **2.** $x = 0, \frac{4}{3}$ **3.** $x = \frac{2}{5}, 4$ **4.** $x = \frac{7}{3}, -2$ **5.** $x = \pm 2\sqrt{3}$ **6.** $x = \pm 5$ **7.** $x = -4 \pm \sqrt{11}$ **8.** $x = -\frac{3}{4} \pm \frac{\sqrt{65}}{4}$ **9.** $x = \frac{3 \pm \sqrt{3}}{2}$ **10.** $x = \frac{1}{4}, x = -2$ **11.** $x = \frac{-4 \pm \sqrt{13}}{3}$ **12.** no real solution

13.

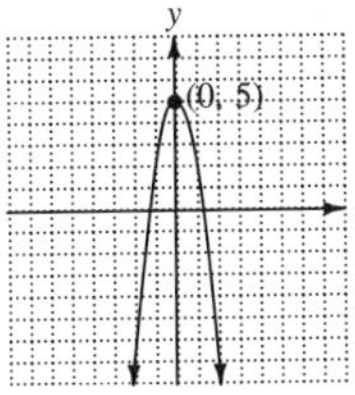

14.

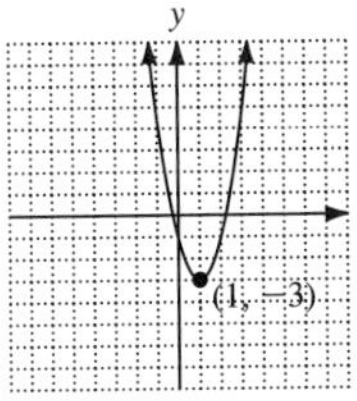

15. base = 6 cm; altitude = 13 cm

10.1 Exercises

1. $a = 1, b = 5, c = 6$ **3.** $a = 7, b = -11, c = -2$ **5.** $a = -12, b = 3, c = -2$ **7.** $a = 3, b = -8, c = 0$ **9.** $a = 1, b = 3, c = -15$ **11.** $a = 1, b = -9, c = 10$ **13.** $x = 0, x = -\frac{1}{2}$ **15.** $x = 0, x = \frac{14}{11}$ **17.** $x = 0, x = \frac{8}{7}$ **19.** $x = 0, x = -3$ **21.** $x = -8, x = 3$ **23.** $x = -\frac{1}{2}, x = 8$ **25.** $x = \frac{1}{3}, x = 4$ **27.** $x = 2, x = 7$ **29.** $x = -10, x = -5$ **31.** $x = -\frac{2}{3}, x = -\frac{5}{2}$ **33.** $y = -3, y = -4$ **35.** $y = \frac{1}{3}, y = \frac{3}{2}$ **37.** $y = \frac{4}{3}$ **39.** $n = 5$ **41.** $n = -4, n = 5$ **43.** $x = 0, x = -\frac{14}{3}$ **45.** $x = -\frac{1}{2}, x = 4$ **47.** $x = -2, x = 7$ **49.** $x = 4$ **51.** $x = 1$ **53.** $x = 0, x = -8$ **55.** $x = \frac{5}{2}, x = -4$ **57.** $x = -\frac{5}{2}, x = 6$ **59.** $x = -2, x = -3$ **61.** $x = 3, x = 5$ **63.** $x = -\frac{10}{3}, x = 4$ **65.** You can always factor out x. **66.** You must have the product of two or more factors equal to 0, not 14. Remember the zero factor property. **67.** $a = \frac{14}{9}, c = -14$ **69.** 231 truck routes **71.** 6 cities **73.** 11 cities **74.** $\frac{(2x-1)(x+2)}{2x}$ **75.** $\frac{5}{22}$ **76.** $\frac{2}{x(x+4)}$ **77.** $\frac{2x^2 - 4x + 2}{3x^2 - 11x + 10}$

10.2 Exercises

1. ± 7 **3.** $\pm 7\sqrt{2}$ **5.** $\pm 5\sqrt{3}$ **7.** ± 5 **9.** $\pm 2\sqrt{5}$ **11.** $\pm\sqrt{3}$ **13.** $\pm 7\sqrt{2}$ **15.** ± 6 **17.** $\pm 2\sqrt{3}$ **19.** $\pm\sqrt{5}$ **21.** $3 \pm \sqrt{5}$ **23.** $x = -8, -4$ **25.** $\frac{-5 \pm \sqrt{2}}{2}$ **27.** $\frac{1 \pm \sqrt{7}}{3}$ **29.** $\frac{-2 \pm 2\sqrt{3}}{7}$ **31.** $\frac{5 \pm 3\sqrt{2}}{6}$ **33.** ± 4.389 **35.** ± 1.600 **37.** $x = 1, x = -15$ **39.** $2 \pm \sqrt{5}$ **41.** $-3 \pm \sqrt{2}$ **43.** $5 \pm \sqrt{29}$ **45.** $x = 0, x = -3$ **47.** $x = 3, x = -2$ **49.** $x = \frac{3}{2}, x = \frac{1}{2}$ **51.** $x = \frac{-5 \pm \sqrt{3}}{2}$ **52.** $x = \pm\sqrt{\frac{c}{a}}$ **53.** $x = \frac{-b \pm \sqrt{28 + b^2}}{2}$ **54.** $x = \frac{-b \pm \sqrt{b^2 - 4c}}{2}$ **55.** $a = -4, b = -4$ **56.** $x = 1, x = 5$ **57.** $\frac{16}{9}$ **58.** $x = -1$

10.3 Exercises

1. $x = -2 \pm \sqrt{3}$ **3.** $x = \frac{5 \pm \sqrt{33}}{2}$ **5.** $x = -1, x = \frac{9}{2}$ **7.** $x = \frac{-1 \pm \sqrt{21}}{10}$ **9.** $x = \frac{5 \pm \sqrt{17}}{4}$ **11.** $x = \frac{3 \pm \sqrt{33}}{12}$ **13.** $x = \frac{-5 \pm \sqrt{65}}{4}$ **15.** $x = \frac{1 \pm \sqrt{3}}{3}$ **17.** no real solution **19.** $x = 6, x = 1$ **21.** $x = \frac{-7 \pm \sqrt{109}}{10}$ **23.** $m = 3, m = -\frac{3}{2}$ **25.** $x = 10.511, x = 6.779$ **27.** $x = 1.541, x = -4.541$ **29.** $x = 1.207, x = -0.207$ **31.** $x = -1.894, x = -0.106$ **33.** $t = \frac{3}{2}, t = \frac{2}{3}$ **35.** $s = 3, s = \frac{1}{3}$ **37.** $p = \frac{1 \pm \sqrt{21}}{2}$ **39.** $y = \frac{3 \pm \sqrt{233}}{14}$ **41.** $x = \pm\frac{\sqrt{21}}{3}$ **43.** $x = 1 \pm 2\sqrt{2}$ **45.** $\frac{3(x + 5)}{x + 2}$ **46.** **(a)** $y = -\frac{5x}{13} - \frac{14}{13}$ **(b)** $5x + 13y = -14$ **47.** $x^2 + 5x + 2$ **48.** $\frac{9p - 20}{3f} = g$

10.4 Exercises

1.
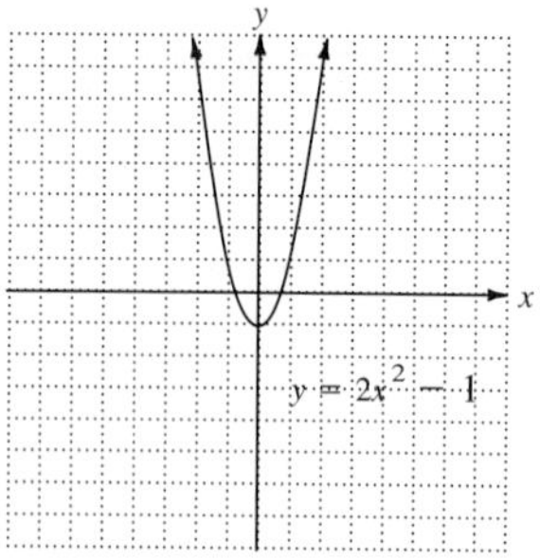

3.
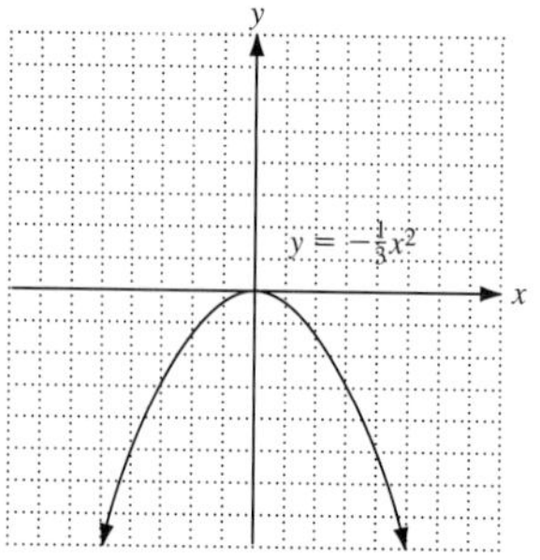

5.
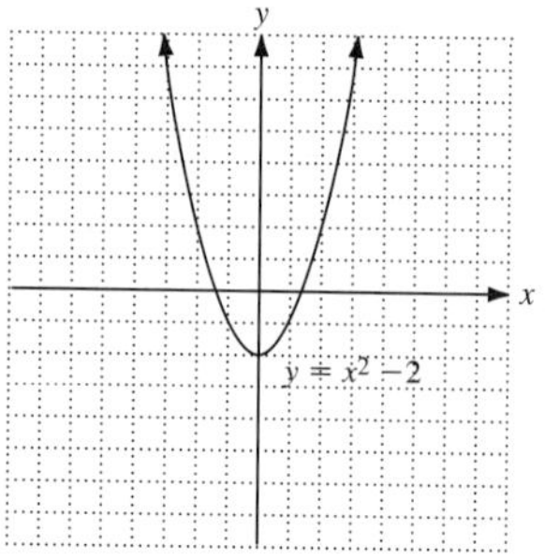

7.
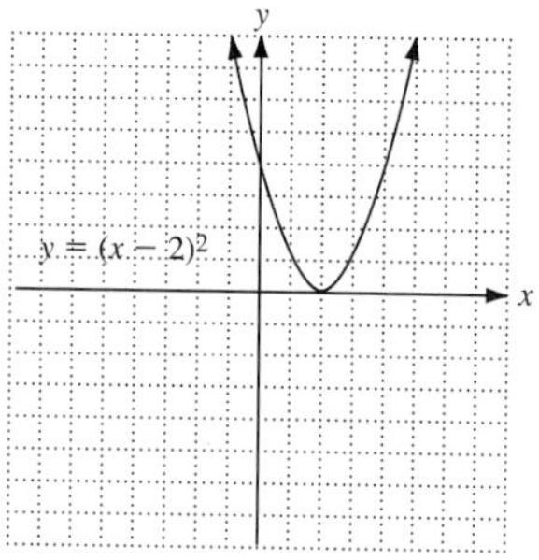

9.
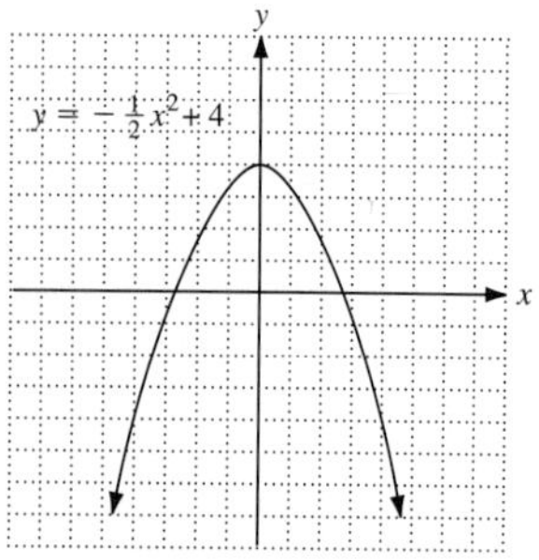

11.
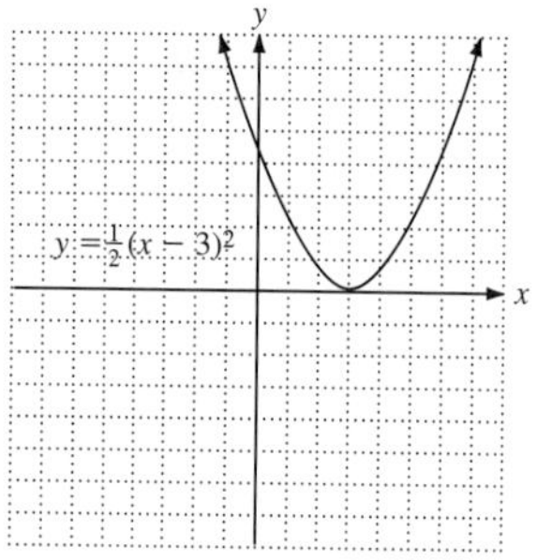

13.
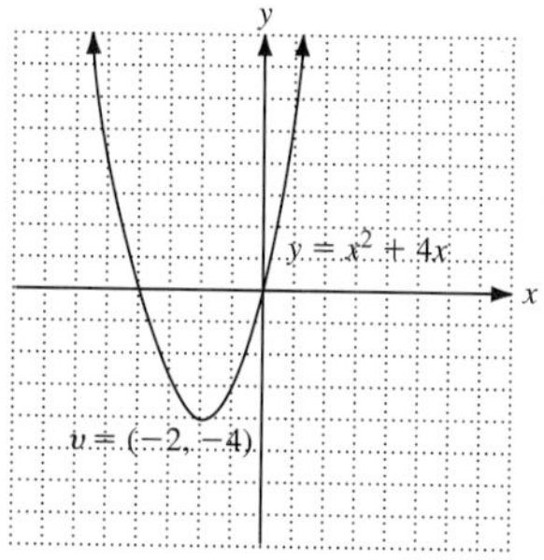

15.
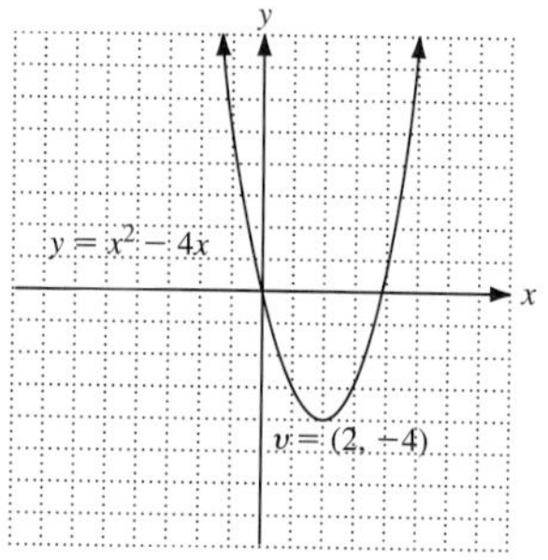

17.
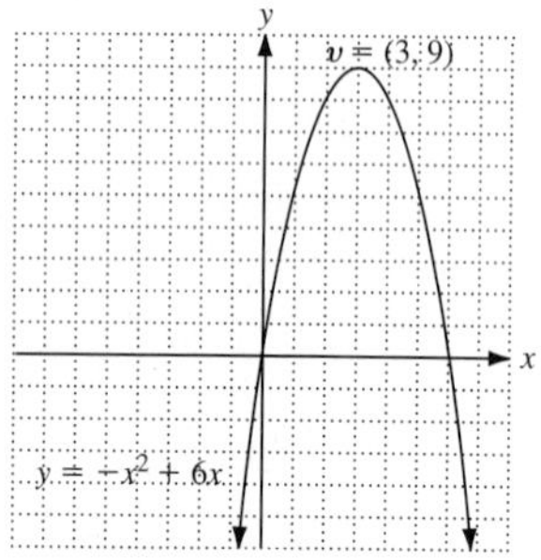

19.
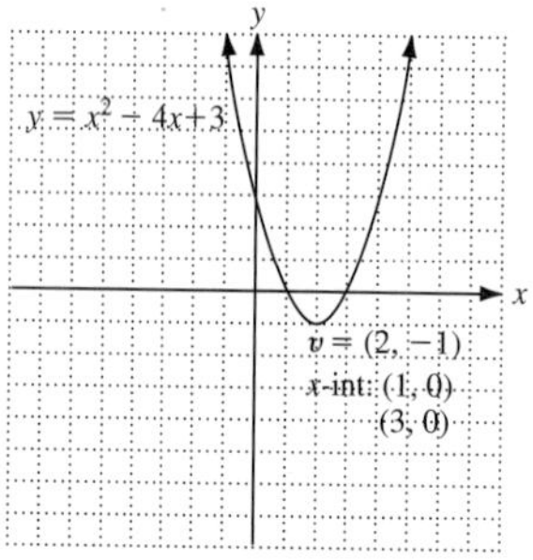

21.
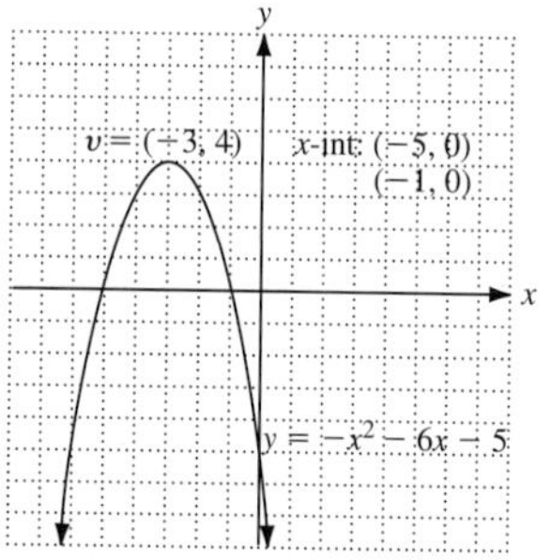

23.
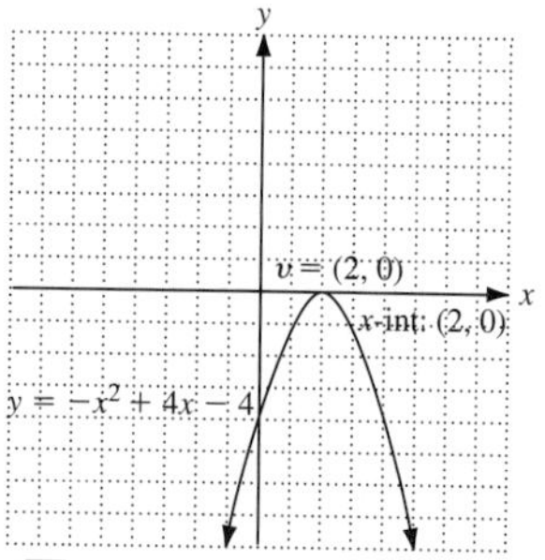

25. **(a)**
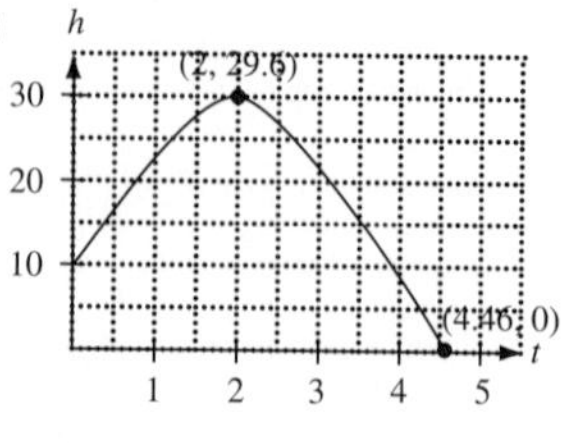

(b) 24.7 meters **(c)** 29.6 meters **(d)** 4.46 seconds **27.** vertex (1, 7), y-intercept (0, 5); x-intercepts $\left(\frac{2 \pm \sqrt{14}}{2}, 0\right)$ **28.** vertex (−1, 3), y-intercept (0, 1); x-intercepts $\left(\frac{-2 \pm \sqrt{6}}{2}, 0\right)$ **29.** $x = \frac{13}{2}$ **30.** $x = 6$ only **31.** $y = 1$ **32.** $y = 75$

Putting Your Skills to Work

1. 11.5 feet **2.** 11.5 feet **3.** 3 yards **4.** 35 yards

10.5 Exercises

1. 32 yards, 24 yards **3.** $L = 12$ m, $W = 5$ m **5.** 21 students originally planned to give the gift. **7.** The school originally planned for 360 students. 450 students actually came to the college. **9.** If the width is 20 feet, the length is 100 feet. If the width is 50 feet, the length is 40 feet. **11.** Jet speed is 600 mph **13.** 13 **14.** 16 **15.** \$1750 **16.** **(a)** 90 boots **(b)** \$1800 **17.**
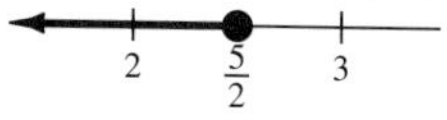

18. $\frac{\sqrt{15} - 2\sqrt{3}}{1}$

Putting Your Skills to Work

1. 16.069 **2.** $5.663 = w$

Chapter 10 Review Problems

1. $a = 5, b = 2, c = -3$ **2.** $a = 15, b = 6, c = 4$ **3.** $a = 1, b = -9, c = -3$ **4.** $a = 3, b = 6, c = -10$ **5.** $a = 5, b = -3, c = 1$ **6.** $a = 4, b = 5, c = -14$ **7.** $a = 3, b = -6, c = 0$ **8.** $a = 12, b = 0, c = -14$ **9.** $-20, -1$ **10.** -5 **11.** $3, -9$ **12.** $\frac{3}{2}$ **13.** $-\frac{5}{6}, -3$ **14.** $\frac{1}{4}, 1$ **15.** $-\frac{4}{3}, \frac{1}{4}$ **16.** $-\frac{1}{3}, -\frac{8}{3}$ **17.** $\frac{2}{3}, -\frac{3}{2}$ **18.** $\frac{1}{3}, -\frac{3}{2}$ **19.** $\frac{1}{5}, -\frac{1}{3}$ **20.** $\frac{1}{4}, -\frac{4}{3}$ **21.** $0, -\frac{16}{5}$ **22.** $0, \frac{9}{2}$ **23.** $0, -4$ **24.** $0, 3$ **25.** $0, -\frac{5}{3}$ **26.** $0, 4$ **27.** $-\frac{2}{5}, 10$ **28.** $7, 2$ **29.** $-\frac{2}{3}, 4$ **30.** $\frac{1}{3}, \frac{4}{3}$ **31.** ± 10 **32.** ± 11 **33.** ± 5 **34.** ± 6 **35.** $\pm\sqrt{22}$ **36.** $\pm\sqrt{39}$ **37.** $\pm 2\sqrt{2}$ **38.** $\pm 2\sqrt{5}$ **39.** $\pm 3\sqrt{2}$ **40.** $\pm 2\sqrt{6}$ **41.** $4 \pm \sqrt{7}$ **42.** $2 \pm \sqrt{3}$ **43.** $\frac{-6 \pm 2\sqrt{5}}{5}$ **44.** $\frac{-8 \pm 2\sqrt{3}}{3}$ **45.** $-1, -5$ **46.** $-5, -3$ **47.** $2 \pm \sqrt{2}$ **48.** $-1 \pm \sqrt{7}$ **49.** $9, -5$ **50.** $6 \pm 2\sqrt{14}$ **51.** $-1 \pm \sqrt{3}$ **52.** $\frac{-5 \pm \sqrt{31}}{2}$ **53.** $-8, 5$ **54.** $6, 3$ **55.** $-2 \pm \sqrt{10}$ **56.** $-2 \pm 2\sqrt{3}$ **57.** $\frac{7 \pm \sqrt{17}}{4}$ **58.** $\frac{-5 \pm \sqrt{73}}{4}$ **59.** $\frac{-2 \pm \sqrt{14}}{2}$ **60.** $\frac{-3 \pm \sqrt{19}}{2}$ **61.** $\frac{5 \pm \sqrt{73}}{6}$ **62.** $\frac{-3 \pm \sqrt{41}}{8}$ **63.** $-\frac{1 \pm \sqrt{2}}{2}$ **64.** $\frac{4 \pm \sqrt{10}}{3}$ **65.** $\frac{5}{2}, 2$ **66.** $\frac{3}{2}, -\frac{1}{2}$ **67.** $\frac{1}{3}$ **68.** $4, \frac{3}{2}$ **69.** $\frac{3 \pm \sqrt{3}}{3}$ **70.** $\frac{7 \pm \sqrt{209}}{10}$ **71.** $\frac{-2 \pm \sqrt{19}}{3}$ **72.** $\frac{-7 \pm \sqrt{29}}{10}$ **73.** $\frac{-3 \pm \sqrt{5}}{2}$ **74.** $-3, \frac{2}{3}$ **75.** $-\frac{1}{2}, -1$ **76.** $1, \frac{1}{4}$ **77.** ± 6 **78.** $\pm 3\sqrt{2}$ **79.** -8 only **80.** -7 only **81.** $\frac{-4 \pm \sqrt{19}}{3}$ **82.** $\frac{-4 \pm \sqrt{2}}{2}$ **83.** $\frac{11 \pm \sqrt{21}}{10}$ **84.** $-4, 2$ **85.** $-3, 5$ **86.** $0, \frac{1}{6}$

87.
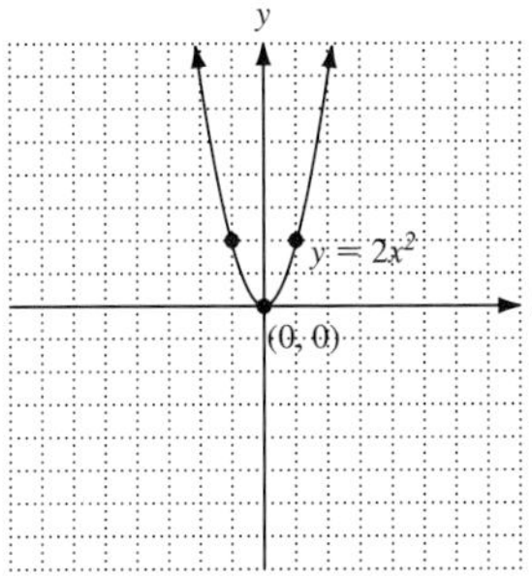

88.
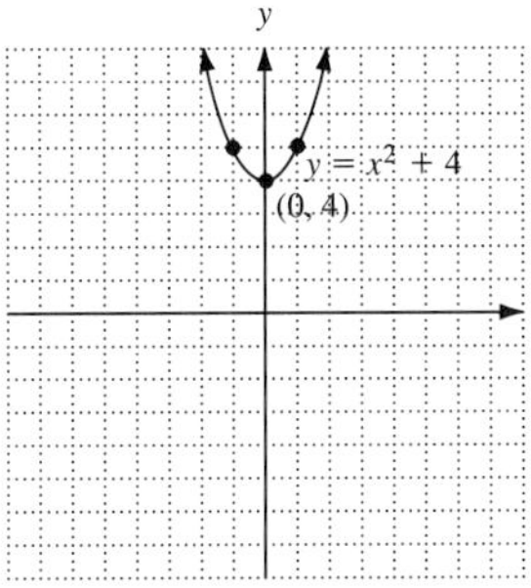

89.
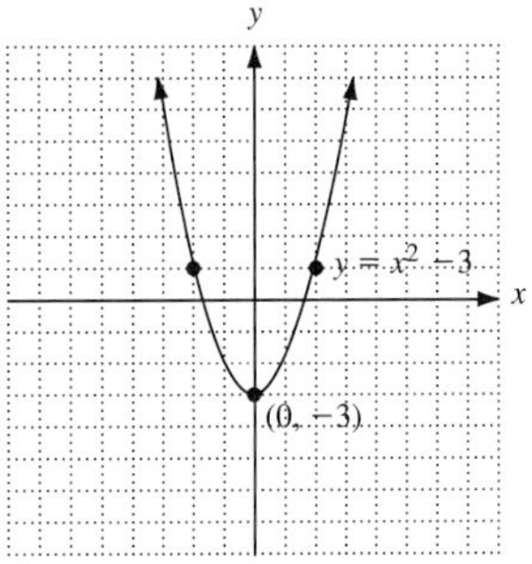

90.
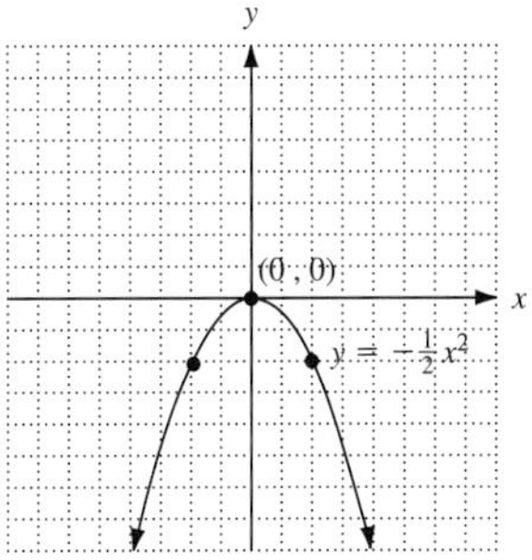

91.
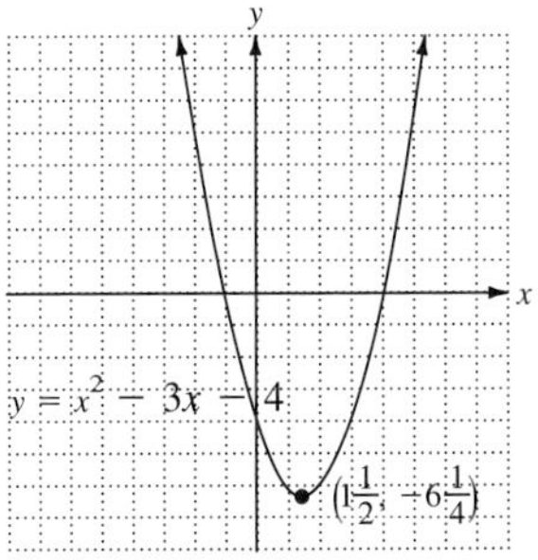

92.
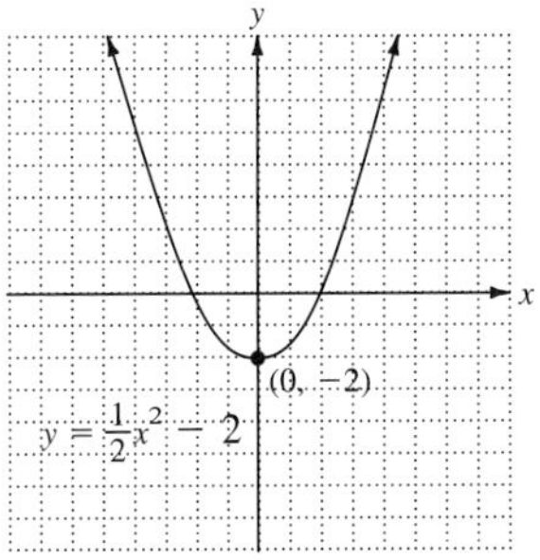

93.
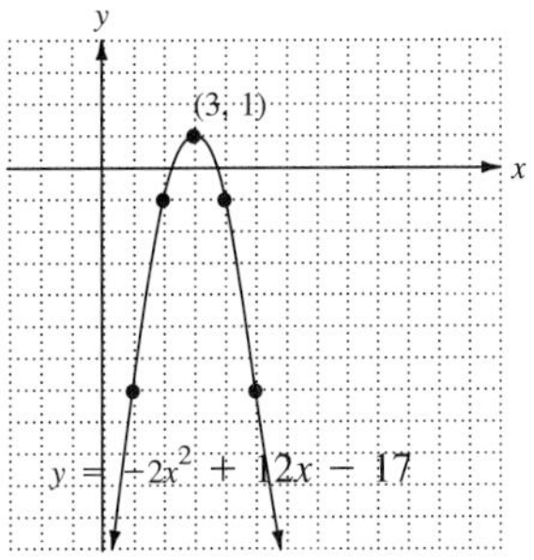

94.
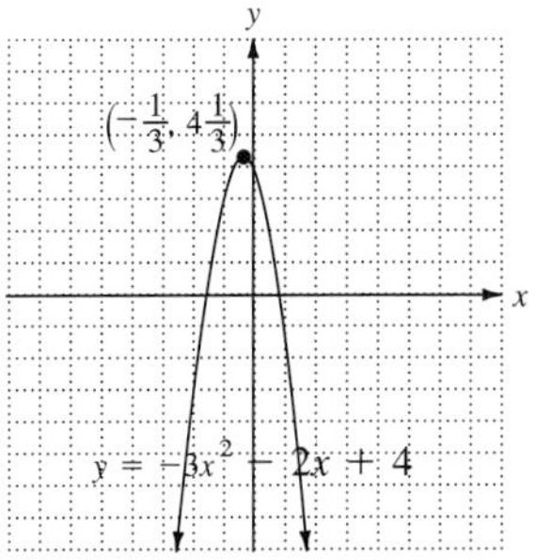

95. $l = 5$ feet, $w = 3$ feet **96.** 16 centimeters, 12 centimeters **97.** 24 members were in the club *last year.* **98.** 5 and 12 **41.** 38.25 ft^2 **100.** going 45 mph, returning 60 mph

Chapter 10 Test

1. $\frac{-7 \pm \sqrt{129}}{10}$ **2.** $-5, \frac{2}{3}$ **3.** no real solution **4.** $-\frac{1}{2}, \frac{5}{3}$ **5.** $-\frac{5}{4}, \frac{1}{3}$ **6.** $\frac{4}{3}$ **7.** $0, 8$ **8.** ± 4 **9.** $-\frac{1}{2}$ **10.** $\frac{3}{2}, -\frac{1}{2}$

11.
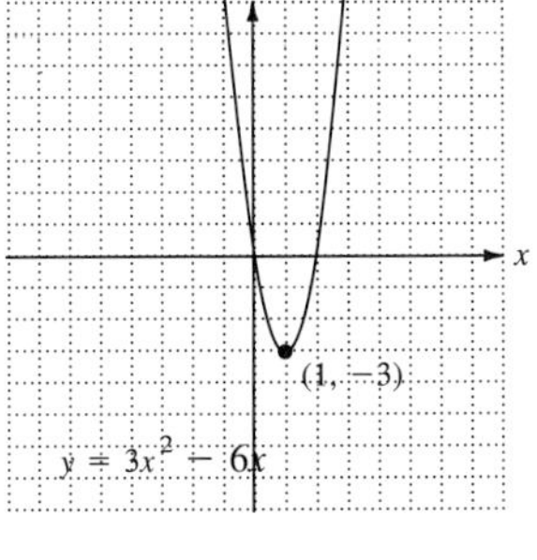

12.
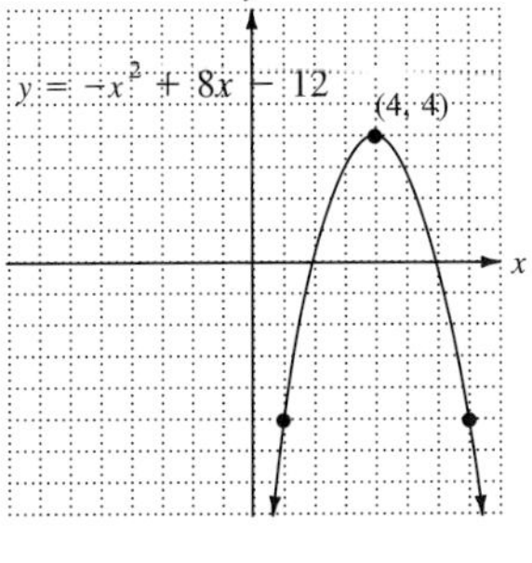

13. 9 and 12 **14.** 7 seconds

Cumulative Test for Chapters 1–10

1. $-27x^2 - 30xy$ **2.** 35 **3.** 2 **4.** $(2x - 1)(2x + 1)(4x^2 + 1)$ **5.** $6x^3 + 17x^2 - 31x - 12$ **6.** $-\frac{12}{5}$

7.

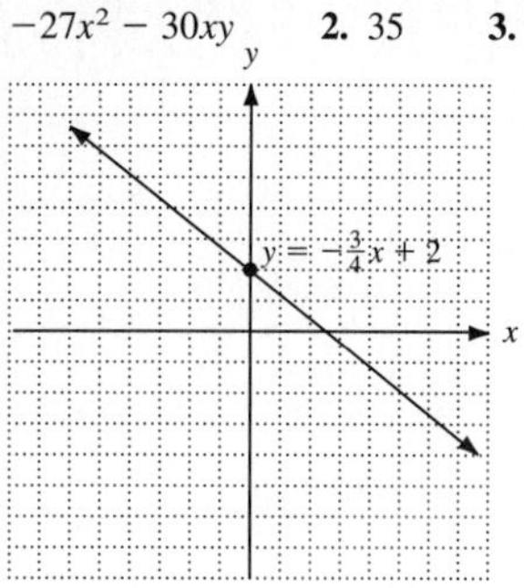

slope $-\frac{3}{4}$; y-intercept $= +2$

8. $-3x + 5$ **9.** $(3, -2)$ **10.** $81x^{12}y^8$ **11.** $3x^2y^3z\sqrt{2xz}$ **12.** $-2\sqrt{6} - 8$ **13.** $-5\sqrt{5} + 11$ **14.** $\dfrac{A \pm \sqrt{A^2 + 3A}}{3}$ **15.** $b = \pm\dfrac{\sqrt{H + 6}}{5}$ **16.** $\frac{7}{2}, -5$ **17.** $-\frac{1}{2} \pm \sqrt{5}$ **18.** $-12, 4$ **19.** $\dfrac{-11 \pm \sqrt{97}}{6}$ **20.** $\dfrac{-1}{7}, -5$ **21.** $\pm 7\sqrt{2}$ **22.** ± 5

23.

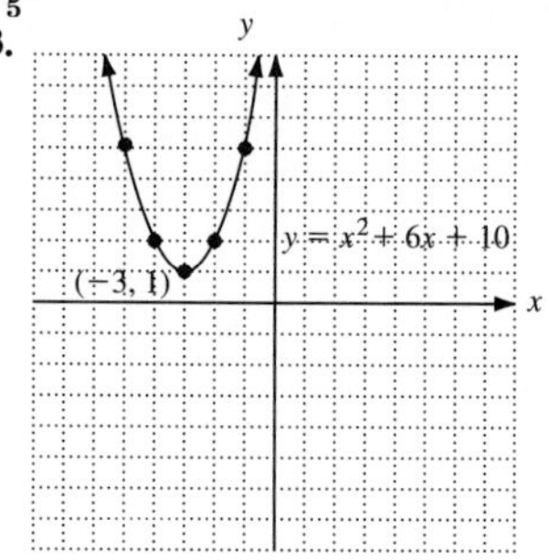

24. 6, 16

Practice Final Examination

1. $-2x + 21y + 18xy + 6y^2$ **2.** 14 **3.** $18x^5y^5$ **4.** $2x^2y - 8xy$ **5.** $\frac{8}{5}$ **6.** $\dfrac{p - 2a}{2}$ **7.** $x \geq 2.6$ (number line: 2, 2.6, 3, 4) **8.** $2x^3 - 5x^2y + xy^2 + 2y^3$ **9.** $2(2x + 1)(x - 5)$ **10.** $25(x^2 + 1)(x + 1)(x - 1)$ **11.** $\dfrac{6x - 11}{(x + 2)(x - 3)}$ **12.** $\dfrac{11(x + 2)}{2x(x + 4)}$ **13.** $-\frac{12}{5}$ **14.** $\frac{5}{2}$

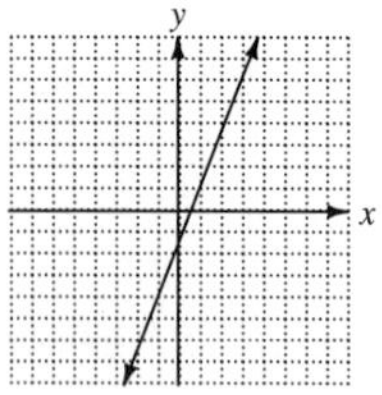

15. 14 **16.** 38.13 sq. in. **17.** $7, -2$ **18.** $-2, -3$ **19.** $8x\sqrt{2}$ **20.** $6\sqrt{3} - 12 + 12\sqrt{2}$ **21.** $-\dfrac{\sqrt{15} + \sqrt{35} + \sqrt{21} + 7}{2}$ **22.** $\dfrac{5}{3}, -1$ **23.** $\dfrac{3 \pm \sqrt{7}}{2}$ **24.** ± 2 **25.** $\sqrt{39}$ **26.** The number is 5. **27.** Width is 7 meters. Length is 12 meters. **28.** \$3000 at 10%, \$4000 at 14% **29.** 200 reserved tickets and 160 general admission tickets **30.** The base is 8 meters. The altitude is 17 meters.

Appendix B

Exercises

1. 34,000 m **3.** 5700 cm **5.** 250 mm **7.** 0.563 m **9.** 29,400 mg **11.** 0.0984 kg **13.** 7000 mL **15.** 0.000004 kL **17.** 1.7 in. **19.** 22.5 km **21.** 3.8 m **23.** 17.8 cm **25.** 1089.6 g **27.** 171.6 lb **29.** 2.8 L **31.** 193,248,000 in. **33.** 0.45 miles per hour **35.** 0.11 yr

Appendix C

Exercises

1. 14 **3.** 52 **5.** 1296 **7.** -3 **9.** $10x - 5$ **11.** 1111811 **13.** 9876549 **15.** 6481100 **17.** 1 6 15 20 15 6 1 **19.** Steps will vary **21.** Steps will vary **23.** Steps will vary **25.** Steps will vary **27.** Ans. will vary

Applications Index

Note: Boldface page numbers highlight particularly significant information.

Index

Note: Boldface page numbers highlight particularly significant information.